Lecture Notes in Production Engineering

Lecture Notes in Production Engineering (LNPE) is book series that reports the latest research and developments in Production Engineering, comprising:

- Biomanufacturing
- Control and Management of Processes
- Cutting and Forming
- Design
- Life Cycle Engineering
- Machines and Systems
- Optimization
- Precision Engineering and Metrology
- Surfaces

LNPE publishes authored conference proceedings, contributed volumes and authored monographs that present cutting-edge research information as well as new perspectives on classical fields, while maintaining Springer's high standards of excellence. Also considered for publication are lecture notes and other related material of exceptionally high quality and interest. The subject matter should be original and timely, reporting the latest research and developments in all areas of production engineering. The target audience of LNPE consists of advanced level students, researchers, as well as industry professionals working at the forefront of their fields. Much like Springer's other Lecture Notes series, LNPE will be distributed through Springer's print and electronic publishing channels.

Indexed in Scopus* *Indexed in EI Compendex To submit a proposal or request further information please contact Anthony Doyle, Executive Editor, Springer (anthony.doyle@springer.com).

Christian Brecher · Manfred Weck

Machine Tools Production Systems 1

Machine Types and Application Fields

Springer

Christian Brecher
RWTH Aachen University
Laboratory for Machine Tools and
Production Engineering (WZL)
Aachen, Germany

Manfred Weck
RWTH Aachen University
Laboratory for Machine Tools and
Production Engineering (WZL)
Aachen, Germany

ISSN 2194-0525 ISSN 2194-0533 (electronic)
Lecture Notes in Production Engineering
ISBN 978-3-662-68119-0 ISBN 978-3-662-68120-6 (eBook)
https://doi.org/10.1007/978-3-662-68120-6

This Springer imprint is published by the registered company Springer-Verlag GmbH, DE, part of Springer Nature.
The registered company address is: Heidelberger Platz 3, 14197 Berlin, Germany

In memory of

Prof. em. Dr.-Ing. Dr.-Ing. E. h. Dr.-Ing. E. h. Manfred Weck, an outstanding engineer, extraordinary pioneer, passionate researcher and teacher.

1937 – 2023

Foreword for the Machine Tools and Manufacturing Systems Compendium

Machine tools are among the most important means of production in the metalworking industry. The current high standard of living in industrialized nations would be inconceivable had this category of machinery not been developed. The Federal Republic of Germany is a world leader in the production of machine tools. In the Federal Republic of Germany, approximately 7.44% of the total mechanical engineering production volume comes from the machine tool industry and 6.96% of those employed in mechanical engineering work in the machine tool industry (German Machine Tool Builders' Association (VDW), as of 2017).

The application areas for machine tools are just as diverse as their structural design and degree of automation. Depending on the technological processes, this extensive field ranges from machine tools for primary forming and forming, machine tools for cutting (including chip-removing machine tools) to joining machines. These machines have varying degrees of automation and greater or lesser flexibility depending on the workpieces to be machined and the batch sizes. As a result, single-purpose and special machine tools as well as universal machines with an extensive range of potential applications are available on the market.

Due to the increasing demands on performance and accuracy, the designers of these machines must ensure there are no compromises in the design of the individual machine components. To do this, the designers must have a comprehensive knowledge of the relationships between the physical properties of the components and of the machine elements. Today, computer-aided design tools are available to designers via a substantial library of programs. The criteria determining performance and accuracy, such as the geometric, kinematic, static, dynamic, thermo-elastic and acoustic properties, can be evaluated through metrological analyses and objective assessment methods. The specific improvements necessary can then be implemented.

The continuing tendency toward automating machine tools has led to a broad spectrum of control system alternatives. In recent years, developments in electrical engineering, electronics and software technology have had a significant influence on machine control systems. Microprocessors and process computers can now provide control-engineering solutions that were previously unthinkable. The mechanization and automation efforts have also included the areas of material transport and machine feeding. Considerations in this area led to transfer lines in mass production and to flexible manufacturing cells and systems in small and medium series manufacturing. Information and communication technologies are becoming more important areas in the machine tool industry with respect to Industry 4.0.

The three volumes that make up this series on the subject of Machine Tools and Manufacturing Systems are aimed at both university students majoring in production engineering and all practicing professionals who must keep pace with the increasingly complex material covering this branch of mechanical engineering. In addition, these volumes will help the reader select suitable machines and relevant control systems. Machine manufacturers are provided with ways of producing the best design of the machine components, drive systems and control systems together with options for specific improvements based on metrological analyses and objective assessment methods.

The content of the complete work is closely based on the Machine Tools lectures and the Mechatronics and Control Technology for Manufacturing Systems at the RWTH Aachen University and is structured as follows:

Volume 1: Machine tools—Machine types and areas of application
Volume 2: Machine tools—Design, calculation and metrological assessment
Volume 3: Machine tools—Mechatronic systems, control engineering and automation.

Christian Brecher
Aachen
September 2019

Foreword to Volume 1

The first volume of this compendium aims to provide machine tool users with a comprehensive overview of the various areas of application and the appropriate machine types. The different types of machine tools for the individual process groups are initially introduced based on the manufacturing process standard DIN 8580. Primary shaping and forming as well as cutting machine tools are taken into consideration as well as material removal machines and equipment and laser beam machining. Various single-purpose machine tools, which are designed for a specific product, are then described before a range of different types of multi-machine systems are discussed. A detailed explanation of various equipment components for machine tools is then given in a separate chapter. The volume concludes with the final chapter of the revised edition. This describes various design examples of selected machine tools in detail to provide the reader with greater depth and illustration of machine design.

The content of this volume is structured as follows:

A short introduction *(Chap. 1)* provides an overview of the significance, history, and basic requirements of machine tools. This is followed by a summary of the classification of manufacturing processes and machine automation.

Chap. 2 covers various machines and systems used in the primary forming stages of machining. In addition to classic primary forming methods such as casting with expendable and reusable molds as well as sintering, explanations are provided on casting machines for reaction resin concrete as well as an introduction to machines for additive manufacturing.

Chap. 3 is a comprehensive explanation of various types of forming and severing machines. In the field of forming machines, this includes various types of presses (e.g., spindle, crank, toggle lever, or wedge presses, etc.) as well as machines for continuous forming processes such as rolling, bending, drawing, and straightening machines. Special guards and safety devices for forming machines are also covered. Shearing machines and cutting presses are the main focus of severing machines.

Finally, *Chap. 4* deals with cutting machines for tools with geometrically defined cutting edges, one of the key application areas for machine tools. In addition to the different types of sawing and drilling machines and the introduction of slotting, drawing, and broaching machines, the main section of the chapter deals in detail with milling machines and lathes as particularly widely used types of machine tools. Different designs and their basic characteristics are explained. In the field of milling machines, for example, a distinction is made between machines with horizontal and vertical tool spindles as well as machines with parallel kinematic structures. The individual descriptions are illustrated through many figures and schematic diagrams.

Cutting machines for tools with geometrically undefined cutting edges also form an important area of application for machine tools in addition to machines for tools with geometrically defined cutting edges. This topic is covered in *Chap. 5*. Honing and lapping machines as well as machines for mechanical deburring and slide grinding machines are also discussed. In addition, there is a comprehensive description of various types of grinding machines (e.g., cylindrical and face grinding machines) and specific devices such as dressing systems and measurement equipment. The last part of the chapter deals with water jet cutting systems.

Chap. 6 provides an overview of material removal machines. Descriptions are provided of machine tools for performing machining processes defined as chip removal manufacturing processes in accordance with DIN 8580. A short, detailed introduction to chemical etching systems and machines for electrochemical and electrical discharge machining is also provided. Systems for chip removal by electron beam and for thermal-chemical deburring are also covered.

Due to the increasing significance in production, laser beam processing machines now have their own separate chapter in this revised edition of Volume 1. Current laser beam sources are initially described in *Chap. 7* followed by options for implementing beam guidance and shaping. This is followed by an explanation of the different types of laser beam processing systems. It includes both laser cutting and laser ablation systems as well as systems for additive manufacturing using laser machining and machines for laser-assisted machining.

Chap. 8 is dedicated exclusively to single-purpose machine tools. These are specialist machines for the manufacture of a specific product. For example, gear cutting machines are used exclusively for manufacturing gears. Due to the large number of different types of gear cutting machines, these are covered in particular detail. Different types of rolling and profiling as well as special and forming gear cutting machines are discussed. However, machines for the manufacture of crankshafts using milling, turning, grinding or rolling operations as well as machines for producing camshafts and ball screws are also covered in this chapter.

The constant demand for increased performance and automation in production is leading to a greater role for multi-machine systems. *Chap. 9*, therefore, deals with the various options for implementing these types of systems. It starts with the basics of machine interlinking to introduce the reader to this topic. Various handling systems such as pick and place units or industrial robots combined with the appropriate end effectors (e.g., grippers) are used as the interface between the individual machines in a multi-machine system are presented in this chapter. Special-purpose machine tools such as rotary or linear transfer machines are then discussed before various types of transfer lines and flexible manufacturing cells and systems are explained.

In addition to the machine tool components with basic functional relevance, there are also many equipment components, some of which are optional, that can also be integrated into the machines. These components are covered in detail in *Chap. 10*. Following a short overview of the various equipment components, peripherals for cooling systems, handling chips, for lubrication and hydraulic units are explained. The details of various tool systems for lathes, drilling and milling machines as well as the mechanical properties of tool interfaces are then covered. The tasks and the various types of workpiece clamping and supply systems are explained at the end of the chapter.

Finally, three design examples of selected machine tools are explained in detail in *Chap. 11* to give the reader a detailed and practical illustration of various types of machines. This chapter is a new addition to the revised edition of this volume and details the design structure and individual functionality of a milling machine, a grinding machine, and a turning and milling machine from different manufacturers using assembly and section drawings.

This ninth edition was revised with the cooperation of my research assistants. My sincere thanks go to all those involved for their enormous commitment. I particularly wish to thank Mathias Dehn M.Sc., Dipl.-Ing. Alexander Hassis and Caroline Kiesewetter-Marco M.Sc. for coordinating and organizing the basic revision of the ninth edition.

I also wish to thank most sincerely the companies who provided the many illustrations for this volume.

Christian Brecher
Aachen
August 2023

Contents

Introduction

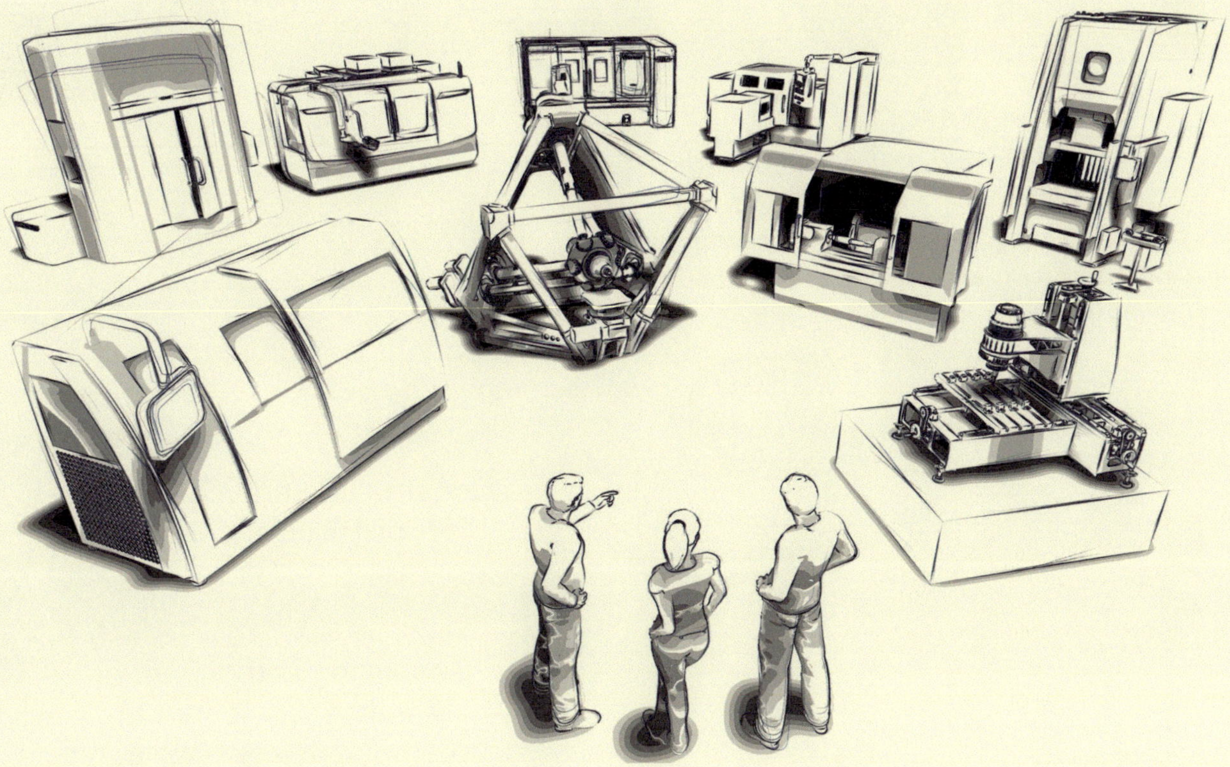

I thank my staff member Benedikt Biernat M.Sc. and Mathias Dehn M.Sc. for their contribution in revising this chapter.

C. Brecher and M. Weck, *Machine Tools Production Systems 1*,
Lecture Notes in Production Engineering, https://doi.org/10.1007/978-3-662-68120-6_1

Contents

1

Tools and production equipment have always played a key role in the economic and social development of society. The mechanization of manufacturing processes and the provision of, initially, mechanical, and later of electrical, energy was at the heart of the Industrial Revolution: This started in the late eighteenth century and brought about fundamental changes in society and hitherto unknown economic growth, initially in England and Europe, later in the USA and Japan, and now in nearly every part of the world. Production equipment is also required to manufacture the necessary manufacturing and energy generation systems. Machine tools represent the most important element of this production equipment in the industrial field. For this reason, the machine tool is known as the "mother of all machines" [CIUP16, KÖHN16].

We will therefore first examine the economic significance of the machine tool industry and the historical development of machine tools before providing a basic overview of the various types of machine tools, their design and their use, in the following chapter. This is followed by an introduction to a few essential terms from the field of mechanical metal processing.

1.1 The Economic Significance of Machine Tools

While the proportion of gross domestic product represented by industrial production has been declining in recent years in most Western countries, in Germany it is currently at the same level as in 1994 at approximately 22% and is therefore at a significantly higher level than most other European countries [BMWI16, STAT16b, STAT16a]. Mechanical engineering is the second largest industrial sector in Germany after the automotive industry [VCI16]. As can be seen from Fig. 1.1, the machine tool industry is a leading player within this sector in terms of both the number of employees and manufacturing volume, which emphasizes the importance of machine tools for industrial production stated above.

Fig. 1.1 Employees and value of production by mechanical engineering sector (as of 2015), according to [VDMA15]

Having a domestic manufacturing base for production equipment—and therefore machine tools—is thus extremely important for a country with a significant element of industrial production. ◘ Figure 1.2 emphasizes this by showing the development of the value of production of the machine tool industry in the ten most important manufacturing countries. Germany and Japan have led the way here for decades and have therefore been more severely affected by economic fluctuations than other countries. In addition to other traditionally important machine tool producing countries such as Italy and the USA, China's growth to become the leading manufacturing country of machine tools today, when measured by the value of production, initially began slowly during the mid-1990s and then continued extremely quickly since the middle of the last decade.

In China, this trend is just one aspect of strong economic growth, which, just as it was in Western countries many decades before, is initially based on industrialization and a significant infrastructure expansion—as well as other factors. Visible indications and other drivers of this development include the acceptance of China into the World Trade Organization in 2001 and China's own desire in recent times to achieve market economy status [DREG16] or China's status as the world's largest economy (as of 2014), which will continue to consolidate in the coming decades [PWC16].

However, national machine tool production alone is not sufficient to cover domestic demand: This is particularly true in China, but also in the USA. In the USA, this is partly due to a trend in reindustrialization [BAIN16] and the subsequent demand for machine tools. In addition to the rising demand for machines in China, which is accompanied by a general increase in industrial production, Chinese manufacturers are not yet able to cover the entire portfolio of machine tools required today. This explains China's and the USA's leading positions in the rankings of machine tool importers shown in ◘ Fig. 1.3.

◘ Figure 1.3 also shows that specifically Germany and Japan export a significant volume of the machine tools they produce. Together with the relevant production volumes and the comparably low imports, this means that the machine tool industry in these countries is not only equipping its domestic industry but is also contributing to export volumes, strengthening the economic significance of this sector even further.

◘ Figure 1.4 provides a more detailed analysis of the German machine tool industry. Cutting machine tools have a significantly greater share of the German market in terms of manufacturing volume compared to forming machines. The reasons for this include a much broader range of cutting machine tools compared to forming machines and the higher demand for cutting machines. This demand can be attributed to the greater distribution of this type of machine and the greater demand for components they produce. Furthermore, the graphics show that the service and spare parts business is an important element of the German machine tool industry, not least because of the longevity of many machine tools as an investment.

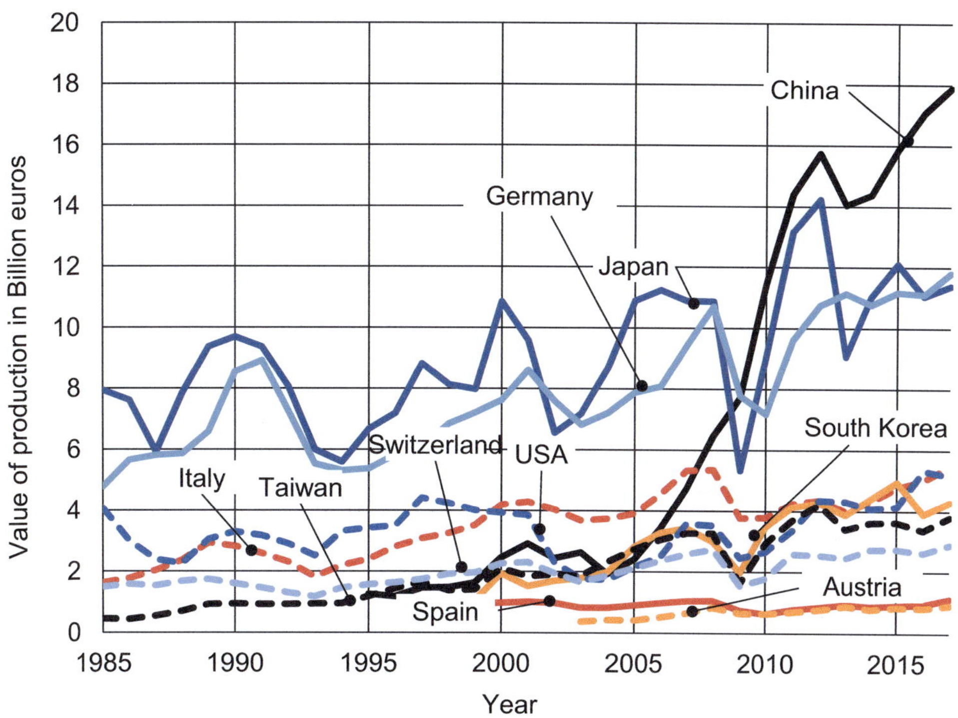

◘ **Fig. 1.2** Historical development of machine tool production in the ten leading manufacturing countries, according to [VDW17]

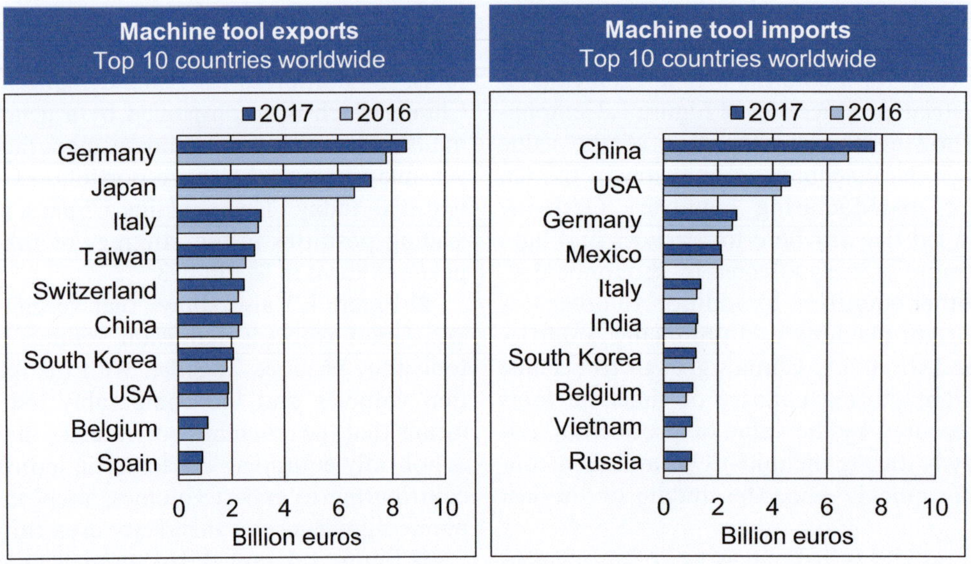

Fig. 1.3 International machine tool trade indices, according to [VDW17]

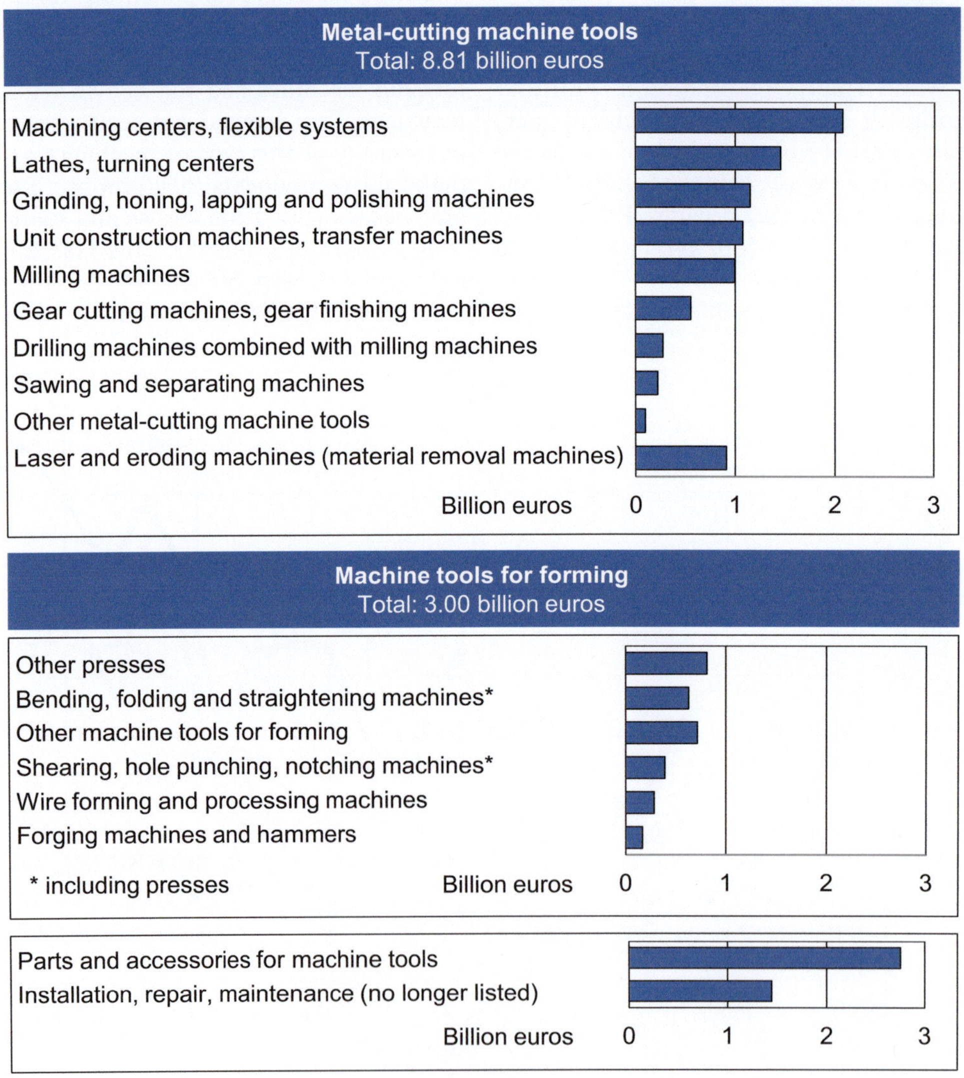

Fig. 1.4 Machine tool production 2016 in Germany by machine type, according to [VDW17]

In view of the economic importance of metal-cutting machine tools in Germany and the great variety in this group, the following chapters focus on the descriptions of the individual machine types.

1.2 Historical Development of Machine Tools

The development of machine tools can be reconstructed from many finds from prehistoric periods, through images and written records as well as historical collections: The first tools were chisels and shaving tools whose efficacy was soon increased through levers and wedges. The oldest drilling tools that have been found originate in the stone age. In these tools, a flint, acting as a cutter, was fixed to the end of a wooden stick. A bow-string was wrapped around the stick allowing the tool to be rotated by moving the bow back and forth [SPRI41].

The first mention of a lathe in its most primitive form can be found in Homer's poems from the eighth century BC. This was also driven by a bow and this remained the most common drive principle until the fifteenth century. After this, the foot-driven rocker gained significance as the drive for processing machines [WITT60].

Overall, mechanical processing machines developed at a slow pace up to the beginning of the industrial revolution, not least due to the lack of suitable materials for the machine components. Leonardo da Vinci's (1452 to 1519) drafts were substantially ahead of their time. As cast iron was still a relatively rarely used material in its time, most of the machines were constructed of wood. This meant that, among other things, they lacked accuracy due to the low stiffness and natural warping [SPRI41].

The real breakthrough in the development of more powerful machine tools, which can machine metal workpieces, came with the further development of the steam engine invented in 1712 by Thomas Newcomen. James Watt optimized the efficiency of the steam engine in the second half of the eighteenth century and thus contributed to the industrial revolution. Machines, which had previously been driven exclusively by muscle power, could now be driven by a motor, increasing their performance many times over.

In 1774, the Englishman John Wilkinson introduced a cylinder boring machine based on an invention by John Smeaton from 1765. This can be designated the first metal processing machine or machine tool. This allowed the working precision to be increased significantly so that this machine bored out the first useful steam engine cylinder for James Watt [WITT60]. In 1776, Matthew Boulton wrote to his partner Watt, "Mr. Wilkinson has bored out various cylinders for us almost without defect and the 50" diameter which we produced in Tipton does not deviate at any point by the thickness of an old shilling" [ROE26].

In 1794, Henry Maudslay developed the first front-bed lathe. It made use of the tailstock guideway to guide the tool carrier slide whose bearing had until now been fixed to the tailstock. This allowed the stiffness of the tool guide to increase substantially. By mechanizing the feed using a lead screw, the working precision and productivity could be increased [WITT60].

A further significant step in the provision of mechanical drive power was the invention by Nicolaus August Otto of the atmospheric engine in 1864 and the 4-stroke engine in 1876. Due to its poor efficiency, the steam engine was replaced by this solution in the following decades in many, primarily mobile, applications. However, the machine shops in production facilities were still dominated by transmission drives which transfer the energy from a central machine to the individual machine tools. It was only with the invention of the three-phase a.c. motor in 1889 that transmission drives slowly began to be replaced by individual drives [WZL86].

The development of the most important types of standard machine tools, such as turning, planing, slotting, drilling and milling machines, was basically completed at the end of the nineteenth century [WZL86].

In addition to the developments in the machine tools themselves, the availability of suitable processing tools was particularly significant for machining workpieces. An important step came with the development of high-speed steel by the American Frederic Winslow Taylor who presented this at the 1900 World Fair in Paris. By alloying 3.8% chromium, 8% tungsten, as well as some vanadium and molybdenum, the tool steel retained its cutting ability up to working temperatures of 600 °C. Compared to pure carbon tool steel, the cutting speed could be increased three to five times [WZL86].

Figure 1.5 shows the historical development of the maximum achievable cutting speed when machining mild steel as a function of the cutting material used. As the cutting material improves, so do the requirements on the machine tools. Increased driving forces, higher working spindle speeds, stiffer guide systems, working spindles, frame components capable of supporting higher loads and improved transmissions were all required. The development of machines and tools has been, and remains, both a mutually beneficial and driving factor in the advancement of mechanical metal processing. This has resulted in cutting and feed rates for some cutting ceramics reaching the performance limits of the spindle bearing arrangements and feed systems currently available. In addition to this, coating

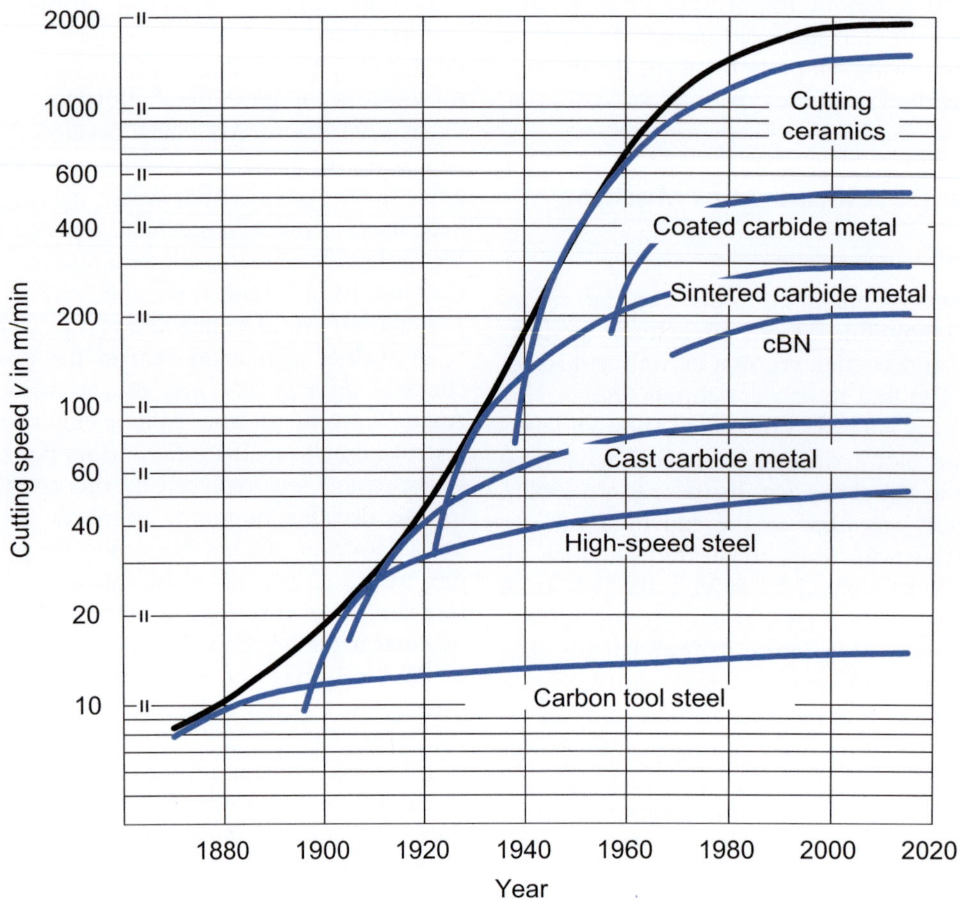

Fig. 1.5 Trend in cutting speeds when turning mild steel

technologies are used to increase the performance of carbide metals.

As machines and tools continue to develop, so do the demands on the achievable manufacturing accuracy, as shown in Fig. 1.6. A logarithmic scale is used to illustrate the substantial spread of manufacturing accuracies. Various significant technical events and inventions are shown on the time axis. The corresponding measuring equipment necessary for verifying manufacturing quality had to be developed in parallel with the improvement in the achievable machine accuracy. This development is shown in the upper section of the chart. The right-hand section of the image shows the component accuracy achieved with commonly available manufacturing processes today. On ultra-precision machines, the physical limit of the achievable accuracy is determined by the crystal structure or the spacing of the atomic lattice of the material being machined.

Figure 1.7 illustrates the development of machine tools over the last two hundred years using lathes as an example. The first screw-cutting lathe made by Maudslay (1797) consisted of a precision-machined, three-foot (approx. 0.9 m) long double prismatic guide system, a headstock with a rotating center and a lead screw driven by the main drive via gears. The feed per revolution could be adjusted by changing the lead screw. Of note is the position of the working spindle on the right-hand side; today, it is located on the left-hand side for ease of operation.

The lathe from 1906 has the same typical basic structure as today's lathes. The concept was derived from the wood-turning lathes used in workshops. A series of pulleys with different diameters is located immediately on the rotating spindle and it was possible to adjust the spindle speed by moving the drive belt between these pulleys. The dependency of the transmission unit and the central working machine tied the machine tool to a specific location.

The 1925 lathe no longer had this disadvantage. It was driven by its own electric motor located in the foot of the frame. A range of spindle speeds could be set via the gearbox. The gearbox was located in the fully enclosed headstock.

The lathe from 1960 had a closed frame design meeting the high stiffness demands. The almost vertical arrangement of the bed allowed unimpeded chip fall.

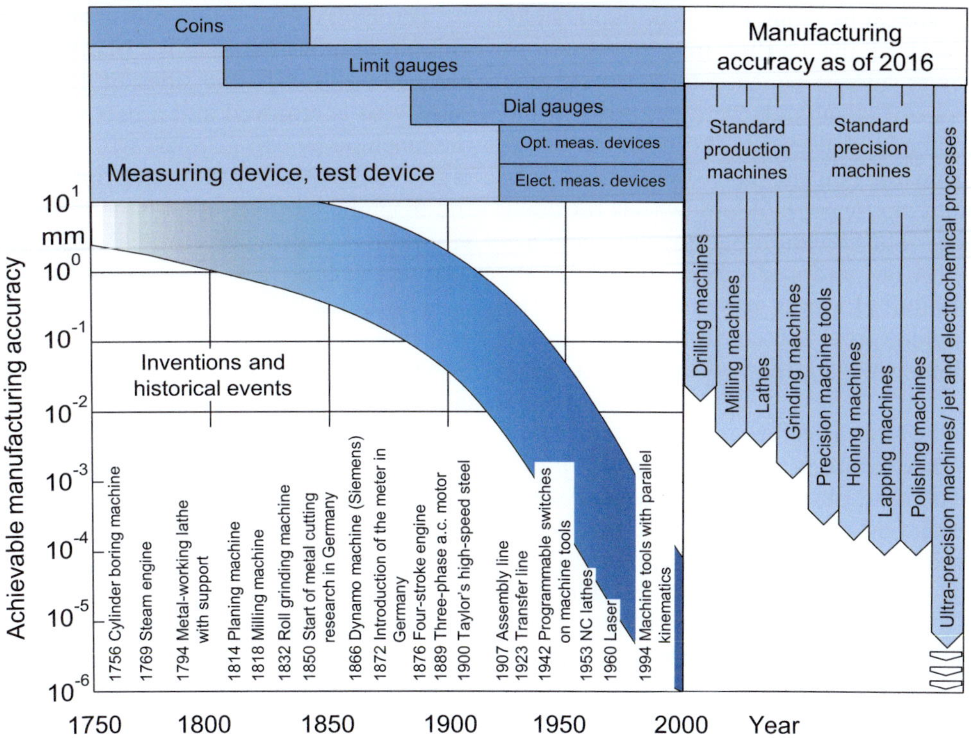

Fig. 1.6 Overview of the progress of development in the manufacturing accuracy achievable by machine tools

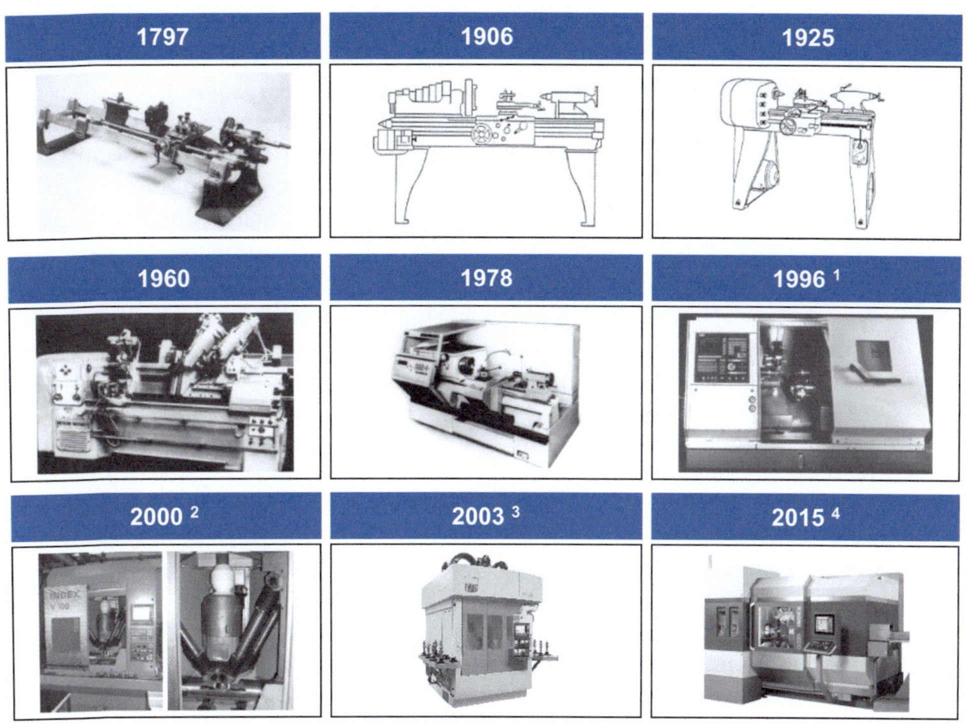

Fig. 1.7 Historical development of lathes. *Source* 1) Traub, 2, 4) INDEX, 3) EMAG

The machine was fitted with a hydraulic postforming device allowing any workpiece contour to be produced automatically. The control commands for the feed motions were produced by scanning a template or a master workpiece. This type of automatic feed control was used successfully in both small and large volume production. Since the start of the 1970s, it has been replaced by flexible numerical control systems.

The Gildemeister lathe from 1978 was numerically controlled. It received the control information from digital data carriers such as punched tape or magnetic cassettes. The control commands were provided in the form of alphanumeric characters. The machine could be controlled from the control panel by pressing buttons. This machine no longer had handwheels for moving the feed slides. The control programs for manufacturing parts could be produced offline at a separate programming station. However, it was also possible to program the machine from the control panel and this could even be done while another workpiece was being machined.

The Traub lathe from 1996 has many structural and control system modifications compared to the previous machines. The most obvious external feature is the complete encapsulation of the machine. It was possible to increase the machine stiffness considerably by using a slant-bed design. At the same time, the slant-bed provided good chip removal which is a necessary requirement for automatic production machines. By using tool turrets and driven tools, it was possible to machine complete workpieces on this machine. The machine was controlled by an NC controller based on a microprocessor which had many additional functions: This allowed the NC programs to be displayed on the control system's screen through graphic machining simulations and to be checked for programming errors. The design and control system concept for the machine show it was intended for use in flexible manufacturing systems. Workpieces, tools, and clamping jaws on chucks could be changed automatically by using handling equipment. The machine had many monitoring and diagnostic functions as well as integrated sensors, which ensured that the automatic production ran without any errors and provided rapid error diagnosis. Options for remote diagnosis and production data acquisition allowed machine status information to be read and evaluated at any time by an external service station which was one of the first steps toward remote machine tool maintenance [HEKE85, ZEPP86].

This development continues today with connecting machine tools to the Internet. This allows, for example, large volumes of data to be transferred with information on the machine's status. This is a condition for implementing predictive maintenance strategies.

The Index V100 machine from 2000 has a vertical main spindle and a new kinematic concept. Using parallel kinematics (shown enlarged) produces a completely three-dimensional movement for the suspended motor spindle. This allows the workpiece to be picked up by the spindle (pick-up spindle) and to be taken to the stationary tools extremely quickly. The short travel distances and the high rapid traverse velocity mean the non-productive times related to the workpiece and the tool are significantly reduced.

The EMAG turning center from 2003 has a vertical pick-up spindle. Storage for blanks and finished parts has been added to the sides of the machine. The finished part is removed and a new workpiece is placed in the machine by two grippers working in parallel, resulting in reduced non-productive times. The base body of the machine consists of polymer concrete. This damps vibrations, improving surface quality and increasing tool service life.

The Index G220 from 2015 is a combined turning and milling machine. In addition to a turret with stationary tools, it has an extra milling spindle for machining workpieces. It also has two main spindles. These allow two workpieces to be machined at the same time and workpieces to be passed from one spindle to the other. This feature allows machining to be completed without any external workpiece handling.

1.3 Basic Demands on Machine Tools

Modern industrial production makes a wide range of demands on machine tools. From an operational perspective, this is primarily concerned with the machine's technical suitability for the processes to be carried out as well as economic aspects such as the piece costs or life cycle costs. At the same time, the state and society make demands on the machines in the form of laws which concern, for example, the user's safety or the environmental characteristics of the machine. As shown in ◘ Fig. 1.8, all these demands affect the construction and design of machine tools.

The displacements which occur at the machining point and which influence quality are themselves affected by the static, dynamic, and thermo-elastic behavior of all the assemblies present within the flux of forces. The manufacturing accuracy, the surface quality of the workpieces, the usable mechanical power output, and the resultant productivity are a function of these machine properties. In addition to this, economical use is determined by the degree of automation. Automation is not merely limited to the completion of the actual machining process but also to the machine feed, i.e., the supply and removal of workpieces, tools, and chips.

Legal framework conditions are largely specifications from professional associations for occupational health and safety or governmental regulations. The European Machinery Directive (abbreviated as EMD, 2006/42/EG) has been particularly important in this respect for some years now. It initially focused on the safe design of machines and systems and has been expanded recently to include regulations on environmental protection.

◘ Figure 1.9 shows that every aspect of the manufacturing tasks to be carried out must be taken

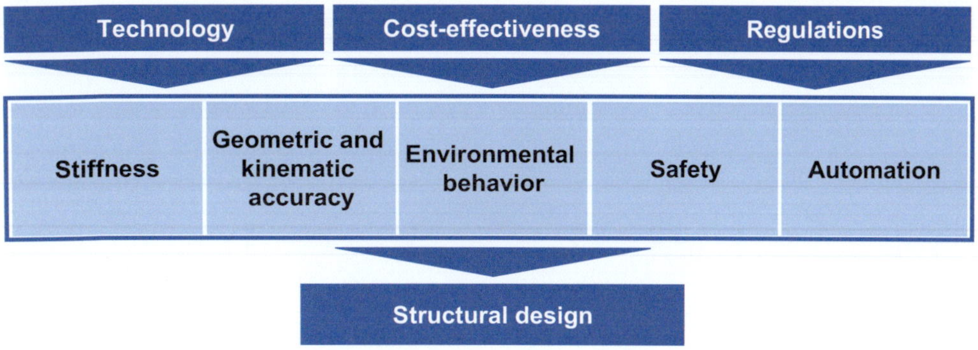

Fig. 1.8 Key demands on machine tools

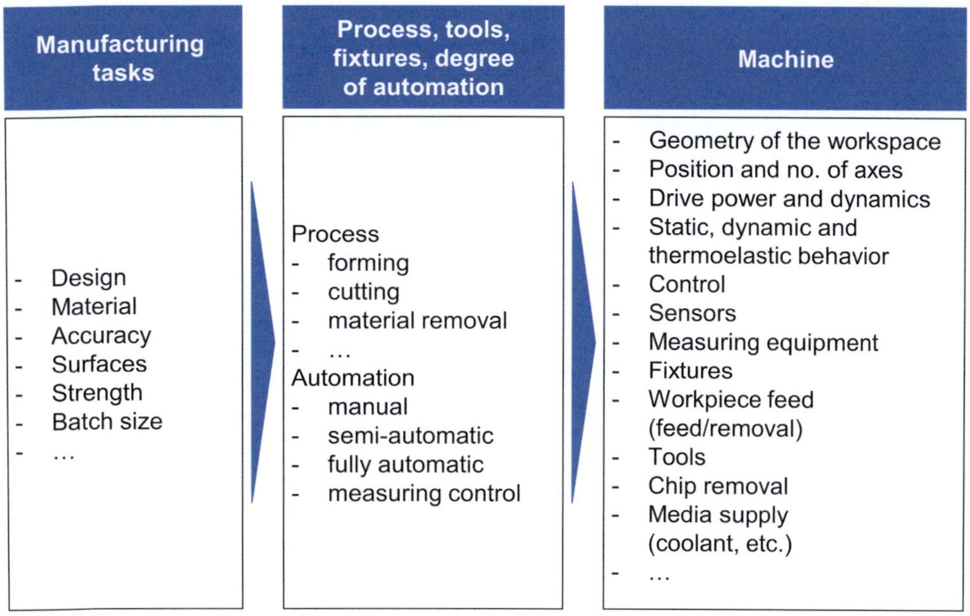

Fig. 1.9 Considerations in the selection and design of machine tools for specific manufacturing tasks

into consideration in the design of a machine tool to achieve the best possible suitability for the processes required. First, the characteristics of the manufacturing process are to be defined. The parameters that determine the working area, such as the number and position of the axes of motion, and the functional characteristics of the machining units (e.g., speed range, torques) are established in accordance with the manufacturing task and the selected process. The required accuracy and surface quality of the workpieces determine the static and dynamic stiffness requirements for the individual assemblies. The design of the machine components with respect to performance and loading depends on the range of tasks. The variation in the workpieces to be machined and the batch sizes determine the degree of automation. In addition to this, all the resources required for production, particularly the clamping tools and measuring and transportation equipment, are to be taken into consideration when designing the machine tool.

1.4 Categorizing Manufacturing Processes and Machine Tools by Description

1.4.1 Subdividing Manufacturing Processes

Machine tools are utilized in various domains/areas, leading to a diverse range of machines. Over the decades of development, it has made sense to systematize and unambiguously define the terms which have evolved over time. This classification and designation of machine tools is based on the terminology for metal processing manufacturing processes defined in the standard DIN 8580, Fig. 1.10 [DIN03].

1.4.2 Categorization of Machine Tools

The classification, categorization, and designation of machine tools were laid down in the withdrawn

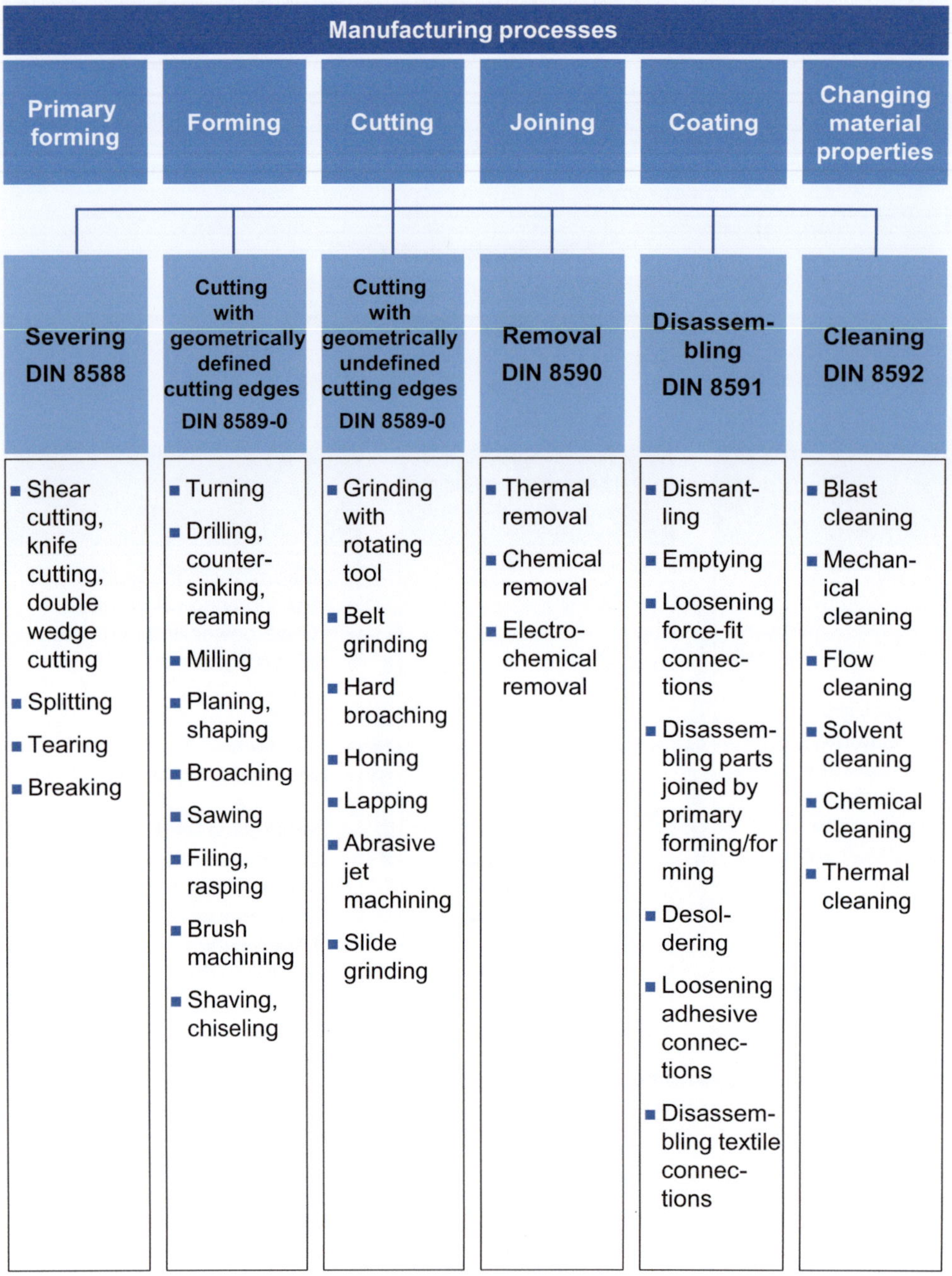

Fig. 1.10 Classification of manufacturing processes according to DIN 8580 [DIN03]

standard DIN 69651 [DIN81]. Figure 1.11 shows the allocation of machine tools within the group of manufacturing systems.

Machine tools are defined as "… mechanized and to a greater or lesser extent automated manufacturing equipment which produces a prescribed form or change to the workpiece through relative movement between the tool and workpiece" [DIN81]. They are understood as part of the production facility.

The categorization of manufacturing equipment is based on the classification of the metal processing manufacturing processes to DIN 8580. As can be seen in Fig. 1.11, the term machine tool is limited to the manufacturing processes of forming, cutting, and joining.

1.4.3 Designation of Machine Tools

According to DIN 69651, machine tools are subdivided into stand-alone machines and multi-machine systems [DIN81]. The designation of machine tools is guided by the manufacturing process and the degree of automation.

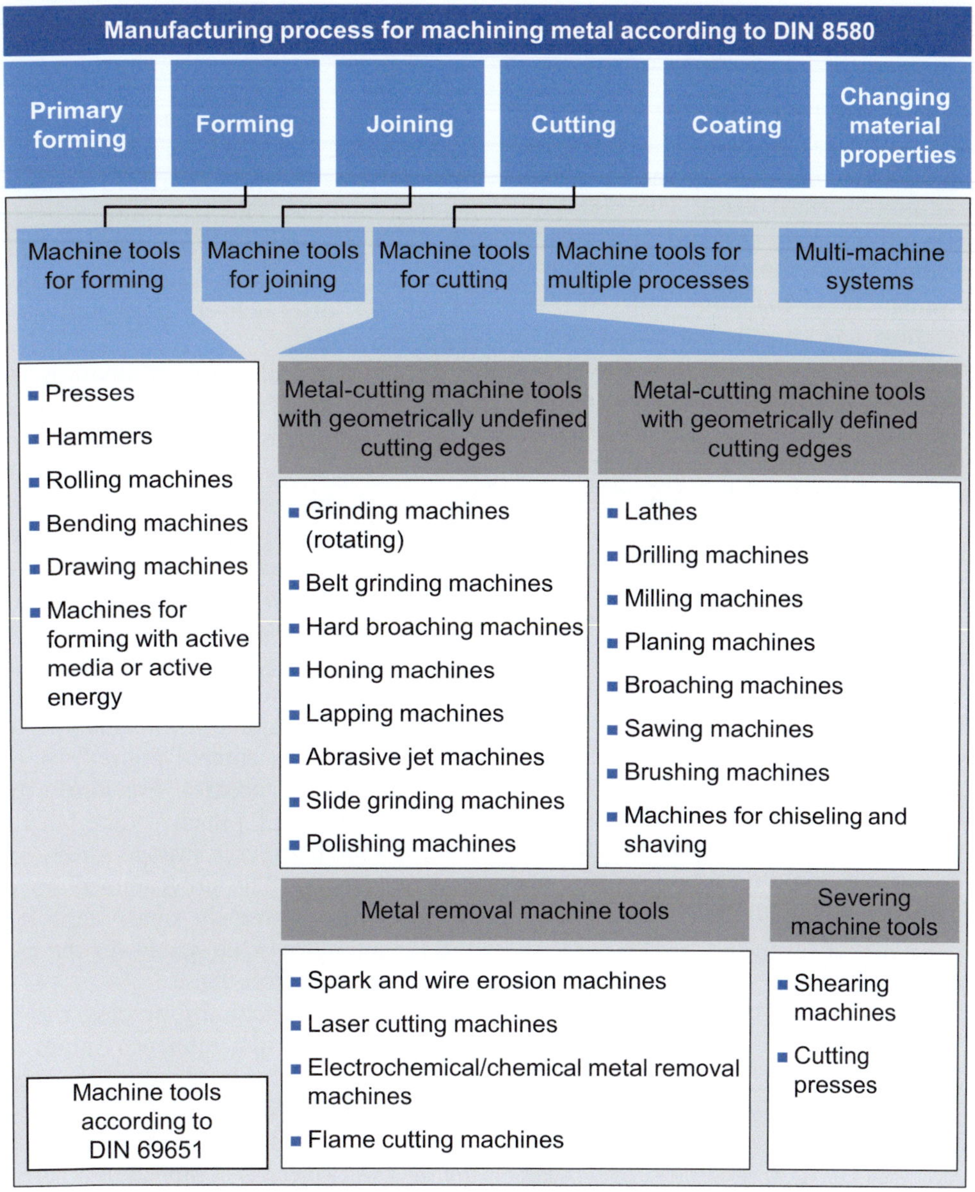

Fig. 1.11 Classification of machine tools according to DIN 69651 [DIN81]

Stand-alone machines are designated by the manufacturing process. If they can execute only one process, they are designated by this process, e.g., lathe.

However, if they can also execute other processes, they are designated by the name of the process for which they are primarily intended and equipped. For example, a milling machine is usually capable of drilling as well.

If a machine can execute several manufacturing processes the term "processing" is placed before the designation and the process group is attached as an attributive supplement, e.g., processing machine for turned parts, prismatic parts, or sheet metal parts.

The single-purpose machines included in the group of stand-alone machines are suitable for one specific machining task only and not for universal use. They are designated according to the workpiece to be produced and the manufacturing process being used, e.g., bevel gear planing machine or saw blade grinding machine.

Machine tools are frequently designated by the design to make a further distinction between them, e.g., knee-type milling machine, portal milling machine, multi-spindle lathe, slant-bed lathe, or double column-mounted drill.

As already mentioned, the more accurate designation of production systems, which includes both stand-

1

alone machines and multi-machine systems, is driven by the degree of automation. ◘ Figure 1.12 illustrates the designations.

The figure has a layered structure. At its core are machines together with their mechanical systems and the drives for producing the working motion (main spindle) and feed motion (feed axes). The next layer contains the NC control system required for the automatic and programmable sequencing of the machine functions. This makes these machines into NC machines. If the machine has an automatic tool change device and a tool magazine system, it is designated a machining center. In a flexible manufacturing cell, a handling system (e.g., a robot) is used to load one or more machining centers from a workpiece magazine. Larger plants in which several machining centers are linked to a material store and sometimes to a tool store by a flexible transport system are known as flexible manufacturing systems (FMS). Automated multi-machine systems, which are frequently less flexible, are used in large volume production, e.g., rotary transfer machines or transfer lines.

The latter consist of interlinked machining systems that today increasingly include NC-controlled machine units or machining centers. These perform one or more subfunctions of a machine tool and are linked together by a guided material flow system. A more accurate explanation of the various multi-machine systems and their application areas can be found in ▶ Chap. 9.

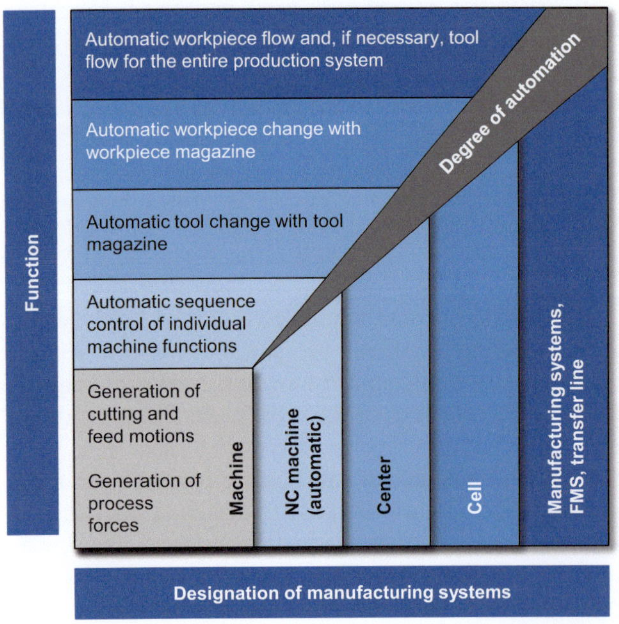

◘ **Fig. 1.12** Designation of manufacturing systems by the degree of automation

1.5 Basic Principles of Machine Tool Automation

Automation technology has made a significant contribution to the increase in productivity, flexibility, manufacturing quality and cost-effectiveness of modern production facilities. This has resulted in shorter throughput times, lower capital commitment costs, lower labor costs and increased utilization of production potential, as well as shorter delivery times and lower manufacturing costs.

In the early days of automation technology, machine tools were fitted with mechanical flow control systems. The individual functions were controlled by levers, cams, or cam disks. The use of hydraulic post-forming devices allowed complex shapes to be machined economically, particularly in larger quantities.

The first numerically controlled machine tool was put into operation in 1953 at MIT (Massachusetts Institute of Technology). NC control systems automatically execute coordinate movements and control commands in production machines. These are provided in the form of alphanumeric information.

Information is entered digitally, decoded and processed in the NC control. Previously this information was in the form of punched tape but today it is from conventional computer storage media or via a connection to a server. The programmed traversing motions are split into extremely small steps by interpolation and the reference value signals for the movement of the axes are created from these.

During the machining process, the travel distances are compared to the reference values in the position control loops and this generates the trigger signals for the drives [WERN83].

The NC information for the automatic manufacturing of a workpiece is compiled in a parts program. The parts program is divided into individual NC lines with a format defined in DIN 66025 [DIN83]. The control data record consists of alphanumeric characters conforming to ISO (International Standards Organization) or EIA code (Electronic Industries Association), ◘ Fig. 1.13.

At the start of the 1970s, the individual sequences for the machine functions were still implemented using hard-wired logical operations. With the development of the field of semiconductor technology, CNC (Computerized Numerical Control) control appeared on the market in the mid-70s. Here, the logical operations were implemented by a digital computer using appropriate software. In these control systems, subprograms can be entered via a control panel, modified, and saved in a memory. As this development

Example of an NC line according to DIN 66025

N1 G54 Z50	Zero point offset
N2 T10 S1000 M06 M07 M04	Tool change, rotational speed: 1,000 min^{-1}, spindle on, coolant on
N3 G46 G90	Absolute dimensioning, Cutting radius compensation "on"
N4 G00 X75 Z45	Positioning
N5 G01 X29 F4000	Planing
N6 G01 Z46	Lifting
N7 G00 Z100	Retraction
N8 M30	End of program

Elements of NC control commands according to DIN 66025

NC – axes	
A, B, C	Angular motion around X-, Y-, Z-axis
D, E	Additional angle
X, Y, Z	Main motion in X-, Y-, Z-direction
U, V, W	2nd motion X-, Y-, Z-direction
P, Q, R	3rd motion X-, Y-, Z-direction

Organizational characters	
%	Program start
N	NC line number
:	Main NC line
/	Skip NC line
LF	End of NC line
DEL	Delete

Technological and geometrical data	
F	Feed
E	2nd feed
D	3rd feed
G	Path condition
I, J, K	Interpolation functions
S	Spindle speed
T	Tool number

Special functions	
H, L	Freely available (e.g., coolant supply, number of runs for a subroutine)
N	Additional function (e.g., direction of spindle rotation)

Fig. 1.13 NC information according to DIN 66025 [DIN83]

continued, the storage media for the control information changed from NC punched tape, magnetic cassettes and floppy disks, to a connection to an external computer using a DNC (Direct Numerical Control) interface [WERN83].

Only control systems based on microcomputers have been offered since the beginning of the 1980s, so a distinction between NC and CNC is no longer necessary. For this reason, the term "NC" is used in the following text for computer control systems in common use today.

In addition to the pure entry and modification of program commands, today's machine control systems offer a series of software-supported auxiliary functions that simplify the production of parts programs. Many functions and programming tools have been adopted from the closely associated fields of information processing so that NC control systems offer more programming convenience than high-level programming languages [PETE01]. Graphic support and macro-programming to generate parts programs are worthy of specific note here.

Figure 1.14 shows the control panel from a modern NC control system with a few geometrical programming support functions.

This development in NC technology was an important condition for modern, automated production. Today, hardly any manually operated machines are built

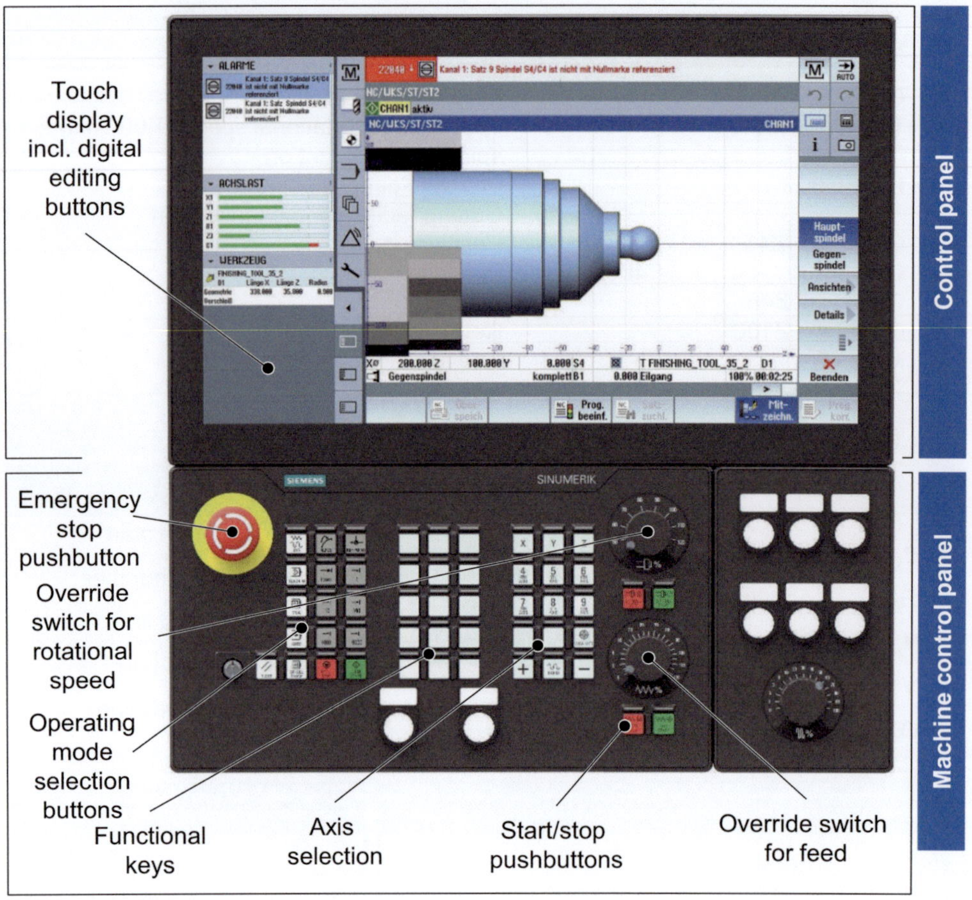

Fig. 1.14 Control and machine control panel on an NC system. *Source* Siemens

in Germany in the field of universal machines. However, they are still used in repair shops.

The use of microprocessors in NC technology opened up completely new possibilities in terms of data processing and integration. As part of the increasing effort to implement electronic data processing throughout all areas of production facilities, the arrival of NC machines made it possible to integrate these units into the plant's data network as equipment that can both understand and implement data. This capability makes NC machines one of the essential building blocks of a modern factory.

The many and varied options provided by control technology and information processing inevitably produced new machine concepts as well. Machine tools suitable for automation were developed and these were ready for use in automated manufacturing systems, see Fig. 1.15. Tool and workpiece magazines, automatic changing devices, and integrated measuring stations now carry out tasks that were previously done manually. Sensors for monitoring machine functions and process states, such as tool wear and tool fracture, ensure manufacturing runs automatically. Figure 1.15 shows the most important components, assemblies, and properties of a milling center.

Flexible manufacturing systems consist of one or more machining stations with a shared workpiece supply, all coordinated by a cell computer.

Figure 1.16 shows the information and material flow in this type of manufacturing system. The production orders to be completed in a flexible manufacturing system are defined during production planning. The order data are available from a production planning and control system and are transferred to the master computer. The master computer is responsible for the automated execution of the production orders. Some master computers also initially optimize the order of the specified production orders to avoid unnecessary setup processes or to make complete use of the remaining tool service lives. To perform the orders, they are broken down into individual operations that are transferred to the control systems of the devices in the manufacturing system according to the timing specifications. Feedback is provided to the master computer from the control system level and this can be evaluated to monitor the progress of the manufacturing process

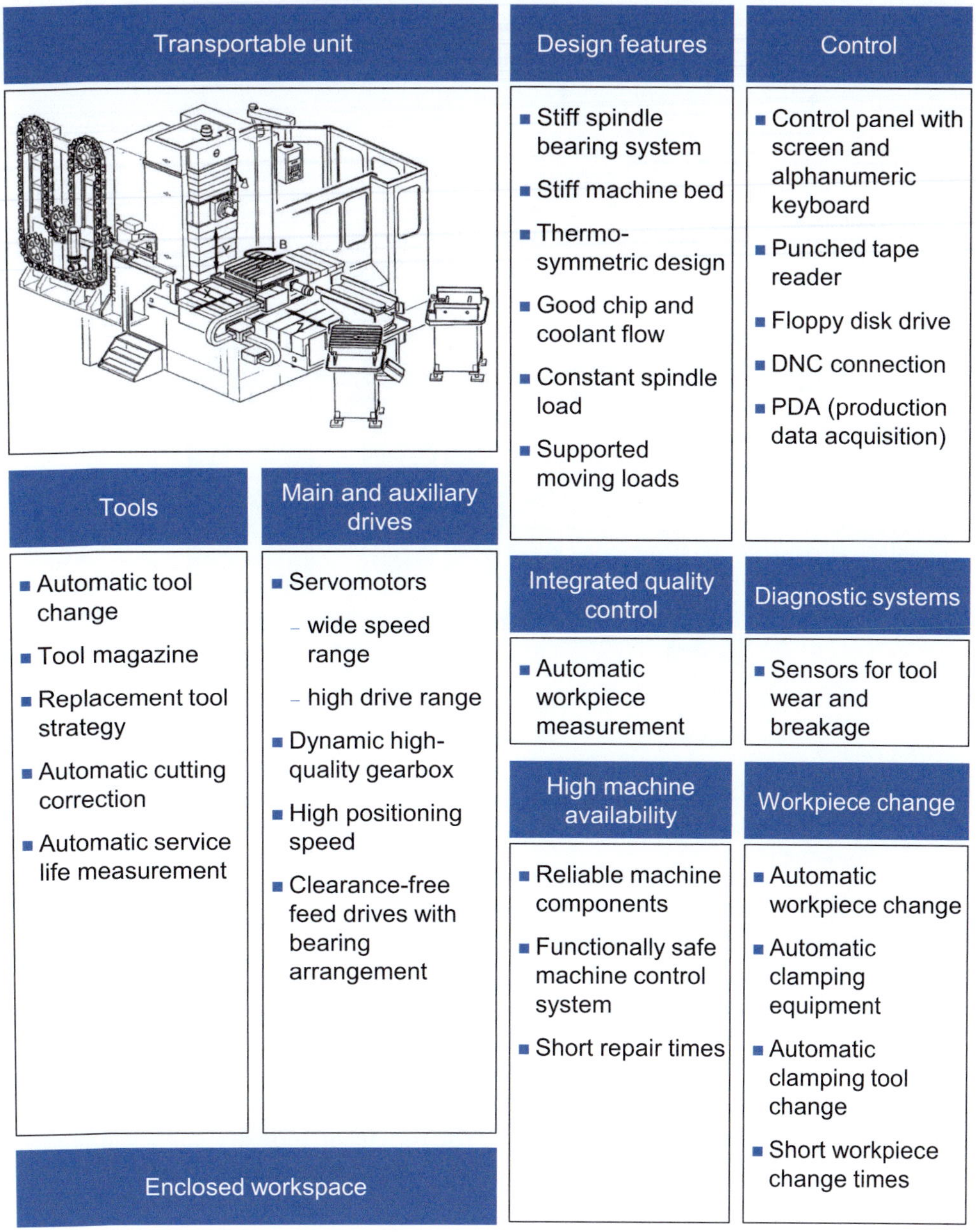

Fig. 1.15 Components and properties of a machining center

and can be made available to the plant operator if necessary.

Flexible manufacturing systems consist of several stations such as flexible manufacturing or assembly cells, parts storage, and transport systems. These are controlled by a central production control computer.

Figure 1.17 shows one possible control structure using the example of a flexible, integrated manufacturing and assembly system (IMAS). The control structure is divided into three distinct layers: the command, cell, and machine levels. At the command level, the overall sequence and the material flows are defined at the order planning stage as is the coordination of the execution of the planned production orders. At the cell level, the part orders received from the command level are autonomously processed. This includes initiating and monitoring the material flow, manufacturing and assembly processes. Machines, robots, and transport devices are interlinked at the machine level. The DNC interfaces for the relevant device controls can be addressed

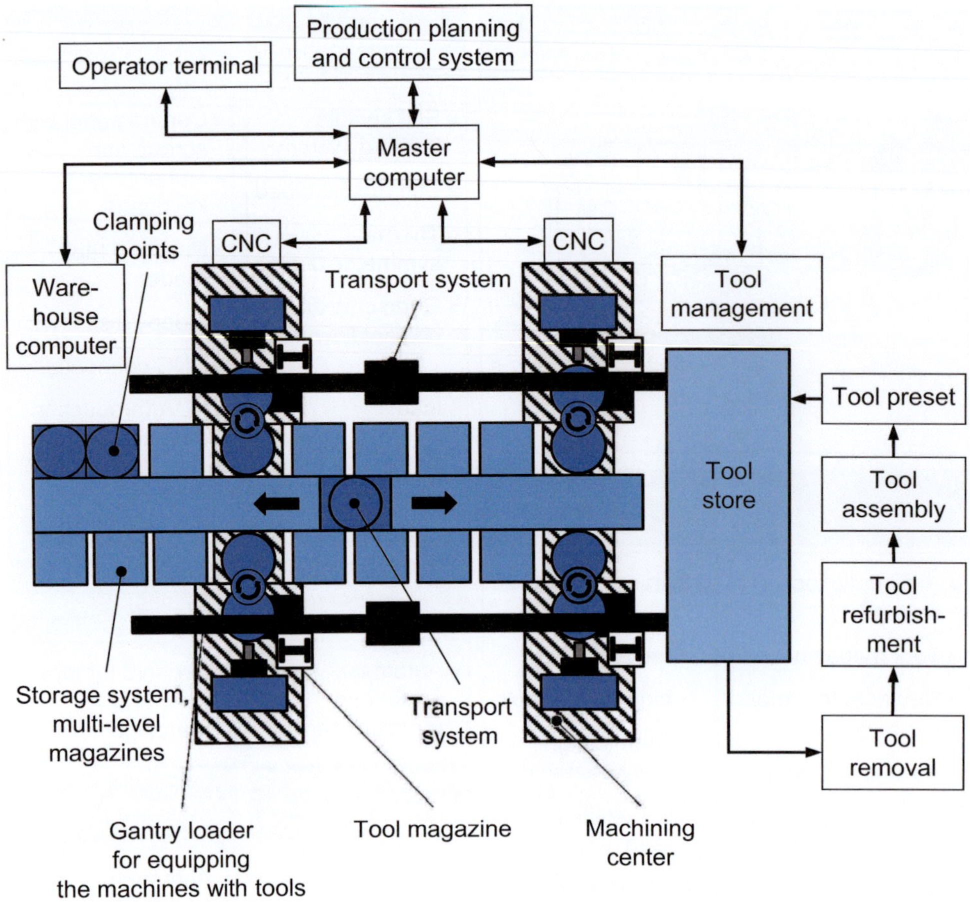

Fig. 1.16 Information and material flow in a flexible manufacturing system

via appropriate driver modules to trigger the necessary physical operations (manufacturing, transportation, handling, etc.) [PETE01].

Planning, decision-making, and actions within every area of a business are all based on information. The use of EDP systems has introduced working methods that efficiently process and manage the information at every level of the business and quickly provide the user with the correct information.

However, the increasing number of stand-alone EDP solutions has led to many disjunctions in the flow of information in the business. Efforts to develop an overarching integration concept for information processing in production companies were summarized in the 1980s under the term Computer-Integrated Manufacturing (CIM). Its primary task was to unite Enterprise Resource Planning with CAx technologies. [BREC14]

However, suitable technology was not available at the time and attempts were made to control all instances centrally. The vision at the time did not consider people as the focal point of a factory. The introduction of new technologies such as the Internet, WLAN, high-performance computing and Big Data has led to fundamentally new possibilities in networking and data processing and to the fourth industrial revolution (Industry 4.0). The benefit of Industry 4.0 lies in highly flexible, adaptable value-added systems which allow value added to be transferred to those parts of the business where it can be most beneficially implemented. This lowers the complexity costs, particularly in individualized production and this generates new economic potential. To this end, all value-added partners, and customers as well, are integrated and networked into the value-added systems with minimal hierarchy. The integration of employees through adaptive, self-teaching human-machine interfaces ensures widespread acceptance of the working system and creates the basis for agile, robust, and learning value-added systems. [BAUE16]

The control aspects of machine tools (NC control) and the manufacturing cells and flexible manufacturing systems (control stations) are dealt with in detail in **Volume 3 of this compendium.**

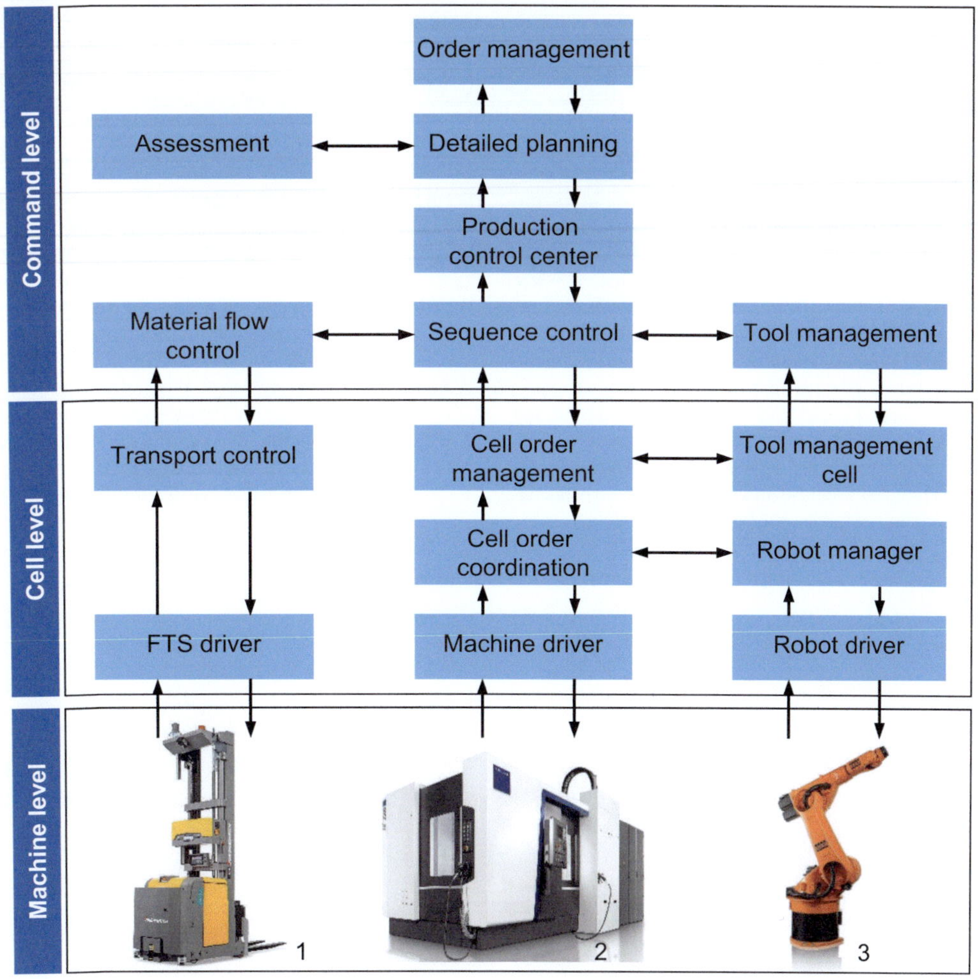

Fig. 1.17 Software modules and control structure in an IMAS. *Source* 1) Jungheinrich, 2) Heller, 3) KUKA

References

[BAIN16] Bain & Company: Go West! Wie Unternehmen von der Reindustrialisierung der USA profitieren können. ► http://www.bain.de/Images/Bain-Studie_Reindustrialisierung_USA_FINAL_ES.pdf. [Stand: 30.07.2016]

[BAUE16] Bauernhansl, T., Krüger, J., Reinhart, G., Schuh, G.: WGP-Standpunkt Industrie 4.0, Wissenschaftliche Gesellschaft für Produktionstechnik (2016)

[BMWI16] BMWi—Bundesministerium für Wirtschaft und Energie: Strukturelle Entwicklungen. ► http://www.bmwi.de/DE/Themen/Industrie/Industrienation-Deutschland/strukturelle-entwicklungen.html. [Stand: 30.07.2016]

[BREC14] Brecher, C., Behnen, D., Brumm, M., Carl, C., Ecker, C., Herfs, W., Klement, R., Königs, M.: Virtualisierung und Vernetzung in Produktionssystemen. In: Brecher, C., Klocke, F., Schmitt, R., Schuh, G. (Hrsg.) Integrative Produktion, S. 35–68, Shaker, Aachen (2014)

[CIUP16] Ciupek, M.: Werkzeugmaschinen auf Wachstumskurs. ► http://www.vdi-nachrichten.com/Technik-Gesellschaft/Werkzeugmaschinen-Wachstumskurs. [Stand: 30.07.2016]

[DIN03] Norm DIN 8580 (September 2003). Fertigungsverfahren - Begriffe, Einteilung

[DIN81] Norm DIN 69651 (März 1981). Werkzeugmaschinen für die Metallbearbeitung. Teil 1: Begriffe

[DIN83] Norm DIN 66025-1 (Januar 1983). Programmaufbau für numerisch gesteuerte Arbeitsmaschinen; Allgemeines

[DREG16] Dreger, C.: Ist China eine Marktwirtschaft? ► https://www.diw.de/documents/publikationen/73/diw_01.c.534441.de/16-21-4.pdf. [Stand: 30.07.2016]

[HEKE85] Hekeler, M.: In Stufen zur flexiblen Drehzelle durch Automatisierung von Werkstück- Werzeug- und Spanbackenhandhabung. TZ für die Metallbearbeitung. 79(9) (1985)

[KÖHN16] Köhn, R.: Ohne die „Mutter aller Maschinen“ gäbe es keine Autos. ► http://www.faz.net/aktuell/wirtschaft/unternehmen/branchen-10-werkzeugmaschinenbau-ohne-die-mutter-aller-maschinen-gaebe-es-keine-autos-1256459.html. [Stand: 30.07.2016]

[PETE01] Peters, A.: Benutzerorientierte Leitsoftware für automatisierte Fertigungssysteme in Komponentenstruktur. Diss. RWTH Aachen (2001)

[PWC16] PwC—PricewaterhouseCoopers: The world in 2050—Will the shift in global economic power continue? ► http://www.pwc.com/gx/en/issues/the-economy/assets/world-in-2050-february-2015.pdf. [Stand: 30.07.2016]

[ROE26] Roe, J. W.: English and American Tool Builders. 2. Aufl. Mc Graw-Hill, New York (1926)

[SPRI41] Springer, W.: Der Weg zur modernen Bohrmaschine. H. Kolb Werkzeugmaschinenfabrik, Köln (1941)

[STAT16a] Statistisches Bundesamt: Japan—Statistische Länderprofile G20 Industrie- und Schwellenländer. ▶ https://www.destatis.de/DE/Publikationen/Thematisch/Internationales/Laenderprofile/Japan2013.pdf?__blob=publicationFile. [Stand: 23.09.2016]

[STAT16b] Statistisches Bundesamt: Anteil der Industrie am BIP seit 20 Jahren nahezu konstant. ▶ https://www.destatis.de/DE/PresseService/Presse/Pressemitteilungen/2015/04/PD15_124_811pdf.pdf;jsessionid=FEDEF99E03095D09A3DF7660A56C5FA5.cae3?__blob=-publicationFile. [Stand: 30.07.2016]

[VCI16] VCI—Verband der chemischen Industrie: Industrieland Deutschland—Wettbewerbsfähigkeit der Chemie stärken. ▶ https://www.vci.de/vci/downloads-vci/top-thema/daten-fakten-industrieland-deutschland-wettbewerbsfaehigkeit-der-chemie-staerken.pdf. [Stand: 30.07.2016]

[VDMA15] VDMA—Verband Deutscher Maschinen- und Anlagenbau e. V.: Statistisches Handbuch für den Maschinenbau—Ausgabe 2015. VDMA, Frankfurt a. M. (2015)

[VDW16] VDW—Verein Deutscher Werkzeugmaschinenfabriken e. V.: Marktbericht 2017 (2017)

[VDW17] VDW—Verein Deutscher Werkzeugmaschinenfabriken e. V.: Marktbericht 2017 (2017)

[WERN83] Werner, P., Schmid, R.: Möglichkeiten zur Lösung von Fertigungsaufgaben mit Hilfe von NC-Steuerungen. (Reihe: Trenn-Kompendium. Bd. 2). Bergisch Gladbach: ETF Edition Technischer Fachinformationen (1983)

[WITT60] Wittmann, K.: Die Entwicklung der Drehbank bis zum Jahr 1939, 2nd edn. VDI-Verlag, Düsseldorf (1960)

[WZL86] WZL: 80 Jahre WZL. Innovation aus Tradition. Impulse für die Produktionstechnik. Greven, Köln (1986)

[ZEPP86] Zeppelin, W.: von: Informationstechnische Einbindung einer NC-Werkzeugmaschine in eine rechnergestützte Betriebsorganisation. ZWF. **81**(11), (1986)

Primary Forming Machines and Systems

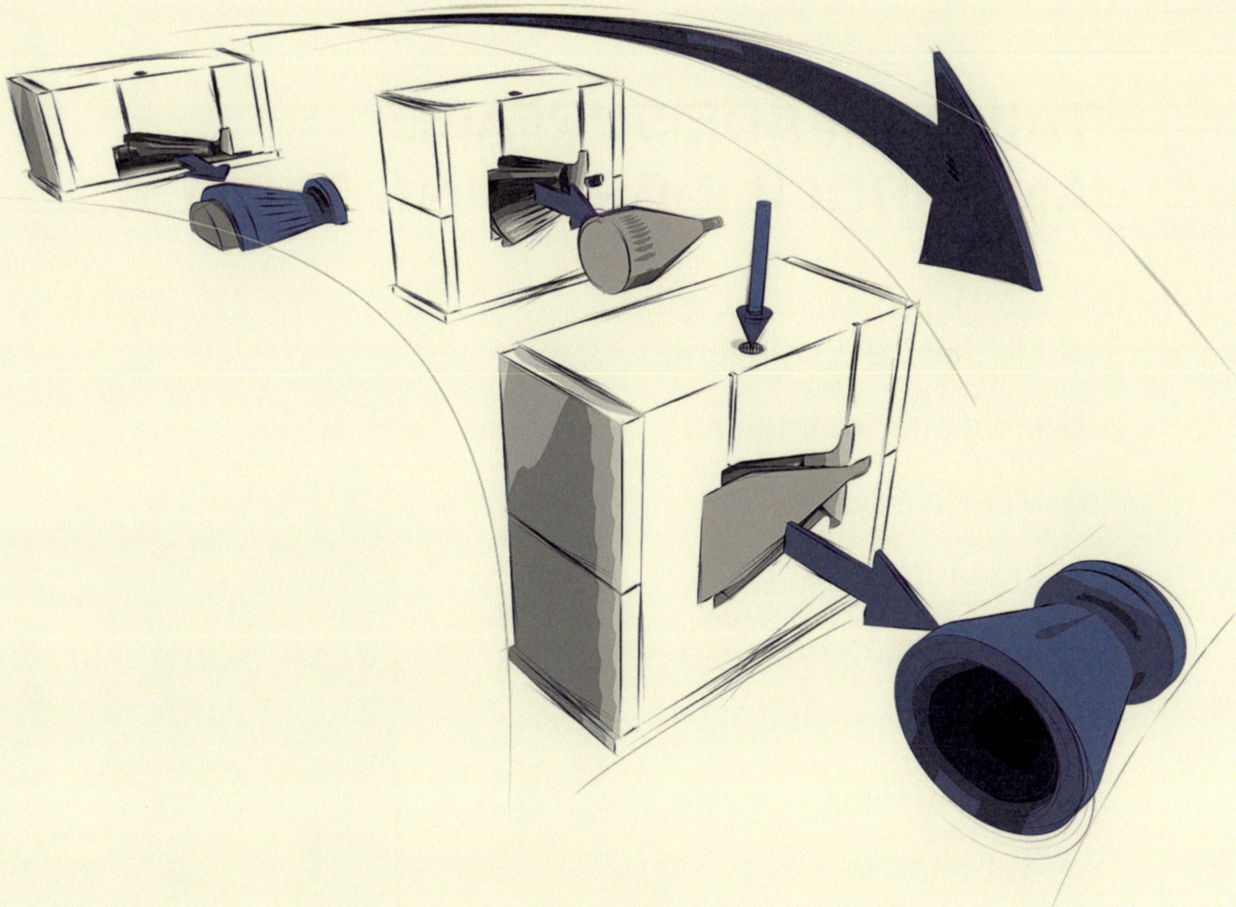

I thank my staff members Gabriel Axelrad M.Sc., Mathias Dehn M.Sc. and Oscar Malinowski M.Sc. for their contribution in revising this chapter.

C. Brecher and M. Weck, *Machine Tools Production Systems 1*,
Lecture Notes in Production Engineering, https://doi.org/10.1007/978-3-662-68120-6_2

Contents

The term primary forming is used to describe the production of a solid body of a specified shape from a material which is present in a formless initial state in liquid or powder form. During the primary forming process, the material changes its state, and it retains its solid cohesion over the entire workpiece volume.

The different primary forming methods classified in DIN 8580 are shown in ◘ Fig. 2.1 [DIN03]. The methods relevant for manufacturing metallic molded parts are assigned to groups "Primary forming from the liquid state" and "Primary forming from the granular or powdery" state. If the shapeless initial state is the liquid phase of a metal and if the cohesion of this material is achieved by the solidification process in the shape-defining mold, this is referred to as primary forming from the liquid state. Commonly used processes are gravity, high-pressure, centrifugal and continuous casting. Another popular primary forming method for metallic

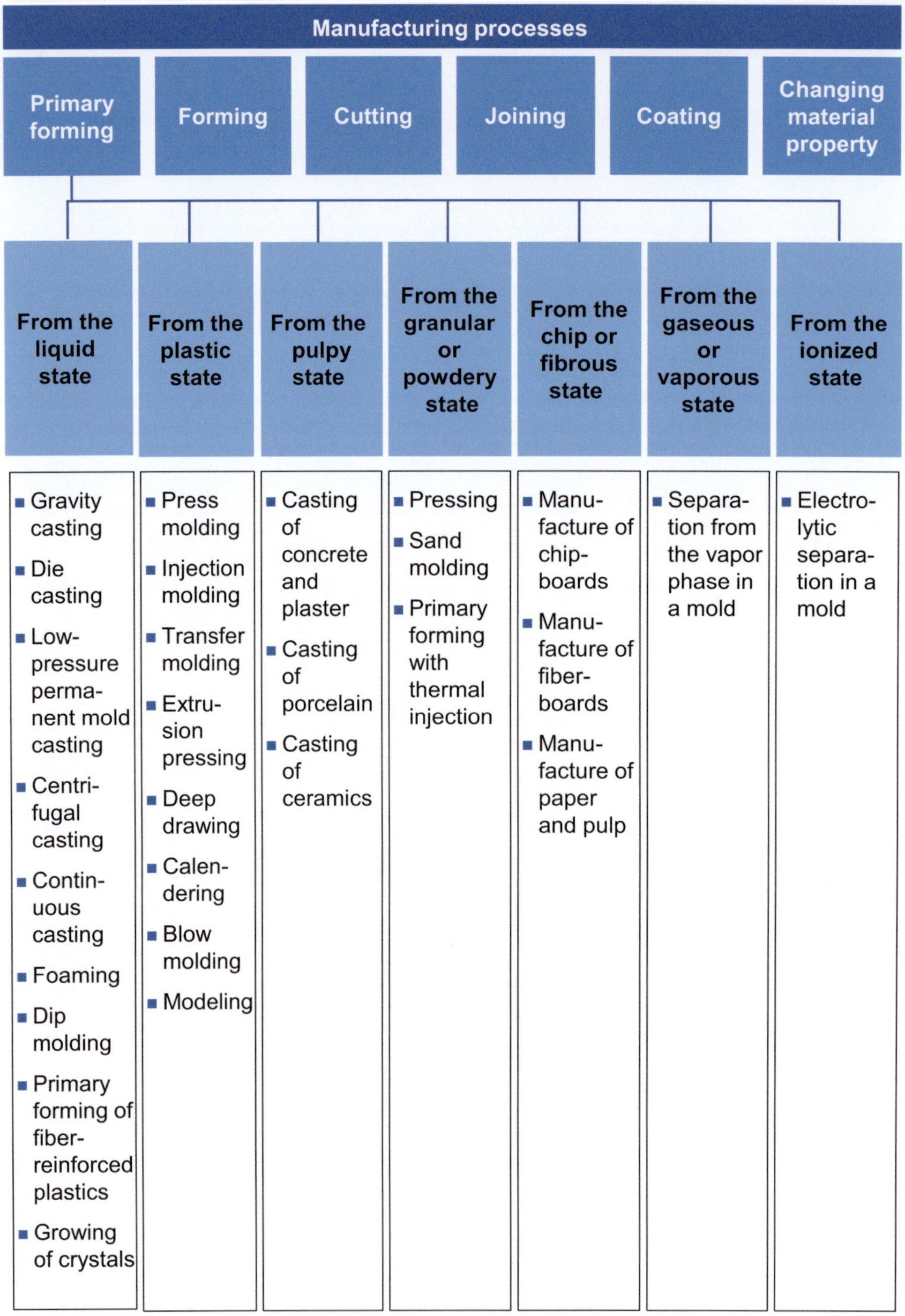

◘ **Fig. 2.1** Categorization of primary forming methods. *Source* [DIN03]

molded parts is pressing. In this process, a molded part is made from granular or powder materials. The powder is pressed into the required shape using high pressure, which activates the grain boundaries. The diffusion and crystallization annealing processes cause the pressed blank (also called green body) to merge and form a high-strength molded part. The subsequent temperature treatment of the green body is called sintering.

Due to the increasing shortage of raw materials and energy, the primary forming processes, as well as other forming processes (see ▶ Chap. 3) are becoming increasingly important. The benefit of primary forming methods is virtually complete freedom in the design and molding of workpieces while producing minimum workpiece waste (such as chips). Applications range from customer-specific single-part production to mass production of automotive components in the millions. This allows the manufacture of components weighing a few grams to several hundred tons for the heavy-duty machinery sector [BÜHR14].

In addition to the categorization of primary forming methods as per DIN 8580, casting methods can be differentiated using the classic casting methods with permanent molds and expendable molds (◘ Fig. 2.2). Expendable molds are only used once. Such a mold is typically made from a granular mold material, a binding agent and additives and is broken up after the casting process. Permanent molds are usually made of metallic materials and can be used for repeated castings [BÜHR14].

The basic process from design to finished cast part is shown in ◘ Fig. 2.3. The process sequence for casting can be divided into the phases of preparation, molding, casting and post-processing. After preparation, the casting mold must be made. The path on the left in ◘ Fig. 2.3 shows the process for producing expendable molds using reusable models. In contrast to manufacture using permanent molds, production with expendable molds is frequently integrated into the process at the foundry (refer to ▶ Sect. 2.2). After the actual primary forming process comprising casting and solidification of the fused material, the cast part is removed, and post-processing is performed. Here, gates, feeders, risers and burrs arising from inaccuracies in the casting mold are removed from the cast part. It is then sandblasted and heat-treated as required. After the final inspection, the cast part is approved for subsequent use or machining.

2.1 Cost-Effective Machining of Cast Parts

Since the end product in the casting process solidifies directly from the liquid melt, casting is one of the most cost-effective manufacturing processes. This presupposes that the finishing process for a cast part involves only little effort. Minimizing the machining allowances for functional surfaces can guarantee a cost-effective cutting process saving time and money.

For complex components, the rough sizes often differ greatly from nominal dimensions because shrinkage prevention when cooling with the cores cannot be predicted accurately despite using the latest computer programs. As a result, an inexpensive, quickly available preliminary model is first deployed for volume production. This model is then optimized for volume production in line with experiences from the production of the initial parts, initial installation at the customer, findings from the market launch and subsequent testing of the product. Among other things, the costs of fettling and machining allowances are minimized systematically, and the required long-term accuracy of the model is

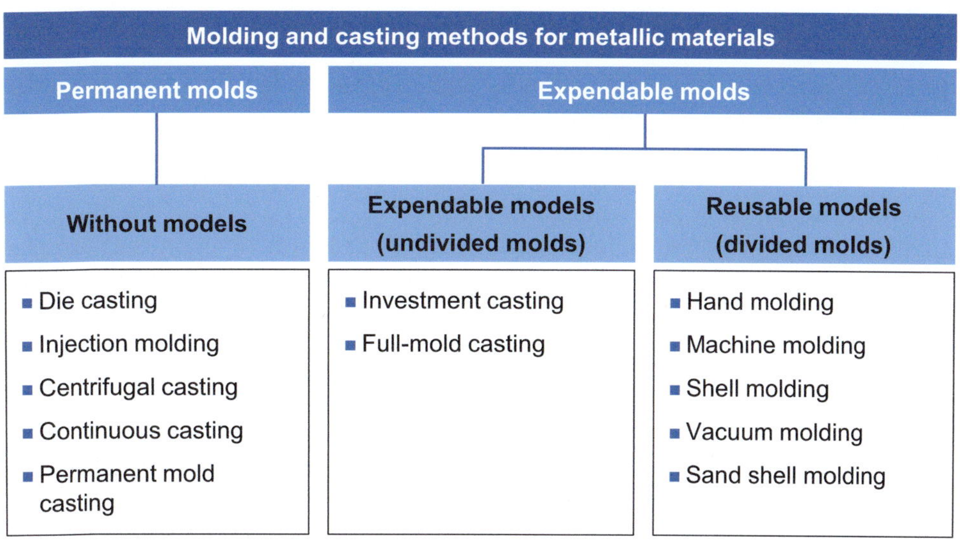

◘ **Fig. 2.2** Forming and casting methods

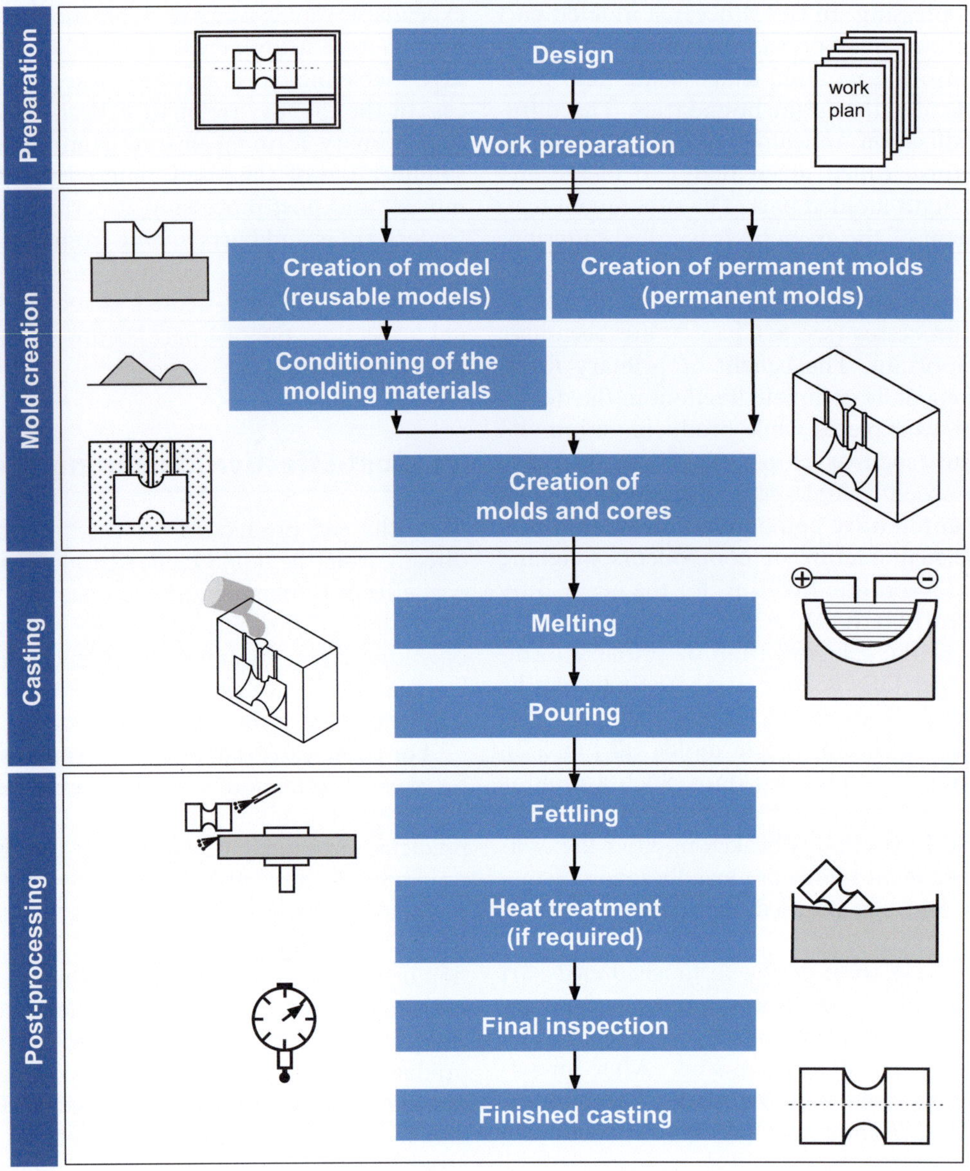

Fig. 2.3 Process sequence for casting

guaranteed. Experience shows that the additional cost for such step-by-step optimization is low compared to the enormous time savings in finishing processes and risk minimization.

Figure 2.4 shows a column for a gear cutting machine and some critical position dimensions. The chart compares the maximum deviations to be met as defined in DIN EN ISO 8062-3 and the actual values currently possible following a model optimization. It is evident that the deviations can be reduced to 20% of the tightest dimensional tolerances for machine-molded and hand-molded parts following the specified DIN standard. The narrow spread of the actual value assures adequate process reliability, which enables machining without supervision.

2.2 Systems for Expendable-Mold Casting

Sand casting is one of the methods for expendable-mold casting. Figure 2.5 shows the manufacture of a split mold for a rotationally symmetric cast part. Both halves of the model are placed in a molding flask and filled with molding material, a mixture of silica sand and a binding agent. The molding material is compacted by means of shaking or pressing. To produce the top half of the mold, additional models for the gating system and feeders are used. The core is inserted into the finished mold halves and the mold is closed. After the casting process, the cast part is removed by breaking the mold.

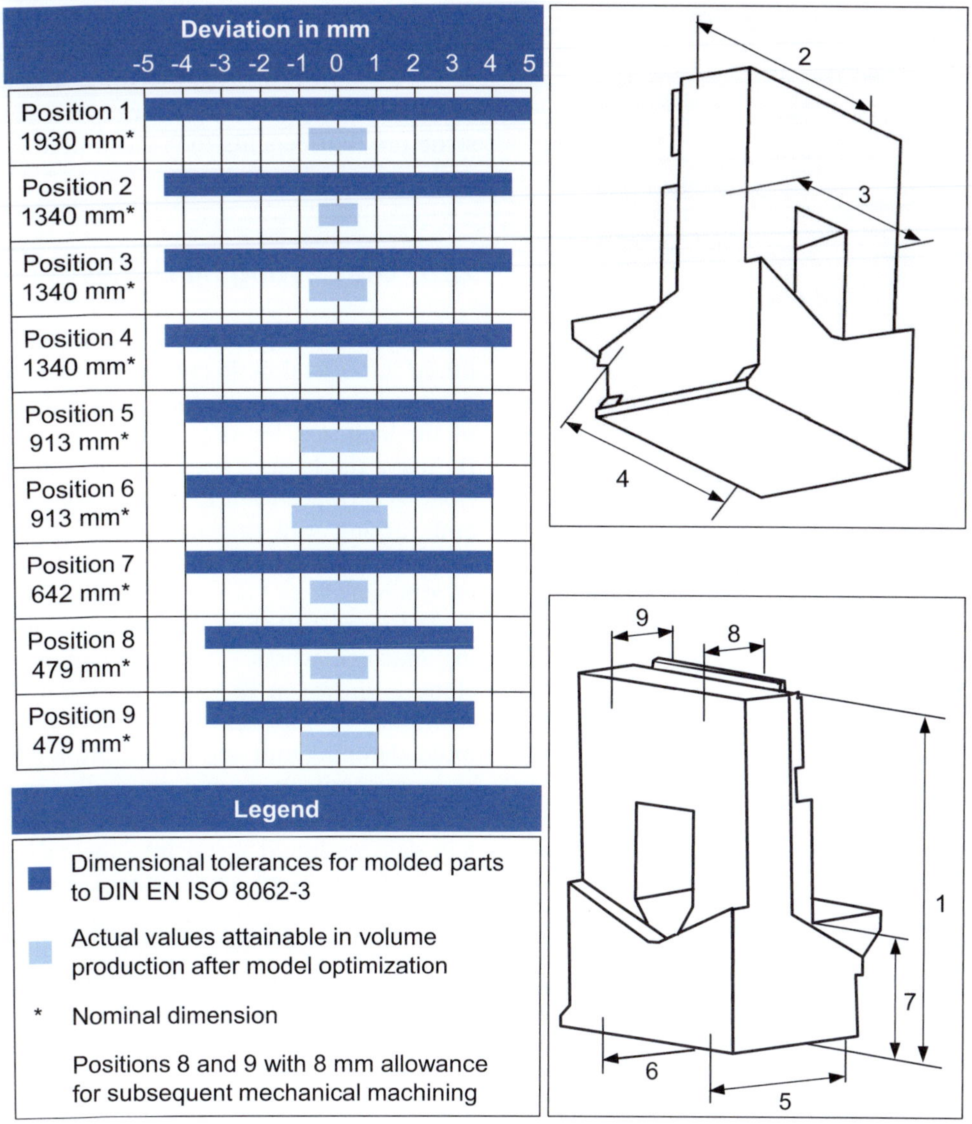

Fig. 2.4 Manufacturing spreads for a column of a gear cutting machine, according to Heidenreich & Harbeck

In the past, foundry production using sand molds was mainly characterized by manual activities. To increase productivity, attempts are being made to broadly automate the handling of molds, cores and molding materials. Such an automated casting machine is shown in Fig. 2.6. Molding machines are used to make the flask shape so that the casting mold is left after the model plate is raised. In terms of the technology used, the material is compacted by shaking, pressing or by using the impulse compaction method.

Cast workpieces are often removed from the mold using an oscillating cylinder (on the left in Fig. 2.7). The molding flasks are emptied into the spring-mounted steel cylinder. Directional exciters positioned perpendicular to the cylinder axis cause the cylinder to oscillate such that the mold breaks away and the cast part is transported to the cylinder outlet by the flow of sand. A downstream grating can be used to separate the cast part from the molding sand. This type of cylinder is suitable mainly for the mass production of smaller parts. These cylinders can separate sand from 5 to 20 metric tons of cast parts per hour. Crusher equipment is used to remove smaller volumes and large cast products from molds (on the right in Fig. 2.7). In the process, eccentric shafts normally cause a separation grate to oscillate.

Sand reconditioning systems are used to recondition molding sand (Fig. 2.8). Used sand, coming from a station for emptying cast parts (such as machine frames), is ground in a spiral crusher and, following separation (dedusting) and cooling, is mixed with new sand to then be used again to create the mold.

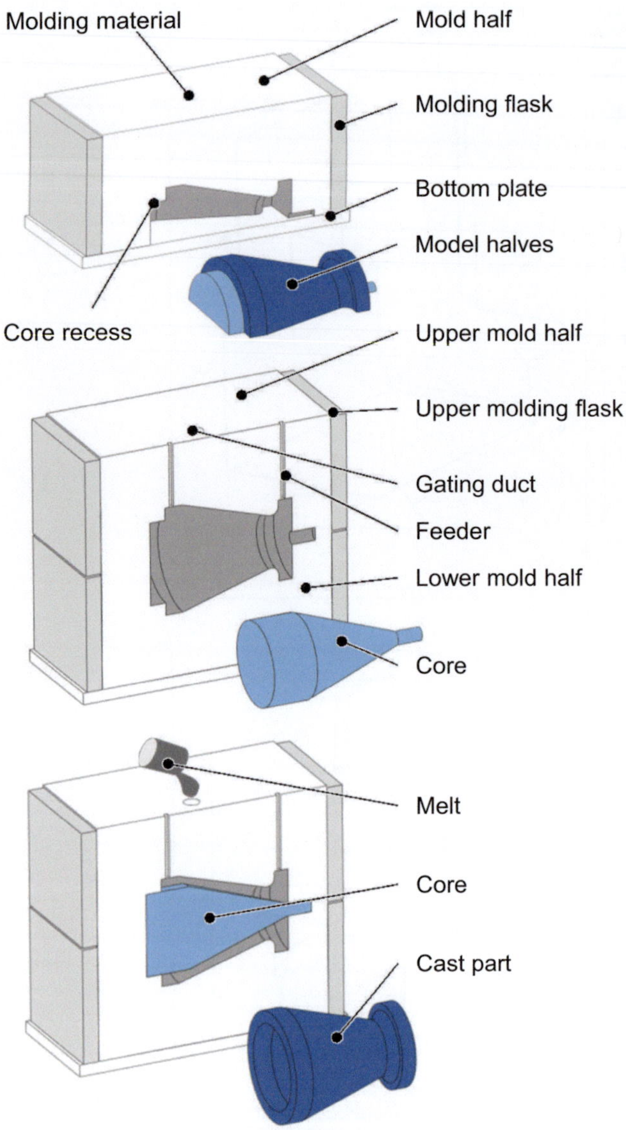

Fig. 2.5 Sand casting manufacturing with expendable molds, according to Aixonix

2.3 Systems for Casting Using Permanent Molds

2.3.1 Centrifugal Casting Machines

The material properties of a workpiece depend, among other factors, on the density of the structure and consequently on the pressure the liquid metal is under during the solidification process. For gravity casting, only the casting height has an influence on the pressure in the liquid metal. For the centrifugal casting of tubular workpieces, the liquid pressure can be increased greatly by the effect of the centrifugal acceleration (i.e., a multiple of gravitational acceleration) of the rotating permanent mold. A centrifugal casting machine with a horizontal axis of rotation and preheated permanent mold, enabling controlled solidification, is shown in Fig. 2.9. In centrifugal casting, the continual effect of the centrifugal force gives rise to a dense and aligned structure that benefits the mechanical properties of the cast part.

2.3.2 Die Casting Machines

In the die casting method, the liquid metal is introduced into metal molds under pressure. The high liquid pressure is maintained during the solidification process of the melt, e.g. by means of a piston-cylinder system. By using the highest possible injection speed, it is possible to produce even angled molded parts with small and varying wall thicknesses. By using smooth mold surfaces, high surface qualities on the workpieces can be attained.

Figure 2.10 shows a schematic of how a die casting machine works. The design of a die casting machine is shown in Fig. 2.11. A die casting machine essentially comprises two functional modules—the mold clamping unit and the injection system. The mold clamping unit opens and closes the casting mold and generates the necessary locking force. The injection system delivers liquid metal into the casting mold at the required speed and applies the casting pressure necessary. The casting pressure is usually transmitted to the casting piston by a hydraulic drive cylinder [BÜHR14].

A distinction is made between hot and cold chamber machines. Using one type or the other is determined by the material to cast. In the cold chamber method in Fig. 2.10, the material to cast is melted in an external furnace and maintained at a hot temperature. A metering unit feeds the amount of molten metal necessary for the casting process to the casting chamber. For a hot chamber machine, the material to cast is stored in the immediate vicinity of the machine in a heated container. The casting chamber is usually located directly in the container filled with melt, meaning the chamber fills by itself when the plunger is retracted. This method can only be used for zinc, tin and lead alloys because the molten metals of these materials are not aggressive with respect to the normal container material (steel or cast iron). Brass, magnesium and aluminum materials are usually processed on cold chamber machines due to their higher melting temperature or their solubility compared to iron and steel. Nowadays, however, the hot chamber method is also being used for aluminum die casting. In such cases, the heavy-duty components of the hot chamber casting assembly coming into contact with aluminum are made of titanium diboride.

The casting pressure, which reaches about 500 to 1000 bar for average workpiece dimensions, attempts to press apart the two halves of the mold. The clamping

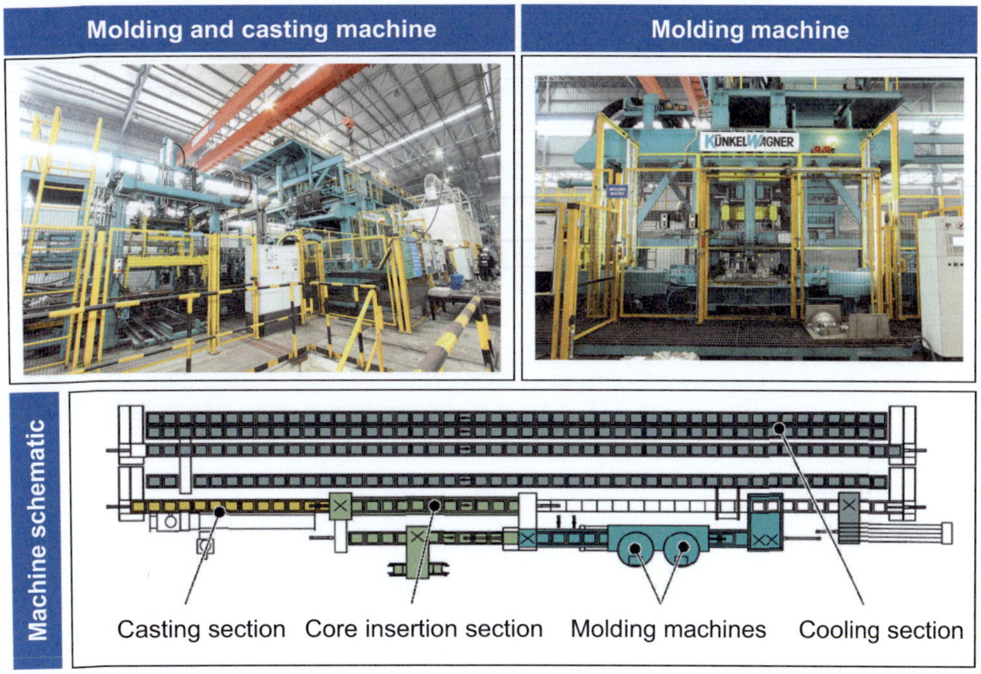

Fig. 2.6 Automated casting machine. *Source* Künkel Wagner

Fig. 2.7 Demolding systems. *Source* Küttner

unit of the die casting machine must absorb the resultant high forces. The closing mechanism on all die casting machines is actuated by a hydraulic closing cylinder. For small die casting machines, the actuator can be used directly for keeping the mold closed in a force-fit manner. For medium and large-sized designs, a transfer of force is necessary—that is usually realized in the form of a double toggle lever (Fig. 2.10). The machine is set in such a way that, when the toggle lever is in the elongated position, the mold halves are under a sufficient preload, meaning the casting pressure is unable to force the mold halves apart. Hot and cold chamber machines are built with clamping forces of up to 1000 MN and 45,000 MN respectively [BÜHR14].

A special form of die casting is thixocasting. Here, the primary material is semi-solid. Some of the material-forming phases are already liquid and other phases are still solid. The primary material thus exhibits thixotropic flow behavior, meaning the viscosity of the primary material decreases with increasing shear loading. There is a distinction between thixocasting and thixomolding. In thixocasting, the primary material is placed into the casting chamber of conventional cold chamber machines and pressed into the casting mold by the plunger. Thixomolding is similar to plastic injection molding. When being transported by an extruder screw, magnesium granulate is partially liquefied by heating and shearing processes [BÜHR14].

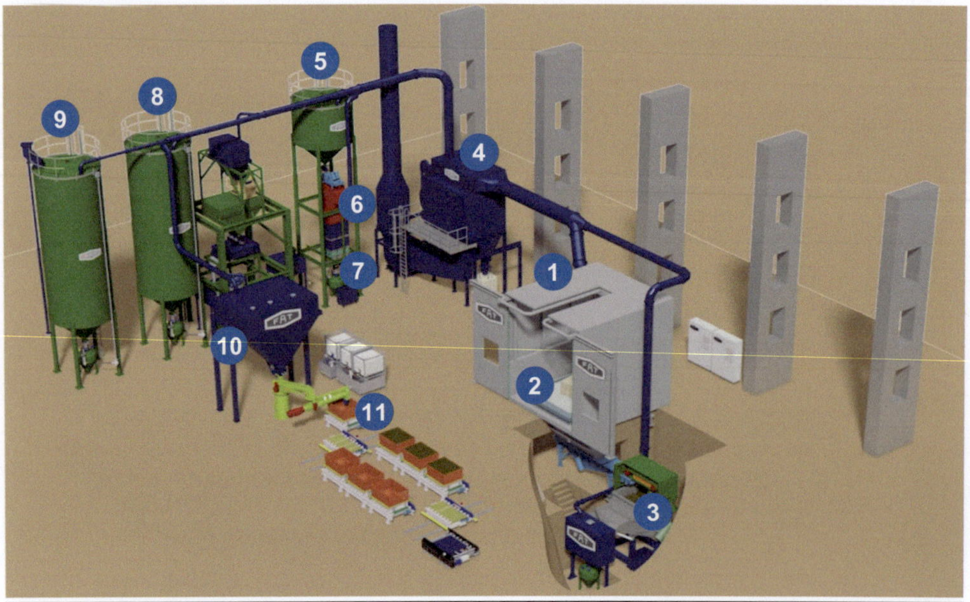

1	Molding flask emptying station	7	Sand cooler
2	Separation grate with spiral crusher	8	Silo for used sand
3	Pneumatic conveyor	9	Silo for new sand
4	Dust extraction system	10	Mixer
5	Silo for hot, used sand	11	Molds
6	Sand sieve		

Fig. 2.8 Mold system with sand reconditioning. *Source* FAT Förder- und Anlagentechnik GmbH

2.3.3 Continuous Casting Machines

For producing ingots and semi-finished sections, ingot casting and continuous casting are used followed by hot rolling. Classic ingot casting has some major downsides, however. These include low productivity and quality deficiencies in workpieces caused by segregation and cavities.

The principle of continuous casting is shown in ◻ Fig. 2.12. Melt is poured continually from the casting ladle via tundish into a water-cooled permanent mold having the cross section profile of the strand casting to be cast. The solidified metal exits the profiled permanent mold downwards and is cooled further by the secondary cooler.

The amount of material poured in, the heat loss in conjunction with the crystallization rate, and the drop-off rate of the solidifying material over time must be carefully synchronized with each other. Larger shrinkage cavities and segregations are prevented by continuous crystallization near the cooled profile permanent mold and with additional equipment, such as electromagnetic stirring (EMS) in the permanent mold and strand area as well as targeted weak thickness reduction (soft-reduction) in the final solidification area [KOBL00]. The drop-off rate of the strand is about 0.1 to 7 m/min depending on the dimensions of the strand cross section and the material.

The original design of vertical continuous casting installations had the subsystems configured in a straight line. Following the tundish are the permanent mold, the support and guide rollers (drivers) and the cutter (on the left in ◻ Fig. 2.12). For these vertical installations, the maximum strand length is restricted to about 12 m for structural reasons. The demand for greater strand dimensions has resulted in curved systems using various intermediate stations. Here, the solidifying or solidified but still ductile material is turned into the horizontal position, either in the permanent mold or after exiting it. As a result, any strand length desired can be generated in this continuous process (on the right in ◻ Fig. 2.12).

There are now processes (inline strip process) in which the rolled end profile can be produced "inline" with near net shape casting, i.e. without interrupting the process. With the CSP technology (compact strip production), a plant concept has been established that enables cost-effective generation of hot-rolled wide strips using a coupled process comprising casting and rolling (◻ Fig. 2.13). With a suitably preadjusted CSP hopper-type permanent mold, slabs ranging from 40 to 90 mm in thickness are produced [ROSE09].

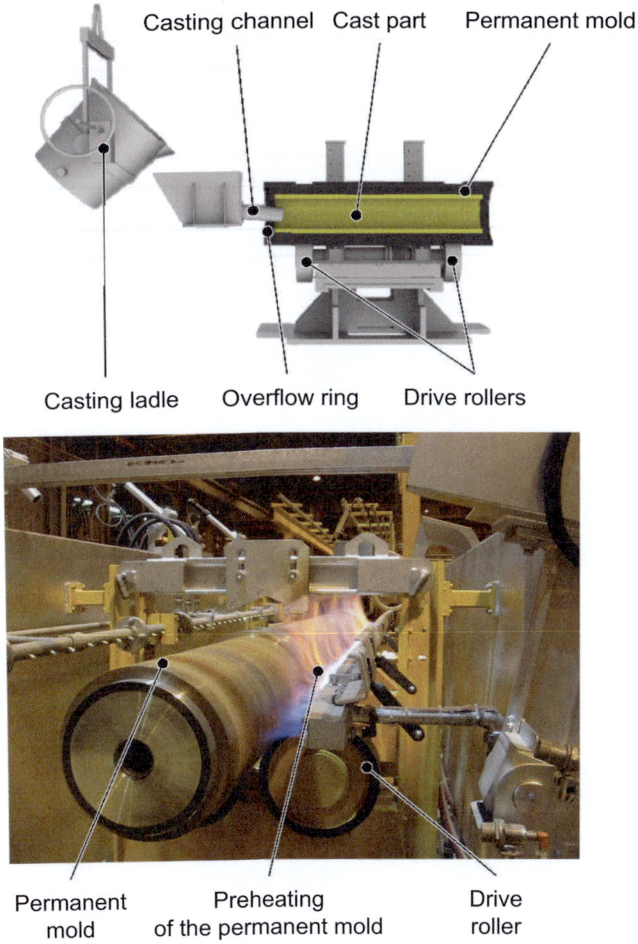

Fig. 2.9 Horizontal centrifugal casting installation. *Source* Küttner

Figure 2.13 shows the principle sketch of a CSP plant in which the thickness of the thin slab is continuously reduced by means of LCR technology (Liquid Core Reduction) to adapt to a corresponding rolling strategy with still liquid core below the mold. After complete solidification, temperature homogenization and final rolling are then performed by an appropriate number of rolling stands. A continuous casting plant is shown in Fig. 2.14. The continuous cast products are then further processed in separate steps, such as heating and rolling, to produce profile and flat products ready for sale. [ROSE09].

2.3.4 Systems for Fettling

To produce usable, dimensionally accurate cast parts, cast bodies need to be finished after the casting process. These finishing processes, grouped under the term fettling, comprise, among other tasks, removing sand from and descaling the cast body and removing gates, feeders, risers and burrs. Fettling is performed mainly by hand. Gates and feeder systems, for example, are removed from cast bodies manually using flame cutters. The cast bodies are deburred using the blasting method. Employees performing this work are exposed to high noise levels and considerable dust emissions. Given the high workpiece weights involved, fettling is often performed under ergonomically unfavorable conditions. The attraction of these types of jobs is low. As a result, turnover of personnel in fettling shops is high and cost-intensive. This is why robots are being used

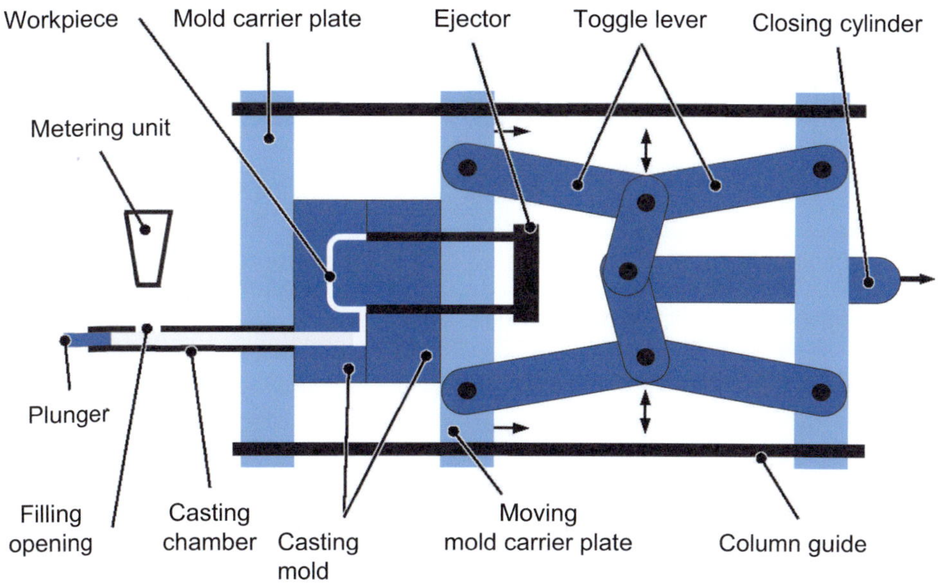

Fig. 2.10 Principle of a die casting machine using the cold chamber method

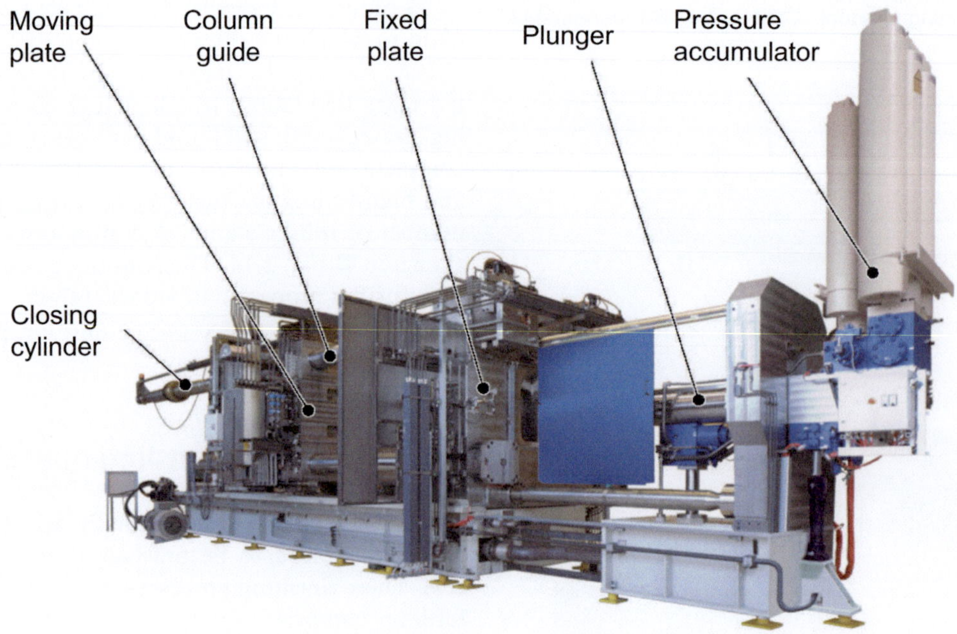

Fig. 2.11 Structure of a die casting machine. *Source* OSKAR FRECH

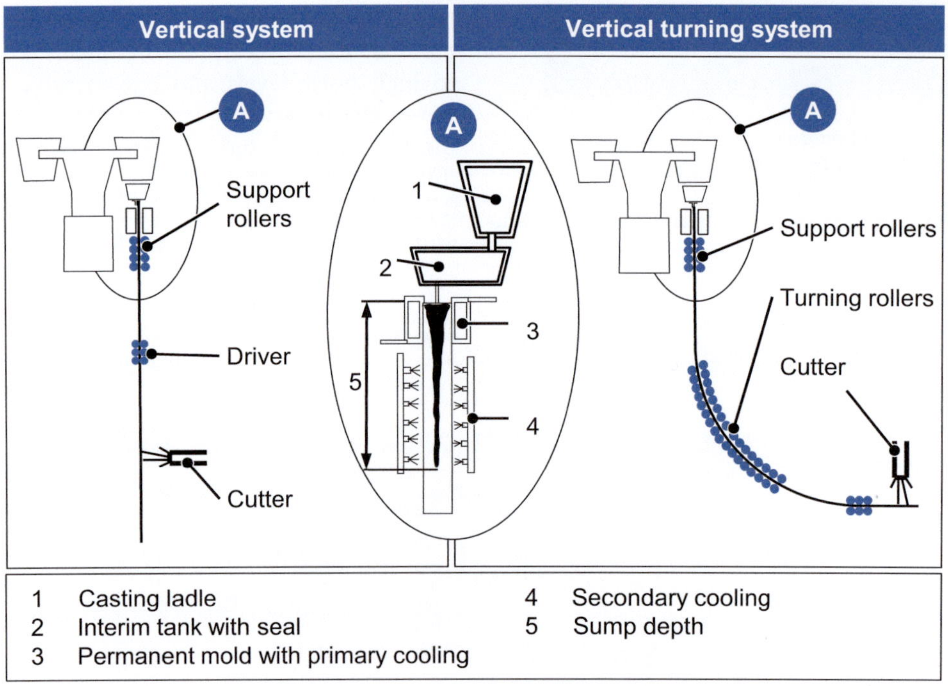

Fig. 2.12 Vertical continuous casting, according to SMS group

more and more for fettling. However, the strongly fluctuating geometric dimensions of burrs, risers, etc., and rough casting tolerances result in considerable fluctuations in cutting forces. This can result in careless burr machining and even tool fractures. Furthermore, the position of the required structure cannot be detected by a machine tool (in contrast to a person). Robots can be converted to machine tools with soft process control or by using cutting force sensors. These measures can correct the preprogrammed path of the fettling tool relative to the cast part. The fact that robots do not exhibit static or dynamic stiffness is reason enough to consider using sensor and numerically controlled machine tools for fettling. This enables, with an essentially

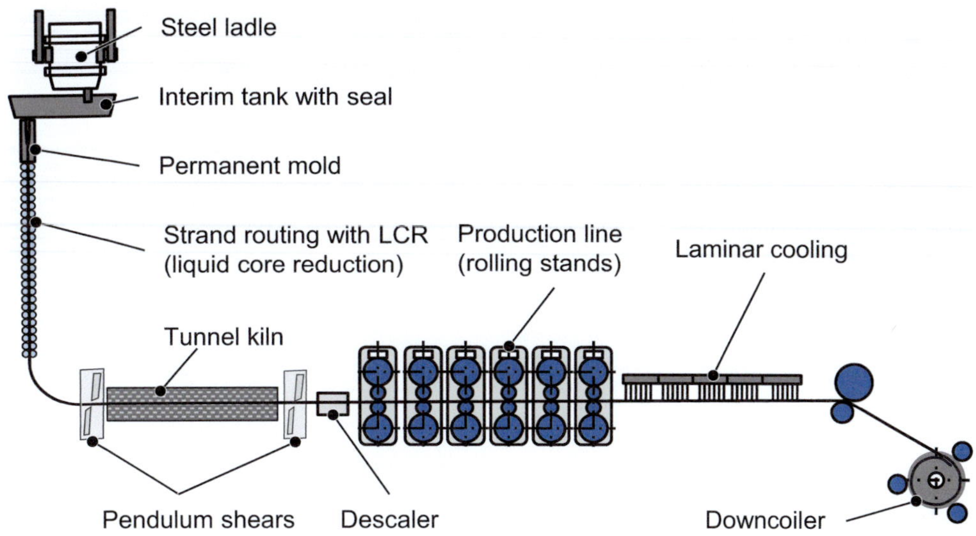

Fig. 2.13 CSP installation, according to SMS group

force-controlled milling machine, cast parts to be fettled automatically, even when they exhibit large tolerances.

2.4 Systems for Sintering

Sintering is a special process step in the powder-metallurgical production of semifinished and finished parts. The entire process sequence for manufacturing workpieces from metallic powders is summarized under the term powder metallurgy. Figure 2.15 shows the individual steps of the manufacturing process. In this process sequence, machines are used to press and calibrate the workpieces.

Increasingly complex forming operations, with a simultaneous increase in strength values, are inevitably resulting in ever-stricter requirements on pressing technology. While standard presses found in forming technology were commonly used originally for powder metallurgy, the trend nowadays is towards machines developed specially for powder metallurgy.

By using several lower punches, a larger component spectrum can be realized. Numerically controlled powder presses enable controlled hydraulic pressing processes with the highest accuracy. The different punches as well as the die and the center mandrel of the sintering plant, can be moved independently of each other in a force-controlled or path-controlled manner. Independent adjustment of the punches (upper and lower punches) and the die enables the material compaction while ensuring optimum space filling. Multi-layer components in particular require complex punch movement sequences. This guarantees that identical section densities, and thus no cracks, are produced during the pressing process.

For pressing the metal powder, it is filled into the annular space between the die and the center mandrel (Fig. 2.16). After the filling process, depending on the component, underfilling is performed in order to achieve the desired density even in thin outer walls. Underfilling is carried out by moving the lower punches accordingly, which provides powder in less densely filled zones. The transfer of powder is completed by closing the filling chamber. In the final phase of compaction, the positions of the different punches are adjusted. To avoid cracks, the green compact is loaded by the upper punch during the extraction movement of the die and the die plate is extracted. The different punch plunge movements and the associated deflections of the lower punches can be compensated during the press force reduction. To prevent the green compact from sticking to the die top, a press-off movement of the upper punch is necessary. The ejection takes place by the stroke movement of a lower punch and retraction of the mandrel, after which the workpiece can be removed from the working area. Figure 2.17 shows a hydraulic powder press with several interacting punch levels.

After the pressing process, sintering takes place at temperatures below the melting point of the metal. To ensure that the diffusion processes are not hindered by oxide formation, sintering furnaces are operated under an inert atmosphere. Once no more sintering is required, the workpiece is calibrated or forged. Figure 2.18 illustrates the complete sintering process by means of an equipment schematic.

Fig. 2.14 Five-segment continuous casting installation. *Source* SMS group

2.5 Systems for Casting Reaction Resin Concrete

A mixture of stones of different grain sizes and reaction resin is called reaction resin concrete. The material was originally used in the construction industry to manufacture precast elements (such as acid-resistant pipes and drainage channels). As a result of its high thermal inertia and good damping behavior and economic interests, the application spectrum of this material has extended to the manufacturing of parts used in general mechanical engineering, such as machine frames and gearbox housings.

Reaction resin concrete consists of a reaction resin as a binding agent and mineral additives. The most important resins are shown with their properties in Fig. 2.19. The mineral additives used primarily are granite and quartz. These are either used in their natural form (sand and gravel) or are obtained by crushing rocks (stone chips, crushed sand and stone dust). In the production of reaction resin concrete, additives of different grain sizes are mixed according to grading curves

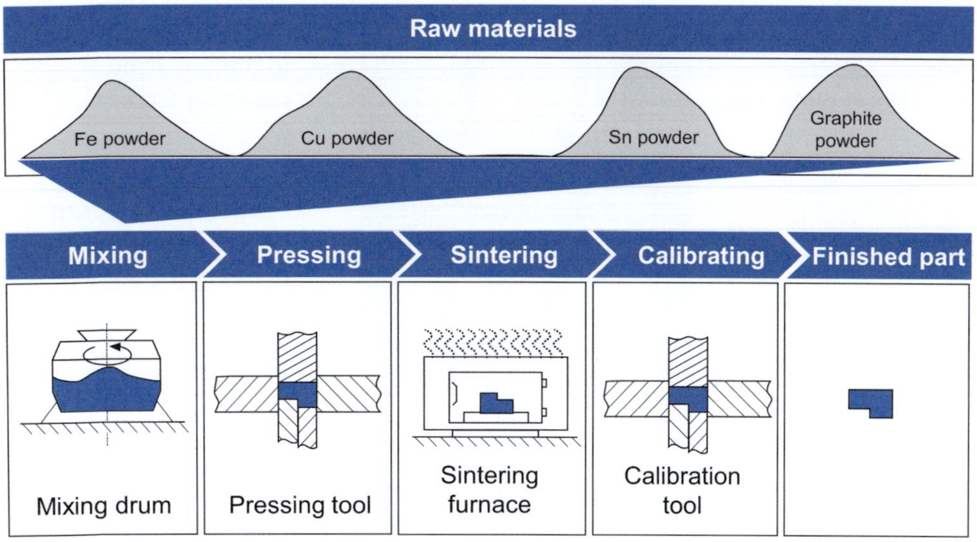

Fig. 2.15 Process steps in powder metallurgy

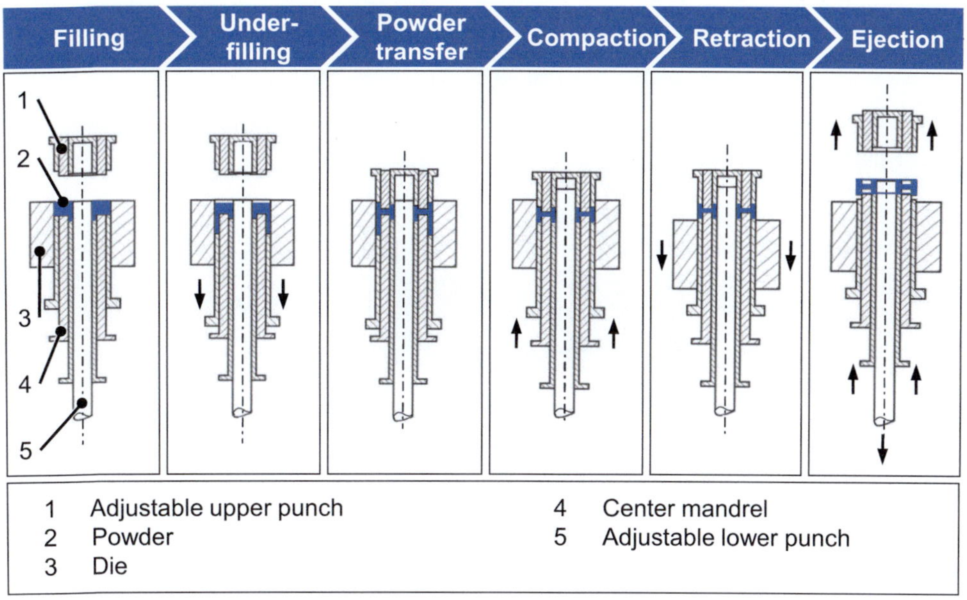

Fig. 2.16 Work steps for sinter pressing, according to SMS group

(bottom right in Fig. 2.19). In this diagram, the ordinate shows the passage by weight through standardized test sieves, the abscissa is the corresponding sieve hole width of the sieves. The interpretation of the grading curve can be additionally seen from the lower table of the diagram, which again indicates the individual weight fractions of the contained grain sizes. The grading curve specified is used for reaction resin concrete, thus meeting the requirement for the smallest possible grain surface to be wetted combined with the highest packing density.

If the reaction resin and the aggregate are mixed with the addition of reactants (such as hardening agents and accelerators), the mixture begins to harden after a certain time (pot life) with a loss of volume. During the pot life, the reaction resin concrete is poured into a mold and compacted by vibration. A plant for the production of components made of reactive resin concrete is shown in Fig. 2.20.

The compulsory concrete mixer is used to mix reaction resin concrete components. The finished, not yet hardened mixture, known as the reaction resin mass, is filled into the mold, sprayed with a release agent and then compacted by vibration. After the reaction resin concrete has hardened, the component is removed from the mold. The casting of mounting plates or strips enables the attachment of add-on parts such as guideways, bearing blocks or similar. Precision functional surfaces

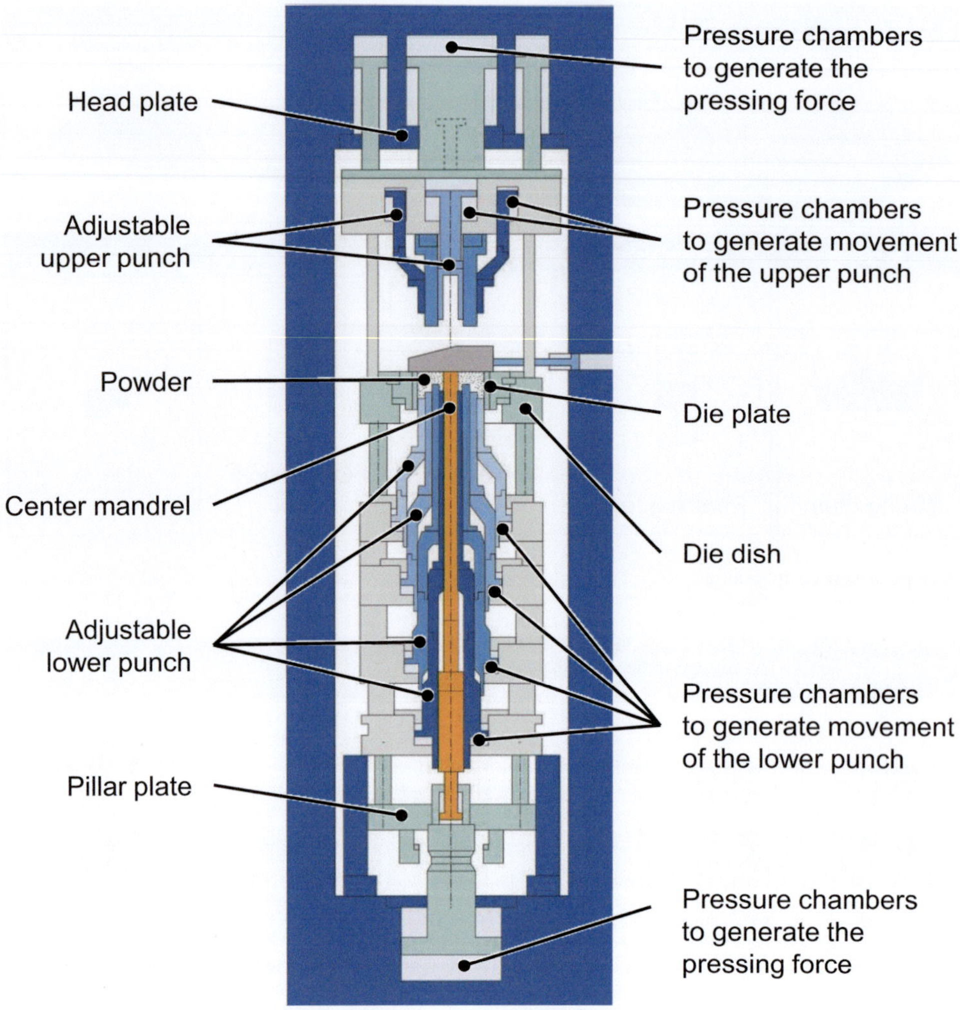

Fig. 2.17 Adapter technology for controlled punch movement in the powder press, according to SMS group

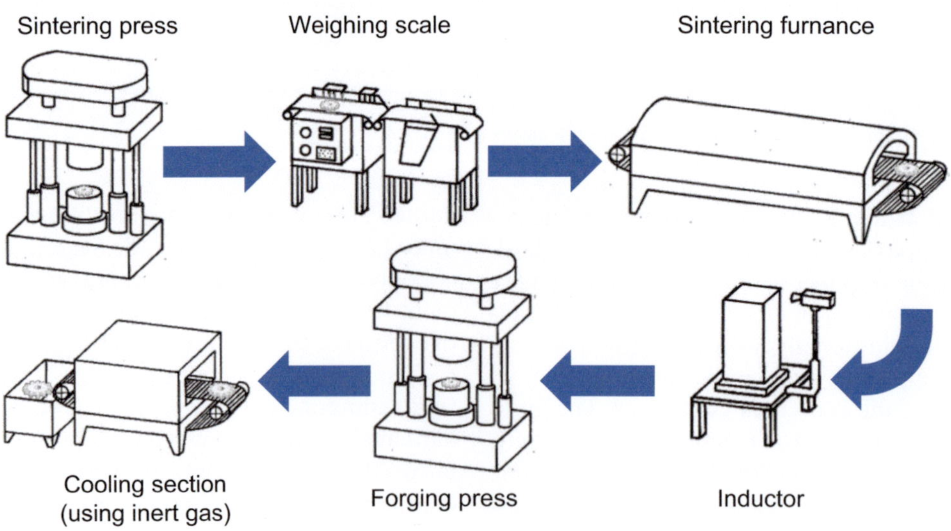

Fig. 2.18 Equipment schematic for sinter pressing. *Source* Sintermetall Krebsöge GmbH & Co.KG

Binding agents: Synthetic resins				
	Properties	Polyester resins	Epoxy resins	Methacrylic resins
Processing	Hardening mechanism	Polymerization	Polyaddition	Polymerization
Processing	Ductility (kg/m/s)	0.03 – 1.7	0.01 - 20	0.001 – 0.17
Processing	Pot life (min)	4 - 120	5 - 240	1 - 40
Processing	Volumetric shrinkage (%)	8 - 12	0.5 - 5	8 - 21
Usage	Tensile strength (N/mm²)	20 - 70	27 - 92	20 - 100
Usage	Young's modulus (kN/mm²)	3 - 5	2 - 4	2.8 – 4.5
Usage	Density (g/cm³)	1.0 – 1.8	1.1 – 1.8	0.9 – 1.7
Usage	Thermal expansion coefficient (µm/m/K)	20 - 90	32 - 90	30 - 80

Additives: Quartz, gravel, basalt, granite

Size of additives (mm)	Additives (in % by weight) as per grading curve
0 - 1	35
1 - 2	5
2 - 3	5
3 - 4	-
4 - 8	-
8 - 16	-
16 - 32	55

Grading curve

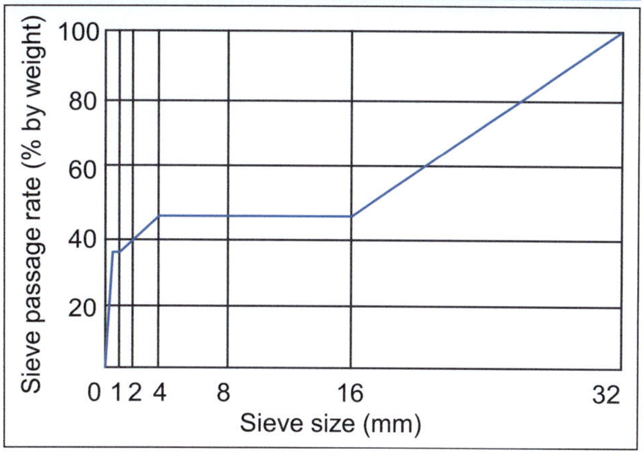

Fig. 2.19 Components of reaction resin concrete, according to [SAHM87]

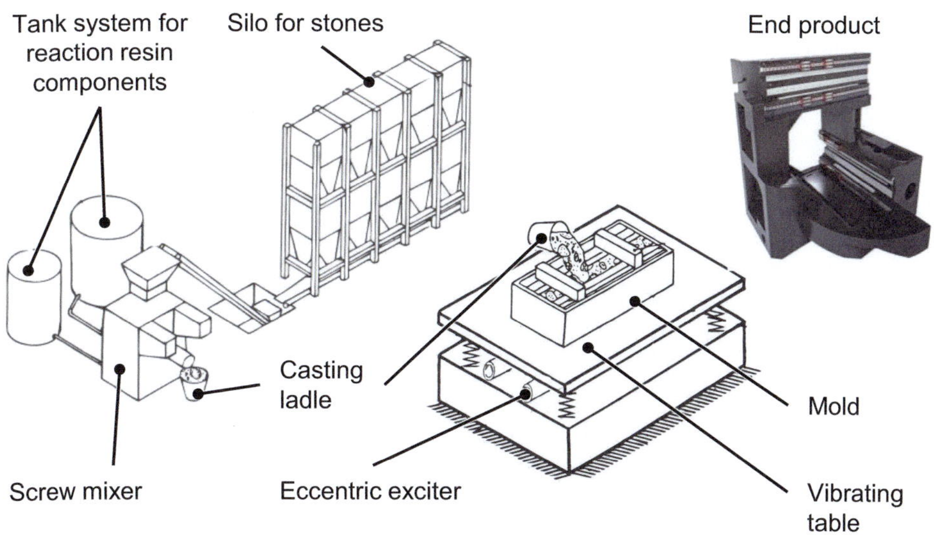

Fig. 2.20 Plant for manufacturing components from reaction resin concrete. *Source* RAMPF Holding

Parameter	Unit	Steel	Casting EN-GJL-			RRC
			200	250	300	
Density ρ	kg/dm³	7.8	7.15	7.2	7.3	2.1 – 2.4
Young's modulus E	kN/mm²	210	88 - 113	103 - 118	123 - 143	15 - 55
Young's modulus in relation to density	10^6 m²/s²	27	14	15.4	18,2	15
Tensile strength R_m	kN/mm²	0.35 – 1.5	0.2 – 0.3	0.25 - 0.35	0.35 – 0.45	0.01 – 0.03
Yield strength R_p	kN/mm²	0.25	0.13 - 0.195	0.165 - 0.228	0.228 – 0.285	0.002
Material damping (log. decrement)	10^3 kg/m/s	0.002	0.003	0.003	0.003	0.02 – 0.04
Thermal expansion coefficient α	µm/m/K	12	10 - 13	10 - 13	10 - 13	10 - 20
Thermal conductivity λ	W/m/K	45 – 50	46 – 50	44.5 – 48.5	41.5 – 45.5	1.3 – 3.0
Thermal capacity C_p	kJ/kg/K	0.45 – 0.5	0.46 - 0.535	0.46 – 0.535	0.46 – 0.535	0.9 – 1.3
Relative material costs	-	1	0.63	0.63	0.63	0.32

Fig. 2.21 Comparison of the material characteristics of steel, casting and reaction resin concrete (RRC). *Source* [WEIß12, ANGU78]

can also be produced by using high-precision casting gauges. These transfer their contour to the cast blank using special coating systems.

The mechanical and thermal properties of hardened reaction resin concrete are shown in Fig. 2.21, compared to cast and steel materials. In terms of the dimensional stability of a machine tool frame under load, the yield strength, i.e., the tension at which the material starts to suffer plastic deformation, is of interest in addition to the Young's modulus.

This parameter for reaction resin concrete is very low compared to steel or cast iron. It must be remembered, however, that higher stress only occurs at force transmission points on conventional machine tool frames, with the exception of presses and hammers. Excessively high stress in the reaction resin concrete can be avoided by means of adapted force transmission elements. The thermal properties of reaction resin concrete also differ greatly from those of ferrous materials. The thermal conductivity is much lower than for ferrous materials, while the specific thermal capacity is much higher. This means that the time constants of the thermos-elastic deformation behavior of machines made of reaction resin concrete are considerably greater than for comparable machines made of steel or cast iron. To prevent bimetal effects at points in the composite material between reaction resin concrete and steel, appropriate compositions of resins and additives are used to align the linear thermal expansion coefficients of the concrete to those of the ferrous materials.

▶ Section 3.5 in the second volume of this book series discusses in detail the structural and production-related special features of machine components made of reaction resin concrete.

2.6 Machines for Additive Manufacturing

Additive manufacturing processes, also known as generative manufacturing processes, create a component geometry by adding volume elements layer by layer without using tools, such as molds and cutting tools. This allows the production of complex structures without geometric restrictions that are necessary in conventional methods. Due to the direct conversions of CAD data into a physical component, additive manufacturing processes are particularly suitable for prototypes, small quantities and customer-specific products. [VDI14].

In addition to the aggregate state of the base material, the classification of additive manufacturing

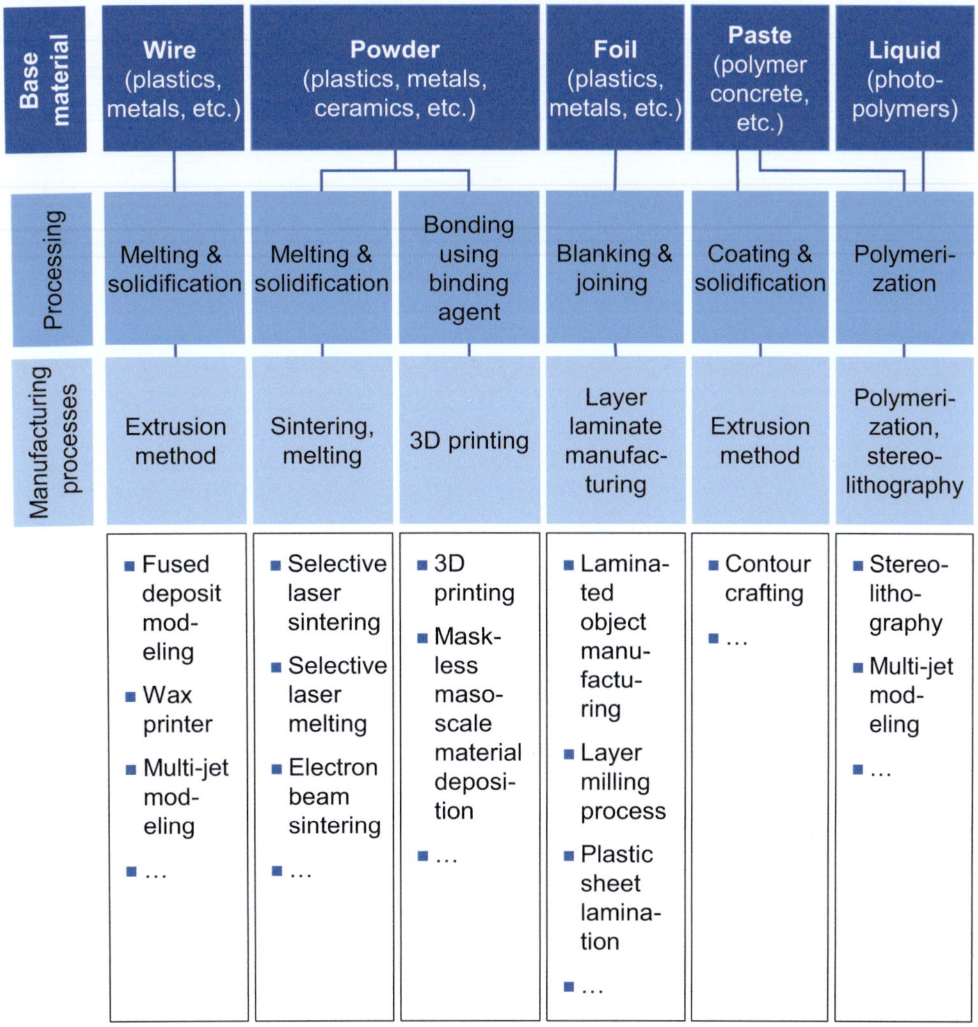

Fig. 2.22 Classification of additive manufacturing processes. *Source* [GEBH07]

processes can be based on the appearance of the base material and the mechanism for naming the method (Fig. 2.22).

▶ Section 7.5 is dedicated to discussing additive manufacturing processes in detail. Additive manufacturing facilities are introduced and additional insights into this production type provided.

References

[ANGU78] Angus, H.T.: Cast Iron—Physical and Engineering Properties, 2nd edn. Butterworths, London (1978)

[BÜHR14] Bühring-Polaczek, A., Michaeli, W., Spur, G.: Handbuch Urformen. (Reihe: Ed. Handbuch Fertigungstechnik). Hanser, München (2014)

[DIN03] Norm DIN 8580 (September 2003). Fertigungsverfahren—Begriffe, Einteilung

[GEBH07] Gebhardt, A.: Generative Fertigungsverfahren, 3rd edn. Hanser, München (2007)

[KOBL00] Koblenzer, H.: Einfluß einer Umformung mit flüssigem Kern auf die Innenbeschaffenheit von im Strang gegossenen Stahlknüppeln. Diss. RWTH Aachen (2000)

[ROSE09] Rosenthal, D., Krämer, S., Klein, C., Geerkens, C., Müller, J.: 20 years of CSP. Stahl und Eisen. **129**(11), 73–89 (2009)

[SAHM87] Sahm, D.: Reaktionsharzbeton für Gestellbauteile spanender Werkzeugmaschinen. Diss. RWTH Aachen (1987)

[VDI14] Richtlinie VDI 3405 (Dezember 2014). Additive Fertigungsverfahren

[WEIß12] Weißbach, W., Dahms, M.: Werkstoffkunde. (Reihe: Studium Festigkeitslehre und Werkstofftechnik). 18. Aufl. Vieweg und Teubner, Wiesbaden (2012)

Forming and Severing Machines

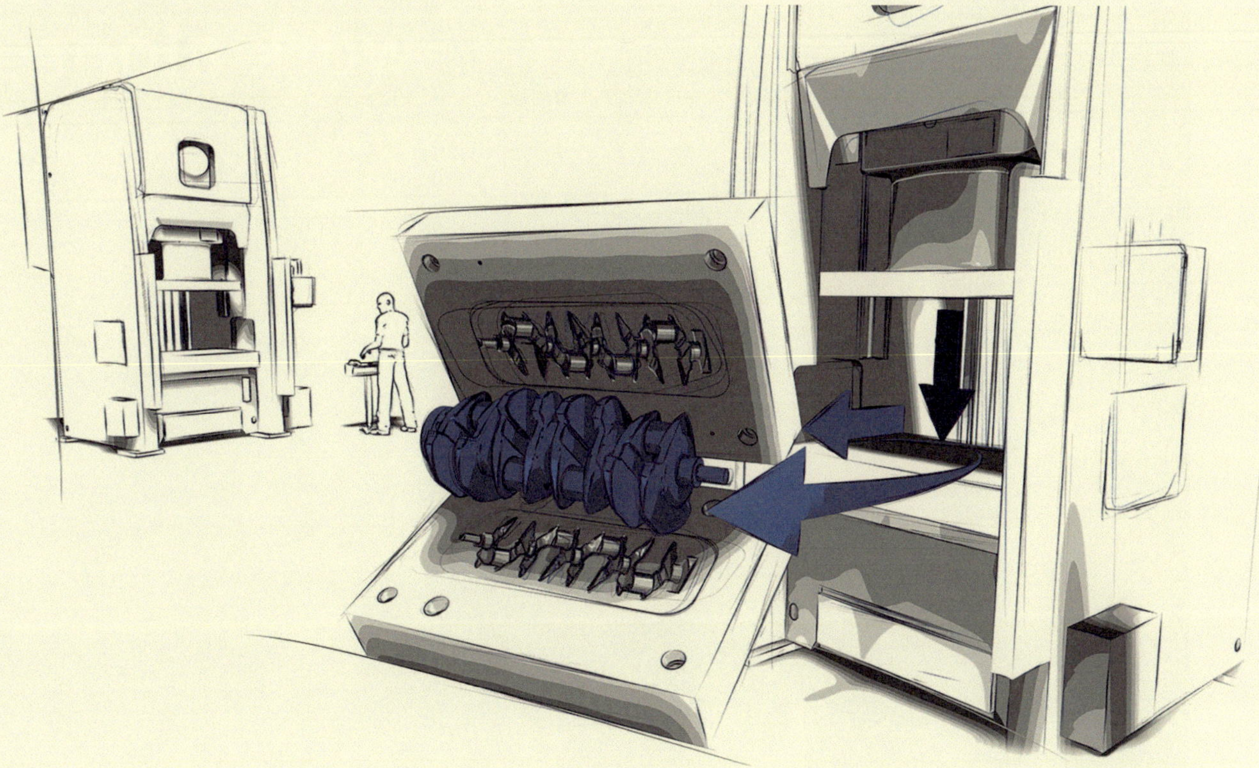

I thank my staff members Mathias Dehn M.Sc., Janis Schäfer M.Sc. and Paul Weiler M.Sc. for their contribution in revising this chapter.

C. Brecher and M. Weck, *Machine Tools Production Systems 1*,
Lecture Notes in Production Engineering, https://doi.org/10.1007/978-3-662-68120-6_3

Contents

3.1 Forming Machines

"Forming", according to DIN 8580 [DIN03f], is understood as the technical manufacturing term for shaping workpieces by ductile (plastic) alteration of the geometry of an object. In this process, both the mass and the cohesion of the material are maintained. However, the partial masses are distributed differently. Forming is therefore a non-cutting manufacturing process, generally characterized by low material losses and short manufacturing times. This process also frequently provides the additional benefits of optimum flow of material fibers in the workpieces and an increase in strength due to work hardening during cold forming.

However, these processes are usually used for large batch sizes so that high tool costs are less of a consideration.

The manufacturing processes used in forming technology are classified in accordance with DIN 8582 [DIN03g] standard based on the main material demands made during the forming process when modifying the geometry. A distinction is made between the five main process groups of forming under compressive, combined tensile and compressive, tensile, bending, and shearing conditions, ◘ Fig. 3.1. These are further categorized in accordance with DIN 8583 to 8587 [DIN03h, DIN03e, DIN03c, DIN03d, DIN03b] depending on the kinematics, (relative motion between the workpiece and the tool) as well as the workpiece and tool geometry, ◘ Fig. 3.2.

Independent of this categorization, a different classification of forming processes has established itself in industrial practice. A distinction is made between sheet metal and massive forming in accordance with the spatial expansion of the workpieces [LANG02]. During massive forming processes (e.g., extrusion, upsetting, forging), there is a three-dimensional distribution of material with large changes in the cross sections and wall thicknesses using spatially distinct blanks. Processes for sheet metal forming (e.g., deep-drawing and bending) are characterized by the formation of three-dimensional components from flat workpieces. Typically, this includes hollow parts with an approximately constant wall thickness, corresponding roughly to the sheet thickness. The machines and tools used in massive forming (forging machines) have a stiffer design as a result of the significantly greater forces required than those used in sheet metal forming. Another way of categorizing the processes is by the forming temperature: A distinction is made between cold, semi-hot and hot forming. In cold forming, the initial temperature of the material is room temperature. In semi-hot forming, the workpiece is in a temperature range between 600 °C and 900 °C. In hot forming, the temperature is generally greater than the recrystallization temperature of the material being formed. According to other definitions (see [DIN03g]), forming is already regarded as semi-hot if the workpiece is warmed beyond room temperature before forming. One benefit of hot forming is the ability to make significant changes to the geometry. Components made through the cold massive forming generally have higher dimensional and geometrical accuracies as well as better surface quality. Semi-hot forming combines the benefits of cold and hot forming. Due to the greater development and tool costs, it is mostly used for very large quantities.

Forming at high temperatures requires less forming energy and a lower forming force due to the greater ductility of the material. The yield strength curve for the material 1.0401 (carbon steel C15) as a function of temperature and the degree of deformation, as shown in ◘ Fig. 3.3 demonstrates this clearly. The high-temperature strength of the tools, the coarse grain formation of the material, and the energy costs limit the maximum temperature for hot forming [KÖNI06].

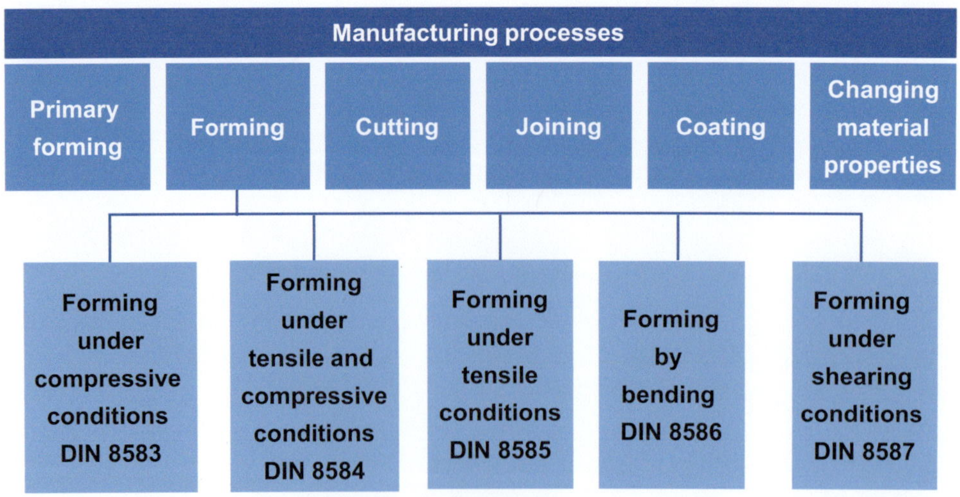

◘ **Fig. 3.1** Classification of forming processes, according to DIN 8580 [DIN03f], DIN 8582 [DIN03g]

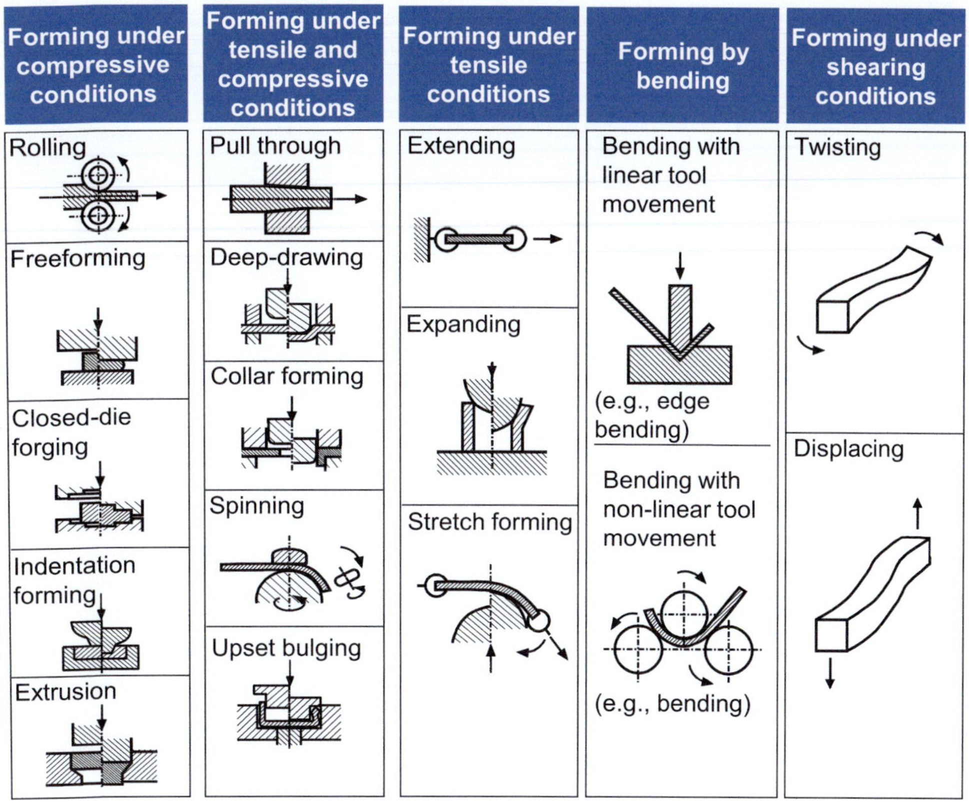

Fig. 3.2 Categorization of forming processes, according to DIN 8583–8587 [DIN03h, DIN03e, DIN03c, DIN03d, DIN03b]

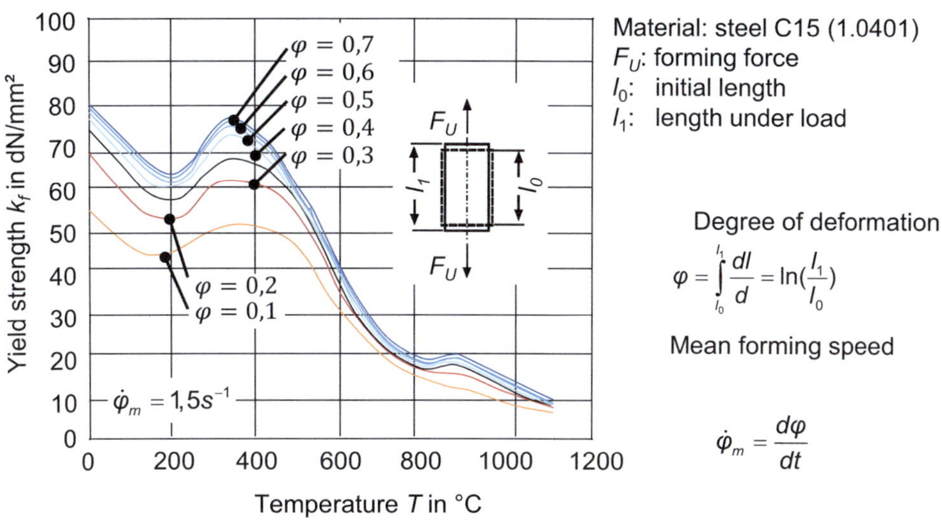

Fig. 3.3 Yield strength as a function of temperature and degree of deformation

The forming machines used in forming processes for manufacturing must supply the necessary forces and energies for the required forming strokes for the process used, and they must guide and control the action of the forming tools accurately.

Forming machines do not necessarily have to be categorized in accordance with the manufacturing processes described above [DIN03h, DIN03e, DIN03c, DIN03d, DIN03b] as several manufacturing processes can be implemented with the same machine kinematics on one machine by using a range of different tools and auxiliary equipment. For this reason, it is possible to categorize by machine type and machine function, as shown in Fig. 3.4. A further breakdown is possible

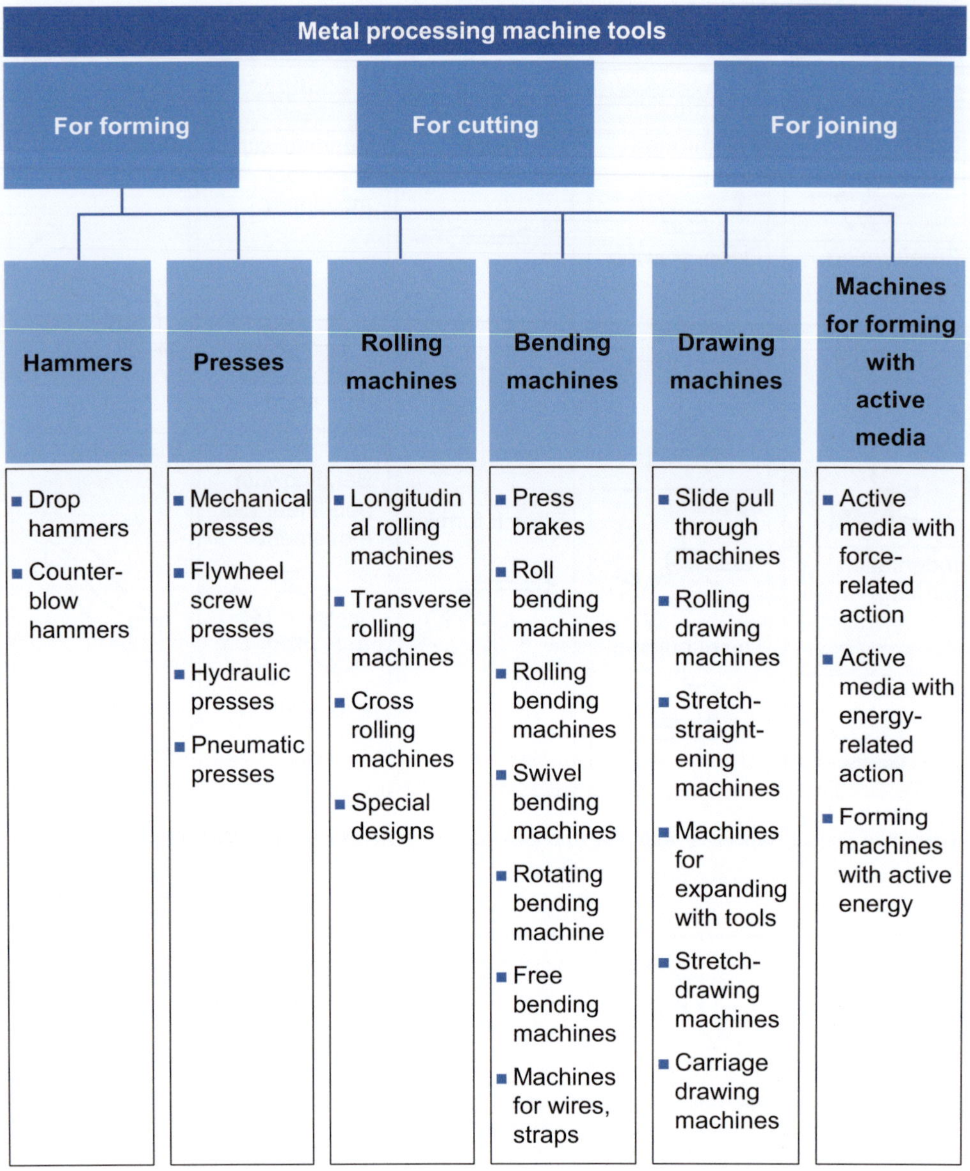

Fig. 3.4 Forming machine types and designs

using the following categories: Functional principle, drive type, kinematics, tool movement, tool and workpiece geometry, scope and application area.

Figure 3.5 provides some insight of a broad brush classification of the functional principles of forming machines and their parameters. Forming machines are initially categorized into energy-controlled machines, travel-controlled machines, and force-controlled machines. Hammers and screw presses are machines which are largely characterized by their work capacity. A certain amount of work capacity is made available for each forming process. This can be influenced and adjusted by the masses and speeds of the moving machine components. The forming process is completed when the kinetic energy of the masses, moving either linearly or rotationally, has been completely converted into forming work (energy-controlled machines). The forces occurring during the forming process are a function of the process. Factors influencing the forming forces are the shape, dimensions, and material of the workpiece as well as the required forming travel.

The ram travel, which is limited by the dead-center positions of the crank drive, is the characteristic feature of eccentric, crank, toggle, and wedge presses (travel-related machines). The magnitude of the usable force on the tool or ram depends on the actual ram position. The nominal press force characteristic for the machine is achieved a few angular degrees before the lower return point depending on the drive type used. Travel-related forming machines include rolling machines where the roller feed, also known as the pass reduction, is the characteristic value.

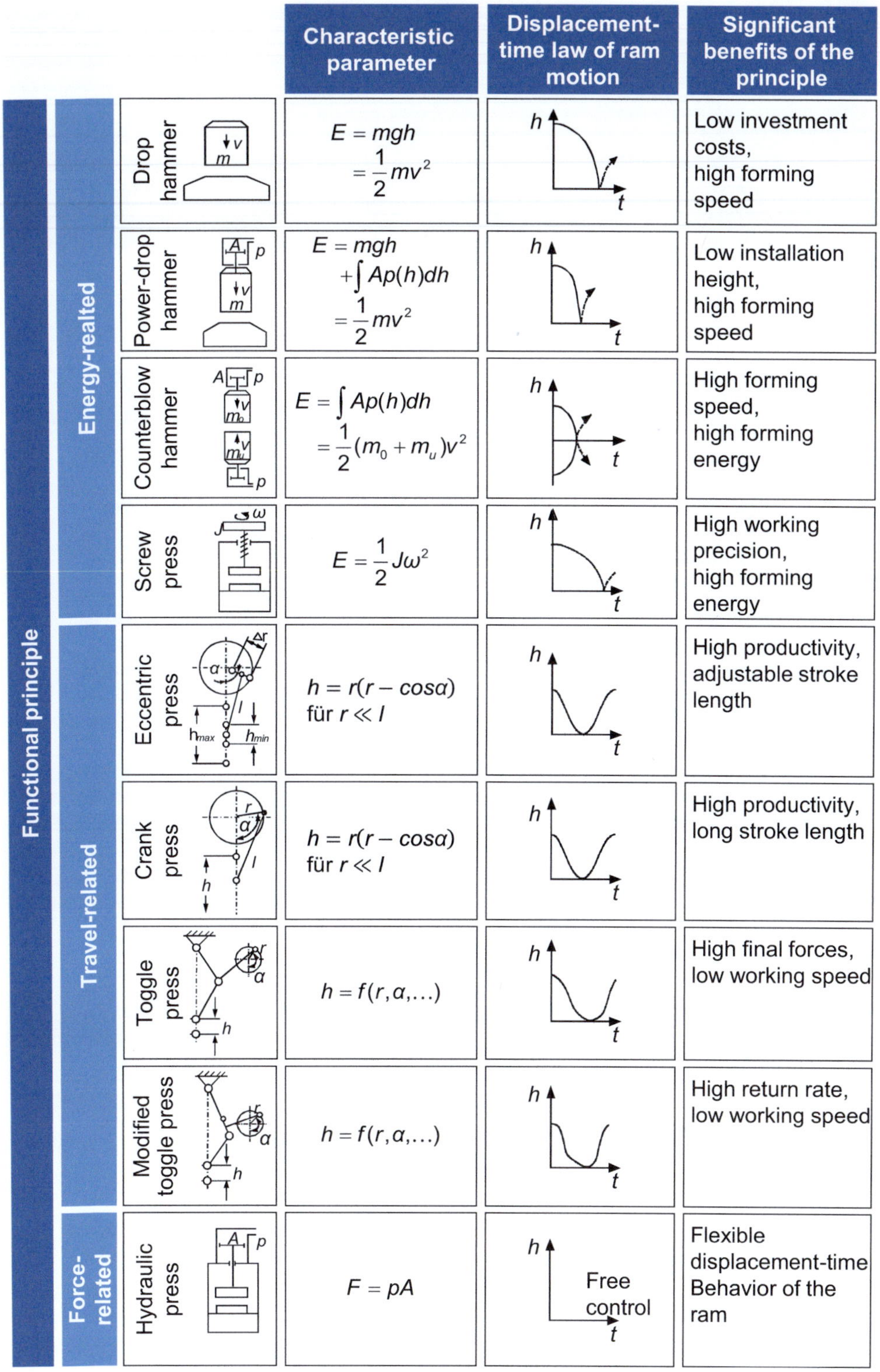

Fig. 3.5 Functional principles of forming machines

The essential parameter for hydraulic presses is the force that can be applied along the ram travel, which results from the system pressure and the piston surface (force-related machines). This means that hydraulic presses can be easily adjusted to the demands of a forming process in terms of the force and working requirements, the speed, and forming travel.

A number of criteria have to be taken into consideration when selecting a forming machine for a specific forming task. Several important considerations are given in Fig. 3.6:

Figure 3.7 provides a compilation of the most important parameters for describing the properties of forming machines. A distinction is made between force,

Criterion	Machine characteristics
Workpiece size	Workspace
Required force / energy	Machine size
Forming speed	Functional principle
Required workpiece accuracy	Geometric accuracy and stiffness of tool and machine

Fig. 3.6 Selection criteria for a forming machine

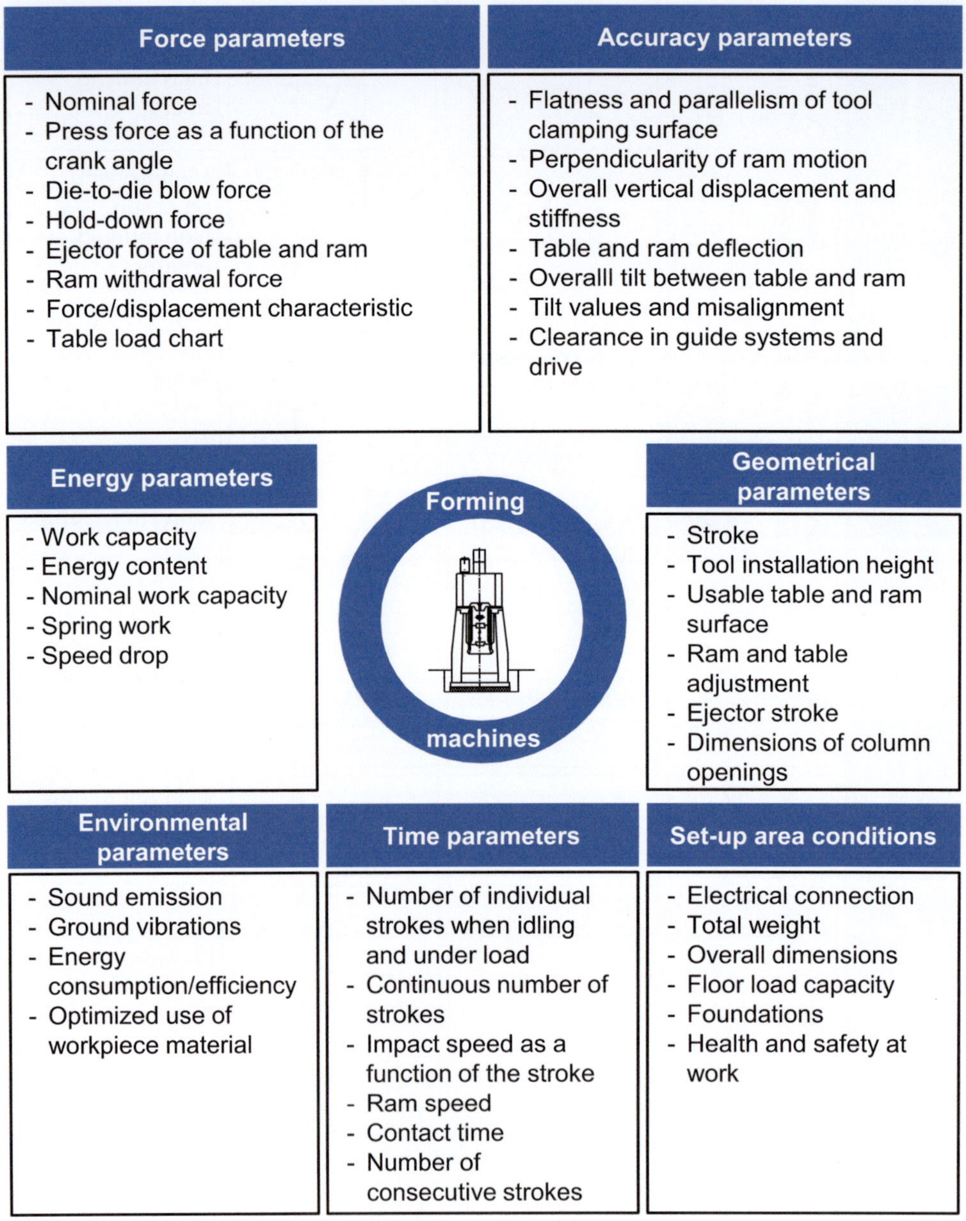

Fig. 3.7 Forming machine parameters

energy, environmental, time, and accuracy parameters as well as geometrical parameters and conditions at the place of installation. By comparing these parameters with the demands of the required forming process, it is possible to select the best machine for the process. The productivity of a press is generally limited by the number of strokes achievable in a given amount of time.

3.1.1 Hammers

Hammers are energy-related forming machines. Due to their straightforward design, they are one of the most economical forming machines for generating very high forces and providing a high work capacity. Simple alterations to the impact strength and sequence allow their impact program to be very quickly adjusted to the relevant process-related machining task. A further benefit of hammers is that they cannot be overloaded, as the hammer frame and drive are not within the flux of forces at the point of operation.

Depending on the type of moving masses (rams), a distinction is made between drop hammers and counterblow hammers, ◘ Fig. 3.8. Drop hammers are characterized by an accelerated ram meeting a fixed anvil bed. The forces have to be absorbed by the anvil bed and the foundation. Counterblow hammers, on the other hand, have an upper and a lower ram of similar mass. These rams meet each other at the same speed such that the externally acting forces theoretically cancel each other out. The vibrations sent into the foundation and the environment when the rams meet are significantly less than in forming processes using drop hammers. As a result, only small foundations are required when using counterblow hammers. In spite of their greater work capacity, the overall design of counterblow hammers can be lighter.

Depending on the design of the moving masses, a distinction is made between purely drop hammers and power-drop hammers. The latter have become common with electrohydraulic hammers predominating. In this design, the ram is accelerated by a hydraulic linear drive during its descent. This increases the work capacity of the hammers or reduces the construction height (drop height) for the same work capacity.

In counterblow hammers, a distinction is made between mechanical (steel strip), pneumatic, or hydraulic ram couplings depending on the coupling between the upper and lower rams. The upward movement of the lower ram is controlled by the ram coupling with the upper ram.

A further subdivision can be made by the type of ram column. A distinction is made between single and double-column designs. Double-column designs are usually used due to the workpiece accuracies generally required. The whole column is usually made as a single piece. ◘ Figure 3.9 shows a power-drop hammer with a hydraulic drive and a double-column frame. Pressure oil is used as the drive medium for the rams; the working pressure can be up to 200 bar with a work capacity of 200 kNm and an impact rate of between 80 and 140 min^{-1}. For the downward movement, the pressure oil is guided from the pressure accumulators through the control block to the top of the piston (see ◘ Fig. 3.10). This produces downward acceleration on the piston rod connected to the ram. The upper

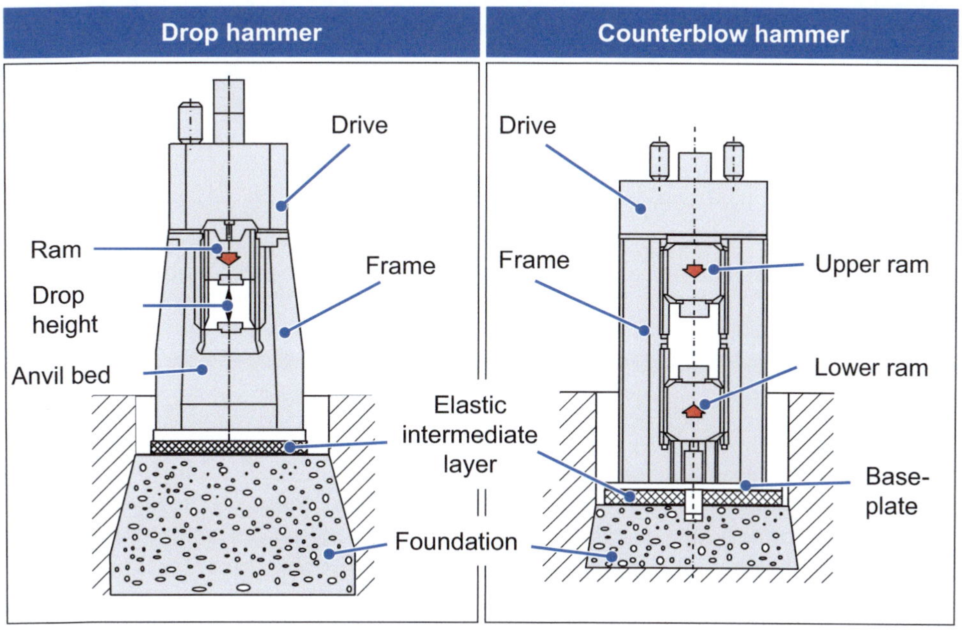

◘ **Fig. 3.8** Schematic structure of forge hammers

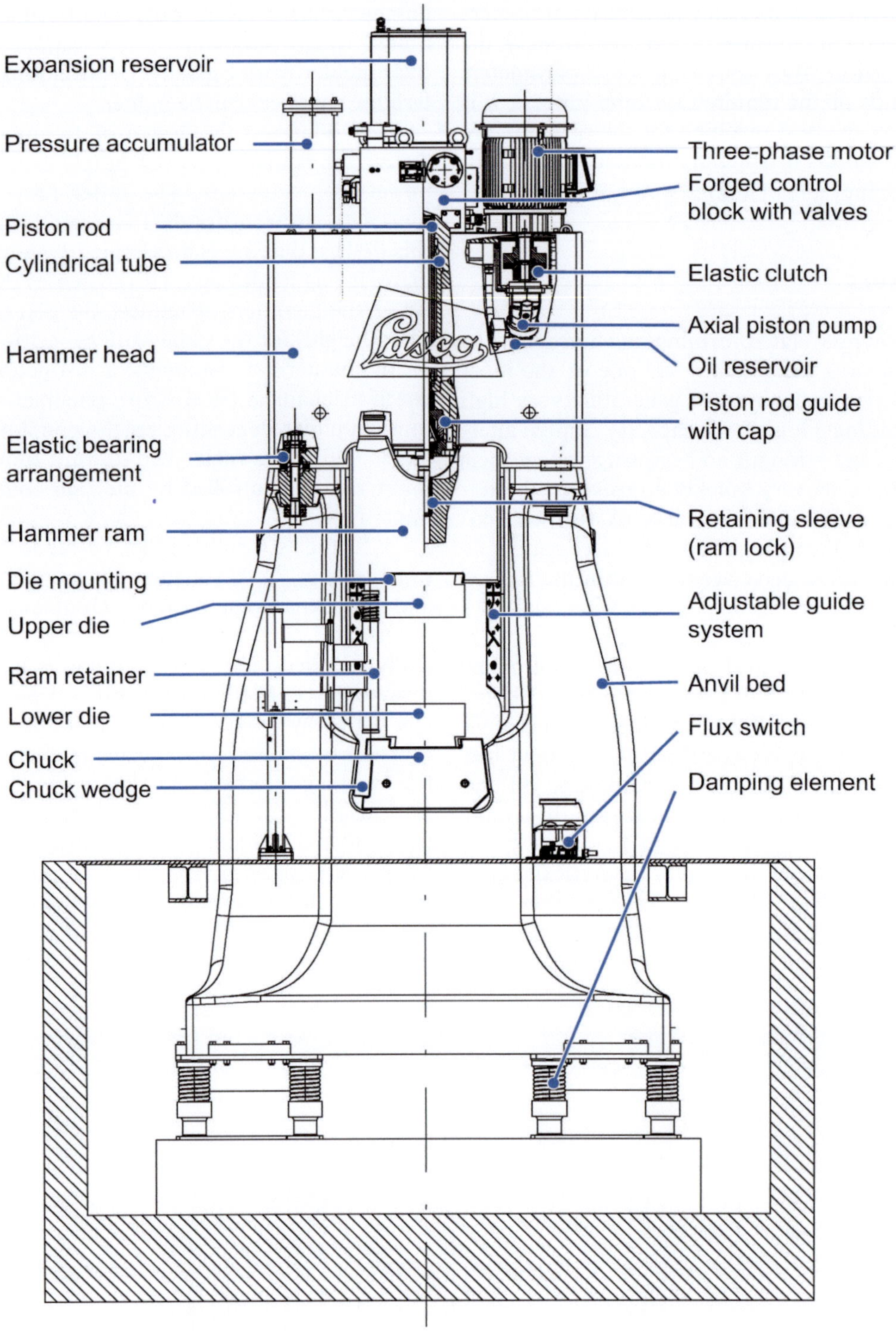

Fig. 3.9 Hydraulic power-drop hammer. *Source* LASCO Umformtechnik

cylinder chamber is then depressurized and the oil pressure constantly applied to the ring surface of the piston underside provides the return stroke.

The anvil bed is made from alloyed cast steel and has a U-shape. The hammer head, configured as an oil reservoir and elastically connected to the lower hammer frame, accommodates all the hydraulic drive units such as the cylindrical tube, axial piston pumps, pressure accumulator, valves, and the three-phase a.c. motors. The bending-flexible, low-mass piston rod is precisely guided and is elastically connected to the ram via a ram lock. High operational accuracy can be achieved as the adjustable ram guidance allows precise adaptation of the guide clearance. The required impact strength and impact frequency can be set via the electrical control.

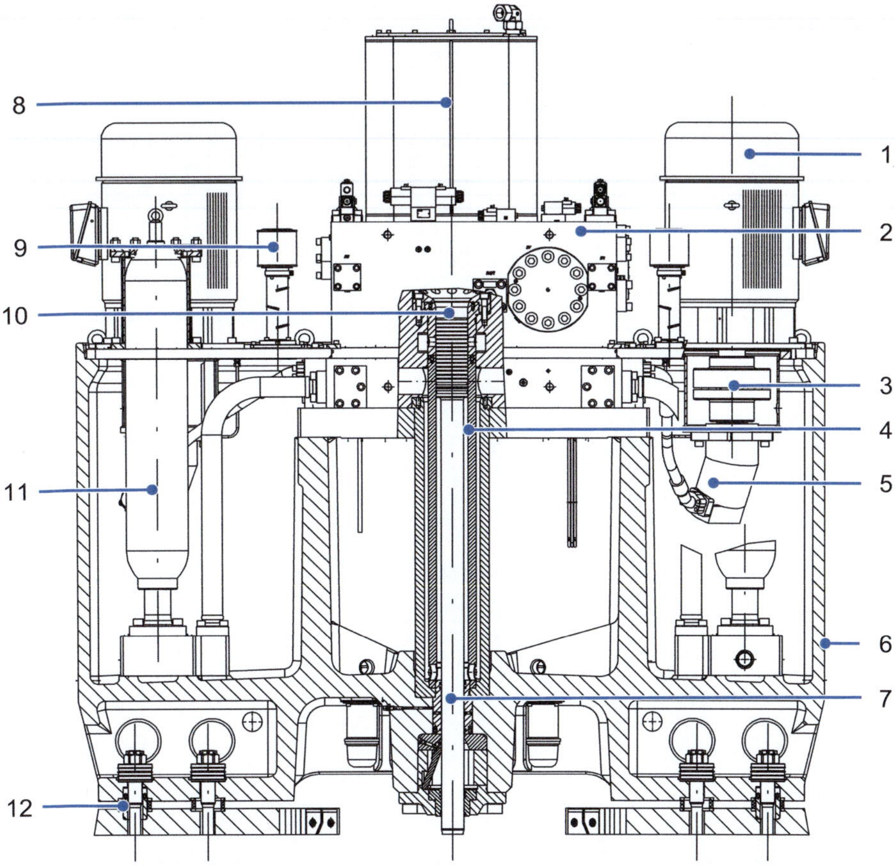

Legend

1 Three-phase motor	7 Piston rod
2 Control block	8 Expansion reservoir
3 Elastic clutch	9 Air filter
4 Ring piston surface for return stroke	10 Piston surface for working stroke
5 Axial piston pump	11 Pressure accumulator
6 Oil sump	12 Rubber buffer

Fig. 3.10 Hydraulic power-drop drive for a drop-forging hammer. *Source* LASCO Umformtechnik

The characteristic values (force, energy and forming rate) can be further increased with a counterblow hammer. Figure 3.11 shows an electrohydraulically driven counterblow hammer in which the movement of the upper and lower ram is connected via a hydraulic ram coupling.

The drive system for the upper ram uses the described power-drop principle, see Fig. 3.12. If the piston speed becomes so high that insufficient oil flows through the inlet valve, the suction valve opens to allow the pressure chamber to fill. The lower ram is indirectly driven by the upper ram via the hydraulic ram coupling. This essentially consists of two clutch pistons connected to the upper ram, which during the impact stroke (downward movement of the upper ram) pressurize the clutch piston on the lower ram via an oil column. These pistons are fixed directly to the ram without any intermediate elastic elements. The mass ratio of the rams is selected such that they return to their initial positions under their own weight. In addition, they are accelerated by oil pressure on the ring surface of the piston underside in order to realize short stroke sequences. The piston rod is connected to the piston tube, which has a deliberately slim design in order to achieve high elasticity. This compensates for possible alignment errors between the piston and the ram. Just before the top dead center, the piston and the piston tube hit a hydraulic end position damper with a conical annular gap. As the oil flows out of the chamber above the piston via this annular gap, which becomes narrower during the upward movement, this slows down the ram movement. In addition to this, the rams are absorbed elastically by

3

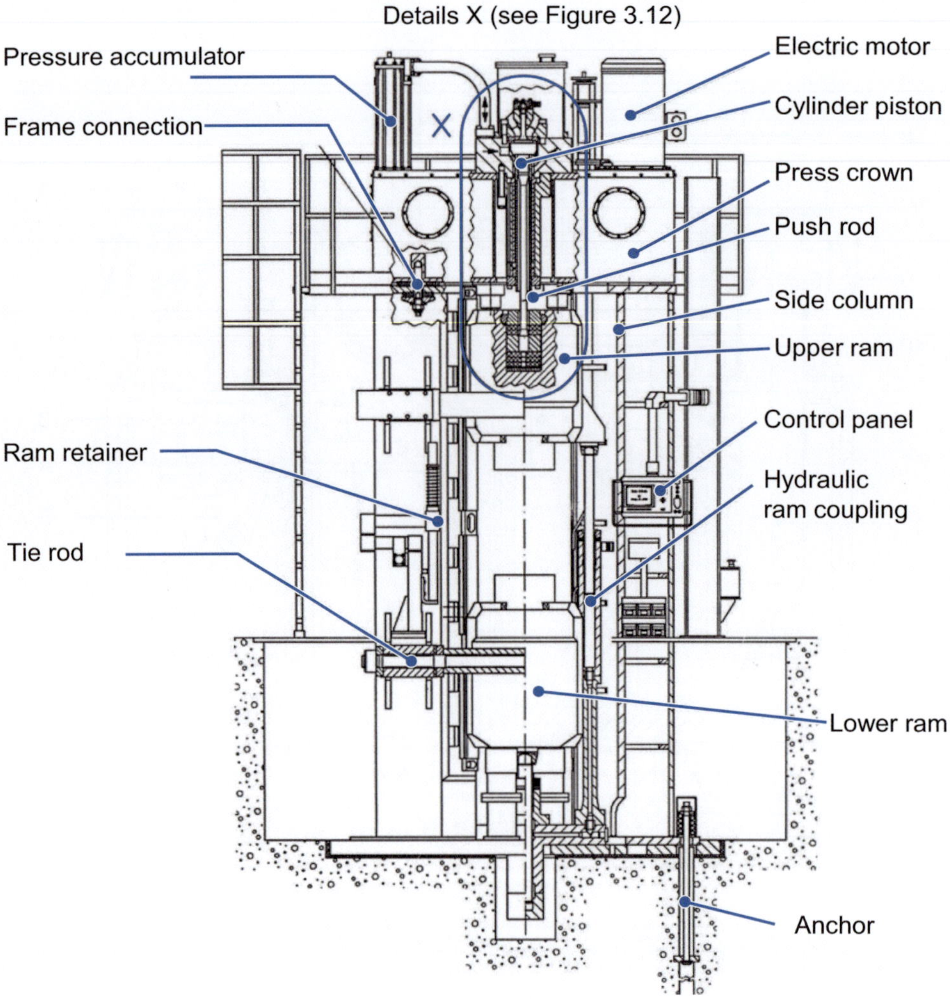

Fig. 3.11 Counterblow hammer with hydraulic ram coupling. *Source* Schuler

end position dampers attached in the upper and lower areas. The hammer frame consists of two box-shaped stands with a stable welded structure. The stands are connected via crossbars and a baseplate to form a precisely preloaded guide system. An asymmetrical arrangement of the clutch pistons on the rams achieves a well-defined contact of the rams on the guide frame. Lateral movements of the upper and lower rams within the ram clearance during the load-free downward movement are avoided. In this way, both halves of the die are precisely aligned as they move toward each other. The movement of the two halves of the die is guided by the tool's integral guide systems during the actual forming process. A high overall manufacturing accuracy is achieved that is comparable to that of drop hammers. The primary application areas for the hammers described here are freeform forging and, above all, drop forging. Small to medium-sized parts (levers, wrenches, etc.) are predominantly produced on drop hammers while heavy and large workpieces (crankshafts, flanges, rings, etc.) are produced on counterblow hammers with high impact energy. Due to the high forming forces required for forged parts with thin wall thicknesses, such as turbine blades, these are produced on hammers with greater impact speeds and shorter pressure dwell times. Figure 3.13 shows a few typical components manufactured by hammer forging.

Process vibrations and noise emissions from operating a hammer can have very deleterious effects on the operating environment. Keeping these emissions at acceptable levels usually requires a great deal of effort to provide a suitable machine installation. Figure 3.14 shows various machine mounting options for reducing ground vibrations. An isolation efficiency of 90% can be achieved with a vibration-isolated mounting arrangement [GERB92]. However, the loud noise and relatively high vibration emissions from these machines will continue to be a problem in the future.

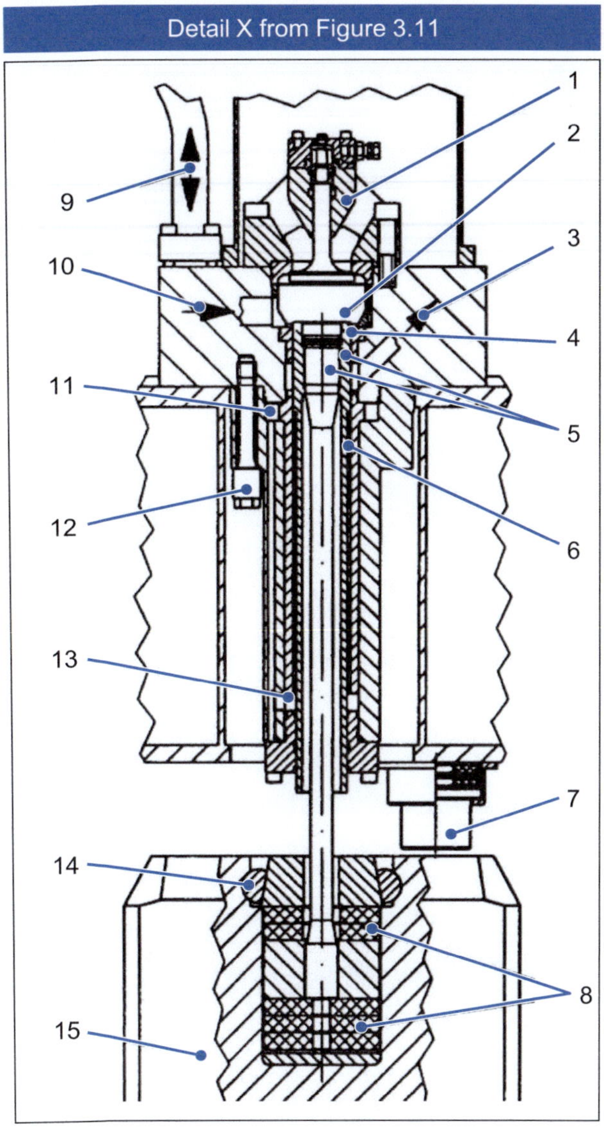

Legend

1 Suction valve
2 Working stroke acceleration surface
3 Tank line (via discharge valve)
4 Brake bushing with conical annular gap
5 Piston rod fixed to piston tube
6 Return stroke ring surface
7 End position damper
8 Piston rod elastic connection
9 Hydraulic accumulator line to inlet valve
10 Hydraulic accumulator line from inlet valve
11 Annular channel for return stroke control
12 Screw connection from crosshead and hydraulic block
13 Oil inlet and outlet line to return stroke ring surface
14 Clamp
15 Upper ram

Fig. 3.12 Electrohydraulic counterblow hammer drive. *Source* Schuler

3.1.2 Screw Presses

Screw presses are energy-related forming machines. They obtain their work capacity from a rotating mass. The rotational speed and dimensions of the flywheel thus determine the magnitude of the work capacity. A vertical ball screw nut drive converts the rotation of the flywheel into a longitudinal motion of a ram guided in the frame. Most ball screws have a multi-thread design and have either a form-fit or force-fit connection to the flywheel, which is driven by a motor.

In terms of their construction, the machines are differentiated by the type of flywheel drive and whether the screw is fixed or can move along the vertical axis. Figure 3.15 shows various drives for screw presses.

Traditional drives are friction wheel drives with three-disk cylinder friction gearing and a screw moving along the vertical axis, Fig. 3.16. Using a flat belt, an electric motor drives the drive shaft and the drive wheels arranged on the shaft. These rotate at a constant speed and can be moved in the axial direction with the drive shaft. Engaging the drive wheels in either direction causes the flywheel to be either raised or lowered by frictional locking. This provides the change of direction for the working stroke or return stroke. The flywheel has a fixed connection to the three-speed ball screw. The screw rotates in the nut, which is fixed in the head of the press frame, and moves vertically together with the rotating mass. This increases the contact radius on the drive disk with the downward movement. As a result, the flywheel is accelerated as required while the drive disk rotates at a constant speed. The transmission ratios are very disadvantageous when raising the ram with the opposing drive disk driving the flywheel

3

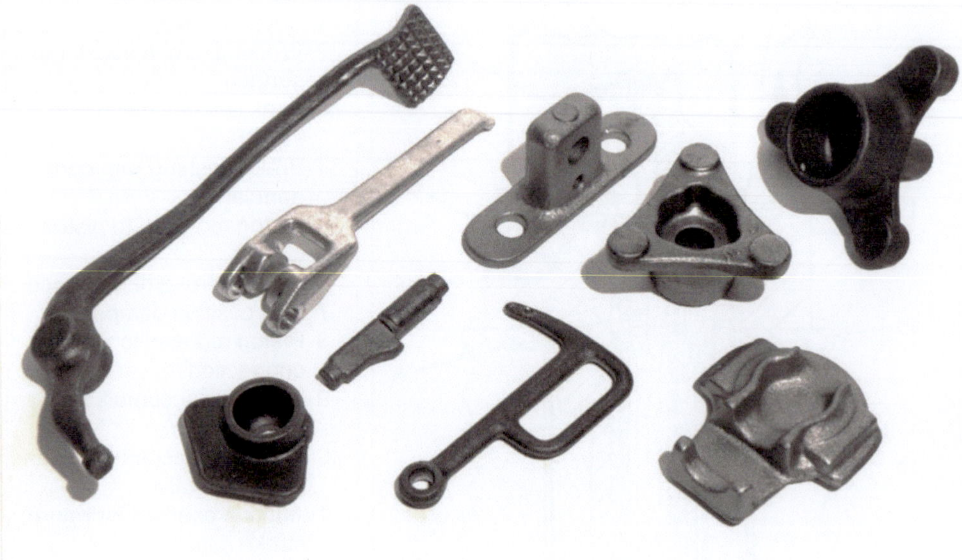

Fig. 3.13 Typical components manufactured by die forging. *Source* Siempelkamp

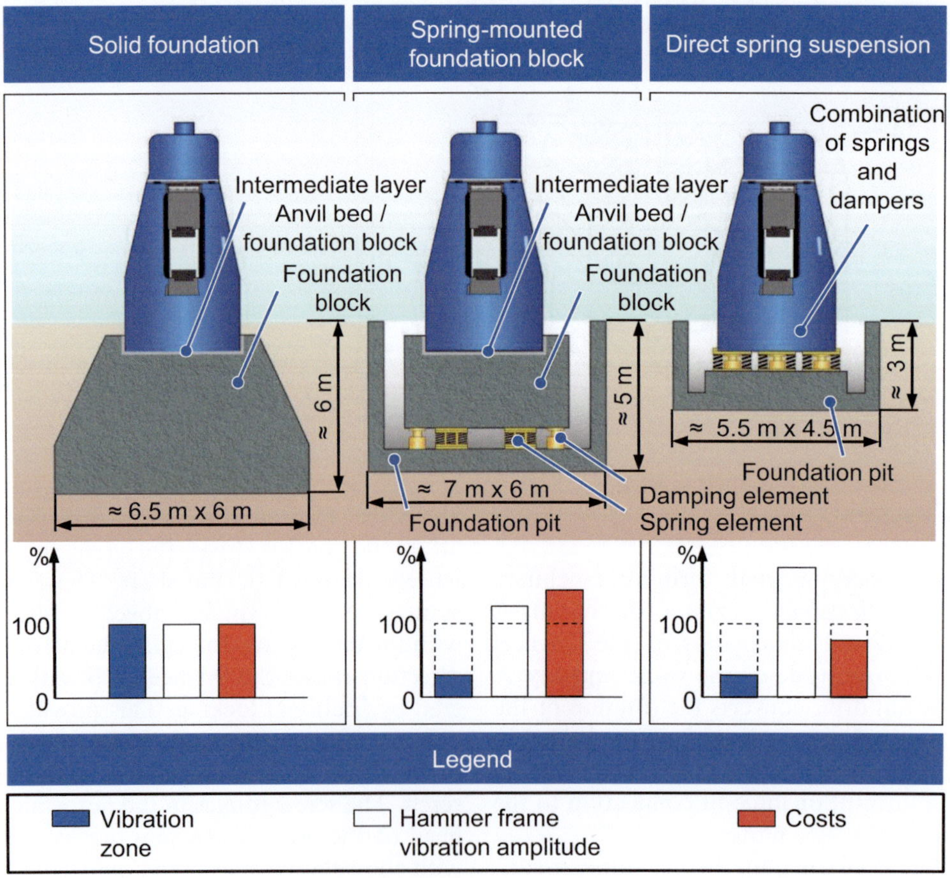

Fig. 3.14 Vibration isolation on forming machines. *Source* GERB Berlin

through frictional locking. The flywheel has to be accelerated from an almost stationary position at the maximum friction radius of the driving wheel. This means the slippage and the associated wear on the friction linings are very high, resulting in very low efficiency.

The disadvantages of conventional three-disk presses can be avoided by using more expensive drives such as the four-disk screw press, ◘ Fig. 3.15. More straightforward drive concepts can be achieved with single-disk screw presses and screw presses with direct drive. Here, the drive motor drives the flywheel directly or the motor's rotor is a part of the flywheel. The disadvantage of these concepts is that the fixed screw has to support both tensile or compressive stresses as well as torsion stresses between the axial bearing in the press crown and the ram, and the nut has to be integrated into the ram. This means a larger screw diameter is required for the same transmissible forming force than the screw on disk screw presses. Here, the screw is subjected exclusively to torsion loads above the fixed spindle nut and exclusively to tensile or compressive loads between the spindle nut and the ram.

Tie rods are frequently used to preload press frames on friction screw presses, ◘ Fig. 3.16. The nuts, which transfer the preload force and the forming force from the frame to the tie rods, are designed as tension nuts. The special design of the tension nuts ensures the first threads are not overloaded. The connection between the screw and the ram is achieved with a hydrodynamic axial bearing. The drive shaft is controlled by a compressed air cylinder mounted directly on the shaft. An electro-pneumatic brake acting directly on the flywheel ensures the ram is brought to a halt as quickly as possible.

Due to the poor energy transfer and high wear on the friction wheel drive, machines with greater forming capacity (>80 MN die-to-die blow force) make use of a pinion flywheel drive in a single-disk design,

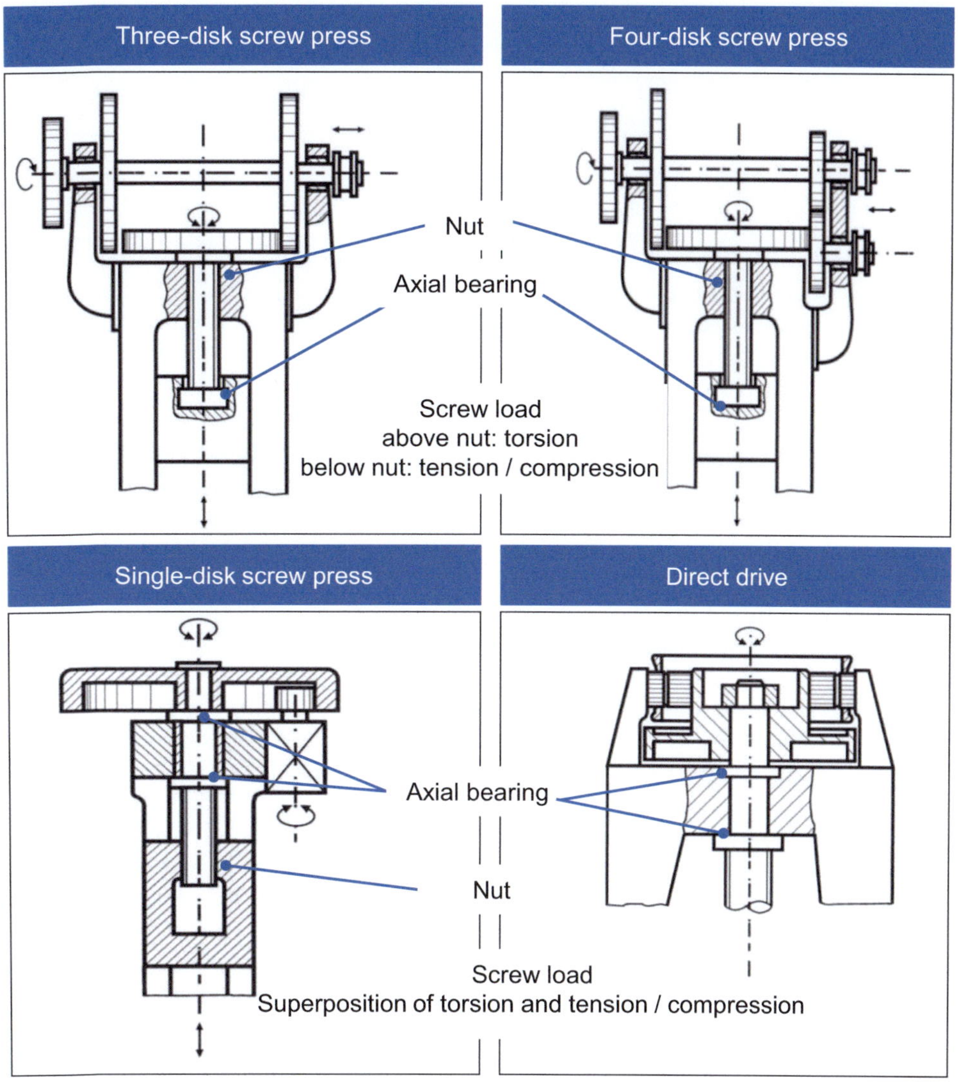

◘ Fig. 3.15 Screw press drives

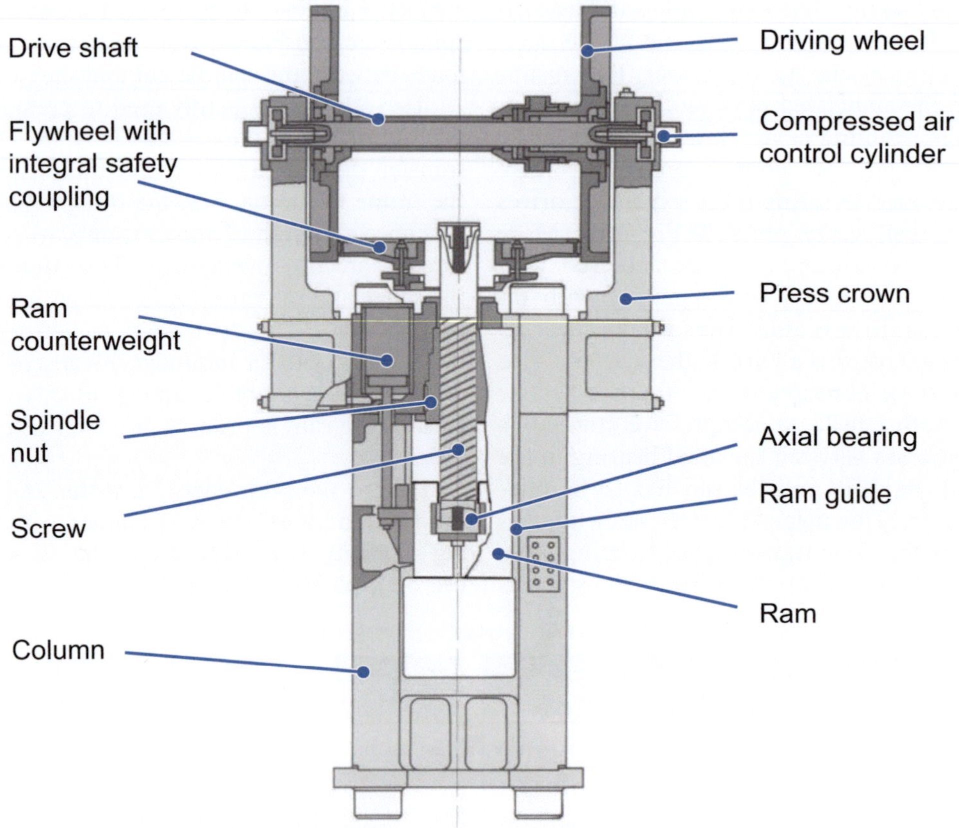

Fig. 3.16 Friction screw press in three-disk design. *Source* SMS group

Fig. 3.17. Here, the flywheel is a helical gear design. The flywheel is driven by pinions connected to asynchronous motors or hydraulic axial piston motors. The flywheel and screw are connected by a slipping clutch in order to limit the pressing force. During its upward and downward stroke, the screw rotates through the nut that is set in the cross member of the body of the press.

In each work cycle, the flywheel and screw are accelerated by the drive motors and move up and down by the stroke length. As the motors are fixed in the body of the press, the pinions slide in the flywheel's gearing in the longitudinal direction of the teeth. The geared flywheel is therefore designed very wide. The minimum width is derived from the stroke of the press plus the width of the pinion. Reversing the rotational direction of the motors provides the return stroke. The pressure supply and control in hydraulic drives are provided by pump units, hydraulic accumulators, and control blocks, which are located outside the machine.

The system of a screw moving longitudinally as opposed to having a fixed screw has the advantage that the screw is subjected only to torsion loading above the spindle nut and only to compressive loading below it. Fixed screws with the same cross section on the other hand are subjected to both compressive and torsion forces. This means that they can only transfer lower pressing forces with the same screw diameter.

This type of press with a fixed screw and a spindle nut located in the ram is shown in Fig. 3.18. A significant feature of this type of forming machine is its direct drive in which the flywheel and spindle are driven by a contactless reversing motor. The drive motor's rotor is connected directly to the flywheel and the screw while the stator is fixed to the crosshead of the press body. The torque required to accelerate the drive masses is produced by the magnetic field between the rotor and the stator.

Compared to the drives described above, the mechanical efficiency of this direct electrical drive is very high at 70 to 80% due to the slippage and other mechanical losses being lower. In addition, the machine's electrical control also allows the work capacity to be set with very high accuracy to match the forming task. A particular feature of the screw press shown here is the hydraulic cushioning integrated in the longitudinal direction of the screw, which prevents the machine from being overloaded. This hydraulic overload protection system performs the function of a slipping clutch.

The press frame consists of a monoblock column. The ball screw bearing arrangement to absorb the downward weight of the screw comprises a rolling element bearing arranged on the upper side of the head piece. The load from the forming process is supported by a pain sliding bearing on the underside of

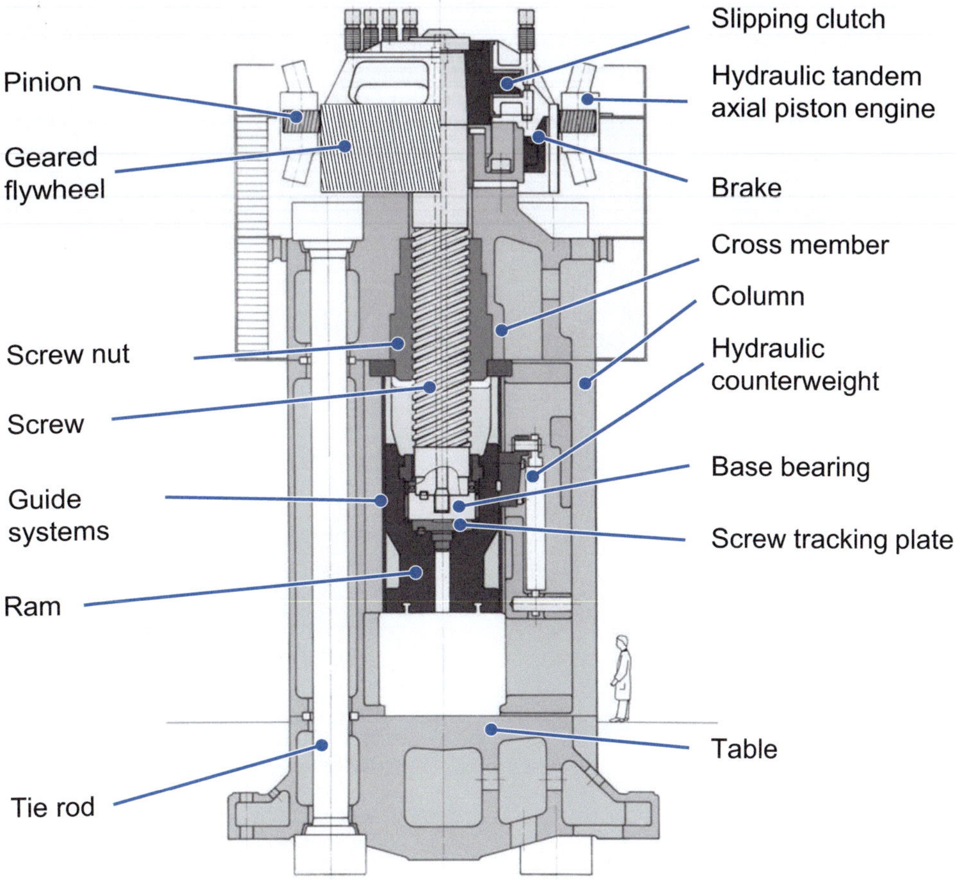

Fig. 3.17 Screw press with hydraulic gear drive. *Source* SMS group

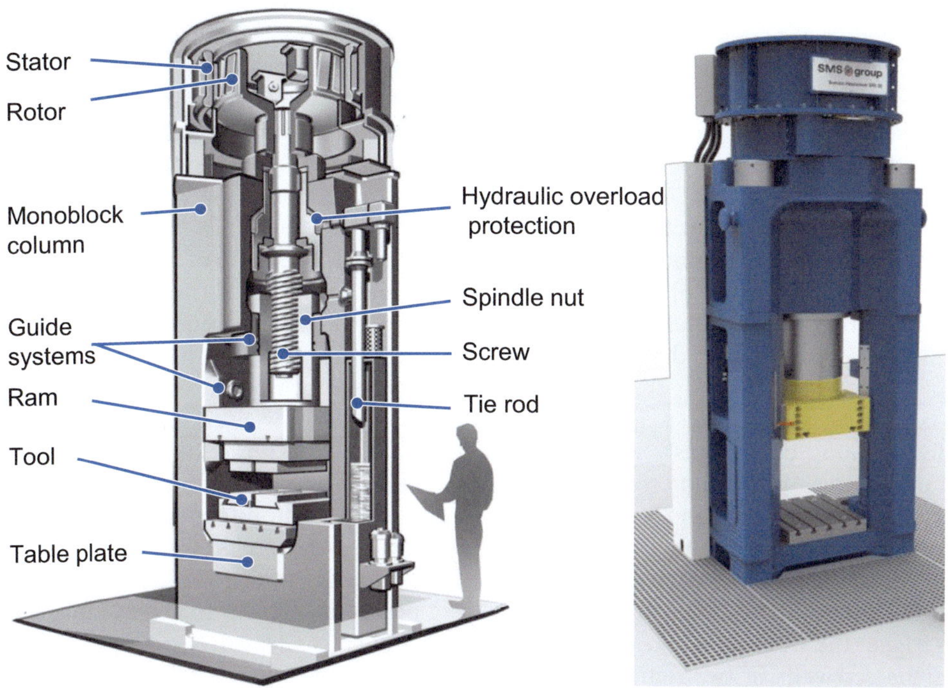

Fig. 3.18 Screw press with direct electrical drive. *Source* SMS group

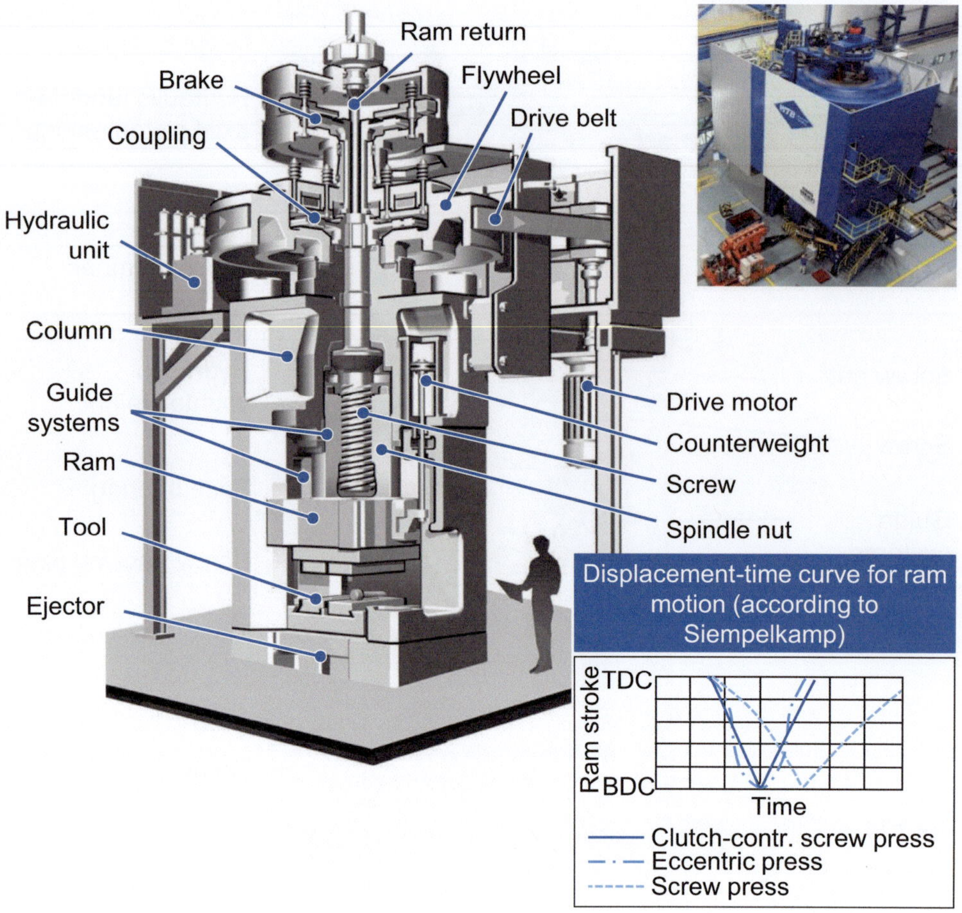

Fig. 3.19 Clutch-controlled screw press. *Source* SMS group

the crosshead. As the pressing force increases, the ball screw moves upwards by the axial clearance in the thrust bearing and comes to rest with its thrust collar against the bronze thrust bearing on the underside of the head piece.

Figure 3.19 shows a screw press with a drive principle in which the disadvantages of screw presses caused by their inherent design such as low stroke rates, limited forming capacity due to the acceleration capacity, and dependency between energy and pressing force no longer apply [GOLF85, SCHI86]. The essential modification compared to conventional screw presses is the connection between the motor, the flywheel, and the screw.

The clutch-controlled screw press is characterized by a flywheel rotating continuously in one direction (similar to crank and eccentric presses). This flywheel is driven by an electric motor and V-belts. As the flywheel is always rotating in the same direction at an almost constant speed, hardly any acceleration is required, meaning the electric motor does not need to be so powerful. To trigger the working stroke, the flywheel is connected to the screw drive by frictional locking via a switchable clutch with several hydraulic cylinders acting on the clutch ring. The drive plate, screw, and ram are accelerated to full rotational speed extremely quickly due to their low mass compared to the flywheel. This means that almost the entire working stroke can be completed at a constant ram speed with a constant forming force limited by the controllable clutch torque.

More broadly speaking, there is no dependency between the pressing force and the work capacity. Even during the actual forming process, this ram speed is maintained except for a small reduction, which is proportional to the speed drop of the flywheel. This means that only the energy required to form the workpiece is extracted from the flywheel during forming, as well as any losses which occur. Only some of the energy stored in the flywheel is used for each forming process. The end of the working stroke is achieved by releasing the coupling. This decoupling process is automatic and depends on a friction torque preset by the oil pressure or when a preset forming travel has been reached. Retraction cylinders return the ram to a preset upper starting position. In order to achieve this, a non-self-locking screw pitch is selected.

Another special design feature of clutch-controlled screw presses is the thrust bearing used to provide the axial bearing arrangements on the screw. It has a spherical design, which avoids the screw bending if eccentric loads occur. A combination of round and diagonal guide systems ensures the ram is precisely and centrally guided on the drive system as well as parallel to the table surface, even with eccentric loads. This design is illustrated in ◘ Fig. 3.20. The ram consists of a single-piece, spheroidal graphite cast iron structure. It is guided by both a round guide system on the head piece of the press column as well as an adjustable flat guide system between the column and the ram. This flat guide system has an O-guide design and is largely unaffected by temperature. The arrangement chosen ensures a very tight guide clearance is set in order to achieve high machining accuracy. For this, the guide rails on the ram can be adjusted using a wedge provided with slots. The required longitudinal motion of the wedge is provided by turning the hollow threaded bolt in the internal thread on the baseplate. The optimized guide rail is then locked using a counternut and screw.

Clutch-controlled screw presses combine many beneficial forming machine features that distinguish them from other designs. These include variable length of stroke, short pressure dwell time, overload protection, high impact sequence, off-center load carrying capacity, high forming energy, high efficiency, and presettable final press force.

The work capacity of the screw presses described here is up to 6,000 kNm with nominal forces of up to 120 MN and maximum impact rates of 30 min^{-1}. Screw presses have a short and hard impact, somewhere between that of a hammer and a crank press. This means they are best suited for solid stamping and forging work.

Typical installation examples are stamping and shaping cutlery and coins as well as forging levers, flanges, steering knuckles, etc. Another important application area is calibrated forging and precision forging of turbine blades [REIH97], which are manufactured with high dimensional accuracy.

3.1.3 Eccentric and Crank Presses

Eccentric and crank presses are travel-related forming machines with a slider-crank drive. An essential feature of this type of drive is the movement of the working ram. This movement is dictated by the kinematics of the slider crank mechanism and the forming force, which is dependent on the ram position.

◘ Figure 3.21 shows the design principle of this type of press drive using a crank press as an example. An electric motor drives a crankshaft via a belt, a pneumatically or hydraulically activated clutch, and a flywheel. This crankshaft moves the ram, which is generally guided vertically in the frame, up and down in a straight line via a connecting rod. The brake is

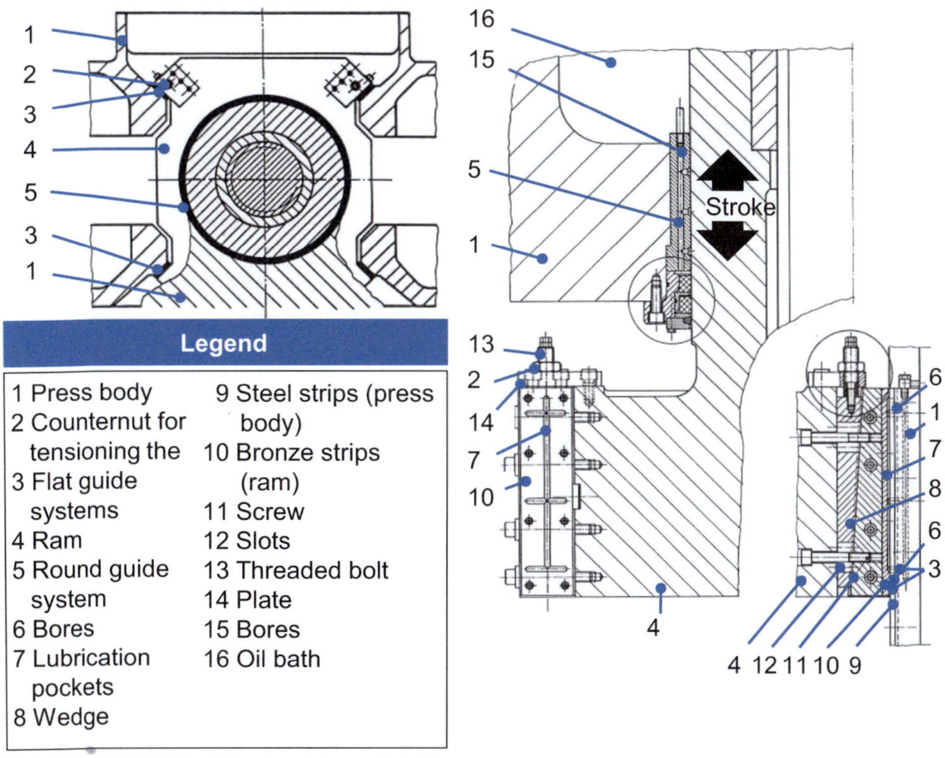

◘ **Fig. 3.20** Ram guide on a clutch-controlled screw press. *Source* Siempelkamp

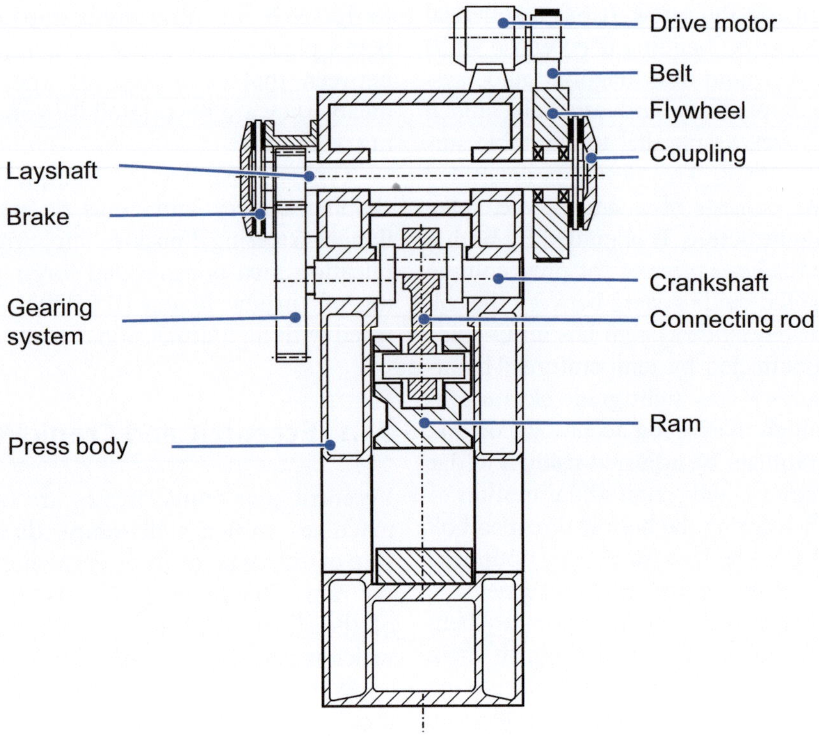

Fig. 3.21 Drive system diagram of a crank press

designed to brake the press ram within a given stopping distance. The crankshaft drive can be direct or via an additional gearing system arranged between the flywheel and the slider crank mechanism as shown here. This allows the rotational speed and work capacity to be adjusted. The flywheel acts as an energy storage medium that provides the energy required for a single forming process.

The press stroke on eccentric presses is produced by an eccentric shaft. A distinction is made between machines with fixed and adjustable eccentrics, Fig. 3.22. The benefit of a fixed eccentric compared to a crankshaft is the stiff and non-weakened shaft design. On eccentric drives with an adjustable stroke, an eccentric disk is mounted on the crankshaft crankpin. The connecting rod incorporates the eccentric disk. The resultant eccentricity of the drive and therefore the stroke can be adjusted via the relative angular position of the pin and the eccentric disk. During operation, the eccentric disk is usually held in its position relative to the crankshaft by a Hirth serration. This position is set by the control system in order to calculate the top dead center of the ram.

The design of crank and eccentric presses is somewhat similar in terms of the eccentric shaft design and the possibility of adjusting the stroke. A distinction is often made between the two designs according to the stroke adjustment options. Compared to crank presses, eccentric presses of the same size generally have a significantly smaller ram stroke, higher stroke rates and a stiffer design due to decreased distances between bearings.

Eccentric presses are mainly used for cutting, stamping and bending work, as long as the smaller ram travels are sufficient. Crank presses are used in preference for forming work where large strokes are required and the forming speed along the ram travel does not have to be constant, i.e., for extruding, drop forging, deep drawing, and bending.

Both machine types can be further distinguished by the type of ram drive and the frame design, Fig. 3.23. In terms of the frame design, a distinction is made between the open, cantilever C-frame, and the closed, O-frame with a two-column design. A further distinction is made between C-frames with single-column and double-column designs, which are mainly used on small to medium-sized presses. The single-column design is characterized by a one-piece vertical column with a rigid box design. Similar to the two-column design, the double-column frame consists of two parallel frame walls directly connected to one another at the press head and the press table by cross connections. For O-frames, the connections in these assemblies are achieved on small machines by using a single-piece machine body. Split machine bodies clamped with tie rods are used on larger machines.

Figure 3.24 shows a single-column eccentric press with an adjustable ram stroke. As described above, the

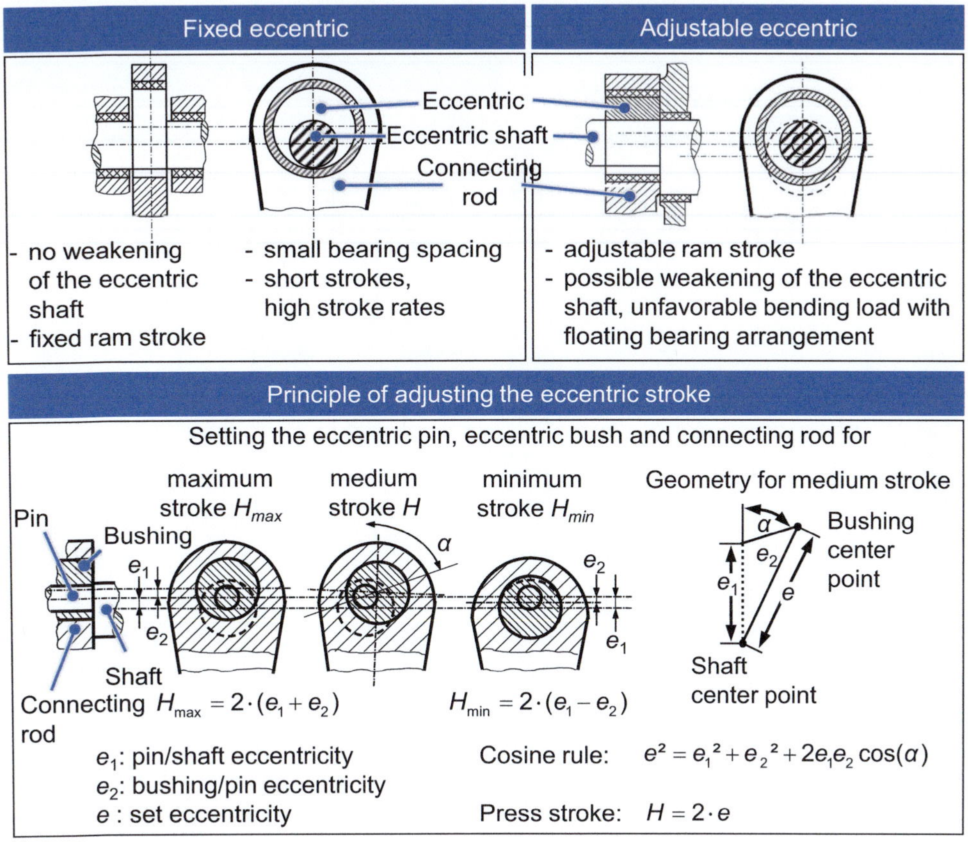

Fig. 3.22 Eccentric shaft drive designs

stroke is adjusted by rotating the eccentric disk relative to the eccentric shaft pin. The connecting rod is in two parts to facilitate height adjustment. The connecting rod length can be modified by a thread adjustment. This allows the position of the ram at the top or bottom dead center to be set and then adjusted to the relevant tool height. A speed-controlled, three-phase a.c. motor rotates the flywheel on the drive shaft of the gearing system via a belt drive. This configuration drives the machine. The coupling/brake combination, which is duplicated for safety reasons, is activated pneumatically. If it is activated, the flywheel is locked to the drive shaft. If there is a pressure drop, the flywheel is decoupled and the drive shaft is braked.

C-frames have the disadvantage that they bend up under forming forces, which means that alignment errors may occur in the tool halves. This is offset by having the workspace accessible from three sides. The deflection and bending of the frame under load can be reduced by adding extra tie rods between the press table and the press head. Closed frame designs are used in medium-sized and larger machines and are always used if the tool requires particularly stiff and precise guiding due to the forces occurring during the forming process. The frames are made from spheroidal graphite cast iron, cast steel, or, increasingly nowadays, in a welded design. As far as the position of the eccentric shaft or crankshaft with respect to the direction of material flow or the frame is concerned, a distinction is made between longitudinal and transverse shaft drives as well as whether the drive is from above or below, Fig. 3.23. Further differentiation is made between single-point, two-point, and four-point systems as defined by the number of connecting rods moving the ram. Two-point and four-point systems are primarily used for wide and large-surface rams, which avoids the ram tilting in the guide systems under eccentric load. Apart from the design of the frame and the drive, the motion accuracy of the ram is largely determined by the type of ram guide. The most significant ram guide designs are shown as diagrams in Fig. 3.25. They are generally based on a slide guide system and are used in all types of forming machines depending on the frame design.

Presses with a C-frame design frequently have dovetail or prismatic guide systems whereas forming machines with a closed frame design usually have two additional prismatic guide systems, also known as diagonal or X-guide and O-guide systems. Machines with increased accuracy requirements for the ram guide are fitted with flat guide systems in a square configuration. A high machine stiffness is also achieved in the horizontal direction using this principle due to the type of ram guide geometry. Rolling element guide systems or

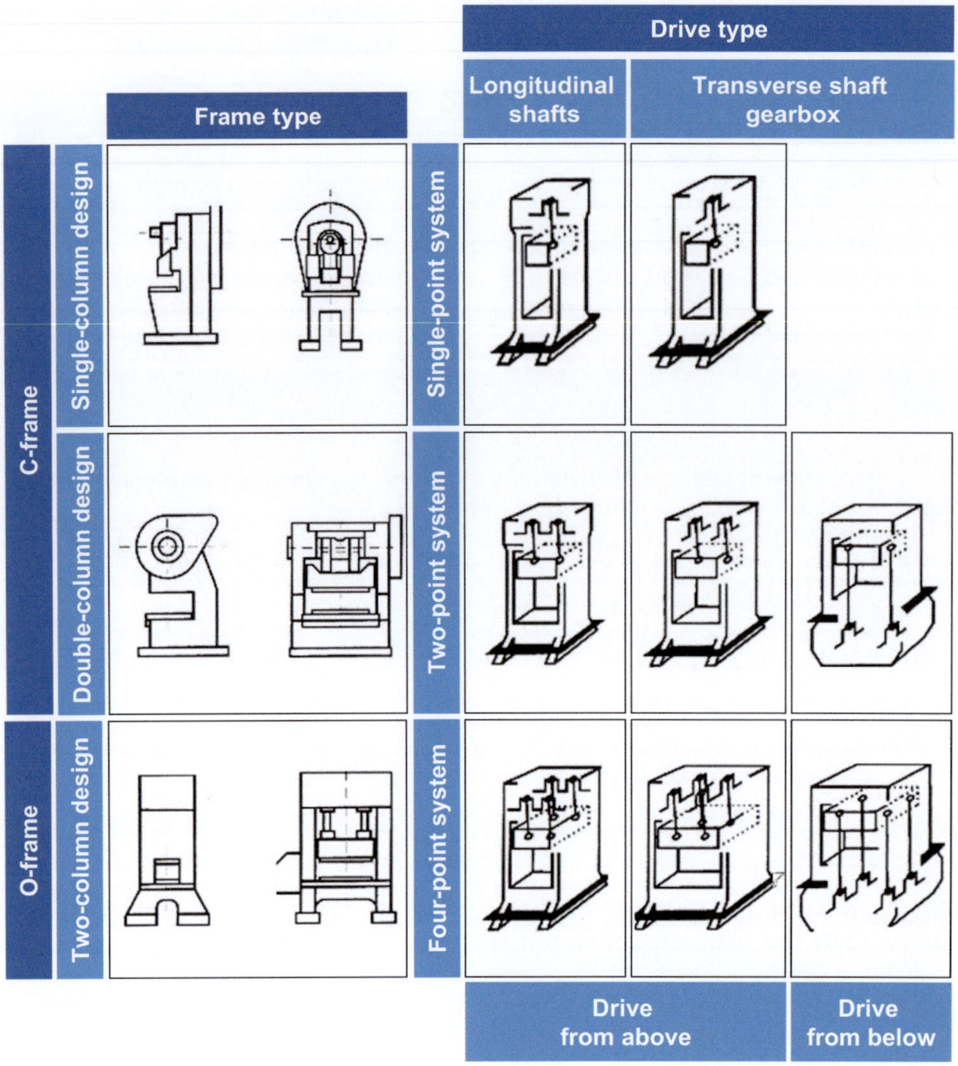

■ **Fig. 3.23** Types of crank and eccentric presses

hydrostatic guide systems are also used in presses today. Round guide systems are frequently used in hydraulic presses driven from below. In addition to the ram guide geometry, the effective guide length and the guide clearance also influence the achievable motion accuracy. The latter can be easily set and adjusted in prismatic and flat guide systems via the guide rails arranged in the press frame using screws, round wedges, or tapered gibs. In terms of manufacturing technology, round guide systems are significantly easier and more economical to produce than flat or prismatic guide systems.

■ Figure 3.26 shows a cross-sectional drawing of a two-column eccentric press with longitudinal shaft drive in a two-point design. The stroke rate is set by the speed controlled, three-phase a.c. motor. The gearing system has helical gearing and runs maintenance-free in an oil bath. The coupling/brake combination has a single-disk design and is pneumatically actuated. The free shaft end of the eccentric shaft is available for driving the material feed and transfer devices. The ram is guided by eight long, flat guide systems. These are designed as either a bronze slide guide system or as a clearance-free, adjustable roller guide system for very high accuracy requirements. Due to the geometry of the guide systems, a high machine stiffness is also achieved in the horizontal direction with this guiding principle. Other significant design features of this type of press are the ram stroke adjustment using an eccentric disk, the motor-driven stroke position adjustment, the hydraulic overload protection and the pneumatic ram counterweight. The ram counterweight allows the press to run smoothly as all drive elements with clearance are preloaded from above. These structural factors mean the forming and cutting presses are universal machines. They are suitable for both forming at a high work capacity (e.g., deep-drawing) and cutting at high stroke rates. They can be used for drawing,

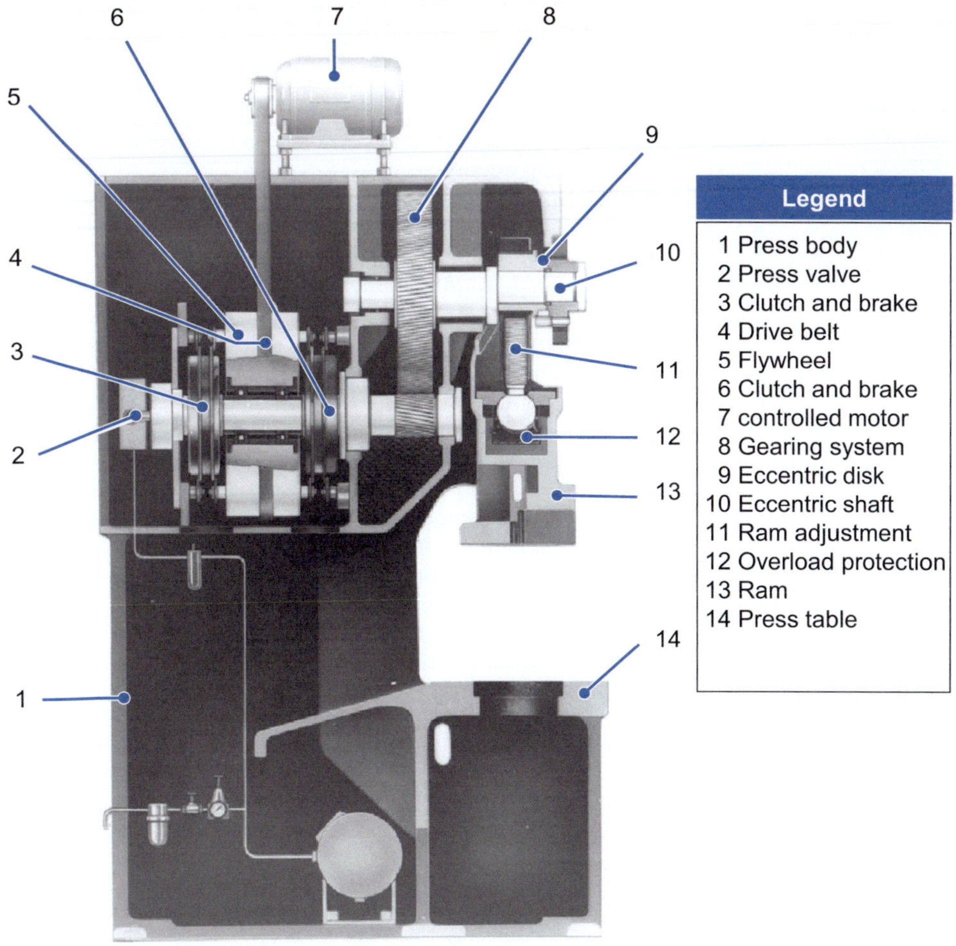

Fig. 3.24 Single-column eccentric press with adjustable ram stroke. *Source* Schuler

cutting and stamping work with single, progressive, or transfer dies. Their further application ranges from large-scale production with an automated material feed to small-scale production of high-quality workpieces with hand-operated machines. Due to their small and symmetrical deflection under load and the precise guidance, this type of machine delivers long tool life and tight workpiece tolerances. The pressing forces of these machines extend up to 6300 kN with stroke rates of between 20 and 250 min^{-1} during continuous operation. Machines with higher pressing forces (10,000 kN) are more frequently designed with a transverse shaft drive. The driving mechanism is particularly stiff due to the short axes of the transverse shafts. Another application area for two-column eccentric presses is the large-scale production of punched parts. Special eccentric presses, so called punching machines and high-speed presses are used here, see Fig. 3.27. These are characterized by an increase in plant efficiency, particularly due to their very high working stroke rates and feed rates [HELL12]. Stroke rates of 2000 min^{-1} can be achieved depending on the press force and the ram stroke.

At these high stroke rates, the oscillating movements of the connecting rod, ram, ram plate and upper die produce very high inertia forces. These forces increase quadratically with the stroke frequency, placing the machine under high dynamic loading conditions. For this reason, high-speed presses are fitted with fully dynamic counterweights that compensate the oscillating and the rotating inertia forces in both the horizontal and vertical directions. The dynamic compensation of the reciprocating masses of the ram and upper part of the tool, shown in red in Fig. 3.27, is achieved by the opposing dynamic force of the counterweights, which is almost equal, shown in yellow in Fig. 3.27. In the machine shown in Fig. 3.27, these compensation forces are produced by the opposite motion of the counterweights produced by the main levers. The horizontal inertia forces of the eccentric shaft are compensated by the movement of the counterweights, which are moved from above by two horizontal linkages. This mass balancing solution automatically accommodates changes in the press stroke and the ram installation position. This drive system with counterweights ensures

System	Type	Design	Properties
Round guide system		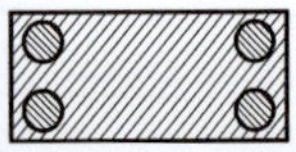	- poor setting and adjustability of the guide clearance - easy and economical manufacture
Rectangular guide system		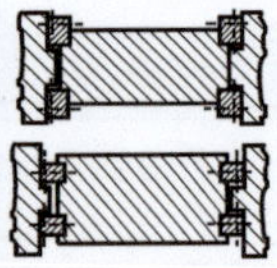	- high guiding accuracy - great horizontal stiffness - easy and accurate adjustability of the guide clearance using screws, round wedges or tapered gibs - complicated to manufacture and adapt
Prismatic guide system	O-guide	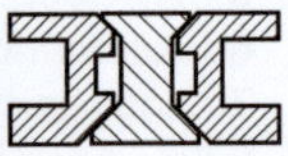	- O arrangement - lines of action of the force result in an O-shape - great horizontal stiffness - good adjustability of the guide clearance - tracking behavior unaffected by temperature
Prismatic guide system	X-guide	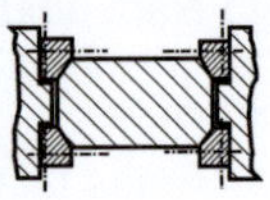	- diagonal or X arrangement - lines of action of the force result in a X-shape - great stiffness, economical - good adjustability of the guide clearance - poor guidance around the vertical ram axis
Prismatic guide system	X-guide		- for C-frame designs - easy setting of the ram clearance - relatively low manufacturing costs
Dovetail guide system			- for C-frame designs - low space requirement, high manufacturing costs - stiffness dependent on the accuracy of the fit

Fig. 3.25 Principles of ram guides

the machine runs almost vibration-free, regardless of the operational settings. The residual vibrations from the machine running and the punching process are absorbed by spring elements in the machine feet. This eliminates the need for expensive foundations.

Other features of high-speed presses are short strokes, stiff frames, fast-actuating coupling/braking systems, extremely small bearing clearances, and high-precision ram guides. As these types of installations are highly efficient when combined with automatic material feed and removal systems, they are frequently used in large volume production for cutting laminations for rotors, stators, and magnets. In addition to this, they can also accommodate automatic tool change equipment to keep downtime to a minimum. An example of this type of high-speed press for punching rotor and stator laminations is shown in Fig. 3.28: This press has a nominal force of 4000 kN and a stepless, adjustable stroke rate of up to 400 min^{-1}.

The main motor on this press has a vibration-isolated mounting so that the drive unit is protected from the punching vibrations caused by the process. The eccentric shaft is arranged in the longitudinal direction of the press. The distance between the two connecting rods minimizes the ram tilt caused by the off-center loading. The connecting rods are fitted with plain bearings, while the eccentric shaft has four rolling element bearings in the head piece of the press. High horizontal stiffness is achieved by using column guides whose rolling elements are arranged in the cutting plane and in a second plane above the ram. The high-speed press has a dynamic counterweight to improve the dynamic machine behavior.

The strip material is fed to the press from a reel using a loop and the strip feed speed at the reel is automatically controlled by the loop sag. A roller feed, which is driven by an NC-controlled servo drive, is used for the strip feed system. The eccentric shaft is also

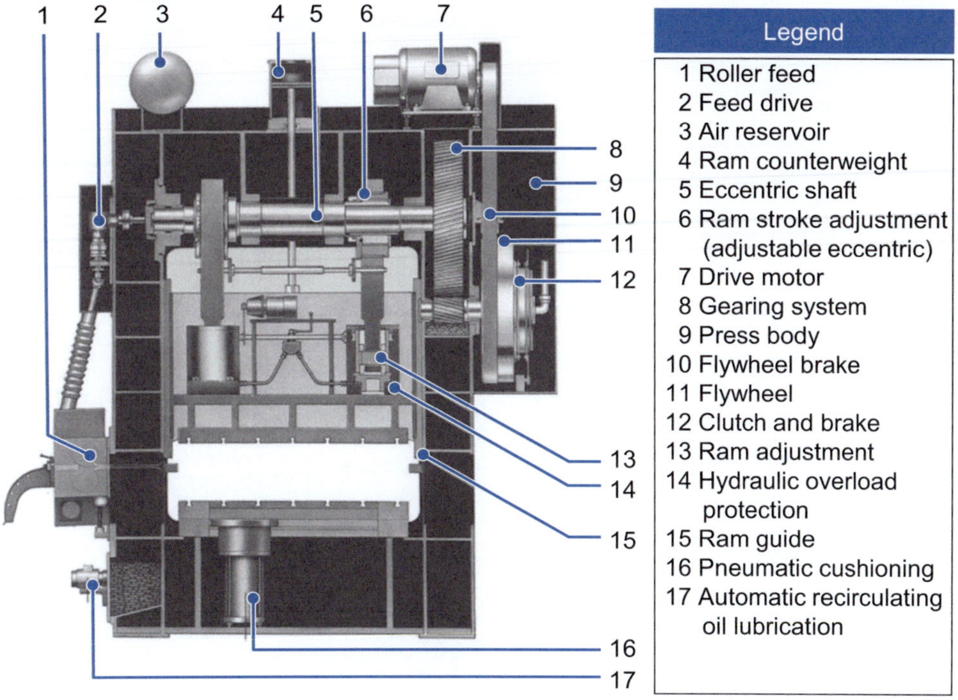

Fig. 3.26 Two-column eccentric press with longitudinal shaft. *Source* Schuler

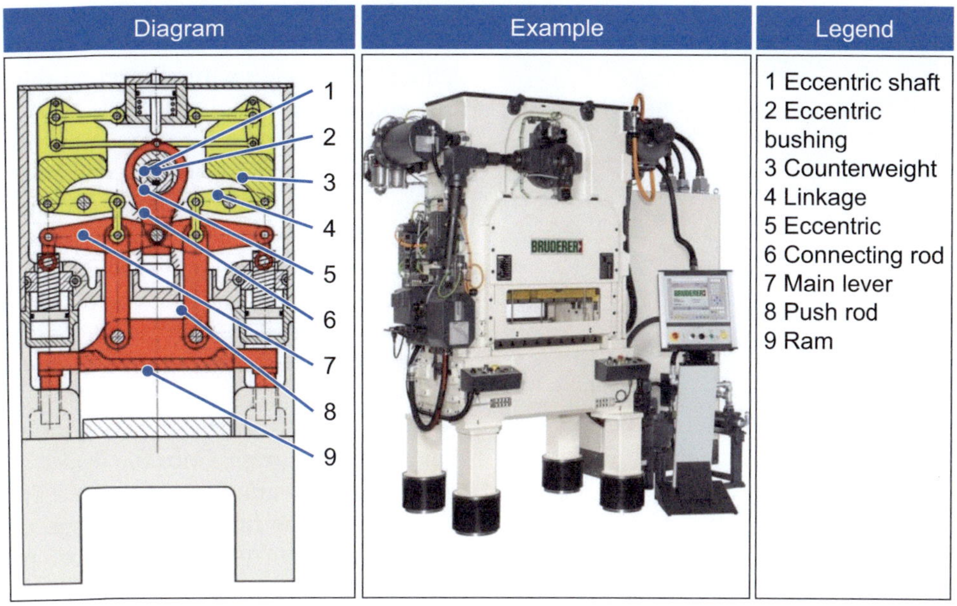

Fig. 3.27 High-performance punching machine with counterweight. *Source* BRUDERER

frequently used to drive the roller feed via a cam mechanism with a form-fit connection. Suitable stacking systems and transport facilities are used in continuous operation to arrange the rotor and stator laminations in accordance with their groove patterns, then to stack them on mandrels and remove them from the press. To reduce noise pollution, the presses are frequently fully enclosed in a noise protection cabin.

Figure 3.29 above shows a typical follow-on cutting tool used to produce the workpieces in a two-stage cutting process. Today, multi-row cutting processes are also used in addition to single-row tools. These offer the benefit of being able to be used on wider and more economical strips.

In addition to this, complete cutting tools are also used where all the cutting operations to manufacture

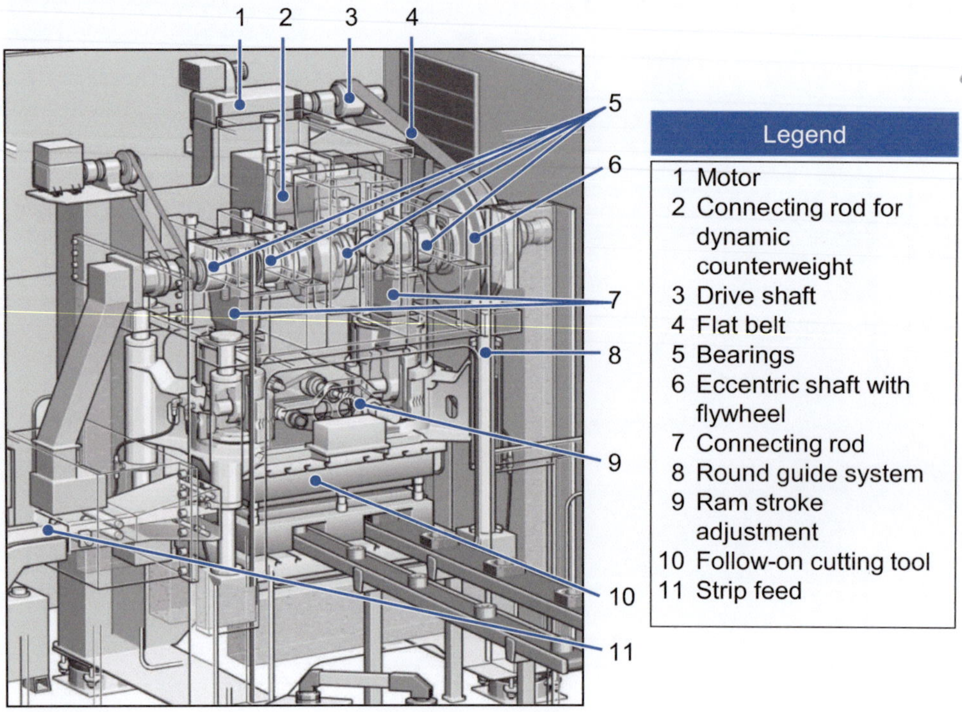

Fig. 3.28 High-speed press for punching rotor or stator laminations. *Source* Schuler

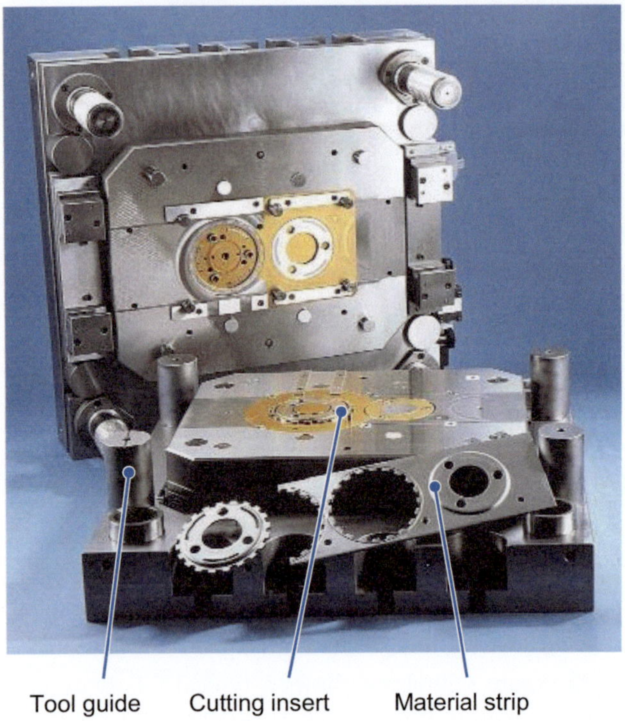

Fig. 3.29 Cutting tool for use in high-speed presses. *Source* Schuler

a finished part are performed simultaneously. These tools are predominantly used for rotor and stator laminations, round blanks, rings, and segments in medium-sized batches. As there is a particular risk of tool offset with complete cutting tools, particularly high demands are made with respect to the guidance accuracy and stiffness of the press. If the manufacture of a workpiece requires more than one working operation and cannot be completed on one universal press, the workpiece has to be processed by individual tools serially. To make the manufacture of sheet metal parts economical, the required individual operations are combined in a special type of two-column press known as a transfer press. The number of tools depends on the number of work stages required to manufacture a step-drawn part.

Figure 3.30 shows the tool mounting area in a typical transfer press. Once the feeder system has taken the workpiece blank from the centering station and inserted it into the first pressing stage, the workpiece is subsequently transported by gripper rails. These gripper rails are controlled in three axes and move the workpieces from one station to the next within the cycle of the installation, Fig. 3.31. An intermediate stage must be provided if the transport distances between the tools vary, for example, due to press columns after the first forming stage. The rails are fitted with pneumatically operated active grippers or paddles as well as interchangeable jaw inserts depending on the geometry of the sheet metal part. For reasons of kinematics, the gripper rails are manufactured in a low-mass box design from hollow profiles. Part transportation is electronically monitored during the process for the presence of parts, the position of the parts

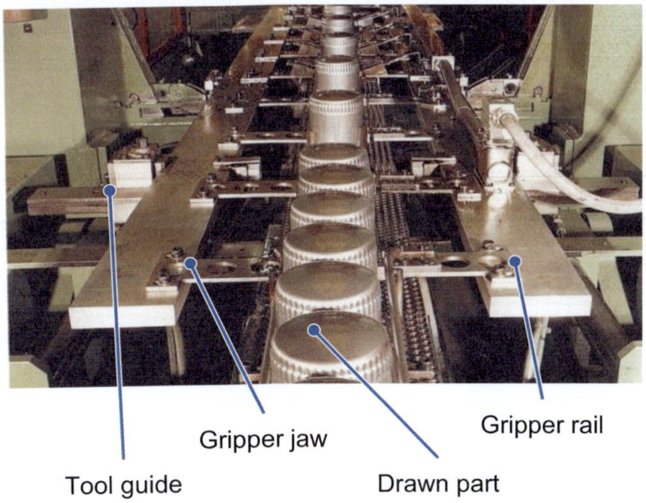

Fig. 3.30 Tool mounting area in a transfer press. *Source* Schuler

and other functions in order to identify process faults as early as possible. Longitudinal and transverse movements controlled by the cam mechanism as well as any lifting movements of the transport equipment are often governed by the main drive and are thus synchronized with the ram motion.

High production output can be achieved with transfer presses because all the manufacturing steps required to produce a workpiece can be performed simultaneously. A part is completed with every ram stroke, regardless of how many workstations are required for production in each individual case. Processes with multiple actions are often required for deep-drawing work. While the tool punch drives the sheet through the die plate, the external edge of the sheet (frame) must be held in place by a hold-down mechanism in order to avoid pleating. The material is drawn into the die plate as the drawing process progresses.

This type of process can be achieved, for example, with presses having a mechanical or hydraulic drive for the hold-down mechanism, Fig. 3.32. The crankshaft synchronously moves a slide vertically in the opposite direction to the drawing ram. A rocker arm with a fixed point of rotation and a push rod provide the movement for the hold-down mechanism. The constant low position of the hold-down mechanism during the actual deep-drawing stroke of the drawing ram is achieved through the gear elements being configured as a gearbox with resting points in that the slider-crank drive and the toggle lever drive are in their return positions at approximately the same time.

One of the main areas of application for mechanical deep-drawing presses is the manufacture of automotive body parts. Here, large-surface blanks are individually processed in work stages on single-action or double-action body presses. Depending on the number of manufacturing steps, the body presses are arranged in sequence with the lead-off press frequently operating as a double-action press and the subsequent presses as single-action. Appropriate feed, transport, and stacking devices link the individual body presses together to form an automated press line. If drawing mechanisms are included, then deep-drawing operations can also be carried out on single-action presses [HOFF91, HOFF92, KOHN01]. A typical drawing mechanism is illustrated in Fig. 3.33. Before the forming process is started, the blank is clamped between the upper part of the tool and the sheet holder. The sheet holder is then raised by plunger cylinders via a lifting element and push rods. During the forming process, the upper part

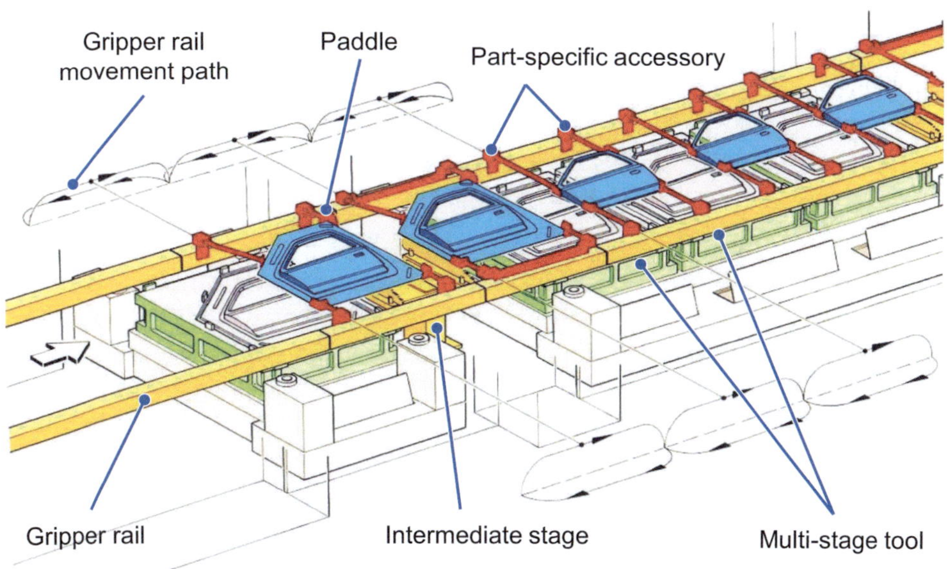

Fig. 3.31 Gripper rail transfer system on a large-panel transfer press. *Source* Schuler

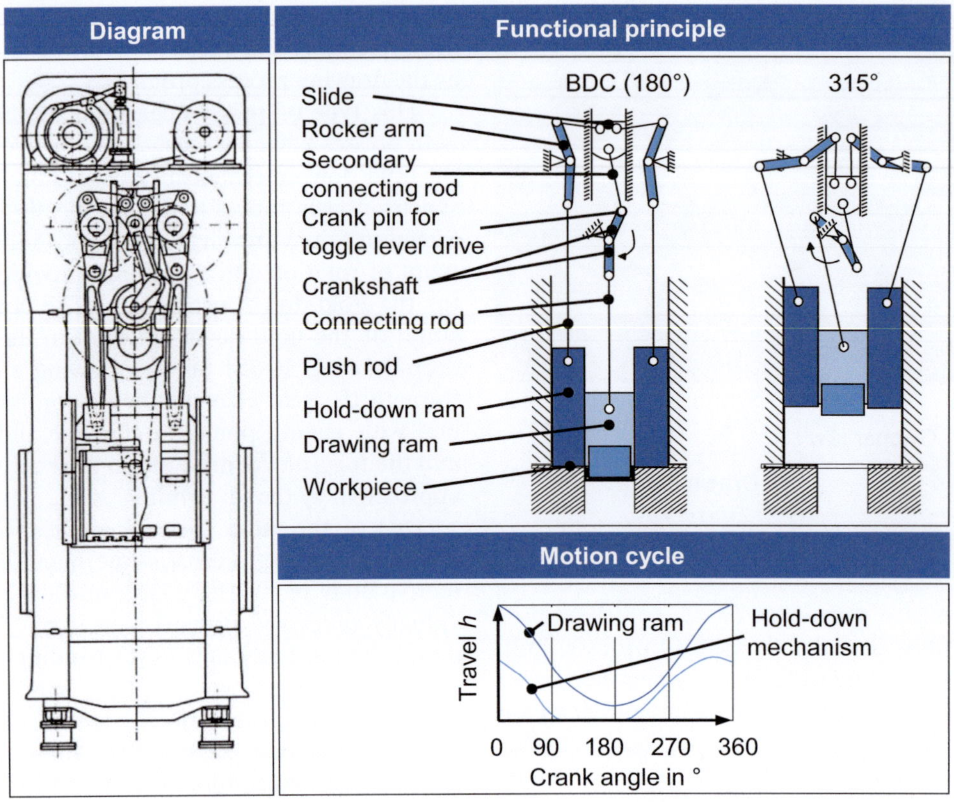

Fig. 3.32 Drive system for a double-action deep-drawing press. *Source* Schuler

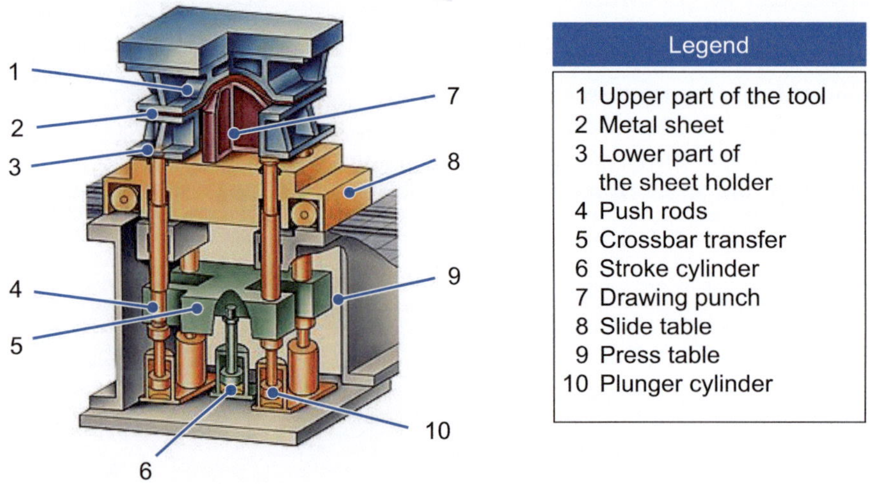

Fig. 3.33 Drawing mechanism. *Source* Schuler

of the tool moves downwards while the sheet holder is also pushed downwards against the sheet holding force. The sheet is formed by the drawing punch while maintaining the sheet holding force, which can be separately adjusted on the four plunger cylinders.

The large-panel transfer press illustrated in Fig. 3.34 was designed for the manufacture of unstable sheet metal parts made from large panels [HOFF89, GEIS96]. In principle, it consists of single-action presses arranged in sequence. It has a modular construction such that an individual ram, head piece, column, and table are available for each of the five forming stations. By having a design that includes a ram for each forming process, the forming force can be individually set as required. It is also possible to avoid influences on the various forming rams as the individual rams are largely centrally loaded.

All the drives and the crossbar transfer are driven via intermediate gears and couplings that are in turn

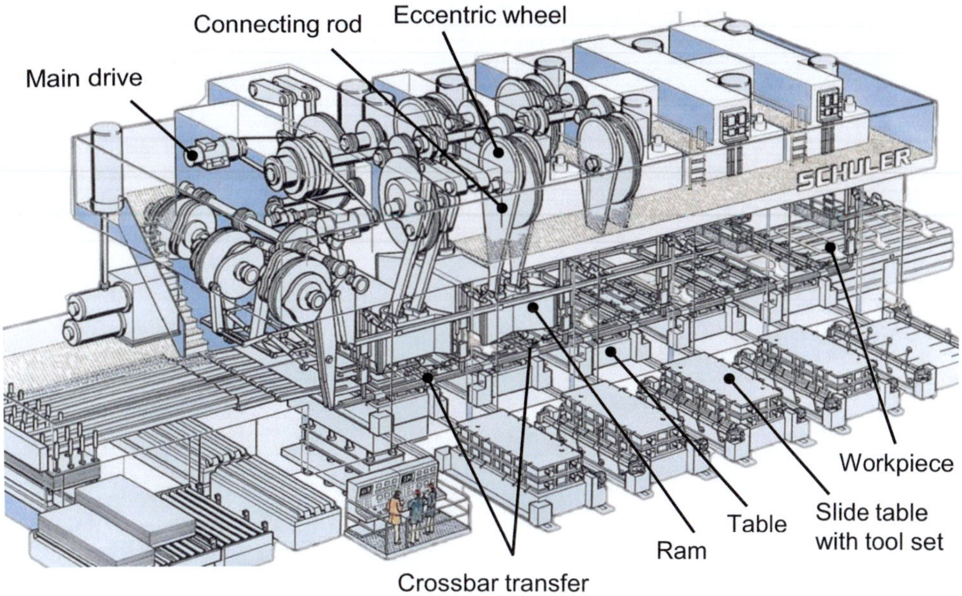

Fig. 3.34 Large-panel transfer press with crossbar transfer. *Source* Schuler

driven by the press crown drive and the transfer drive [HOFF92]. The eccentric shafts have a transverse arrangement to the head piece and drive the rams at four points via eccentric wheels, guiding links, and connecting rods. In this instance, the joint kinematics for the first stage have a special design. One benefit of this modified toggle kinematics is a very low, almost constant, ram speed at the working area. This drive also has high feed and return speeds, Fig. 3.5. The kinematic properties of this joint drive are matched to the specific requirements of the deep-drawing process. The other forming stages have eccentric drives. In order to minimize the ram skewing in the event of off-center loading, each press has four connecting rod pressure points. Slide tables ensure fully automatic tool changes.

Figure 3.35 shows a typical joint drive for use in the initial stage of transfer presses. It shows an eight-part joint drive which is activated by two eccentric systems. The flywheel (2) on the first stage is driven by the main drive shaft of the transfer press (3) and an intermediate wheel (1). The flywheel then drives two symmetrically arranged wheels (5) on the eccentric shafts via additional intermediate wheels. Both eccentric shafts transfer their rotational motion to the ram (15) via two double-arm levers (7, 8) with further coupling levers. Unlike conventional toggle lever systems where the position of the external attachment point of the toggle (11) is fixed, the double-arm lever, which is hidden in the illustration (7, shown on the left-hand side with a dashed line) and acts as a controlling connecting rod, drives an actuator shaft (9), which alters the position of the attachment point of the toggle (11) during the stroke. This means that, for drawing operations, the design of the toggle motion of the lever system that transfers the power, comprising the double-arm lever (9), toggle (12) and connecting rod (13), can be even more favorable with respect to its speed as it approaches the bottom dead center.

Due to the inherent instability of large sheet blanks, gripper rails cannot be used on large-panel transfer presses. On this type of transfer press, the workpieces are conveyed using a crossbar transfer, Fig. 3.36. By using longitudinal beams that move in a vertical direction, the workpieces, which are held by suction cups, are lifted off or lowered onto the dies. The crossbeams, which are directed by control rods, are moved horizontally in pairs in order to load or unload the stations. The workpiece supports between the individual forming stations are moved alternately into a placement or removal position by a control program. In contrast to gripper rails the rail return cannot be realized during the forming process as the crossbars (with the suction cups) are between the tools. So, they are brought to a park position outside the tool area. Figure 3.37 shows a crossbar transfer cycle.

Using the press principle mentioned, a large-panel transfer press was built with a nominal press force of 52,000 kN. Here, the first station is intended for 20,000 kN, the second and third for 9000 kN each and the fourth and fifth for 7000 kN each. The total weight of the press is approximately 2200 metric tons and it is approximately 40 m long, 11 m wide, and extends 9 m above ground level and 6 m below. With a transport step of 2000 mm, the tool clamping surface per tool station is approximately $4500 \times 2000\ mm^2$. This corresponds to a maximum part size of $4000 \times 1750\ mm^2$. The production stroke rate is up to 16 strokes/min. Eccentric and crank presses are used in the field

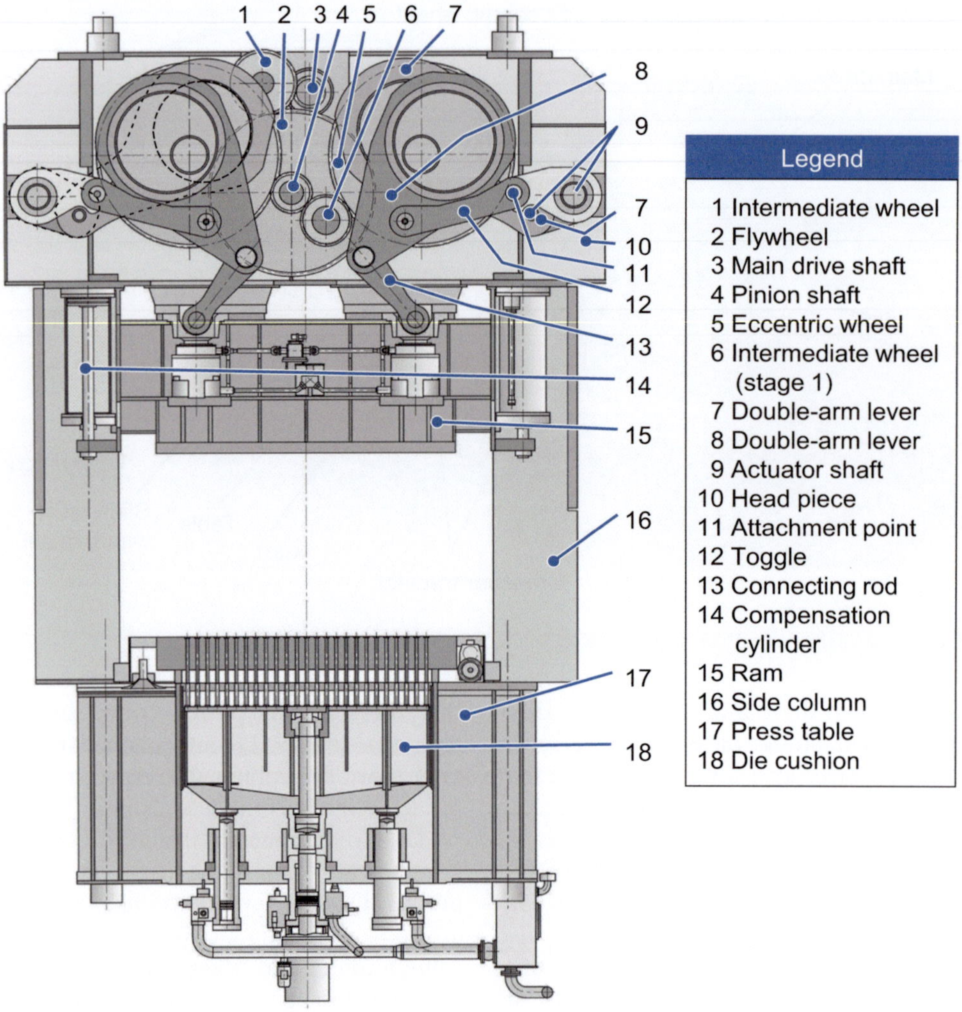

Fig. 3.35 Cross-section of the first stage of a large-panel transfer press. *Source* Schuler

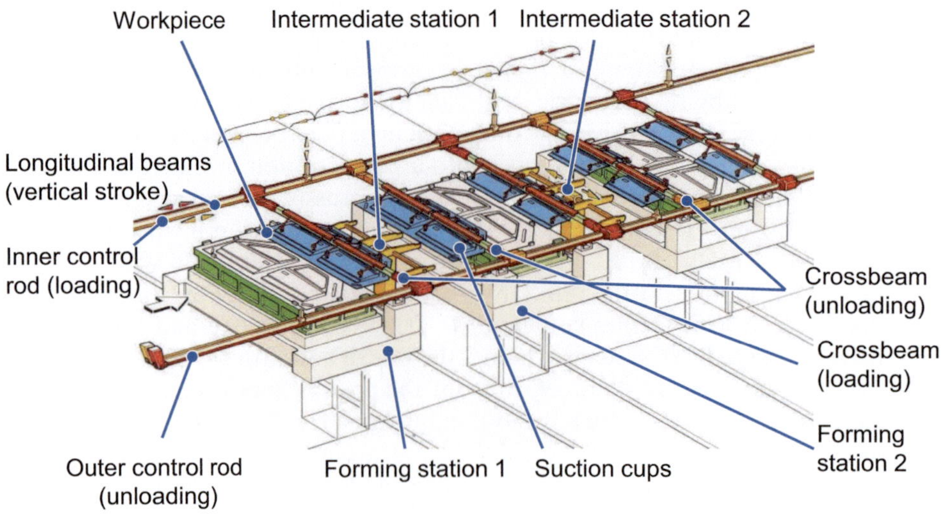

Fig. 3.36 Crossbar transfer on a large-panel transfer press. *Source* Schuler

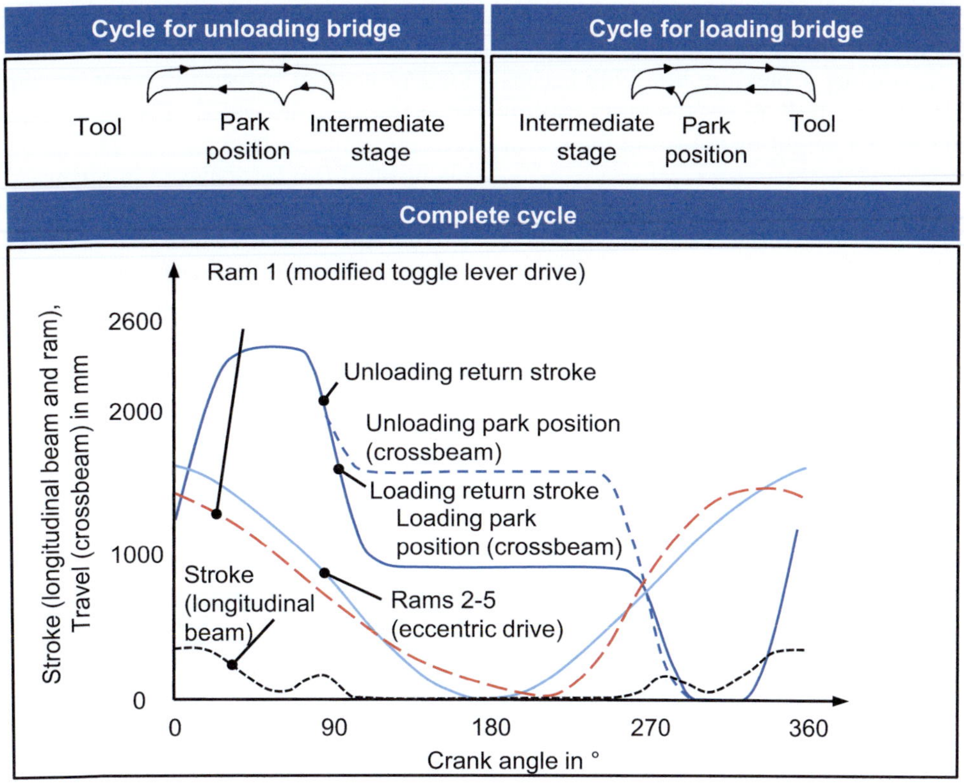

Fig. 3.37 Crossbar transfer cycle work sequence diagram. *Source* Schuler

of massive forming as well as forging presses, see Fig. 3.38, with nominal forces up to 160,000 kN. The stiff and solid construction of the frame and the drive allows for the machine type to accommodate the working precision required under extreme forging loads. The press column is designed as an O-frame with a solid cast steel construction with an extremely small column width for pressing forces of up to 40,000 kN or a frame with tie rods is selected for higher pressing forces. The main drive shaft is constructed as a rigid, forged

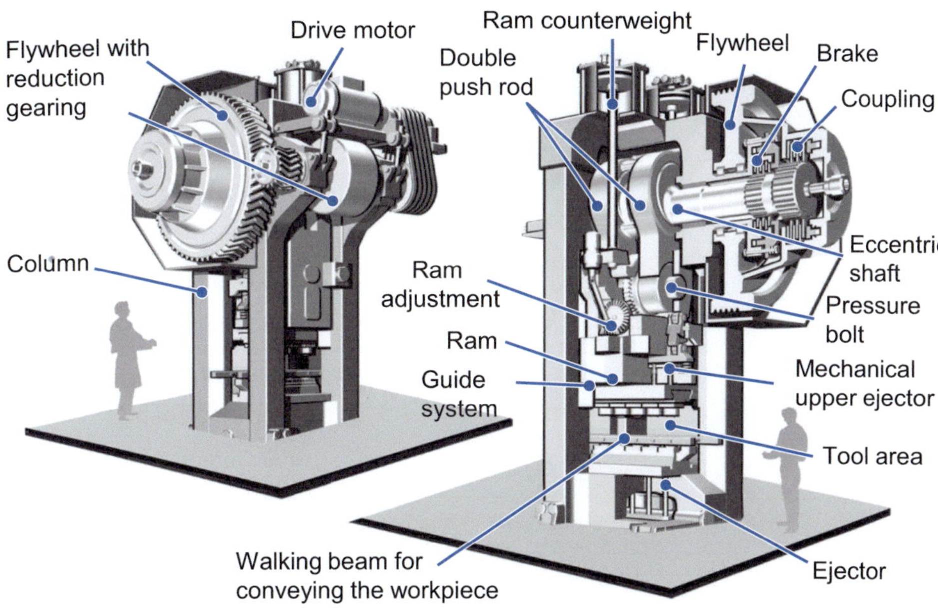

Fig. 3.38 Eccentric forging press. *Source* SMS group

eccentric shaft made from alloyed steel with a particularly wide eccentric. The lower part of the wide connecting rod presses against a pressure bolt with eccentric support. This pressure bolt allows the ram position to be adapted to the tool height by adjusting the worm gearbox.

Pneumatic cylinder/piston systems in the press head constantly hold the ram against the connecting rod and this against the eccentric shaft ensuring that the required preload exists. The rear compensation cylinder is also used to clamp the eccentric pressure bolt for the ram adjustment. When combined with the extremely solidly designed ram guide, which has a diagonal structure, very high accuracies are achieved in the vertical and horizontal directions even with relatively large off-center loading. A friction clutch and brake are arranged directly on the eccentric shaft and have an electro-pneumatic control. The clutch also provides overload protection because it starts to slip if the permissible torque is exceeded. The production process for a steering knuckle is shown in ◘ Fig. 3.39 as an example of what can be achieved in terms of manufacturing with eccentric presses and multi-stage tools.

In addition to the crankshaft and eccentric drives in extensive use, the space crankshaft joint has been developed for high-speed presses, ◘ Fig. 3.40. The machine shown here works at stroke rates of up to 2000 min^{-1} without a counterweight. The whole drive is mounted on a yoke with a swivel support around its horizontal axis. An electric motor drives two drive shafts acting in opposite directions via a belt drive and a flywheel. The upper spherical heads of two connecting rods are

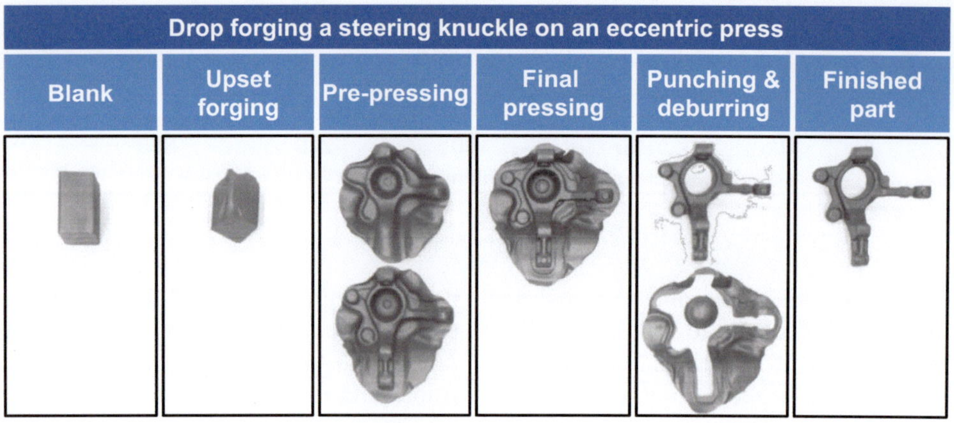

◘ **Fig. 3.39** Example of drop forging on an eccentric press. *Source* SMS group

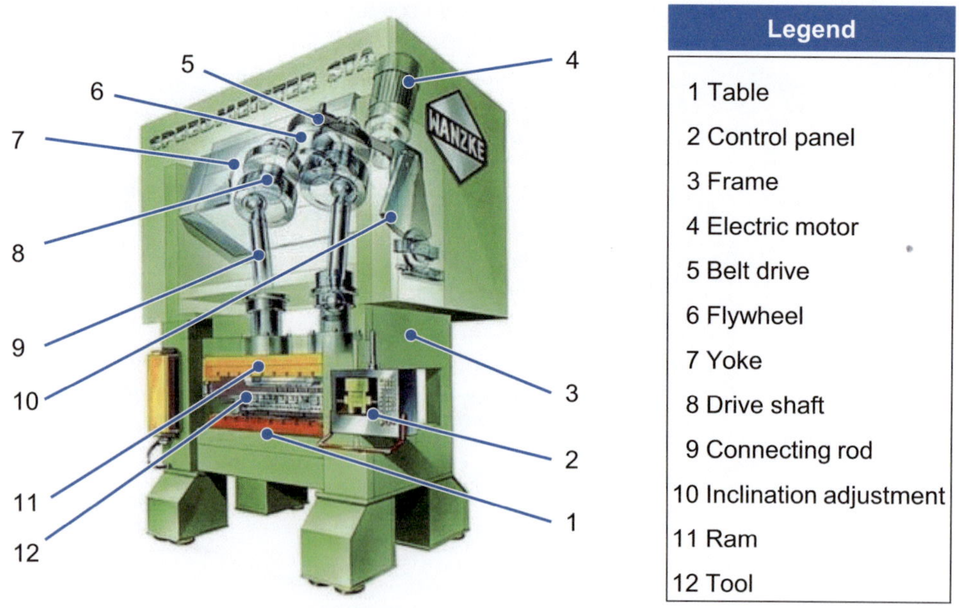

◘ **Fig. 3.40** High-speed press with space crankshaft joint. *Source* Wanzke Umformtechnologie

mounted eccentrically in the shafts. The lower connecting rod heads are mounted in the ram. By adjusting the inclination of the yoke, the press stroke can be steplessly set without displacing the bottom dead center of the ram. This changes the nominal force and the ram speed, but the work capacity is not affected. The guide system on the press is decoupled from the frame so that deformations of the frame caused by forming forces do not act on the ram and the tool. Other benefits of this design are that the bearing loads are more evenly distributed to the upper connecting rod heads as compared to the upper connecting rod ends in eccentric presses and the lateral forces from the connecting rod thrust are more than 50% less than with eccentric drives.

The horizontal multi-stage press shown in ◘ Fig. 3.41 was specially designed for massive forming with wire as the base material. The ejector draws the wire off the coil and conveys the start of the wire into the press after changing a coil. The vertical and horizontal straighteners then align the wire. The feed unit, whose movement, like all the other machine components, is controlled by the central drive unit of the main drive, finally conveys the wire at the machine cycle rate until it comes to the shearing station stop. The wire pieces are cut to length and then conveyed by the transfer system to the first punch drive. In total, this multi-stage press has five punch drives. These are extremely ergonomically configured one below the other in the ram so the user can set the tools. They have five opposing die plates. Ejectors in the punch and the die plates ensure that the components can be gripped by the transfer system after each forming process and moved to the next operation. Just like the transfer system, they are also activated by separately adjustable cam disks. The transfer system can be swung out of the way to change the tool.

3.1.4 Toggle Presses

The drive principle used on toggle presses differentiates them from crank and eccentric presses. The forming force is essentially transmitted by the toggles to the ram and to the head piece of the machine frame. The toggles are driven by hydraulic cylinder/piston systems or crank drives, ◘ Fig. 3.42. Due to the high force transmitted by the toggles, only a fraction of the nominal press force is applied to the upstream drive components, e.g., the connecting rod as well as the crankshaft and its bearing arrangement.

Compared to eccentric and crank presses, toggle presses have a different force/stroke path characteristic. Their disadvantage lies in the fact that the nominal force is only available over a short ram travel distance. The main application areas are minting coins and medals, stamping cutlery and jewelry, calibrating and cold extrusion of high accuracy workpieces (tolerances around IT5/IT6).

A distinction is made between the two working methods of direct extrusion where the tool movement and material flow are in the same direction and indirect

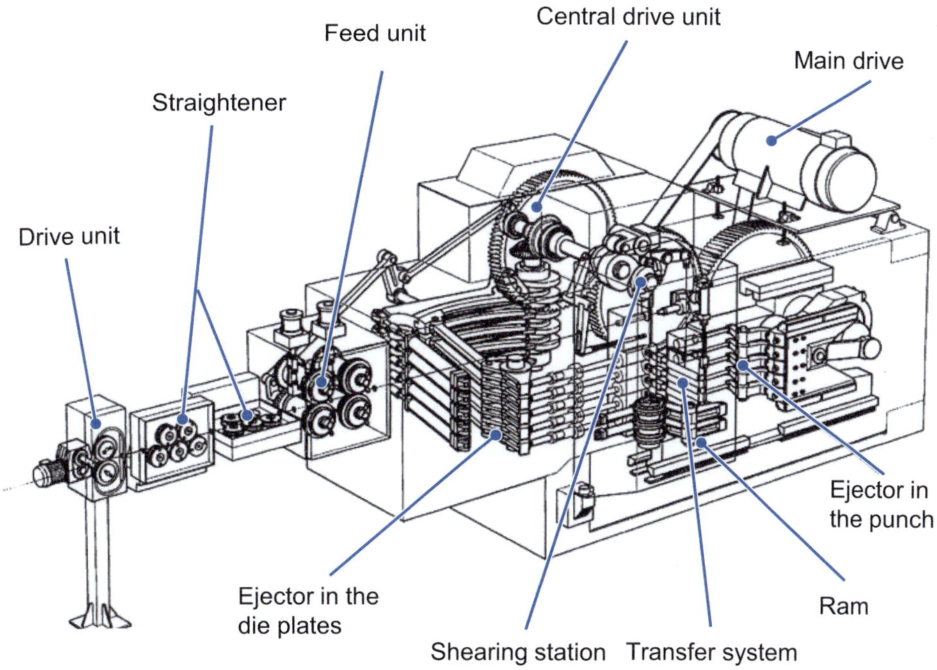

◘ **Fig. 3.41** Horizontal multi-stage press. *Source* Schuler

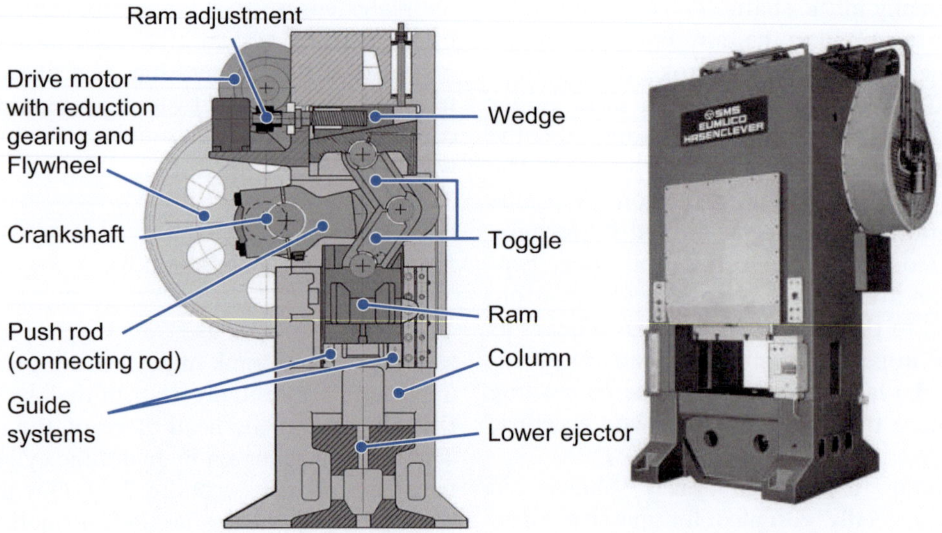

Fig. 3.42 Press with toggle lever drive. *Source* SMS group

extrusion where the workpiece material and tool move in opposite directions. The latter is mainly used to manufacture thin-walled hollow pieces and tubes. There are often technological limits to increasing the working strokes. Increases in the forming speed are generally associated with increased forces. Furthermore, impermissible high dynamic vibration amplitudes can occur in the tool leading to greater wear.

An increase in the stroke rate is therefore only possible if the drive elements can be designed such that the unproductive time segments in a working stroke (e.g., the return stroke, passage of idle travel before starting the forming or cutting process) are shortened while the actual machining process runs slowly. The drive diagram for a servo-mechanical fine blanking press with a double toggle lever drive is shown in Fig. 3.43.

The use of servo technology has almost doubled the production rate of this press compared to conventional fine blanking presses. The drive uses two controlled servo motors directly connected to the double toggle lever. Due to their dynamics, they provide reliable tool protection even at high cycle rates. Unlike mechanical presses, the movement of the ram is not fixed but can be flexibly programmed and controlled. The servo mechanics of the machine allow the forming speed to be partially reduced without any limitation on the lower speed. With this, it is possible to vary the movement and force curves during the pressing process and to adjust them precisely to the particular application.

Thanks to the servo technology, fine blanking presses can run at stroke rates of up to 140 min^{-1}. The machine exploits both the flexible travel profile of hydraulic presses and the high working speed of mechanical presses.

3.1.5 Wedge Presses

Wedge presses are mechanically or hydraulically driven presses for high pressing forces. The ram is moved by a wide wedge located between the upper part of the column and the ram, Fig. 3.44. The wedge has a lead angle of approximately 30°. Ignoring frictional forces, this corresponds to a transmission ratio between the ram force and the wedge driving force of approximately 1.7:1. In this mechanical eccentric wedge press, the wedge is moved by an eccentric drive located on the rear of the press. It has holding strips so that the wedge and ram cannot fall if the counterweight cylinder fails. The wide contact surface between the wedge and the ram and the stiff design of the press frame mean that this type of press is largely unaffected by off-center loading. The ram does not tilt. Wedge presses are the preferred option for drop forging high-precision forged parts and for processes with unavoidable off-center machine loads. The construction height is lower than that of a crank press by configuring the drive on the rear of the machine. As shown in Fig. 3.44, mechanical wedge presses are usually driven by a crank or eccentric drive.

Mechanical wedge presses are designed for nominal forces of 20 to 120 MN and have stroke rates of 20 to 70 min^{-1}. Pressing forces of up to 800 MN can be achieved using hydraulic wedge presses. A typical manufacturing scenario of forging a crankshaft is shown in Fig. 3.45.

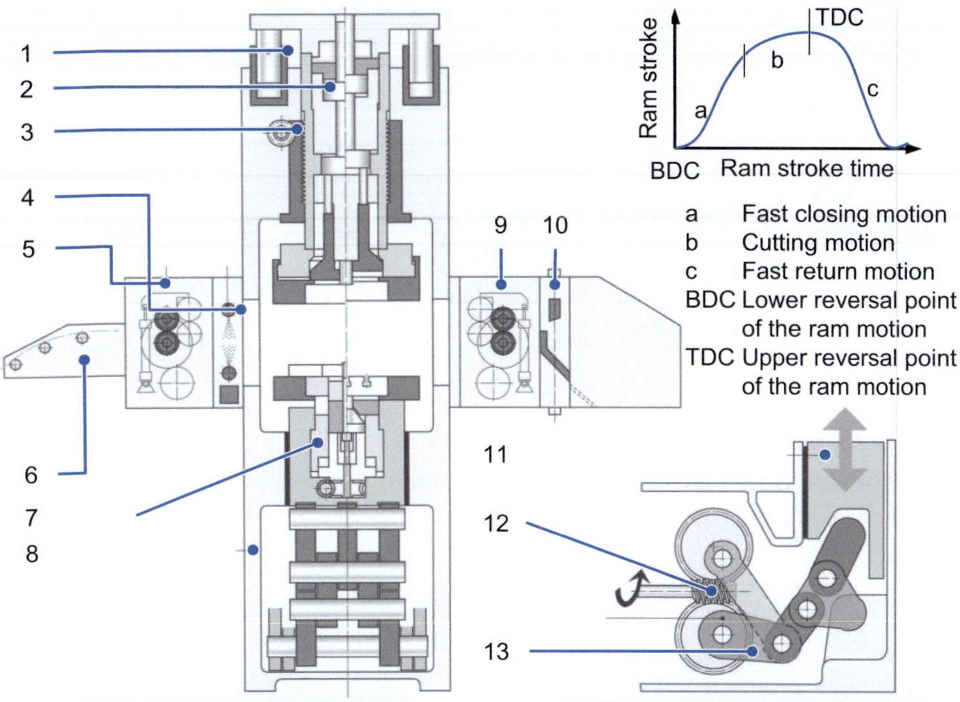

Machine elements	
1	Top spindle lock
2	Ring tooth piston
3	Tool height adjustment
7	Opposing retaining piston
8	Four-column press body
11	Ram
12	Gearbox
13	Double toggle

Elements of the feed system	
4	Spraying device
5	Inlet feed
6	Roller basket
9	Outlet feed
10	Waste separator

Fig. 3.43 Fine blanking press with double toggle lever drive. *Source* Feintool (Technologie AG

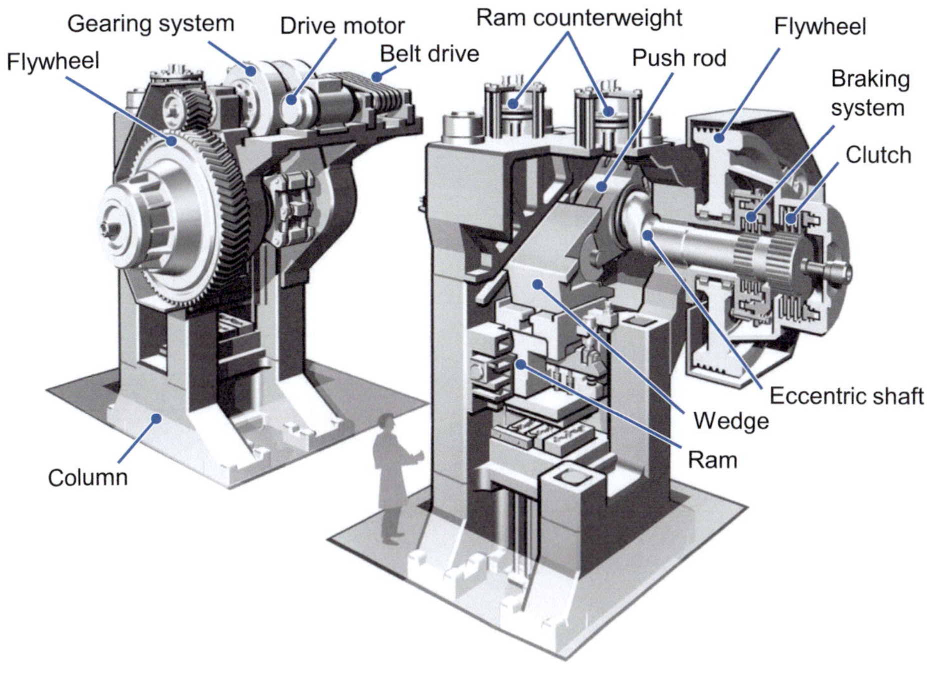

Fig. 3.44 Forging press with wedge drive. *Source* SMS group

3

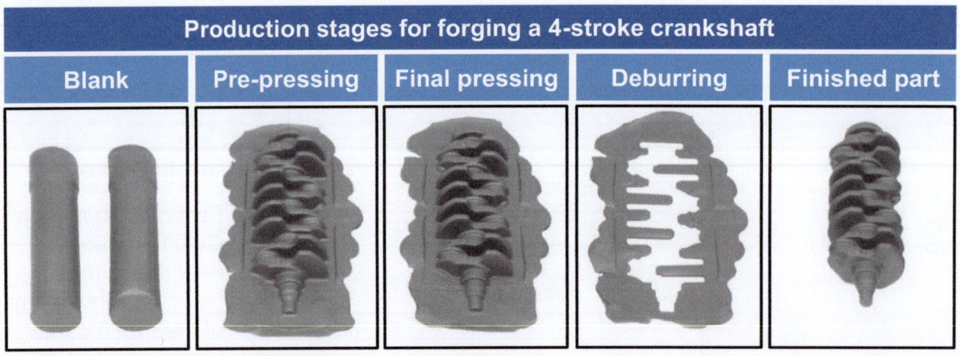

Fig. 3.45 Example of forging on a mechanical wedge press. *Source* SMS group

3.1.6 Hydraulic Presses and Forming Machines Using Hydraulic Principles of Operation

Hydraulically actuated presses are force-controlled machines. They are characterized by the fact that a force up to the nominal force can be applied at any point of the ram stroke, regardless of the tool position. For this reason, hydraulic presses have proven themselves predominantly in those areas of forming technology where the force has to be either constant throughout the ram travel or controllable as required by the process.

The basic physical principle of the working method of hydraulic presses is the hydrostatic principle in which the pressure in a fluid is equally distributed in all directions and produces a force on the adjacent surfaces proportional to the pressure and surface area. Within a cylinder/piston system that is directly coupled to a ram, the high pressure energy in the pressure medium is thus converted into mechanical work to produce the working motion. The essential parameters of the hydraulic drive are the pressure and the flow rate of the pressure medium. By using suitable pumps and control equipment (power, pressure, and zero stroke controllers), both these parameters can be set appropriately for the relevant operation. Further information on the basic principles of hydraulics can be found in Volume 2 of this series.

A fundamental distinction is made between direct pump drives and accumulator drives. However, they are rarely used in their pure form. Instead, they are frequently adapted to the relevant requirements. Figure 3.46 shows the schematic for a hydraulic circuit demonstrating the principle of a direct pump drive. Effectively, the drive pump volume is controlled such that it provides the working medium (oil or water) on a demand-oriented basis. To achieve high drive dynamics and compensate for pressure variations, a hydraulic accumulator is used (similar to accumulator drives). This can briefly provide a specific volume of the medium at

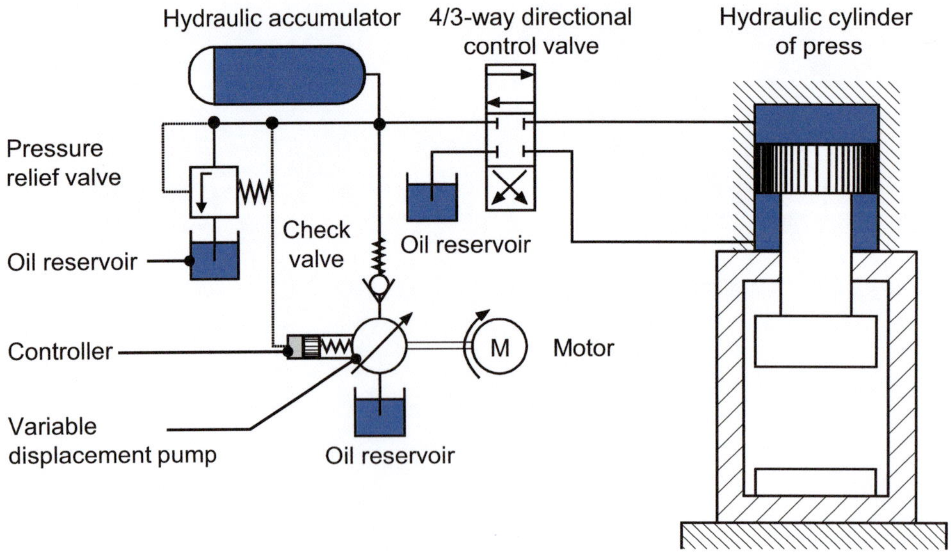

Fig. 3.46 Hydraulic press drive

the set pressure. The medium acts on the main working cylinder or on the differential piston, which is subjected to the large surface to generate the pressing force and to the small surface to lift the ram. The pump and drive motor do not have to be designed for the maximum specific power requirement of the press, as is the case with the pure form of direct drives. The main advantage over accumulator drives is retained and the controlled pump operates on a demand-oriented basis making the system more energy-efficient. The pressure regulator ensures that the pump volume is governed by the quantity of oil required. The operating pressures for hydraulic presses are between 200 and 300 bar.

Hydraulic presses use a similar frame design to that of mechanical presses (see ◘ Fig. 3.23). However, they usually have a two-column, welded design. The machine columns on presses with greater pressing forces and larger column widths are constructed from multiple components (table, head piece, side columns) that are braced against each other with tie rods.

The hydraulic drive elements are relatively easy to install in the machine frame. Several hydraulic drives can be readily installed in a machine for combined forming and cutting operations (drawing, hold-down, cutting, ejection, etc.) and the required movements can be very easily coordinated. A distinction is made between single-, double- and triple-action presses: The different movement options are shown in the drawing tools illustrated in

◘ Figure 3.47. While the main ram moves only once in single-action presses, in double-action and triple-action presses, one or two additional movements are included that allow the functions of a die cushion, ejector, or hold-down device

One example of a single-action press is a freeform forging press. Freeform forging is used to preform or manufacture large forgings (up to 500 metric tons). The benefits of this process lie in the formability of large workpieces, that require a large stroke and great forming force. In addition, the quick handling of tools can lead to greater efficiency in small batches. Unlike in drop forging, in freeform forging no enclosing dies are used. The shape results from guiding the workpiece between the dies whereby the die geometry is either flat or the shape of the reference geometry. Workpieces can be handled by forging robots (up to 300 kg component weight) or manipulators (up to 250 metric tons component weight). The disadvantages are limited achievability of tight tolerances and limited options for manufacturing larger batches compared to drop forging. Typical application areas are in the manufacture of crankshafts, rollers, disks, rings, or blocks from alloys of steel, aluminum, titanium, nickel, or cobalt.

There are various fundamentally different designs of freeform forging presses: Overhead or underground design, two- or four-column design, hydraulic drive using oil or water. The two-column freeform forging press shown in ◘ Fig. 3.48 can produce finished forgings or preforms for subsequent drop forging processes. A process force of 25 MN is achieved during forging and 30 MN during upsetting. The press stroke is 2400 mm. This stroke is required for forming large structural components. The 6 m

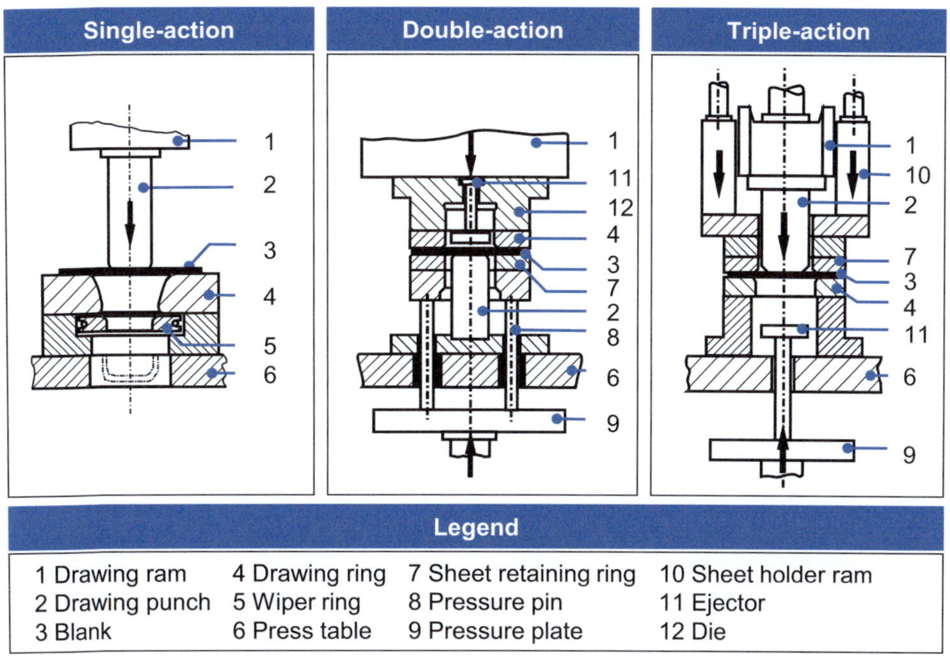

◘ **Fig. 3.47** Working principle for hydraulic forming presses. *Source* Schuler

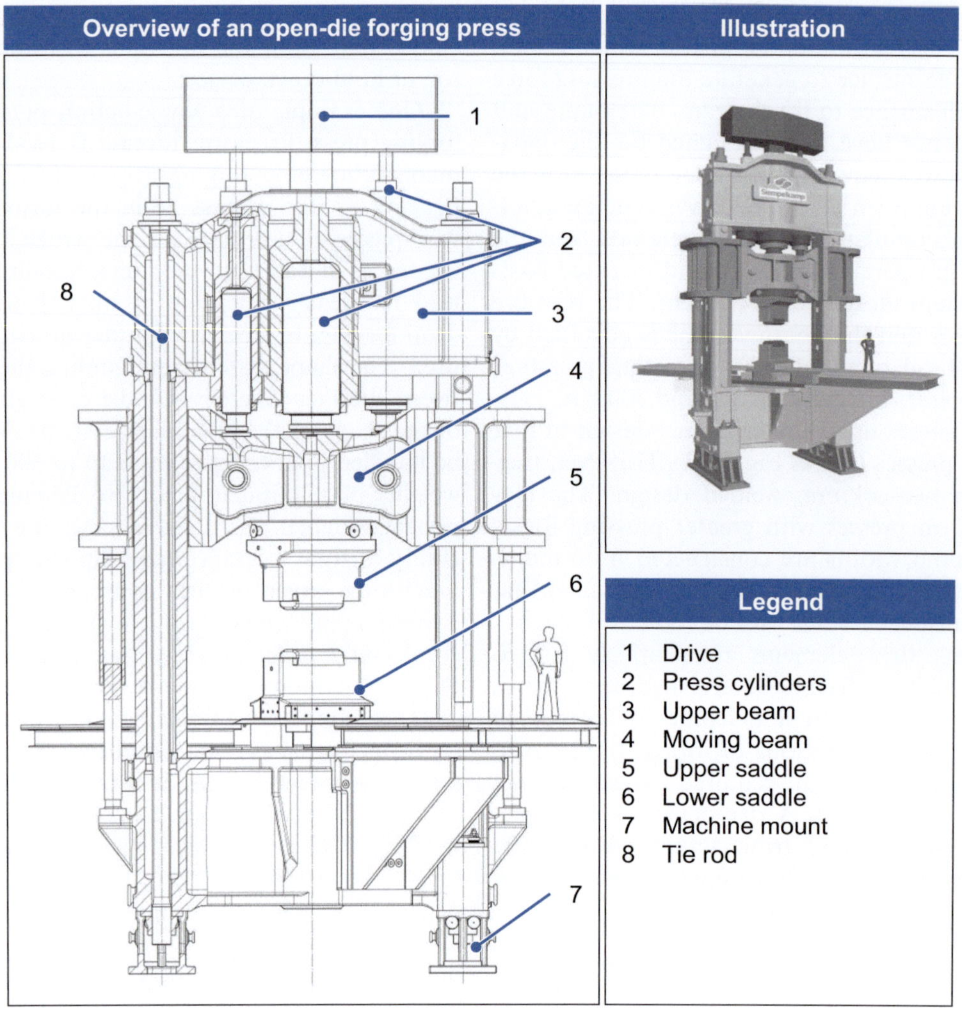

Fig. 3.48 Structure of an freeform forging press. *Source* Siempelkamp

long press table can be extended by 6.5 m to change the workpiece. This allows workpieces to be produced while preparing the next workpiece at the same time. The central press cylinder is integrated into the upper beam and its piston is mortised into the moving beam. The combination with the X-guide system ensures precise guidance of the moving beam throughout the press stroke as well as in the event of high eccentric loads. The weight of the moving beam is less than that of a surrounding moving beam meaning the forging frequency can be increased. The multi-part structure is preloaded using tie rods to prevent the structure from bending open.

Figure 3.49 shows a triple-action hydraulic deep-drawing press with two of the rams acting from above and one from below. The upper internal ram is the actual drawing ram, which slides in the guide systems of the external sheet holder ram. The drawing ram is driven by the main cylinder with the differential piston and the sheet holder ram is driven by four cylinders with plunger pistons. In this case, the die cushion arranged beneath the table can also act as an ejector.

During the work process, drawing ram and sheet holder ram both move down until the sheet retaining ring in the die contacts the blank, as shown in Fig. 3.49 (right). Forming now begins through the sole continued movement of the drawing punch. It pulls the blank downwards through the drawing ring. The die cushion remains in its lowest position during this process and only acts as an ejector once forming is completed.

Fine blanking also requires a triple-action press such as shown in Fig. 3.50. The force is transmitted from the electric motor to the pressure accumulator via an axial piston pump. The pumped pressure oil moves to the working cylinder via controlled valves. For the fast closing motion, oil is conveyed out of the pressure accumulator into the rapid closing cylinders causing the main working piston to draw oil out of the tank. The area of the cutting path requires the pressure oil to be switched to the main working cylinder. The required ram speed is set by the pump flow rate. At the top dead center of the movement, the ram runs up against a fixed stop. The main piston and the rapid closing piston are depressurized, switched and cause the fast return movement. The ring tooth force and the opposing force are also applied hydraulically. It must be possible to set these independently of each other. In the ma-

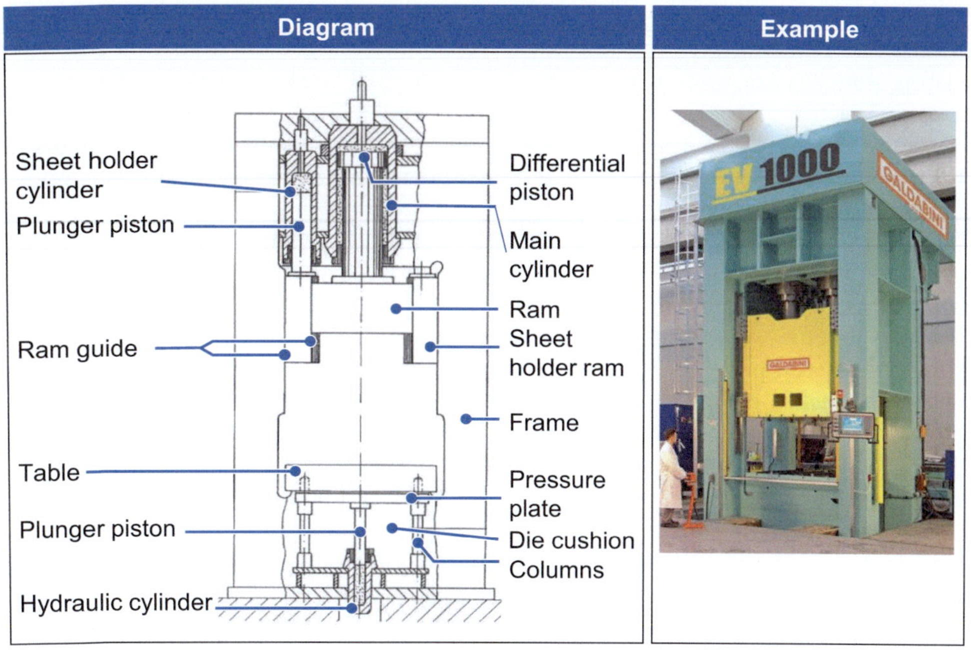

Fig. 3.49 Triple-action hydraulic deep-drawing press. *Source* Schuler, Galdabini [GALD]

Machine elements

1 Four-column press body
2 Main working cylinder
3 Fixed ram stop
4 Actuator
5 Ram guide
6 Rapid closing cylinder
7 Opposing retaining cyl.
8 Central supports
9 Tool clamping table
10 Ring tooth cylinder

Feed drive elements

11 Inlet feed
12 Strip spraying device
13 Automatic cutting mechanism
14 Strip inspection
15 Feed height adjustment
16 Outlet feed
17 Waste separator

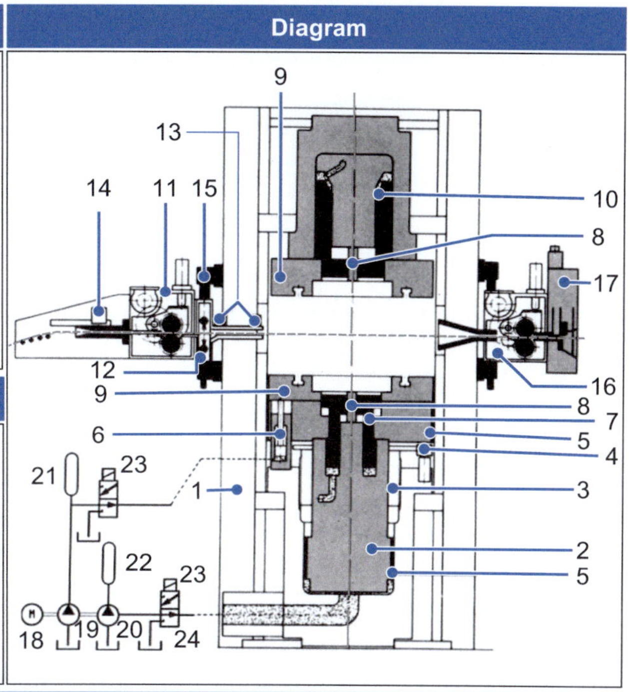

Main drive elements

18 Drive motor
19 Low pressure pump
20 High pressure pump
21 Low pressure accumulator
22 High pressure accumulator
23 Valve
24 Tank

Fig. 3.50 Fine blanking press with hydraulic drive. *Source* Feintool (Technologie AG)

jority of applications, tools with fixed cutting punches are used in hydraulic fine blanking presses while tools with movable cutting punches are used in mechanical fine blanking presses. The use of tools with fixed cutting punches allows fine blanking of thick and large components. The structure and the function of a typical tool during a ram stroke are shown in ◘ Fig. 3.51 and ◘ Fig. 3.52.

The fixed cutting punch (3) is placed on top of a pressure plate (2) and the upper frame of the cutting tool (1) and is screwed together with them. Holes and the internal forms of the workpiece are cut into the main cutting punch and pushed out by the ejector (6). The force acts by way of the ring tooth hydraulics (11) and the guide plate with ring teeth. Pressure bolts (5) press the ring teeth of the guide plate (6) into the workpiece. The active part of the tool thus supports the ring teeth and is guided by the columns on the tool frame. Before cutting starts, the cutting insert (7) and the guide plate are locked using locking pins (8).

The first step in fine blanking is to bring the two tool halves together (state a, ◘ Fig. 3.52) until the tool is closed. The main working cylinder on the press handles this action. The workpiece is then clamped inside and outside the cutting line by means of the ring teeth (upper hydraulic piston) and the opposing force (lower hydraulic piston). The material is cut in the next step (state b). To achieve this, the lower tool side is moved upwards by the main press drive. The applied cutting force, opposing force and retaining force from the ring tooth cylinder are controlled by the press control. In the next process step (state c), the ram stroke is ended by the main drive and the component is completely cut. The waste cut out from the sheet is notched into the main cutting punch for the internal forms, while the cut-out component is notched into the cutting insert. Once the ring tooth force and the opposing force are deactivated, the two die halves are opened by the main drive of the press (state d). By activating the pressure control valve for the ring tooth cylinder, the wiper force now acts instead of the ring tooth force to eject the in-

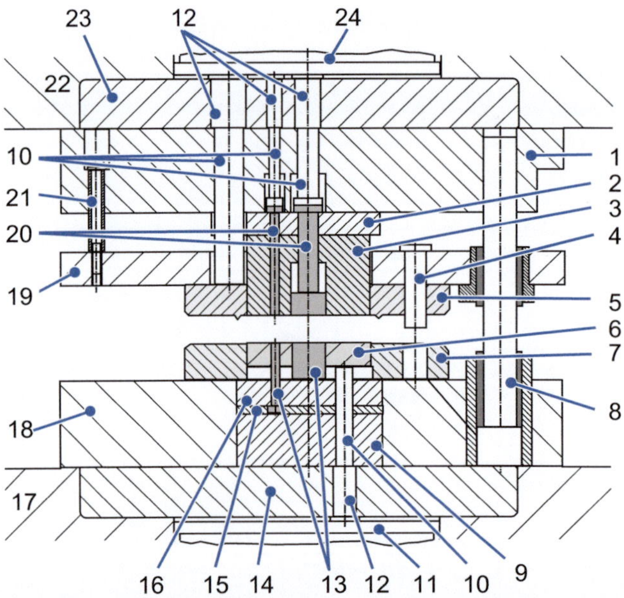

Legend		
1 Upper part of the tool	11 Opposing retaining pist.	20 Ejector (upper part)
2 Pressure plate	12 Pressure bolt in insert ring	21 Retaining screw (in the tube)
3 Cutting punch	13 Hole or form punch	22 Upper machine table
4 Locking bolt	14 Insert ring (lower part)	23 Insert ring (upper part)
5 Guide plate with ring teeth	15 Spacer plate	24 Ring tooth piston
6 Ejector (lower part)	16 Hole punch retaining plate	
7 Cutting insert	17 Lower machine table	
8 Tool guide	18 Lower part of the tool	
9 Pedestal	19 Guide plate	
10 Pressure bolt		

◘ **Fig. 3.51** Structure of a fine blanking tool. *Source* Feintool (Technologie AG)

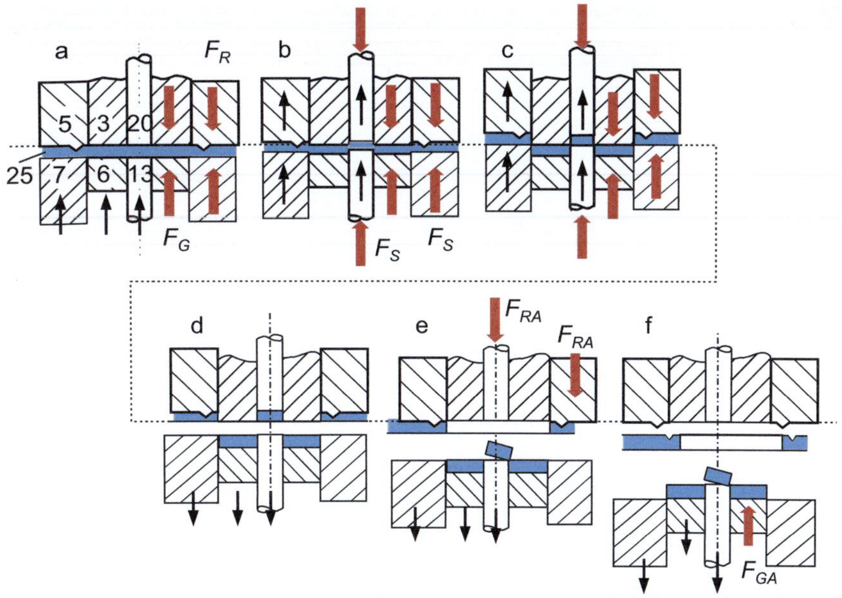

Legend

3 Cutting punch
5 Guide system
6 Ejector
7 Cutting insert
13 Internal form punch
20 Ejector
25 Sheet metal strips
F_S: Cutting force
F_R: Ring tooth force
F_G: Counterforce
F_{RA}: Wiper force
F_{GA}: Ejector force

↓ Movements (completed)
⬇ Forces

Fig. 3.52 Workflow in a fine blanking tool. *Source* Feintool (Technologie AG)

ternal form waste and to wipe the punching grid (state e). Subsequently, the opposing retaining cylinder is also pressurized causing it to eject the finished cut workpiece due to the lack of an opposing force (state f). All waste material and the cut component can be removed and the punching grid can be advanced via the roller feed before the cutting cycle starts again.

The essential advantages of presses with hydraulic drives compared to mechanical presses are better adjustment of force and speed to suit operations, easier setting of the stroke height, the option for interrupting the working stroke, and protection against overloading. These advantages are particularly evident in modern NC-controlled presses where, for example, force/distance curves can be measured and precisely controlled. Hydraulic drive systems are also used in cold and hot bulk metal forming, in GFRP processing and notably in rubber cushion stamping presses. However, in cutting and forming machines with high stroke rates, mechanical drives are obviously superior to hydraulic drives.

Figure 3.53 shows a diagram for active hydromechanical deep-drawing and a machine suitable for this process. The sheet holder cylinders move the sheet holder ram with the sheet holder downward until the sheet holder contacts the blank and thereby sealing the space in the trough underneath. Following the movement cycle of the sheet holder drive, a defined pressure (20–30 bar) is built up in the area under the sheet holder which is formed as a trough or water tank. This internal pressure stretches the blank over the entire component surface until it contacts the punch drive. The die is now moved downward by the ram cylinder, ram and connecting columns until it reaches the stop, which can be adjusted by the press control. The vertical distance between the sheet holder and the die is set by the adjustable stops in the forming ram (shown in the illustration in two different planes). The die is fixed to the forming ram by connecting columns. The pretensioned blank is formed by the downward movement of the tool (ram). During this movement known as the back pull, emulsion is forced out from the water trough. The pressure in the tank can be controlled by the control system. Once the drawing process is completed by the ram, the pressure in the tank is built up again in order to place and shape the blank optimally

3

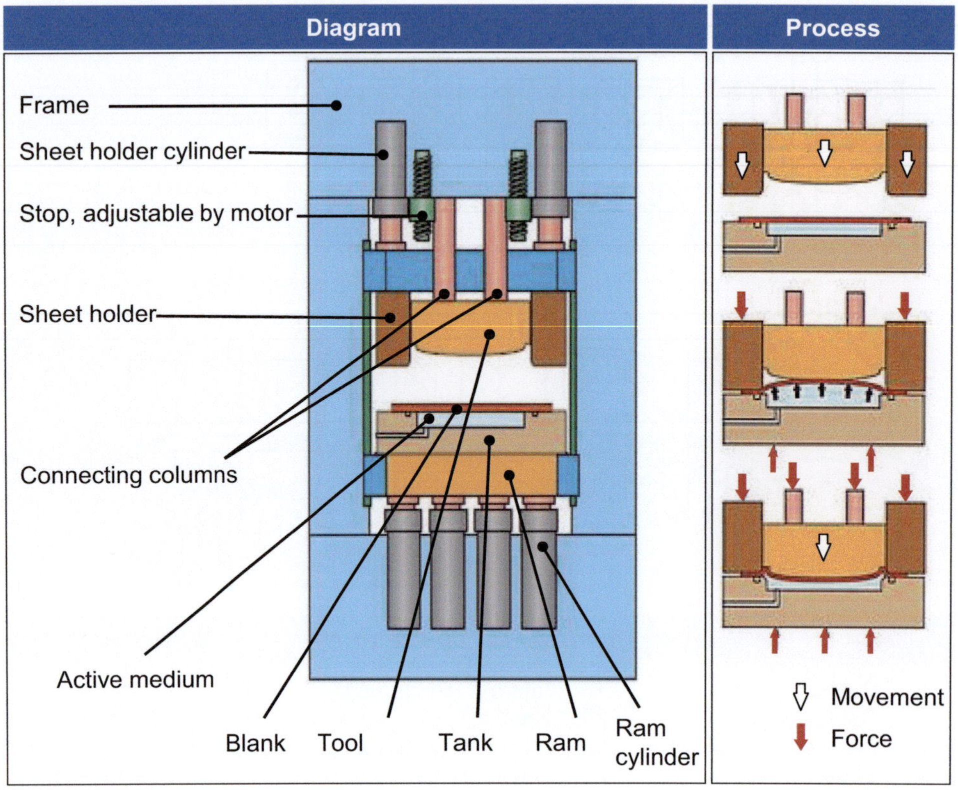

Fig. 3.53 Active hydromechanical deep-drawing. *Source* Schuler

according to the contour of the punch drive [AUST03]. This process has several benefits. The work hardening achieved by pre-stretching allows for increased buckling strength, particularly in large surface area components. Tool costs are also reduced as only one mold half is required to create a shape. At the same time, longer tool service lives are achieved as wear is kept to a minimum due to forming using an active medium. Lastly, components can be produced with extremely high-quality surfaces, particularly on the side facing the active medium.

Similar to active hydromechanical deep-drawing, the process of internal high-pressure forming (IHPF) is a cold forming process which is used to manufacture hollow components with complex geometry from tubular or flat blanks. It is based on flaring the hollow bodies, compressing them in axial or radial direction, and finally expanding them against the tool wall through calibrated pressure.

In IHPF, a basic distinction is made between components which, permit a continuous axial material flow (ASE components) due to specific conditions, particularly the component geometry and friction, or do not permit axial material flow due to axial upsetting (AE components). ASE components are characterized by a high degree of deformation over the entire die contour due to the axial material flow. They are flared, upset and expanded during the process. Due to the axial material flow, more significant deformation can be achieved than with radial forming. In the case of AE components, axial feeding cannot be achieved for component-specific reasons. This process is therefore a calibrated process. Forming occurs only in radial direction. An example of internal high-pressure forming is shown in Fig. 3.54 illustrating the forming of a T-piece. A special hydraulic press is equipped with a two-piece multi-cavity die, two sealing punches (horizontal cylinder) acting axially on the tube ends, and one vertically acting opposing retaining punch. The cylindrical blank is placed into the lower die and the die is closed. The ends of the tube are sealed by the axial punches, and the blank is filled with the pressure medium (water/oil emulsion). During the actual forming process, the punches compress the blank while it is simultaneously expanded by the active medium and contacts the die contour. Additionally, the opposing retainer controls the material flow of the forming T-piece. At the end of the process, the calibrated pressure ensures the workpiece is shaped to match the contour of the die cavity. Then, the die is opened and the component ejected.

Presses for internal high-pressure forming have a special design. Their component parts are adapted to the particular requirements of the process. They are either four-column presses or frame presses (Fig. 3.55). Plunger and differential cylinders, as well as position, stroke and diameter of the horizontal and

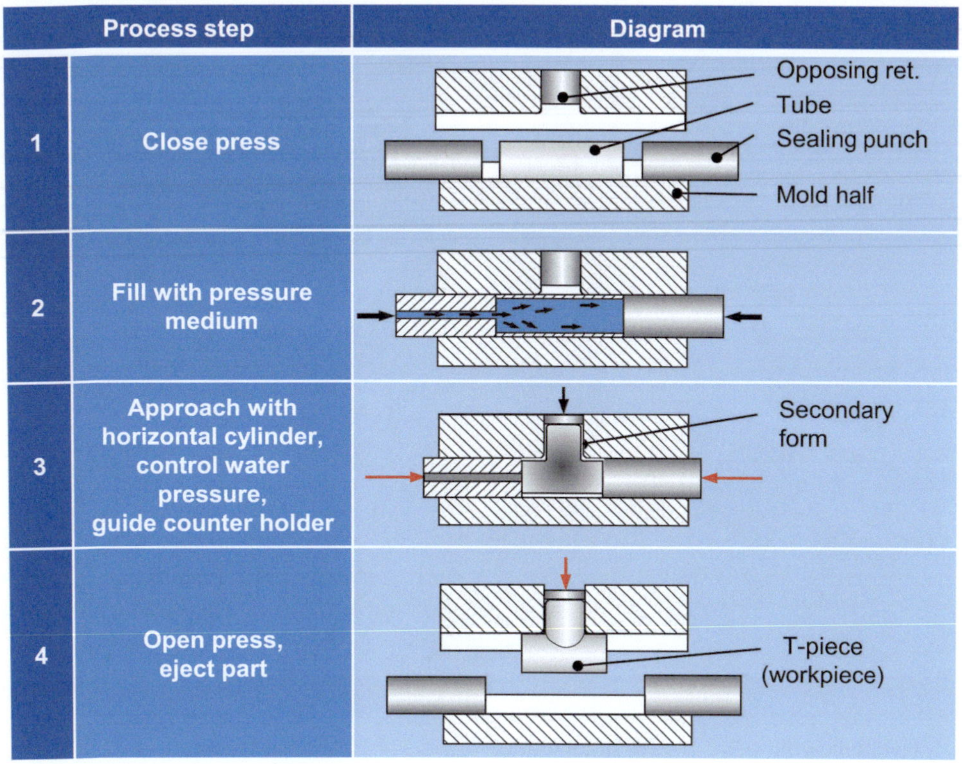

Fig. 3.54 Internal high-pressure forming process. *Source* Schuler

opposing retaining cylinders are determined based on the workpiece. Controllable pumps fill the workpieces with a pressure medium up to a defined initial pressure. Higher pressures are generated by pressure intensifiers [VDI00].

Another hydraulically driven machine for massive forming is the short-stroke front-loading extrusion press in Fig. 3.56. These machines are used for profile manufacturing, where a pre-heated block (15) of the base material is precisely positioned in the center of the press by the linear ingot loader (8) and centrically clamped between the press punch (18) and the forming die plate (14). In the second step, the ingot container (16) moves towards the opposing beam (11) and

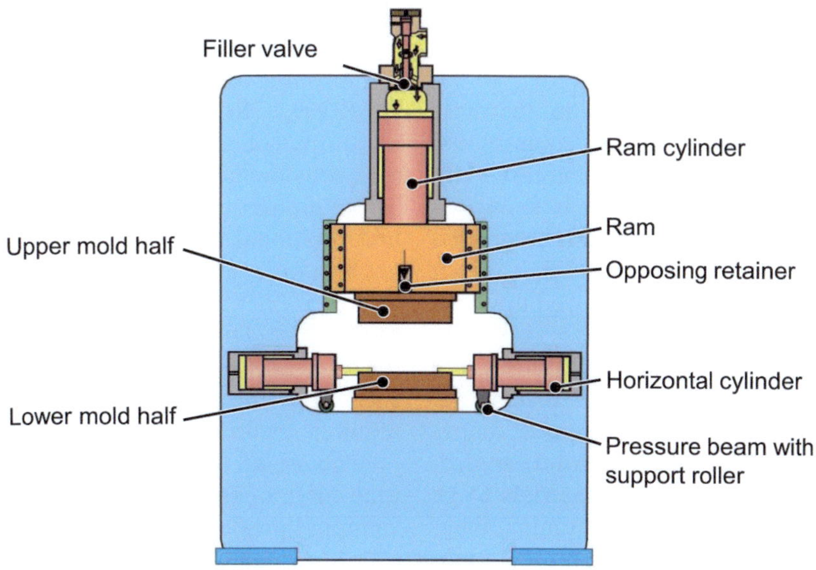

Fig. 3.55 Press for internal high-pressure forming. *Source* Schuler

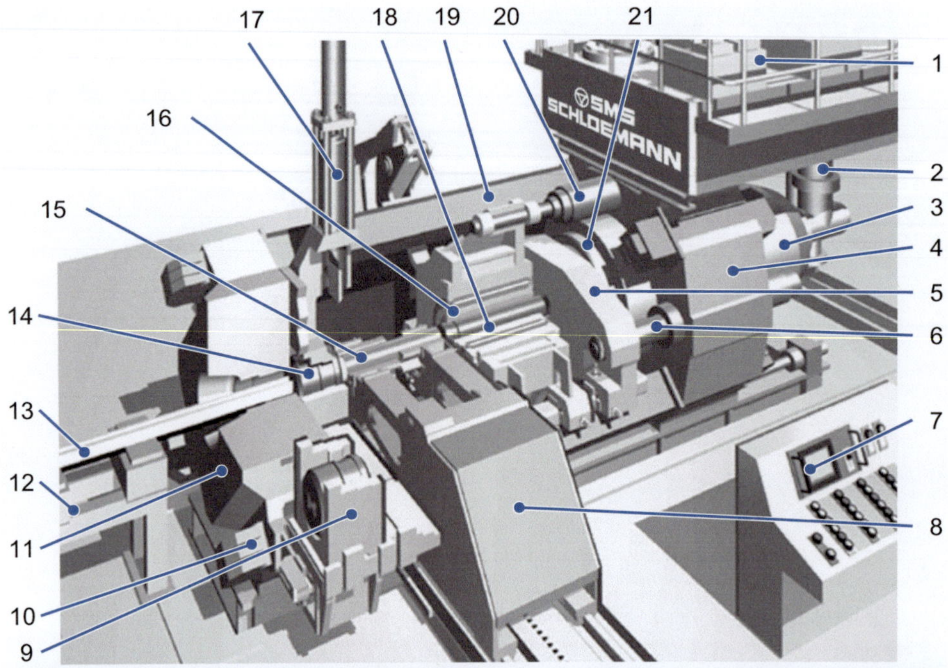

Legend

1	Hydraulic unit	8	Ingot loader	15	Press block
2	Filler valve	9	Cassette displacement	16	Ingot container
3	Main cylinder	10	Tensile lamella	17	Press rest shear
4	Cylinder beam	11	Opposing beam	18	Press punch
5	Moving beam	12	Start of outlet	19	Pressure boxes
6	Feed/ return cylinder	13	Extruded section	20	Holder displacement cylinder
7	Control panel	14	Die	21	Main piston

Fig. 3.56 Short-stroke front-loading extrusion press. *Source* SMS group

encloses the block. Then, the main cylinder (3) presses the block through the die plate with high pressure. Due to the high pressure, the material starts to flow and can thus be pushed through the complex die plates. The result is a long product that is cooled and cut in length in the outlet area.

To increase the stiffness of the machine, the press frame is pretensioned with pre-stressed columns using a tensile lamella/compressive box structure (10). A machine for concentric cold forming is shown in Fig. 3.57. With this machine, fittings can be pressed onto hydraulic hoses, for example. To achieve this, a hydraulically driven prismatic guide is moved upwards so that the concentrically arranged base jaws are moved towards each other. When the prismatic guide is retracted, the base jaws are returned to their initial position by springs. A broad range of profile shapes can be produced by extrusion such as solid, angled, hollow and tubular profiles. The base materials to be processed by extrusion are usually aluminum, magnesium and copper alloys, with steel being less commonly extruded.

3.1.7 Servo Presses

In recent years, a new press technology known as the servo press has become established on the market. These press types are becoming increasingly important for forming technology due to their specific benefits [BERG09, GROC09]. Torque motors with an individual output of up to 500 kW, also known as servo motors, provide the drive for this type of press [BEHR11]. These motors generate very high torques at relatively low rotational speeds and are highly dynamic. They are also quiet and have low maintenance requirements. The fundamental design of a servo press is shown in Fig. 3.58. The ram motion is produced directly from the servo motor via a gearbox. Mechanical components such as flywheels and transfer elements are no longer required resulting in a press with reduced maintenance.

The use of torque motors allows the displacement–time curve of the ram to be adjusted as required. Conventional presses are usually limited by their drive concept, e.g. toggle lever mechanism, to a single form of movement. In servo presses, the ram travel can be opti-

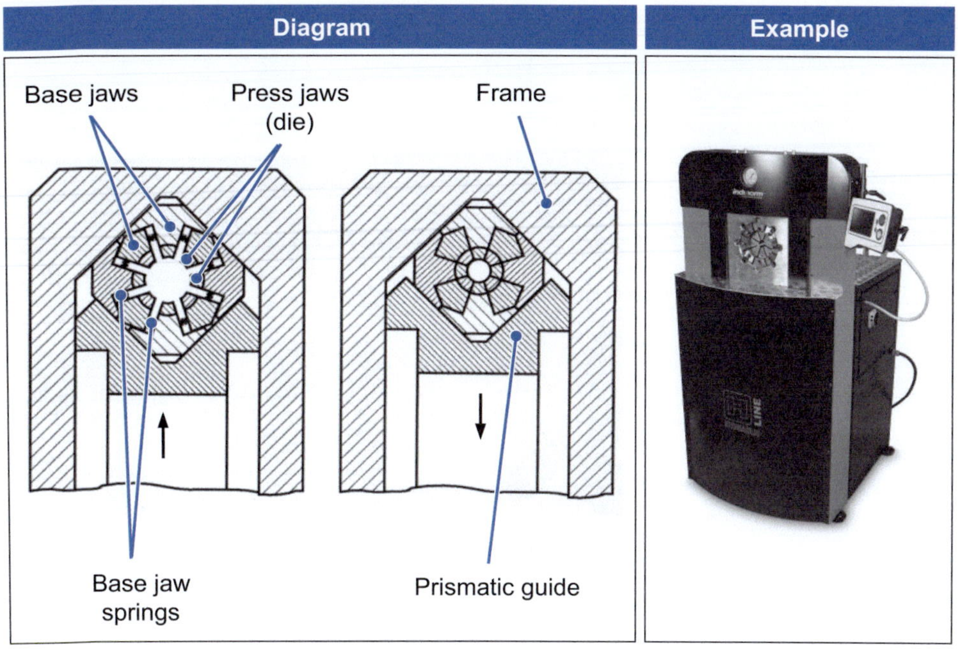

Fig. 3.57 Press for concentric cold forming. *Source* Indunorm Hydraulik

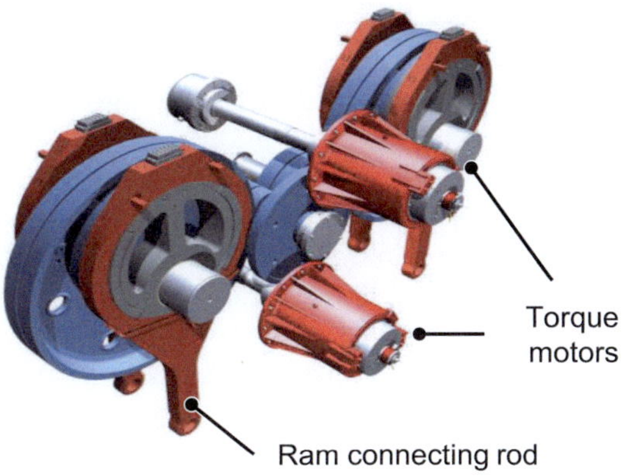

Fig. 3.58 Typical structure of a servo press. *Source* Schuler

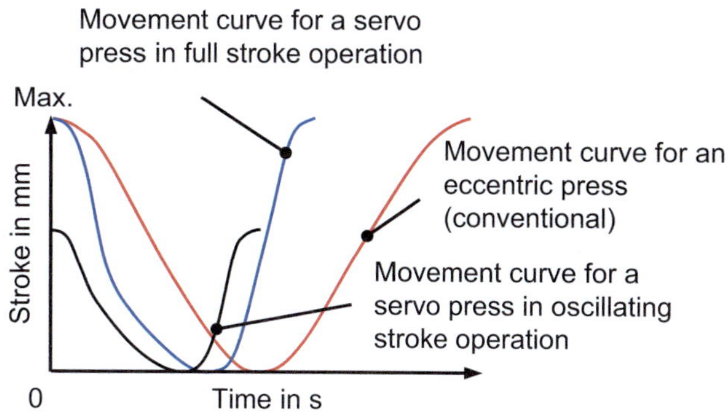

Fig. 3.59 Displacement-time motion for a conventional press and a servo press. *Source* Schuler

mally adapted to the relevant forming task [BEYE10]. This allows them to be used for a wide range of forming tasks. For example, the ram speed can be increased significantly during the infeed and return motion. This feature minimizes production time per component, leading to higher productivity. Figure 3.59 shows a possible displacement-time curve for a servo press compared to a mechanical press with a sinusoidal movement of the ram. During the forming process, it is possible to reduce the forming speed allowing greater degrees of deformation. The return stroke of the servo press takes less time compared to a conventional press.

Thus, the benefits of servo presses compared to conventional presses are higher productivity, increased component quality due to optimized forming speed and reduced operating costs.

One disadvantage of servo technology is the high drive power required. In addition, the systems are more expensive than conventional presses due to the complex control system. The stated benefit of flexibility only applies if it is required. If components are mass-produced, the free displacement-time curve does not offer any advantage so that conventional presses can be used instead. Another disadvantage of servo presses is the limited forming force set by the power of the servo motors. However, recent research results indicate that higher degrees of deformation and a simultaneous reduction of the forming force can be achieved by using superimposed vibrations [BRÜN09, GROC07]. This allows components requiring extremely high forming forces on conventional presses to be produced on servo presses with a lower forming force. Figure 3.60 shows typical ram movement curves that can be implemented using servo technology. Servo technology can basically be used in every area of forming technology such as in sheet metal processing or in bulk metal form-

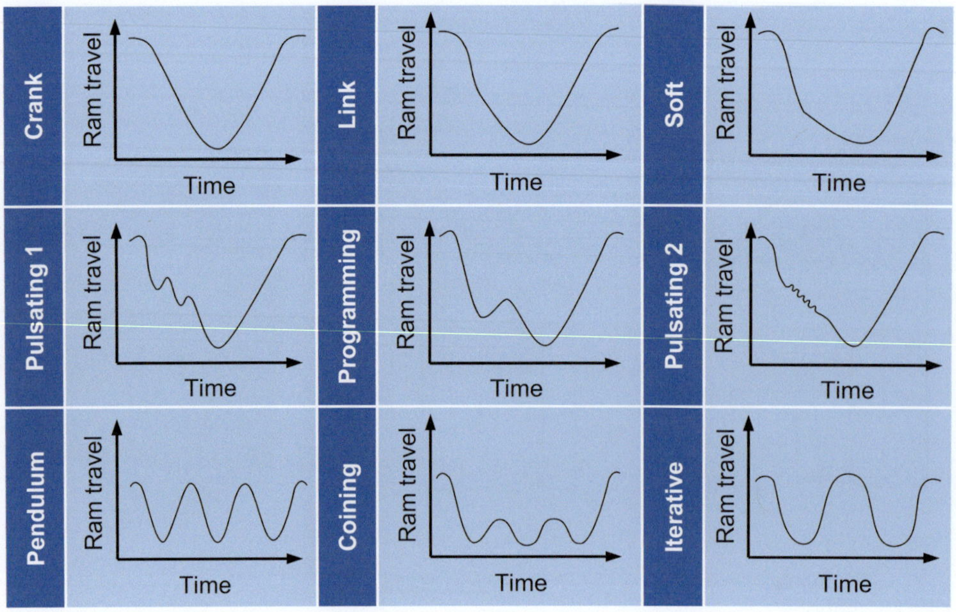

Fig. 3.60 Typical servo press movement profiles. *Source* AMADA

ing. Two different servo presses are shown below as examples.

3.1.7.1 Servo Presses in Monoblock / Tie Rod Design (Direct Drive)

The monoblock press body is realized as a welded construction, Fig. 3.61. Eccentric wheels with double helical tooting are used for axial guiding and reducing noise levels. The ram roller guide system is preloaded to be clearance-free. Forming forces of up to 8000 kN can be achieved with today's monoblock servo presses or 32,000 kN using a tie rod design.

3.1.7.2 Toggle Presses with Servo Drive

The proven toggle concept already provides very high component accuracy due to the characteristic ram motion as the forming speed is reduced before the bottom dead center is reached. In addition to this, the force in toggle mechanisms is introduced on the external sides of the ram. This provides the ram with very high tilting rigidity, Fig. 3.62. An established process can be further optimized in terms of flexibility and productivity by combining this concept with servo technology. Component-specific parameters such as the forming speed can be set even more optimally and with less

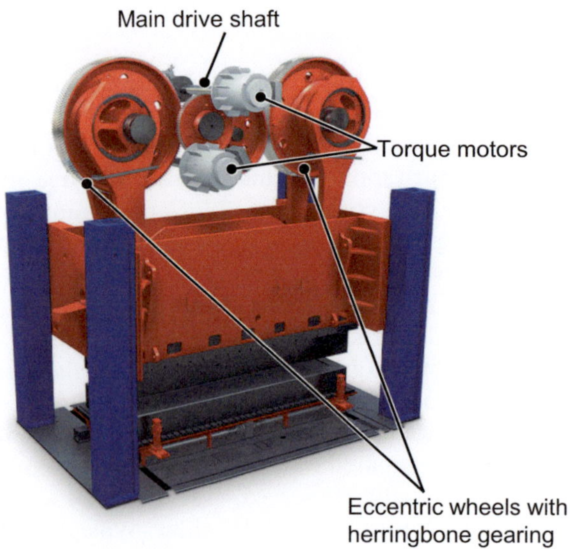

Fig. 3.61 Servo press in tie rod design. *Source* Schuler

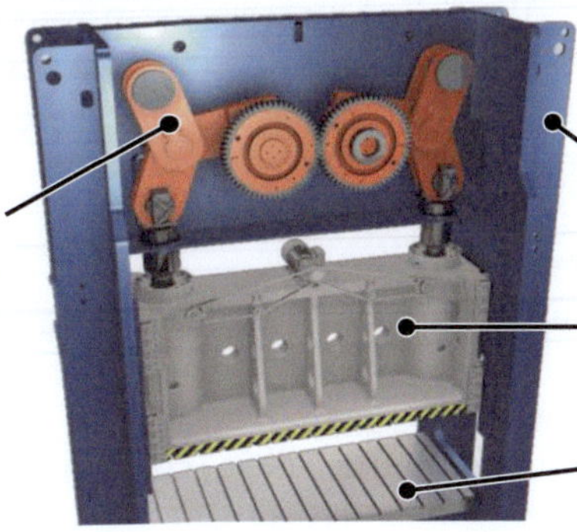

Fig. 3.62 Toggle press with servo drive. *Source* Schuler

effort. This reduces incidental costs while increasing component quality.

3.1.8 Radial Forging Machines

The hydraulically driven radial forging machine, which is shown in Fig. 3.63, was designed for forging slim components with a low diameter/length ratio.

Four radially acting rams are used simultaneously so that the central axis of the workpiece handled by one or two manipulators does not move. The cylinder units are tensioned to the frame components of the forging unit by tie rods. The cylinder chambers above the pistons are continuously supplied with pressure oil. During the return stroke or when the press is stationary, the oil flows off via the axial holes in the piston. To trigger a press stroke, the holes in the piston are closed by valve spools. The valve spools are driven by servo motors which allows the travel and force of the ram to be controlled. Pressure oil continuously feeds the annular grooves below the piston so that the return stroke

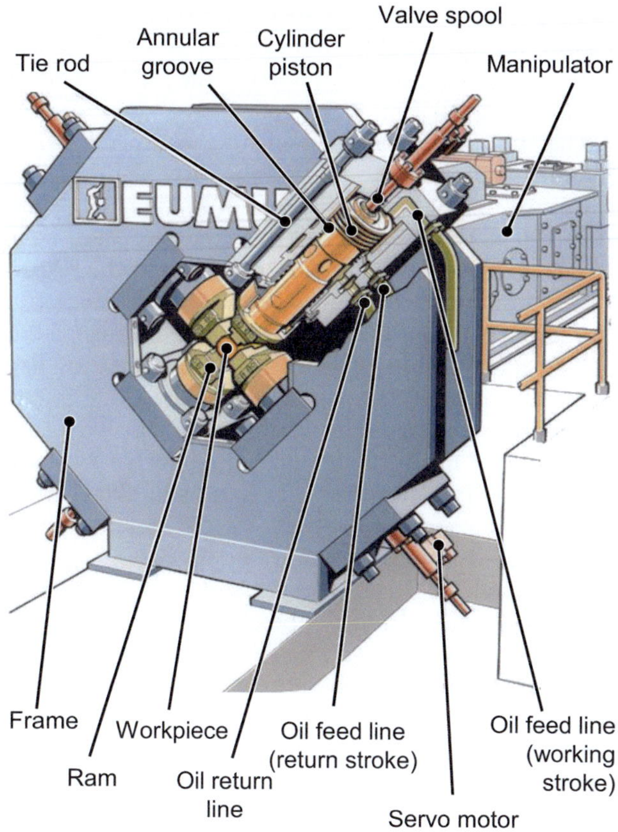

Fig. 3.63 Radial forging machine. *Source* SMS group

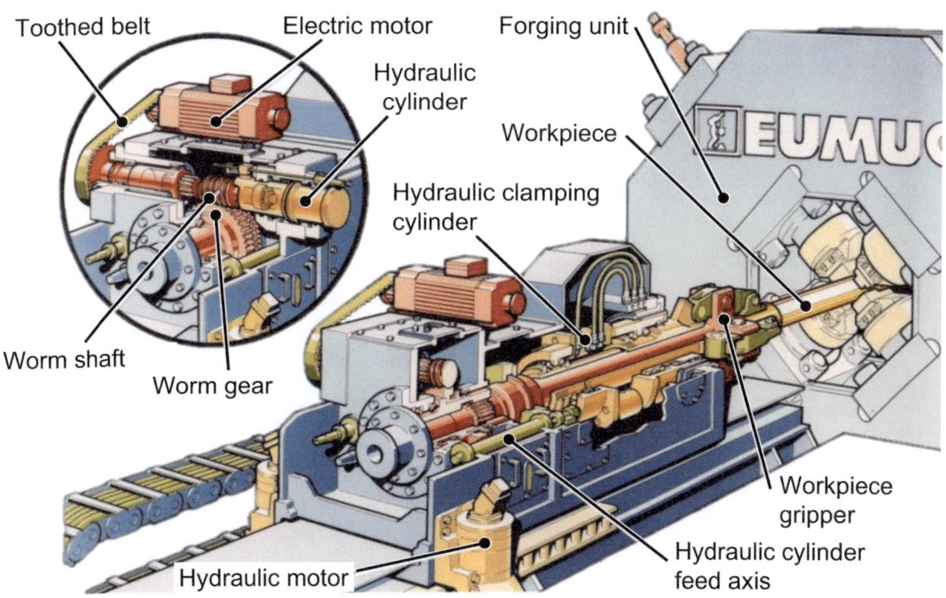

Fig. 3.64 Manipulator for radial forging machine. *Source* SMS group

occurs as soon as the valve spools are retracted and the pressure oil flows out of the cylinder chambers above the piston through the axial holes in the piston. This allows stroke rates of up to 300 min^{-1}.

The dual-sided handling of the workpiece in the forging machine is managed using two identical manipulators, one of which is shown in ◻ Fig. 3.64. These rotate the workpieces, realize the axial feed, and compensate changes in workpiece length during the respective forming process. To achieve this, the manipulators are equipped with a rotary and a linear axis which can both be controlled in two ways. The main rotation axis comprises a worm drive with a worm shaft that is uniformly driven by an electric motor. The rotary axis can also be moved in a highly dynamic manner. In this case, the worm shaft is moved axially by a hydraulic cylinder during the forming process so that the rotational motion of the workpiece is briefly interrupted.

The linear feed axis is implemented with two rack and pinion drives. Both of these are uniformly driven by one hydraulic motor each. To ensure that the workpiece and the forging box are relatively stationary and the changes in length of the workpiece are compensated during the forming process, an additional dynamic drive was installed. The grippers with the clamped workpiece can be moved by hydraulic cylinders in a force-controlled manner along the workpiece axis.

3.1.9 Rolling Machines

Rolling machines are machines with one or more rotating rollers or differently moved tools. The relative motion between the tool and the workpiece is a rolling motion. The geometry and/or surface structure of the workpieces is changed by forming under compressive conditions. A further distinction is made between longitudinal, transverse and cross rolling machines, as well as special designs, as per ◻ Fig. 3.4. In longitudinal rolling, the rolled material is moved through the roll gap perpendicular to the roller axes without any inherent rotation. A distinction is made here between flat and profile rolling.

3.1.9.1 Stretch Rolling Machines

Stretch rolling is preferably used in forging lines to preform drop-forged workpieces. The principle of stretch rolling is shown in ◻ Fig. 3.65.

Roller segments, arranged in pairs on the rollers and easily interchangeable, act as the die that is the negative form of the workpieces. In order to roll in several forming steps, the necessary pairs of roller segments are arranged next to each other on the rollers. The workpieces are fed by an automated handling device whose transport movements are mechanically coupled to the roller movement via a crank-rocker mechanism, ◻ Fig. 3.66.

The hydraulically actuated workpiece gripper sequentially guides the preheated workpieces into the various roller segment dies and can rotate a workpiece between the rolling steps through 90° if required. As roller segments with a maximum contact angle of 180° are used in automatic handling, half a roller rotation is available for retracting, rotating, lateral displacement, and feeding the workpieces between the rolling steps. The final rolled part is ejected between the rollers. ◻ Figure 3.67 shows the preformed blanks produced by stretch rolling and the finished parts produced by drop forging.

3.1.9.2 Cross Rolling Machines

Cross rolling, ◻ Fig. 3.68, is a kinematic reversal of stretch rolling. The workpiece rotates around its own axis and the dies are moved in a straight or curved path in opposite directions to each other. This means that only workpieces with rotational symmetry can be produced.

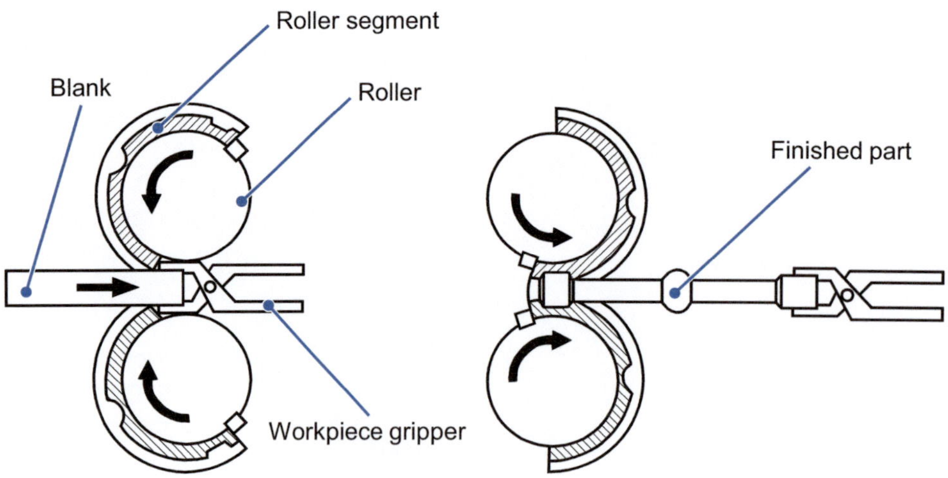

◻ **Fig. 3.65** Principle of stretch rolling. *Source* SMS group

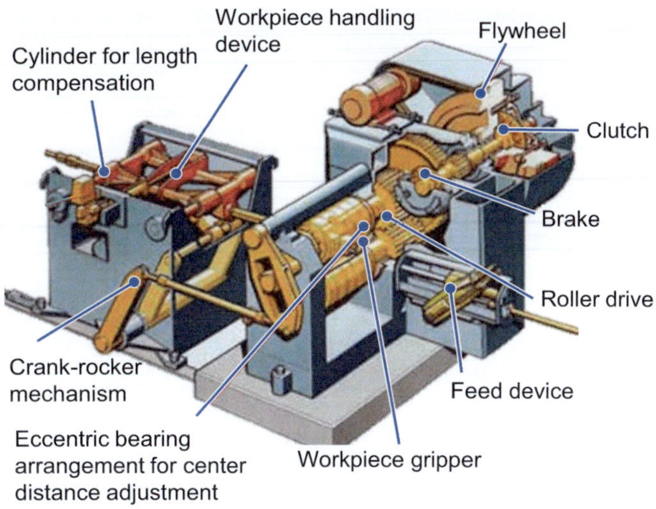

Fig. 3.66 Stretch rolling machine with automatic workpiece transport. *Source* SMS group

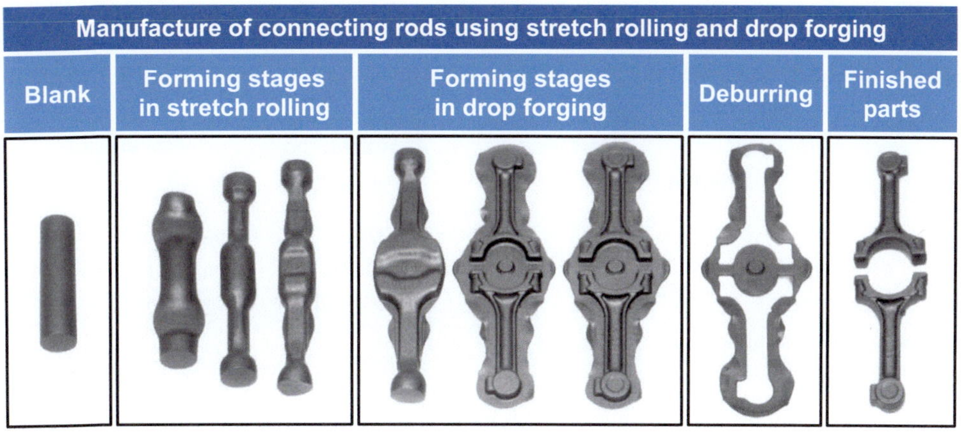

Fig. 3.67 Manufacture of connecting rods by stretch rolling and drop forging. *Source* SMS group

The geometry of the cross rolling tools is transferred to the rolling workpiece by the opposing linear movements in flat jaw rolling or rotational movements in round jaw rolling of the tools. A round jaw cross rolling machine is shown in Fig. 3.69.

The two rollers can be seen, one of which is arranged on a slide for fine adjustment of the rolling gap. The opposing movement of the tools in the rolling gap is produced by worm gearboxes which are coupled together by a shared worm shaft. Both rollers have the same direction of rotation. The part of the roller circumference (approx. 180°) not covered by the tool is partially used to support the handling devices (e.g., guide rails). Once the processed workpiece has been removed from below the rolling gap, the next unfinished workpiece falls into the rolling gap from the guide rail. After the rolling process, the workpiece, which is still hot, is processed further in a drop forging plant.

The cycle time to produce a preform for forging is generally a few seconds. If shorter cycle times are required, several cross rollers can be operated in parallel in order to increase the output [SMS03].

3.1.9.3 Ring Rolling Machines

A radial-axial ring rolling machine, Fig. 3.70, is used to roll seamless, ring-shaped workpieces from preheated ring-shaped blanks produced by pressing or hammering. Workpieces can be, for example, blanks for large rolling bearings. The wall thickness of the ring to be rolled is set in the radial rolling gap, consisting of a main roller and a radially movable mandrel roller, while at the same time the height of the ring is reduced in the axial rolling gap by the two conical axial rollers. During the process, the ring is automatically centered by two centering rollers that hydraulically follow the changing ring geometry. The workpiece is driven by the main and axial rollers which are driven by controllable electric motors.

The axial rolling speed is controlled as a function of the centering force differences in order to prevent lat-

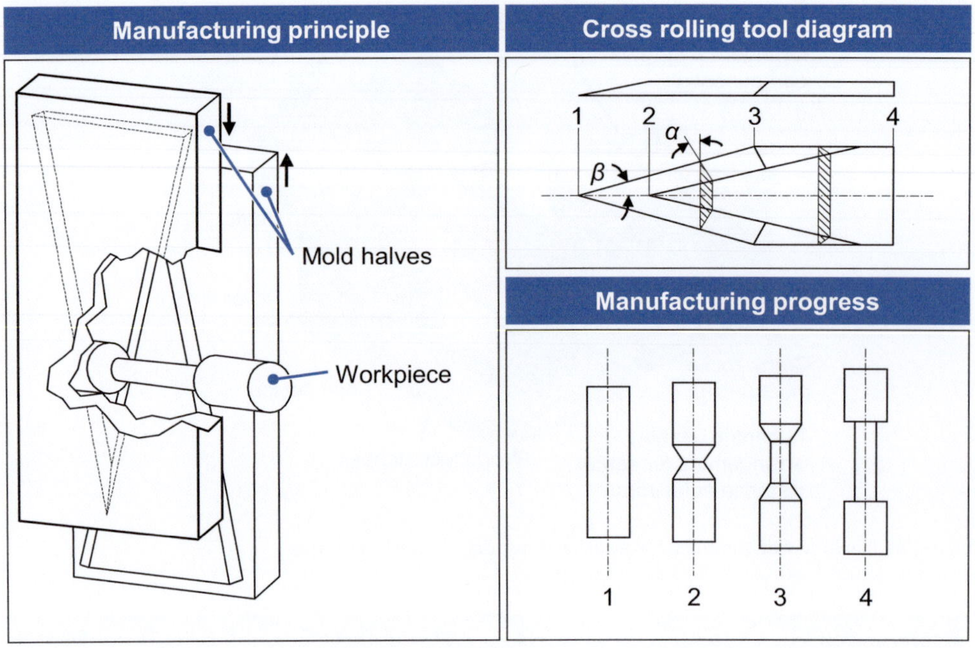

Fig. 3.68 Principle of cross rolling. *Source* Schuler

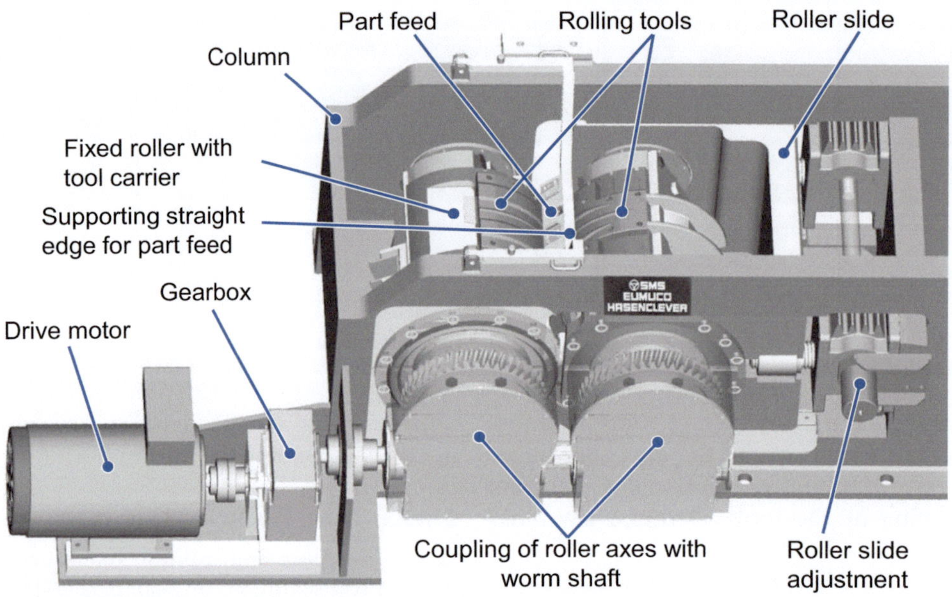

Fig. 3.69 Round jaw cross rolling machine. *Source* SMS group

eral deflection of the ring. The infeed motions are realized via position control loops. The external ring diameter is continuously monitored with the aid of a laser measurement system. An adaptive power controller ensures that the rolling feed is adjusted accordingly to avoid exceeding the maximum rolling forces or motor torques.

3.1.9.4 Closed-Die Rolling Machines

The principle of ring rolling is based on the partial forming of a workpiece. The closed-die rolling process makes further use of the benefits of closed-die technology with respect to accuracy, geometrical design and repeatability. Figure 3.71 shows an axial closed-die rolling machine. The lower die rotates around its vertical axis and holds the blank to be formed. The upper die, which rotates around an inclined axis, rolls the material into the engraving with an axial infeed. A hydraulic ejection system is located in the lower rotating unit. The figure also shows the kinematic relationships in the forming process. Due to the small size of the locally restricted forming zone A*p* and the correspondingly low

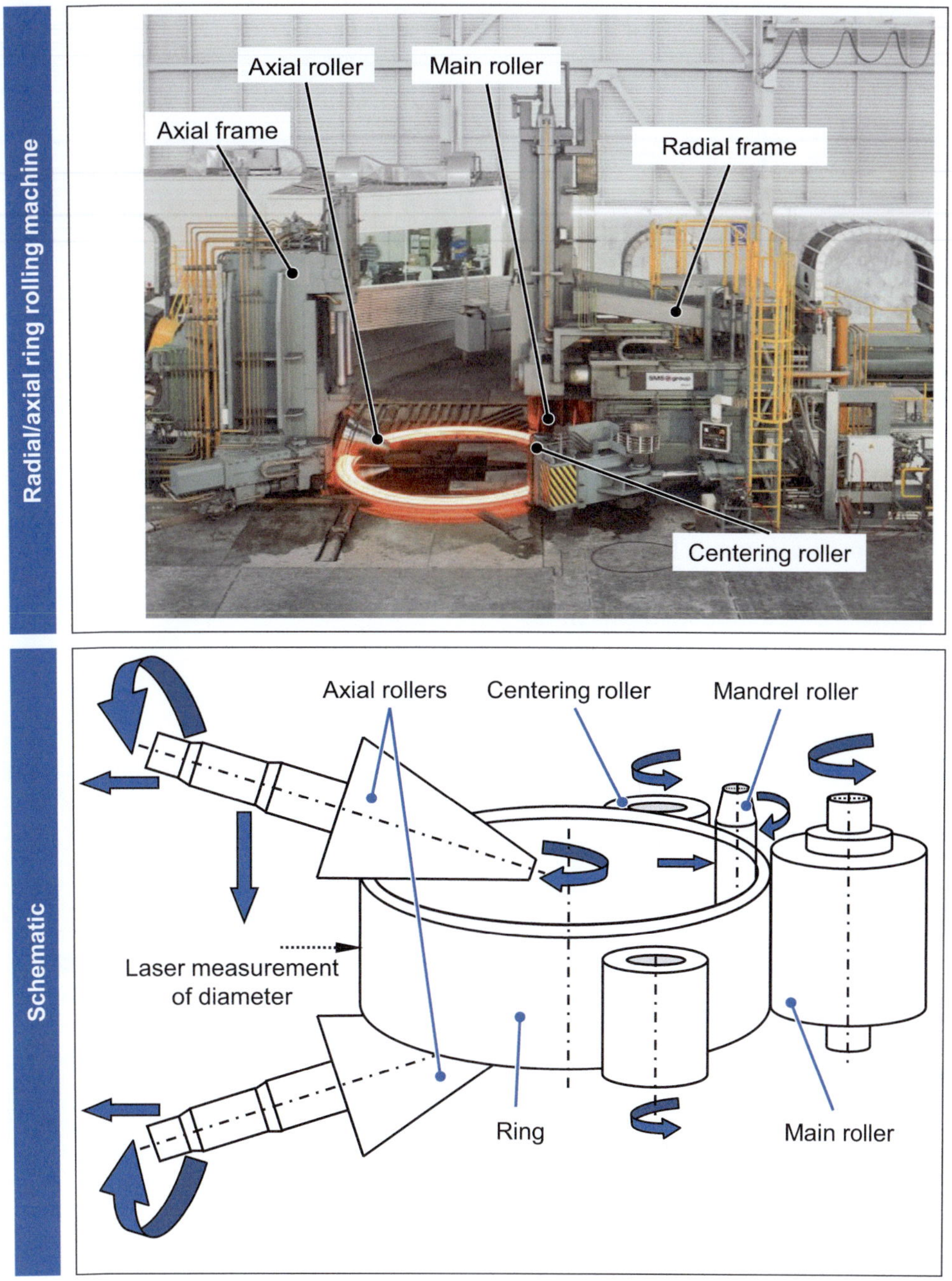

Fig. 3.70 Radial-axial ring rolling machine. *Source* SMS group

axial rolling force F, the closed-die rolling machine can be significantly smaller compared to a forging press for comparable forming tasks. Typical workpieces for axial closed-die rolling are train wheels and aluminum rim blanks.

3.1.9.5 Pressure Rolling Machines

The process of pressure rolling, Fig. 3.72, is used for the chipless manufacturing of rotationally symmetrical hollow components [LEIF10]. Here, the material in its initial shape as a tube or a round sheet metal blank is made to flow between a rotating mandrel and the radially and axially set pressure rollers. In synchronous pressure rolling, shown in the upper section of Fig. 3.72, the axial feed of the pressure rollers and the flow of the material have the same direction. In reverse pressure rolling, lower section of the figure, the movement of the material in the direction of the axial feed motion of the pressure rollers is prevented by a stop. This causes the material under the pressure rollers to flow in the opposite direction of the feed direction.

Fig. 3.73 shows a pressure rolling machine for forming hubs on pulleys or geared drive wheels. A diagram of the process is shown in Fig. 3.74. The in-

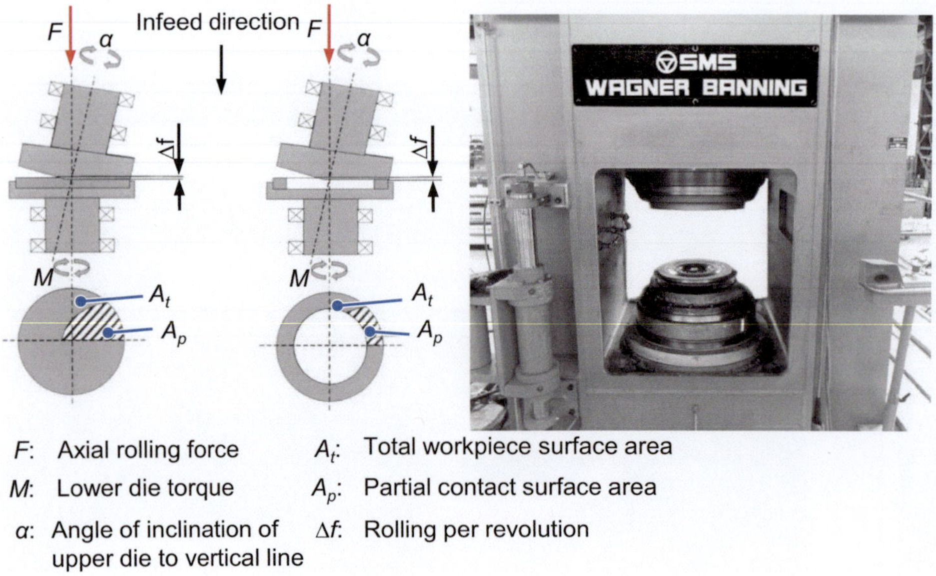

Fig. 3.71 Axial closed-die rolling machine. *Source* SMS group

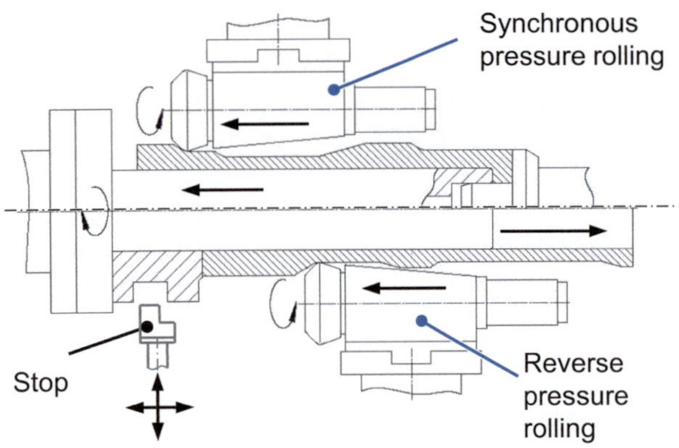

Fig. 3.72 Principle of pressure rolling. *Source* Leifeld Metal Spinning

itial round blank initially lies flat on the rotating tool. It is centered by a mandrel and fixed in place axially by the tool mandrel. A specially-shaped pressure roller is pushed into the surface of the round blank once the tool has started and then moved radially towards the center of the round blank [WFM03]. The material is pushed from the flat surface of the round blank towards the tool mandrel. The hub is raised between the tool mandrel and the pressure roller. The internal and external features of the hub are defined by the shape of the tool mandrel (internal) and the shape of the pressure roller (external).

There are economic and technical benefits for forming hubs this way compared to welding or machining solutions:

- The heavy cold forming of the hubs ensures greater strength.
- Pressure rolling of hubs requires less time.
- The material distortion expected with welding does not occur.
- The component can be designed as a single piece.

3.1.9.6 Reverse Pressure Rolling Machines

Reverse pressure rolling machines, Fig. 3.75, are special-purpose machines for the manufacture of cylindrical hollow shapes with large dimensions such as the housing for the solid-fuel motor (booster) of the Ariane 5 rocket.

Figure 3.76 shows the principle of reverse pressure rolling. The workpiece is clamped on the rotating table of the machine. The forming process is carried out by four undriven pairs of pressure rollers, which roll on the driven workpiece. All pairs of rollers are simultaneously moved downwards by the feed drive. During the resulting forming process, the wall thickness of the tube is uniformly reduced by continuous rolling on the outer and inner walls while the mean diameter is maintained. The material displaced by the pressure rolling process flows upwards in the opposite direction to the feed direction of the pressure rollers (reverse process, see Fig. 3.72). This not only reduces the wall thickness of the tube but also increases its length.

The pairs of rollers are offset by 3 mm in height. This means they engage sequentially and produce the progressive reduction in wall thickness due to the decreasing radial distance between the rollers from pair to pair. As the rollers are arranged on a 90° pitch on an ex-

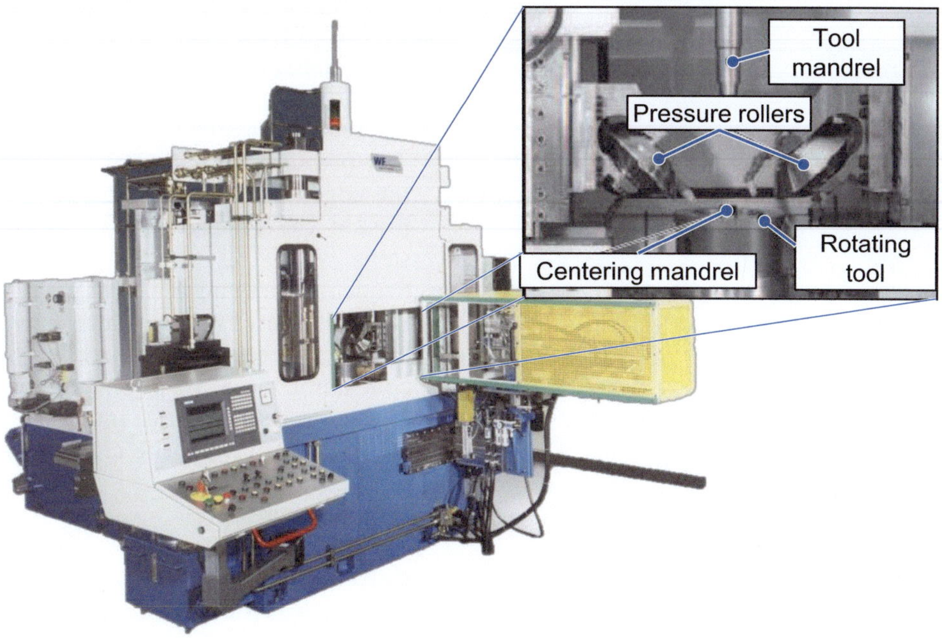

Fig. 3.73 Pressure rolling machine. *Source* WF Maschinenbau und Blechformtechnik

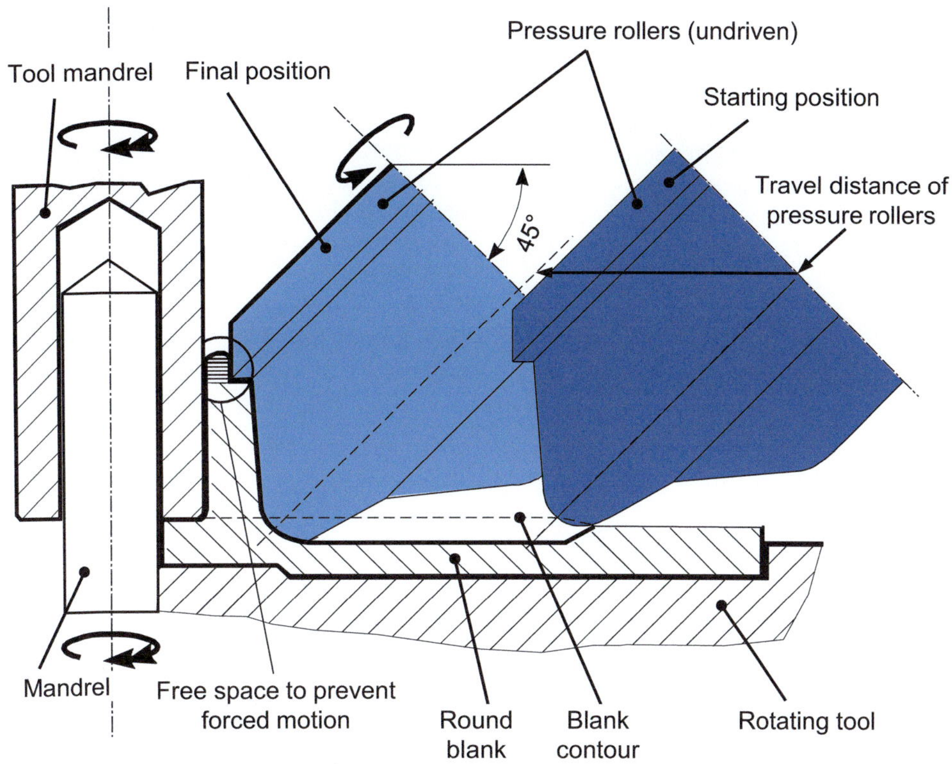

Fig. 3.74 Producing hubs using pressure rolling. *Source* WF Maschinenbau und Blechformtechnik

ternal and internal pressure roller carrier, the high radial forces are largely balanced out. This results in a closed force flow, avoiding the need for a massive machine structure. To ensure the machine is easy to feed and to prevent the machine from limiting the maximum possible height of the finished parts, there are no connections on the head side between the external and internal pressure roller carriers. When cold forming the seven

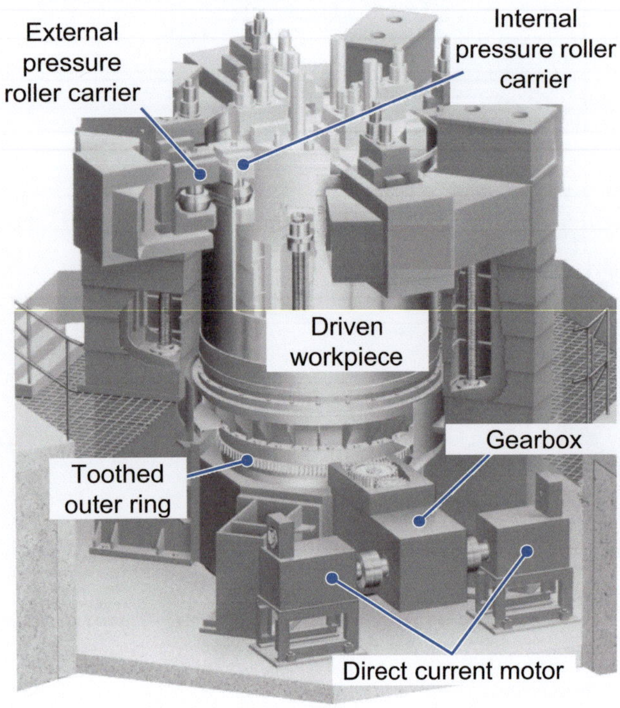

Fig. 3.75 Reverse pressure rolling machine. *Source* SMS group

individual segments for each booster of the Ariane 5 rocket, the wall thickness of the tube segments was reduced from a maximum of 48 mm to 8 mm and the height of the workpiece increased from 1 m to 3.50 m.

3.1.9.7 Spin Forming Machines

Spin forming machines are used to form flat round sheets (round blanks) into rotationally symmetrical hollow components. The round blank is clamped between a rotating spinning chuck and the pressure head and rotated. By using a pressure roller guided in three axes (X, Z, B), the sheet metal is adapted to the shape of the spinning chuck, Fig. 3.77.

Spin forming uses the following strategies:

- Sequential pressing
  The pressure roller describes curved movements to the edge of the rotating blank and back and thus presses the material onto the spinning chuck in steps.
- Sequential drawing
  If a workpiece is to be drawn into radii and/or bevels, it is necessary to work with two-piece axial chucks for pieces open on both sides or with radially split partial chucks, internal rollers or "in air", i.e. without an opposite chuck, for parts closed on one side.
- Projection pressing
  The pressure roller moves parallel to the spinning chuck, whereby the blank material moves axially and the wall thickness decreases. The unprocessed part of the round blank remains perpendicular to the axis of rotation. This process produces polishable surfaces with low tolerances.

The flat bed spinning machine shown in Fig. 3.78 has basically the same design as a lathe. It has a main spindle drive with rotational speeds up to 2500 min^{-1}. It can process round blanks with a diameter of up to 1000 mm. The variously shaped pressure rollers are accommodated in a 6-way tool turret with clampable quick-change holders. Other tools can be various flanging and profiling rollers as well as cutting rollers or a turning chisel for finishing. Some options for additional processing equipment used in the machine are shown in Fig. 3.77, in addition to the spinning techniques.

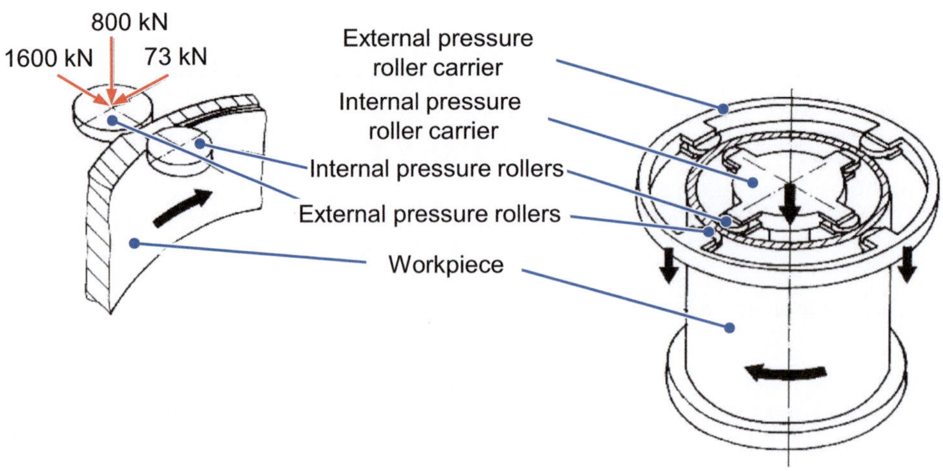

Fig. 3.76 Principle of reverse pressure rolling. *Source* SMS group

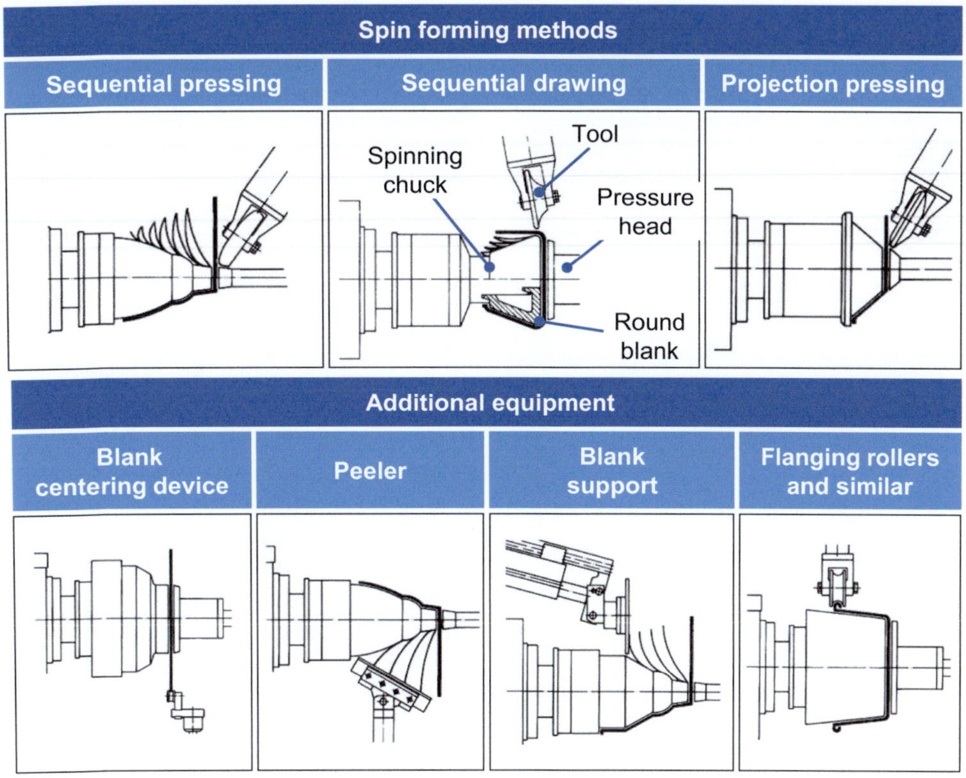

Fig. 3.77 Principle of spin forming and additional equipment. *Source* Leifeld Metal Spinning AG

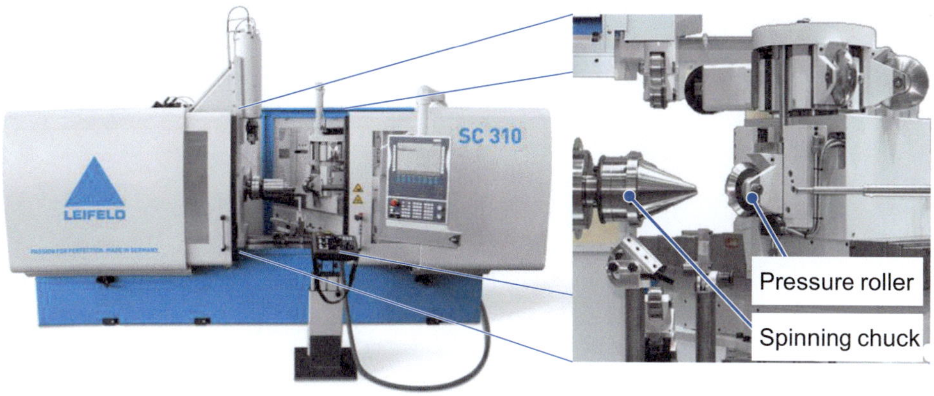

Fig. 3.78 NC spin forming machine. *Source* Leifeld Metal Spinning AG

3.1.10 Bending Machines

According to DIN 8586, a bending machine is defined as a forming machine tool in which discontinuous or continuous bends are made on workpieces around straight bending axes using tools moved in a straight or rotary line [DIN03d]. Bending machines are further subdivided by their kinematics.

3.1.10.1 Press Brakes

A bending machine with a straight tool movement is the hydraulically-driven CNC press brake, Fig. 3.79, which bends flat sheet metals in an open die with sharp edges or any corner radii. Depending on the machining task, the press punch and die can be changed.

The main design requirements for such machines are the bending stiffness of the upper beam with the press punch and the table with the die and sheet support as well as the frame stiffness against bending open. The essential components of a press brake are shown in Fig. 3.80. Possible types of bending on this kind of machine and a working example are shown in Fig. 3.81.

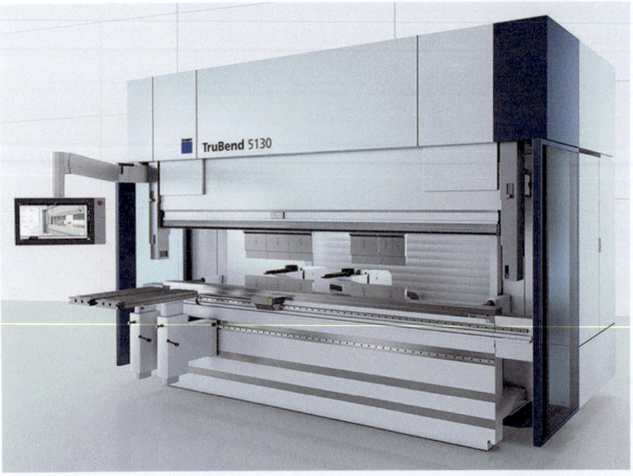

Fig. 3.79 CNC press brake with hydraulic drive. *Source* TRUMPF-Gruppe

As lightweight materials such as magnesium, titanium, or aluminum have gained in importance in recent years, research is in progress on developing a hybrid machine concept consisting of a press brake and a laser unit for local heating of a component [KUHN09]. Local heating to 150 °C to 300 °C along the bending line increases the material's elongation at break multiple times so that the component can be formed without being damaged. Laser-supported sheet metal forming is still partially in the development phase but it is exhibiting huge potential for the future.

3.1.10.2 Automatic Bending Cells

Bending cells consisting of a press brake and an industrial robot have become an established design for the flexible, automatic manufacture of bent parts. Figure 3.82 shows a typical example.

During the sequence of various bending operations, the workpiece is placed in the required position relative to the bending tool by a robot, so that manual handling is not necessary. Special software is used for the programming that synchronizes the movements of the bending machine and the robot. It automatically calculates a collision-free path for the robot. The process is graphically simulated before it is executed so that a functional and collision check can be completed. Sensor technology built into the pincer or suction grippers allows the workpiece to be detected and correctly positioned. This eliminates the need for positioning stations that are usually used. The robot tracks the workpiece during the bending process. The bending cell can also be expanded due to its flexible design.

An online angle measuring system for detecting the sheet bending angle ensures machining accuracy during the bending process, Fig. 3.83. The system consists of two sensor plates with different diameters. The plates are movably integrated into the upper tool. Dur-

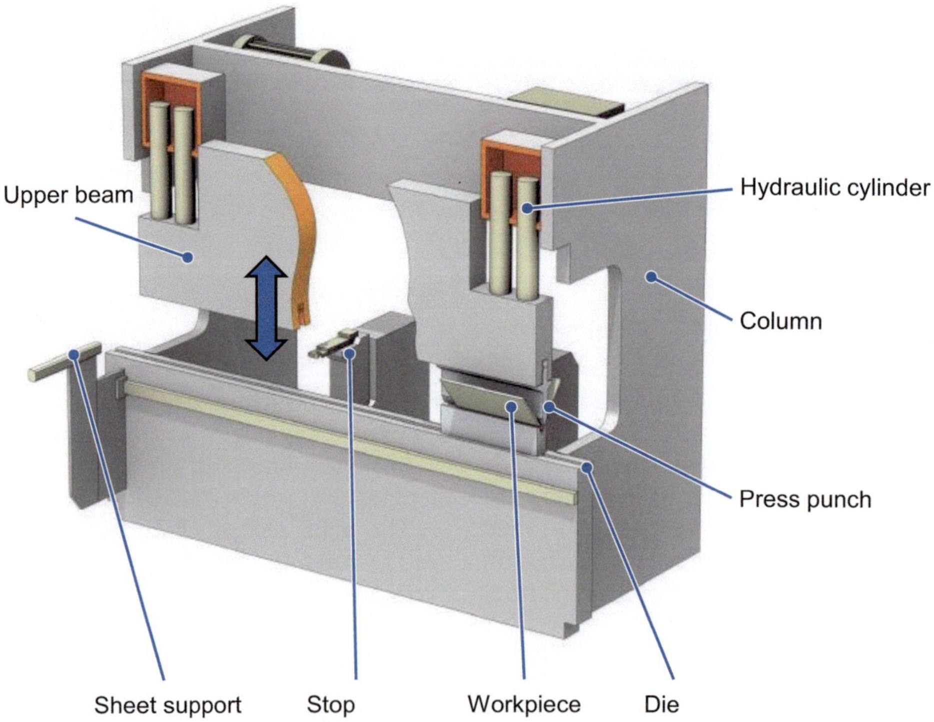

Fig. 3.80 Principle of a press brake. *Source* TRUMPF-Gruppe

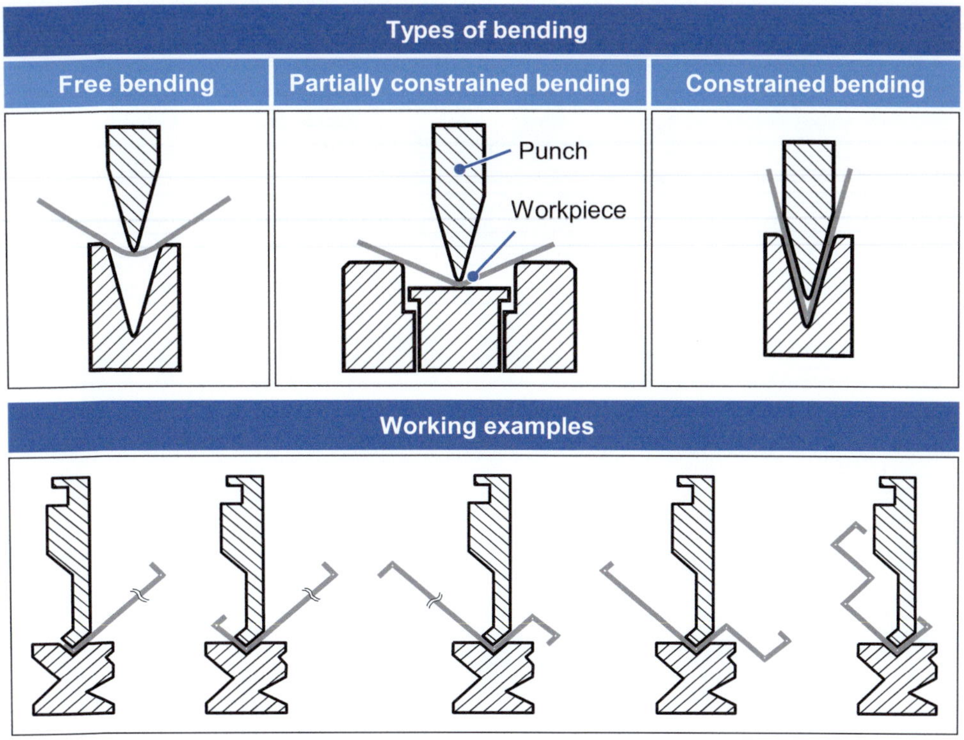

Fig. 3.81 Types of bending and working examples. *Source* TRUMPF-Gruppe

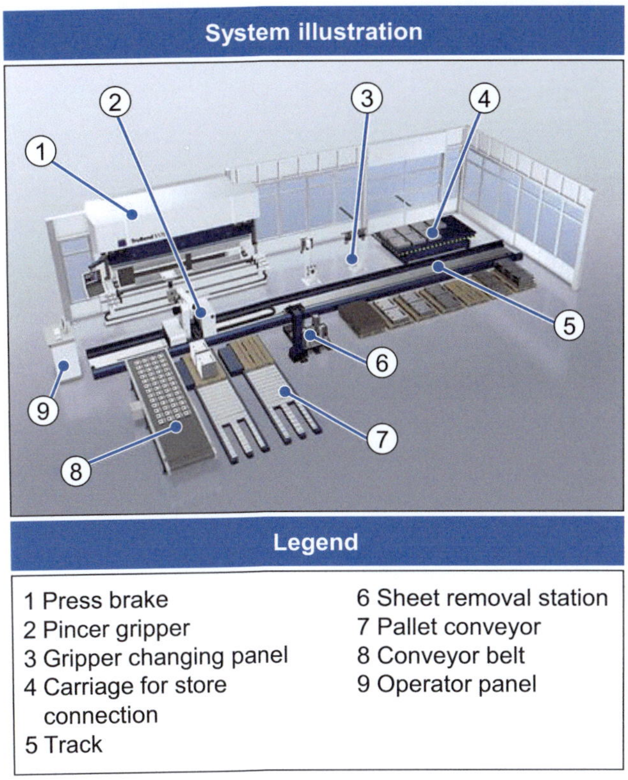

Fig. 3.82 Automatic bending cell. *Source* TRUMPF-Gruppe

ing the bending process, the plates center themselves between the bending legs so that each plate is in contact with the workpiece at two points. The current bending angle is calculated from the distance of the center points of the plates, which increases with increasing bending angle. If necessary, the control system alters the insertion depth of the upper tool. This allows the springback of the sheet and a corresponding height adjustment of the upper tool to be automatically determined during an identification phase at the start of production and to be taken into consideration during the production phase without any time delay. If deviations between the reference and actual bending angle occur during production, this is again automatically adjusted in the program [JANS11].

3.1.10.3 Sheet Round Bending Machines

Sheet round bending machines are used to manufacture tubes, troughs, containers, coils, and conical parts. The workpieces are formed from a flat metal sheet in a continuous working process. The achievable degree of accuracy depends on the type of initial sheet bending. Four-roller machines have been developed for volume production as sheets cannot receive the initial bend automatically, or only with a great deal of difficulty, using conventional three-roller machines. The principle of four-roller sheet round bending is shown in Fig. 3.84.

Here, the workpiece is initially fed in until it touches bending roller B, step 1 in Fig. 3.84. The workpiece is then clamped between rollers C and D and positioned for the following rounding (step 2). During rounding, step 3, roller A is advanced to bend the sheet. For final rounding, the sheet is drawn in by

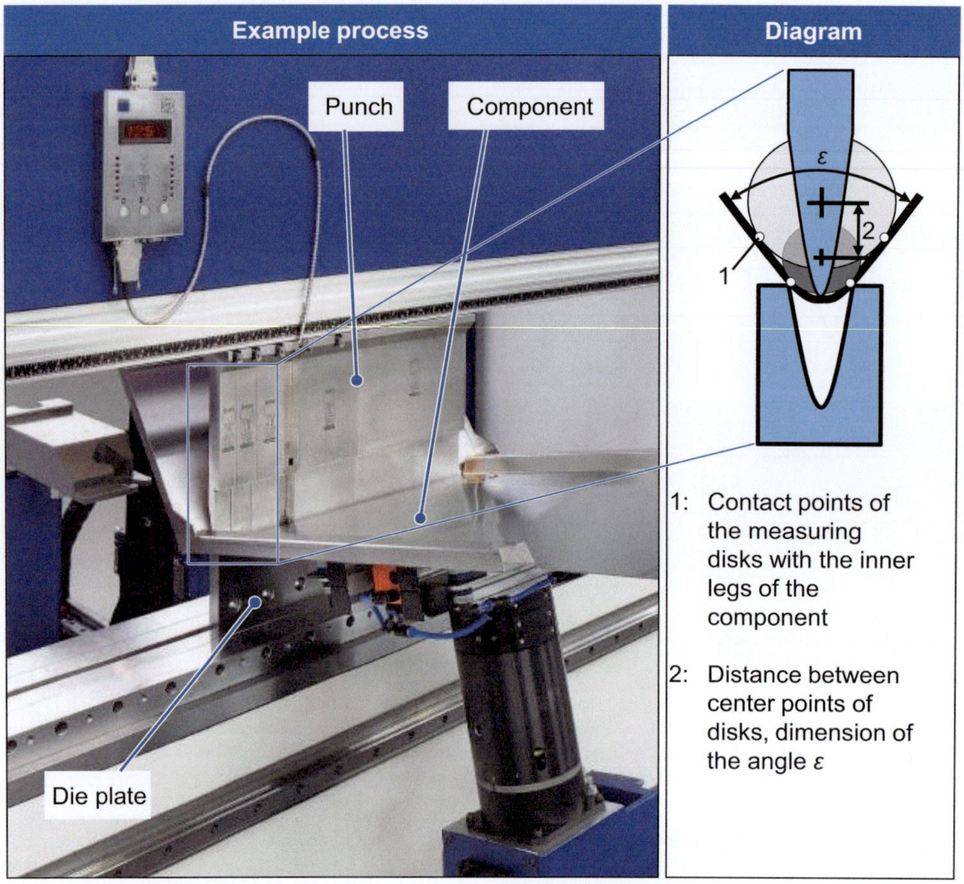

Fig. 3.83 Tool with bending angle measuring system. *Source* TRUMPF-Gruppe

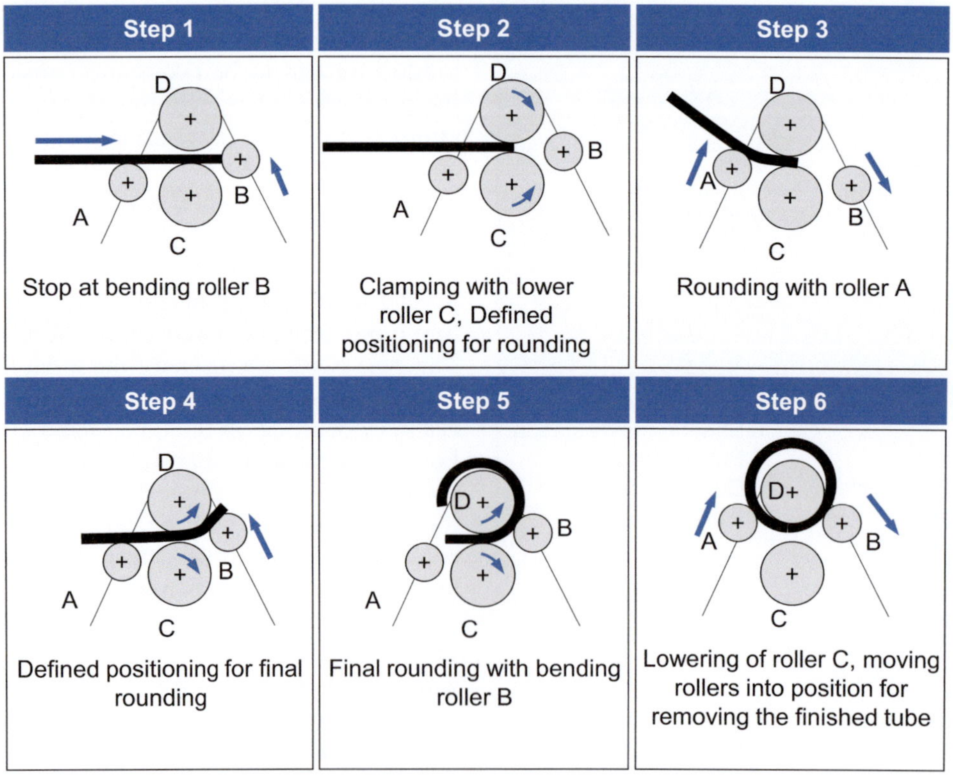

Fig. 3.84 Principle of four-roller round bending

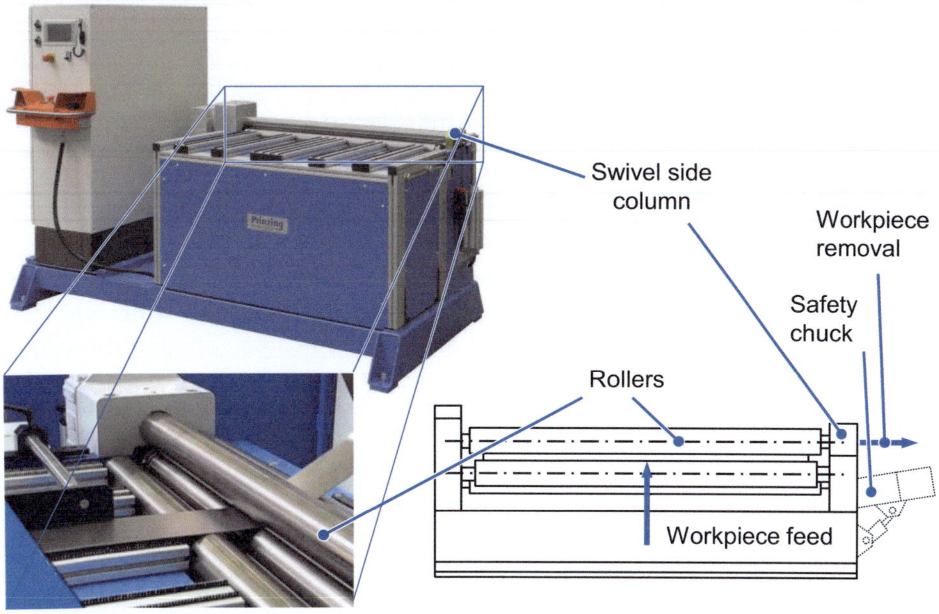

Fig. 3.85 Four-roller round bending machine. *Source* Peter Prinzing

rollers C and D (steps 4 and 5), roller A is retracted and roller B is advanced accordingly. The tube is now bent around roller D. In the final manufacturing step, roller C is retracted. The tube can then be removed and welded (step 6). Figure 3.85 shows a four-roller sheet round bending machine. The rollers are positioned using stepless motors and provide the feed, the bending angle and the angular adjustment of the sheet.

Four-roller sheet round bending machines are particularly suitable for volume production as the bending process can be fully automated once the sheet has been inserted. With three-roller machines, the sheet has to be re-inserted a second time for rounding. However, the benefit of three-roller machines is a better bending quality when rounding the sheet. In addition, the procurement costs for three-roller machines are significantly lower than the corresponding machines with four rollers.

3.1.10.4 Swivel Bending Machines

Bending machines with rotating tool movement are known as swivel bending machines, Fig. 3.86. The machine shown here can be used to bend sheets up to 6 mm thick. A clamping force of up to 200 metric tons is achieved with a three-axis servo drive. Sharp folds with various bending angles as well as various radii can be produced using appropriate hold-down and bending rails. For adapting the thickness of the sheet to be bent, for precise round bending, and for using special rails in exceptional bending operations, the position of the pivot point and the bending cheek can be adjusted in the movement radius with respect to the pivot point. If the pivot point is adjusted upwards and the bending beam is adjusted downwards by the same amount, the bending tool remains at the same height as the sheet support of the lower beam in its starting position. However, the pivot point is higher, and the movement radius of the bending cheek is correspondingly larger.

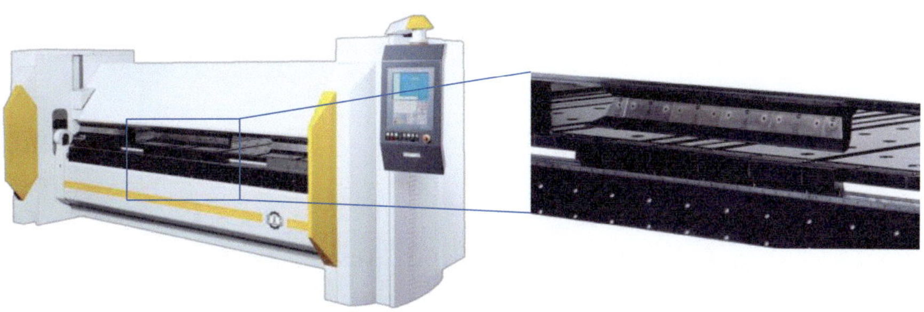

Fig. 3.86 Swivel bending machine. *Source* RAS Reinhardt Maschinenbau

As a result, the bending tool slides along the underside of the sheet during the swivel movement, thus forming a round shape.

To compensate the deflection of the bending cheek under the bending force, a crowning can be inserted between the bending beam and the bending tool. For this purpose, two long wedges, which can be moved against each other, are pushed together under the bending tool so that it is slightly raised in the center. Under the load occurring during the bending process, the tool is then almost straight due to the opposing deformation. An integral sensor measurement system determines the deflection of the bending cheek and the intelligent crowning system automatically compensates the measured deformation. ◘ Figure 3.87 shows the functional principle for a swivel bending machine.

3.1.11 Drawing Machines

Drawing machines are mainly used to manufacture or rework wires, tubes, bars or profiles. ◘ Figure 3.4 shows the categorization of the different machine designs.

◘ Figure 3.88 shows a continuous carriage drawing machine for reducing the cross-section of tubes, rods and folding material.

The material is drawn alternately through the die plate (drawing ring) located in the drawing block by the two reciprocating drawing carriages. Drawing forces of up to 450 kN and drawing speeds of 150 m/min can be achieved. The movements of the drawing carriages are synchronized in such a way that the material moves continuously. Mechanically connected axial cam disks realize the drive and the synchronization

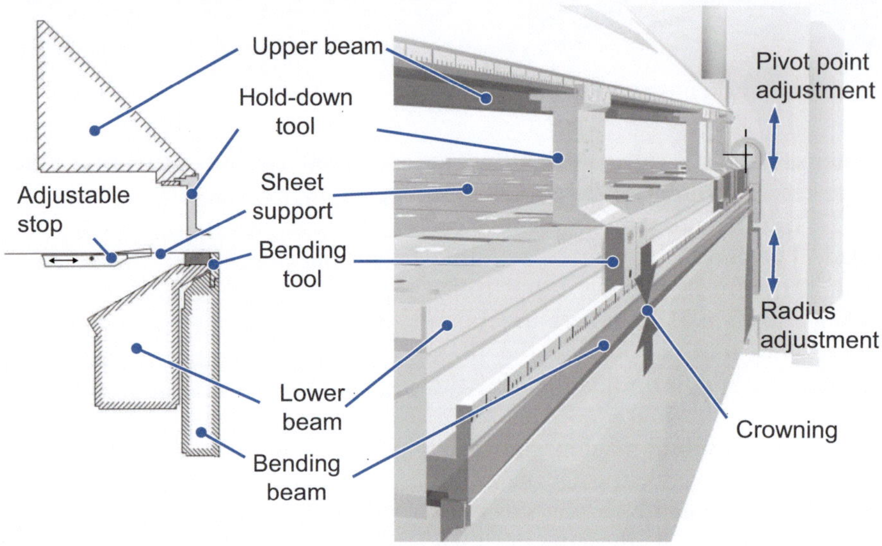

◘ **Fig. 3.87** Functional principle for a swivel bending machine. *Source* Schröder-Fasti Technologie

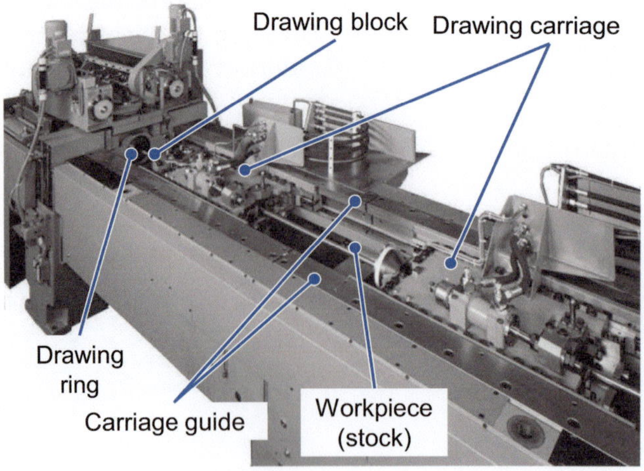

◘ **Fig. 3.88** Continuous carriage drawing machine. *Source* SMS group

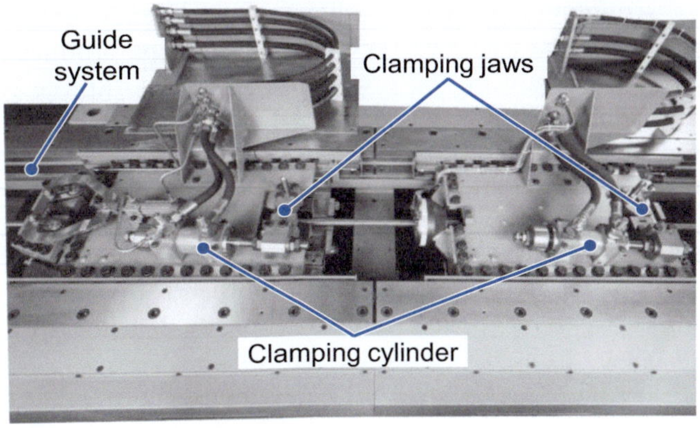

Fig. 3.89 Drawing carriage on a drawing machine. *Source* SMS group

of the drawing carriage movement. This achieves high drawing performances. The main shaft, which supports the cam disks, is located in the body of the machine beneath the drawing carriages. The clamping jaws in the two drawing carriages, Fig. 3.89, which serve to grip the material during the working stroke, are actuated hydraulically by a plunger control system whose hydraulic cylinders are controlled by cam disks mounted on the main shaft. At the start of the drawing process, the material is either pointed by rotary swaging machines or stretch rolling and pushed through the drawing ring or pressed through the drawing ring with high force by a hydraulic pusher.

One option for achieving higher drawing speeds is shown in Fig. 3.90. This shows a caterpillar drawing machine that transfers the continuous movement of its drawing chains to the workpiece. The driven chains with attached clamping pieces (block link chains) provide for the transport of the material. As the clamping pieces transport the material through friction, a high press force is required. This force is transferred by the roller chains located behind the clamping pieces. The roller chains are supported by connectors between the chain wheels. This allows drawing speeds of up to 500 m/min to be achieved.

In addition to the actual drawing units and the pusher shown here, there are many more devices associated with drawing machines for pre- and post-processing of the material. These include decoiling and straightening machines for pre-processing as well as devices for inspection and cutting to length for post-processing. The manufacture of copper tubes includes cascades in which up to three connected drawing machines work in line with correspondingly high productivity.

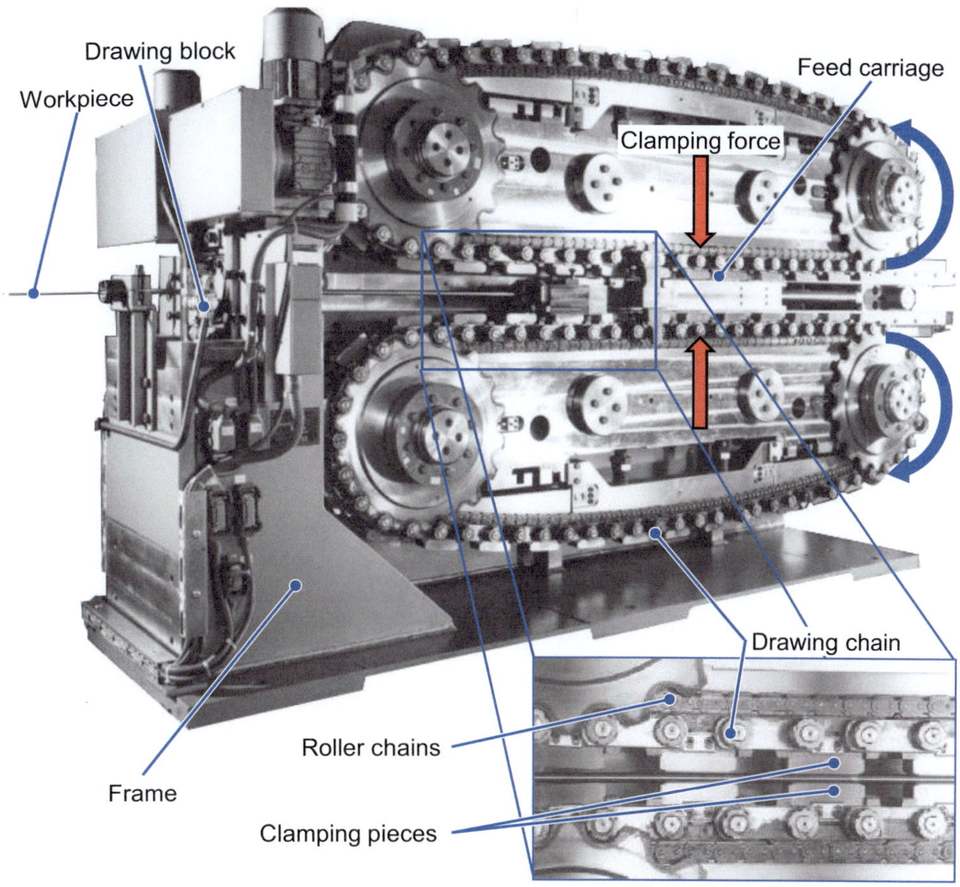

Fig. 3.90 Caterpillar drawing unit of a drawing machine. *Source* SMS group

3

3.1.12 Straightening Machines

Machining as well as heat treatment usually causes workpieces to warp. Straightening processes which can be traced back to the manufacturing process of 'forming by bending' are used to straighten these workpieces or semi-finished parts such as sheets, bars, wires and tubes. This means that the straightening process is carried out by deforming the workpiece past the elastic range. By doing this, the desired plastic deformation is achieved.

A basic distinction is made between two straightening processes that use this type of forming: Bend straightening and roll straightening. ◘ Figure 3.91 and ◘ Fig. 3.92 provide an overview of possible variants of these two processes that can be categorized into discontinuous and continuous processes.

The discontinuous processes include bend straightening processes involving straightening using straightening or roller blocks or a straightening press. These processes are mostly used for straightening individual workpieces before final grinding. A typical application is straightening guideway gibs or rails after hardening and before final grinding. Hydraulic straightening presses with a regulated stroke have proven themselves as straightening presses, allowing accuracies within 1/1000 µm, ◘ Fig. 3.93. The change in shape produced by the pressure of the press ram on the workpiece is displayed on dial gauges as shown in ◘ Fig. 3.93. Here, the measuring sensors are constantly in contact with the workpiece by lever transmission.

Self-learning processes are also often used in large-scale production. The machine control system learns from previous bending results which bends or bending forces are to be applied to the workpieces depending on their curvature.

◘ Figure 3.93 shows the straightening process for a bevel gear shaft. The runout is determined using the dial gauges by rotating the workpiece, which is supported at both ends. The stroke movement of the press ram is then synchronously controlled using a manual lever. The result can then be read directly off the dial gauges. Alternatively, the radial runout deviation can be checked using an illuminated column display. The ram speed is also proportional to the manual lever movement as well as the straightening stroke. The ram stroke is limited to avoid over-straightening and thus the risk of workpiece fractures. Several straightening strokes are usually required to straighten the workpiece [DUNK98].

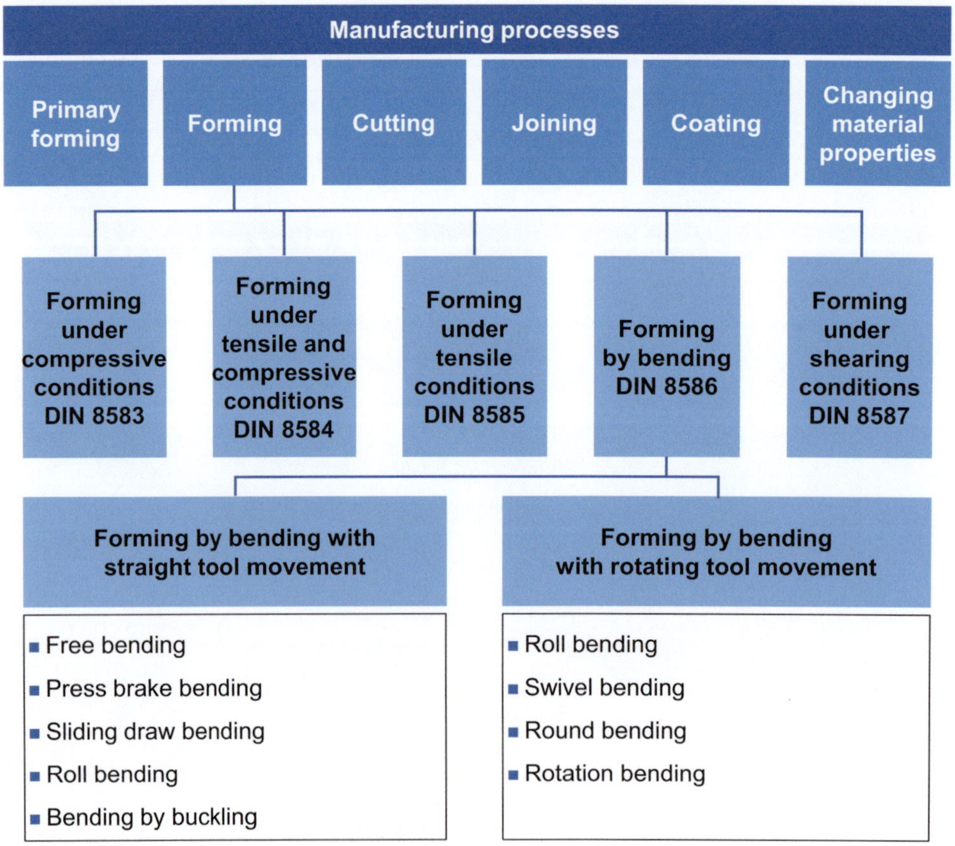

◘ **Fig. 3.91** Classification of the manufacturing processes 'bend straightening' and 'roll straightening', according to DIN 8580 [DIN03f], DIN 8586 [DIN03d]

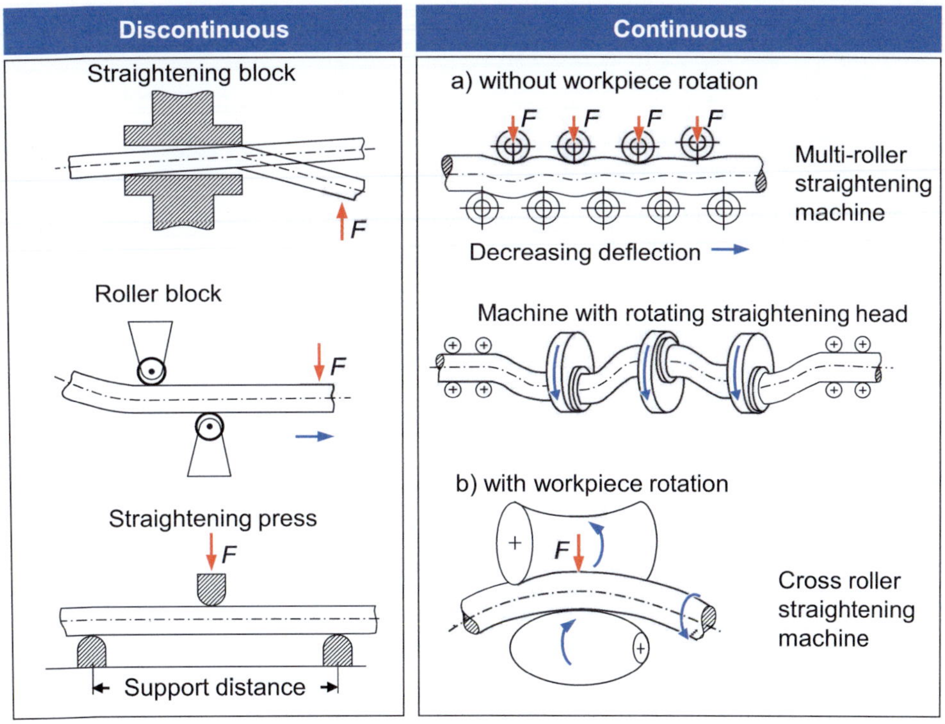

Fig. 3.92 Straightening process for round steel bars. *Source* SMS group

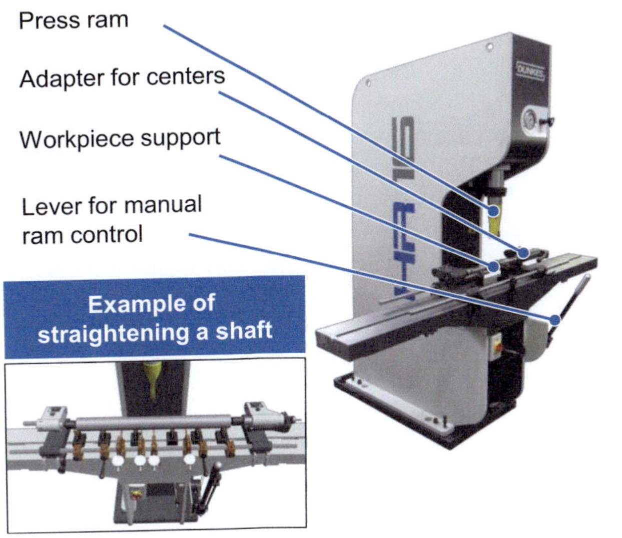

Fig. 3.93 Hydraulic straightening press with regulated stroke. *Source* DUNKES

The continuous roll straightening processes shown in Fig. 3.92 are broken down into processes with and without workpiece rotation. Multi-roller straightening machines, for example, operate without workpiece rotation. The processes with workpiece rotation are generally known as cross roller straightening. A typical processing task for cross roller straightening, which is only suitable for workpieces with a circular cross-section due to the process (workpiece rotation), is straightening bright steel round bars and tubes as semi-finished parts. Steel round bars are used, for example, to manufacture springs and screws. There are high demands on the straightness of the bar material. These cannot be achieved with cutting processes (bar peeling or centerless grinding) of the rolled raw material alone. The required straightness is only achieved in the downstream straightening process.

Figure 3.94 shows the principle of a two-roller straightening process (cross roller straightening). Due to the position and shape of the straightening rollers, the workpiece is bent to such an extent that it is plastically deformed in the center of the rollers. The magnitude of the plastic deformation must be selected such that the maximum curvature in the unstraightened bar can still be compensated. The required maximum deflection f is set by the two rolling angles, i.e. by the skew between the two straightening rollers and the bar as well as the roller infeed. If the bending process is repeated with a decreasing bending torque, the residual curvature of the straightened bar decreases until a technically useful value is achieved after only a few passes. The third important control variable apart from the roller angles is the distance between the rollers. By adjusting this distance, it is possible to straighten different diameters of steel bars with the same pair of rollers. By setting the distance between the rollers appropriately, a stamping force is applied in the center of the rollers in addition to the straightening force (bending force). A sufficiently high stamping force will smooth

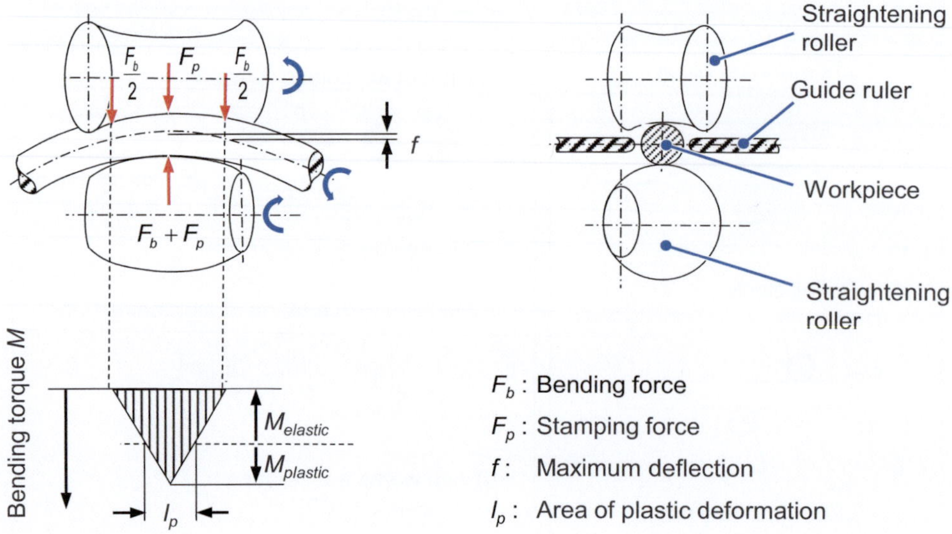

Fig. 3.94 Two-roll straightening process

the surface of the bar such that roughness depths of up to Rz = 2 µm can be achieved with good pre-processing quality. A hydraulic overload protection system is used to avoid workpiece fractures due to over-straightening.

Figure 3.95 shows the construction of a two-roller straightening machine. The single-piece, welded body forms a stiff structure and absorbs the high rolling forces. The adjustable straightening rollers are driven by shafts with universal joints. Bar stock with a diameter between 3 and 300 mm can be straightened on these machines. A straightening accuracy of up to 0.1 mm/m straightness deviation can be achieved for bright material. Straightening speeds (feed rates) of up to 120 m/min are possible for material with a diameter of 10 mm.

3.1.13 Guards and Safety Devices on Forming Machines

Forming machines are fitted with mechanical and electronic devices which provide the greatest possible protection for operators, the machine (overloading), the tool, the workpiece and the environment. Detailed information on the guidelines for machine design is given in ▶ Chap. 8 of Volume 2 in this series.

To protect operators from injury, gratings or protective windows are used to screen the tool area during the working stroke to prevent manual access. Contactless protection devices such as light barriers and light curtains are also suitable for protecting operators from inadvertently starting the machine during manual place-

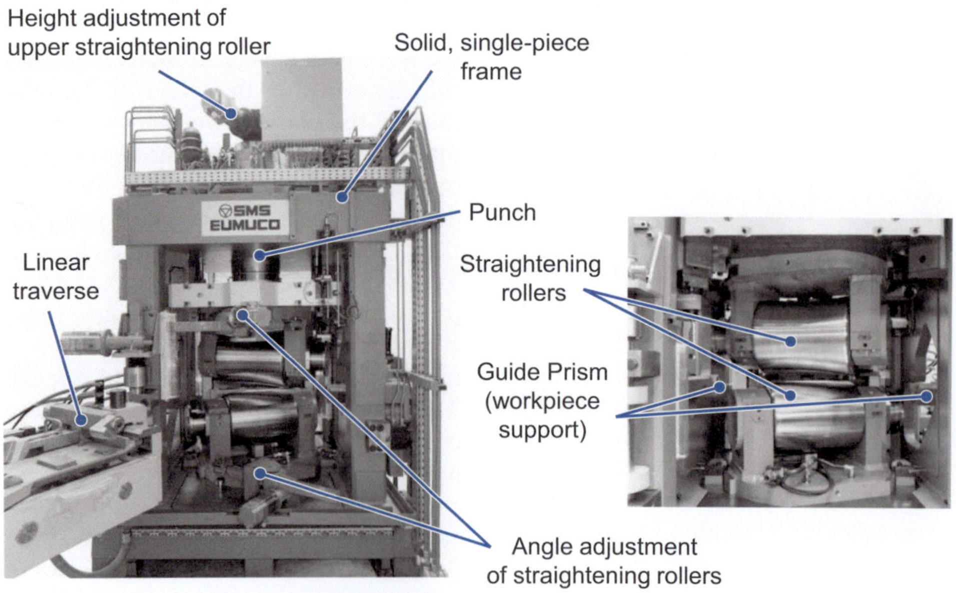

Fig. 3.95 Two-roller straightening machines. *Source* SMS group

ment of workpieces. Appropriate legally binding regulations have been issued via the German Federation of Professional and Trade Associations in Industry [DIN09, DIN11, DIN12, DIN14, SCHU08].

Another, but not necessarily adequate protection for operators is two-hand controls on forming machines. It should only be possible to initiate the working stroke once the switches have been activated with both hands at the same time. The switches must remain activated throughout the forming process. A further working stroke after loading the machine can only take place once the two-hand controls have been released and activated again [SCHU08]. Other important elements of the safety equipment on mechanical presses are overrun monitoring and run-through protection. Overrun monitoring detects impermissible extensions of the braking process to lock the press control if required and thus to prevent the machine from operating with a non-functioning brake. Run-through protection shall, in the event of a brake failure after a stroke has begun, prevent the ram from stopping too late, that is when the tool starts to close or even after it has closed. Run-through protection usually requires a second operational brake that operates with the normal braking process but which is also able to bring the machine safely to a halt on its own [SCHU08]. This means, the coupling and the brake have a significant influence on the safety of a press. According to the accident prevention regulations, only friction clutches and brakes may be used for manual loading with two-hand controls or light barriers. Pneumatically actuated single-disk couplings and brakes are being used here.

Overall, the machine control must also be designed such that no accidents can be caused by malfunctions. The function of additional safety equipment can only be guaranteed by a safe control system. Mechanical predetermined breaking points and hydraulic systems are used as overload protection on forming machines. As changing mechanical safety devices has a relatively high mounting effort, and therefore a high time and cost requirement, crank and eccentric presses now largely use hydraulic overload protection, ▪ Fig. 3.96.

In the system shown, the valve plate is loaded by pressure oil from below. It is supported by the housing via the valve seat. The pressing force acts on the valve plate via the connecting rod, pressure piece and the adjustment sleeve. If the pressing force exceeds the preload force set by the oil pressure, the valve plate lifts off the valve seat, the force fit between the connecting rod and the ram is interrupted and the oil flows back into the free space above the valve plate under zero pressure. The press is turned off by the pressure switch. Compared to mechanical systems, the hydraulic overload protection has the advantage that the maximum

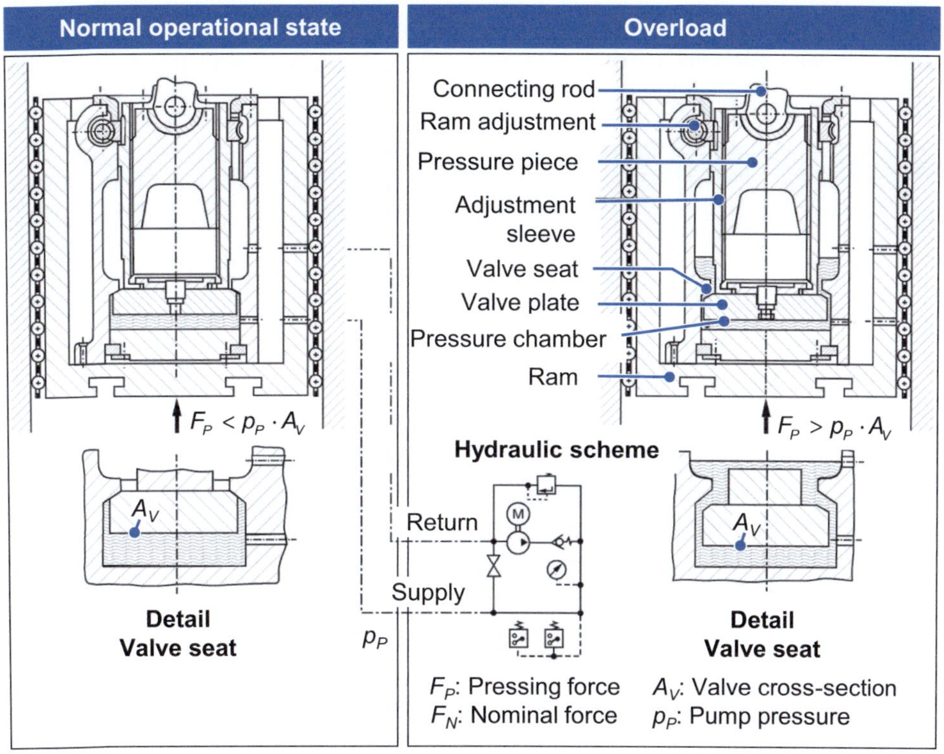

▪ **Fig. 3.96** Hydraulic overload protection in the press ram

transferable pressing force can easily be adjusted by the oil pressure. The press is available again just a few minutes after the overload protection has been activated.

To ensure the torque is not exceeded, e.g., in large eccentric or screw presses, friction clutches are usually used in the press drive chain. If the components within the flux of forces are overloaded, the clutch slips. Different coefficients of friction and natural wear have a significant influence on the time consistency of the set limit load in this type of safety element so that sufficient safety is only provided when the torque is exceeded by a significant amount.

The key safety element in hydraulic machines is the press safety valve placed in the hydraulic circuit between the drive (pump) and the cylinder. Once the maximum adjustable permissible system pressure has been reached, this valve opens and prevents the frame or tool from being overloaded.

Tool damage can typically occur if workpieces or cutout slugs are not completely removed from the tool working area before a new blank is inserted. The working area is monitored by multiple styluses or electronic sensors integrated into the tool. An alternative method of ensuring the safety of the tool is path-dependent force measurement: High-speed clutches and brakes immediately bring the tool to a halt if there is an unplanned, for example premature, increase in the force due to the presence of a tool. Hazards and negative environmental impacts from forming machines are mainly caused by vibrations and noise emissions [HUBE12]. These negative environmental impacts are prevented by various means such as cutting impact damping [FAHR05], noise protection cabins, vibration-isolated machine setup and specific technological measures.

3.2 Severing Machines

Severing machines are particularly important when processing sheet metal as they provide an economical means of producing semi-finished parts and finished workpieces. The severing manufacturing process is defined as the mechanical cutting of materials without forming chips. ◘ Figure 3.97 shows the classification of severing processes, starting with the generic term 'cutting' from DIN 8580 and subdivisions of cutting processes. According to DIN 8588, severing processes can be subdivided into shear cutting, knife cutting, wedge-action cutting, splitting, tearing and breaking.

While shear cutting is used in many manufacturing technology applications, the other severing processes are less significant in industrial metalworking. Most severing machines use the shear cutting principle as shown in ◘ Fig. 3.98.

The shear cutting process can be split into four phases. Setting the upper and lower knives onto the workpiece produces an elastic deformation in the workpiece. The infeed motion between the upper and lower knives then leads to plastic deformation and flow in the workpiece. In the third phase, cracks start forming in the workpiece, originating from the cutting of the upper and lower knives, and this results in the workpiece being cut in the fourth phase. The cut quality in shear

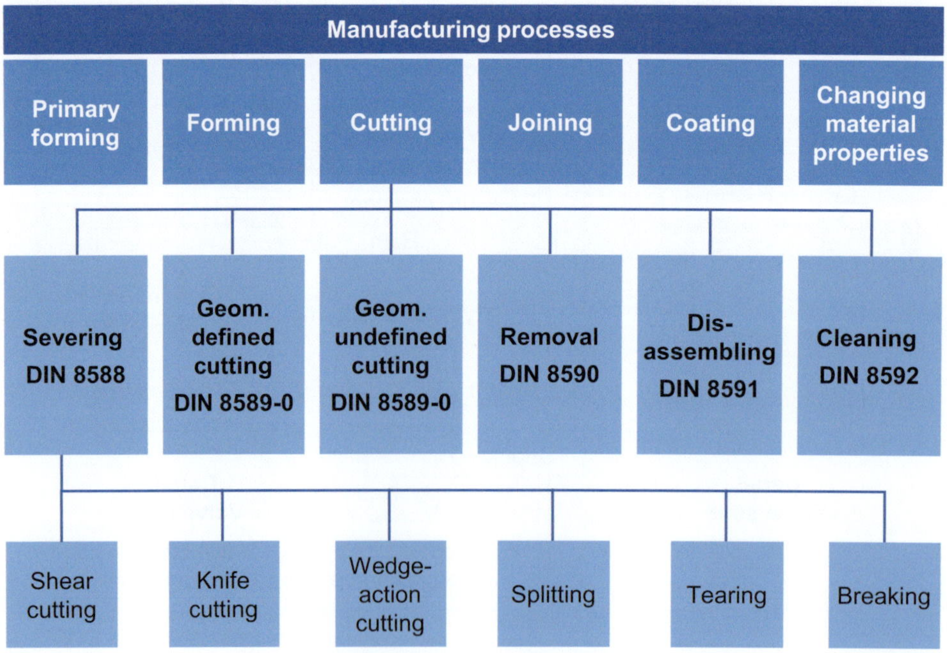

◘ **Fig. 3.97** Subdivision of the severing manufacturing processes, according to DIN 8588 [DIN03a]

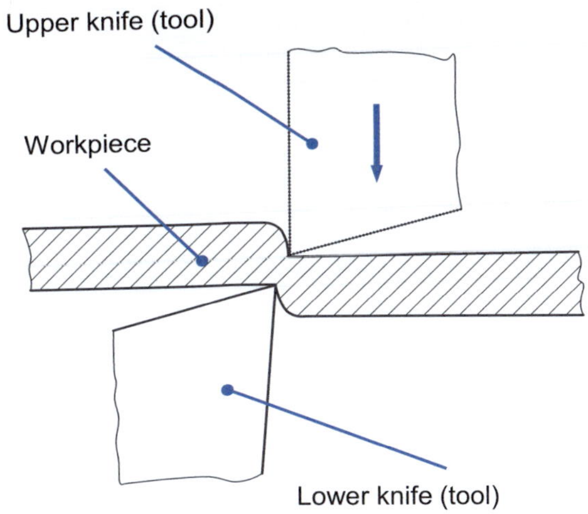

Fig. 3.98 Principle of shear cutting, according to DIN 8588 [DIN03a]

cutting is determined by the thickness of the workpiece and the material as well as the cutting gap. Furthermore, the cut quality decreases as the upper and lower knives wear. The upper and lower knives engage continuously.

3.2.1 Shearing Machines

Shearing machines are used for separating sheets, bars, profiles and hollow elements. A characteristic feature of shearing machines is that the tools (knives) are an integral part of the machine. They are bolted to the components of the machine structure and can be easily replaced. Shearing machines are subdivided into guillotine and circular knife shearing machines, depending on the design of the tools (knives). Guillotine shearing machines are predominantly used for separating sheets with a thickness of approximately 2 mm. The translational shearing motion of the knives is produced by a crank drive or a hydraulic cylinder. Hand lever shearing machines are also used for thinner sheets.

In guillotine shearing machines, the lower knife is fixed to the machine frame while the upper knife executes the guided translational stroke movement. Figure 3.99 shows an example of a guillotine shearing machine. The detail shows the arrangement of the upper and lower knives and the hold-down device for the workpiece and other equipment such as the sheet stops and the hand guard.

In circular knife shearing machines, the cutting tools consist of two or more circular knives arranged in pairs. The knives are driven by motors and cut the material while rolling between the upper and lower knives. Figure 3.100 shows a circular shearing machine for cutting out circular blanks and rings.

3.2.2 Cutting Presses

Cutting presses are universal machines for punching, cutting or trimming metal sheets. The majority of this type of machine includes an automatic tool change and a CNC-controlled sheet movement. These machines are characterized by small strokes and high stroke rates with accurate punch guidance. The sheet cutting operations are generally performed by crank and eccentric presses with the latter being used more frequently as the stroke length can be adjusted while maintaining high stroke rates. Electrohydraulically driven machines are now seeing wider use. The various drive concepts are described in ► Sect. 3.1.

Figure 3.101 shows a cutting press with peripheral equipment and automation components. The sheet blank is positioned numerically controlled on the table.

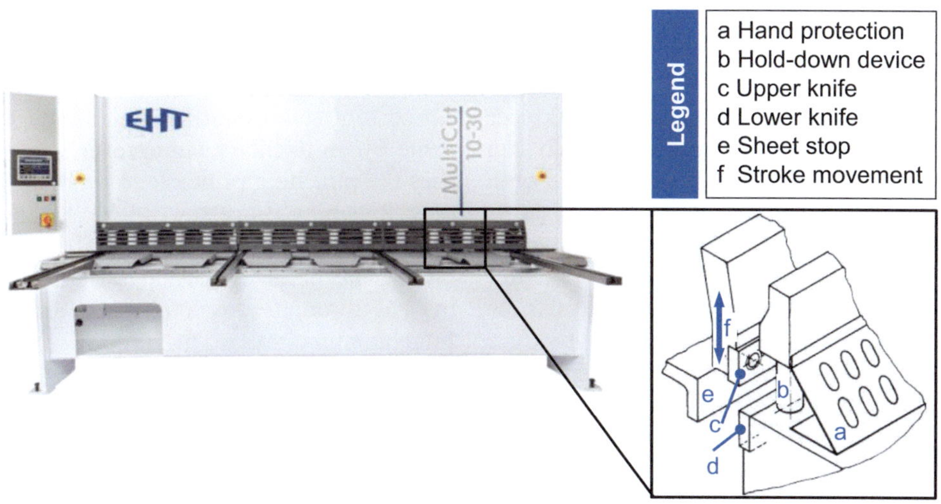

Fig. 3.99 Guillotine shearing machine. *Source* TRUMPF-Gruppe

3

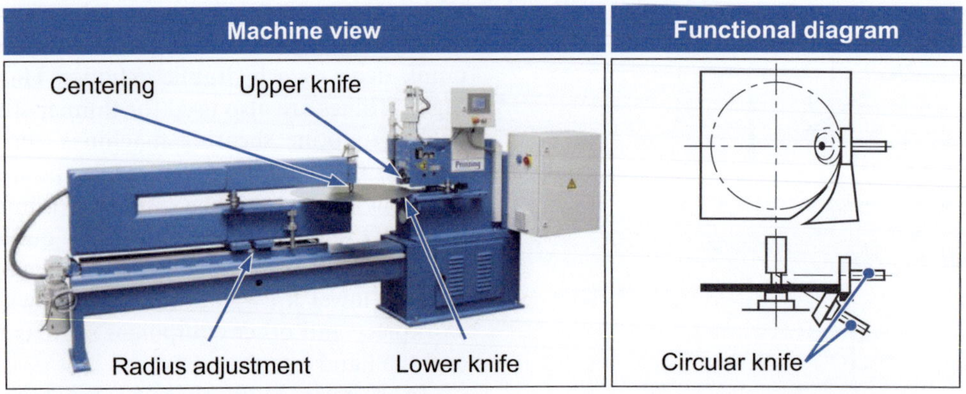

Fig. 3.100 Circular shearing machine. *Source* Peter Prinzing

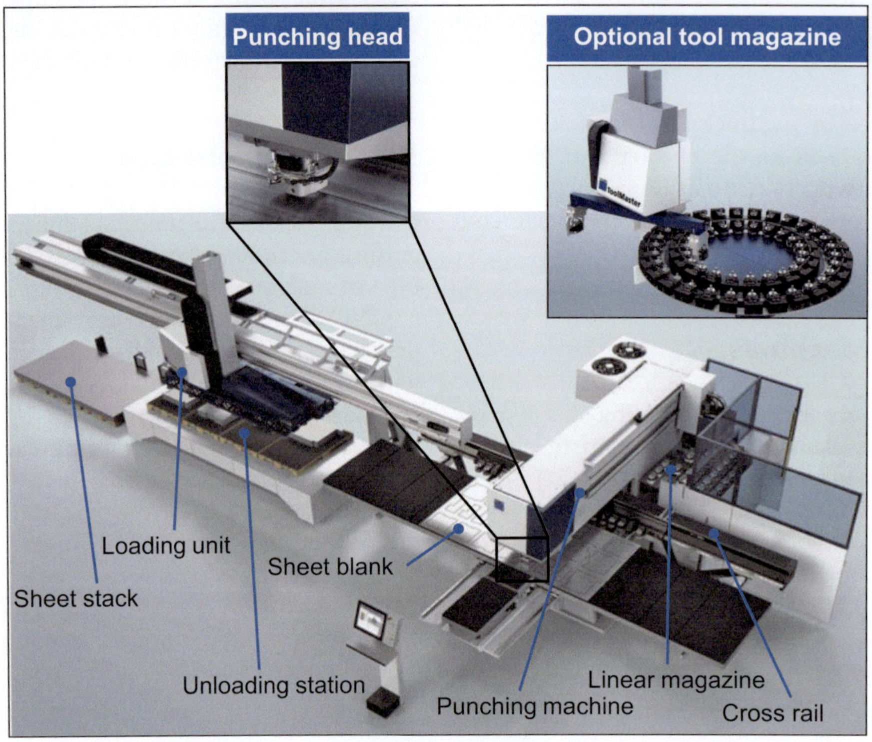

Fig. 3.101 Cutting press with peripheral equipment. *Source* TRUMPF-Gruppe

To achieve this, the sheet is held in place by workpiece grippers, which are fixed to the cross rail which can move in the x and y-directions, and is then moved with the cross rail. The machine is loaded with blank sheets and the finished parts are removed by a loading/unloading unit. Small parts are removed via a parts chute underneath the punching head. The cutting head can be replaced using an automatic tool changer to produce a range of machining contours in the sheet. The tools for machining sheet metal are stored in a linear magazine on the machine and moved to the ram position using the positioning drive and automatically changed there. If the linear magazine does not have enough slots, an additional external tool magazine can be installed. These tool sets are added to the machine's tool magazine by an ancillary changeover system. The machine uses the punching principle to produce edges and openings. Punching processes can be classified as shear cutting and are used in sheet metal processing to manufacture flat parts. This is characterized by the sheet being cut through completely on each individual stroke. Punching processes do not produce straight cuts. Due to the high forces, the material is cut in the upper area only while the material breaks off in the lower area. Figure 3.102 shows a microscopic image of the cross-section of a punching process with the characteristic areas.

Punching can be subdivided depending on the application. By using tools with the same contour as the

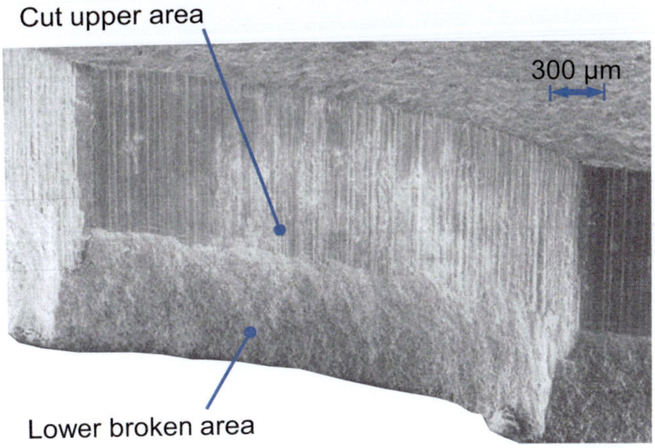

Fig. 3.102 Cross-section of a punching process. *Source* TRUMPF-Gruppe

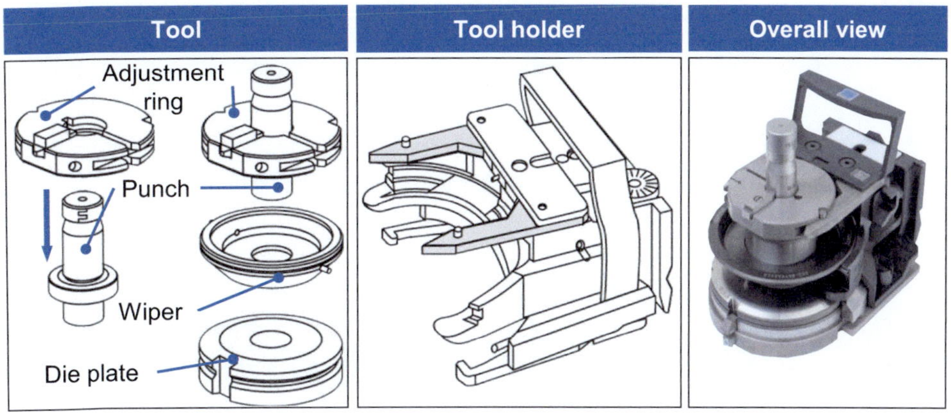

Fig. 3.103 Forming punching tool of a cutting press. *Source* TRUMPF-Gruppe

required component, it is possible to produce finished workpieces in a single stroke. This allows both small and large components to be mass-produced. Figure 3.103 shows a forming punching tool as used in cutting presses. In principle, punching tools comprise a punch and a die plate. The metal sheet to be processed is positioned between the punch and the die plate. To produce the required part, the punch is moved through the die plate; the edges of the punch and the die plate pass by each other in parallel and separate the sheet. Wipers are often used in addition to the punch and the die plate to ensure the punch can return from the die plate unobstructed.

Another type of punching is nibbling. This process is defined as a periodic series of individual overlapping punching processes to produce sheet parts with an extremely wide range of contours. Figure 3.104 shows nibbling tools and the contours they produce. This process is used in small batches or one-off production where creating special cutting tools for specific workpiece shapes is not cost-effective. Nibbling is a particularly economical solution for cutting tasks with large, complex cutting edge lengths.

Due to the diverse range of applications for punching processes in sheet metal processing, there has been an increased focus in recent years on improving the process. One option for optimization is to use the principle of laser-supported shear cutting, Fig. 3.105 where the area of the sheet to be processed is heated using a laser and then the cutting process is performed. The aim is to improve component quality by increasing the cutting component and to reduce the die roll and the overall process forces. In addition to this, the range of products can be expanded as thicker sheets and stronger materials can be processed. Detailed information on the structure and function of lasers can be found in ▶ Chap. 7.

A further option for extending cutting presses is shown in Fig. 3.106. This sheet metal processing center integrates a laser cutting unit in the machine. This allows laser beam cutting as well as punching. The benefits of laser cutting compared to punching are most apparent when processing thin sheets (sheets less than 5 mm thick) and are characterized by higher cut quality and significantly greater flexibility when machining complex and detailed workpiece geometries.

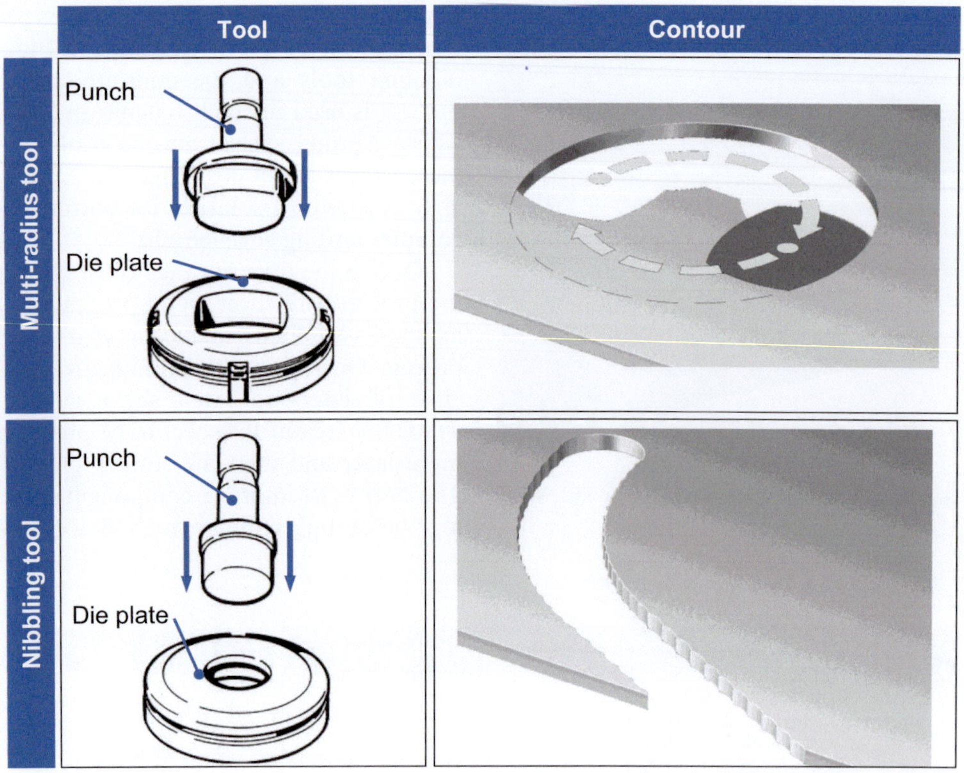

Fig. 3.104 Multi-radius tools and nibbling tools with their emerging contours. *Source* TRUMPF-Gruppe

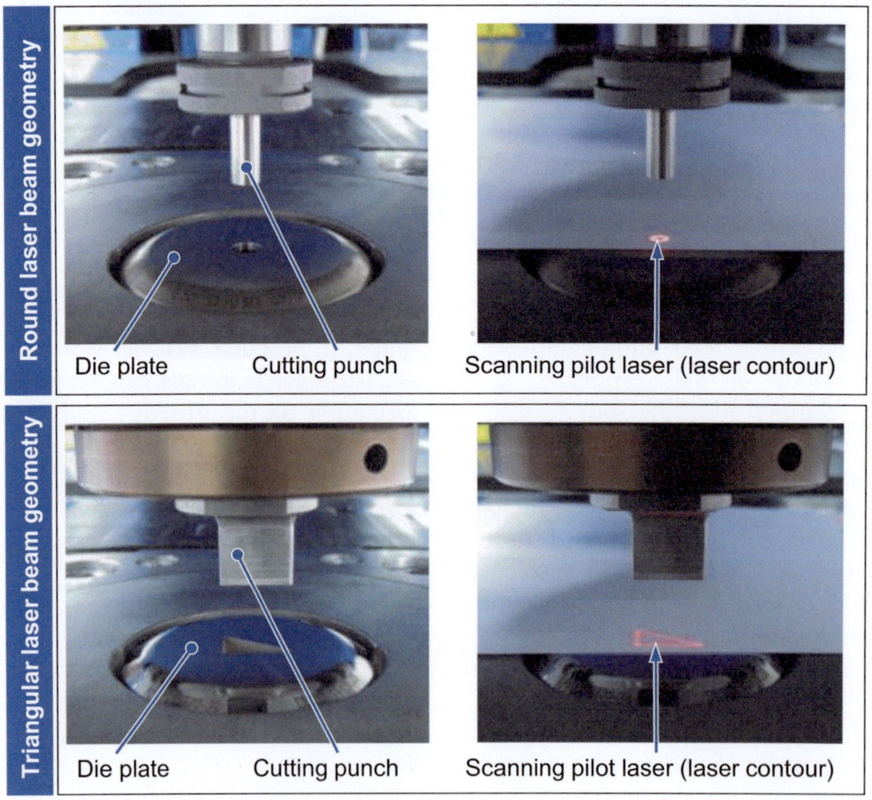

Fig. 3.105 Laser-supported shear cutting (Punching)

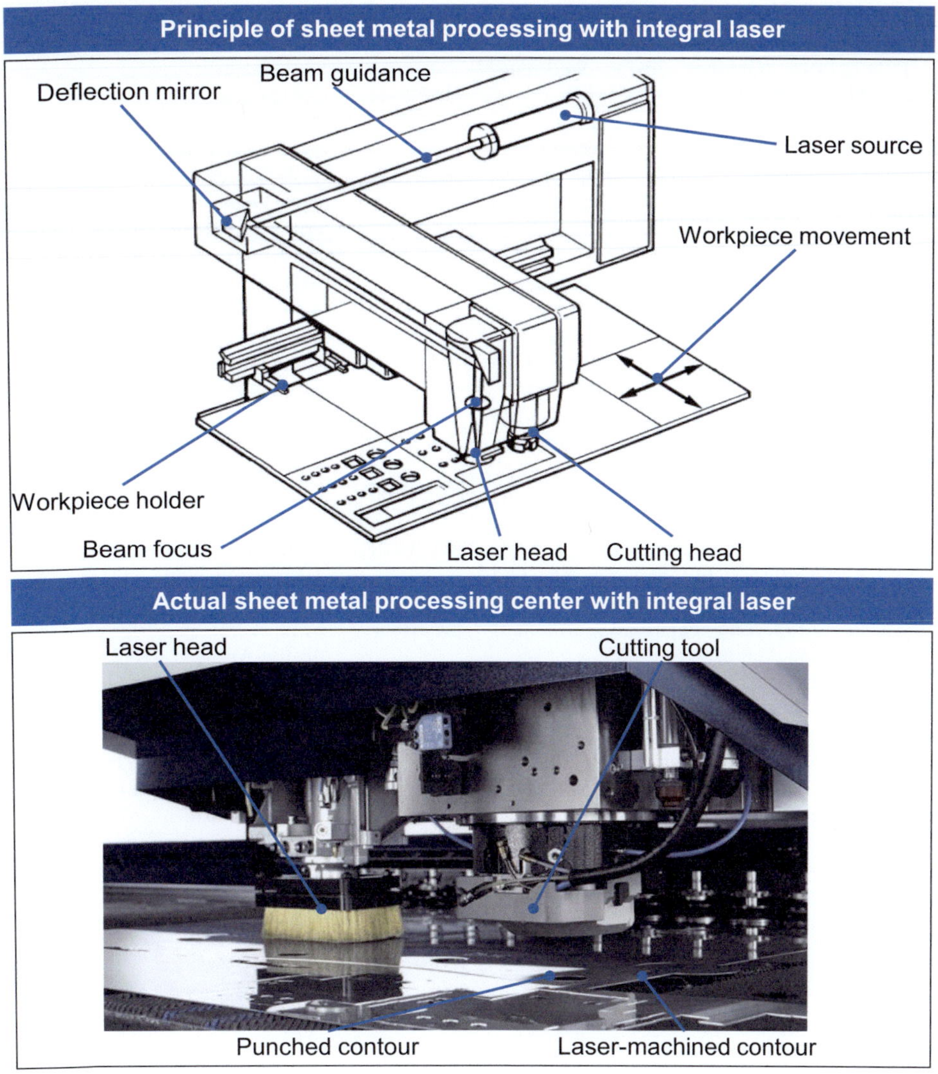

Fig. 3.106 Sheet metal processing with integrated cutting laser. *Source* TRUMPF-Gruppe

References

[AUST03] Aust, M.: Hydromechanisches Tiefziehen von Karosserieteilen. Universität Stuttgart, Diss (2003)

[BEHR11] Behrends, P.: Elektrische Maschinen. (Reihe: Vogel-Fachbuch: Die Meisterprüfung). 14. Aufl. Vogel, Würzburg (2011)

[BERG09] Berger, J.: Man muss die Servotechnik vernünftig sehen. Blech Rohre Profile. **51**(4), 30–31 (2009)

[BEYE10] Beyer, J.: Energieeffiziente Servopressen - Energieeinsparung durch ServoDirekt-Technologie. In Neugebauer, R. (Hrsg.) Energieeffiziente Produkt- und Prozessinnovationen in der Produktionstechnik, S. 139–156. Wiss. Scripten, Auerbach (2010)

[BRÜN09] Brüninghaus, G.: Servopressen. In: Servopressen und Werkzeugsysteme zur Blechverarbeitung. Bad Böll, 10.–11. März 2009. Europäische Forschungsgesellschaft für Blechverarbeitung (Veranst.). S. 11–17

[DIN03a] Norm DIN 8588 (September 2003). Fertigungsverfahren Zerteilen—Einordnung, Unterteilung, Begriffee

[DIN03b] Norm DIN 8587 (September 2003). Fertigungsverfahren Schubumformen—Einordnung, Unterteilung, Begriffe

[DIN03c] Norm DIN 8585 (September 2003). Fertigungsverfahren Zugumformen—Tiefen; Einordnung, Unterteilung, Begriffe

[DIN03d] Norm DIN 8586 (September 2003). Fertigungsverfahren Biegeumformen—Einordnung, Unterteilung, Begriffe

[DIN03e] Norm DIN 8584 (September 2003). Fertigungsverfahren Zugdruckumformen—Tiefziehen; Einordnung, Unterteilung, Begriffe

[DIN03f] Norm DIN 8580 (September 2003). Fertigungsverfahren—Begriffe, Einteilung

[DIN03g] Norm DIN 8582 (September 2003). Fertigungsverfahren Umformen - Einordnung; Unterteilung, Begriffe, Alphabetische Übersicht

[DIN03h] Norm DIN 8583 (September 2003). Fertigungsverfahren Druckumformen—Eindrücken; Einordnung, Unterteilung, Begriffe

[DIN09] Norm DIN EN 13736 (November 2009). Sicherheit von Werkzeugmaschinen—Pneumatische Pressen

[DIN11] Norm DIN EN 693 (November 2011). Werkzeugmaschinen—Sicherheit—Hydraulische Pressen
[DIN12] Norm DIN EN 692 (Oktober 2012). Werkzeugmaschinen—Mechanische Pressen—Sicherheit
[DIN14] Norm DIN EN 12622 (Februar 2014). Sicherheit von Werkzeugmaschinen—Hydraulische Gesenkbiegepressen
[DUNK98] Die hubgesteuerten Präzisions-Richtpressen von Dunkes. Kirchheim. Firmenschrift (1998)
[FAHR05] Schutzrecht DE102005053350 A1 (07.11.2005). Fahrenbach, J.: Presse mit Schnittschlagdämpfung
[GALD] Galdabini SPA: stamping-deep-drawing. ▶ http://www.galdabini.eu/metalforming/stamping-deep-drawing
[GEIS96] Schutzrecht DE 19642684 C2 (16.10.1996). Geisendörfer, E., Reinkunz.H.: Großteil-Transferpresse
[GERB92] Schwingungsisolierungen. Essen, 1992. Firmenschrift
[GOLF85] Golf, K.H.: Kupplungsspindelpresse. Industrie-Anzeiger. **107**(4) (1985)
[GROC07] Groche, P.: Oszillierende Kaltmassivumformung. wt Werkstatttechnik online. **97**(1), 741–744 (2007)
[GROC09] Groche, P.: Konstruktion und Steuerung von Servopressen—Ein Überblick. In: Servopressen und Werkzeugsysteme zur Blechverarbeitung. Bad Boll, 10.-11.03.2009. Europäische Forschungsgesellschaft für Blechverarbeitung (Veranst.). S. 19–28
[HELL12] Hellwig, W., Kolbe, M.: Spanlose Fertigung Stanzen, 10th edn. Vieweg & Teubner, Wiesbaden (2012)
[HOFF89] Hoffmann, H.: Wirtschaftlich fertigen auf Großteil-Stufenpressen. Werkstatt und Betrieb. **122**(5/6), 489–491 (1989)
[HOFF91] Hoffmann, H.: Vergleichende Untersuchung von Zieheinrichtungen. In: Bänder Bleche Rohre. **32**(12) (1991)
[HOFF92] Hoffmann, H.: Prozeßgesteuerte Zieheinrichtung paßt Blechhaltekraft flexibel an. Bänder Bleche Rohre. **33**(1) (1992)
[HUBE12] Huber, W.: An diesen Elementen bleibt der Schall sicher hängen. Industrie-Anzeiger. **122**(18), 46–47 (2012)
[JANS11] Schutzrecht EP 2 529 850 A1 (30.05.2011). Janson, M.: Biegewinkelsensor und Verfahren zum Messen eines Biegewinkels
[KOHN01] Kohnhäuser, M.: Entwicklung und praxisgerechte Verifizierung von CAE-Methoden zur Optimierung des Karosserieteiletransfers in Grosspressen. (Reihe: Utg-Forschungsberichte. Bd. 14). Hieronymus, München (2001)
[KÖNI06] König, W., Klocke, F.: Fertigungsverfahren. (Reihe: VDI. Vol. 4). 5th edn. Springer, Berlin (2006)
[KUHN09] Kuhn, D.: Der Laser ist ein elementarer Baustein der Umformtechnik. MM Maschinenmarkt. **115**(50), 24–25 (2009)
[LANG02] Lange, K. (ed.): Umformtechnik, 2nd edn. Springer, Berlin (2002)
[LEIF10] Leifeld Metal Spinning: Drückwalzen in neuen Dimensionen. Blech Rohre Profile. **55**(3), 54–57 (2010)
[REIH97] Reihle, M.: Spindelschlagpressen zum Schmieden von Turbinenschaufeln. In: Industrie-Anzeiger. **107**(86) (1997)
[SCHI86] Schiller, W.: Spindelpressen mit Schaltkupplung. Industrie-Anzeiger. **108**(23), 20–22 (1986)
[SCHU08] Schulte, M., Volkmar, E., Liedtke, H.: Unfallverhütung an Pressen, 3rd edn. Erich Schmidt, Berlin (2008)
[SMS03] Querwalzen. Mönchengladbach. Firmenschrift (2003)
[VDI00] Richtlinie VDI 3146 (April 2000). Innenhochdruck-Umformen—Maschinen und Anlagen
[WFM03] WF Maschinenbau und Blechformtechnik: Nabenanformung im Kaltrollverfahren. Sendenhorst. Firmenschrift (2003)

Metal Cutting Machines for Tools with Geometrically Defined Cutting Edge (Chip Removal)

I thank my staff members Mathias Dehn M.Sc., Franziska Plum M.Sc., Alexander Steinert M.Sc. and Anton Strachkov M.Sc. for their contribution in revising this chapter.

C. Brecher and M. Weck, *Machine Tools Production Systems 1*,
Lecture Notes in Production Engineering, https://doi.org/10.1007/978-3-662-68120-6_4

Contents

The manufacturing processes of machining with geometrically defined cutting edge belong to the main group of cutting. Their categorization into manufacturing processes according to DIN 8580 [DIN03f] is shown in ◘ Fig. 4.1.

The term "chip removal" is defined in standard sheet DIN 8589-0: Chip removal is a "cutting process where the cutting edges of a tool mechanically remove material layers of a workpiece in the form of chips" [DIN03e, p. 4]. When machining with geometrically defined cutting edges, a tool is used "where the number of cutters, the geometry of cutting wedges and the positions of cutters in relation to the workpiece are defined for the process." [DIN03e, p. 4] The principle of chip removal with geometrically defined cutting edge is shown in ◘ Fig. 4.2 using an example of a single-cutter tool with translational motion.

The cutting forces exerted on the cutter can be calculated using different cutting force models. The Kienzle model, or the Kienzle formulas[KLOC08], sees the most widespread use. The cutting force parameters are specified in ◘ Fig. 4.3 for the most important material groups [KÖNI82].

The force flux between tool and workpiece is closed by the machine structure. The power required for the main cutting motion is provided by the main drive. The auxiliary drives (feed drives) determine the relative position between tool and workpiece, and thus the workpiece geometry. Rotating three-phase a.c. motors (synchronous and asynchronous machines) are common drive types for the main and auxiliary drives [BREC16]. In recent years, electrical linear direct drives have also been used for auxiliary drives. These direct drives offer advantages for high-speed machining due to their fast accelerations and travel speeds. Assistance functions of a machine tool for cutting with a geometrically defined cutting edge include cooling lubricant supply, automatic tool and workpiece change, and chip removal.

Machines for tools with geometrically defined cutting edges are classified according to the manufacturing processes used (see also ▶ Sect. 1.4). The following sections introduce sawing, drilling, milling, slotting, drawing, and broaching machines.

4.1 Sawing Machines

Sawing machines are used to cut materials such as steel bars, plates and sheet metal. The material is cut with a circular or straight cutting motion using a multi-toothed tool having a short cutting width. The cutting motion is on the tool side [DIN03a]. The process is very similar to milling, as shown by circular sawing in particular. The thin saw blade with a large diameter acts like a side milling cutter. For hacksaws and band saws, the miller diameter takes on a theoretically infinite value, resulting in a translational cutting motion (shown in ◘ Fig. 4.4). While band and circular sawing machines work with continuous cutting motions, hacksaws use discontinuous motions.

In addition to the specification of cross-section-dependent feeds, automated band and circular saws enable the cutting speed to be adjusted depending on the material. Correspondingly, powerful workpiece

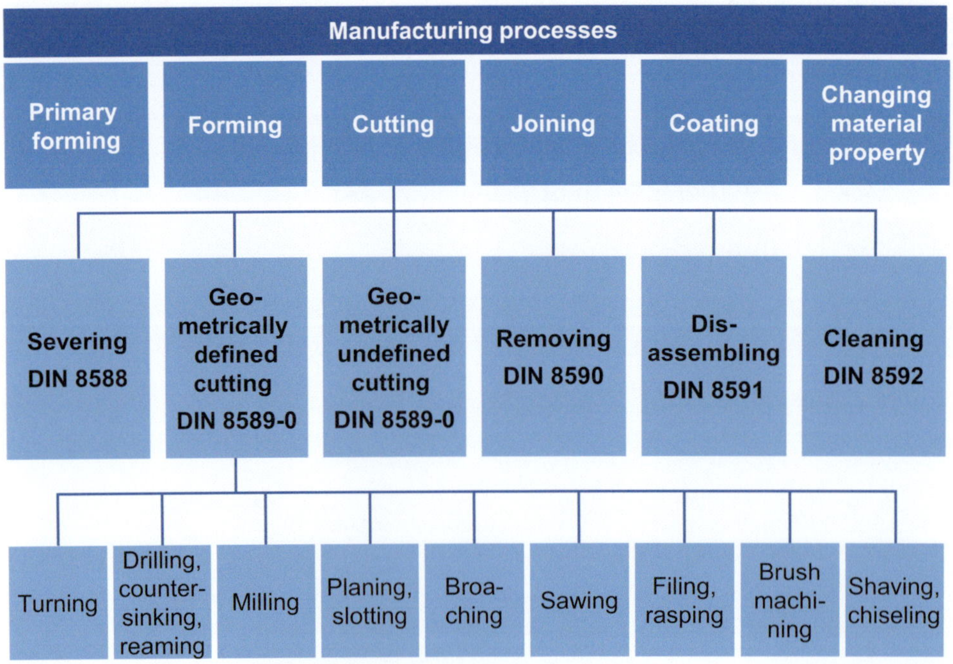

◘ Fig. 4.1 Categorization of chip removal manufacturing processes, according to [DIN03f]

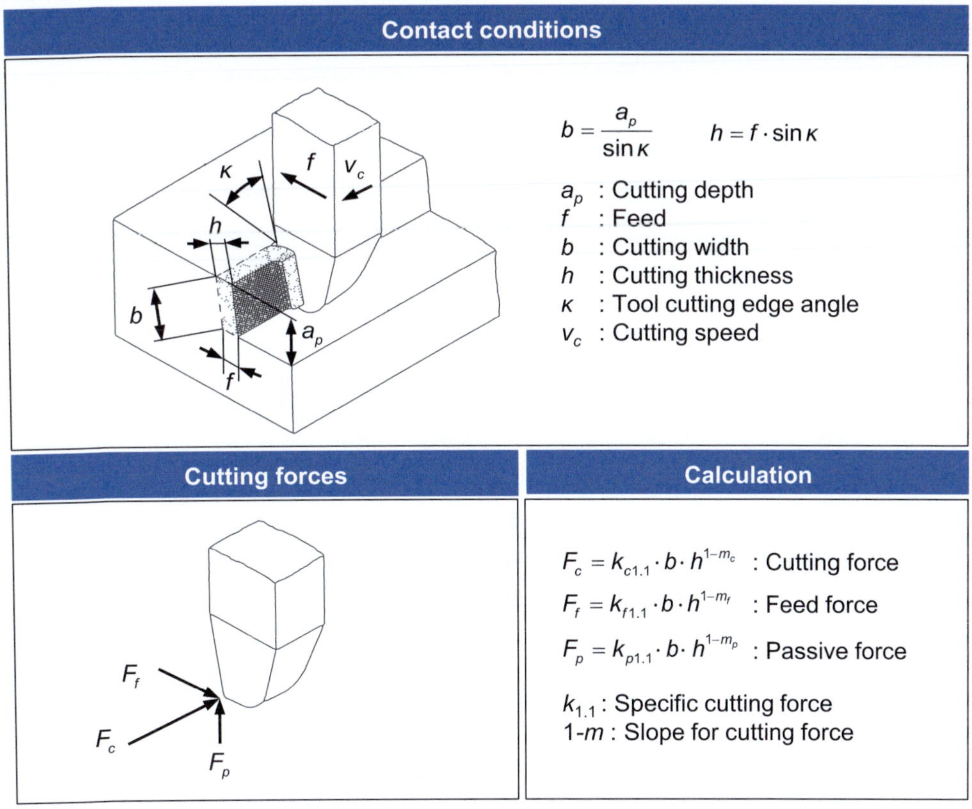

Fig. 4.2 Parameters of chip removal with geometrically defined cutting edge, according to [KLOC08]

Material	Specific cutting force $k_{c1.1}$ [N/mm²]	Slope $1\text{-}m_c$	Specific feed force $k_{f1.1}$ [N/mm²]	Slope $1\text{-}m_f$	Specific passive force $k_{p1.1}$ [N/mm²]	Slope $1\text{-}m_p$
GCI 30	897	0.6308	156	0.2806	124	0.2901
St 70-2	1,547	0.6983	454	0.3835	311	0.5067
C 45	1,320	0.7224	302	0.3534	209	0.4926
42CrMo4	1,601	0.8262	268	0.3120	247	0.6066
9SMnPb36	1,303	0.9198	250	0.3669	207	0.5823
Cast AlMg5	574	0.7635	159	0.4224	120	0.5065

Cutting geometry and technological data

Clearance angle :	$\alpha = 5°$	Corner angle :	$\varepsilon = 90°$
Cutting angle :	$\gamma = 6°$	Corner radius :	$r = 0.8$ mm
Inclination angle :	$\delta = 0°$	Cutting speed :	$v_c = 200$ m/min
Tool cutting edge angle :	$\kappa = 70°$		

Fig. 4.3 Cutting force values for metallic materials, according to [KÖNI82]

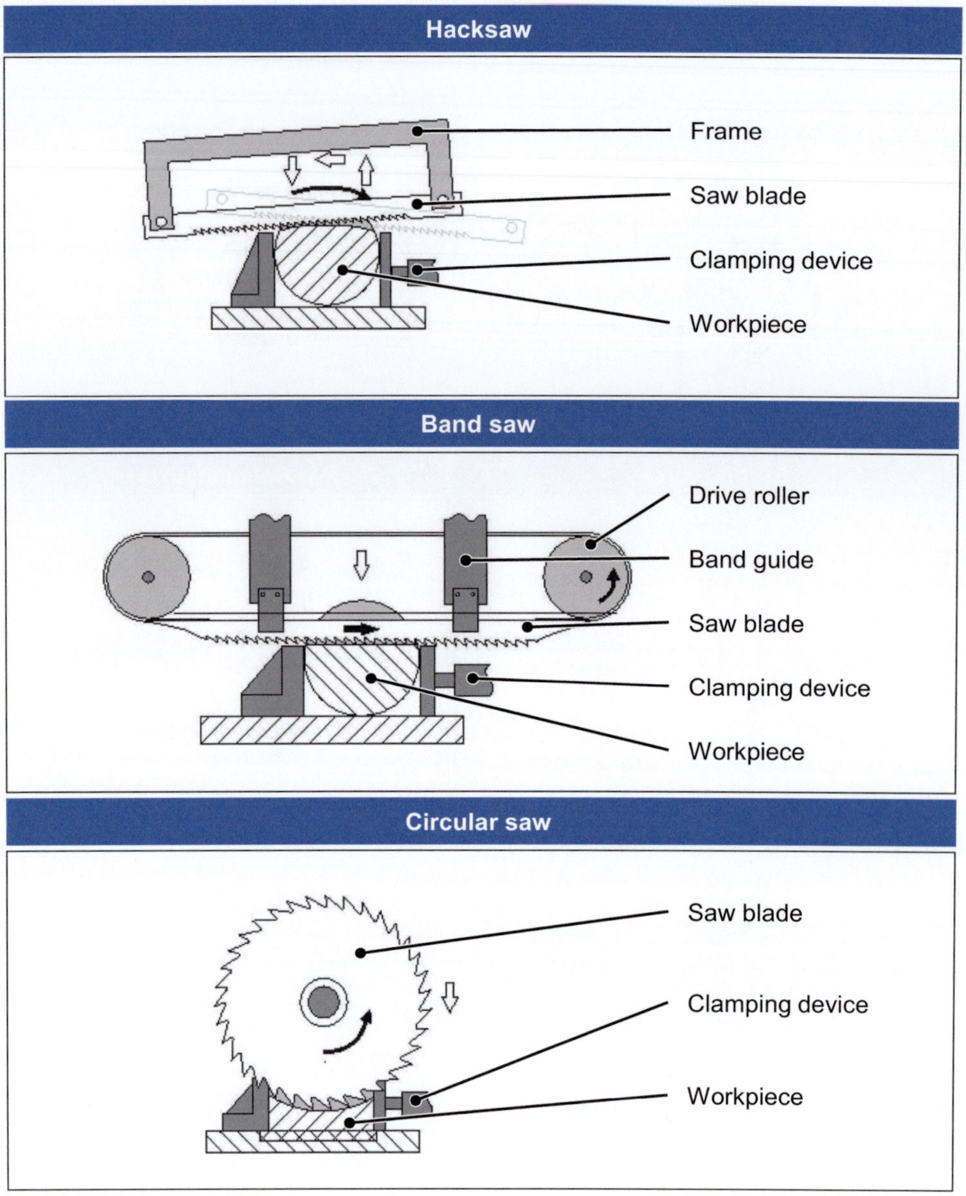

Fig. 4.4 Sawing techniques

handling peripherals are important for working efficiently with saws. Upstream flat magazines, for example, enable larger orders to be compiled. These can be worked on by the saw without operator intervention. Fully automatic supply units and removal/sorting devices control the material flow. Dynamic clamping devices with pressure-displacement restriction guarantee clamping without deformation, also of thin-walled tubes and profiles. The reduced need for cooling lubricants contributes to preventing costs and harming the environment. Minimum quantity lubrication for saw blades is available to address this. These micro spray systems disperse a fluid, creating a fine lubricant film on the tool surface. This type of spray head for band and circular saws is shown in Fig. 4.5.

4.1.1 Hacksawing Machines

Hacksawing machines cut workpieces within a working range of 200 to 500 mm. A slotted link mechanism generates the reversing main motion during the curved cut. The saw blade works in a thrusting manner, and on the return stroke is lifted mechanically or hydraulically from the workpiece, as shown in Fig. 4.4. The contact zone shifting along the saw blade due to these kinematics also enables the cutting of thin-walled parts with high power, and results in good service life utilization of the saw blade. The discontinuous cutting profile for hacksawing means however the cutting performance is lower than for continuous sawing techniques, enabling hacksawing to be used when capacity

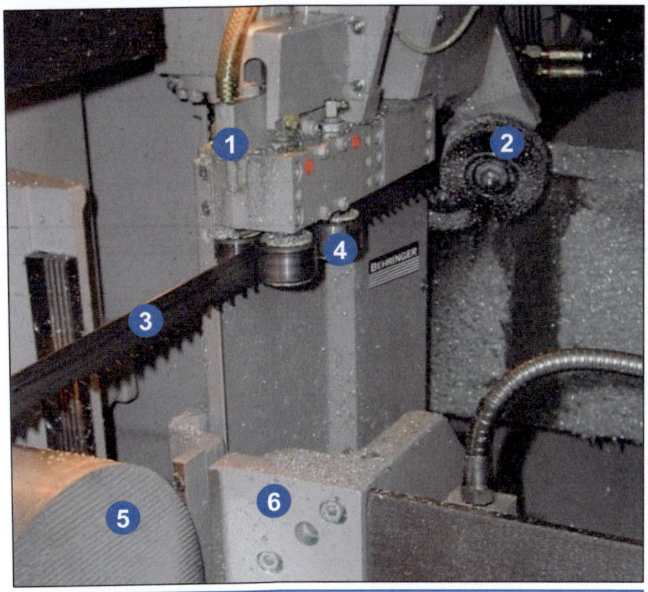

1 Mist generator head	4 Guide roller
2 Chip brush	5 Workpiece
3 Saw blade	6 Chuck

Fig. 4.5 Saw blade roller guide with minimum quantity lubrication. *Source* Behringer

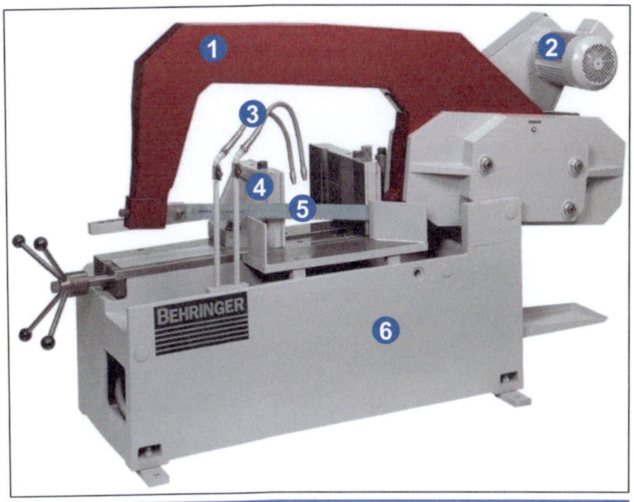

1 Frame	4 Chuck
2 Drive motor	5 Saw blade
3 Cooling lubricant supply	6 Bed

Fig. 4.6 Hacksawing machine. *Source* Behringer

requirements are less strict. Their low tool costs make them suitable as sturdy machine tools (Fig. 4.6).

4.1.2 Band Sawing Machines

Band sawing machines are available for the most diverse degrees of automation and are particularly suited to cutting larger workpiece diameters. Due to the low chip removal per saw tooth of about 0.01 mm, the drive power required is low. Smaller workpiece diameters, as used in mass production, can be cut cost-effectively with layer or bundle cutting. Horizontal band sawing machines may have a swiveling frame design but those with the stiffer two-column design are more common. Figure 4.7 shows such a CNC large band sawing machine. Machines of this size are suited for workpiece dimensions in the range of 600 to 2000 mm.

The band guides prevent the saw band from deviating and adjust their spacing to the workpiece cross-section. Deviation of the band also represents the service life criterion for the tool. For this reason, the path of the saw blade is monitored electronically, and the technical data are adapted automatically to the wear level of the band and the machine is switched off when the end of service life is reached. Cut-to-length accuracies of 0.1 mm per 100 mm of workpiece height can be attained [STOL96].

Bimetal saw bands with a carrier band made of alloyed tempering steel in combination with cutters made of high-performance high-speed steel are used as tools for sawing unhardened steel. To prevent vibration, saw bands have variable teeth spacings. The machining of hardened steels makes it easier to manufacture carbide-tipped saw bands with teeth that alternate between pre-cutting and finishing cutting edge geometries. These saw bands are also suitable for machining non-ferrous metals (such as aluminum) with high cutting capacities [KULL94, OV94].

Two band saws for making miter cuts with variable angles are shown in Fig. 4.8. The design with the saw band running vertically (on the left) requires a smaller installation space and is used, for example, for cutting profiled supports in steel construction. In the horizontal design (on the right in the figure), a rotary table enables the swiveling frame to be turned to both sides. The pivot point corresponds to the point of intersection of the saw blade contact line and material support edge. Here, dimensional accuracy is assured for different miter squares.

Miter band saws are commonly also integrated into modular, automated sawing systems. The example shown in Fig. 4.9 is a saw that can make miter cuts up to 550 × 700 mm (H × W) at a 45° miter angle, and 550 × 460 mm (H × W) at a 60° miter angle. Miter squares are NC-controlled and positioned fully automatically, assuring the reproducibility of miter cuts. Various equipment is available for material flow and subsequent automation. They are, for example on the feed side, roller segments and roller conveyors. In these, the roller segments are integrated directly into the machine and can be raised and lowered hydraulically. The same applies to the removal side where simple handling is possible. An NC-controlled measuring stop with a

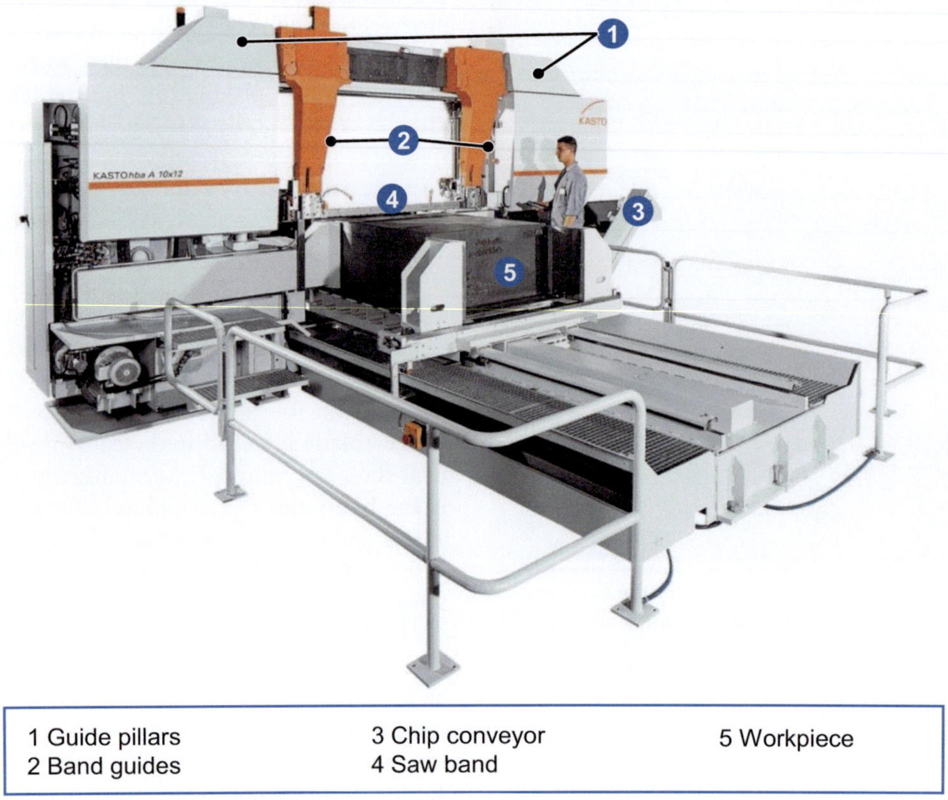

Fig. 4.7 Large band sawing machine. *Source* Kasto

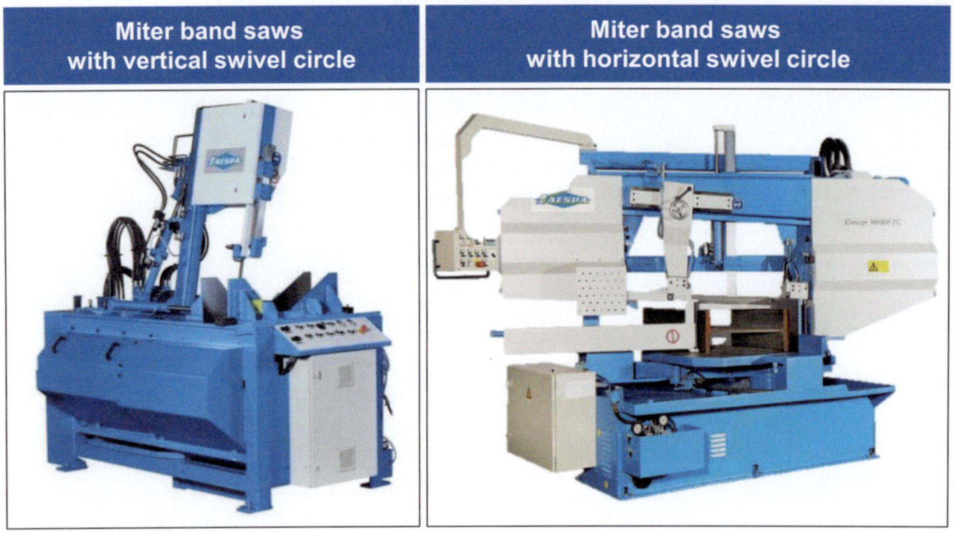

Fig. 4.8 Miter band saws. *Source* Jaespa

measuring range of 60 to 16,000 mm is used as another material flow module. [KAST04].

4.1.3 Circular Sawing Machines

Circular sawing machines can be used as universal saws with highly automated peripherals because they reach maximum cutting capacities when using carbide-tipped saw blades. The circular saw blade, compact compared to the saw band, permits automated tool changes depending on the level of wear or workpiece material. Using automatic saw blade monitoring with electromagnetic sensors, broken teeth or damaged cutting edges can be detected and displayed. Machines designed specially for the use of vibration-sensitive carbide saw

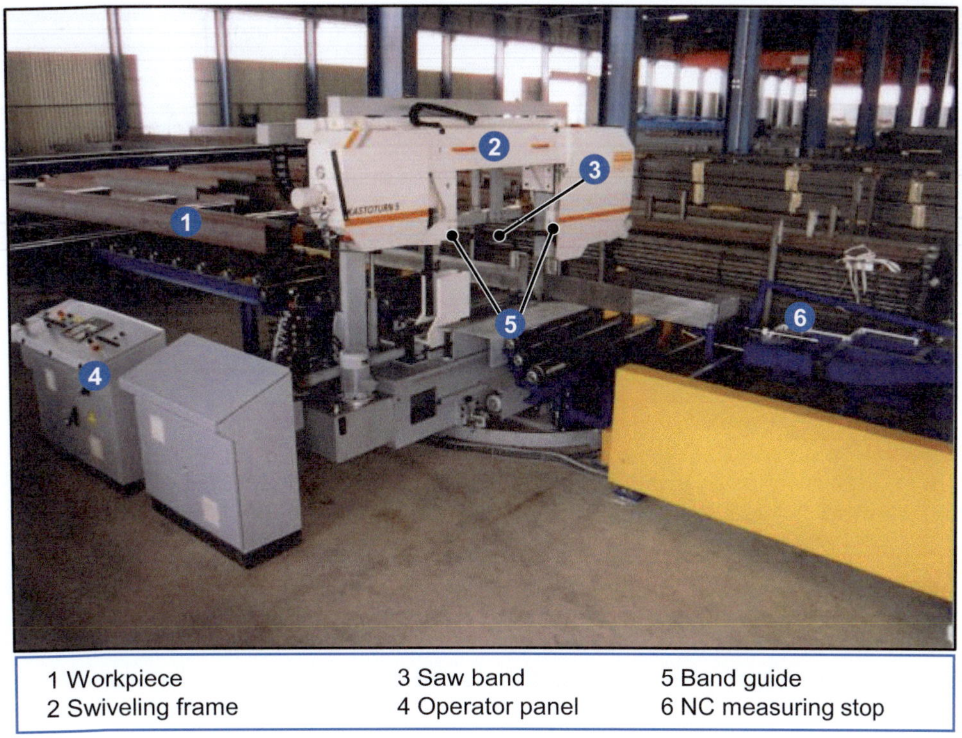

Fig. 4.9 Modular miter band saw. *Source* Kasto

blades feature rigid structural designs and attain high degrees of intrinsic damping from the mineral cast filling of the machine body. One third of the saw blade diameter can be deemed the limit for the cuttable workpiece diameter. This is why working ranges over 500 mm are assigned to band saws. For every 100 mm of workpiece height, length tolerances of $\pm$0.1 mm are attained (even $\pm$0.02 mm for fine sawing with solid carbide blades) [STOL96]. Carbide-tipped, thin-cutting saw blades with cut widths of 2.2 mm are appropriate for dry machining. They add only a little energy into the workpiece at low feed and cutting speeds up to 1200 m/min [OV96a, OV96b]. Designs are differentiated between circular saws with the saw blade coming down from above (see Fig. 4.10) and machines with the saw blade located under the table (see Fig. 4.11). These machine types are commonly provided with an additional swivel circle. This means that miter cuts can also be performed and the machine can be used universally.

Figure 4.10 shows a universal miter circular sawing machine for large workpieces that has a saw

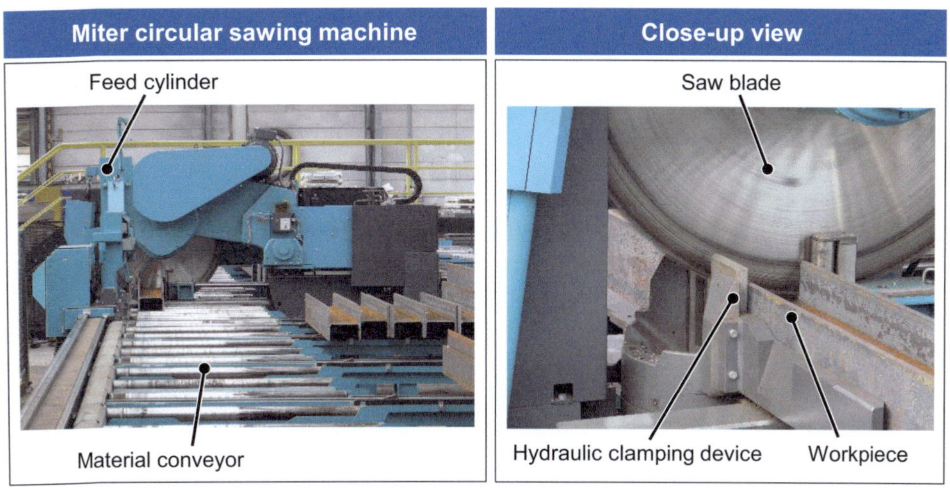

Fig. 4.10 Universal miter circular sawing machine for large parts. *Source* Kaltenbach

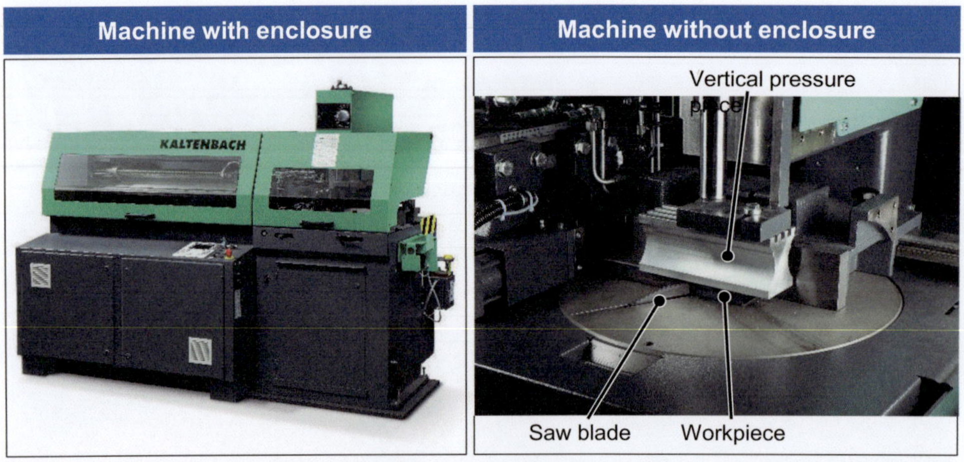

Fig. 4.11 Universal cold circular sawing machine. *Source* Kaltenbach

blade diameter of 1.30 mm. The feed cylinder shown in Fig. 4.10 supports the heavy weight of the saw blade and heavy-duty gearbox. This allows control of the feed rate at which the saw blade penetrates into the workpiece under its weight.

For the universal cold circular sawing machine shown in Fig. 4.11, with a saw blade coming up from underneath, the saw blade feed uses a hydraulic or electromechanical drive with rapid return to shorten cycle times. A material supply gripper guarantees automatic material supply and material clamping. The material is also clamped by a vertical pressure piece and the cutoff gripper. At the same time, the cutoff gripper provides for automatic removal of good parts as well as cutoff and residual pieces. The cutoff pieces are automatically distributed to preselectable positions by corresponding removal devices.

Circular saw operations with smaller saw blade diameters are also performed in milling and turning centers. For this, the saw blade to be exchanged is in the tool magazine. Long-cutting cold circular saws, where the saw blade can move in the cutting direction, are used for cutting steel plates. Large cutting lengths can be achieved in this way.

4.1.4 Sawing Centers

Powerful sawing in combination with a cantilever and/or cassette rack, known as sawing centers, permit flexible and automated material handling. The bar stock deposited at the storage input and output stations is sorted by an NC-controlled rack feeder inside the storage blocks, fetched from there as required and fed to a saw via the quick-change station [OV96a]. With this, material change times of around 30 s (chip to chip) can be reached. A cutoff sorting device picks the sawn material in accordance with the individual orders, while the residual material is transferred back into stock by the rack feeder. In the central control station, a storage computer manages workpiece qualities, dimensions and corresponding quantities stored. Figure 4.12 shows such a sawing center.

4.2 Drilling Machines

In the drilling process, chips are formed by the rotatory cutting motion of the tool. The feed motion comes either from the workpiece or the tool, directed along the tool rotation axis (Z-axis). In addition to drilling with twist drills, the machining types shown in Fig. 4.13 can be performed on drilling machines. These are "drilling with boring rod," "countersinking," "reaming" and "thread cutting." The cutting and feed motions of these machining types correspond to those of the drilling process.

4.2.1 Pillar Drilling Machines

Drilling machines with a vertical spindle arrangement are generally selected for use in metalworking workshops. Their ease of use and low maintenance effort make pillar drilling machines the most widely used in this field.

Pillar drilling machines are bench-mounted or free-standing machines depending on the drive power needed. Figure 4.14 shows an NC-controlled high-performance pillar drilling machine used for drilling holes with diameters of up to 60 mm into steel, depending on the motor power installed. The table height can be aligned to the workpiece size. For the example machine shown, an NC-controlled feed drive provides the feed motion of the drill automatically. However, frequently the feed is manually operated using

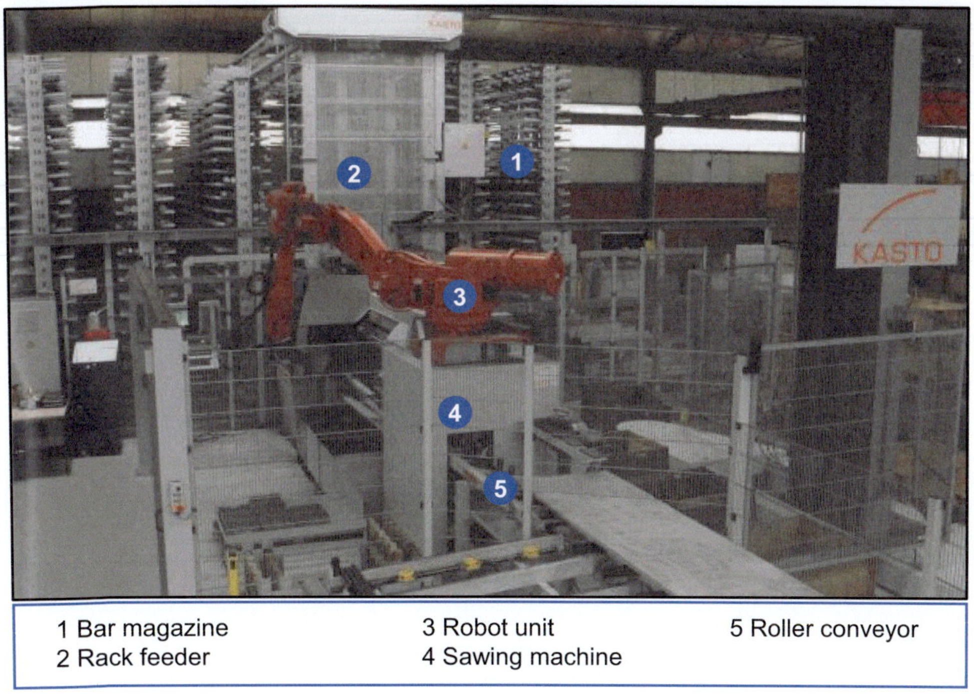

Fig. 4.12 Sawing center with bar stock storage, sawing machine and cutoff orientation. *Source* Kasto

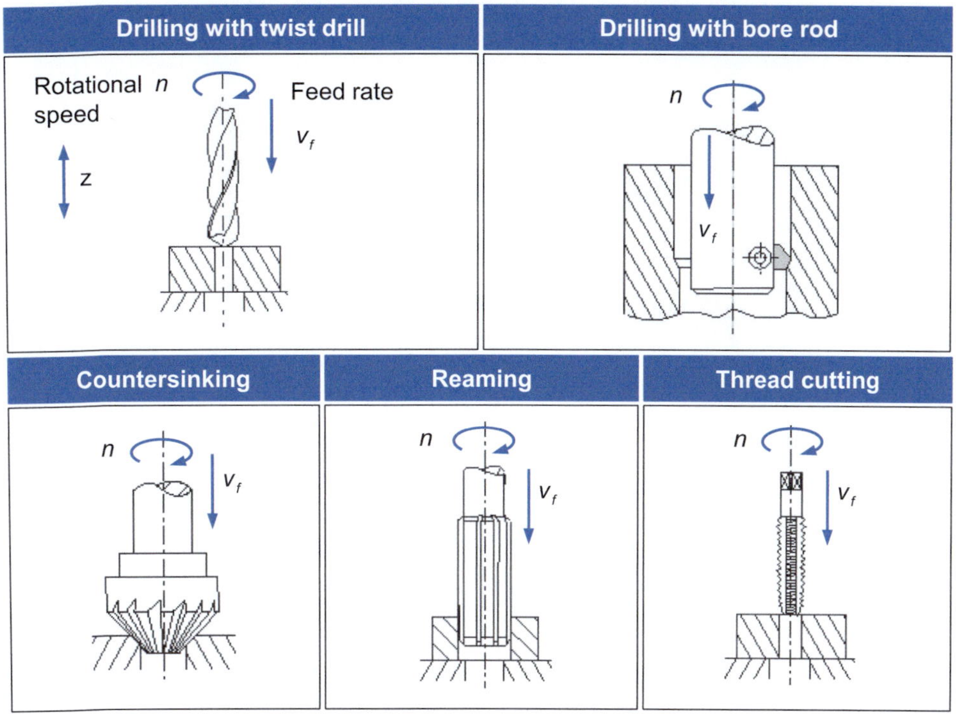

Fig. 4.13 Machining types possible on drilling machines

a feed lever attached to the machine housing (see Fig. 4.15). In NC operation, coupling tool rotation to the feed motion enables threads up to M42 to be cut without a compensating chuck.

To increase efficiency, drilling machines can be adapted perfectly to the machining task using different drill heads. The turret drill head used on the left in Fig. 4.15 permits a complete machining cycle to run

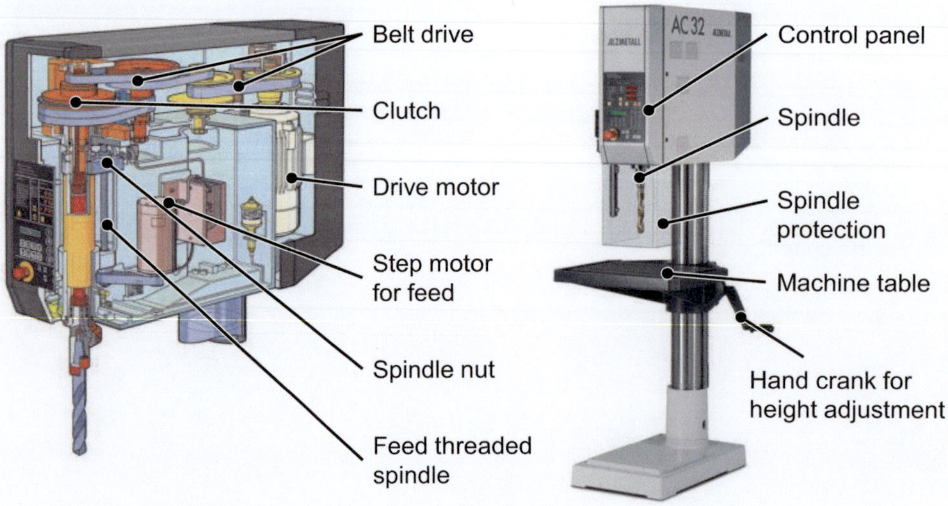

Fig. 4.14 Pillar drilling machine, according to Alzmetall

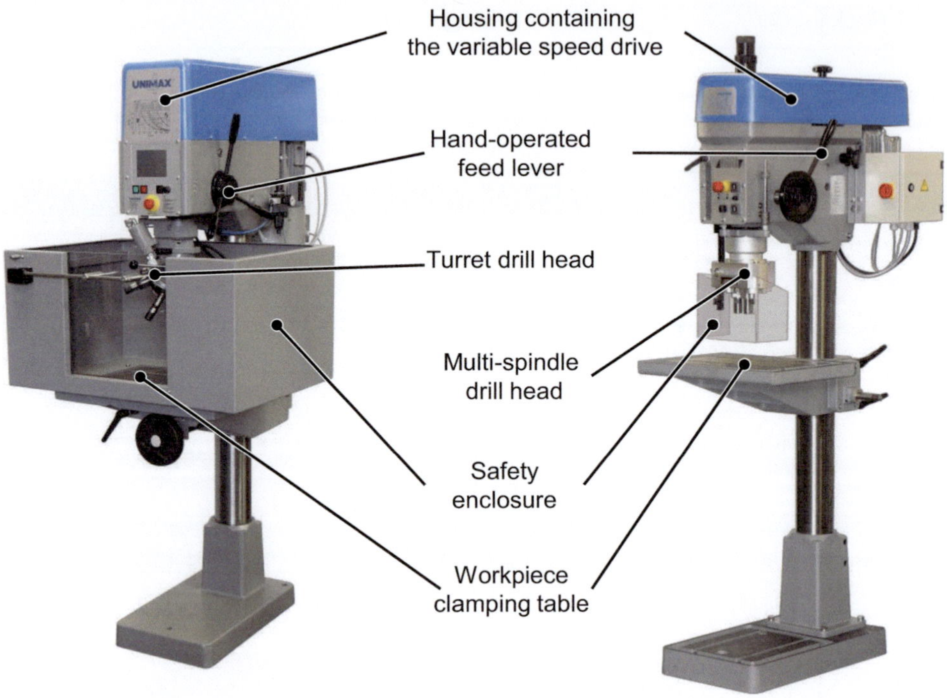

Fig. 4.15 Pillar drilling machine with turret and multi-spindle drill head, according to Maxion

fully automatically because the machining parameters (such as rotational speed, feed and drilling depth) can be programmed for all six tool spindles. The multi-spindle drill head on the right enables multiple bores and/or threads to be sunk simultaneously in one machining cycle.

Adjustable multi-spindle drill heads are used in addition to conventional multi-spindle drill heads with a fixed hole pattern. Figure 4.16 shows a detailed view and the function of such a drill head. The distance between the drilling spindles is continuously adjustable by hand because of the eccentric bearing of the drilling spindles. As a result, it can be adapted to different bore spacings. The drilling spindles with rolling element bearings are driven by the machine spindle using gears. The drill head is held by the machine spindle using a steep taper interface. A pin on the side prevents the housing from twisting. Adjustable multi-spindle drill heads are used first and foremost to increase productivity in low-volume production. They can be adapted to drilling and milling machines over standardized tool interfaces.

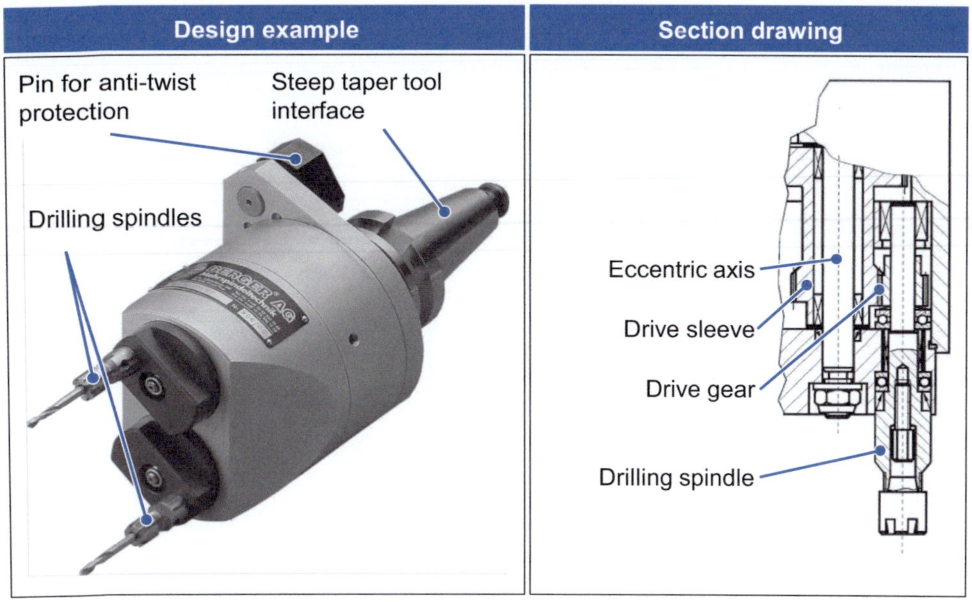

Fig. 4.16 Adjustable multi-spindle drill head, according to Berger

Directly coupling a motor to a multi-spindle drill head creates standalone drilling units used in special-purpose machines (Fig. 4.17). Using them, several holes can be bored into a workpiece simultaneously with one infeed motion. The relative tolerances of these holes can also be very low for short machining times.

4.2.2 Cantilever Drilling Machines

For drilling larger workpieces, cantilever drilling machines, also called radial drilling machines, are used in shop floor areas (Fig. 4.18). The drill head, attached to a cantilever that swivels by 330° and can move radially, is ergonomic to use because aligning and reclamping the workpiece are not required. The cantilever can be moved in the vertical direction and clamped to align with the workpiece height. The rotation and radial axes can be clamped hydraulically and are moved by applying a minor force (approx. 20 N of frictional force). Furthermore, the machine has a digital position display on the three axes.

Figure 4.19 shows a NC-controlled high-speed radial drilling machine. In this machine design, manual and 4-axis CNC operation modes are combined. In addition to the axes for carrying out the pillar and quill strokes, the rotational axis and radial axis, optional, additional axes can be moved (such as an integrated, NC-controlled rotary table). The aims of this variant are a marked reduction in the non-productive time and consistent production quality. The workpiece zero point (even of unaligned components) can be determined quickly by making contact three times using an automated calibration method. Programming involves a computer-aided dialog with the operator. Several cycles are available for selection. Automatic positioning replaces marking out and center-punching and any necessary drilling templates. For automated tool change, the CNC-controlled high-speed radial drilling machine is equipped with a tool magazine (as shown in Fig. 4.19), which, like two other workstations, is located on the rear of the machine.

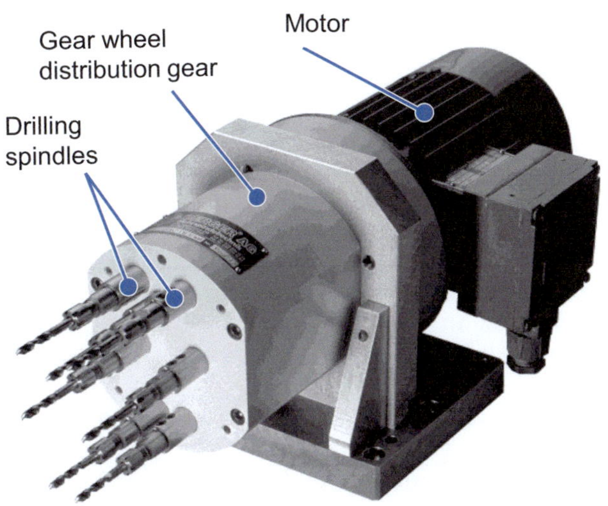

Fig. 4.17 Directly driven multi-spindle drill head, according to Berger

4.2.3 Deep Drilling Machines

Deep drilling refers to the process of creating bores where the ratio of drill depth (l) to borehole diameter

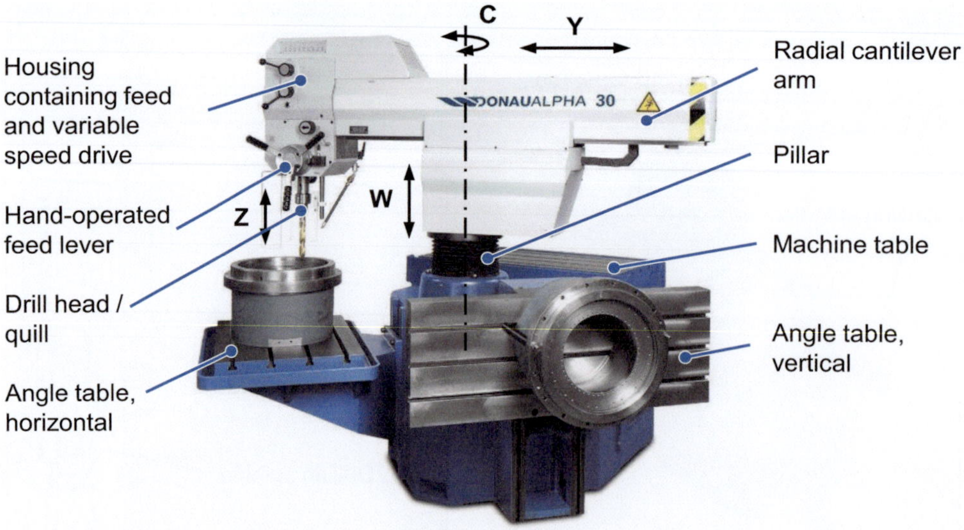

Fig. 4.18 High-speed radial drilling machine, according to DONAU

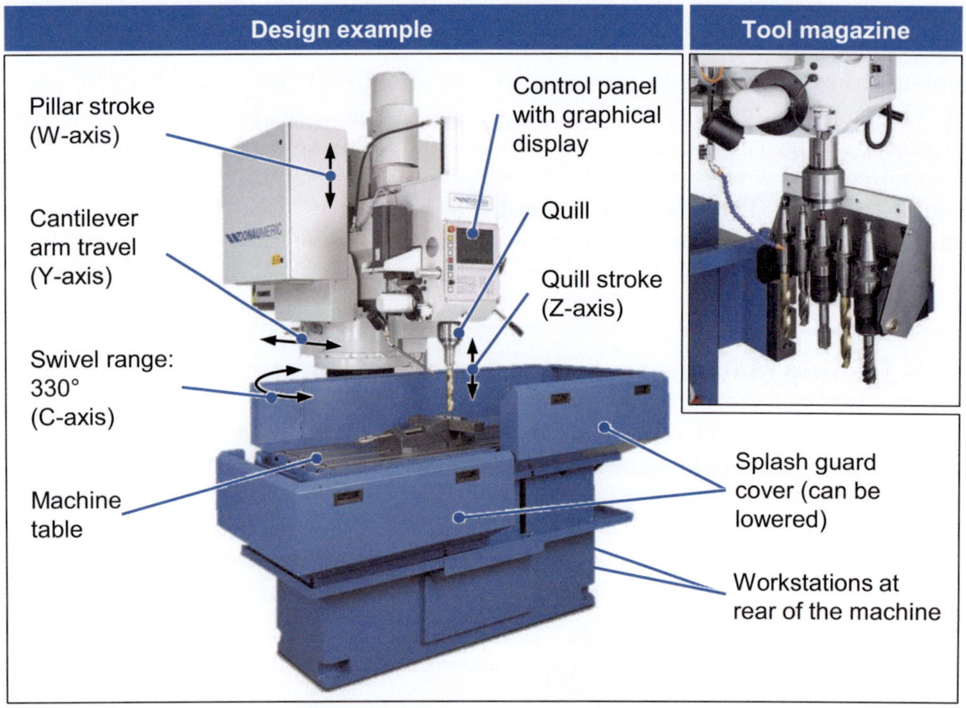

Fig. 4.19 CNC-controlled high-speed radial drilling machine, according to DONAU

(*D*) is greater than 10. While bores with a *l*/*D* ratio of about 5 can be generated in drilling using normal drilling machines, ratios up to 150 can be realized using deep drilling machines. The deep drilling principle is characterized by high cutting volume as well as high precision with regard to the straightness, dimensional accuracy and surface quality of bores. [SAND12].

In the deep drilling process, chips are washed out of the bore by the increased cooling lubricant pressure along the tool. A distinction is generally made between three drilling systems.

The single-lip drilling system has the simplest design (Fig. 4.20). Single-lip drills are used for a diameter range of 1 to 40 mm. They comprise a single-cutter carbide drill head or a drill head fitted with exchangeable carbide cutting inserts. For chip discharge, the drill shaft has a longitudinal groove (see Fig. 4.21, left and center). Guide rails center the drill in the drill hole. These rails are ground in or attached to the drill head. Their number and configuration depend on the drill variant. The cooling lubricant is routed under high pressure through one or two ducts inside the drill

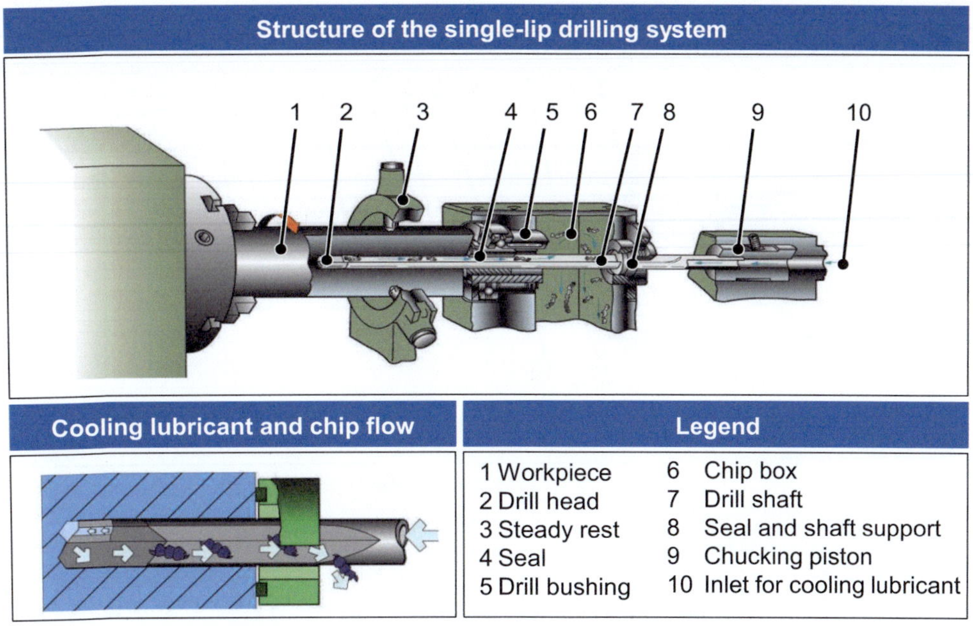

Fig. 4.20 Structure and function of the single-lip drilling system, according to Sandvik Coromant

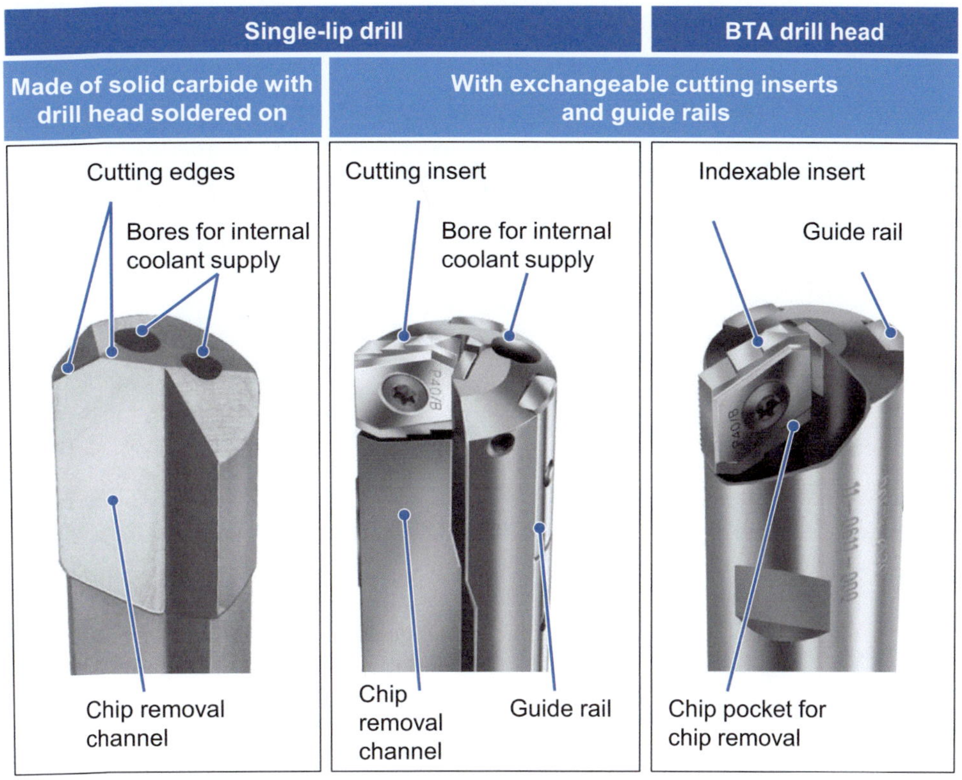

Fig. 4.21 Deep drilling tools, according to Botek

to the cutting edge. From there, the chips are removed by the cooling lubricant through a V-shaped groove between the drill shaft and the borehole wall. The mixture of cooling lubricant and chips flows through the sealed drill bushing into the drill bushing holder (chip box). From here, the mixture is routed to a cooling lubricant filter where chips and fine dirt (abraded particles) are separated.

In the BTA drilling system (BTA: Boring and Trepanning Association), in contrast to the single-lip drilling system, the directions of cooling lubricant flow and chip transport are reversed (Fig. 4.22). The cooling

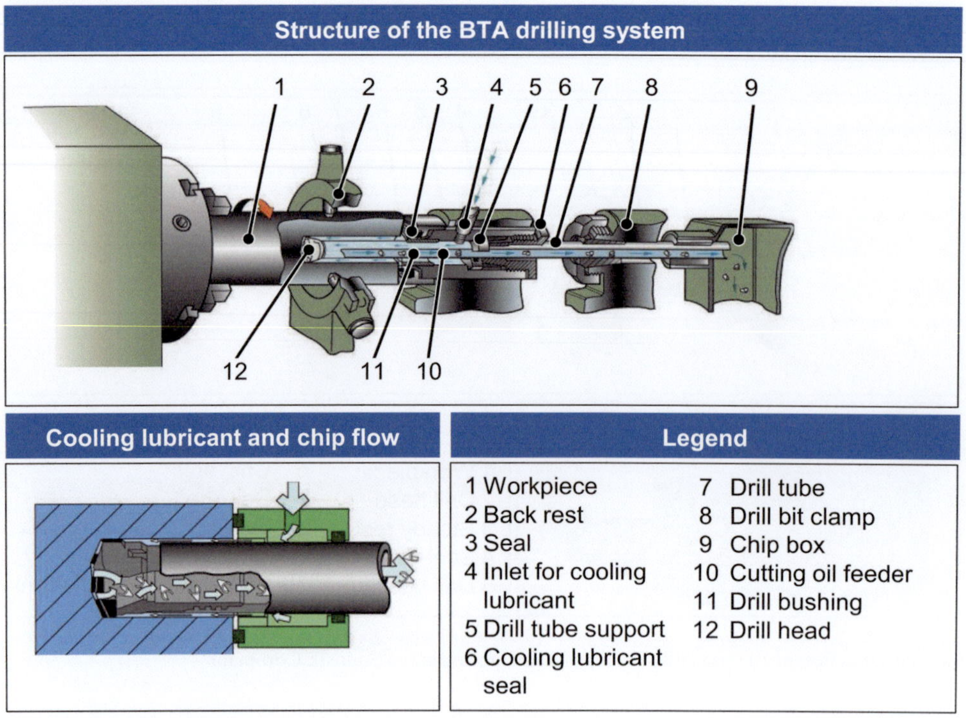

■ **Fig. 4.22** Structure and function of the BTA drilling system, according to: Sandvik Coromant

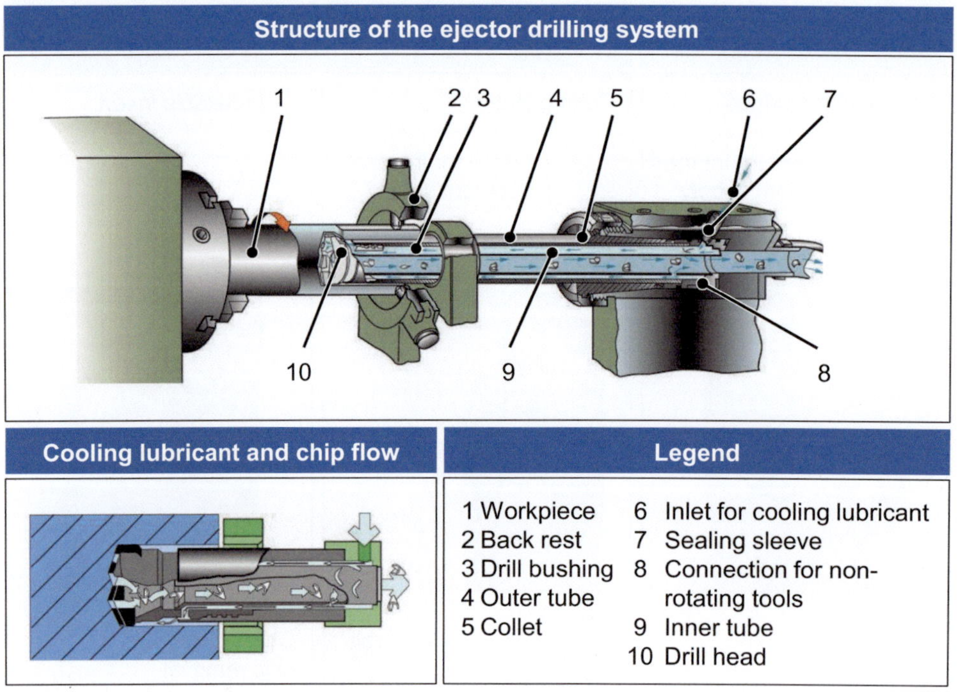

■ **Fig. 4.23** Design and function of the ejector drilling system, according to: Sandvik Coromant

lubricant is supplied from outside under high pressure by a cutting oil feeder that holds the drill bushing. The cooling lubricant flows through the drill bushing via the annular space between the borehole wall and the drill tube to the tool cutters. Together with the chips, the cooling lubricant flows back through the chip pocket in the drill head and the drill tube. The high fluid speed ensures efficient chip conveyance. The BTA drill head (see ■ Fig. 4.21, right) is fitted with one or more cutting inserts and is guided radially through rails in the drilling hole (like the single-lip drill). With its round cross-section and associated higher tool stiffness,

BTA drilling has higher productivity than single-lip drilling [SAND12]. The BTA drilling system is used for a diameter range of 6 to 200 mm and for multi-stage boring up to 2000 mm.

The ejector drilling system, a variant of the BTA drilling system, feeds cooling lubricant supply in an annular space between the drill tube and an inner tube (Fig. 4.23). Just behind the drill head, the cooling lubricant escapes from the side of the drill tube, cleanses the drill head and flows back into the inner tube together with the chips. Some of the cooling lubricant is routed into the inner tube via a ring die. The low pressure generated at the chip pocket aids the backflow of the mixture of cooling lubricant and chips (ejector effect) [VDI99]. In contrast to the BTA drilling system, there is no seal preventing cooling lubricant from escaping at the drill bushing. With this, the system can be used on conventional machine tools. The ejector method is only used on special deep drilling machines when the sealing required for BTA drilling is not possible. The ejector drilling system is used for drilling for a diameter range of 18 to 180 mm and for reboring to 250 mm.

While axial feed motion is always provided by the tool for machines developed especially for deep drilling, the rotation required for chip formation can be generated in three different ways [VDI14]:

- Fixed workpiece and rotating tool, suited to general-purpose use for all workpieces.
- Rotating workpiece and non-rotating tool, only suitable for rotationally symmetric workpieces. Low hole straightness deviation is possible for long drilling depths.
- Counter-rotating workpiece and rotating tool, for rotationally symmetric workpieces, has the lowest hole straightness deviation from the workpiece axis.

Figure 4.24 shows a single-spindle long-bed deep drilling machine for machining rotationally symmetric components, as used primarily for BTA drilling for larger borehole diameters. The machine has a fixed workpiece spindle housing and a moving drilling spindle housing. The position of the cutting oil feeder can be adjusted to the length of the workpiece to machine using the moving guide carriage. To prevent longer workpieces from deflecting, they are supported with a roller steady rest. The same applies to long tools that are supported by a steady rest. The rotation required for chip formation comes from the tool and/or workpiece. A worm gear drive provides the feed motion of the drilling spindle housing. This drive can generate a maximum feed force of 60 kN. The maximum nominal drilling depth for the machine shown is 15,000 mm for a borehole diameter of up to 200 mm for drilling and 400 mm for reboring.

The deep drilling machine shown in Fig. 4.25 is suited to deep drilling and conventional machining types, such as twist drilling, thread cutting and milling. Cubic workpieces with maximum weights of 10,000 kg can be accommodated on the workpiece rotary table that moves in the X-direction and has a clamping surface of $1500 \times 1000\ \text{mm}^2$. The drilling-milling unit is

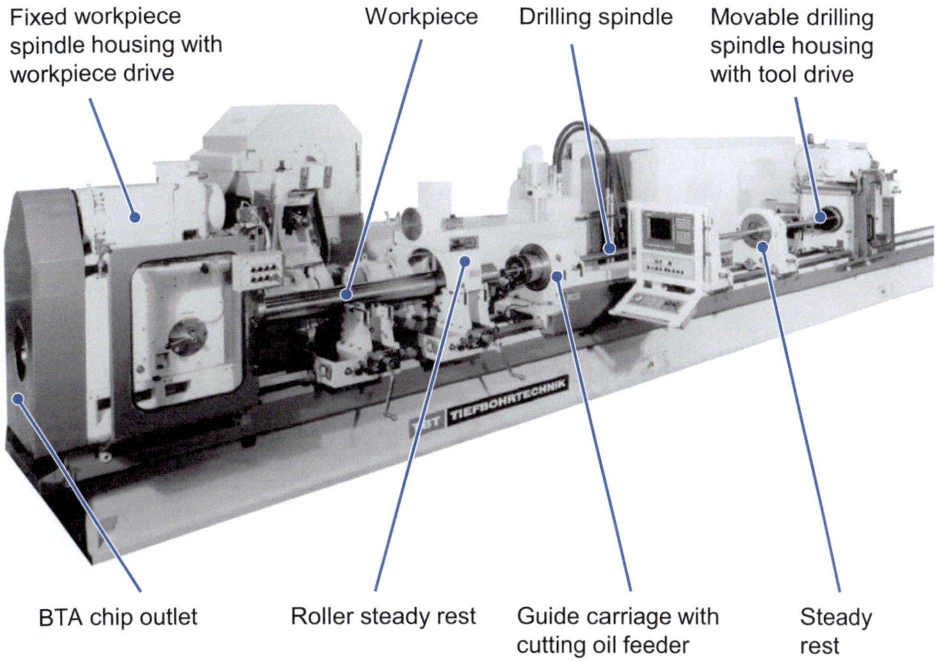

Fig. 4.24 Long-bed deep drilling machine, according to TBT Tiefbohrtechnik

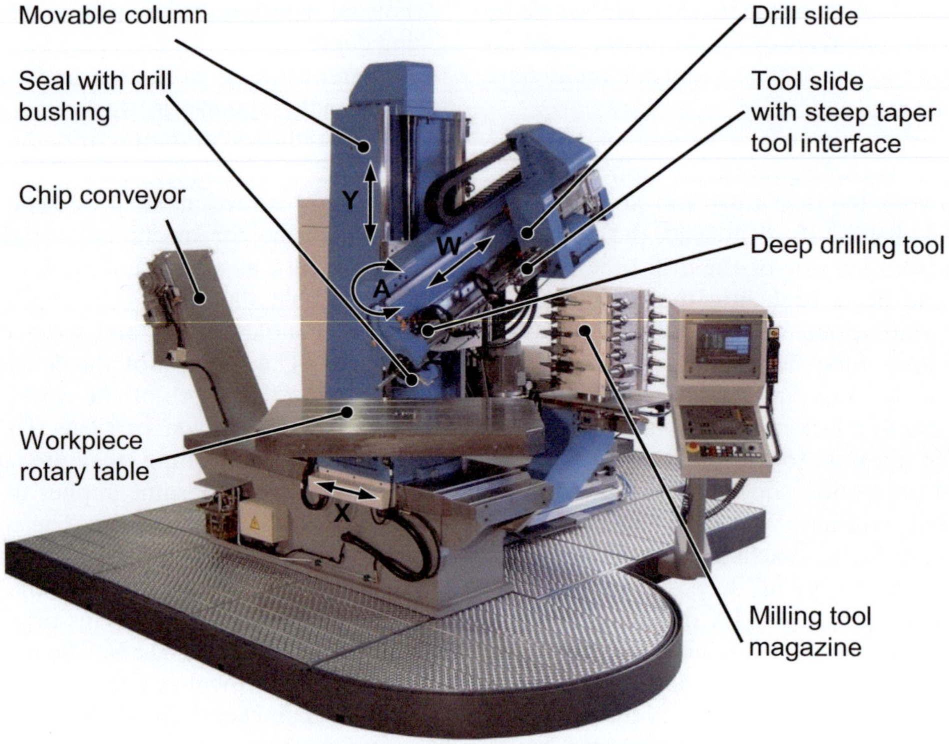

Fig. 4.25 Deep drilling machine, according to Ermafa Auerbach

routed in the Y-direction at the column that can move in the Z-direction. Furthermore, the drilling-milling unit can be lined up at an angle in the range of −25° to + 30° by the integrated swivel axis A. The drill slide, realized as a W-axis, provides the linear feed motion of the deep drilling tool. All tools are accommodated by the tool spindle over a standard steep taper interface. For conventional milling, there is a semi-automated tool change in which initially the complete boring rod, including front seal with drill bushing, is removed manually. Then the drill slide moves to the front W-axis position so that, for the horizontal drilling-milling unit, the milling tools can be replaced automatically.

A deep drilling mill center for 4-side machining of cubic workpieces in one clamping process is shown in Fig. 4.26 (without the workpiece clamping table). The main feature of this machine concept is the parallel kinematic rocker system producing the vertical and swivel movements of the spindle housing. On a cross carriage that produces the traversing movements in the X and Z-directions, three traversing carriages determine, depending on their positions in the rocker system, the position in the Y-direction and about the A-axis. There are two front rockers to better support the machining forces acting in the Z-direction. On the other hand, there is only one rear rocker because only the tilting moments generated need to be supported. In the single-lip drilling method, the machine can be used to a maximum drilling depth of 1500 mm. A milling unit is integrated into the machine in addition to the deep drilling unit. When a milling unit is exchanged, the boring rod is moved back with the drive. Then the drilling unit is swiveled out by the automatic turret head and the milling unit is swiveled in. Then the drive moves again to the front and is coupled with the milling spindle. The automatic turret head makes very short change times between drilling and milling operations possible. Furthermore, the programming of machining positions is easier because drilling and milling use only one axis and thus there is no risk of collision.

An example of a fully automatic tool changer that can exchange up to five deep drilling tools is shown in Fig. 4.27. For BTA drilling, the complete, ready-to-use tool units with the tool, boring rod, cutting oil feeder, vibration damper and clamping element can be exchanged from the turret into the machine.

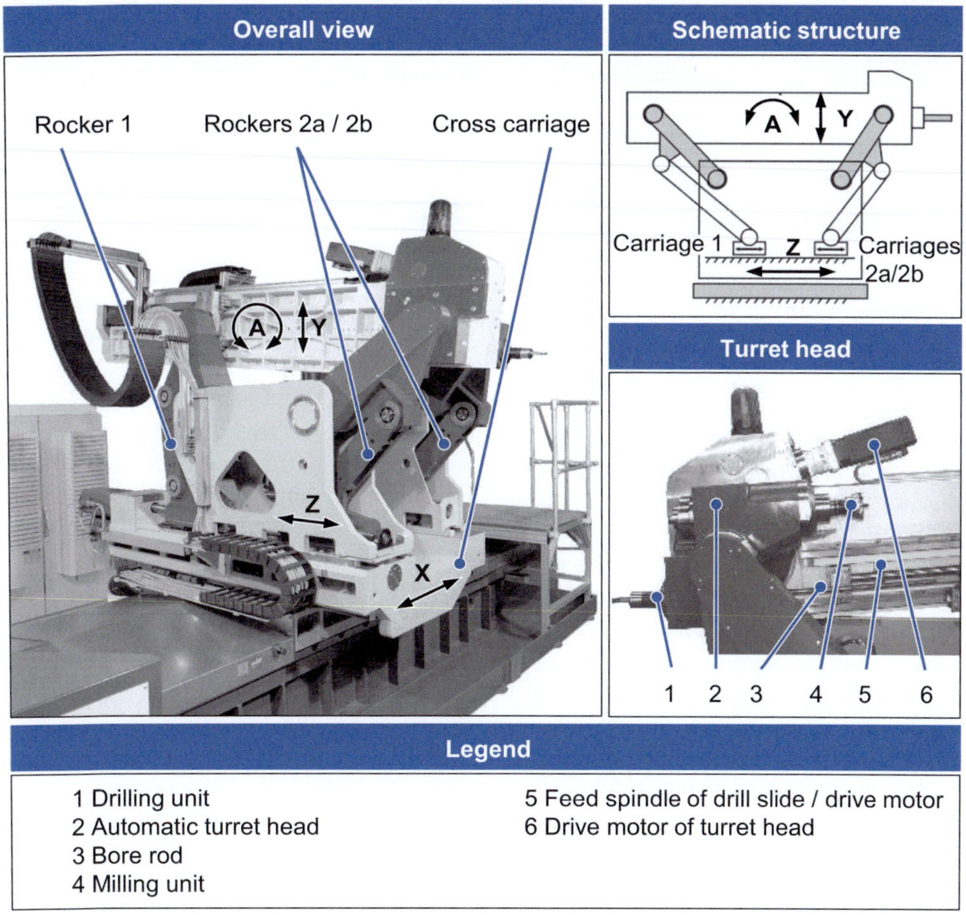

Fig. 4.26 Deep drilling mill center with parallel kinematics, according to TBT Tiefbohrtechnik

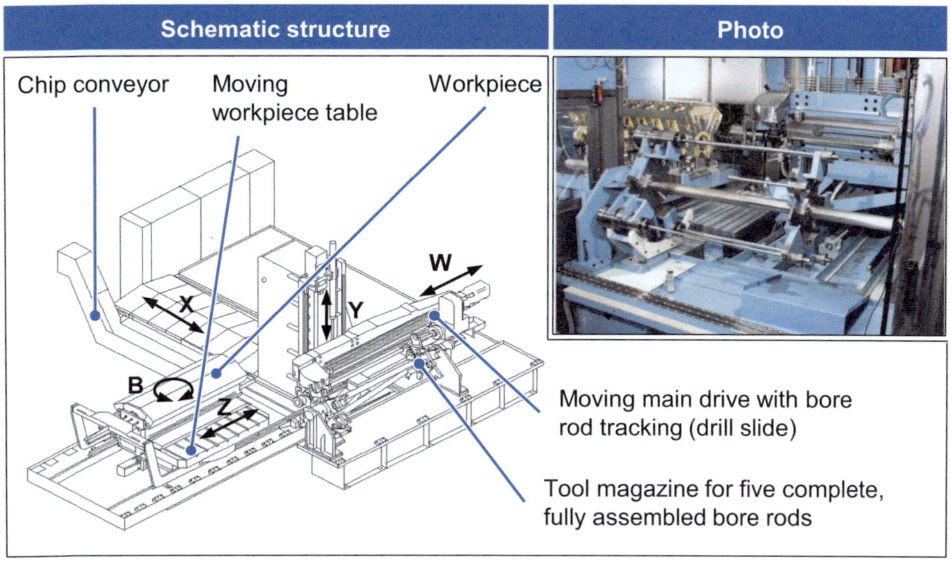

Fig. 4.27 Deep drilling center with turret tool changer, according to TBT Tiefbohrtechnik

4.3 Milling Machines and Machining Centers

Milling is a machining method very similar to drilling where the cutting motion is also generated by rotation of the tool. In contrast to drilling, the feed motion here is mainly perpendicular to the rotation axis of the tool, whereby for some process variants, several feed motions coupled with each other superimpose.

Figure 4.28 provides an overview of the most important milling processes [DIN03b]. Flat surfaces can be created using peripheral milling, face milling or face-peripheral milling, a combination of the two. These processes are also used to generate round bodies. Instead of the linear feed used to manufacture flat surfaces, the tool and the workpiece rotate in circular milling. Rotation of the workpiece can also be replaced by path-controlled movement of the tool on the plane perpendicular to the tool rotation axis. Milling with a helical feed motion to create spiral surfaces, such as cylindrical worms and threads, is known as helical milling. The long-thread milling and thread whirling processes can create lengths that are, in theory, unlimited. In contrast, the lengths of threads achievable with short-thread milling are restricted to the width of the multi-profiled thread milling cutter. Hobbing is used to manufacture gears. Here, the profiled tool performs a rolling movement during the cutting process simultaneous to the feed motion with the workpiece (see ► Sect. 8.1.1.1). In profile milling, the profile of the tool is reproduced on the workpiece. In addition to flat and rotationally symmetric profile surfaces, profiled holes can also be made. Drilling operations can normally also be performed on milling machines and machining centers regardless of the milling processes shown.

Figure 4.29 and 4.30 provide an overview of the most important designs of drilling and milling machines and machining centers with a horizontal and vertical tool spindle. The positioning and feed motions are performed by the tool holder (column, spindle housing) and/or workpiece carrier (table). The design principles are shown schematically in the figures in line with the subdivision of the axes of motion.

Due to the vertical workpiece table movement during which the table drive must bear the workpiece weight, the knee-and-column design is only used for smaller machine tools. The main area of application for these universal machines are tool and fixture making and production of small and medium-sized batches. Bed-type milling machines are used to machine heavy workpieces. In contrast to the knee-and-column design, the workpiece table is on a fixed machine bed in this design. For bed designs, the term cross-table design is used when the workpiece carrier performs two

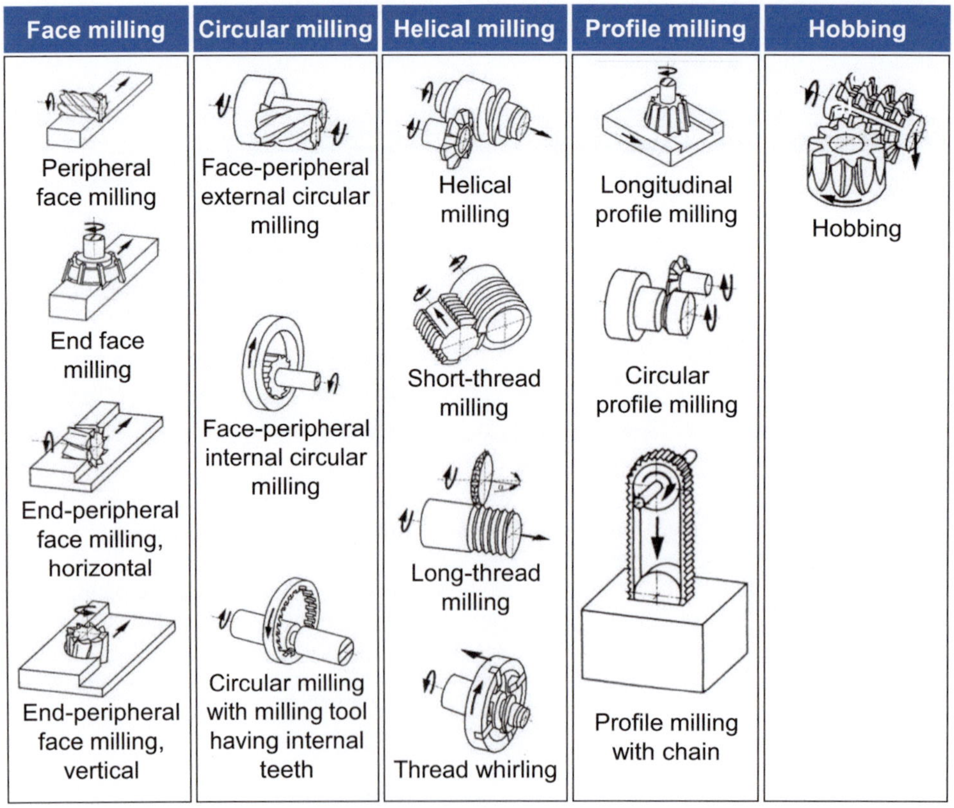

Fig. 4.28 Different milling processes

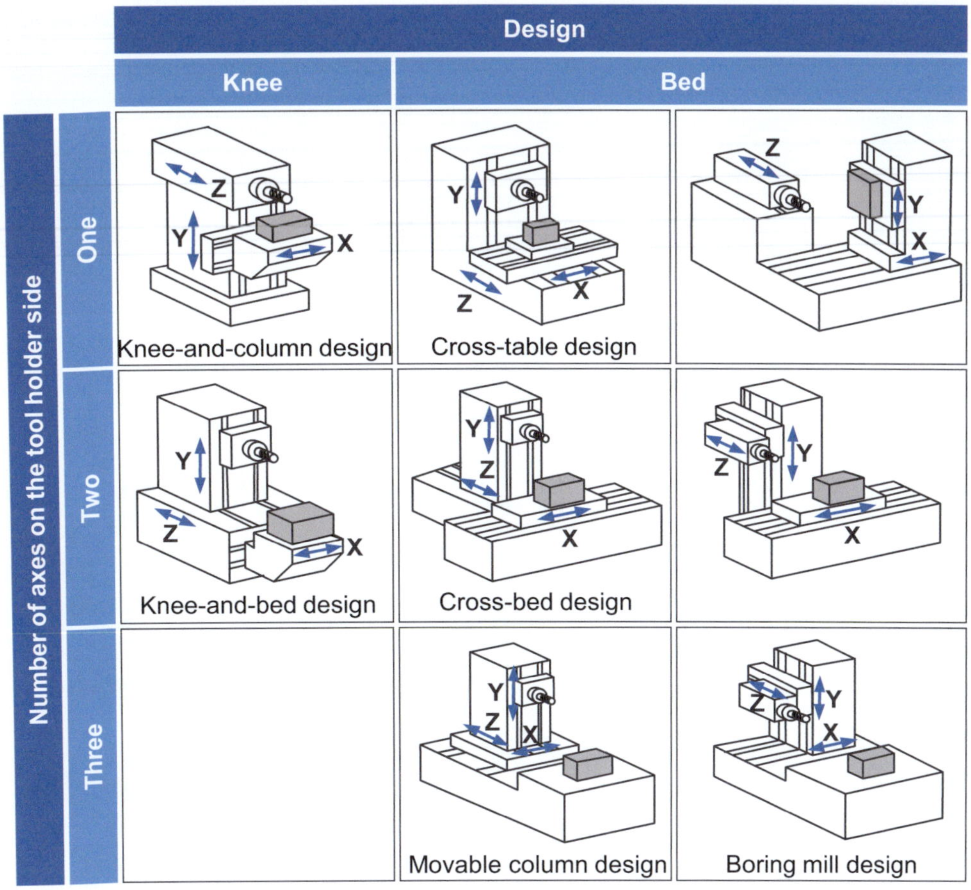

Fig. 4.29 Designs of horizontal drilling and milling machines

directions of motion that are perpendicular in relation to one another. The cross-bed design denotes those variants where two perpendicular feed motions are used on the bed. One motion is for the tool-holding assembly (usually the column) and the other for the workpiece-holding assembly (usually the table). Machine designs where all linear axes of motion are assigned to the tool side are called moving column or boring mill design depending on variant. In contrast to the moving column design, for a boring mill the main spindle is normally positioned horizontally and is attached to the column so as to traverse vertically. The headstock contains here the complete drive and feed unit for the main spindle, often designed as a quill. One design that has only been used in machines with a vertical tool spindle is the portal design, also called the two-column design with a crosshead. In the table-type design, two axes of motion are on the tool side and one axis of motion is on the workpiece side. In the gantry design, all motions are performed on the tool side.

Four-axis and five-axis machine systems are increasingly being used to machine complex components. As all three translational directions of motion are already implemented in machine tools with Cartesian design, the fourth and possibly fifth axes are added as rotational axes. These axes are implemented on the tool or workpiece side depending on application. If the workpieces have a comparatively low mass, workpiece-side rotary tables for four-axis machining and rotary swivel tables for five-axis machining are often used. Rotary and rotary swivel tables enable a workpiece to be positioned around one or two additional axes (normally the A- and B-axes) (Fig. 4.31).

When very large workpieces are machined on a portal milling machine, for example, the axes required in addition can be realized on the tool side. Attachment heads are used to realize a fourth axis on the tool side. These attachment heads can be built with angle heads with a fixed angle or with angle heads that can assume different angle positions, for example, with Hirth serration at fixed locking positions. Hydraulically clamped angle heads can be locked into place at any angle. There are also angle heads that are not locked into place but that can be moved during the process. If a fifth axis is required on the tool side, it can be realized as a C-axis. Thus the angle head can be rotated around the Z-axis. There are also different variants here, where the C-axis is moved during the machining process or

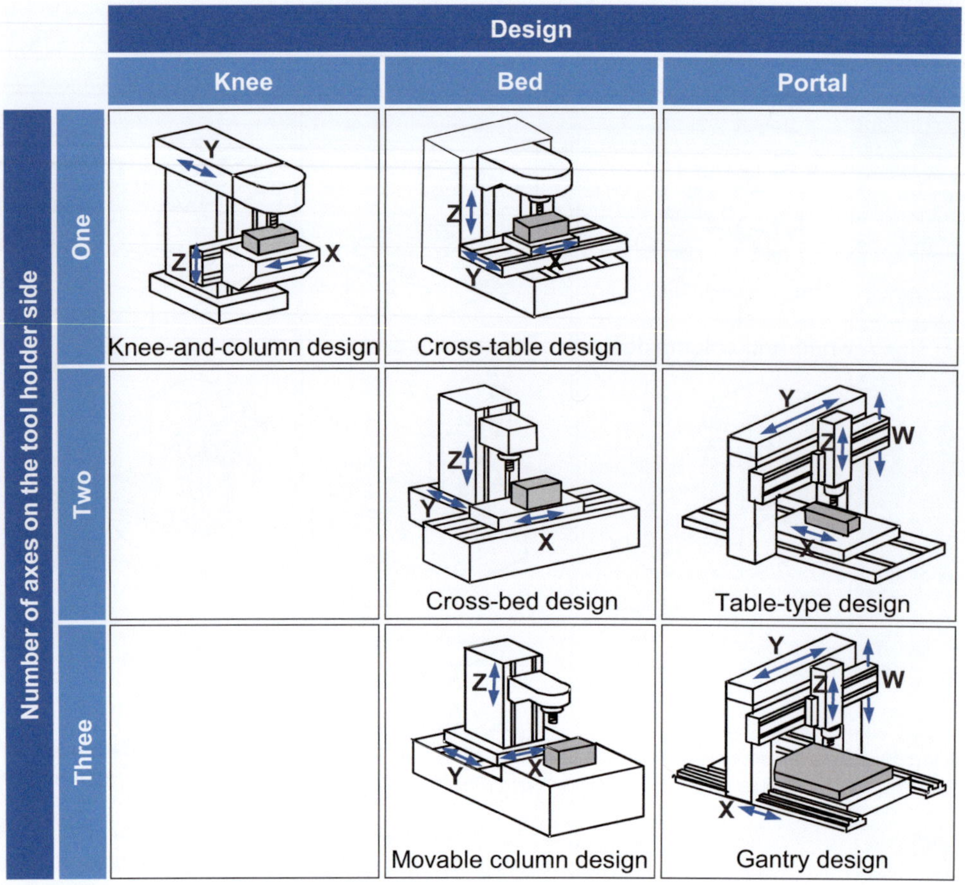

Fig. 4.30 Designs of vertical drilling and milling machines

can be clamped at fixed locking positions at any angle. HV heads (horizontal-vertical heads) or fork heads enable the fourth and fifth axes to be traversed during the machining process. For very high drive powers in particular, attachment heads are very complex in terms of design because all of the drive torque needs to be diverted in the attachment head clearance-free using complex gearboxes.

In addition to being able to implement both axes on the tool and workpiece sides for five-axis machining, there are also mixed forms. Boring mills for example often have a fourth axis in the form of a rotary table. An additional fifth axis can be provided as required on the tool side using an angle head.

In addition to the conventional designs shown in Fig. 4.29 and Fig. 4.30, machine tools exist that have parallel kinematic structures. In contrast to conventional machine tools with generally Cartesian design for which the individual axes are positioned serially, that is, one after another, for parallel kinematic machines all axes of motion act directly upon a movable platform that carries the main spindle with tool. To perform a platform movement, the general requirement for parallel kinematic machines is to move all drives at the same time, that is, in parallel with each other. Parallel kinematic machines are presented and described in ▸ Sect. 4.3.3.

The following subsections introduce and explain some example milling machines with different, conventional designs. The machines are grouped according to the position of their tool spindle.

4.3.1 Machines with Horizontal Tool Spindle

Introduced first in this subsection are machines with a horizontal tool spindle. Then in ▸ Sect. 4.3.2, machine systems with a vertical tool spindle are explained.

4.3.1.1 Machines with Cross-Bed Design

The most common design of milling machines with a horizontal tool spindle is the cross-bed design. As can be seen in Fig. 4.30, the traversing motion of machines having this design uses two axes on the tool side (often the X- and Y-axes) and one axis on the workpiece side (often the Z-axis). The machine can be expanded to a 4-axis machining center by adding a rotary table to the workpiece side.

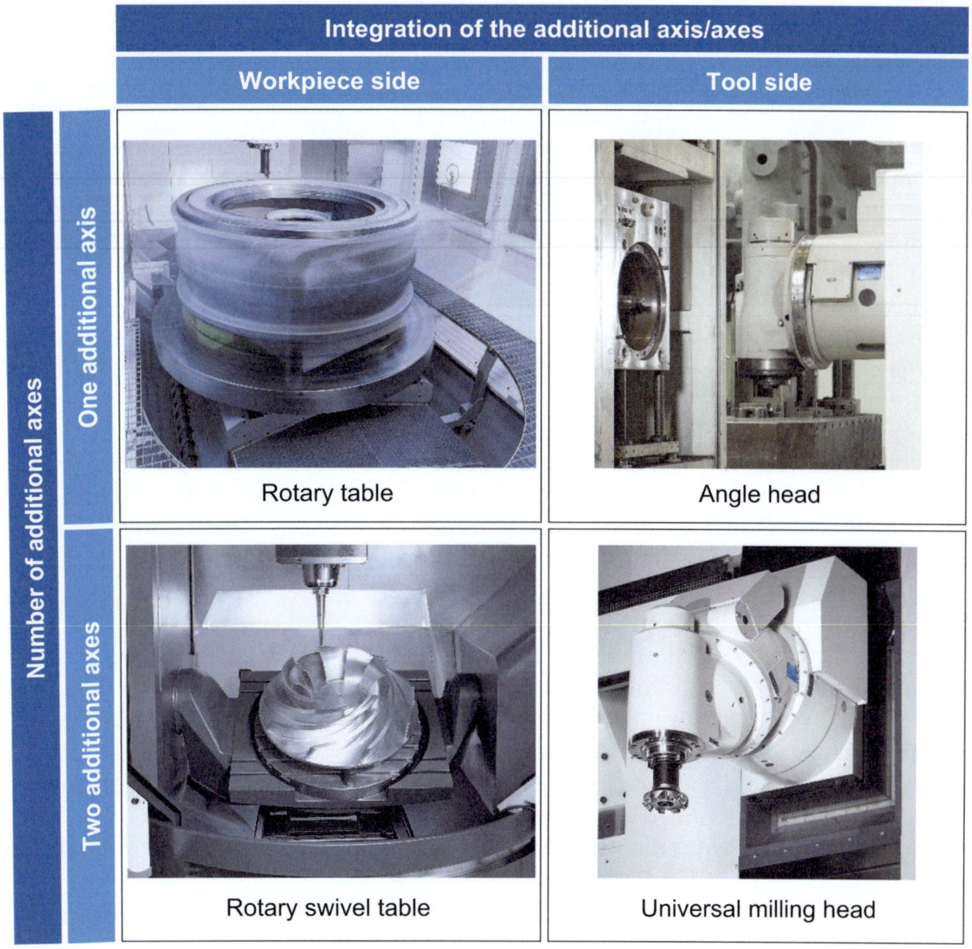

Fig. 4.31 Possibilities for realizing fourth and fifth machine axes. *Sources:* Starrag Technology (Scharmann division), Union Chemnitz

An example of such a machining center is shown in Fig. 4.32. The guides of the rotary table and the thermo-symmetric gate column are attached to the T-shape machine bed made of gray cast iron. The machine bed is optimized as regards high dynamic stiffness for low mass and is also made of gray cast iron, like the machine column providing the X-axis motion. Given the statically and dynamically stiff design, this machine concept is preferred in the automotive industry for heavy-duty machining of engine components.

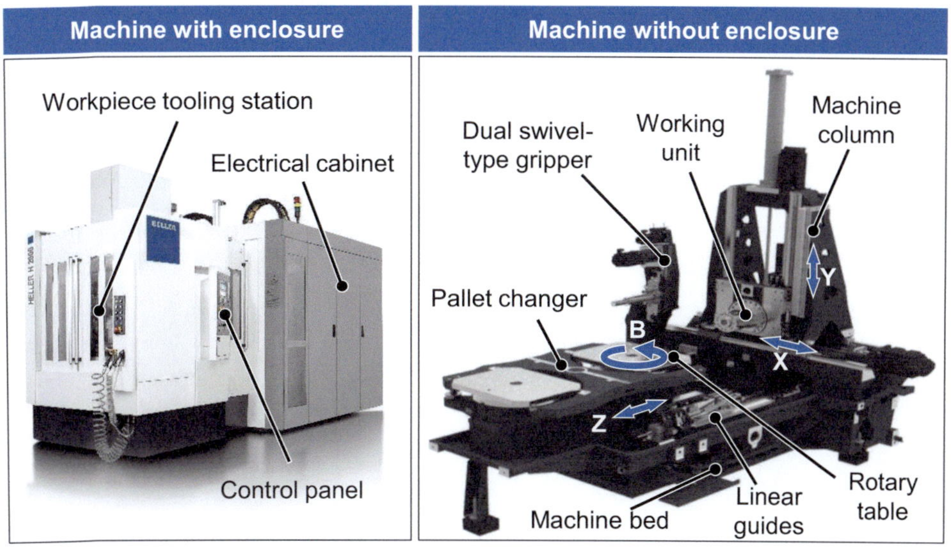

Fig. 4.32 Machining center with cross-bed design. *Source* HELLER

The linear traversing motion in the Z-direction and the rotational motion around the B-axis are realized by the rotary table. The B-axis enables here a maximum rotational speed of 40 rpm at a maximum tangential torque of 5000 Nm and maximum breakdown torque of 8000 Nm. The maximum speed reached in rapid motion is 65 m/min and 90 m/min for Power/Speed drive variants. In three configurations of the working unit with direct spindle drive, speeds of 10,000 rpm to 24,000 rpm and torques of 40 Nm to 242 Nm are possible. In these situations, oil-air lubrication of the spindle bearings and water-cooled spindle neck minimize displacements caused thermo-elastically.

To reduce non-productive times due to component handling and clamping during primary machining time, a lifting-pivoting pallet changer with a maximum loading on both sides of 1000 kg is integrated in the 4-axis machining center (◘ Fig. 4.33). In this case, the maximum component weight on one loading side is 800 kg. As an alternative to the manually rotating set-up station, the next index position of the rotating set-up station can be approached automatically using a spur gear drive and servo gear motor with absolute encoder. When the safety door of the set-up station is closed, the rotational motion is at a rapid speed. When the guard door is open, safety for the operator is guaranteed by a light barrier. For hydraulic workpiece clamping, pallet change time is 10.5 s, done manually, it is 0.5 s quicker.

To build up a larger work list or for parallel retooling of fixtures, pallet storage can be used in combination with a pallet management system for multiple clamping. The system, together with the machine, can then hold six pallets.

A 2-axis servo tool changer is used to shorten chip-to-chip time. An automatic tool change is performed using a dual swivel-type gripper controlled over two servo axes. The magazine type used is a chain or rack-type. A chain-type magazine holds up to 240 tools with a maximum diameter of 160 mm. The rack-type variant of the magazine can hold up to 409 tools with a maximum diameter of 188 mm.

Furthermore, the 4-axis machining center enables online compensation of the Z-spindle growth on the basis of diverse sensor values regardless of whether they are caused by the process or the environment. Also, expansion to a flexible production system is realized with the option of incorporating external pallet automation systems, robot loading and an automatic set-up process.

Side views of another 4-axis machining center with cross-bed design are shown in ◘ Fig. 4.34, with and without enclosure. ◘ Figure 4.35 shows more detailed views of the different machine features of a 5-axis machining center. The infeed motion in the Z-direction of both machining centers uses the moving rotary table, in the Y-direction by the moving column and in the X-direction by the spindle housing. Linear direct drives, or alternatively ball screw drives, provide the translational movements. The rotary table shown in ◘ Fig. 4.34 also has a rotational B-axis and can expand the machine to a 5-axis machining center with the addition of another rotational A-axis (as shown in ◘ Fig. 4.35).

One way of automating production processes is to fit the machine tool variant shown with a pallet changer that can hold up to 12 pallets, depending on the design, on the workpiece side. On the tool side, the machine has a modular wheel magazine with 40 tool slots for tools having an HSK-A63 interface. It is possible to expand the magazine to five wheels having up to 303 tools. From the second magazine wheel, primary machining time and non-productive time can be implemented with parallel set-up on a set by set basis.

Shown on the left in ◘ Fig. 4.35 are the cooling provisions for the machine system (for variant with linear direct drives). In addition to cooling the linear direct drives for the translational movements in the X, Y and Z-directions, a major part of the cooling capacity is used by the spindle system. The drives of the turn-swivel unit are cooled in addition. The cooling provi-

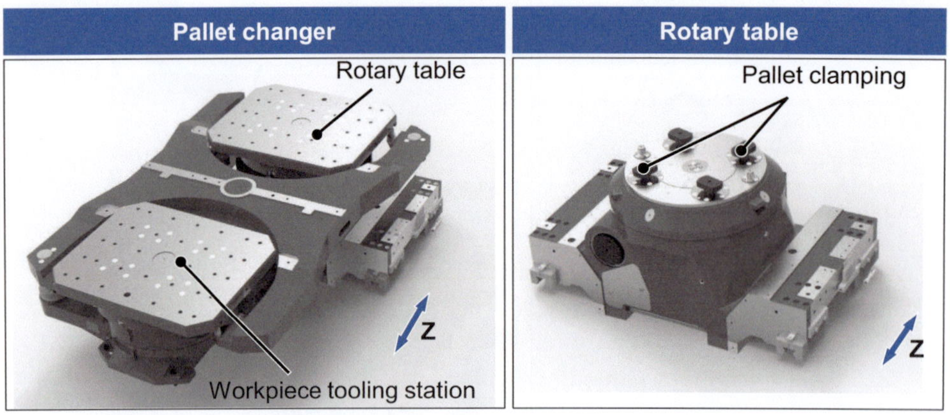

◘ **Fig. 4.33** Pallet changing equipment and rotary table. *Source* HELLER

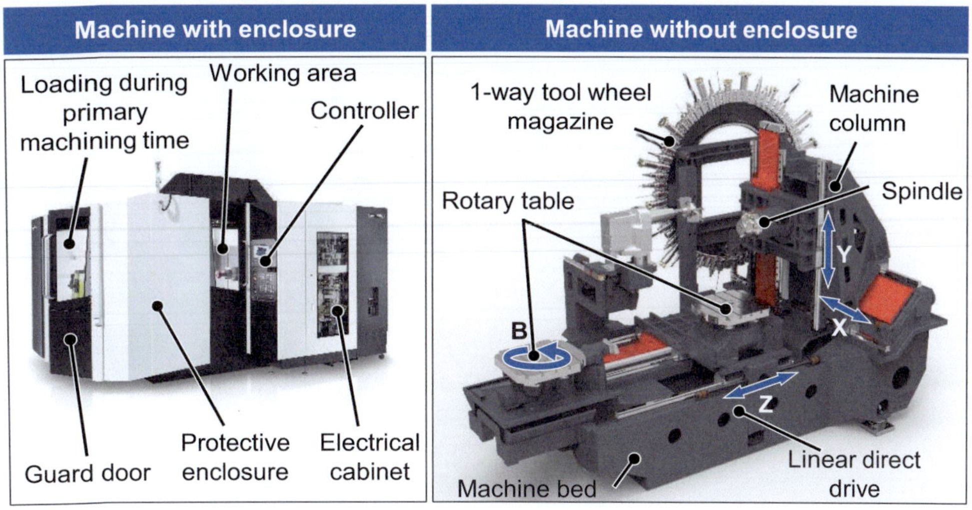

Fig. 4.34 Machining center with cross-bed design. *Source* DMG MORI

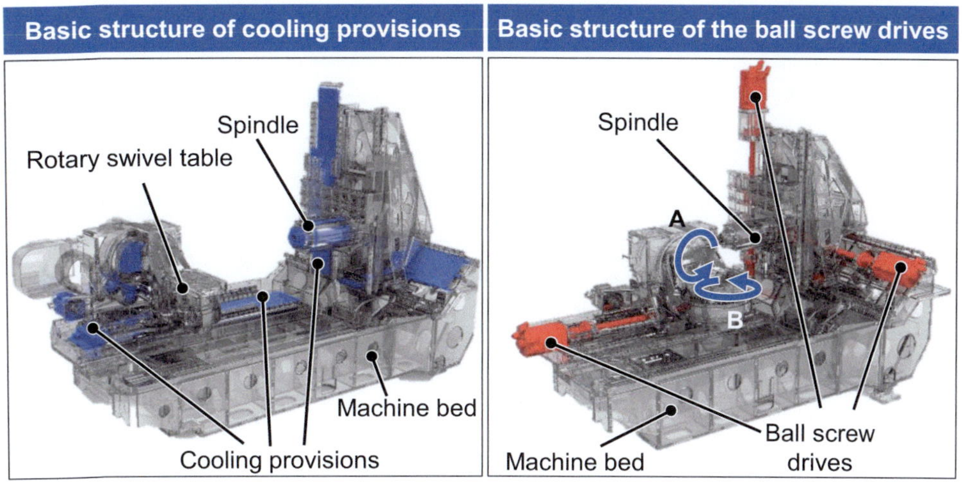

Fig. 4.35 Machine bed with cross-bed design. *Source* DMG MORI

sions are thus concentrated in areas where heat is generated from energy lost in the drives. The linear guide systems also generate heat from friction for traversing movements and do not have additional cooling in the machine concept introduced. However, there are also machine designs incorporating cooling of the guides (see for example Fig. 4.42).

The machine system variant with ball screw drives is shown on the right in Fig. 4.35. Using ball screw drives instead of linear direct drives means this drive concept is far more cost-effective, but the loss of traverse dynamics is a downside.

Figure 4.36 shows an inexpensive alternative to the high-performance machining centers presented. Due to international cost pressure, particular attention was paid in the development phase to low manufacturing costs combined with a high level of reliability. The horizontal machining center with cross-bed design, made primarily using standard components, is in the medium performance range. With ball screw drives for the individual axes, rapid motion rates of 50 m/min at accelerations of 7 m/s2 can be achieved. The chip-to-chip time for the magazine chain shown is about four seconds for 40 tools. Using the pallet swivel changer, workpiece pallets weighing up to 500 kg can be exchanged in seven seconds. By designing the machine in a systematic and cost-effective manner using proven components and technologies, the investment total for this machine could ultimately be reduced by 50% compared to conventional high-performance models from the manufacturer. This expands market opportunities in contract manufacturing in particular.

Shown in Fig. 4.37 as another example of a special form of cross-bed design is a machine tool realized as a boring machine with a horizontal tool spindle for heavy-duty machining. The spindle housing

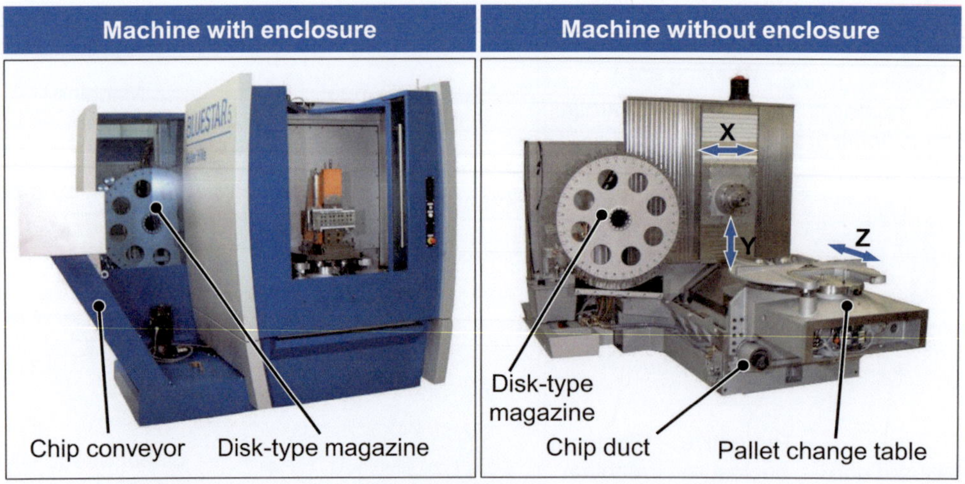

Fig. 4.36 Horizontal machining center. *Source* MAG IAS/Zuse Hüller Hille

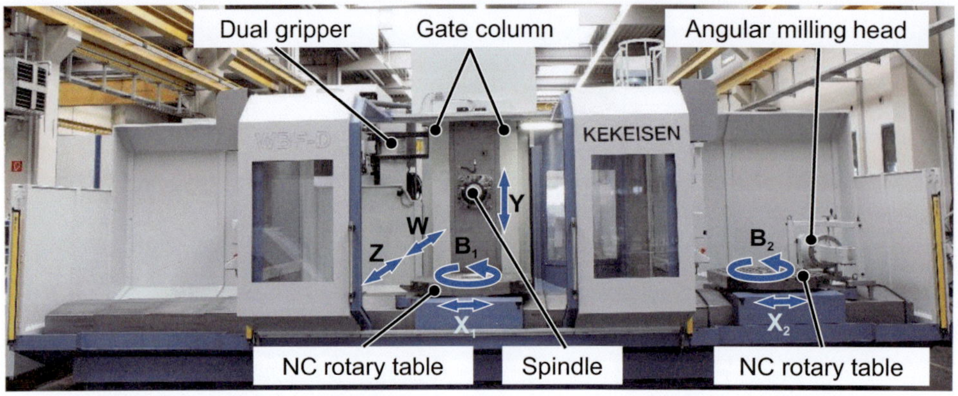

Fig. 4.37 Boring machine with cross-bed design. *Source* Kekeisen

in the gate column provides the infeeds in the Z- and Y-directions. The gate column ensures symmetric load distribution and thermal balance. Traversing the NC rotary tables effects movements in the X-direction. The configuration shown with two NC rotary tables allows the set-up of large components during primary machining time. The maximum workspace is 3500 mm × 1800 mm × 1300 mm on the three linear axes for a maximum table clamping area of 1250 mm × 1400 mm. In addition, the drilling spindle (with its main bearing lying more in the front) can be extended by 500 mm on the W-axis, thereby enabling machining near the center of the table. A two-speed gearbox and powerful main drive rated to 37 kW enable rotational speeds up to 3500 rpm and high torques. Oil cooling for the spindle housing is standard and ensures thermal stability.

The tool magazine can hold 100 tools. Tool changes can be performed automatically with a dual gripper. In addition, to increase the degree of automation, a pallet change system with up to three deposit stations and automatic door opener can be installed.

The application spectrum is extended by means of different vertical milling heads or universal milling heads, with or without Hirth serration, with manual or automatic adjustment. Furthermore, for difficult machining tasks and when the W-axis is extended a long way, various attachment support bearings of differing lengths can be used.

To increase productivity in volume production, dual-spindle machining centers with two main spindles arranged next to each other are deployed in addition to corresponding high-speed machines. Figure 4.38 shows such a machining center with two horizontal main spindles having a cross-bed design and a modified moving column with frame column design. The column, like the pallet change unit, is securely connected to the machine frame. A cross carriage of motion link design with X- and Y-axes is routed in the column. Traversing the table unit, which integrates the A-axis and both B-axes, provides the infeed in the Z-direction. The column and cross carriage are made of steel profiles that are weight and topology-optimized. They form a stiff and low-mass welded construction.

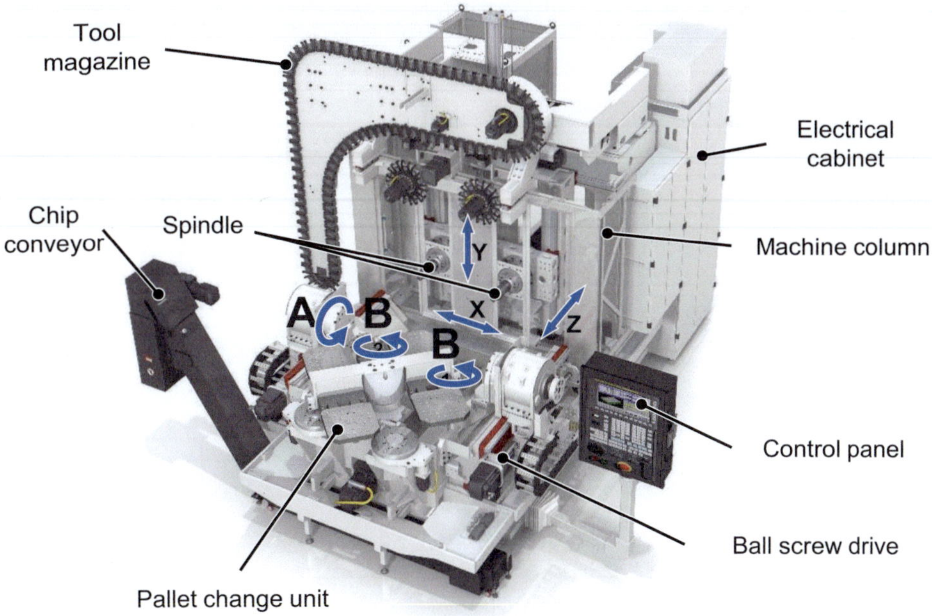

Fig. 4.38 Dual-spindle, high-speed machining center with cross-bed design. *Source* MAG IAS

The low masses to move mean that feed rates of up to 120 m/min are possible on the linear axes with linear drives. The tool spindle is a motor spindle to accommodate tools with the HSK-A63 or HSK-100 interface. Maximum speeds of 18,000 rpm, maximum power of 52 kW and a maximum torque of 378 Nm are possible depending on the motor spindle variant. The L-chain magazine shown holds 72 tools with a maximum tool diameter of 230 mm, maximum tool length of 550 mm and maximum weight of 3.5 kg each. The table unit shown, with a combination of A- and B-axes, can clamp workpieces with a maximum weight of 2×290 kg. With the tool magazine and pallet change unit, the machine shown forms a standalone 5-axis machining cell. The modular machine design also allows the individual units to be interlinked to form flexible manufacturing cells or transfer lines.

4.3.1.2 Machines with Moving Column Design

The column with frame column design for the machine shown in **Fig. 4.39** differs considerably from the

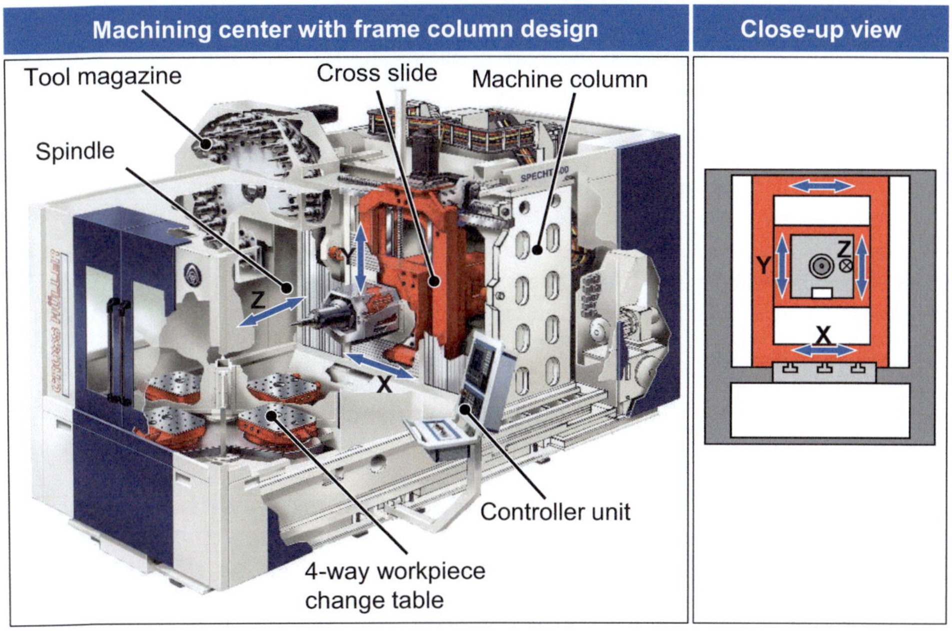

Fig. 4.39 High-speed machining center with frame column design. *Source* Cross Hüller

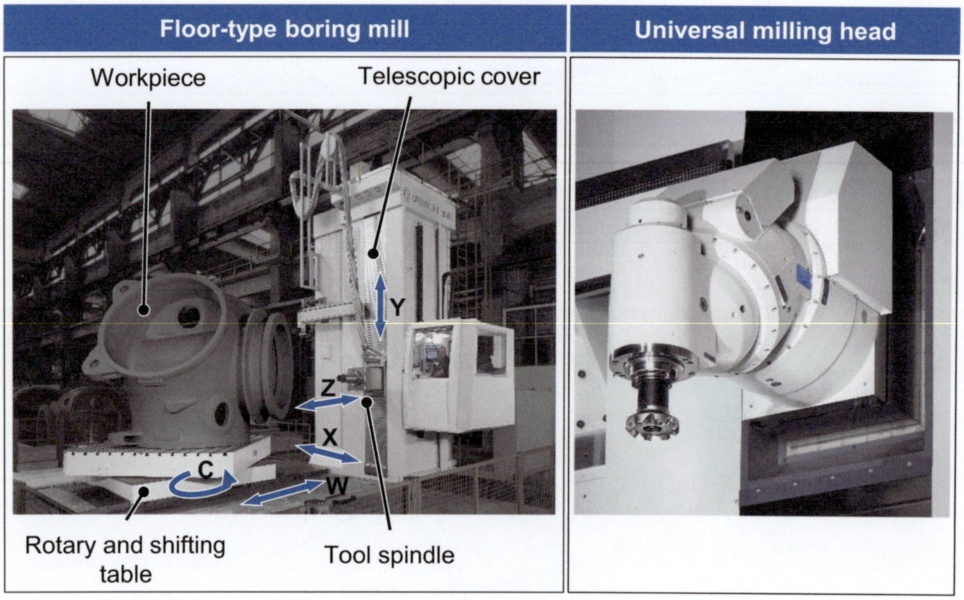

Fig. 4.40 Boring machine with moving column design. *Source* Union-Chemnitz

classic moving column design (Fig. 4.43). The column, like the table/pallet change unit, is securely connected to the machine frame. A cross carriage with a motion link design and linear X- and Y-axes is routed in the column. The Z-carriage is implemented as a quill, carries the tool spindle and is attached to the Y-slide. The column and cross carriage are made of steel profiles that are weight and topology-optimized. They form a stiff and low-mass welded construction. The low masses to move mean that feed rates faster than 60 m/min are possible on all the linear axes with the machine. The tool spindle is a motor spindle that is fitted with hybrid rolling bearings (steel rings with ceramic rolling elements). The maximum speed is 16,000 rpm. Spindle speeds up to 24,000 rpm are possible when the bearings have oil/air lubrication. With a tool magazine, here designed as a disk, and pallet changing equipment, the machine forms a standalone machining cell.

Figure 4.40 shows another example of a machine with a moving column design and a horizontal tool spindle — a boring machine with its linear axes of motion on the tool side, in a box column that moves crosswise and is vibration-resistant. The column unit is routed in the X-direction on a long bed of up to 40 m. A backlash-free dual pinion drive with rack is used as the drive. The guides are controlled hydrostatically or designed as pretensioned linear compact roller guides depending on the model. The boring machine shown has multi-compensation. The girder bend and column inclination are compensated in addition to the spindle slack. The additional integration of a fully automatic tool changer extends the functionality of the boring machine to a machining center. Up to 300 tools are available for tool changes using pick-up stations. Fixed or transportable rotary, shifting or tilting tables provide workpiece clamping and fast-turning tables for carousel machining, designed with a slide guide, rolling element guide or hydrostatic guide. The universal milling head shown on the right in Fig. 4.40 can be used to add a fourth or fifth axis to a boring machine on the tool side.

Besides boring machines for large machine components, there are also compact machining centers in a moving column design. A horizontal, dual-spindle machining center with moving column design is shown in Fig. 4.41. The two spindles are 600 mm apart. Depending on the design, the spindles can traverse autonomously or coupled in the Z-direction. On the tool side, the three linear axes ensure workpiece-independent dynamics. Two workpieces with a maximum diameter of 600 mm and height of 500 mm can be clamped simultaneously on the two-axis rotary swivel table. Loading and unloading of the workpiece spindles, during primary machining time, use a turret workpiece changer with maximum load of 1000 kg per side. The chip-to-chip time using the pick-up technique is about 2 s for the magazine chain shown, for 2 × 42 tools. The machine shown reaches rapid motion rates of 60 m/min and can be operated with motor spindle speeds up to 20,000 rpm. Also, the machining center requires a compact installation space of just 3045 mm × 4,820 mm.

4.3.2 Machines with Vertical Tool Spindle

After the machine systems with a horizontal tool spindle, this subsection introduces milling machines with a vertical tool spindle. The classification is structured according to the designs shown in Fig. 4.30.

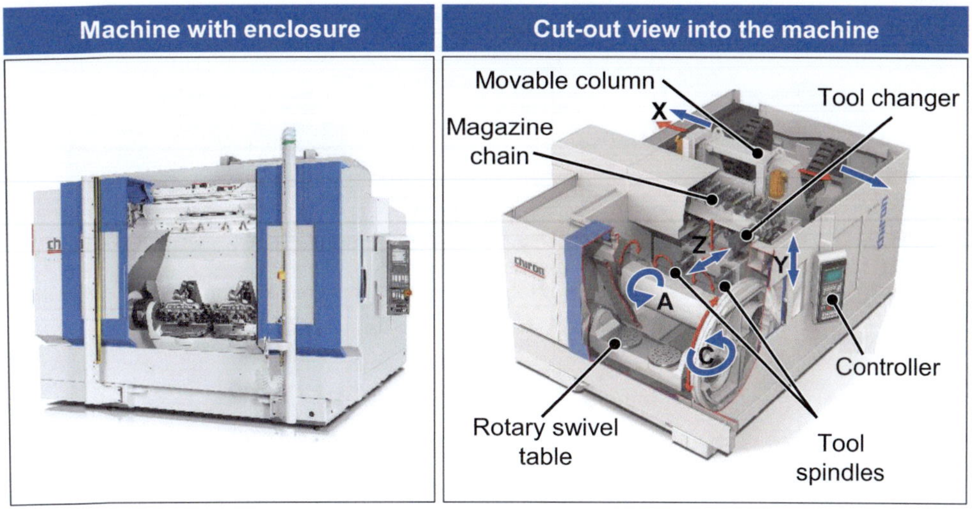

Fig. 4.41 Dual-spindle, high-speed machining center with moving column design. *Source* CHIRON Werke

4.3.2.1 Machines with Moving Column Design

Figure 4.42 shows a machining center with a vertical tool spindle and moving column design. The machine has three linear and optionally two rotational axes in the workpiece carrier. Integrated machining with milling and turning is possible with a rotational speed of 1200 rpm at a torque of 1000 Nm. This compactly designed machine tool requires just 7.5 m^2 of floor space.

It is possible to select from two table variants, a fixed table and a rotary swivel table with a drive system on one or both sides. Using a rotary swivel table with a drive system on both sides, simultaneous 5-axis machining is possible for workpieces having a maximum diameter of 840 mm and weighing a maximum of 1000 kg. A fixed table can handle workpieces weighing up to 3000 kg with maximum dimensions of 1000 mm × 840 mm × 560 mm.

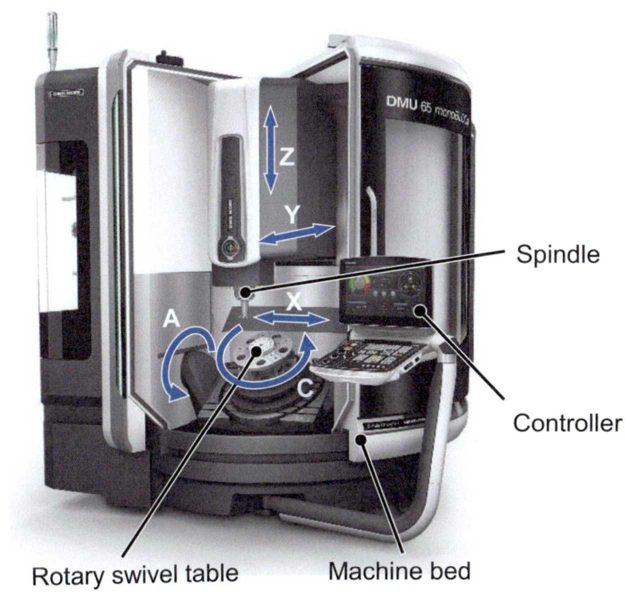

Fig. 4.42 5-axis machine with moving column design. *Source* DMG MORI

To load the machine, a workpiece can be transported by a crane from the top to above the center of the table. The workspace and tool magazine are also fully accessible for automation cases. The tools are loaded in a vertical chain-type magazine with room for 30 tools (expandable to 180). The dimensions and weight of a tool are limited by the maximum tool diameter of 160 mm, tool length of 315 mm and tool weight of 8 kg. The short chip-to-chip time of 4.9 s is achieved by cam-controlled dual grippers.

The machining center with moving column design shown on the right in Fig. 4.43 depicts the implementation of all three infeed and feed axes on the tool side. The column base slide positioned behind the fixed workpiece table moves on the X-axis on the machine bed. The tool column that can move in the Y-direction is on this slide. The tool spindle housing attached to the front of the column performs the Z-infeed. Furthermore, the machine is equipped with a NC rotary swivel table with a swivel range of ± 100°, meaning 5-side machining of complex workpieces is possible in one clamping process. One particular feature of the machine is the basket tool changer, shown on the left in Fig. 4.43. The tools (a maximum of 20) are positioned around the tool spindle in a star shape above the spindle nose and are also moved when the spindle unit traverses. Every tool has its own tool gripper, that continually grasps the tool when it is replaced and when it is in the magazine. This principle means the tool change time by itself is reduced to about 0.9 s and the chip-to-chip time is only about 1.9 s.

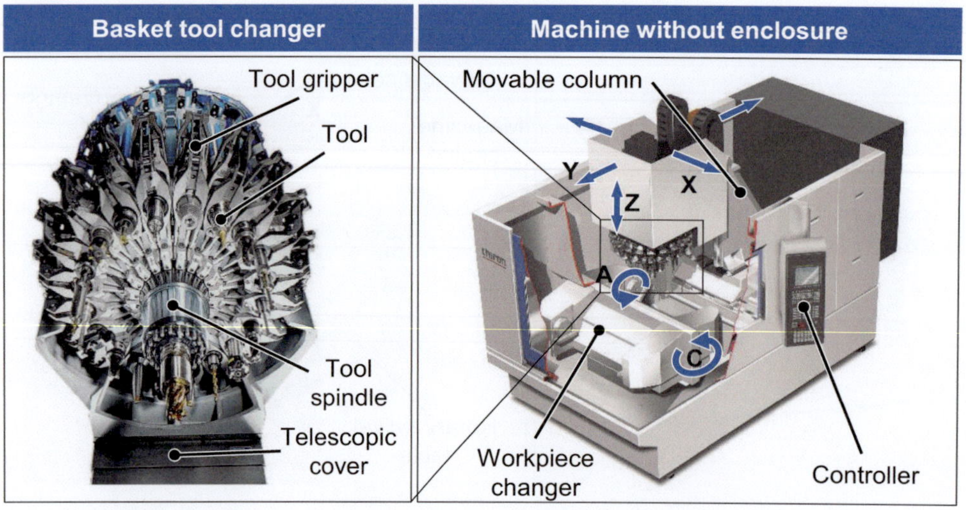

Fig. 4.43 Movable column machining center with basket tool changer. *Source* CHIRON Werke

The movable column machine shown in Fig. 4.44 is a long-bed version that can be configured with a maximum traverse path in the X-direction of 10 m. The 5-axis machining center is mounted on a base frame made of mineral cast. It exhibits vibration-damping and precision mounting properties and allows improved design options. Housed in each of the two tool magazines are 64 tools. Additional magazines can be added to extend this number to 200 tools.

In addition to the machining of long workpieces, operation using the “swing method” is also possible. For this, a partition is mounted in the center of the workpiece table for separating the machining area from the clamping area. This splits the workspace into two separate clamping surfaces. This means loading and unloading can take place during primary machining time. On the one side, a workpiece can be machined, while at the same time on the other table side, the new workpiece is already being clamped. Two independent moving columns make possible 5-axis integrated machining here. Furthermore, the machine can be fitted with an automatically opening roof construction to allow the loading and unloading of heavy workpieces and fixtures using a crane.

4.3.2.2 Knee-and-Column Design Machines

The milling machine shown in Fig. 4.45 with a vertical tool spindle has a knee-and-bed design and two linear axes of motion in the tool and one axis in the workpiece. The workpiece table implemented as a rotary table moves vertically in the Z-axis on the machine bed. The machine bed is an inherently rigid solid block made with reaction resin concrete technology and has high thermal stability and good damping characteristics. The motor spindle unit is in front of the Y-quill and has an additional CNC-controlled swivel axis (B-axis). This means that, in combination with the C-axis of the rotary table, 5-side machining is possible. A tool ring magazine, designed as a pick-up system, complements the flexible machining options. Tools are changed here without the aid of an additional gripper. The tool spindle takes a tool directly from, or places it directly in, the magazine (pick-up technique).

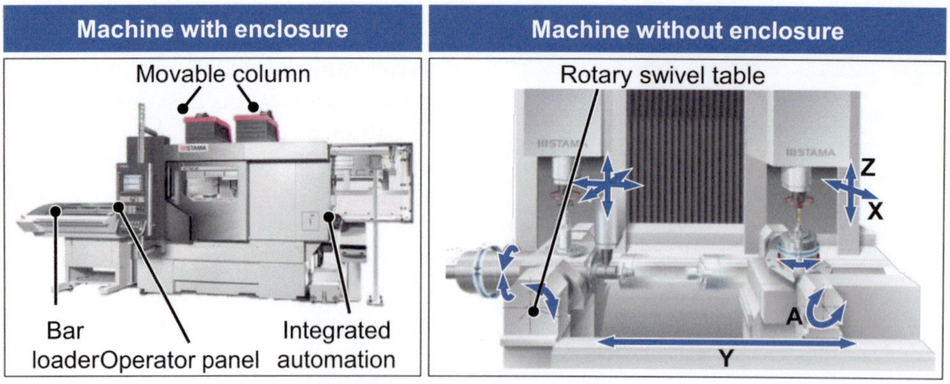

Fig. 4.44 Machining center with moving column design. *Source* STAMA Maschinenfabrik

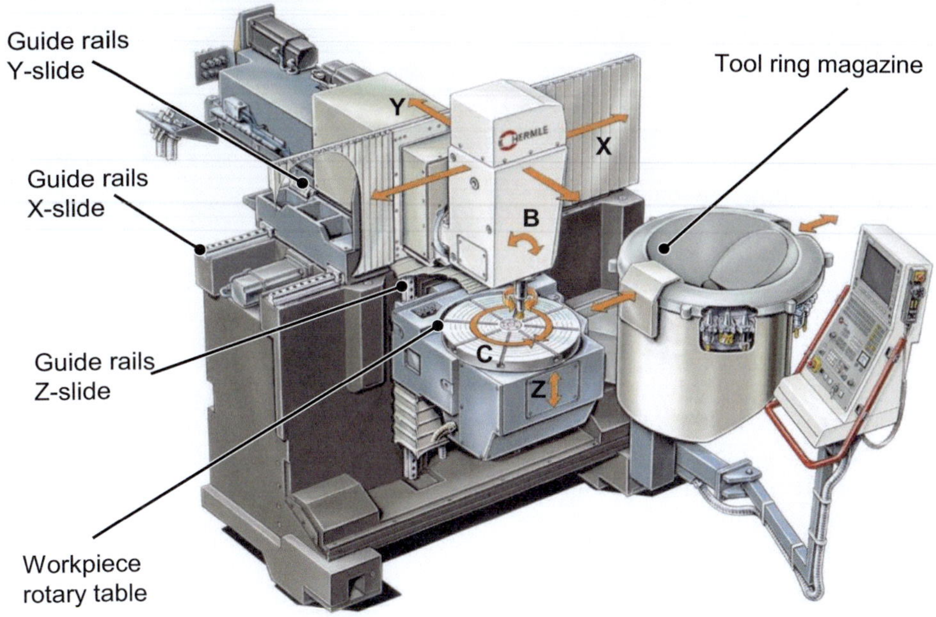

Fig. 4.45 Machining center with knee-and-column design. *Source* Maschinenfabrik Berthold Hermle

4.3.2.3 Portal Milling Machines

Machines of portal design are particularly well suited for the machining of workpieces that are heavy, have large surface areas and require high cutting capacity. Figure 4.46 shows two variants of this type. The table-type design, on the left in Fig. 4.46, is fitted with a table that can be moved on the X-axis. For every point of the workpiece table to be reachable with the tool, the machine bed must be twice as long as the table. All other infeed and feed motions in the direction of the machine coordinate axes are performed by the tool. The crossbeam on which the Z-slide (also called Z-ram), traversable vertically in the Y-direction, is secured can be moved in the vertical direction (W-axis). Its working height is aligned to the height of the workpiece. To keep the extension length of the flexible Z-slide as short as possible, the crossbeam is positioned as low as possible, that is, just above the upper workpiece edge. The working motion in the Z-direction is usually performed by the Z-slide. The milling

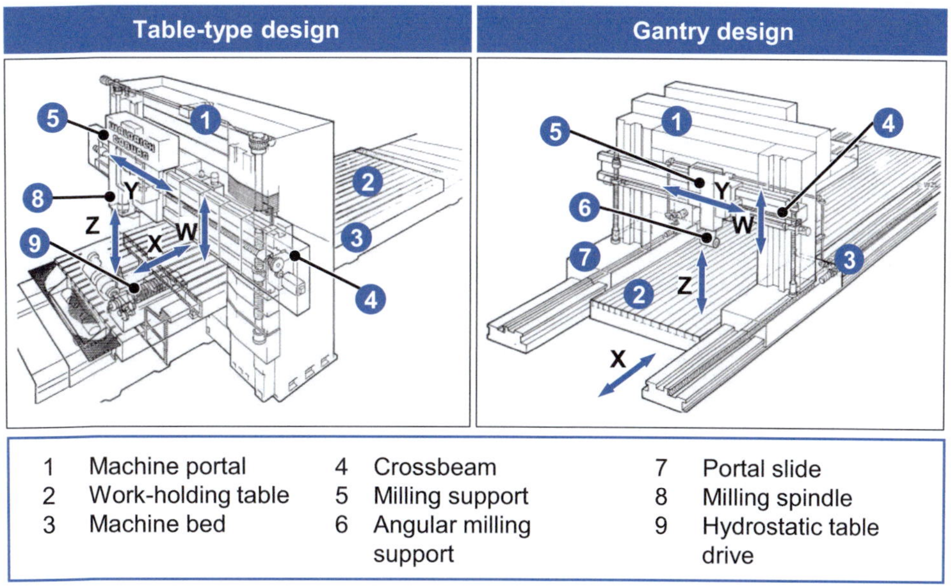

Fig. 4.46 Portal machines with table-type and gantry design. *Source* WALDRICH COBURG

support is integrated on the lower end of this slide. The portal machine with gantry design shown on the right in ◘ Fig. 4.46 has the same design with the crossbeam, milling support and milling spindle. In this type, however, the work-holding table for the workpiece is fixed. For feed motions in the direction of the X-axis, the entire portal is moved. The benefit of the gantry design is that the machine bed need only be as long as the longest workpiece to be machined or the clamping plate. Thus, for the same machining task, a machine with table-type design is built twice as long.

◘ Figure 4.47 shows a portal machine with gantry design for machining large parts. A special feature of this machine is that it has a work-holding table as well as a rotary table (hidden in the picture) that enables turning of rotationally symmetric workpieces. The sliding milling unit has a thermo-symmetric structure and an integrated C-axis. As encapsulating the large portal machines is too complicated, partitions are used to protect operating personnel from flying chips. A head exchange interface enables the milling head exchange.

◘ **Fig. 4.47** Portal machine with gantry design. *Source* Starrag Technology (Droop + Rein division)

One variant of a portal milling machine is the crossbeam machine with gantry design (portal machine with the gantry on top), ◘ Fig. 4.48, where only the crossbeam moves in the X-direction. The crossbeam is mounted on the guideways, located on two stationary, raised sides. The masses moved in the X-direction can be reduced and this improves the dynamic properties of the machine. The X- and Y-movements are each performed by a pretensioned rack and pinion drive. Only the Z-slide (spindle carriage) moves in the Z-direction. A conventional ball screw drive with mass balance is used here. Different machining heads with a maximum of two swivel axes, which can reach rotational speeds of 40,000 rpm, can be adapted for the end of the support. The sturdy sides made of mineral cast provide the machine with excellent damping and high static stiffness. The sides are also used to encapsulate the workspace to protect the immediate vicinity from flying chips. The integrated windows, made of safety glass, enable the process to be viewed by the machine operator.

4.3.2.4 Machines with Modified Gantry Design

Gantry machines are not only used to machine large, heavy workpieces. ◘ Figure 4.49 shows, for example, a compact machine with modified gantry design for

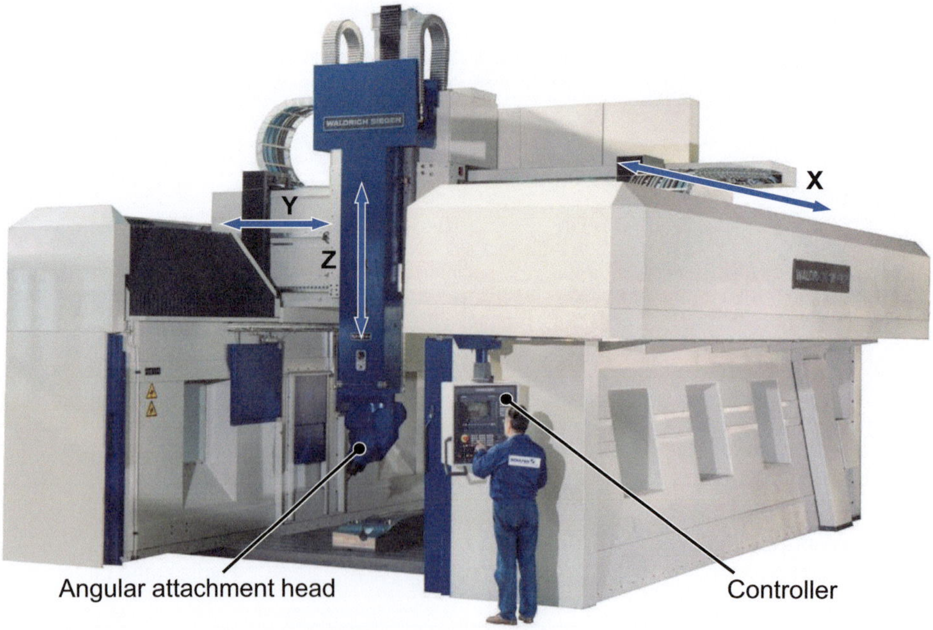

◘ **Fig. 4.48** Portal machine with gantry on top. *Source* Waldrich Siegen

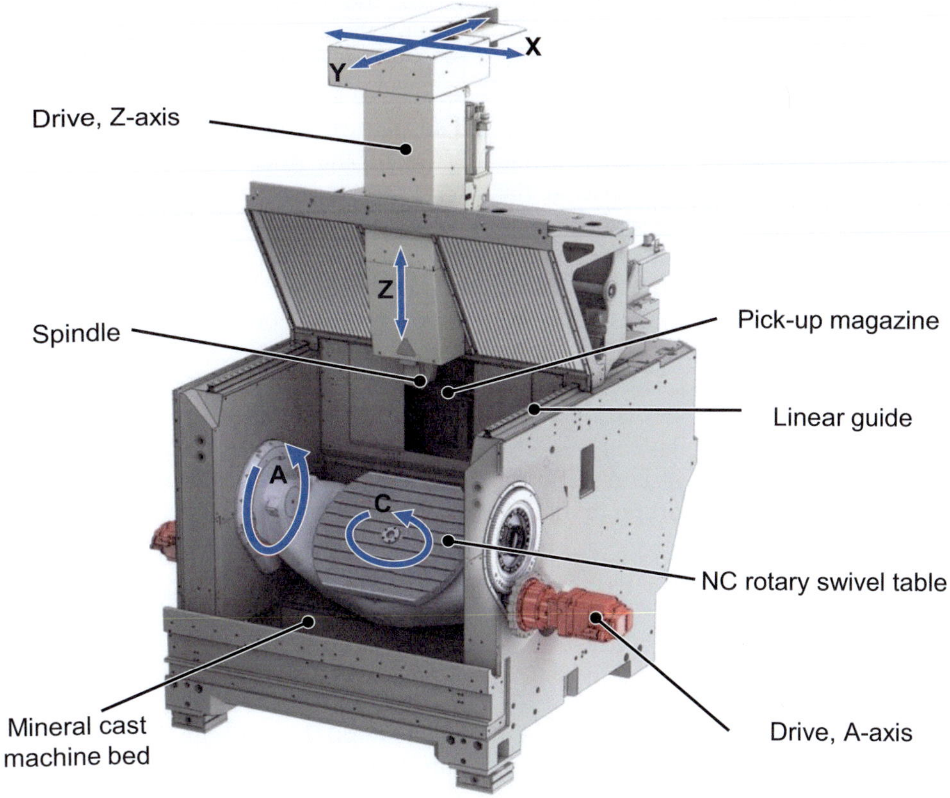

Fig. 4.49 Machining center with modified gantry design. *Source* Maschinenfabrik Berthold Hermle

machining smaller workpieces. The machine bed is a solid block made with mineral casting technology and has high thermal stability and very good damping characteristics. Its semi-open, cuboid-shaped design lends it a very high level of stiffness. In contrast to conventional gantry machines, the X-axis is not perpendicular to the crossbeam but along its longitudinal direction. Also, to attain improved tilting rigidity about the X-axis, the guide rails of the X-slide are not on the Cartesian XZ-plane, but positioned parallel to the Y-axis on the crossbeam with an offset (refer to Fig. 4.50). Linear rolling element guide systems are used as linear guides. The crossbeam, here designed as a Y-slide, and the four carriages are mounted on two guide rails located on the raised sides of the base frame. Movement in the Z-direction is realized by

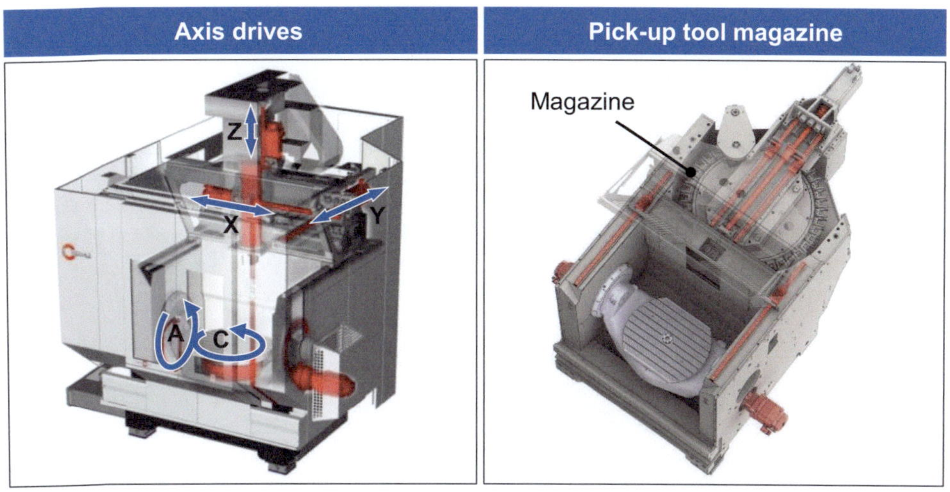

Fig. 4.50 Axis drives and tool magazine of a machining center with modified gantry design. *Source* Maschinenfabrik Berthold Hermle

vertical traversing of the workpiece table. The table is guided at the machine sides. The design of the bed permits different workpiece clamps to be used. In addition to the simple machine table shown in ◘ Fig. 4.49, an NC clamping cube can also be used enabling continuously variable rotation of the workpiece about the B-axis. Up to four workpieces can be clamped at the same time, corresponding to its four cube surfaces. The machine has two rotational axes in the workpiece in addition to the translational axes. Travel speeds to 60 m/min can be achieved on all three linear axes. The NC rotary swivel table shown here can be designed with either a worm drive or a torque drive. The maximum load is 850 or 1400 kg depending on the variant. The rotational speeds of the A-axis and C-axis can reach 55 and 65 rpm, respectively. A third table variant is a fixed table with a payload of 2000 kg and workpiece dimensions of 1050 mm × 805 mm.

The machine can be fitted with two spindle variants for different machining tasks: a 2-part spindle or a compact spindle, each with five speed ranges and an HSK-A63 interface. The maximum speed is 25,000 rpm at a power of 29 kW and torque of 100 Nm.

The tool ring magazine for 42 tools, implemented as a pick-up system, adds to the flexible machining options (see ◘ Fig. 4.50). Tools are changed here without the aid of an additional gripper. The tool spindle takes a tool directly from or places it directly in the magazine (pick-up technique). The tool magazine can be extended to 160 tool slots using additional magazines. For crane loading, the spindle moves into the magazine, just like when a tool is changed, to maximize the space available for part handling.

In the conventional gantry design, the bridges or crossbeams are set on the sides of the machine [STEI01]. This means the sides bend when forces, such as cutting and acceleration forces, are applied to them. The deformation is dependent on the physical properties of the sides (area moments of inertia, Young's modulus and cantilever length). The linear guide elements, arranged in one plane, act like swivel joints. Due to their relatively poor torque resistance, they cannot markedly reduce bending of the sides and so geometry errors occur on the workpiece. The new gantry concept shown in ◘ Fig. 4.51 was developed to remedy this situation and to increase static and dynamic stiffness. Here, the guide elements are not on the sides but inside (◘ Fig. 4.52). Furthermore, the guide elements are not positioned in one plane, but two planes, resulting in tensioning of the cross member. The same concept led to the arrangement of guide elements for the carrier of the Z-slide. It is in the center of the crossbeam. As a result, forces in the Z-direction generate no torsional torque around the longitudinal crossbeam axis. The special guide arrangement means two opposing bending loads are introduced into the sides of the frame when lateral forces occur. This doubles the stiffness and halves the displacement Y. The machine shown is suitable for rapid motion rates of 60 m/s at an acceleration of 1 g, and can be operated with motor spindle speeds up to 24,000 rpm.

4.3.2.5 Special Machining Center Designs

◘ Figure 4.53 shows a special design of a machining center. It is a machine for the 5-axis integrated machining of bar-shaped materials combining the milling,

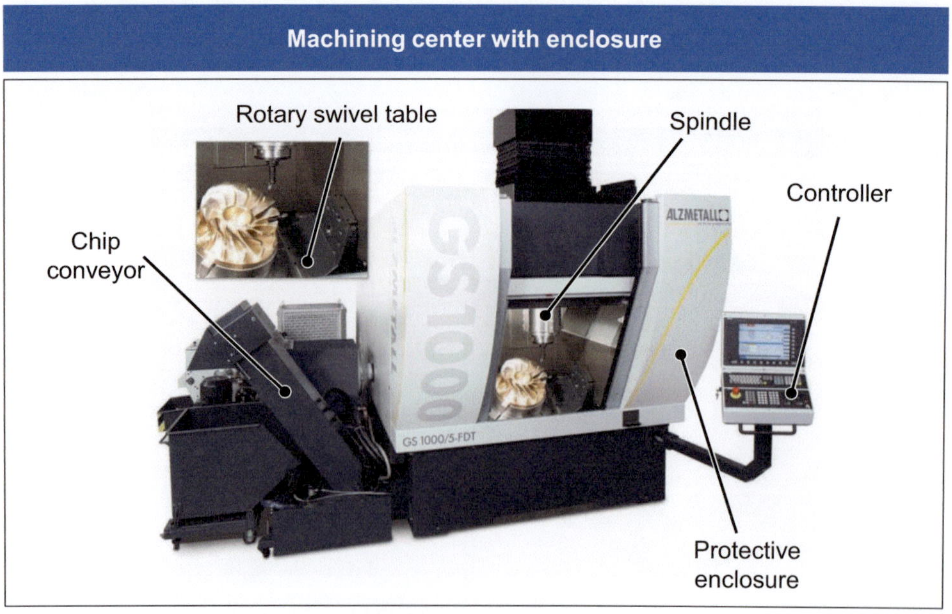

◘ **Fig. 4.51** Machining center with gantry design and coordinate carriage. *Source* ALZMETALL

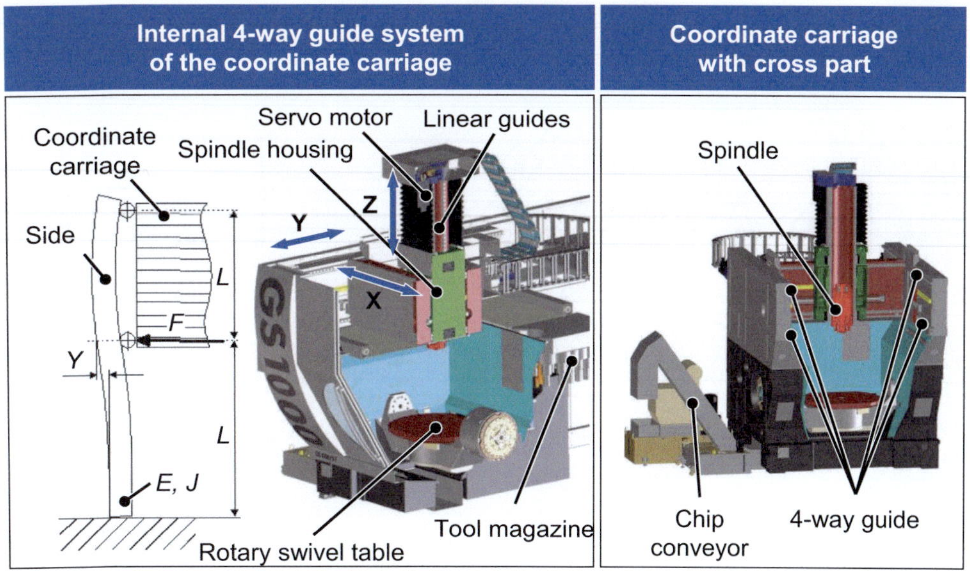

Fig. 4.52 Machine frame with gantry design with coordinate carriage. *Source* ALZMETALL

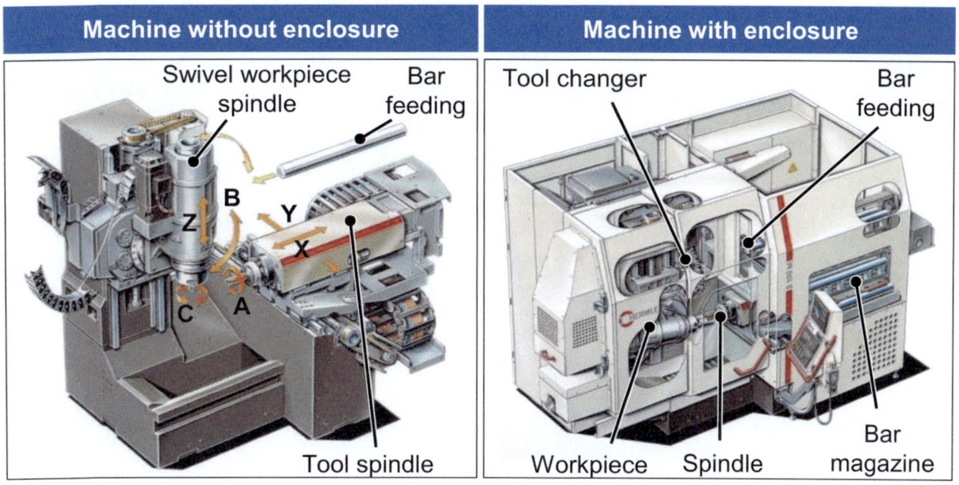

Fig. 4.53 Bar machining center. *Source* Maschinenfabrik Berthold Hermle

drilling, threading and turning manufacturing technologies. The tool and workpiece rotation axes are aligned at 90° to each other. On the tool side, a cross carriage provides the horizontal movements in the X and Z-directions. On the workpiece side, infeed is in the vertical Y-direction. The workpiece spindle can also be swiveled about the X-axis. Workpiece spindles are loaded by a 7-way loading magazine (for round and profiled material bars with lengths to 1.1 m) that is integrated into the machine. For this, the workpiece spindle is swiveled to the horizontal position and traversed to the relevant transfer position. The workpieces are clamped by an automated three-jaw chuck that can be operated at a speed of 30 rpm at a maximum nominal torque of 510 Nm. The maximum rotational speed of the tool spindle is 16,000 rpm at a maximum torque of 139 Nm.

4.3.3 Machines with Parallel Kinematic Structures

Another approach in generating relative motions between tool and workpiece are machine tools with parallel kinematic structures. They have been in use since the mid-1990s (see Fig. 4.54). However, most of the machines that have proven their worth are hybrid solutions, that is, only some of the feed axes are parallel axes. The remaining axes are serial.

Figure 4.55 shows a concept for a purely parallel kinematic design for movement. For these 3-axis parallel rod kinematics, the spindle platform is held by six rod couplers with fixed lengths. The arrangement of two each of these couplers on the three adjuster carriages that move in the Z-direction defines the three

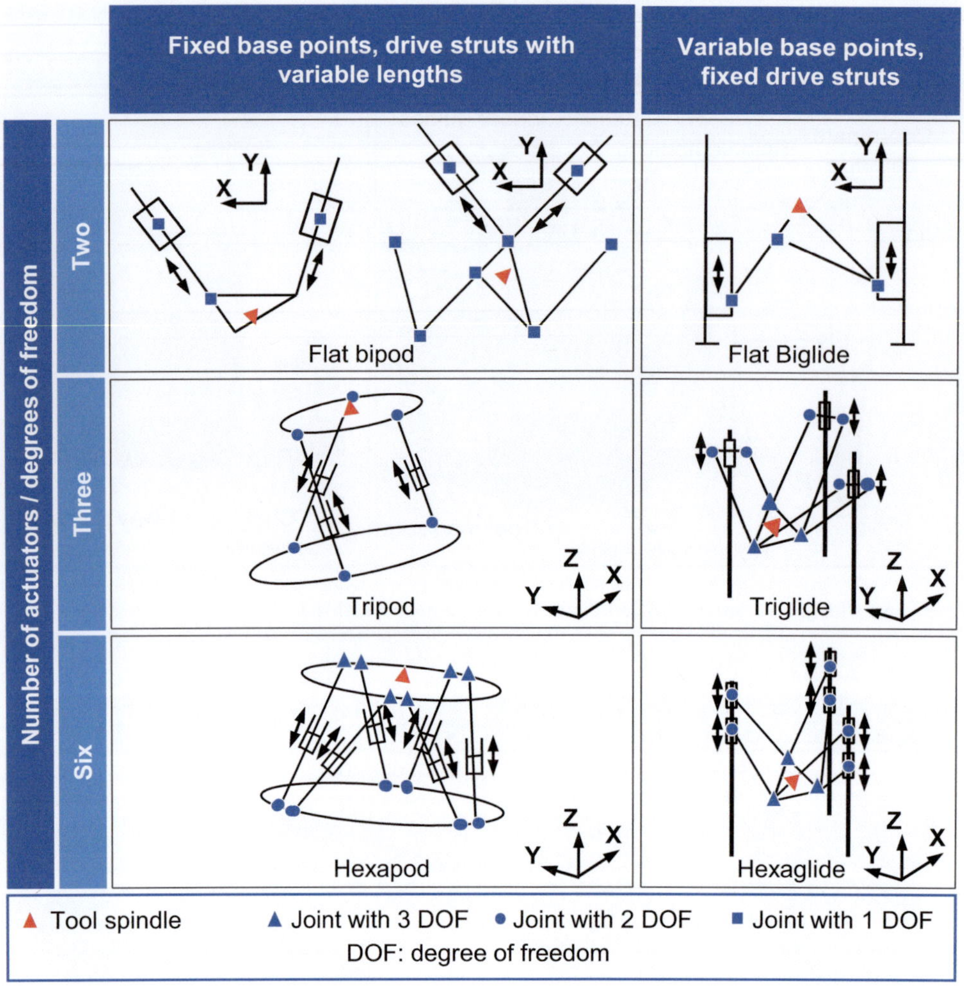

Fig. 4.54 Abstract designs of parallel kinematic milling machines

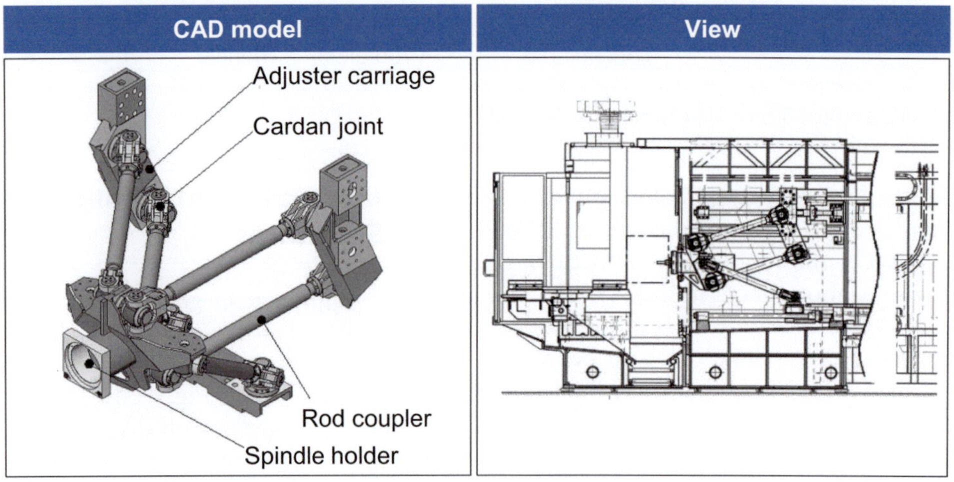

Fig. 4.55 Machining center with parallel kinematics. *Source* Krauseco

rotational degrees of freedom of the spindle platform. As a result, only the three translational movements in coordinate directions X, Y and Z are possible. Another approach for parallel kinematic generation of relative motion between the workpiece and tool are machine tools working to the hexapod principle. These machines are based on the Stewart platform described by Stewart in 1956 and named after him. The platform

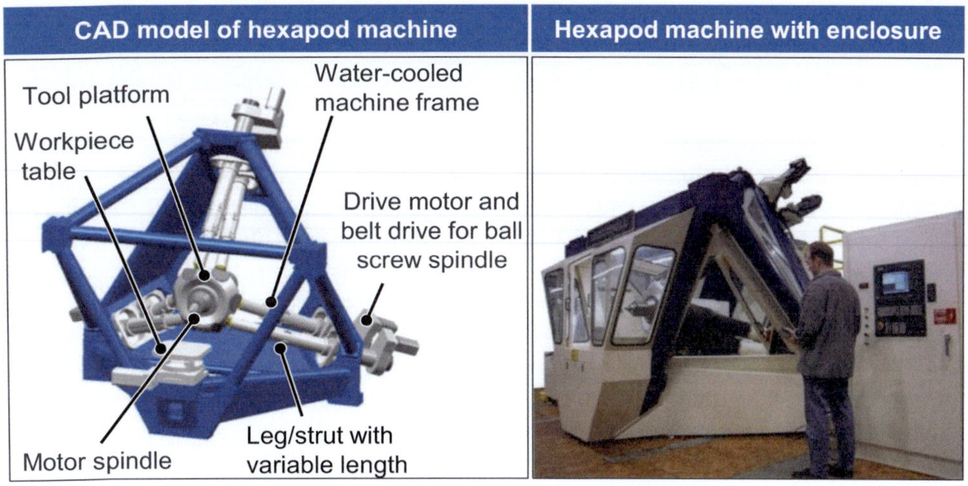

Fig. 4.56 Machining center with hexapod design and horizontal tool spindle. *Source* Ingersoll

is carried by six struts with variable lengths. When the strut lengths change, the position and orientation of the platform (6 degrees of freedom) can be changed.

Figure 4.56 shows an example of such a hexapod machine with horizontal tool spindle. The kinematic machine structure is made up of six struts that are crossed in pairs. They are secured with cardan joints to the tubular frame of the machine and with ball joints to the moving platform that carries the milling spindle with drive motor. The lengths of the telescopic struts are changed by motor-driven, integrated ball screws. The horizontal table is fixed and is used as the workpiece clamp. The tubular half-timbered structure of the machine frame, as triangles, assures a high level of machine stiffness. The joints of the struts are integrated into the lattice nodes. This way, the struts and the lattice supports are only subjected to tension/compression. As there is virtually no bending load, the machine has excellent static and dynamic compliance behavior. To prevent thermo-elastic expansions of the machine structure, temperature-controlled (cooled) water can flow through the tubular supports such that the machine exhibits good thermo-elastic accuracy behavior.

In the field of feed drives, hexapod machines are made of identical parts, simplifying machine manufacturing and assembly. No specific adjustment or alignment work is necessary when the machines are assembled. The positions of the pivot points, which must be known for precise axis control, are determined by a calibration process after assembly. For the linear movements in the direction of the Cartesian machine coordinate axes, all six struts must be activated every time. A model stored in the controller transforms the coordinates from the global Cartesian axes to the machine axes (strut lengths). The 5-axis capability of hexapod machines is restricted by the limited swivel angle of the platforms. Thus for 5-side machining, an additional rotation and swivel axis is normally used on the workpiece side. Lastly, the specific benefits and downsides of hexapod machines are listed in Fig. 4.57.

Benefits	Downsides
✓ Inherently rigid structure ✓ Low moved masses ✓ All drives (struts) identical; repeat parts ✓ Struts only subjected to tension/compression; prevention of bending components ✓ Large dynamic range	✗ Great control effort because coordinate transformations must be performed for all six axes ✗ Complicated leg bearing system and costly position measuring system to guarantee stiffness and accuracy ✗ Long reaches of thermal effects ✗ Unfavorable ratio of workspace to machine volume ✗ Limited swivel angle for 5-axis machining

Fig. 4.57 Benefits and downsides of the hexapod structure

The many drawbacks to the hexapod structure, in particular the intricate cooling and relatively large space requirement compared to the workspace, have resulted in a very low level of acceptance of this machine concept in industry. Hybrid kinematics were developed to make use of the benefits of this machine concept nevertheless, and thus to increase the level of acceptance. Introduced briefly below are the hybrid kinematics, the high-speed machining center with V-shaped parallel kinematics, the bipod machine concept with linkage, the bipod machine concept with linear direct drives, and the milling center with 3-axis attachment milling head.

Figure 4.58 shows a compact high-speed machining center that performs movement on the XY-plane using V-shaped parallel kinematics (PKM), a flat bipod structure, with linear direct drives. Every point on the XY-working plane can be reached using a corresponding extension length for the two actuators. Movement in the Z-direction on the other hand is conventional using a traversing spindle housing hinged to the front spindle platform. The machine concept shown provides a marked reduction in the moved masses yet at the same time increases stiffness. This allows a rapid motion rate of 120 m/min and axis acceleration to 3 g with a chip-to-chip time of 1.2 s. Both the quickly accelerating and decelerating motor spindle with up to 40,000 rpm and the pick-up tool changer with a change time of 0.5 s ensure a very short chip-to-chip time.

Another flat bipod machine concept is shown in Fig. 4.59. This machine comprises a Z-quill that can traverse on the XY-plane with a motor spindle. Movement of the spindle in the XY-direction and its stiffness perpendicular to it (the Z-axis) are ensured by a

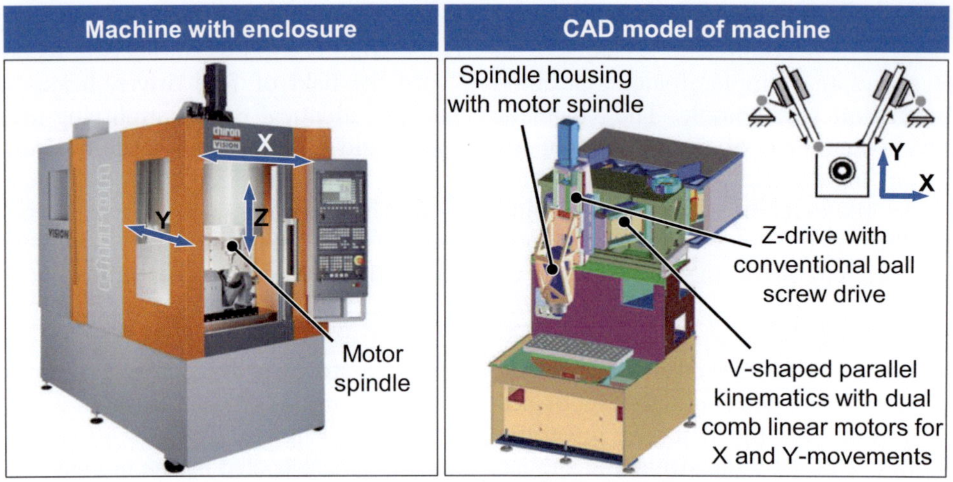

Fig. 4.58 High-speed machining center with V-shaped parallel kinematics. *Source* CHIRON Werke

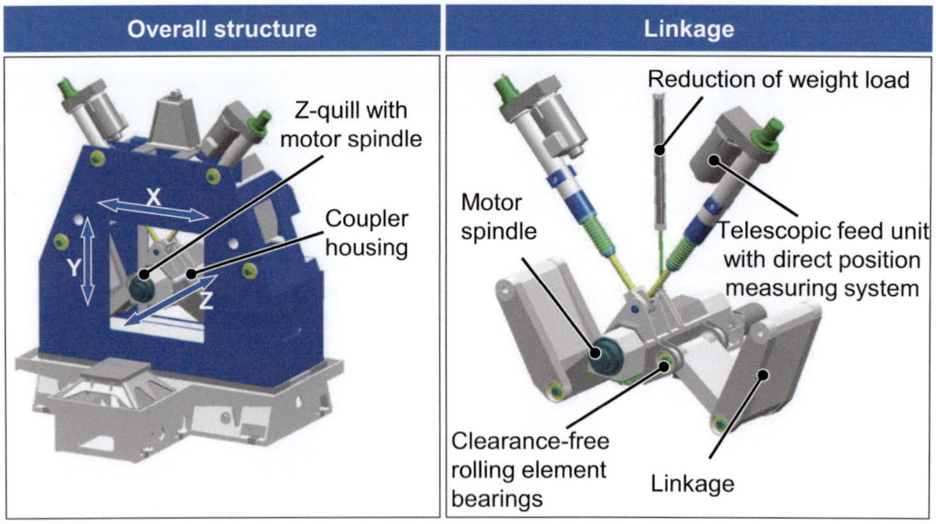

Fig. 4.59 Bipod machine concept with linkage

W-shaped linkage. In the joints of the linkage, rolling element bearings adjusted for zero-backlash ensure a stiff connection with low friction. Two telescopic ball screw spindles drive the linkage. The telescopic spindles are coupled on the one side to the tool spindle, and on the other in a guide frame suspended so it can swivel. Telescopic spindles with a total length of 1450 mm are required for travel distances of 500 mm in the X and Y-directions. The motor spindle implemented as a quill provides the infeed in the Z-direction also on the tool side. Consequently, this machine concept can be used to realize a standalone machine as well as a machining module inside transfer lines.

A variant with spindle unit fixed in the Z-direction is also possible as an alternative, where the infeed on this axis is by a linear motion of the workpiece pallet. ◘ Figure 4.60 shows a machine with parallel kinematics for the XY-plane. The scissor-like coupling kinematics of this machine are very similar to the kinematics in ◘ Fig. 4.59. Movement on the XY-plane comes from two parallel and vertically acting carriages that are driven by linear motors. The base points of the couplers are on the carriages guided at the frame-shaped machine column. These structural features and the integration of linear direct drives increase the stiffness and dynamics of the machine. The maximum travel speed is 120 m/min with an acceleration of 1.5 g.

◘ Figure 4.61 shows another application example of the use of parallel kinematic machine concepts: a high-speed machining center for the 5-axis

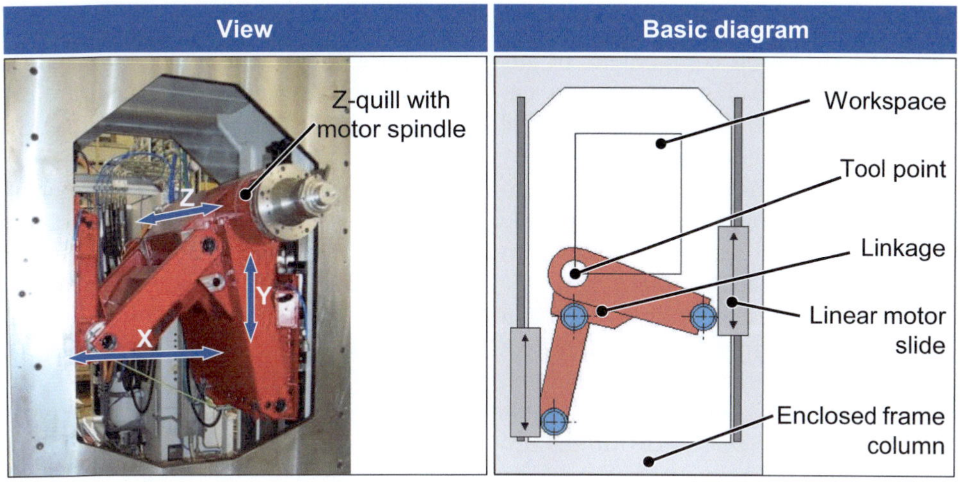

◘ **Fig. 4.60** Bipod machine concept with linear direct drives. *Source* MAG IAS

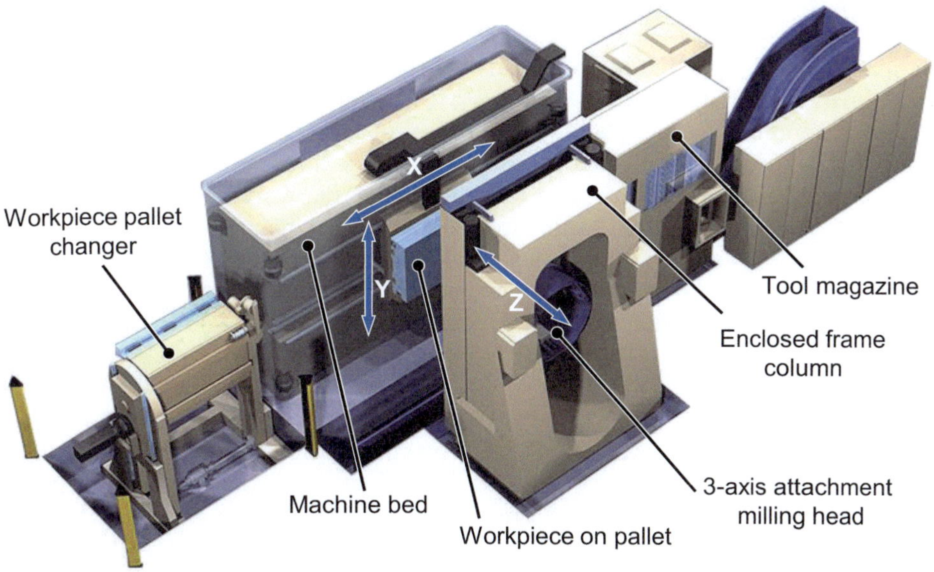

◘ **Fig. 4.61** Milling center with 3-axis attachment milling head. *Source* Starrag Technology (Scharmann division)

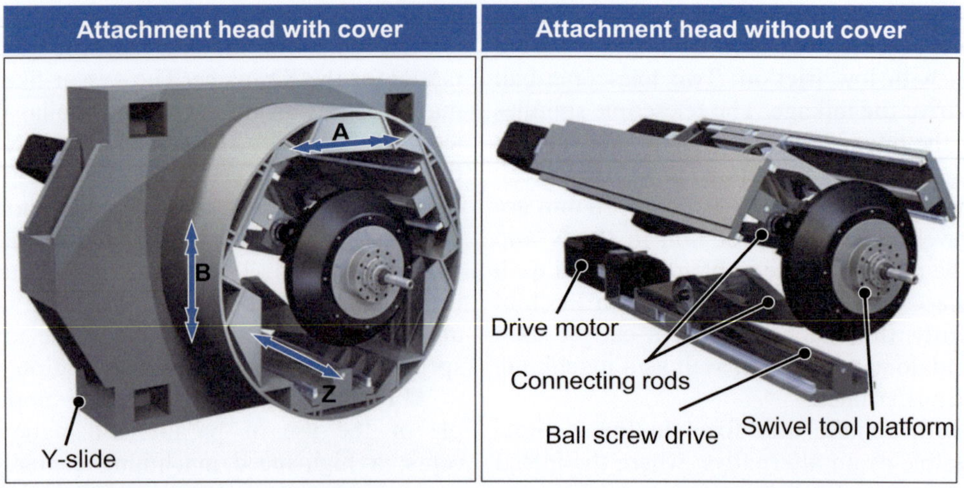

Fig. 4.62 Parallel kinematic 3-axis attachment milling head. *Source* Starrag Technology (Scharmann division)

machining of mid-sized structural components in aircraft construction. The machine has a stationary, enclosed frame column. This column guides a parallel kinematic 3-axis attachment milling head, moved by two ball screw drives, in the Y-direction. The machining head carries, guides and moves the motor spindle mounted on a platform, using three rod couplers with fixed lengths (3-rod kinematics), Fig. 4.62.

The three base points of these couplers, which can move in the Z-direction, are traversed by carriages driven with ball screws. These enable translational (Z-axis) or swivel (A/B-axis) motions of the spindle platform depending on the travel distances of the appropriate coupler base points. The swivel range of the platform for the simultaneous A and B-axes is $\pm 40°$ for a maximum axis velocity of 80°/s and axis acceleration of $685°/s^2$. Movement in the X-direction is realized via an electrically preloaded dual pinion drive with two feed motors, on the workpiece side, by means of the traversable machine table. For the translational movements on all three coordinate axes, the maximum axis velocity is 50 m/min at an acceleration of 1 g. On the pivoting dual pallet station located on the side of the machine, a pallet measuring 3 m × 2 m can be loaded horizontally. It is then transferred in the upright position to the machine table. This means set-up can happen during primary machining time. More than 100 machines with the parallel kinematic 3-axis attachment head are now in industrial use.

4.4 Lathes

Lathes are machine tools used mainly to produce rotationally symmetric workpieces. The cutting motion is generated predominantly by rotating the workpiece. Many lathes nowadays are fitted with additional functions to provide milling and drilling operations parallel to, perpendicular to and even at any angles to the workpiece axis, as well as back side machining. Tool turrets and tool magazines with tool changers allow automated use of a large number of different tools one after another. The plane of the main feed motions (XZ-plane) is perpendicular to the cutting speed vector at the contact point of the turning tool. In this situation, the Z-feed axis is by definition in the direction of the rotation axis of the workpiece.

As Fig. 4.63 shows, the following turning operations are differentiated in accordance with the feed motion: longitudinal turning (axial feed), face turning (radial feed) and contour turning (simultaneous axial and radial feed). The numerically controlled synchronization of the X and Z-axes with the rotational motion of the main drive spindle (C-axis) can create, in combination with driven tools, complex workpiece geometries that previously could only be generated with milling machines.

The most important structural components and designs of lathes are summarized in Fig. 4.64.

Lathes differ mainly in terms of the spatial configuration of the workpiece axis (horizontal/vertical). The design of the machine bed and the position of the spindle are important classification criteria here. Horizontal flat-bed lathes are always used wherever great machine stiffness is required. The slant-bed design enables speedy removal of the hot chips and cooling lubricant. This greatly reduces the risks of thermal deformation of the machine bed compared to other designs. Long workpieces are supported by a tailstock with a center. Front-bed lathes are particularly well-suited for machining chuck parts with an automated workpiece change system. A design with a vertically positioned and suspended spindle offers the benefits of favorable workpiece change (the chuck can grip the workpiece

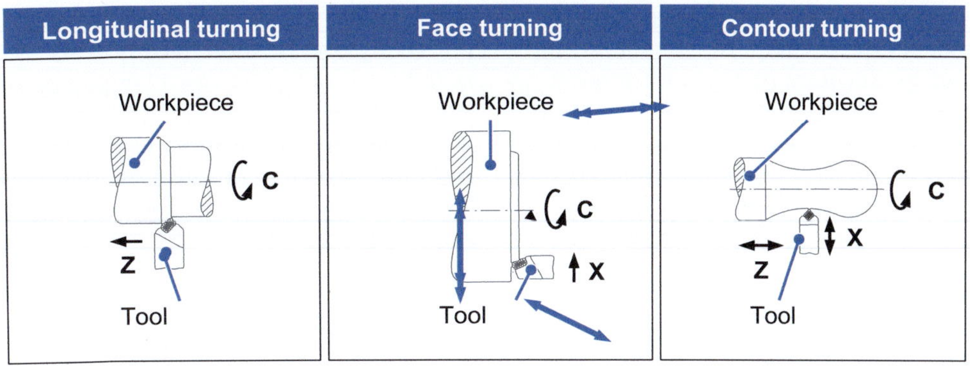

Fig. 4.63 Turning operation types

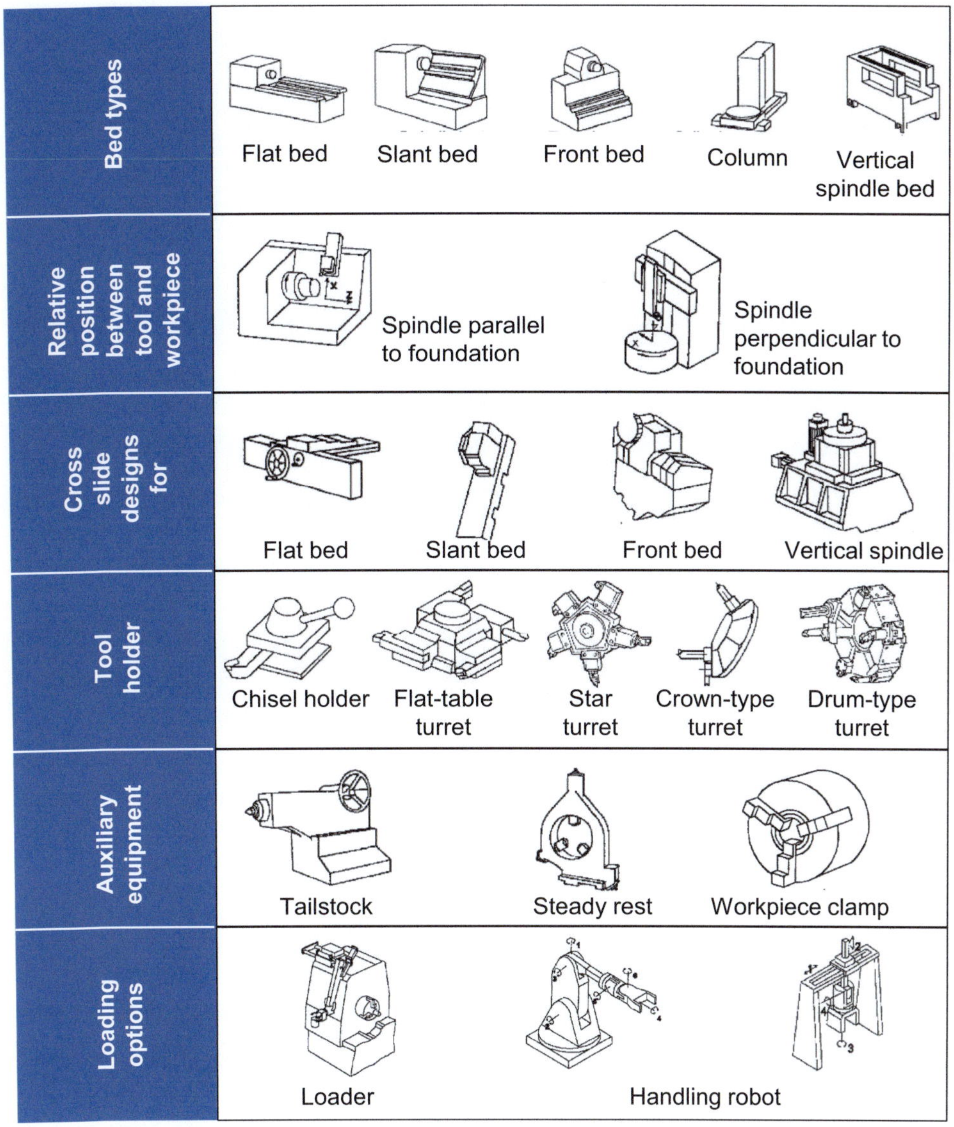

Fig. 4.64 Components and designs of lathes

directly from a pallet and place it back after the machining process) and favorable chip removal. Vertical lathes with column design can accommodate large-diameter, thin-walled components on their rotating workpiece table (face plate). Inherent workpiece deformation due to gravity is eliminated to a large extent.

Different variants are also available for the structure of tool holders (see ► Chap. 10). In the arrangement of tools and specification of the work process, attention must be paid to ensuring there are no collisions between the workpiece and tools (this is dependent on the machining task). Two workpiece clamping types are generally possible for lathes. Long workpieces are normally held by two lathe centers (center lathe). When the slenderness ratio is high, the turned part is also supported by one or more steady rests to prevent deflection. Short workpieces are clamped on one side into the chuck (chuck lathe) or, when the parts are not rotationally symmetric, on a face plate (facing lathe). For the bed types shown in ◘ Fig. 4.64, the structure of typical lathes and the names of key components are summarized in ◘ Fig. 4.65.

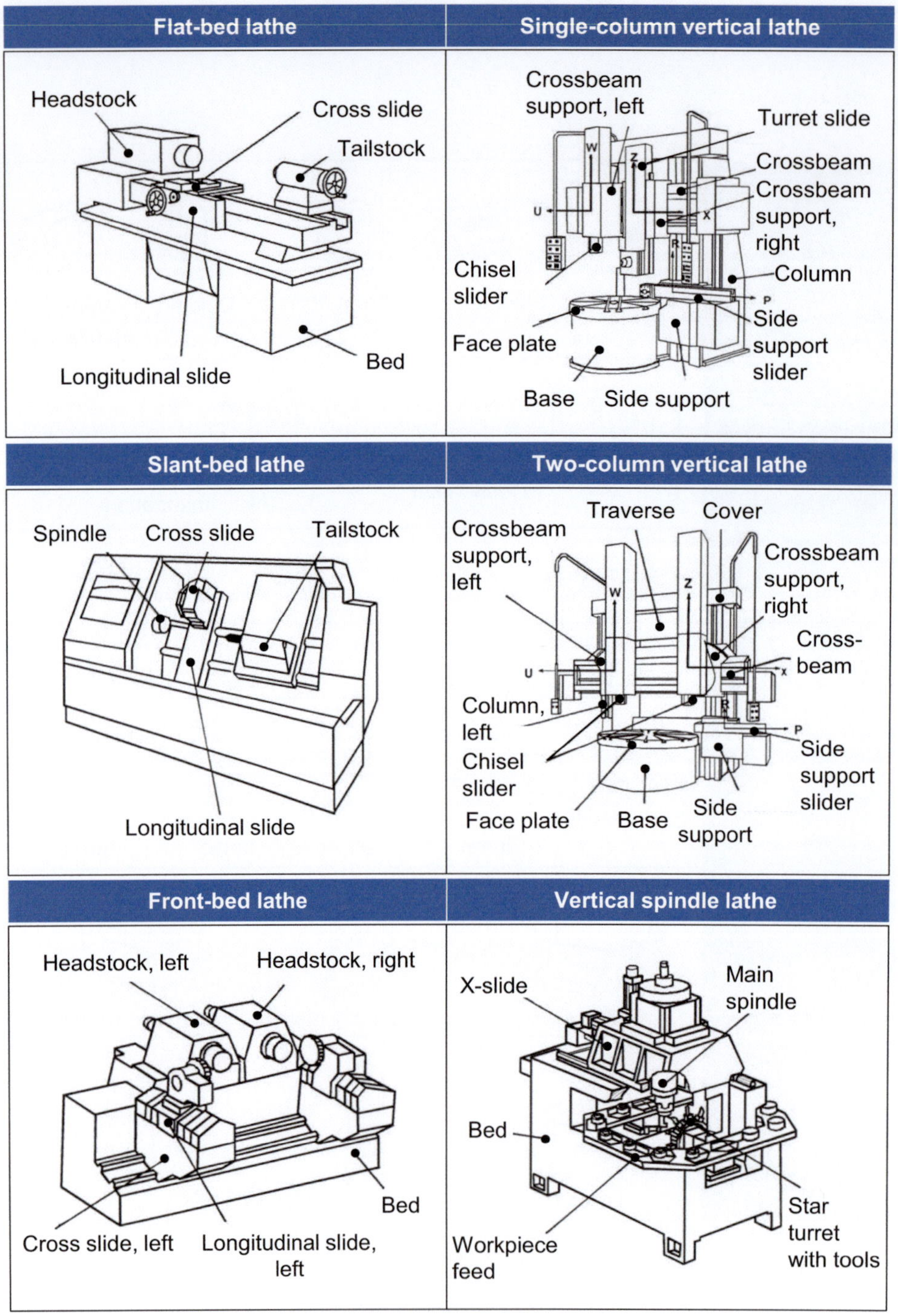

◘ **Fig. 4.65** Classification of lathe designs

4.4.1 Universal or Workshop Lathes

Figure 4.66 shows a hand-operated universal lathe with automatic cycle control for performing lathe work, such as outside turning and internal turning, contour turning and axial drilling and thread cutting work.

This machine offers all the functions of a workshop lathe. In contrast to conventional machines, however, handwheel movements are transferred electronically to the feed drives. This enables automatic functions such as joystick control, automatic cycle control and numerical control (NC) to be added to the lathe. The synchronization of workpiece speed and feed for thread cutting, performed by lead screw and clasp nut for previous mechanically controlled lathes, is also handled here by a numerical controller.

The machine has a flat-bed design. The guideways for longitudinal slide and tailstock are a combination of dual prism and flat guide systems. Large guiding surfaces with low surface pressure guarantee high guide system reliability and low wear, even when there is a large projection outside the center of rotation (for example, for the turning of larger workpieces). For movement of the cross slide, dovetail guide and flat guide systems are used. These systems also prevent the carriage from lifting when high torques generate a load.

4.4.2 Automatic Lathes

Tailored to the requirements of small volume production, numerically controlled automatic lathe with high production rates are in use today. All functions such as cutting speed, feed rate, feed path and tool change being controlled by numerical information is characteristic for this machine type. Workpiece supply and removal can also be automated using handling devices (see ▶ Chap. 9).

Figure 4.67 shows the components of a numerically controlled automatic lathe for chuck and tip work. Its modular design allows the machine to be configured in line with each customer's needs. A tailstock can be used for tip machining and a counter spindle for back side machining. This machine can also be fitted with a swiveling tool holder and milling unit in addition to a conventional tool turret. Another turret can be added for machining the workpiece in a counter spindle or for machining with multiple tools in use simultaneously. Steady rests prevent workpieces from deflecting and provide support for longer workpieces. In the arrangement of tools and specification of the work process, attention must be paid to ensuring there are no collisions between workpiece and tools.

While for chuck lathes the workpieces are automatically or manually taken from a magazine and placed into and removed from the chuck, for automatic bar

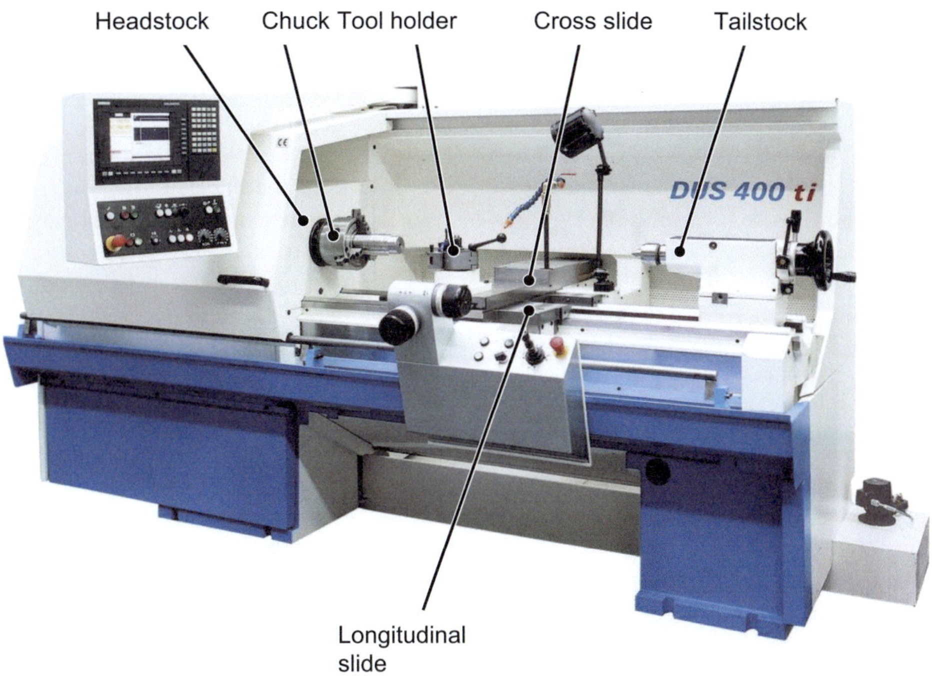

Fig. 4.66 Flat-bed lathe, according to Boehringer

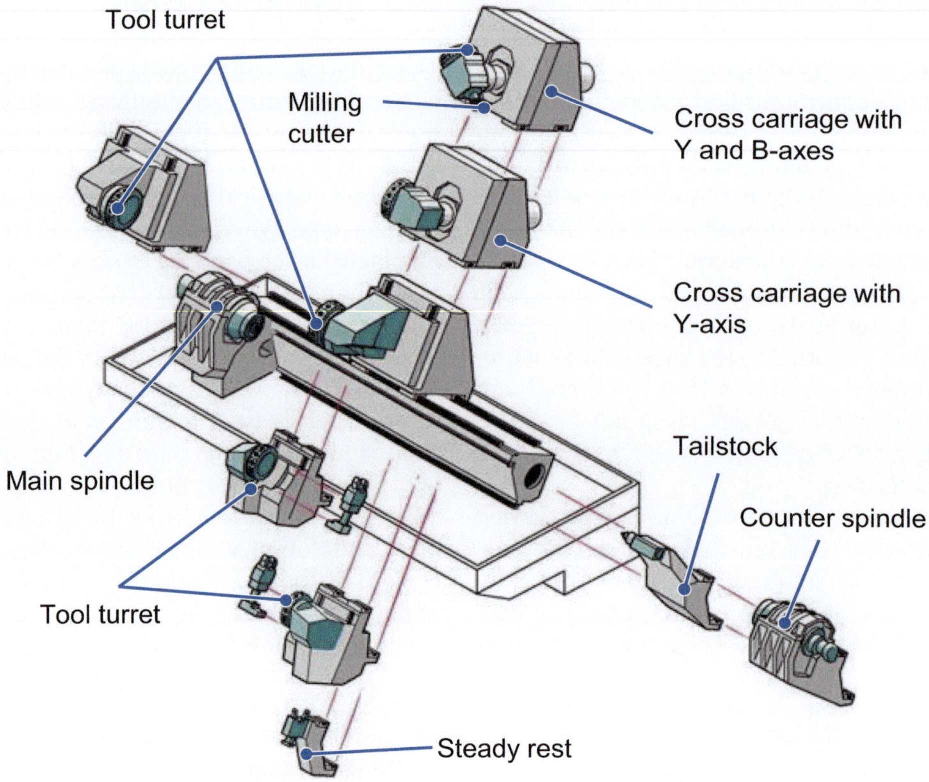

Fig. 4.67 Components of a NC slant-bed lathe, according to Index

lathes, the bar stock is supplied by the spindle. The parting process here is the last machining process on the main spindle. The workpiece is separated from the bar. Figure 4.68 shows a bar turning center. Given the appropriate configuration of such machines with a counter spindle and additional axes (such as with Y-axis and B-axis as shown in Fig. 4.67), as well as driven tools, a number of different drilling and milling operations can be performed beyond the usual turning operations.

After the automatic transfer of the workpiece to the counter spindle, back side machining can be performed. This way, all sides of a component can be machined and one machine can manufacture it automatically without manual re-clamping.

The machine shown in Fig. 4.68 has two independently controllable tool turrets with the upper one also carrying a counter spindle (Fig. 4.69). So with the relevant configuration of tools, several work steps can be performed at the same time. In Fig. 4.69 for example, a contour turning and drilling process at the front of the workpiece in the main spindle and a drilling process at the back of the workpiece in the counter spindle are performed simultaneously.

The bar stock is pushed through the main spindle into the workspace by a hydraulically operated feeder on the bar magazine in accordance with the workpiece length. Figure 4.70 to 4.72 show the diversity of machining options on appropriately equipped automatic

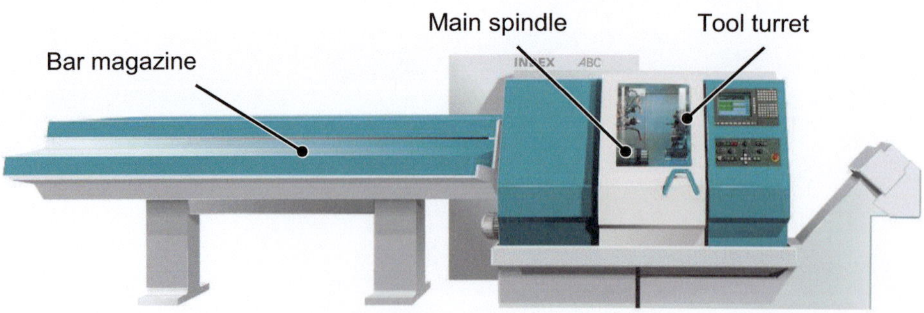

Fig. 4.68 Lathe with bar magazine, according to Index

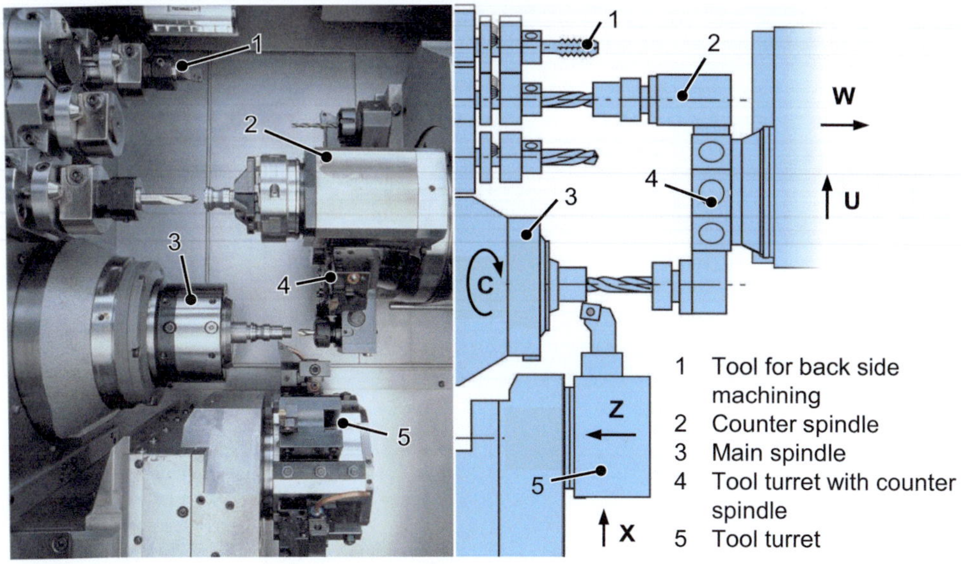

Fig. 4.69 Workspace of the lathe in Fig. 4.68, according to Index

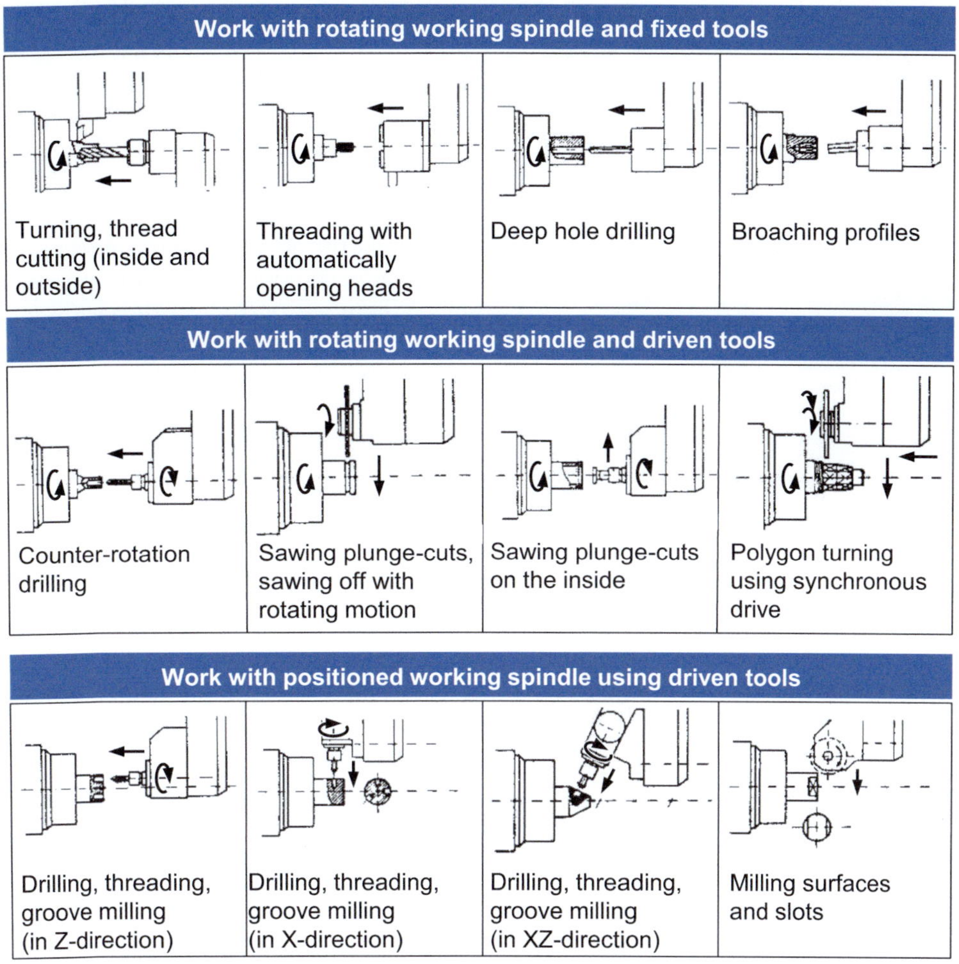

Fig. 4.70 Machining capabilities, Part 1, according to Traub

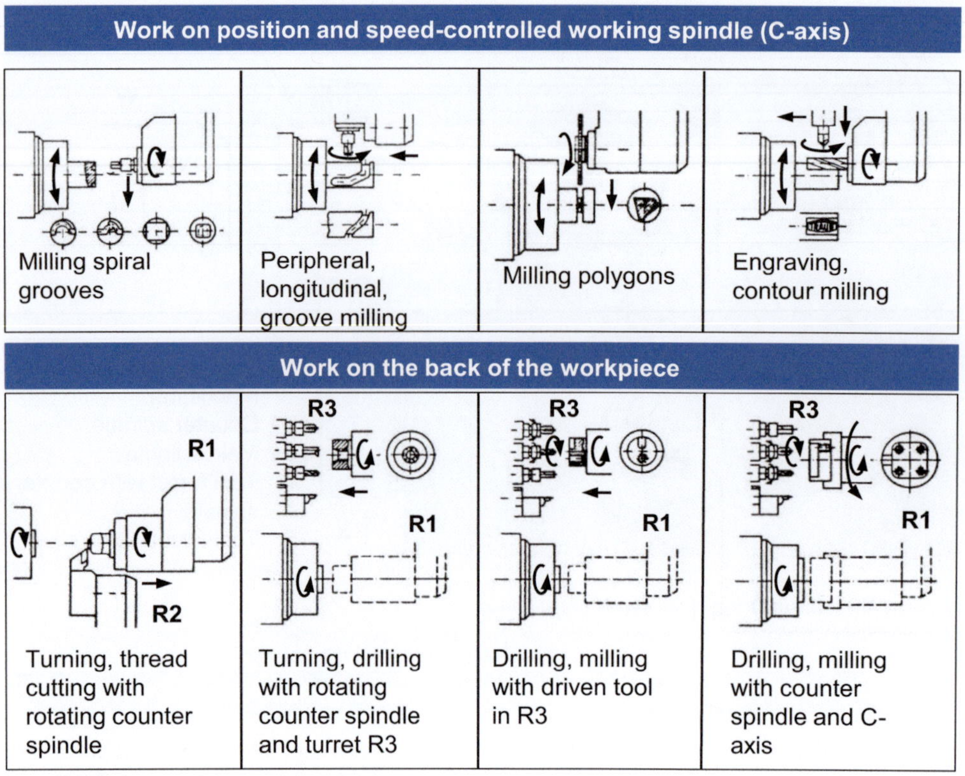

Fig. 4.71 Machining capabilities, Part 2, according to Traub

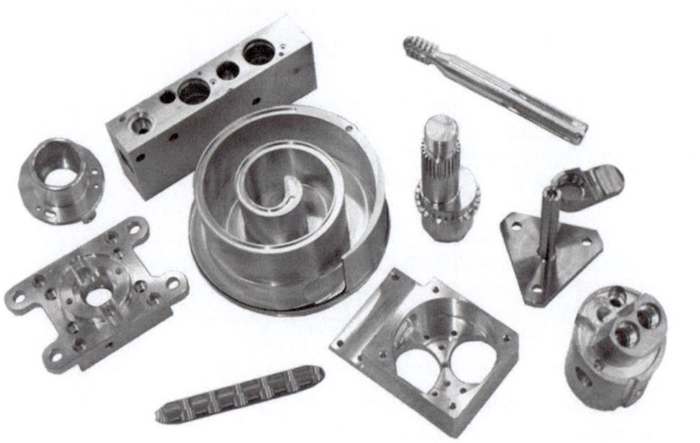

Fig. 4.72 Workpiece spectrum for turning, according to Index

chuck and bar lathes as well as an overview of the workpiece spectrum possible.

4.4.3 Sliding Headstock Lathe

Automatic lathes machine workpieces from bar materials with small and medium diameter sizes. Often, thin workpieces are produced which deflect due to their low stiffness and the forces acting during the machining process. This may lead to dimensional errors and shape defects that are outside the tolerances permitted. Long-part turning represents a suitable process variant where the machining forces, due to the kinematics, always act directly at the chuck or guide bush and thus only cause small deformations at the cutting point. Typical workpieces here are parts from precision engineering (such as the watch/clock industry) and toolbuilding. There are essentially two different design principles for sliding headstock lathes. In the first variant, the workpiece performs the Z-movement using the spindle. A fixed guide bush (slide or rolling element guide) is used to keep the workpiece in the radial direction. The rotating workpiece bar is pushed through the guide bush from behind by the traversing workpiece headstock in the axial direction. The tools are positioned directly in front of the fixed guide bush and are only fed in the X-direction (Swiss principle). In the second variant, the guide bush can also be moved together with the tools relative to the fixed headstock (Offenbach principle).

Figure 4.73 shows the workspace of the Swiss-type lathe and the arrangement of rotation and feed axes. The spindle moving in the Z-direction provides the rotational drive of the workpiece and the feed. The radial infeed motions are implemented by the tool slides. Machining workpieces on the cut-off side is possible using the counter spindle integrated in the front unit.

Figure 4.74 shows the sectional view of the guide bush of a sliding headstock lathe. The workpiece is guided as a bar blank from behind through the spindle and sliding sleeve to the collet chuck. The collet chuck

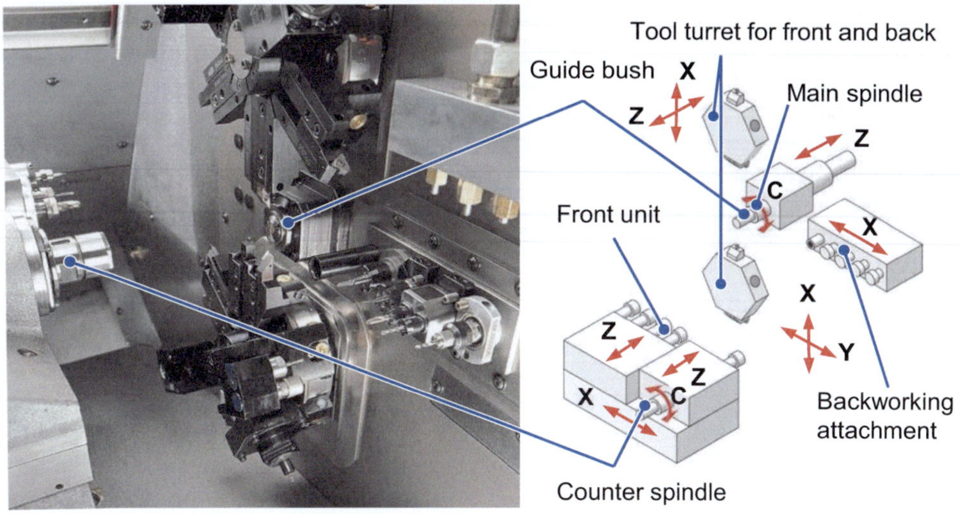

Fig. 4.73 Workspace of a Swiss-type lathe, according to Traub

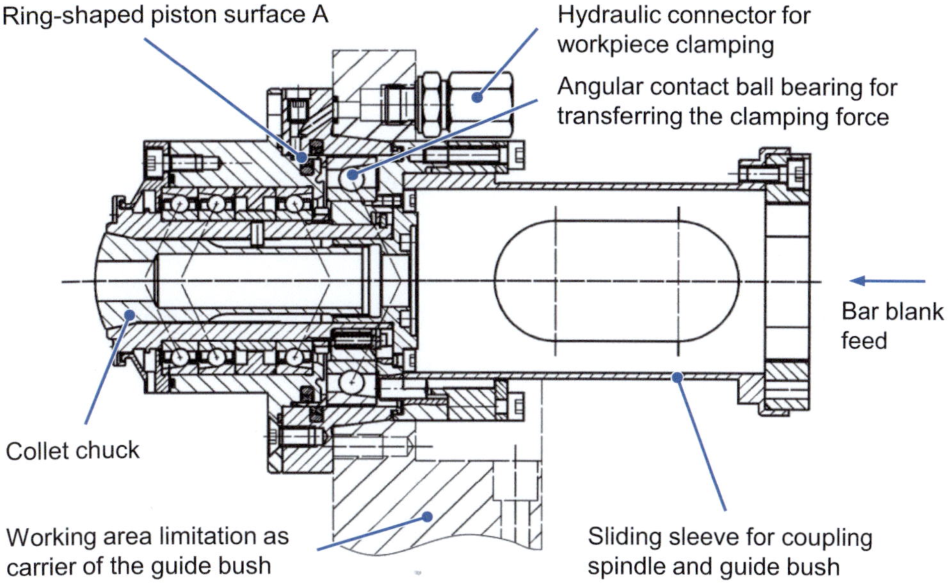

Fig. 4.74 Sectional view of the guide bush of a sliding headstock lathe, according to Traub

must apply the clamping force in a particularly precise and uniform manner because too high a clamping force would impede the axial movement of the workpiece. Too loose a clamping force however would reduce the effectiveness of the guide. For precise actuation of the collet chuck, the relatively small piston surface A, located in a ring shape in the fixed part of the guide bush, is subjected to high oil pressure from a hydraulic connector. The clamping force generated this way is transferred to the rotating collet chuck via an angular contact ball bearing. If the machining process does not require any axial displacement of the workpiece, the collet chuck can also be closed completely with a higher clamping force. An axial groove in the sliding sleeve transfers the rotation of the workpiece spindle from the spindle shaft that moves in the axial groove to the guide bush.

4.4.4 Multi-Carriage Automatic Plunge-Cut Lathes

For larger production volumes, machine productivity is a key condition for cost-effectiveness. It can be raised with higher rotational speeds and cutting rates, but also by using multiple tools that are in use simultaneously. For this, multi-carriage automatic plunge-cut lathes are used where multiple tools cut into the workpiece radially at the same time.

The molding tools mounted on the carriages fitted radially around the main spindle have the contour to be generated on the workpiece. Given that there are large cutting widths in plunge-cut machining, and thus large cutting forces occur, sufficiently stiff dimensioning of the machines is a requirement. If the designed stiffness is not high enough, unwanted chatter vibrations can occur during the machining process and damage the surface being machined. The required stiffness is achieved structurally by placing the cutting carriages in the immediate vicinity of the spindle bearing arrangement.

In a turret located axially in front of the workpiece, tools for axial and additional radial machining processes for the workpiece can be set up on some machines (such as for drilling, milling and face turning). After machining the inner and outer contours, the workpiece can also be held by a counter spindle that is affixed to and moves synchronously with the turret. It is also separated from the bar by a parting tool mounted on one of the radial carriages. Then machining of the back side of the workpiece is possible (such as by introducing bores or threads). Here, the turret head usually pivots the counter spindle with the workpiece to another position where fixed or driven tools can complete the machining process.

4.4.5 Multi-Spindle Automatic Lathes

Multi-spindle automatic lathes have multiple working spindles (usually two to eight). On these machines, multiple operations are performed simultaneously on one workpiece at a time. The spindles are mounted in a spindle drum. They are advanced from one tool station to the next. After every indexing step, one turned part has passed the last work step and is finished. Spindle drum indexing is either derived directly from the control shaft mechanically or is performed by hydraulic or electric drives. While previously cam plates and mechanical gear ratios assumed sole control of such multi-spindle machines, numerical controllers are now being used in some modern multiple-spindle machines for controlling and monitoring machine functions.

An array of tools and additional equipment are accommodated in the workspace of these automatic lathes and, as a result, very different machining tasks can be performed on workpieces. The tools for axial workpiece machining are on longitudinal slides positioned around the rotational axis of the spindle drum, while the tools for radial machining are on cross slides. All of the tools necessary for part production work simultaneously in just one automatically running work cycle, with the time-intensive manufacturing operation determining the single part production time.

▫ Figure 4.75 shows the structure of a cam-controlled multi-spindle automatic lathe. On this machine, the feed motions are controlled by cam plates. Lever mechanisms are used to adapt the stroke paths of the tools to the machining task. A detailed description of the mechanical control of automatic lathes is in Volume 3 of this book series.

Setting up such a machine for a new production task takes from a few hours to days, meaning its use is only cost-effective for large quantities. Change gears are used to adjust the spindle speeds and feed rates. To increase the flexibility of these kinds of machines, it is possible to drive the cam plates not from the central drive via a splitter gearbox with change gears, but to drive the main spindles and cam plates separately. This simplifies the adaptation of tool movements to difficult machining tasks.

While cam disk controls are used in conventional multi-spindle lathes, numerically controlled axes are increasingly being deployed in modern multi-spindle automatic lathes. This increases flexibility above all because converting from one workpiece to another requires relatively short retooling times. In contrast to conventional multi-spindle automatic lathes with cam disk controls where the feed paths need to be adjusted with lever settings and cam disks replaced, conversion of the part spectrum for numerically controlled multi-spindle lathes requires reading a new NC program. The use of conventional multi-spindle automatic lathes is restricted to large-scale production, while numerically controlled multi-spindle lathes with their much shorter retooling times can be deployed cost-effectively even for medium-sized batch sizes. Despite the major benefit of the increased flexibility of NC-controlled machines, due to the numerical control of axes, conventional multi-spindle lathes in large-scale production are continuing to be deployed primarily because they also feature more accurate travel at high speeds and are generally less susceptible to faults.

▫ Figure 4.76 shows the workspace and internal structure of a numerically controlled 6-spindle automatic bar lathe with counter spindles. In this concept, there is no guide block for separate axial carriages. Instead, the radial tool slides move in the X as well as Z-direction as well as around the R-axis. Synchronizing the C-axis of the main spindle and the parallel R-axis of the tool holder accommodates positioning of the tool perpendicular to the X and Z-directions relative to the workpiece. This means off-center radial bores can be realized and minor milling operations performed. On this machine, the feed motions of the individual axes are coordinated by the numerical controller. The spindle drum (▫ Fig. 4.77) comprises six motor spindles and is positioned by Hirth serration after indexing by one spindle division. In addition to driving

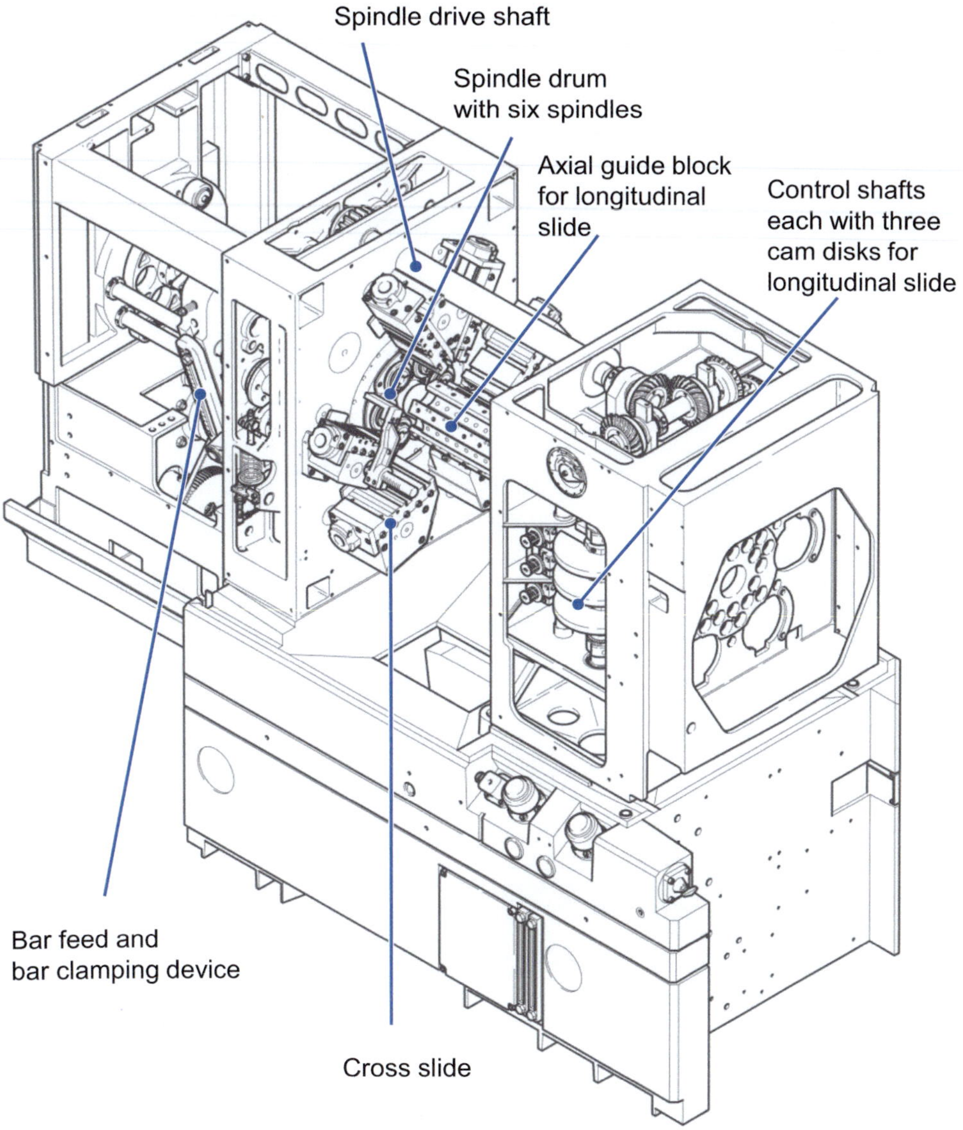

Fig. 4.75 Multi-spindle automatic lathe, according to Schütte

the spindles, the counter spindle drum allows the extension of the spindles in the Z-direction so the workpieces can be transferred from the opposite spindles at every workstation. An automatic lathe with a modular structure as in Fig. 4.76 can, depending on the customer requirement, be expanded to a configuration offering up to 70 NC-controlled feed axes.

The standard mounting of a bar magazine (see ▶ Chap. 10) makes the multi-spindle automatic lathes introduced particularly well-suited to the machining of small to medium-sized turned parts from the bar. For chuck parts, that is, parts needing to be inserted into the spindle chuck individually before machining, integrated handling robots are used (see ▶ Chap. 10).

Good chip fall as well as the capability of gripping chuck parts with the working spindle, and thus greatly simplifying workpiece handling, have led to machines with vertically suspended spindles (see ▶ Sect. 4.4.6). Figure 4.78 shows a vertical multi-spindle automatic lathe. The photo shows the view into the workspace of the machine with the clamping brackets for the tool assemblies. The spindles are located above the tools. Every spindle is mounted on its own machining unit together with its drive and a controller/supply unit. The machining units can, driven by linear motors, move independently in a circular path around the vertical machine axis. Also, the spindles can be moved in the direction of the spindle axes (Z-direction), meaning they are suitable for picking up workpieces from a feed conveyor without the use of additional equipment (pick-up spindles). If an additional Y-axis is required for milling, the corresponding equipment can be placed onto the tool bracket provided, together with the tool (as shown schematically in Fig. 4.78). For the purposes

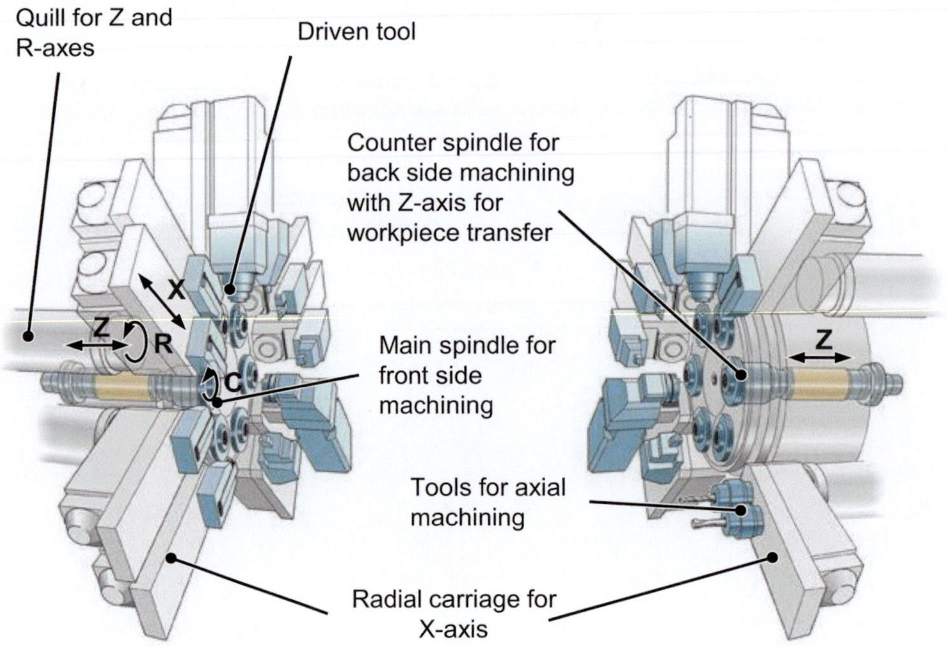

Fig. 4.76 Internal structure of a NC-controlled multi-spindle automatic lathe, according to Index

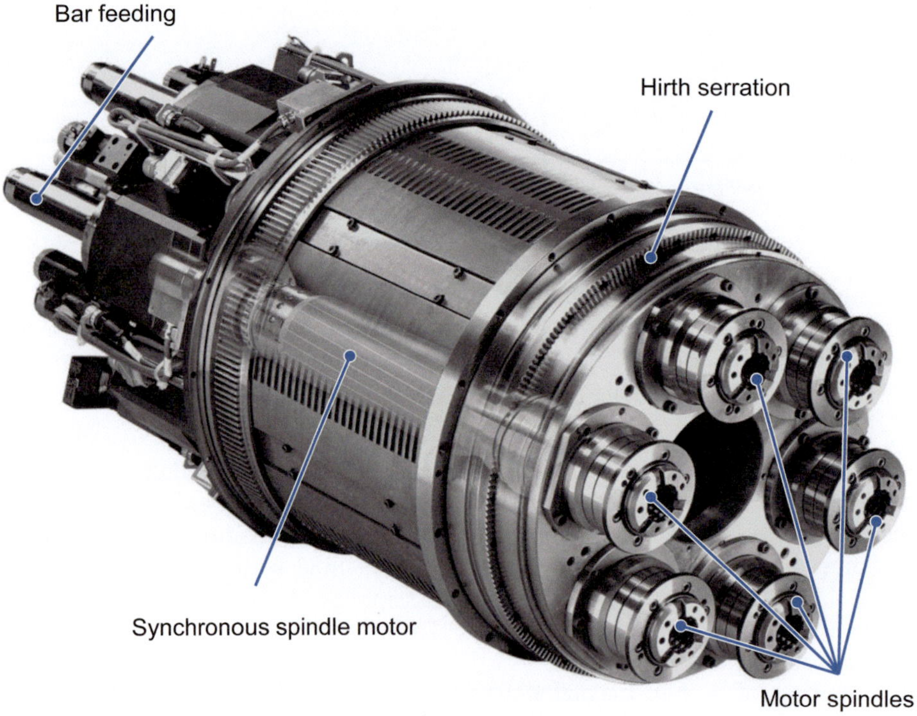

Fig. 4.77 Spindle drum of a multi-spindle automatic lathe, according to Index

of clarity, only one of the movable machining units is shown here. The machine can be fitted with up to eight of these units, however.

The time-independent indexing of the machining units permits, while ensuring no collisions occur, the sequencing of machining stations with different machining times. This means unproductive time can in many cases be reduced (compared to a multi-spindle machine with spindle drum). The cycle time continues however to be determined by the longest machining operation.

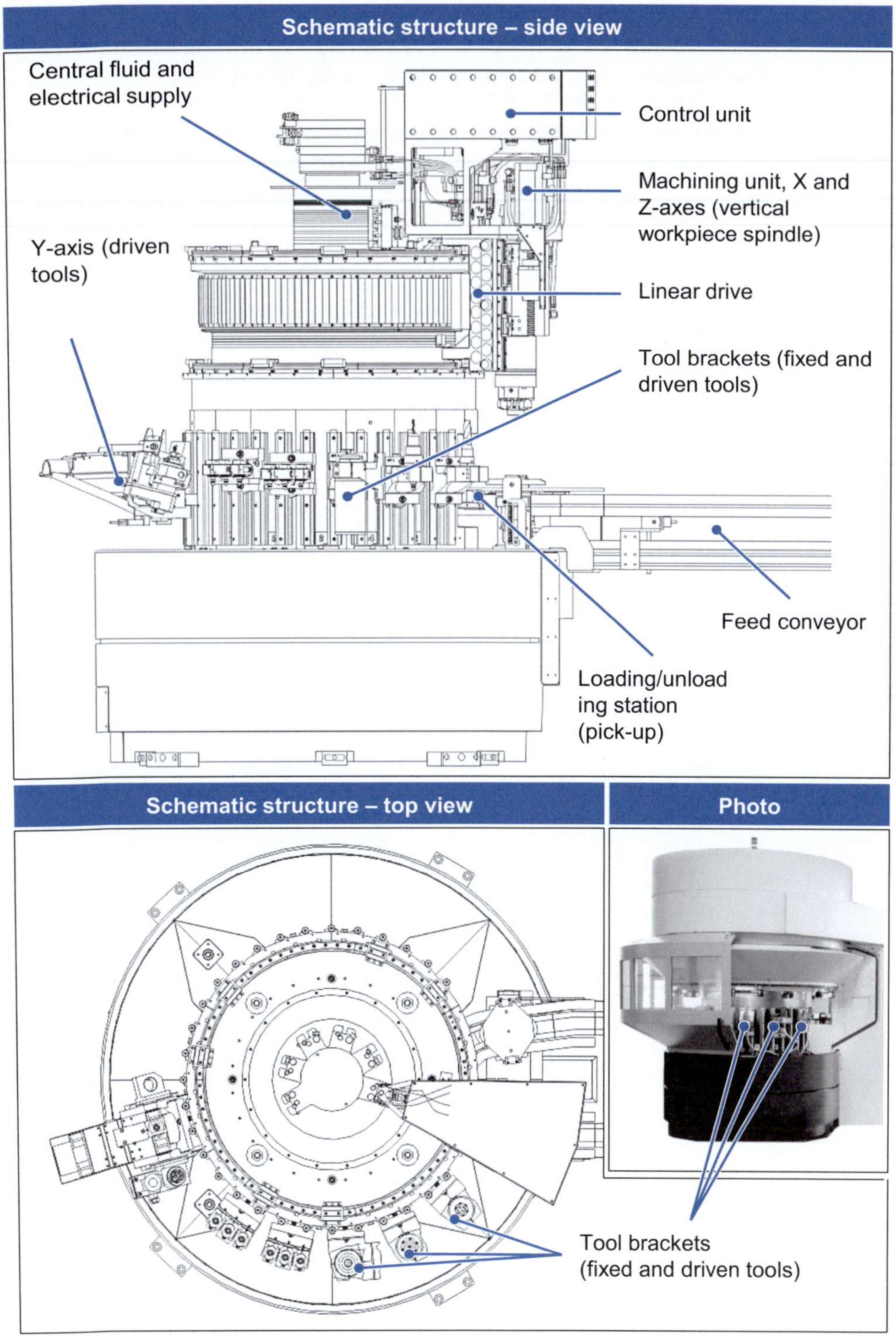

Fig. 4.78 Multi-spindle automatic chuck lathe, according to Schütte

4.4.6 Lathes with Vertical Spindle

In contrast to lathes with horizontal spindles, for lathes with spindles positioned vertically suspended, gravity causes the chips to fall unobstructed into the disposal device, and thus they only stay in the machining area for a short time. The often hot chips have no thermal effect on the machine structure to a large extent. The loading and unloading of workpieces are other criteria for machines used for volume production. Additional handling devices are often required here. These come in the form of robots or gantry loaders that convey workpieces to/from the machining position or between machining cycles or stations. For vertical lathes on the other hand, the axis motion of the main spindle is generally used for workpiece handling. Additional handling devices are thus usually unnecessary.

An example of these lathes is shown in ◘ Fig. 4.79. The figure shows a vertical lathe where the spindle is moved on the three linear feed axes by the parallel kinematics (refer to ► Sect. 4.3.3 for an explanation of parallel kinematic structures on machine tools). Here, three dual struts join the spindle housing to the three vertical carriages. These carriages are moved in the vertical direction on the three machine columns by ball screw drives. The parallel kinematic structure can result in very high machine stiffness enabling, for example, high cutting volumes in heavy-duty machining.

◘ Figure 4.80 shows the workspace of the machine. For turning, the workpiece is moved past the tools fixed to the table. Driven tools for drilling and milling can also be attached to the table plate. The great freedom of motion in the horizontal plane means the machine is able to load chuck parts moving past the workspace on a conveyor belt and to place finished parts back onto the belt. An additional workpiece handing system is not required.

Lathes with vertical spindles are commonly used to machine workpieces on both sides. These machines have two spindles that machine a workpiece, one spindle at the front and one at the back. Two examples are shown in ◘ Fig. 4.81 and 4.82. The machine shown

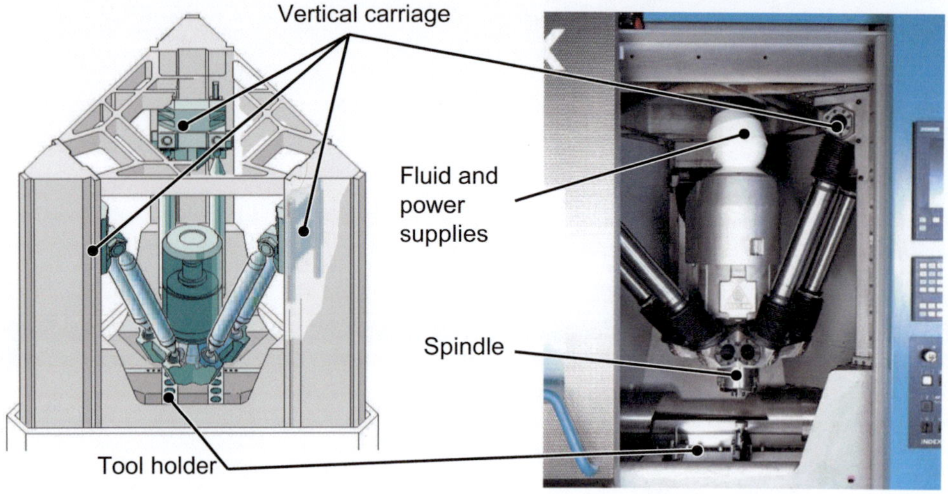

◘ **Fig. 4.79** Parallel kinematic vertical lathe, according to Index

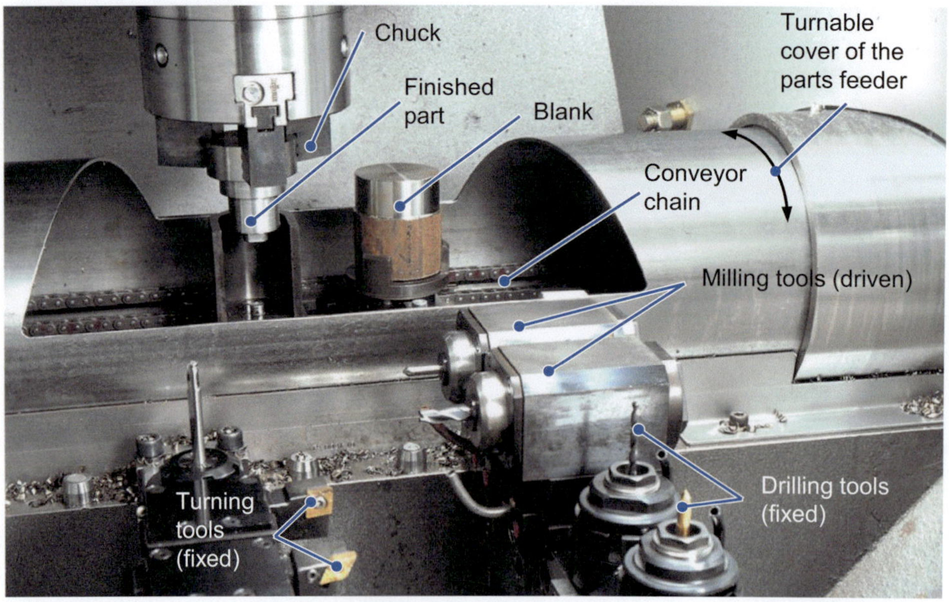

◘ **Fig. 4.80** Workpiece change with pick-up spindle, according to Index

in ◘ Fig. 4.81 has two spindles on opposite sides. The chuck for the spindle movable in the X and Z-axis directions and in the suspended position (left) grips the workpiece from a conveyor. Once the workpiece has been machined on the fixed tool turret, it is transferred to the second fixed spindle linked to the controller. The first spindle is traversed such that it stands opposite the chuck of the second spindle and passes the workpiece. In the chuck of the fixed spindle, the back of the workpiece is machined by the tool slide movable in the X and Z-directions (on the right in ◘ Fig. 4.81). A gripper mounted on the second turret slide takes the finished part from the spindle chuck and places it on the right-hand conveyor. The positions of the two spindles mean the two sides of the workpiece can be machined without needing to turn it. Up to 12 driven tools can be used in each of the turrets of the two machining positions.

◘ Figure 4.82 shows a similarly designed machine but with two suspended spindles. Turning the workpiece around is required here and is handled by a turning fixture outside the workspace. A conveyor moves the workpieces to the spindles and turning station. A major benefit is the modular machine design. This allows the manufacturer, with comparatively little construction overhead, to also build a 1 or 3-spindle system. The user also has the option of using both spindles for front side machining. Between the workspace and conveyor, it is also possible to fit measuring tools. Any dimensional deviations coming from tool wear and thermal drift can be detected immediately and compensated by the controller. External cooling quickly dissipates the heat built up during the machining process from the machine. This keeps the machine bed made of reaction resin concrete, the motor spindle, the hydraulic unit and the control cabinet at a constant temperature. This results in improvements in manufacturing accuracy. The disk turrets located at the bottom in the machining chamber carry up to 12 tools (that can also be driven tools). The machining process can be viewed by the machine operator through a window at the front.

4.4.7 Large-Scale Lathes

4.4.7.1 Horizontal Large-Scale Lathes

Large-scale horizontal lathes (such as roller lathes) machine workpieces up to 30 m in length and 6 m in diameter. They are used for example to manufacture rollers for rolling mills and paper machines, drive shafts for ships, and turbine rotors.

◘ Figure 4.83 shows a turbine rotor clamped on a horizontal large-scale lathe. The lathe is designed for a maximum workpiece diameter of 4.5 m and maximum workpiece length of 20 m. The maximum workpiece weight that can be carried between the centers is 350 tons. The center height is 2.3 m above the guideways of the workpiece bed.

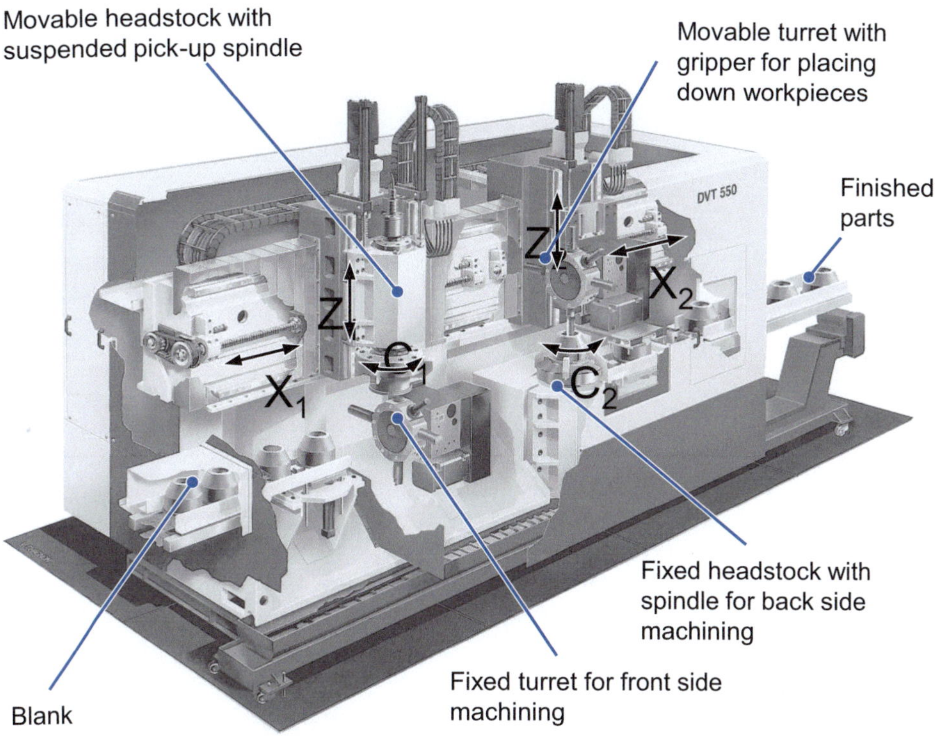

◘ **Fig. 4.81** Dual-spindle vertical lathe, according to FFG Werke

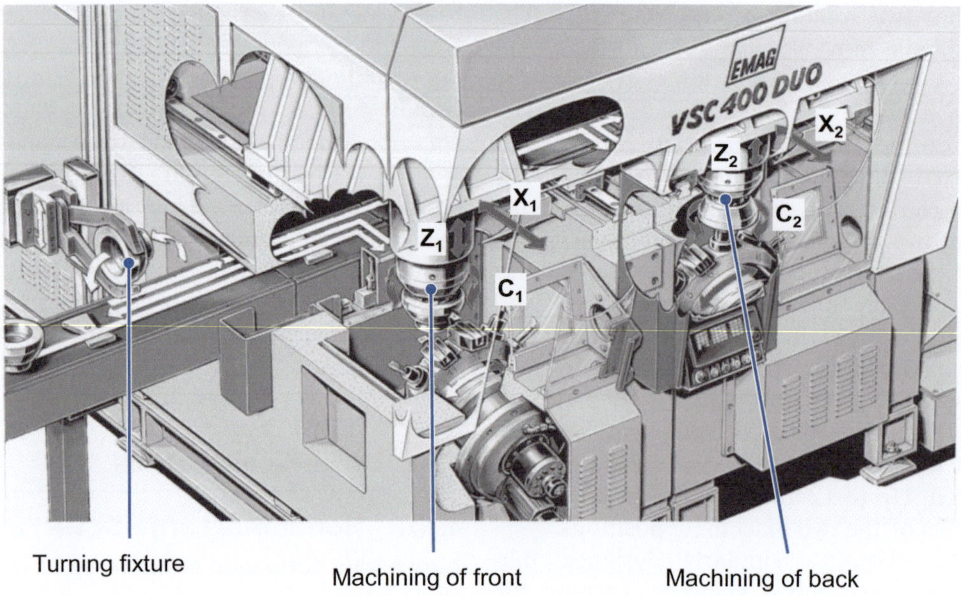

Fig. 4.82 Dual-spindle vertical lathe, according to EMAG

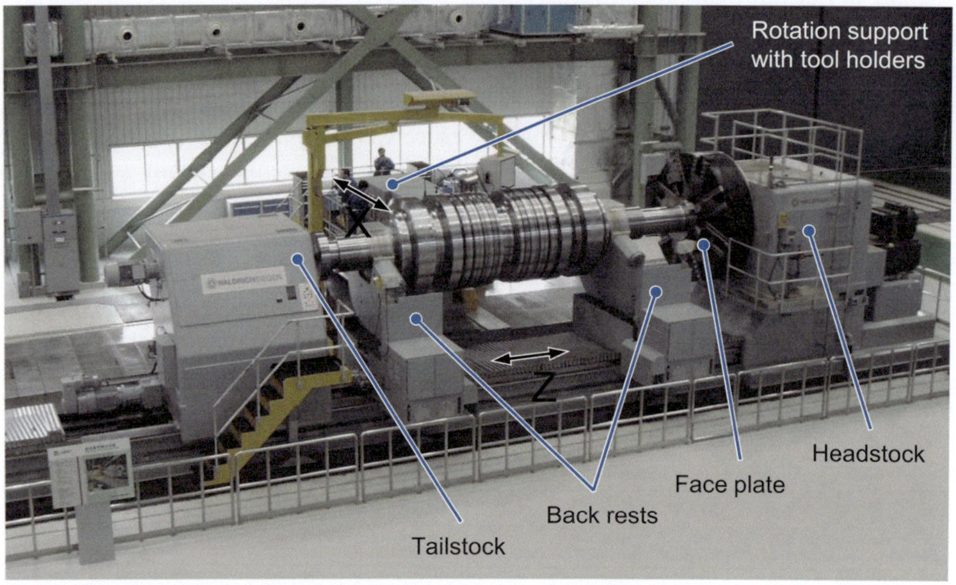

Fig. 4.83 Horizontal large-scale lathe for turbine rotors, according to WaldrichSiegen

On the workpiece side on the right is the headstock, that uses the face plate to drive the workpieces with a drive power of 225 kW and maximum torque of 250,000 Nm. On the opposite side on the left is the traversable tailstock that takes the workpieces using a quill that can move axially, centers and clamps them. Between the headstock and tailstock, additional steady rests are integrated that hydrostatically support the workpieces at the respective bearing points.

On the back of the workpiece is the rotation support movable in the Z-direction. On the rotation support are solid tool holders that can be pre-positioned and secured in different positions. Workpieces are machined with tool cassettes affixed to the tool holders. For precision tool infeed, the support can be traversed towards the workpiece in the X-direction using a pre-tensioned ball screw drive.

Figure 4.84 shows a horizontal large-scale lathe center used for turning and other machining operations. The objective is to finish machining large and heavy workpieces on one machine.

By integrating a milling slide attached to a pivot-mounted tower, machining operations such as milling, drilling and grinding can be performed. The

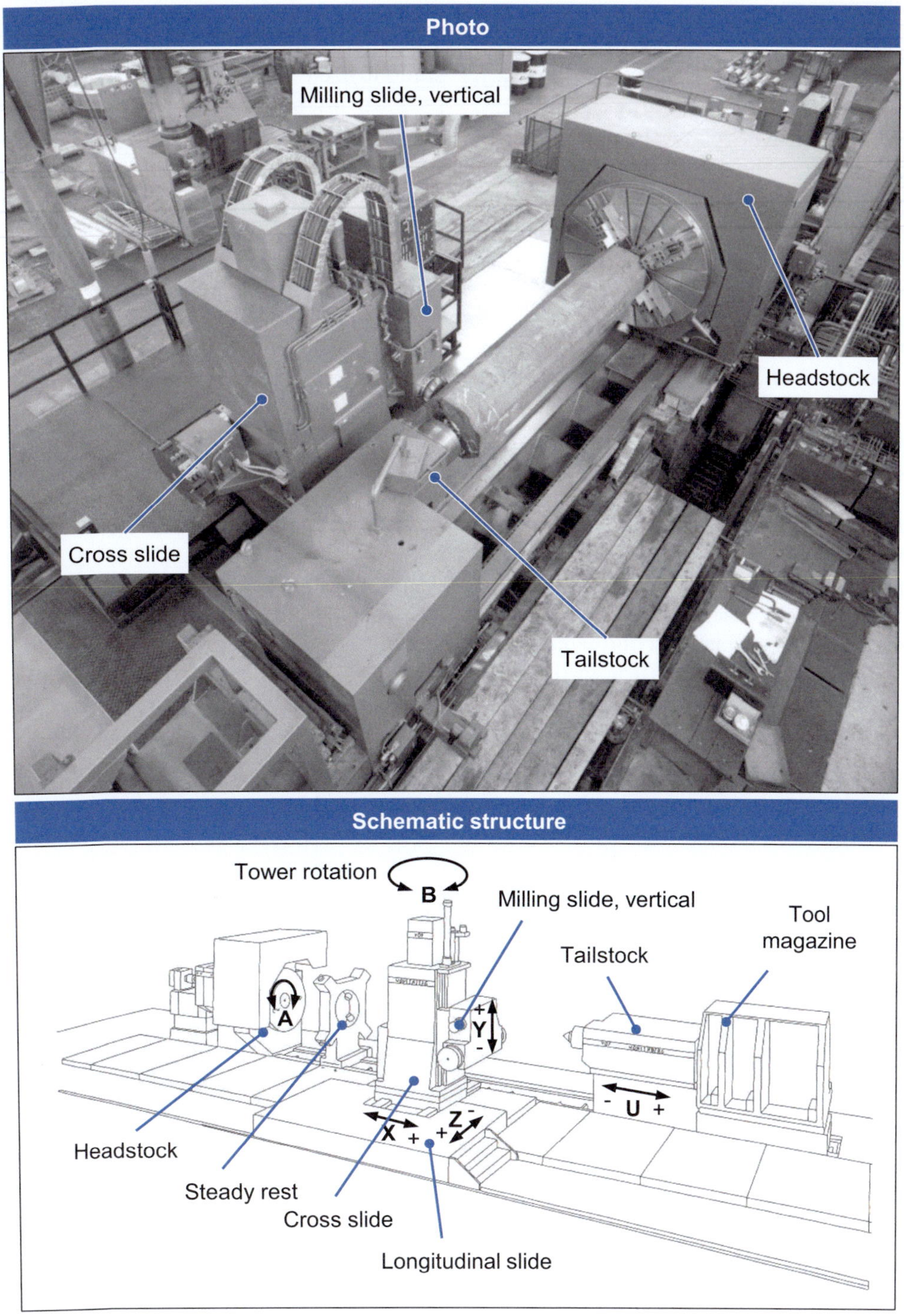

Fig. 4.84 Horizontal large-scale lathe center, according to Wohlenberg

different tools and fixtures are located in a magazine next to the tailstock. They can be replaced by the milling head using the pick-up technique (see ▶ Sect. 10.4.3.3). This means that tool usage is possible at any angle and height position in relation to the workpiece axis over the entire traverse path. With the integration of processes on one machine, it is possible to also finish machining large workpieces in one clamping operation. This eliminates repeated retooling and alignment on different machines. This reduces non-productive and set-up times and eliminates the need for transportation and temporary storage. A large portion of the machine components always remains unused however, something that often calls into question the cost-effectiveness of such systems.

All nine axes of the machine shown in Fig. 4.84 are numerically controlled. Steady rests that prevent workpieces from deflecting are used to machine long workpieces.

4.4.7.2 Vertical Large-Scale Lathes

Vertical large-scale lathes, also called vertical turret lathes, offer the benefits of the ability to accommodate the workpiece weight and straightforward clamping options for unstable workpieces having large diameters. ◘ Figure 4.85 is the basic structure of vertical large-scale lathes with single-column and portal designs. The face plate is driven by a gearbox. Torques to 125,000 Nm and powers exceeding 200 kW are common. Workpieces having diameters to 20 m can be machined on these kinds of machines.

To keep set-up times short and prevent re-clamping, finish machining of the workpiece with only one clamping process is the aim, if possible, for large-scale production in particular. For this, the turning tool holder (◘ Fig. 4.85, bottom left) can usually be replaced today with different grinding, drilling and milling heads (at bottom of ◘ Fig. 4.85, center and right). Automatic tool and workpiece change equipment is also expanding heavy-duty machine tools to make flexible manufacturing cells capable of machining a wide range of different, large workpieces without interruption of the set-up process. While the turning tools are held in a fixed clamping device, the rotating milling and drilling tools are accommodated by a spindle with standardized steep taper or hollow taper shank interface. To increase cutting capacity, two tools are commonly used at the crossbeam on independent Z-carriages (slides) that are in use simultaneously during the turning operation. ◘ Figure 4.86 shows a two-column vertical lathe with two machining stations and a circumferential diameter of 10 m. The diameter of the face plate is 7 m. Both machining stations perform turning operations on the gear ring.

4.4.7.3 Bar Peeling Machines

Bar peeling machines are used primarily in large-scale production for the cost-effective manufacture of high-quality cold drawn steel and for the abrasive peeling of thick-walled tubes and rolled or cast round steel bars. The primary materials for forming (for example, forged parts) or metal-cutting (for example, drive shafts) manufacturing processes are obtained this way. Bar peeling is a manufacturing process derived from turning (see ◘ Fig. 4.87). It is a kinematic reversal of longitudinal circular turning with large feed, using a

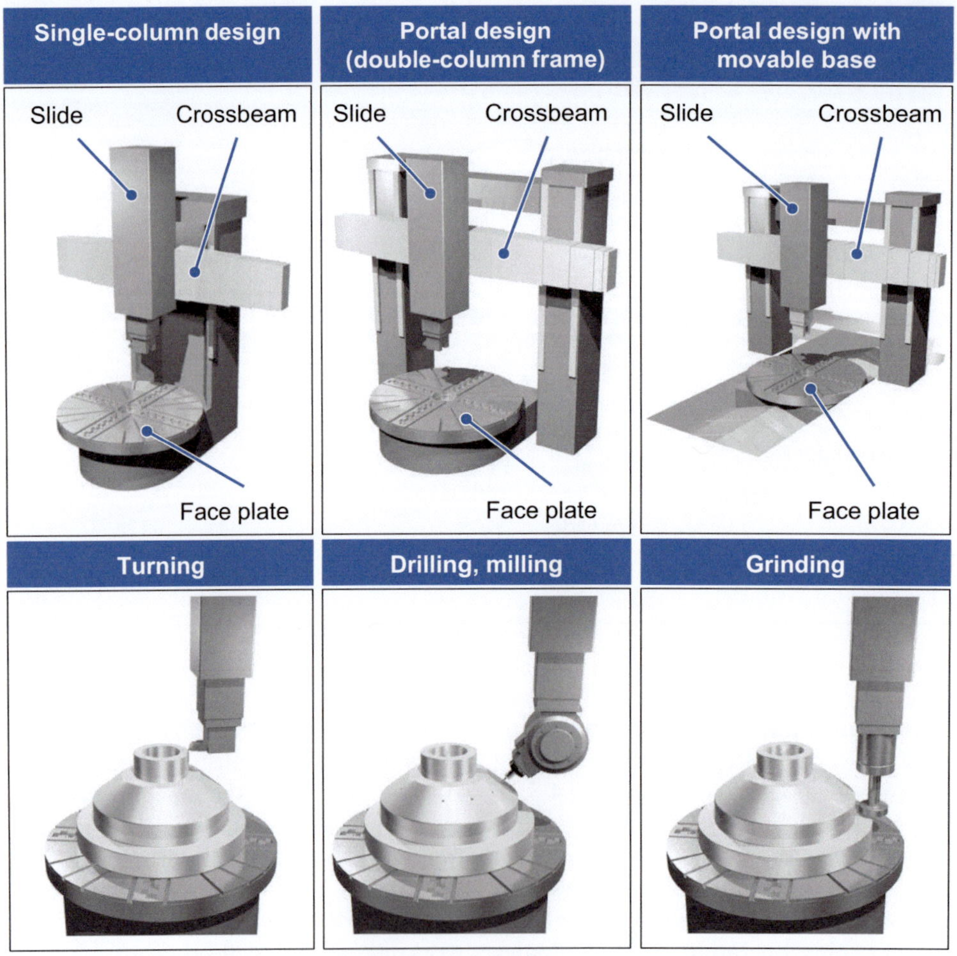

◘ **Fig. 4.85** Types and typical applications of vertical large-scale lathes, according to Dörries Scharmann Technologie

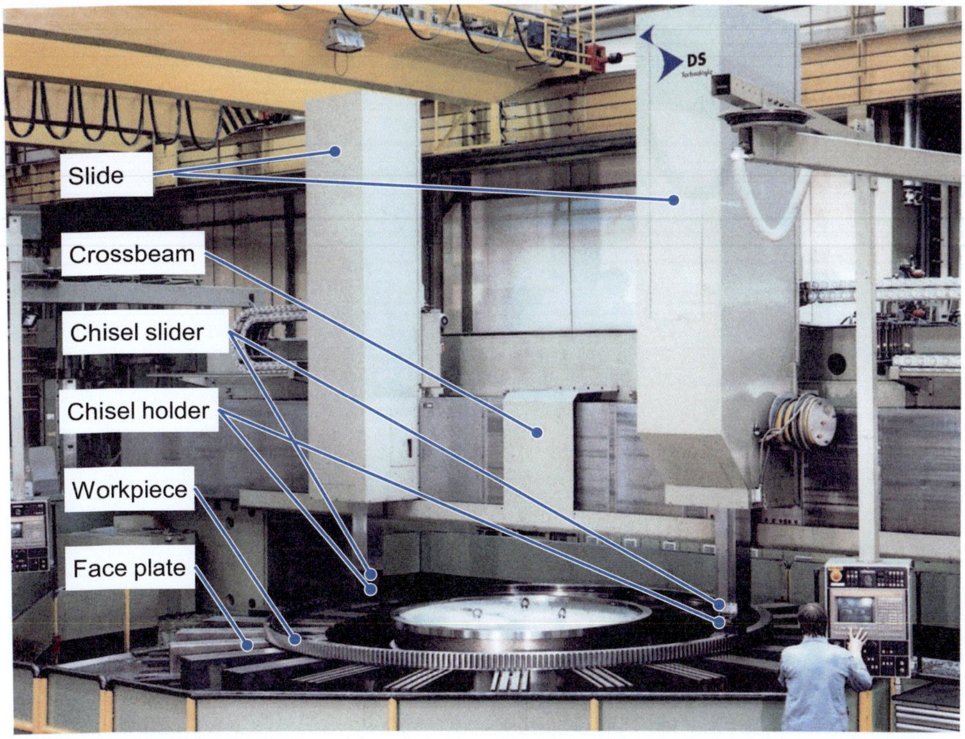

Fig. 4.86 Two-column vertical large-scale lathe with portal design. *Source* Starrag Technology (Dörries division)

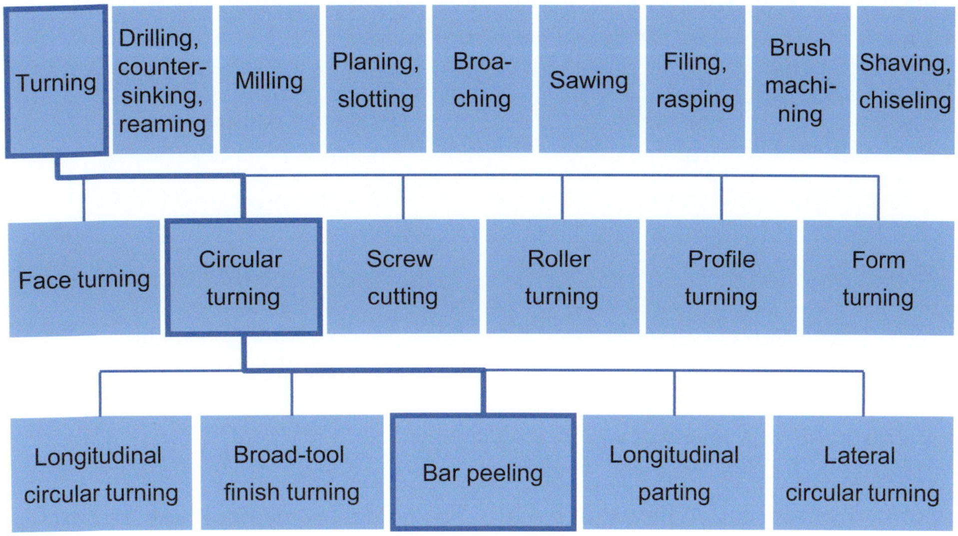

Fig. 4.87 Categorization of the bar peeling manufacturing process, according to [DIN03c]

rotating tool through which the material is pushed or pulled concentrically.

Figure 4.88 shows a schematic of the process sequence in machining coil and bar material. The peeling process can be adapted to the base material by means of a combination of additional cutting and straightening operations. Other tailoring operations follow the cutting processes depending on the end product required.

The most important components of a peeling machine can be seen in Fig. 4.89. The rollers of the in-feed unit affect the axial feed of the bars by the rotating head tipped with carbide tools, and absorb the cutting forces generated in the process. The peeling process generates internal stresses in the material that, for small bar diameters in particular (to 40 mm), can cause uniform curvature of the bar material. This makes handling the bars more difficult. To reduce this effect, the

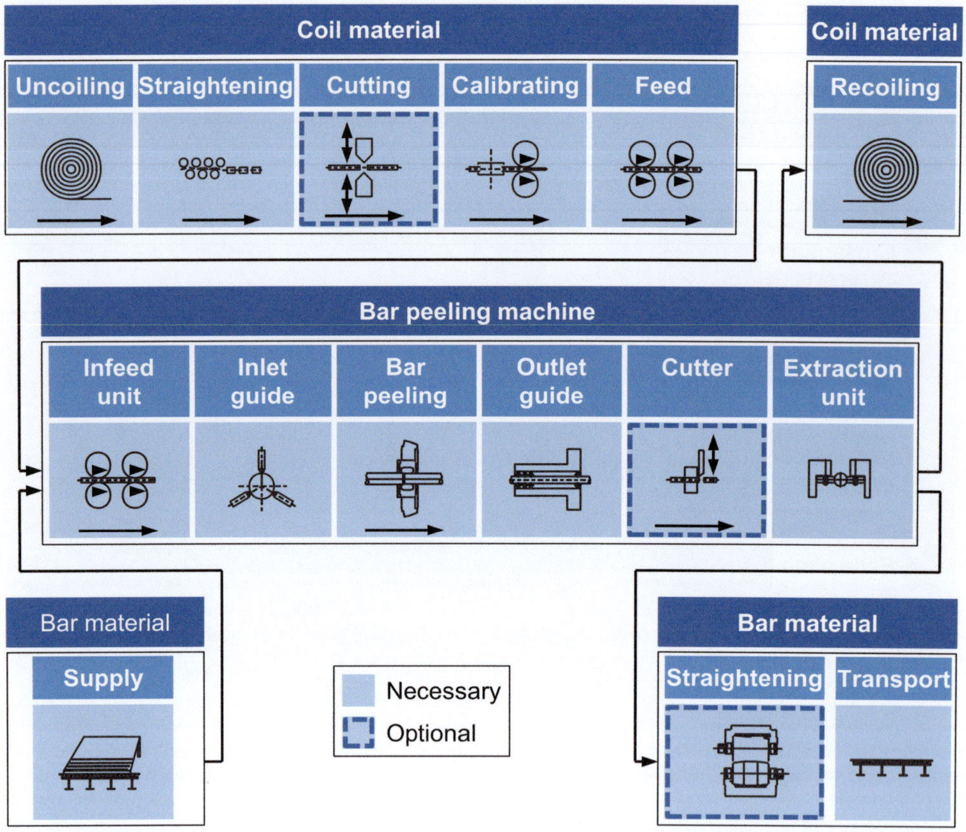

Fig. 4.88 Process chain for machining coil and bar material, according to Schumag

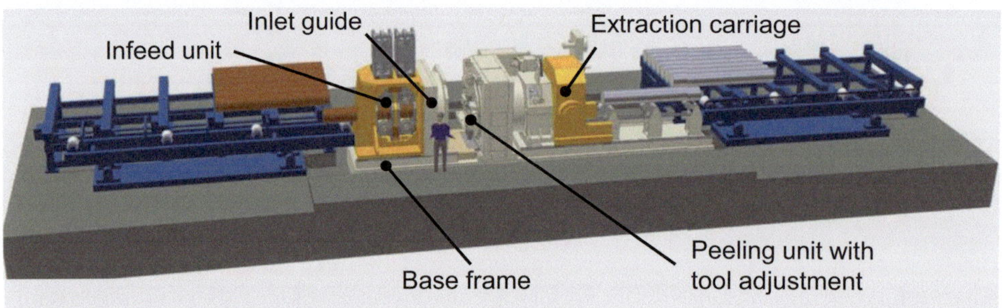

Fig. 4.89 Peeling machine with its main assemblies, according to SMS Meer

infeed rollers are offset from the infeed direction by up to 2.5°. This transports the bar to be peeled through the cutting plane while rotating. The infeed material is subjected to preliminary deformation along the infeed direction, meaning that, after the peeling process, just a slightly spiral residual deformation remains. A wobble axis with a plain bearing is used to adjust the infeed rollers.

The guide rollers immediately in front of and behind the rotating head ensure the material is centered. The workpiece is enclosed by three, four or five guide rollers depending on the machine type. For alignment of the workpiece, the guide roller holders can each be advanced individually by a separate ball screw drive. The holders sometimes have a cartridge design to simplify maintenance and replacement. Each guide roller is engaged radially using disk springs and is also dampened by a hydraulic clamping bush. The extraction carriage provides the axial feed as soon as the workpiece has moved from the infeed rollers. To prevent ledges and marks on the surface, the drives of the infeed rollers and extraction carriage are synchronized.

In the machining of bar material, guiding the bar is particularly problematic for the initial and final cuts (start and end of the bar) and increased tool wear can occur. To prevent this effect and to increase cost-effectiveness, bars are placed end to end. For this, the following material is placed early into the driven roller conveyor. The faster speed of the conveyor rollers causes the bar to catch up with the previous bar.

The rotating head designed for a specific diameter range has multiple internal cutting inserts that are mounted on carriages and are adjustable radially. The integration of tool cassettes into the carriages means the cutting inserts are easier to exchange and the use of different cutting insert geometries and dimensions is possible (see ◘ Fig. 4.90).

Characteristic for the geometry of tools used is the long minor cutting edge, with a setting angle in range 0° to 2°. As a result, there is additional peel cutting at the minor cutting edge in addition to cutting at the major cutting edge. This means surface qualities of $R_z = 15\ \mu m$ and $R_a = 2.5\ \mu m$ are attained. Tolerances to ISO h8 can be achieved.

◘ Figure 4.91 provides good insight into the layouts of a rotating head, hollow drive shafts and material

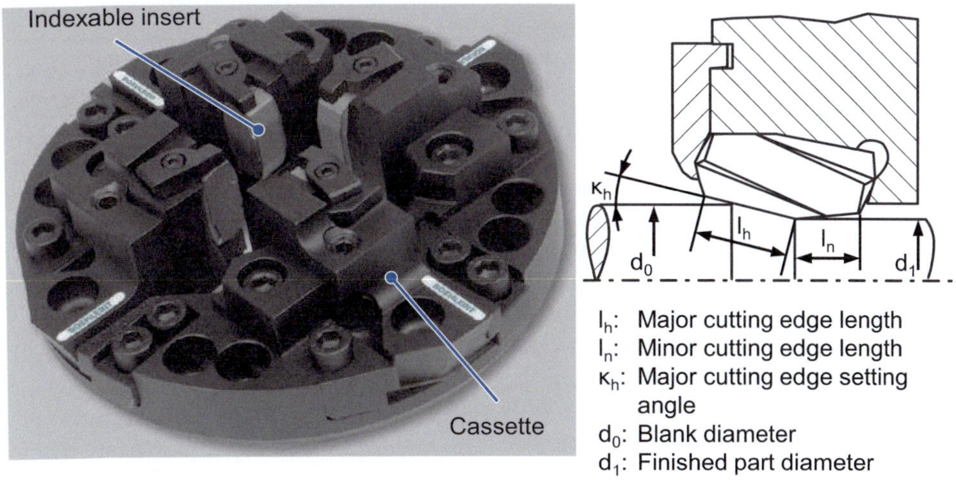

◘ Fig. 4.90 Rotating head design with cassettes, according to Boehlerit, Saar-Hartmetall und Werkzeuge

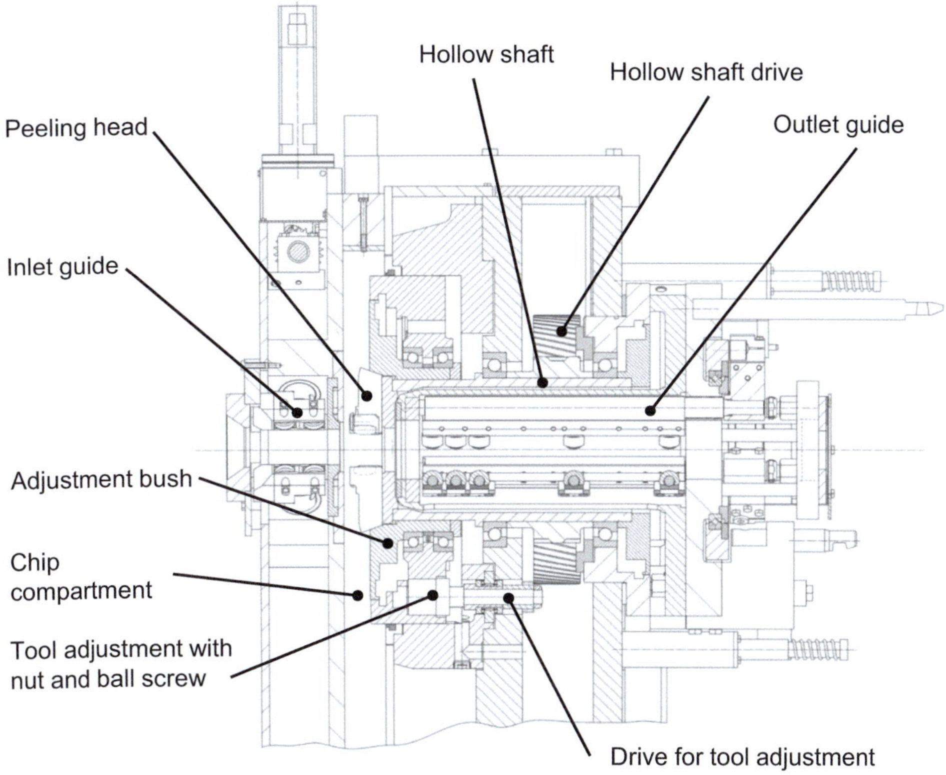

◘ Fig. 4.91 Peeling unit on a bar peeling machine, according to SMS Meer

guiding. For diameter adjustment and for tool wear compensation, the tool can be advanced radially during machining work. This is achieved by the axial adjustment of a conical sleeve, within the hollow shaft, that acts upon the beveled ends of the radially moving tool slide in the rotating head.

Given that peeling is used first and foremost for shape and surface improvement, the cutting depth is generally very small. The feed for every cutter is specified by the long length of the minor cutting edge. Steel cutting enables feeds up to 30 mm per revolution. Standard cutting speeds for carbide cutters depend on the steel quality and material diameter and are between 30 m/min and 160 m/min [SMS98].

The principle of internal tube peeling shown in ◘ Fig. 4.92 is a variant of bar peeling described above. The rotating head with cutters arranged on the outer circumference is located on the front end of a tool bar. The free end of this bar is secured by a collet chuck. It is used to transfer the drive torque to the tool bar. For the machining process, the tube to be peeled is pushed onto the tool bar that is then re-clamped. Now the peeling process starts. The tube is pulled away from the conveyor rollers axially above the rotating head with the carbide tools in contact. During this process, the position of the rotating head remains in the center area of the feed rollers that guarantee best guiding and support of the tube.

4.4.8 High-Precision and Ultra-Precision Lathes

A differentiation can be made in finishing machining between precision, high-precision and ultra-precision turning (◘ Fig. 4.93). In precision turning, steels and non-ferrous metals are cut using various cutting materials and infeeds or feed rates in the 1/10 mm range within ISO tolerances to IT7.

In high-precision turning, cubic crystalline boron nitride (CBN), monocrystalline diamond (MCD), carbide metal (CM) and cermets are used as cutting materials for the same spectrum of materials. With feeds and infeed rates in the 1/100 mm range, ISO tolerances between IT2 and IT6 are achieved.

Non-ferrous metals are machined primarily in ultra-precision turning. Monocrystalline diamond is used as the cutting material, with infeeds and feeds in the range of a few micrometers enabling workpiece tolerances of IT1 bis IT01 to be attained. Ultra-precision lathes are used for polish turning to establish optical surfaces (such as for lenses and mirrors) with direct turning. Subsequent polishing or lapping is unnecessary. The geometrical deviations attained in ultra-precision cutting range from 0.1 to 0.5 µm for a mirror diameter of 100 mm. The mean roughnesses attained (R_a values) depend on the material machined and lie between 2 and 5 nm. To achieve the geometrical

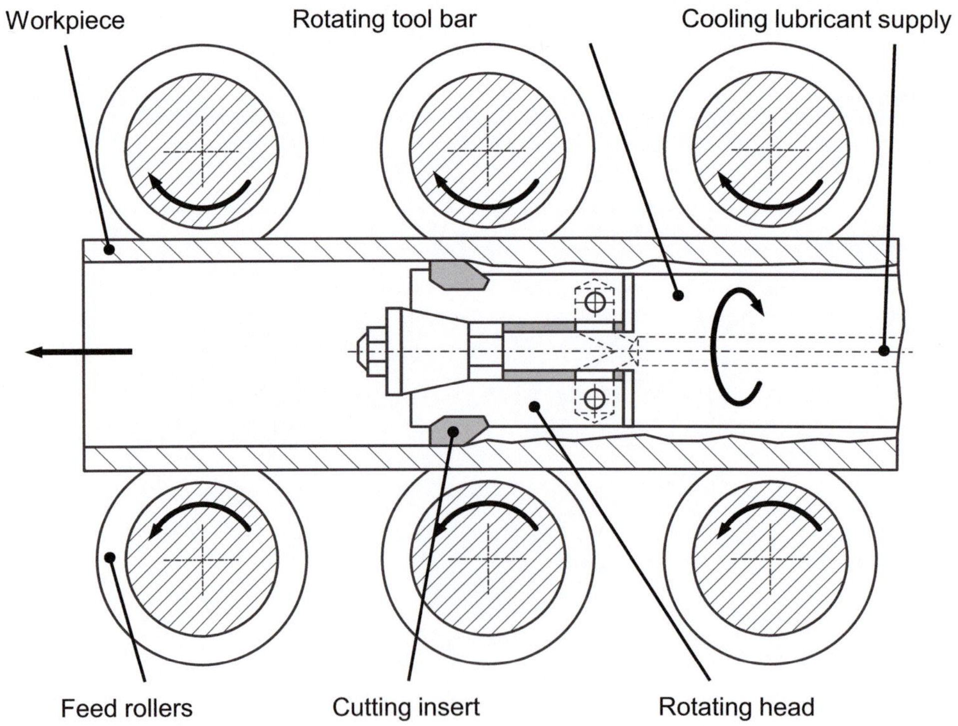

◘ **Fig. 4.92** Principle of internal tube peeling

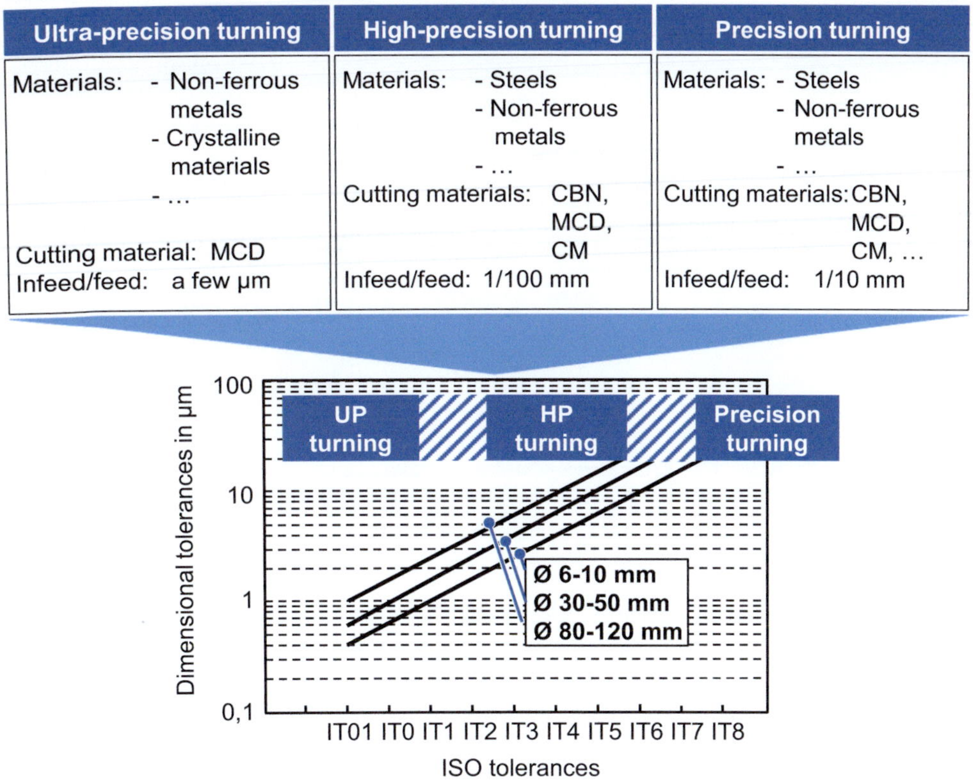

Fig. 4.93 Application areas and attainable accuracies for ultra-precision (UP), high-precision (HP) and precision turning, according to [DIN10]

accuracies required, hydrostatic axes with direct drive, and hydrostatic as well as aerostatic spindles with torque motors are used. To improve long-term thermal stability, the machine bed, mounted on a vibration-isolated base, is placed on granite. The machines are set up in air-conditioned rooms having a maximum temperature difference of ±0.1 K.

Figure 4.94 shows a compact ultra-precision lathe. The main area of application is the manufacture of complex freeform optics and microstructures with monocrystalline diamond tools (average surface roughness $R_a < 10$ nm). For the feed motion, the machine shown has a hydrostatically mounted X-axis with direct drive with positioning accuracy in the single-digit na-

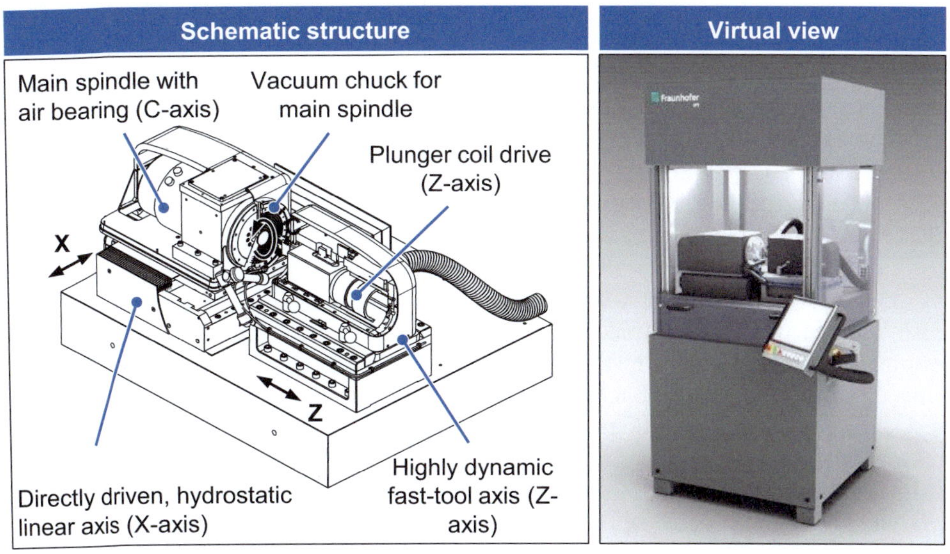

Fig. 4.94 Ultra-precision lathe, according to Fraunhofer IPT

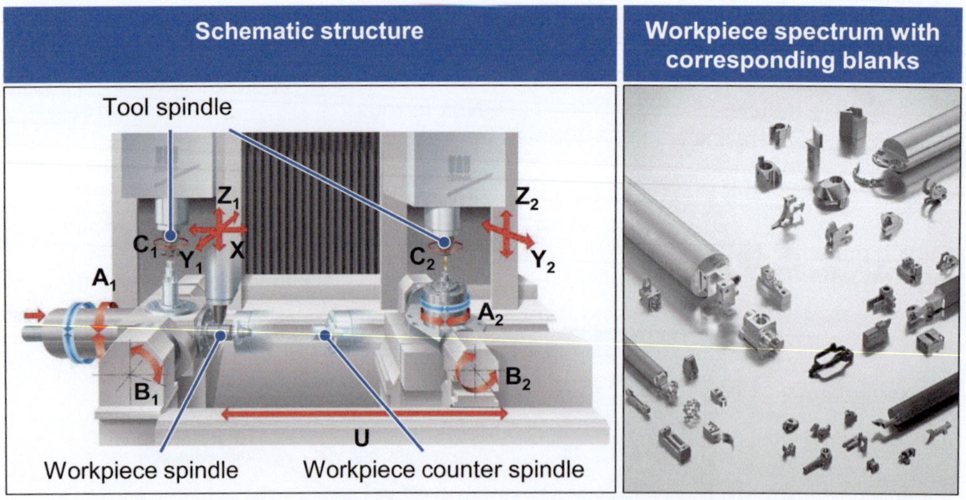

Fig. 4.95 Turning-milling combination machine, according to Stama Maschinenfabrik

nometer range. A glass scale is used to optically measure the carriage movement. The main spindle (C-axis) has an air bearing and permits maximum concentricity accuracies at up to 5000 rpm. The Z-axis has two stages. The stage driven with a plunger coil drive handles low dynamic traversing movements. The highly dynamic infeed motion required to manufacture free forms and microstructures is generated by a feed stage, also with air bearing, and called the fast-tool Z-axis.

4.4.9 Combination Machines

Although integration capabilities are commonly provided on modern lathes for driven tools or additional axes for milling and drilling, the trend is to offer special combination machines that can use different machining methods to the same extent. Two examples are presented below.

4.4.9.1 Milling-Turning Center

Figure 4.95 shows a machine designed for the integrated machining of workpieces from a bar blank. Using 12 NC-controlled axes, it is possible to finish machine a component on all sides. The workpiece bar is held in the spindle chuck with three rotating jaws. With a tool holder that can hold fixed tools for the turning process as well as milling tools, only the first side of the workpiece is finish machined. The workpiece is now passed to the counter spindle and cut from the bar by means of grooving or sawing. Then the back is finish machined. The swivel options of the spindle permit any positions of bores and surfaces.

So, the machine can be reconfigured for another component very quickly in many cases without having to perform retooling work. Shown on the right in Fig. 4.95 is the spectrum of possible workpieces with the corresponding blanks. Figure 4.96 shows some of the possible machining operations on the machine presented.

4.4.9.2 Turning-Grinding Machine

The turning-grinding combination machine in Fig. 4.97 is usually deployed for the hard turning/grinding process combination for the technologically optimized machining of holes, plane surfaces and synchronizing cones/synchronous surfaces on gear wheels and similar workpieces. As the pick-up spindle, the main spindle performs workpiece handling. The workspaces for grinding and turning are separated by a partition. As a result, chips and auxiliary materials for the processes can be disposed of separately. Tools can be positioned quickly and variably on the vertical T-slot plate. A dressing spindle for the grinding tools is attached to the headstock. The work steps necessary for manufacturing a ratchet wheel on the machine introduced are shown in Fig. 4.98 as an example.

4.5 Slotting and Drawing Machines

An overview of different slotting and drawing machines and their features is shown below. A usually straight-line, repeated cutting motion of the tool for step-by-step infeed of the tool cutter perpendicular to the cutting direction characterizes the kinematics of groove slotting and keyseating [DIN03d]. Figure 4.99 illustrates each of the movement sequences.

In the keyseating method, the tool is clamped in a tool holder. This holder slides up and down in the tool guide together with the feed bar. The top end of the feed bar is wedge-shaped and can be traversed by a feed

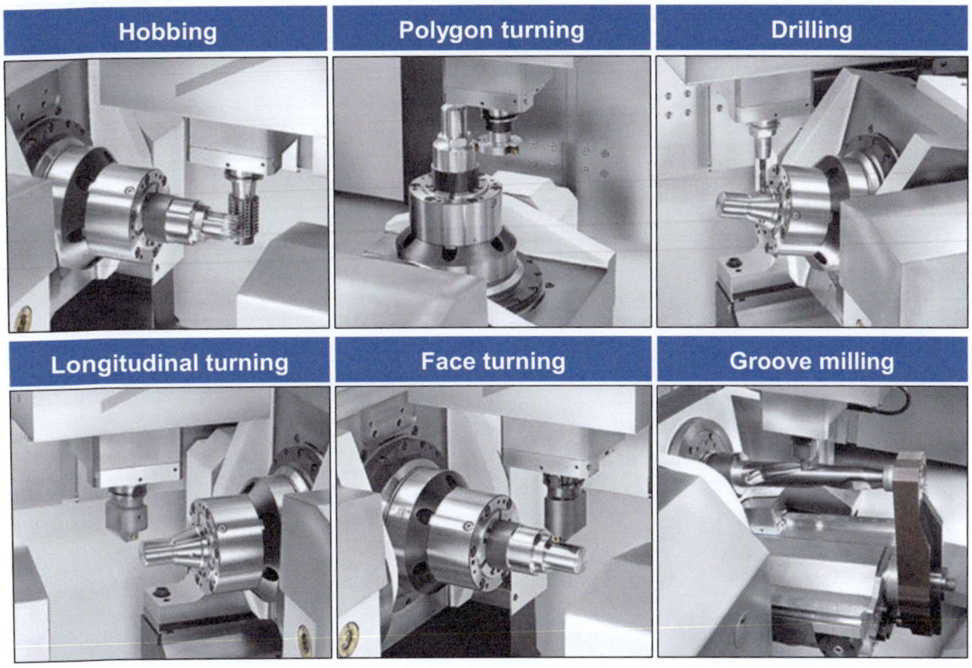

Fig. 4.96 Machining options on a milling-turning machine, according to Stama Maschinenfabrik

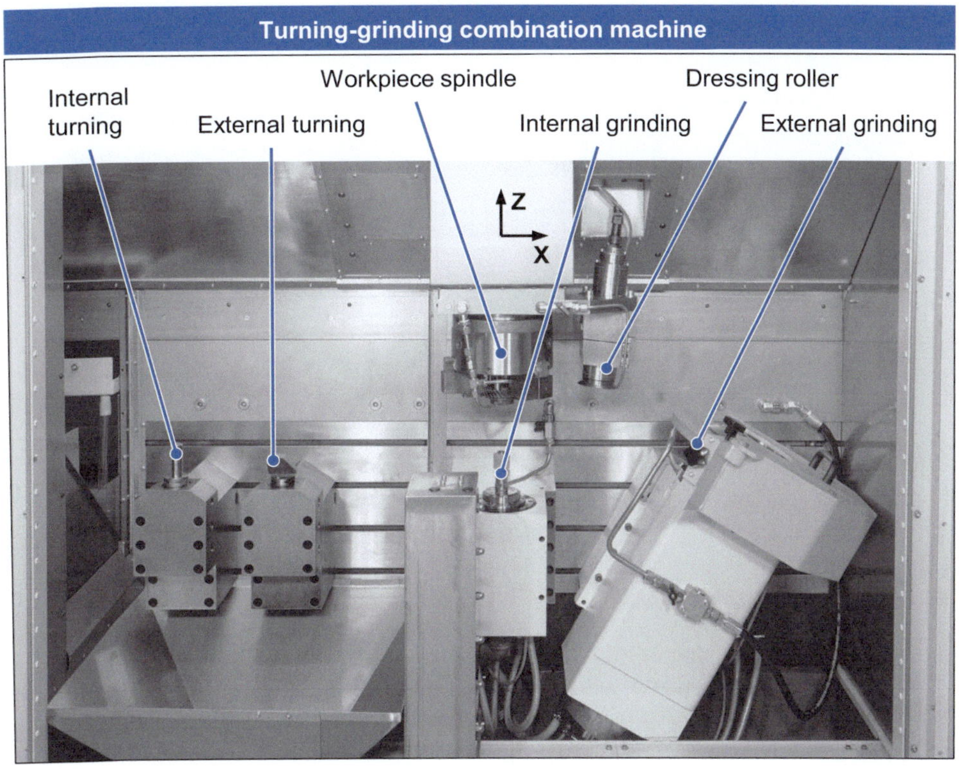

Fig. 4.97 Turning-grinding combination machine, according to Schaudt Mikrosa

slide relative to the cutter holder/stroke carriage. As a result of this, the inclination of the wedge advances the draw cutter in the direction of the groove depth. After a working stroke, the tool (to protect the cutter) is lifted from the base of the groove in an appropriate way on the return stroke. This double-stroke motion is repeated until the programmed groove depth is reached.

For slotting, the working and return strokes work in opposite directions. This machining type is thus particularly suited to profiling blind holes with little tool

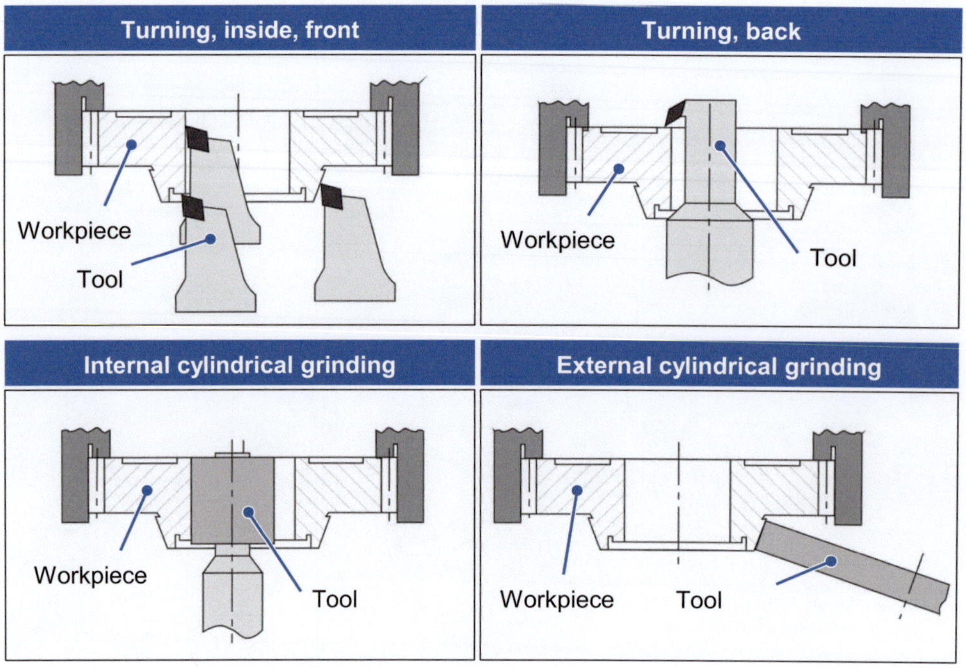

Fig. 4.98 Integrated machining of a ratchet wheel, according to Schaudt Mikrosa

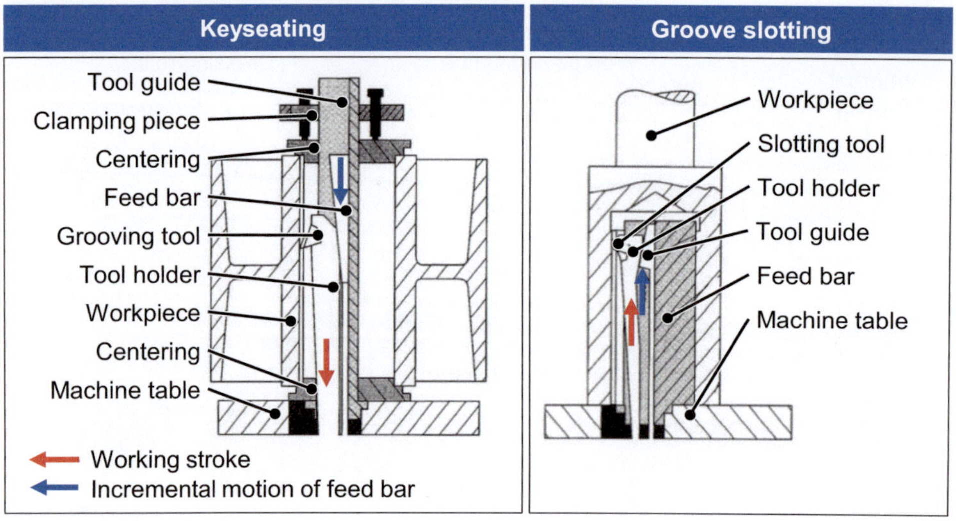

Fig. 4.99 Grooving techniques

runout. Workpieces with complex shapes and sometimes machining points that are difficult to access are typical for slotting.

Both methods require an intensive supply of cutting oil to guarantee correct lubrication of the tool and effective chip removal. Figure 4.100 shows a grooving and profiling machine used for slotting, drawing and broaching operations.

The size of the motor-driven indexing table corresponds to the grooving machine. Using the indexing table allows workpieces to be turned about the vertical axis. This permits grooves with very different pitches to be sunk. The vertical movements of the tool and of the feed slide are electromechanical using two independent ball screw drives. Each drive is connected to its own servo motor via a toothed belt. This enables speedy stroke and exact positioning movements. Infeeds in increments ranging from 0.01 mm to 0.1 mm can be achieved. By synchronizing indexing table rotation, tool and feed movements using a CNC continuous path control, helical and contoured grooves can also be created (as can be seen on the right in Fig. 4.100 using a workpiece as an example). Implementing the indexing table as a combined cross-rotary table means that the grooves

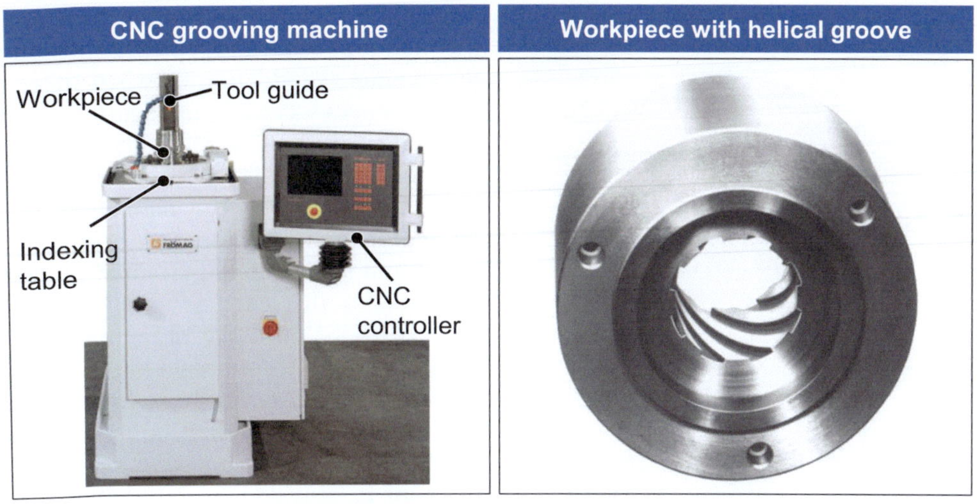

■ **Fig. 4.100** CNC grooving machine with electromechanical drive. *Source* Frömag

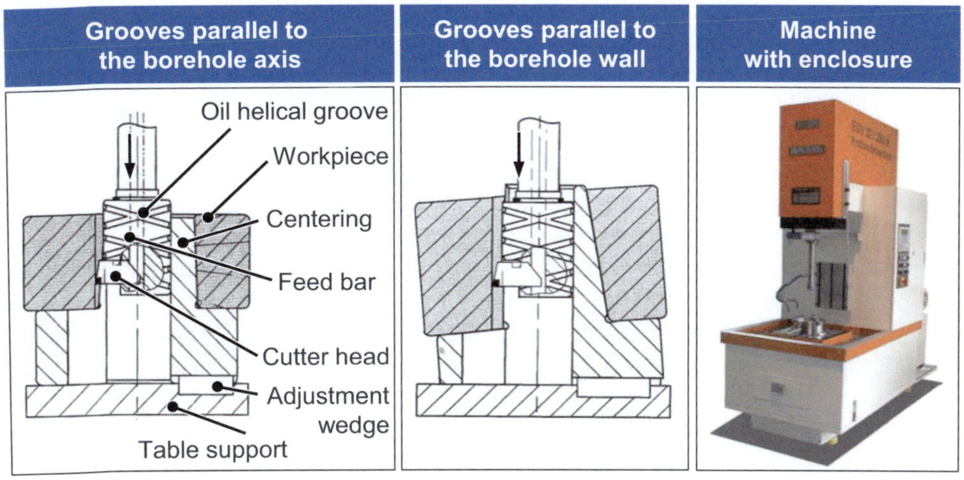

■ **Fig. 4.101** Hydraulic groove system with self-centering cutter head. *Source* Balzat

can be sunk at different positions in the workpiece with one clamping process. The grooving machine can be fitted with a tool changer to increase its flexibility. Using grooving machines available on the market, grooves with a maximum length of 1800 mm and a maximum width of 150 mm can be created. The tensile force can be up to 100 kN depending on the size of the drawing machine.

The slotting tool shown in ■ Fig. 4.101 makes grooves on the circumference of a bore without the workpiece having to be rotated using an indexing device. The tool performs the rotational movement in line with the individual part steps from groove to groove. The guide of the cutter head during the machining process is centered in the bore and is lubricated by the oil helical grooves located on the circumference. The parallel arrangement of such tools permits simultaneous machining of multiple bores. This multi-groove technique is used, for example, in the manufacture of motor housings on transfer lines.

Lastly, a typical part spectrum for slotting and drawing is shown in ■ Fig. 4.102. This type of grooving technique is primarily used whenever enclosed cavity grooves need to be sunk into workpieces, or grooves into enclosed components. Furthermore, slotting and drawing techniques are always used for open components when the cost-effectiveness of manufacturing an expensive broach cannot be justified.

4.6 Broaching Machines

Broaching machines allow integrated machining with one multi-toothed tool in one working stroke. ■ Figure 4.103 shows the function of an internal circular broach. The cutters of the broach are positioned behind each other and are staggered by the appropriate cutting thickness *fz* for roughing/smoothing. This replaces the radial infeed [DIN03g]. In the calibra-

Fig. 4.102 Typical workpieces for slotting and drawing. *Source* Balzat

tion part, however, the gradient is zero. When the tool is resharpened, the entire tool profile shifts towards the calibration part by one tooth. This means the previous first calibration tooth becomes the last finishing tooth and the expensive tool can be resharpened 15 to 20 times depending on the number of calibration teeth. A wear-reducing coating of titanium nitride (TiN) prolongs the tool life of broaching tools. However, due to the high tool costs, broaching is only suitable for use in high volume production or when there are many identical grooves. Typical application examples are the broaching of the fir-tree profiles of turbine rotors for the aviation industry and for power plant construction. The machining of high-strength materials having high-temperature strength results in high broaching forces and represents a particular challenge to the construction of the machine structure (stiffness) and the wear behavior of tools. Figure 4.104 shows a schematic of the different broaching techniques.

Internal or external broaching machines, with vertical or horizontal design, are used depending on the machining task. The cutting motion of the broaching tool is provided by hydraulic or electromechanical drives. Here, the higher stiffness of electromechanical drives reduces the tendency of the tool to vibrate along its longitudinal axis, something that can occur as a consequence of alternating cutting forces [HAßL92]. Although hydraulic drives represent the most cost-effective alternative, eliminating hydraulic fluid and the energy savings from electromechanical drives are becoming more attractive in the face of increasingly restrictive environmental legislation [OV94].

4.6.1 Internal Broaching Machines

Figure 4.105 shows the section drawing of a vertical internal broaching machine and a detailed photo of

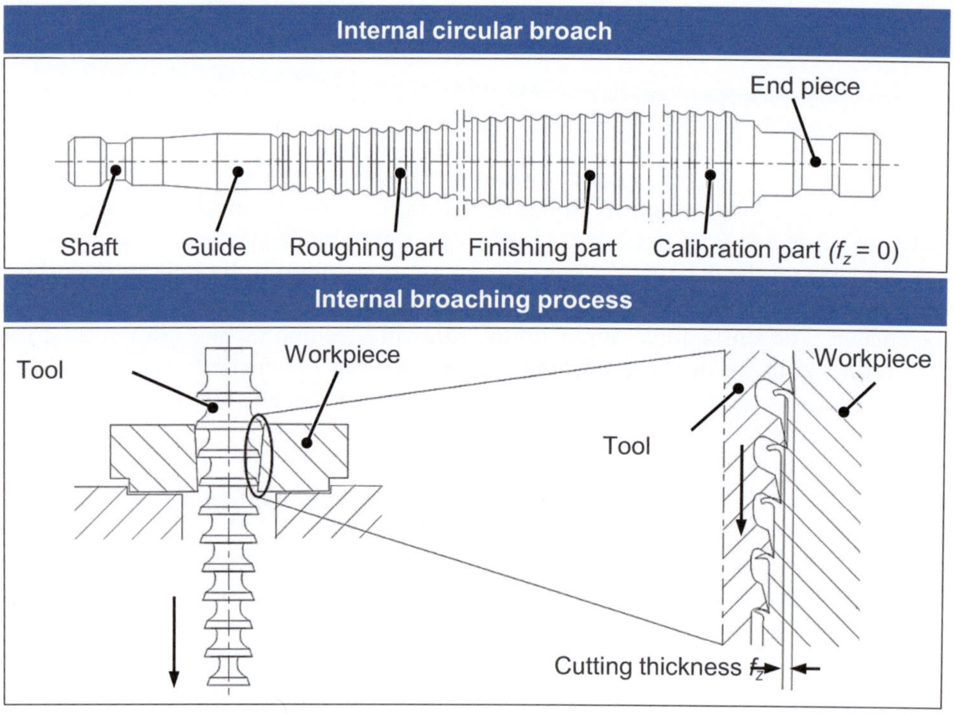

Fig. 4.103 Broaching tool with calibration part

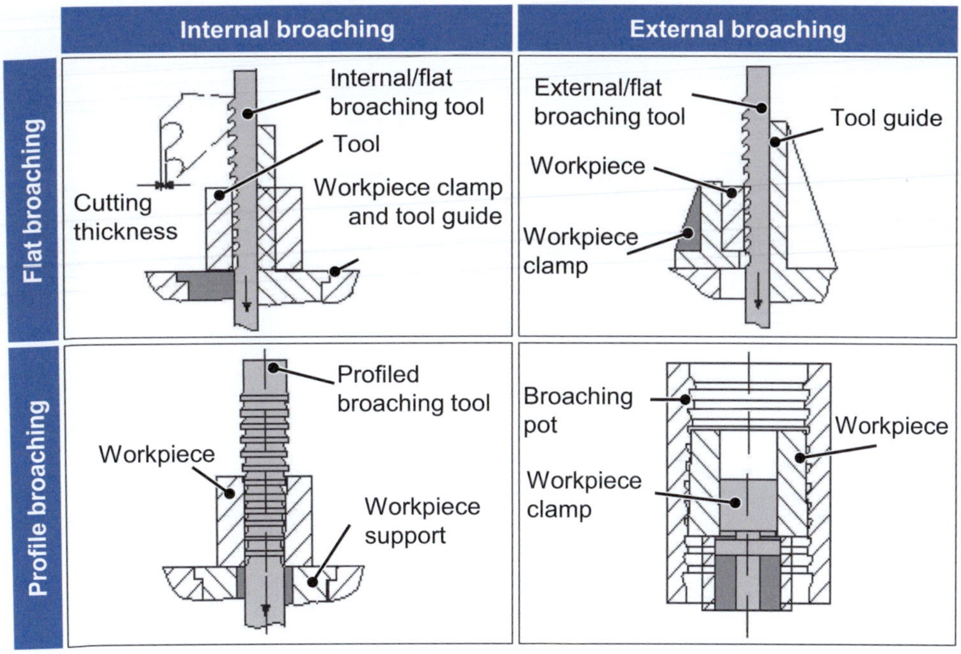

Fig. 4.104 Broaching techniques

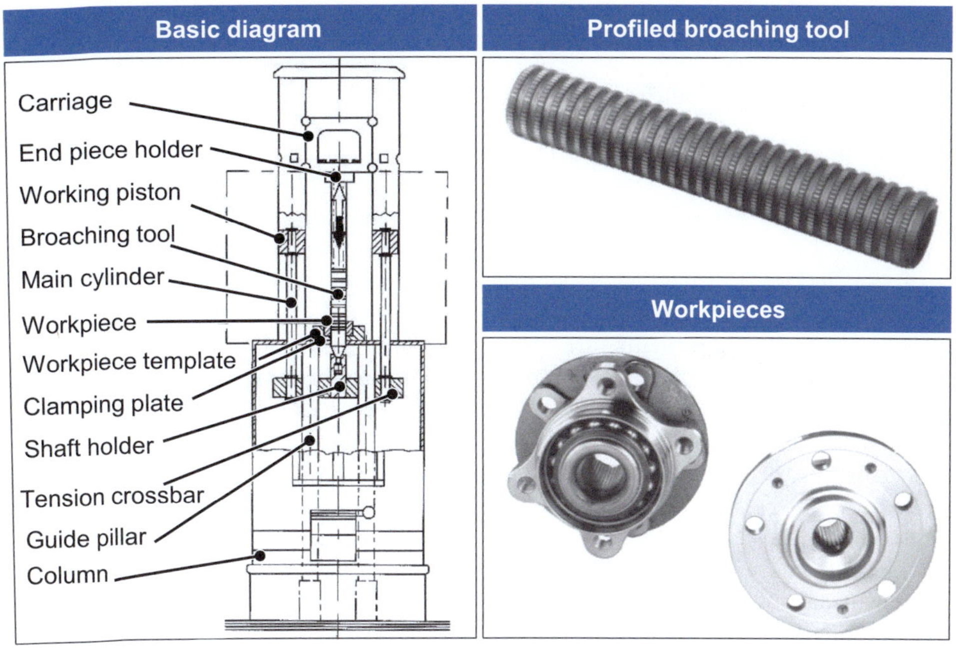

Fig. 4.105 Internal broaching machine. *Source* Forst

a profiled broaching tool with two typical workpieces. For the machining process, the workpiece is placed onto the workpiece template and centered. The broaching tool hanging in the end piece holder is inserted into the workpiece bore and locked in the shaft holder underneath the workpiece. The two hydraulic pistons transfer the force to the tool via the connected tension crossbar and shaft holder and pull the tool through the workpiece bore. The tool performs the last part of the broaching stroke released from the end piece holder, without the top guide. The broach is pulled through the workpiece completely and the machined workpiece is removed from the machine. The shaft holder then moves to the top again with the tool where it is locked in the end piece holder. Once the shaft holder has released the broaching tool at the lower end, it can be

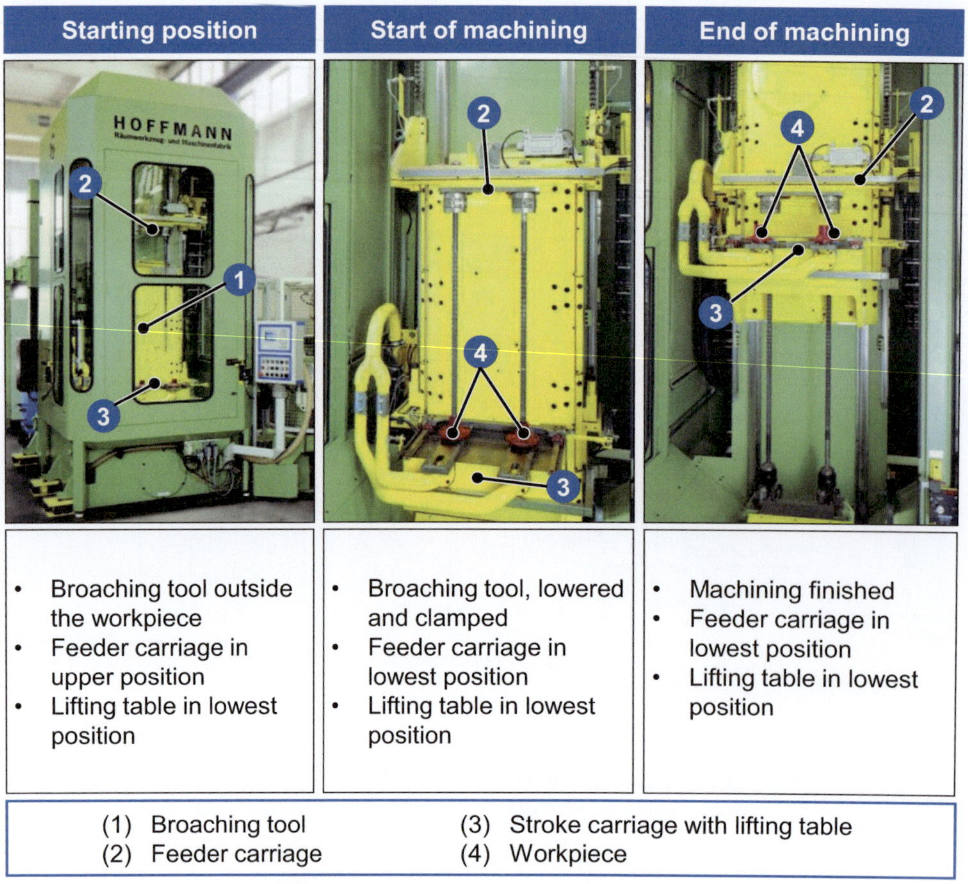

Fig. 4.106 Broaching process for a broaching machine with lifting table. *Source* Hoffmann Räumtechnik

raised into the starting position for entry of the next workpiece.

One variant of an internal broaching machine is the broaching machine with lifting table shown in Fig. 4.106. Here, several components are machined simultaneously by multiple broaches. It is not the broaching tool that performs the working stroke but the lifting table with the workpieces [HOFF03]. This enables the installation height to be reduced significantly compared to conventional vertical internal broaching machines. The broaching tools remain clamped on the shaft and end piece over the entire working stroke ensuring smooth machine running, high surface qualities and long tool lives. The stroke carriage and feeder equipment have joint, pretensioned linear rolling guides. The broaching force is 100 kN for a maximum broaching length of 1250 mm and broaching speed v_f of 25 m/min. [DIN03g].

With conventional broaching tools, only open grooves can be created, that is, grooves having profiles extending from the tool entry level to the tool exit level of the workpiece. For closed groove shapes where the profile does not extend to the appropriate workpiece entry/exit level, only the more time-consuming slotting and drawing techniques were options. For this reason, an internal broaching tool with movable cutting edges was developed. It has closed cavity grooves that can also be machined in one tool stroke. Figure 4.107 shows a cross-section through such a tool and possible groove shapes that it can produce.

During the downward movement of the shaft, the moving cutters are first in the initial position inside the contour of the shaft, until the control unit for the cutters touches the template. During the linear movement of the shaft, the cutters penetrate the side of the workpiece following the template contour and remove the material. The cutter lengths increase from the bottom up, meaning the length difference between two adjacent cutters corresponds exactly to the cutting thickness just like conventional broaches. This allows contoured grooves to be created in internal grooves and on the outside of a workpiece in just one working stroke.

Another special method of internal broaching with a lifting table is the helical broaching technique. In this method, a Cyclo gearbox is used to rotate the tool about the Z-axis during the broaching process (see Fig. 4.108). This means the internal broaching technique can also be used to create inclined grooves. Planetary roller screws or ball screw drives generate the drive along the Z-axis. The helix kinematics can be pro-

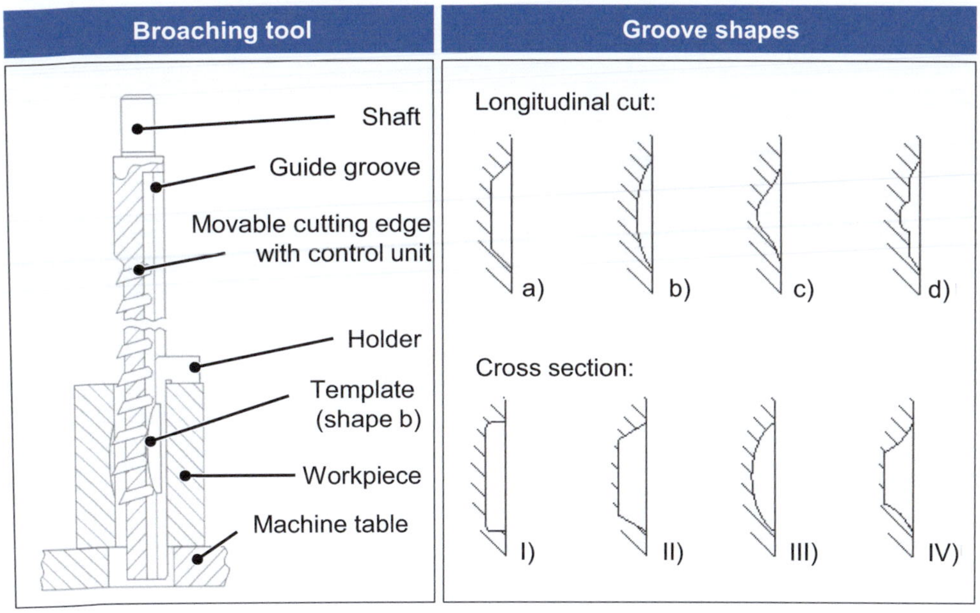

Fig. 4.107 Broaching tool with movable cutting edges. *Source* Balzat

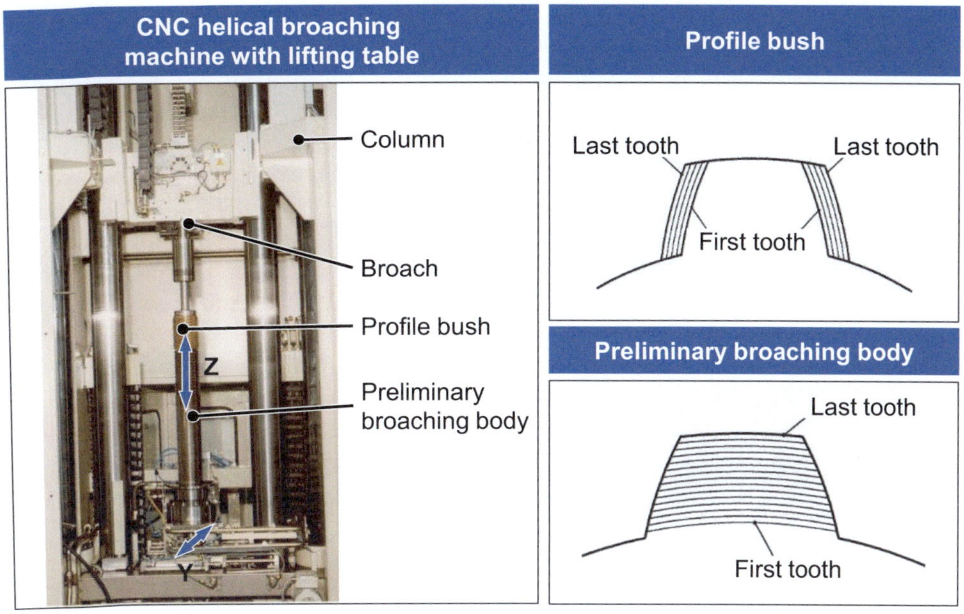

Fig. 4.108 Helical broaching machine. *Source* KARL KLINK

grammed. Clockwise and counterclockwise rotations are possible for different helix angles. For this, the helix unit has a gearbox that is clearance-free, torsionally stiff and can withstand torques up to 10,000 Nm. To traverse the workpieces between the loading and unloading positions, the slide table on the lifting table is moved along the Y-axis. The double-column design permits the installation of automatic loading and unloading from the rear of the machine. At the same time, access is free from the front for setup, tool change, maintenance and manual loading/unloading.

The broaching tool can be divided here into two different cutting parts. These are the preliminary broaching body with offset in depth for preliminary broaching and the flank-cutting profiled broaching bush with lateral offset for the finishing of the broaching process. During the finishing of the broaching process, the cutting pressure is minimized to attain the maximum gear teeth quality. The clamping pressure for the clamping device can be adjusted on the lifting table.

The lifting table (see Fig. 4.109) is driven by a gantry assembly. The electromechanical drive trains on both

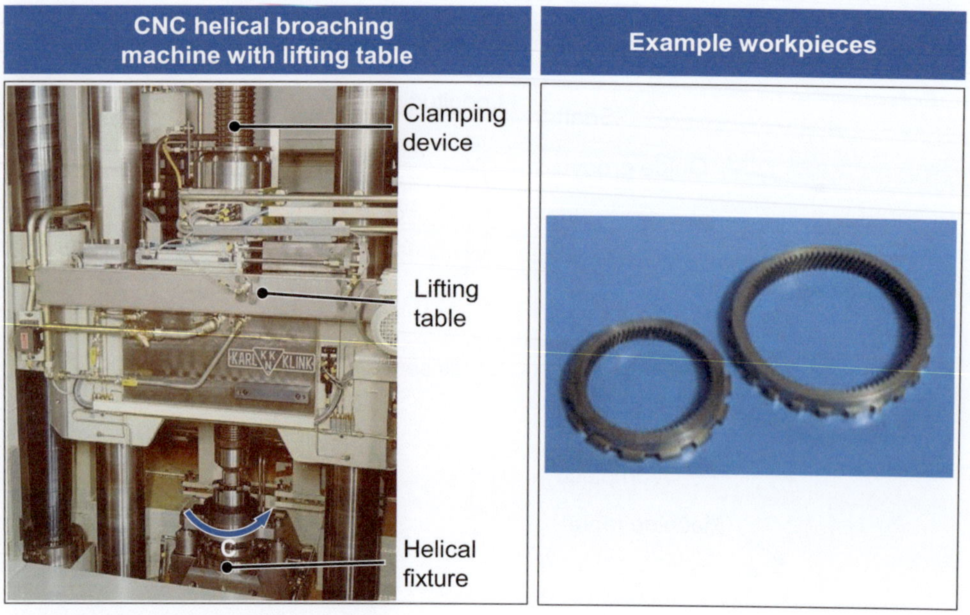

Fig. 4.109 Helical broaching machine. *Source* KARL KLINK

sides of the lifting table are coupled electronically. As thermally and mechanically related geometrical deviations can occur for the drive spindles, there is a longitudinal scale on each side that measures the errors in real-time, and thus enables compensation. This means the mechanical components are protected from one-sided overloading.

4.6.2 External Broaching Machines

For machining tasks requiring very long broaching tool lengths, vertical external broaching machines are used for example (as shown in Fig. 4.110) for machining fir-tree grooves in a turbine compressor disk. Splitting the broaching tool, up to 22.5 m in length, into a maximum of 10 adjacent automatically exchangeable tool cassettes, keeps the machine dimensions comparatively compact. The broaching machine is fitted with an electromechanical drive and has up to seven NC-controlled axes. The benefits of the vertical machine design are lower space requirement, prevention of the broaching tool deflecting under its own weight, improved cooling lubricant effect and ease of integration into transfer lines.

In addition to vertical broaching machines, horizontal variants are also used for longer tool lengths. The

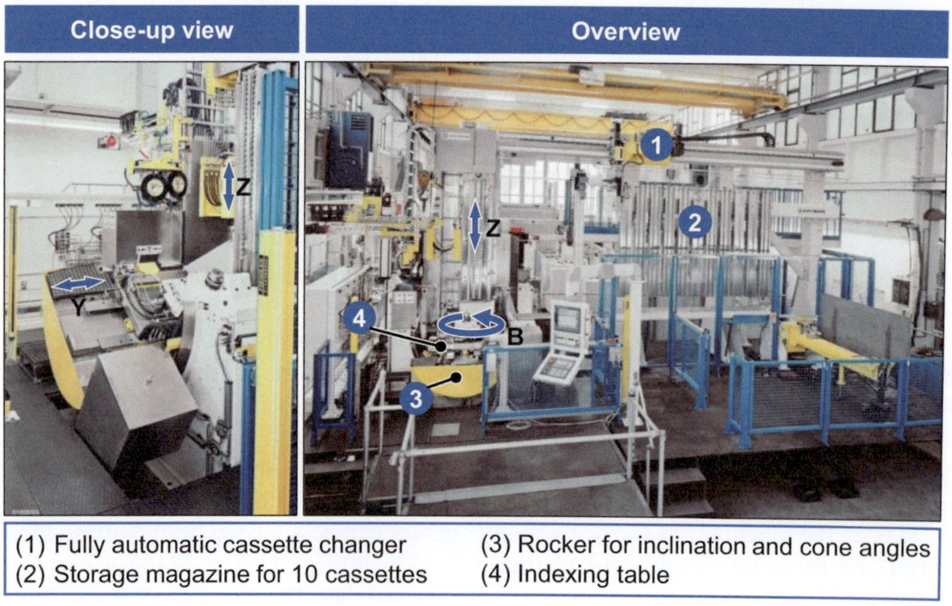

Fig. 4.110 Vertical external broaching machine. *Source* Hoffmann Räumtechnik GmbH

benefits of a horizontal broaching machine are a considerably lower installation height and the fact that no pit foundation is required. Also, it is easier to load heavier workpieces. These benefits must be weighed against the much greater installation space required, fewer options for integration into transfer lines, poorer cooling lubricant effect and a potential for the broach to deflect under its own weight.

In the same way as for vertical broaching machines, the workpiece is clamped on a table and positioned above the machine axes. There are tables with A and B-rotation axes for turning and tilting to execute inclined or tapering grooves. This type of table is shown on the right in ◘ Fig. 4.111. There are two types of feed kinematics: Either the workpiece is guided past the fixed broach or the workpiece is fixed and the broach is moved. A machine type with a moving tool is shown on the left in ◘ Fig. 4.111. The benefit of this concept is that the accuracy requirements placed on the foundation and machine structure are far less strict because the process always takes place in a fixed location (refer also to the benefits and downsides of portal milling machines in ► Sect. 4.3.2). The downside is the large installation space because twice the broach or broaching ram length is required. The broach length is usually 6.3 m. Machines with broach lengths of three to ten meters are also available.

The broaching machine shown in ◘ Fig. 4.111 is designed for the production of turbine disks having diameters in the range of 150 mm to 1400 mm for airplane engines or turbochargers. It also has an electromechanical drive with asynchronous servo motor and worm gear (having a power rating between 80 and 100 kW depending on the variant). The maximum broaching force that can be sustained at a broaching speed of 60 m/min is 250 kN. Variants with broaching forces up to 320 kN are available. The chip thickness is generally 0.01 mm. The guides of the machine system shown are hydrostatic to guarantee high stiffness. No numerical controller is required for path control. Given their functionality, however, numerical controllers are used for tool and program management and flexible programming. An automatic tool changer normally complements the machine system. As an alternative to a tool changer, multiple broaches can be clamped next to one another on a large broaching ram. In industry, however, there is a clear tendency towards tool changers. This is because of the risk of collision with parallel broaches on one carriage and the difficulty of achieving a high overall stiffness of a wide broaching ram with multiple paths. Workpiece clamping is normally manual due to the relatively long machining times on the machine. To shorten retooling times, there are machines that have quick-change systems for fixtures. This means it is possible to load the next workpiece into a fixture during primary machining time and then to clamp the fixture with the workpiece on the machine.

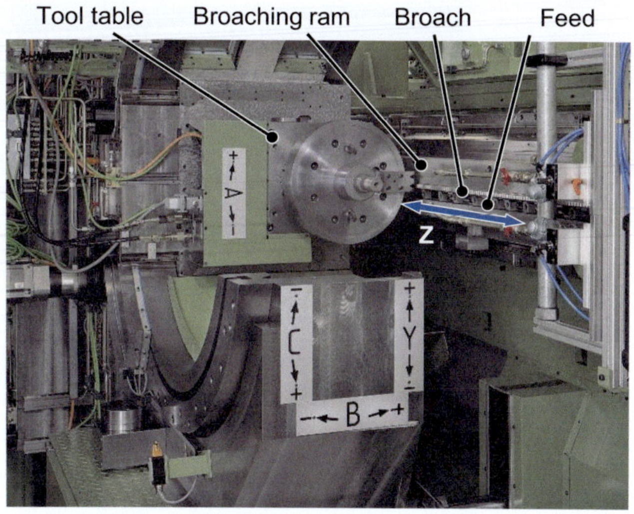

◘ **Fig. 4.111** Horizontal external broaching machine. *Source* Hoffmann Räumtechnik GmbH

Chain broaching is a special type of external broaching with a fixed tool and workpieces with continuous translational motion. This method is used only very rarely today. The clamping devices attached to the rotating broaching chains move with the workpieces in the tool tunnel, past the fixed broaching tool such that the required surfaces can be machined. The loading and unloading of broaching devices with workpieces is usually automated for chain broaching machines.

For broaching with a pot broaching machine, as shown in ◘ Fig. 4.112, the broaching pot surrounding the workpiece generates the outer contour required. The tool is mounted on a linearly guided carriage with spindle drive and is pushed by a mandrel above the workpiece secured to the push rod. Workpiece changes are automated by a conveyor and handling device.

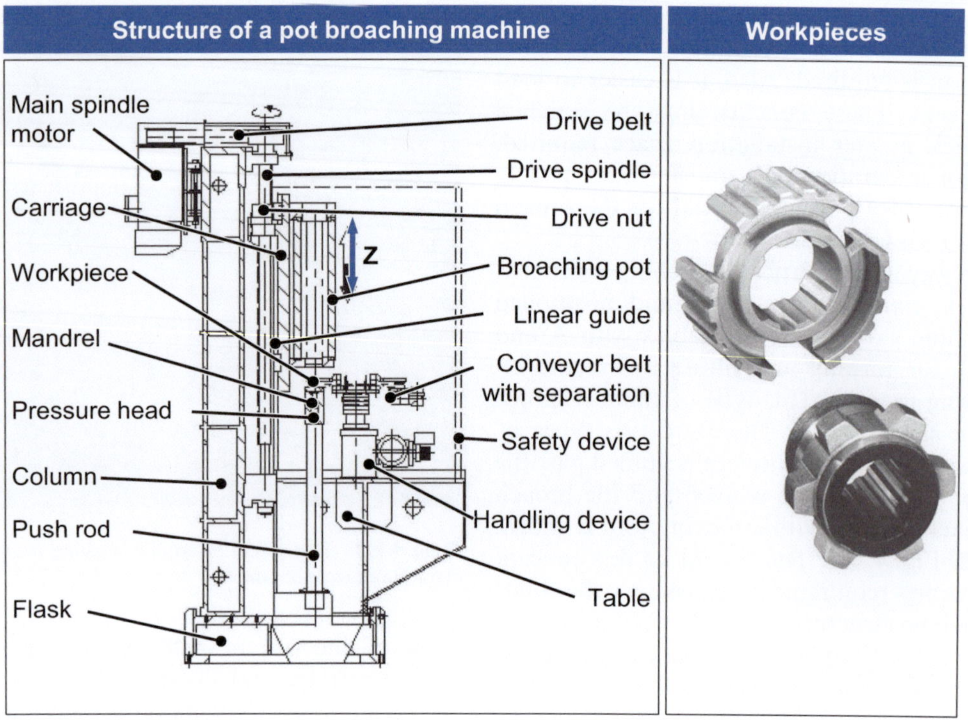

Fig. 4.112 Pot broaching machine. *Source* Forst Technologie

References

[BREC16] Brecher, C.: Werkzeugmaschinen Fertigungssysteme. (Reihe: Werkzeugmaschinen Fertigungssysteme. Bd. 2). 9. Aufl. Springer, Berlin (2016)

[DIN03a] Norm DIN 8589-6 (September 2003). Fertigungsverfahren Spanen—Teil 6: Sägen—Einordnung, Unterteilung, Begriffe

[DIN03b] Norm DIN 8589-3 (September 2003). Fertigungsverfahren Spanen—Teil 3: Fräsen—Einordnung, Unterteilung, Begriffe

[DIN03c] Norm DIN 8580 (September 2003). Fertigungsverfahren—Begriffe, Einteilung

[DIN03d] Norm DIN 8589-4 (September 2003). Fertigungsverfahren Spanen—Teil 4: Hobeln, Stoßen—Einordnung, Unterteilung, Begriffe

[DIN03e] Norm DIN 8589-0 (September 2003). Fertigungsverfahren Spanen—Teil 0: Allgemeines; Einordnung, Unterteilung, Begriffe

[DIN03f] Norm DIN 8580 (September 2003). Fertigungsverfahren—Begriffe, Einteilung

[DIN03g] Norm DIN 8589-5 (September 2003). Fertigungsverfahren Spanen—Teil 5: Räumen—Einordnung, Unterteilung, Begriffe

[DIN10] Norm DIN EN ISO 286-1 (November 2010). Geometrische Produktspezifikation (GPS)—ISO-Toleranzsystem für Längenmaße – Teil 1: Grundlagen für Toleranzen, Abmaße und Passungen

[HAßL92] Haßlach, K.H. (Hrsg.): Elektromechanische Antriebe an Räummaschinen. Karlsruhe (1992)

[HOFF03] Hoffmann Räumtechnik GmbH: Hubtisch-Räummaschine COMPACT. Pforzheim. Firmenschrift (2003)

[KAST04] Kasto Maschinenbau: Rationelles Gehrungssägen großer Querschnitte. Achern-Gamshurst. Firmenschrift (2004)

[KLOC08] Klocke, F., König, W.: Fertigungsverfahren 1, 8th edn. Springer, Berlin (2008)

[KÖNI82] König, W., Witte, K., Essel, L.: Spezifische Schnittkraftwerte für die Zerspanung metallischer Werkstoffe. Stahleisen, Düsseldorf (1982)

[KULL94] Kullmann, J.H.: Profilen Zähne zeigen. In: NC-Fertigung. **2** (1994)

[OV94] o. V.: Sieben in einem Schnitt. In: NC-Fertigung. **6** (1994)

[OV96a] o. V.: Flexibles Sägezentrum für Teile in kleiner und mittlerer Losgröße. MM Maschinenmarkt **8** (1996)

[OV96b] o. V.: Kaltkreissäge ermöglicht Trockenschnitte. MM Maschinenmarkt **25** (1996)

[SAND12] Sandvik Coromant: Tiefbohren. Düsseldorf. Firmenschrift (2012)

[SMS98] SMS Eumuco GmbH: Blank und Edel: Maschinen und Anlagen zur Herstellung von Blankstahl höchster Qualität. Firmenschrift (1998)

[STEI01] Steinbusch, G.: Sicher konstruieren mit Mechatronik-Simulation. Werkstatt und Betrieb. **134**(9), 163–166 (2001)

[STOL96] Stolzer, A.: Wirtschaftliches Zuschneiden und Komissionieren. Technica **11** (1996)

[VDI14] VDI-Richtlinie VDI 3208 (April 2014). Tiefbohren mit Einlippenbohrern

[VDI99] VDI-Richtlinie VDI 3209-1 (November 1999). Tiefbohren mit äußerer Zuführung des Kühlschmierstoffes (BTA- und ähnliche Verfahren)

Metal Cutting Machines for Tools with Geometrically Undefined Cutting Edges (Chip Removal)

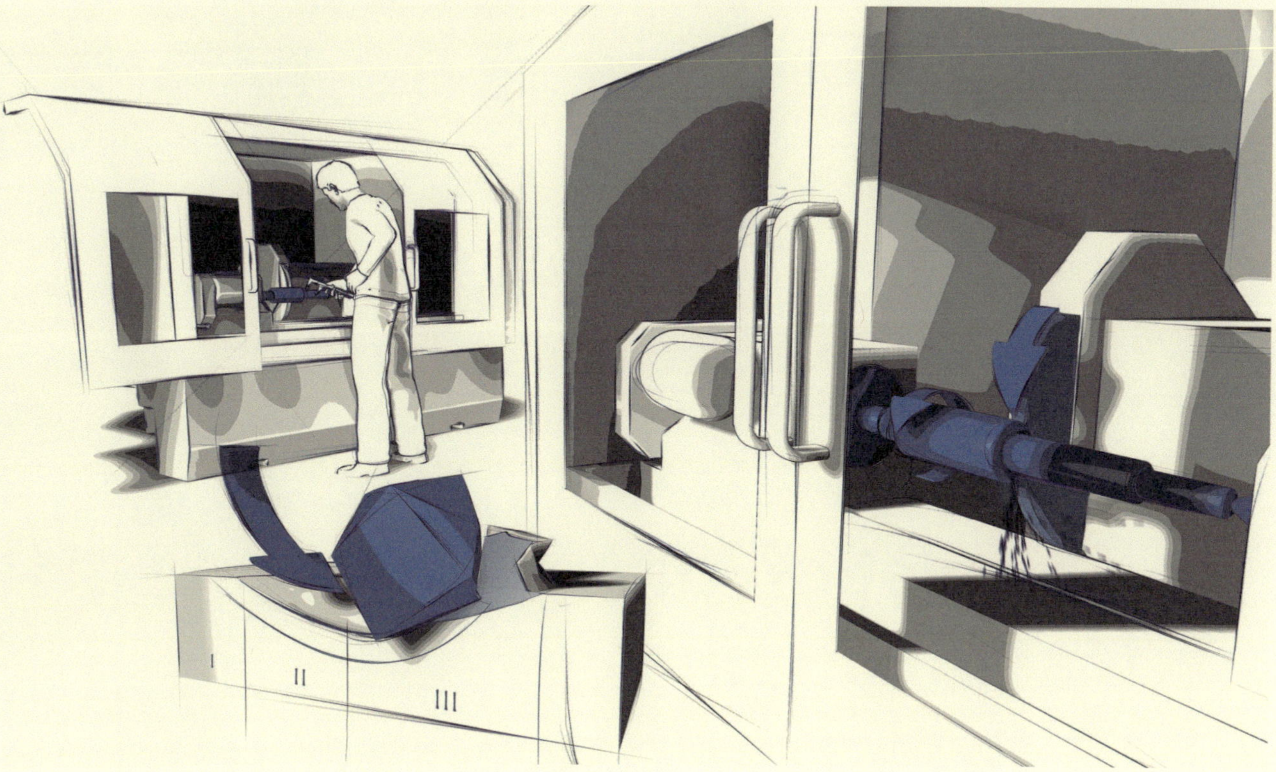

I thank my staff members Mathias Dehn M.Sc., Ralph Klimaschka M.Sc. and Dipl.-Ing. Stephan Neus for their contribution in revising this chapter.

C. Brecher and M. Weck, *Machine Tools Production Systems 1*,
Lecture Notes in Production Engineering, https://doi.org/10.1007/978-3-662-68120-6_5

Contents

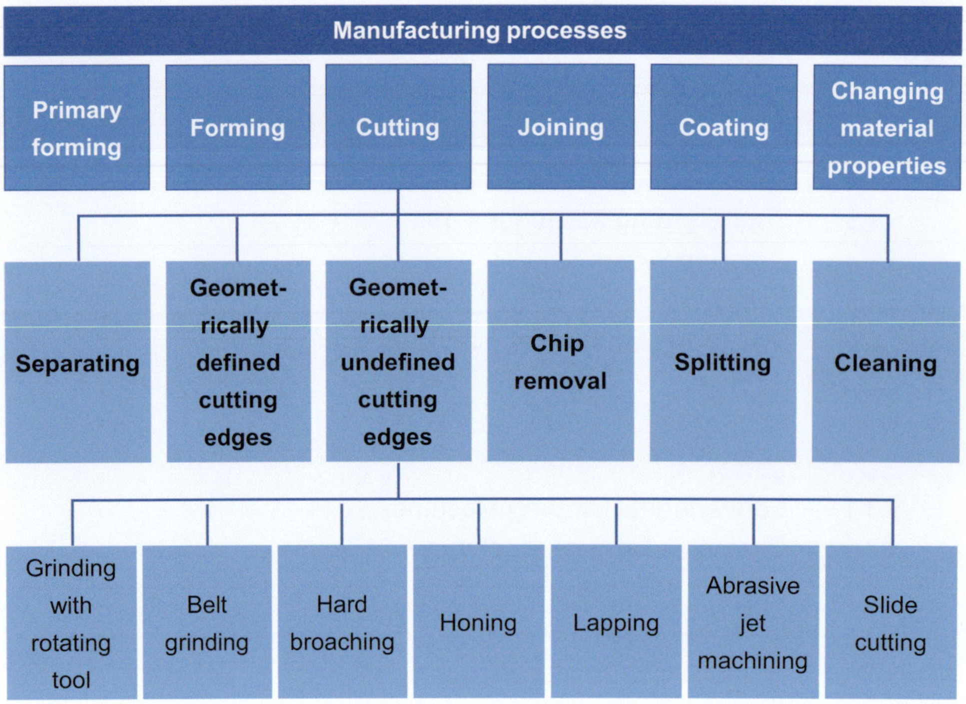

Fig. 5.1 Machining with geometrically undefined cutting edges—categorization of manufacturing processes. *Source:* [DIN03b]

The cutting manufacturing process "Machining with geometrically undefined cutting edges" includes processes which use tools whose number of cutting edges, geometry of the wedges, and position of the cutting edges with respect to the workpiece are undefined [DIN03c]. The classification of the manufacturing processes to [DIN03c] and the various subcategories are shown in Fig. 5.1. Figure 5.2 shows material removal using a tool (in this instance, a grinding wheel) with cutting edges that rotates at a high cutting speed relative to the workpiece. The cutting speeds range of 18–80 m/s for conventional corundum and silicon carbide grinding wheels and of 80–180 m/s for high-performance diamond and CBN grinding wheels.

The main application area for this process is usually precision machining. However, grinding is used nowadays for material removal at high rates. These precision machining processes are generally used if the required geometrical and dimensional accuracy and the surface quality of the workpieces cannot be achieved with other cutting processes. When grinding, the workpiece can be pre-machined and finished in a single clamping process by roughing and then finishing machining. An important area of application for grinding is the machining of workpieces from hardened steel or even carbide metal.

New requirements are being placed on the further development of high-performance grinding wheels, for example, through minimum quantity lubrication. In principle, the cutting speeds and chip removal rates generally expected from milling are possible with CBN grinding.

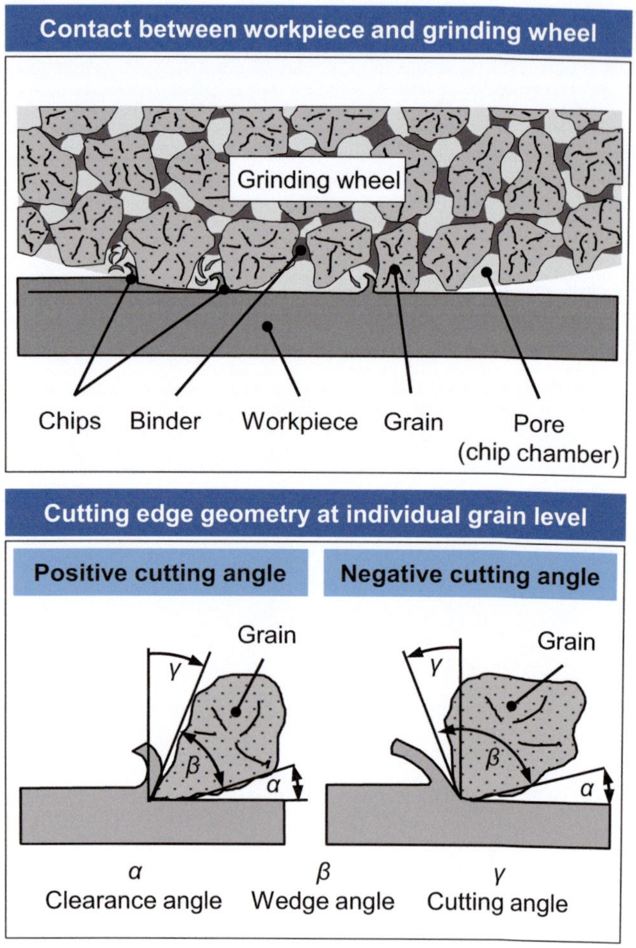

Fig. 5.2 Contact conditions between workpiece and grinding wheel

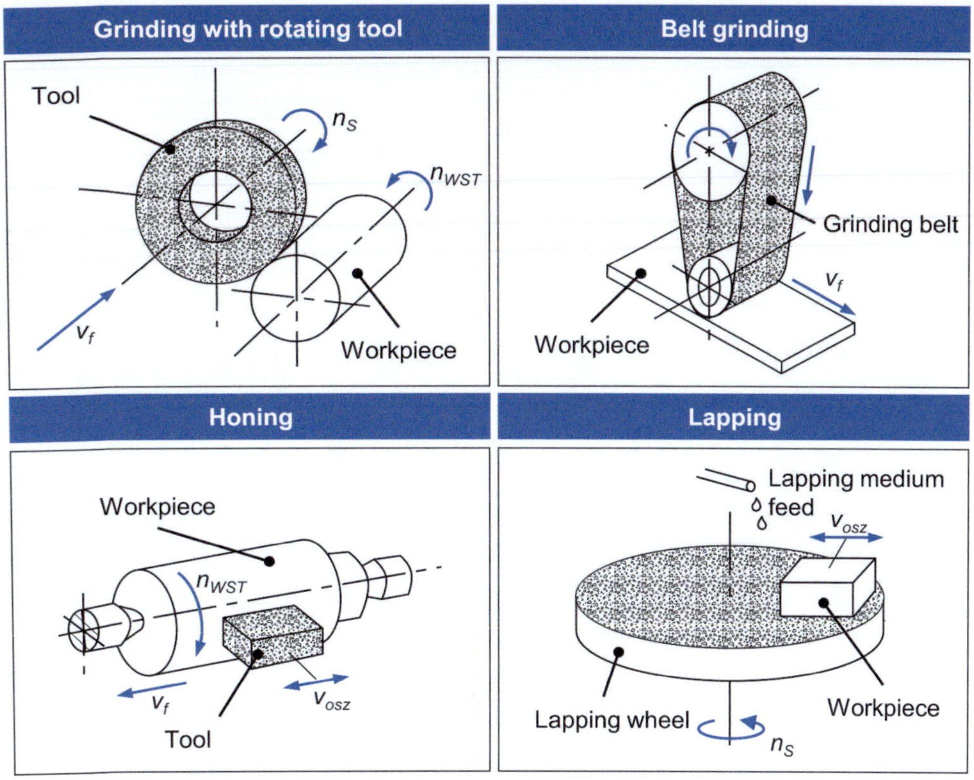

Fig. 5.3 Machining with a geometrically undefined cutting edge: process examples

Cutting processes with geometrically undefined cutting edges are subdivided according to the type of tool movement. Figure 5.3 provides an overview of the most significant processes. The shape and type of tool is a fundamental determinant of the basic subcategory. The most common process is grinding with rotating tools, defined as grinding in the following text. Here, the tool is a disk whose circumference or end face is used to achieve the cutting motion. In belt grinding, the abrasive chips are incorporated into a continuous, revolving belt.

Honing machines are generally subdivided into long-stroke and short-stroke machines. In this variation, material removal comes from a tool with geometrically undefined cutting edges (honing stone). However, unlike grinding, the cutting motions are relatively slow. Honing is mainly used to improve surface quality and dimensional accuracy. However, compensating for errors in shape (improving the dimensional accuracy in more than one coordinate direction) is only possible to a limited extent and with considerable expenditure on technical equipment as the honing stones are pressed onto the workpiece elastically and have no forced guidance.

Lapping machines are basically used for surface lapping. Like honing, lapping is most suitable for reducing the surface roughness and thus increasing the dimensional accuracy of a workpiece. In lapping, material is removed from the workpiece by means of a relative motion between the workpiece and the lapping wheel while a lapping medium comprising fine lapping grains and oil is introduced. In shape or fit lapping, the two workpiece components to be matched are moved against each other while a lapping medium is introduced (e.g., gear lapping, lapping guide pins in drilled holes).

5.1 Grinding Machines

In grinding machines, the cylindrical tool rotates to provide the cutting motion. One or more feed motions and the infeed are superimposed on the cutting motion. The shape of the workpieces to be ground determines the process and the axes of motion, i.e., the type of machine. Figure 5.4 shows the different types of grinding machines and their kinematic relationships with a machining example [DIN03b, DIN81].

The process options for grinding can be categorized by elements such as the tool's contact surface (circumferential grinding—side grinding) or the relative direction of the cutting motion to the main feed motion (down, up, or reciprocating grinding).

Grinding machines come in many different designs due to the extremely diverse range of tasks and re-

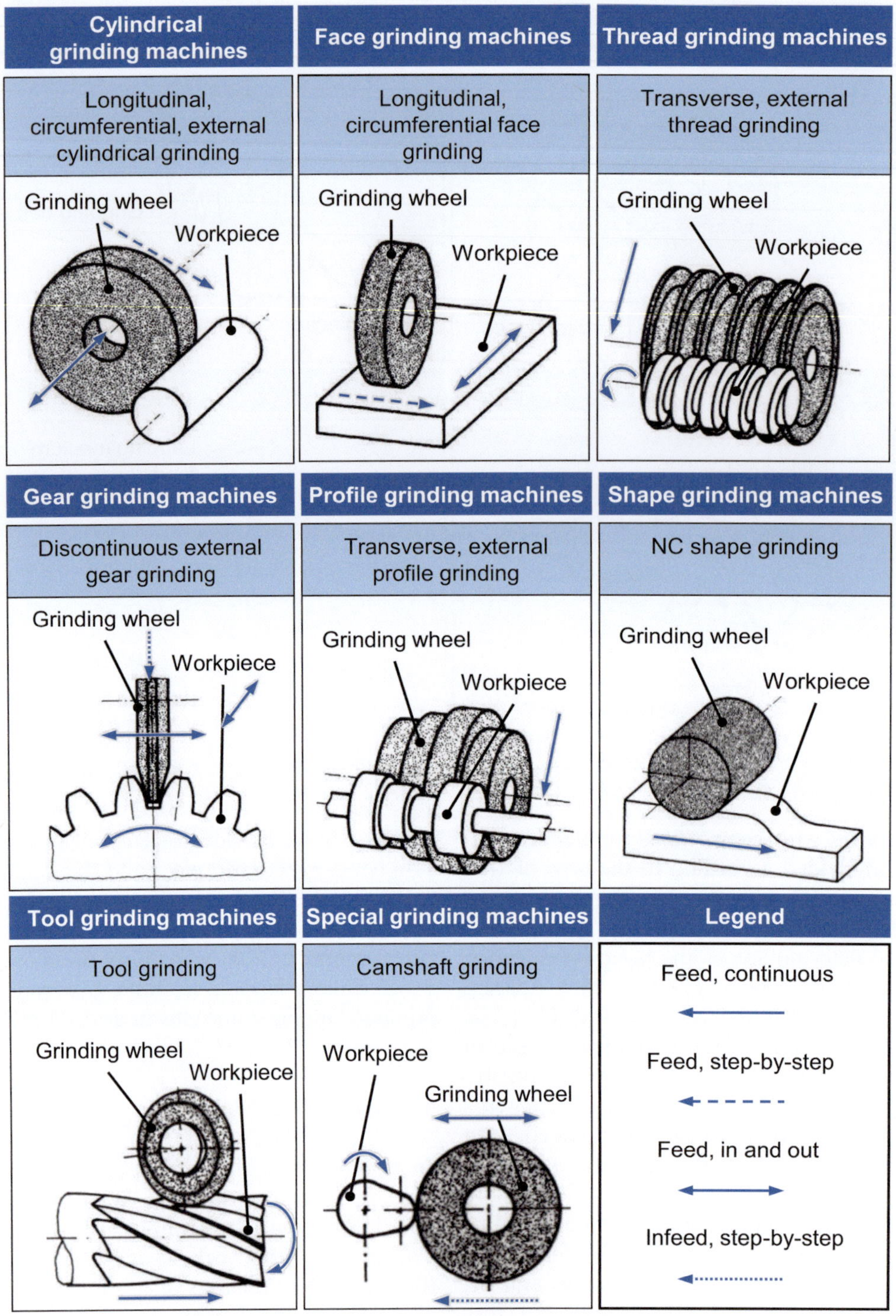

Fig. 5.4 Working principles in different types of grinding machines, according to [DIN03c]

quirements. The most significant machine types are discussed in the following sections together with the relevant technological tasks. Special-purpose machines for machining crankshafts and camshafts and gear grinding machines for machining gears are dealt with separately for editorial reasons (▶ Sect. 8.2).

5.1.1 Cylindrical Grinding Machines

Cylindrical grinding machines are characterized by one component of the feed motion being provided by the round, rotating workpiece. The second feed motion and the infeed act in the longitudinal or radial direction of the workpiece.

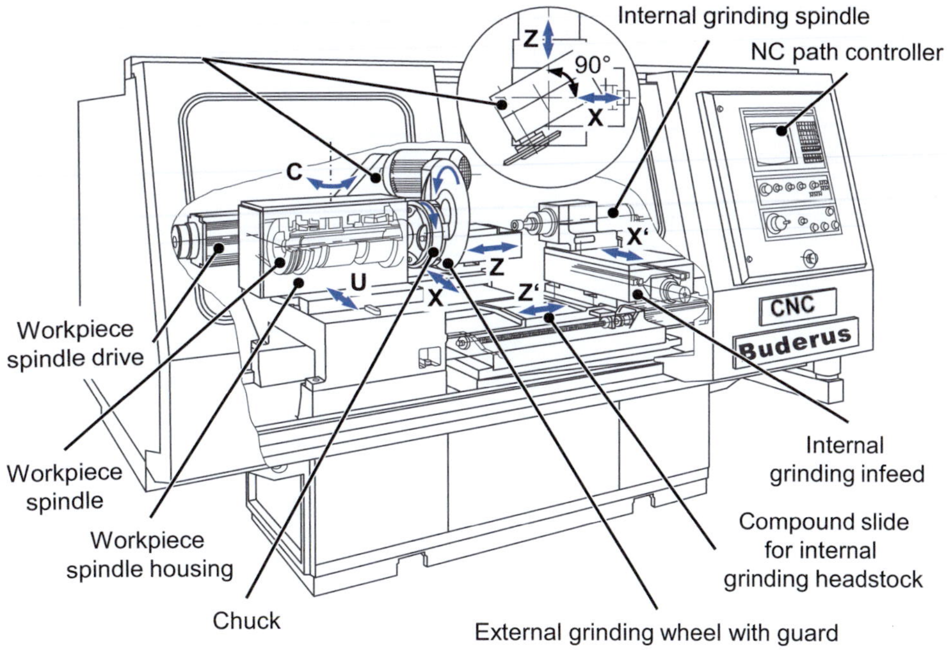

Fig. 5.5 Universal cylindrical grinding machine. *Source* Buderus

Figure 5.5 shows a universal cylindrical grinding machine, i.e., a machine used for external and internal machining as well as facing of the workpiece end faces.

The external grinding unit is supported by a compound slide (X and Z axes) on which the grinding headstock is configured in a swivel arrangement. In longitudinal grinding, the feed is via the Z-axis and the X-axis undertakes the step-by-step infeed; in transverse or side grinding, the feed and infeed directions are reversed. Bevel grinding is possible with this slide arrangement due to the corresponding angular position of the grinding spindle and the coordinated movement in both axes.

The internal grinding unit consists of a separate compound slide to support the grinding spindle. The longitudinal and transverse feeds are provided via the Z' and X' axes. The internal grinding unit also allows face grinding using a suitable grinding wheel (cup wheel).

The workpiece is held in the workpiece spindle housing chuck and is driven by a 3-phase servo motor to achieve the circular feed. The workpiece spindle housing can be swiveled and moved in the x-direction to execute tapered grinding.

Figure 5.6 shows an alternative design for a universal cylindrical grinding machine. The external and internal cylindrical grinding spindles on this machine are arranged on a turret grinding headstock and can be swiveled into the working position. The swivel axis (B-axis) allows angle adjustment of the grinding wheel for longitudinal bevel grinding or angular plunge grinding. Unlike the machine shown in Fig. 5.5 the longitudinal feed is provided by the workpiece table here. Hydrodynamic flat prism guide systems are used in both the X and Z-axes.

Slim components are usually machined between two centers. Both the workpiece spindle housing and the tailstock can be manually adjusted on the Z-slide to adapt to the workpiece length. The workpiece is clamped by a spring-mounted quill in the tailstock. As an alternative to clamping the workpiece between centers, the machine can be equipped with a workpiece chuck.

Long, heavy workpieces such as rollers for industrial applications are machined using equipment where the workpiece is fixed and the grinding headstock performs both the longitudinal feed and the radial infeed motions. Figure 5.7 shows an example. This example is for machining workpieces up to 6 m in length and 0.8 m in diameter weighing 6 metric tons. To ensure extremely high accuracy, the machine is fitted with an in-process measuring system to determine the diameter. The redundant infeed and swivel axes integrated into the headstock are a special design feature of the machine. These allow extremely accurate fine adjustment when hollow grinding and crown grinding rollers.

Machines designed only for internal cylindrical machining are actually relatively rare and are primarily used in volume production. Otherwise, the preferred

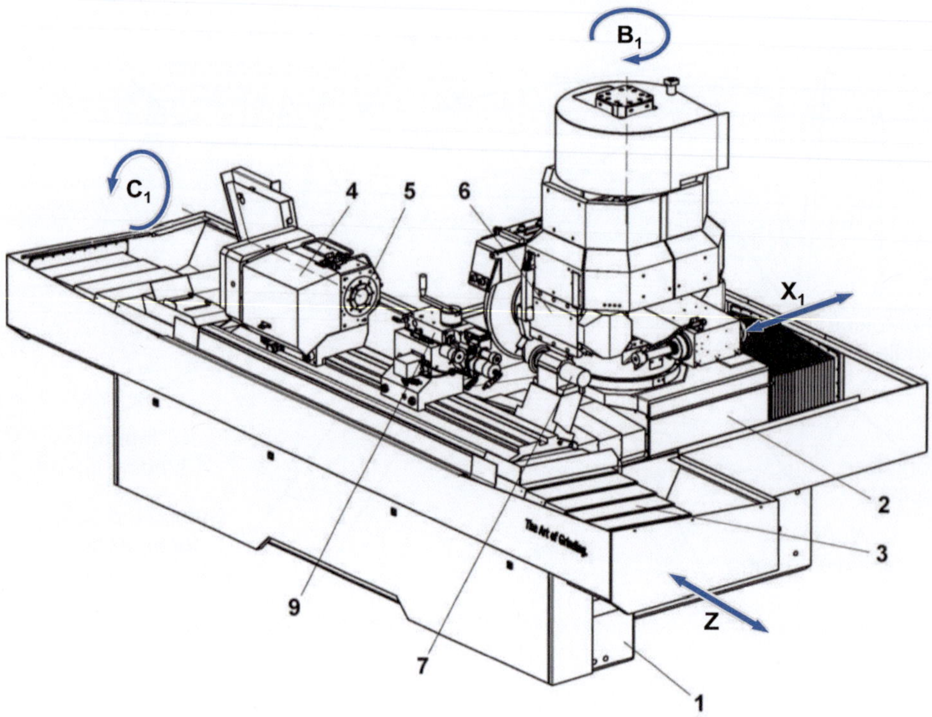

1	Machine bed	6	Grinding spindle
2	Transverse axis (X)	7	Dressing spindle
3	Longitudinal slide (Z)	8	Dressing device (including optional U-axis, not shown here)
4	Workpiece headstock (C)		
5	Workpiece clamp	9	Tailstock

Fig. 5.6 Universal cylindrical grinding machine. *Source* Schleifring, Studer

option is a universal cylindrical grinding machine (Fig. 5.5). Figure 5.8 shows an internal cylindrical grinding machine with two main spindle units. The design with two grinding spindles allows various grinding operations to be carried out on one machine without wheel change. In practice, this can be used for pre-machining and finishing work or for a range of grinding processes such as longitudinal and angular plunge grinding.

The grinding spindles are each arranged on a separate compound slide unit supported on a common guideway for the x-direction. A hydrodynamic flat prism guide system is used in the x-direction while rolling element guide systems are used in the z-direction.

Both the grinding wheel diameter and the diameter of the grinding spindle are generally significantly smaller for internal cylindrical grinding than those used for external cylindrical grinding. This results in heavy radial wear and high static and dynamic compliance. Tool geometries and grinding spindles specifically adapted to the relevant process are therefore often used in internal cylindrical grinding to generate optimum conditions.

The workpiece is held between a support rail and a regulating wheel in centerless external cylindrical grinding, Fig. 5.9. The driven regulating wheel that rotates the workpiece is made of synthetic resin or hard rubber. The more rapidly rotating grinding wheel attempts to accelerate the workpiece, but the greater friction between the regulating wheel and the workpiece leads to the wheel and workpiece having the same circumferential speed except for some slight slip. The material removal

(1) Tailstock
(2) Grinding wheel
(3) Workpiece
(4) Workpiece headstock
(5) Grinding headstock
(6) Measuring system

Fig. 5.7 Roll grinding machine. *Source* Waldrich Siegen

through grinding is achieved by the relative motion produced between the grinding wheel and the workpiece.

The workpiece axis is above the line connecting the grinding wheel and regulating wheel axes, see Fig. 5.9 lower right. This is necessary to avoid or reduce deviations in workpiece circularity [KLOC05]. The workpiece support must be made of wear-resistant material due to the frictional contact. In some supports, the workpiece runs on an oil cushion. The support surface is inclined at an angle of φ with respect to the regulating wheel to increase the contact force of the workpiece on the regulating wheel.

In centerless external cylindrical grinding, both longitudinal and transverse grinding are possible. In transverse grinding, the feed is produced by reducing the distance between the grinding wheel and the regulating wheel or by adjusting the infeed of the support rail. Fixing the axial location in transverse grinding is implemented by swiveling the regulating wheel axis through a slight angle to produce a force in the longitudinal workpiece direction. This is absorbed by a stop.

The angle φ is increased for longitudinal grinding (Fig. 5.9 top). The longitudinal feed rate v_L is approximately.

$$v_L = d_R \cdot n_R \cdot \sin\varphi \tag{5.1}$$

where d_R is the mean regulating wheel diameter, n_R is the regulating wheel rotational speed, and φ is the angle between the workpiece axis and the regulating wheel axis. To ensure the linear contact between the workpiece and the regulating wheel is maintained in these instances, the regulating wheel must be dressed with a rotationally hyperboloid form. The transverse feed in throughfeed grinding is achieved from the relative increase in diameter (cone shape) of the grinding wheel and the regulating wheel in the workpiece's axial feed direction. This corresponds to the reduction in diameter of the moving workpiece to be achieved.

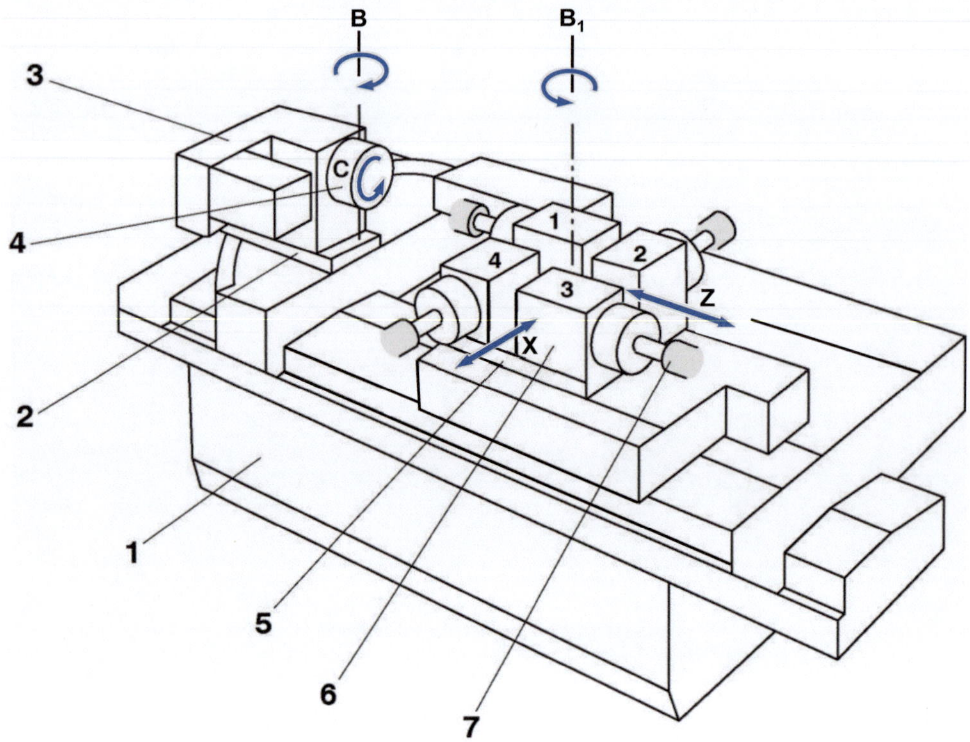

1	Machine bed for the compound slide machine (X/Z)	5	Grinding table with maximum 4 tool positions
2	Workpiece table (swivel table)	6	Grinding spindle carrier
3	Workpiece headstock	7	Grinding wheel
4	Workpiece clamp		

Fig. 5.8 Internal cylindrical grinding machine. *Source* Schleifring, Studer

This process is generally very well suited to large-scale production as the output is considerable due to the high removal rate and the short non-productive times (throughfeed grinding). However, compared to standard external cylindrical grinding machines, the tooling costs are many times higher.

Figure 5.10 shows some examples of the highly varied range of parts that can be machined in centerless grinding machines.

The structure of a centerless external cylindrical grinding machine is shown in Fig. 5.11. A distinction is made here between two different types. In one, as shown in the figure, the grinding wheel and regulating wheel headstocks are each arranged on a different slide and can be moved separately. The workpiece support is height-adjustable. In this arrangement, the workpiece is fixed in place. This makes automated workpiece feed easier. This type is particularly suitable for throughfeed grinding of bars and for automatic workpiece handling. In the other type, the grinding wheel is fixed in place and the regulating wheel and support are configured together on a slide.

In the machine shown in Fig. 5.11 the two slide units for the grinding spindle and regulating spindle units are guided by linear rolling element guide systems on a mineral cast machine bed. The spindle housing on the regulating wheel can be swiveled with respect to the slide to set a defined angle. As is usual in centerless grinding machines, the grinding wheel and the regulating wheel are both supported by bearings due to the large wheel width. The machine is designed with two separate dressing units for the grinding wheel and the regulating wheel. The dressing units are each arranged on the back of the grinding wheel and regulating wheel unit and can be moved on compound slides so that complex geometries can be produced such as the hyperbolic contour of the regulating wheel on throughfeed grinding (Fig. 5.12).

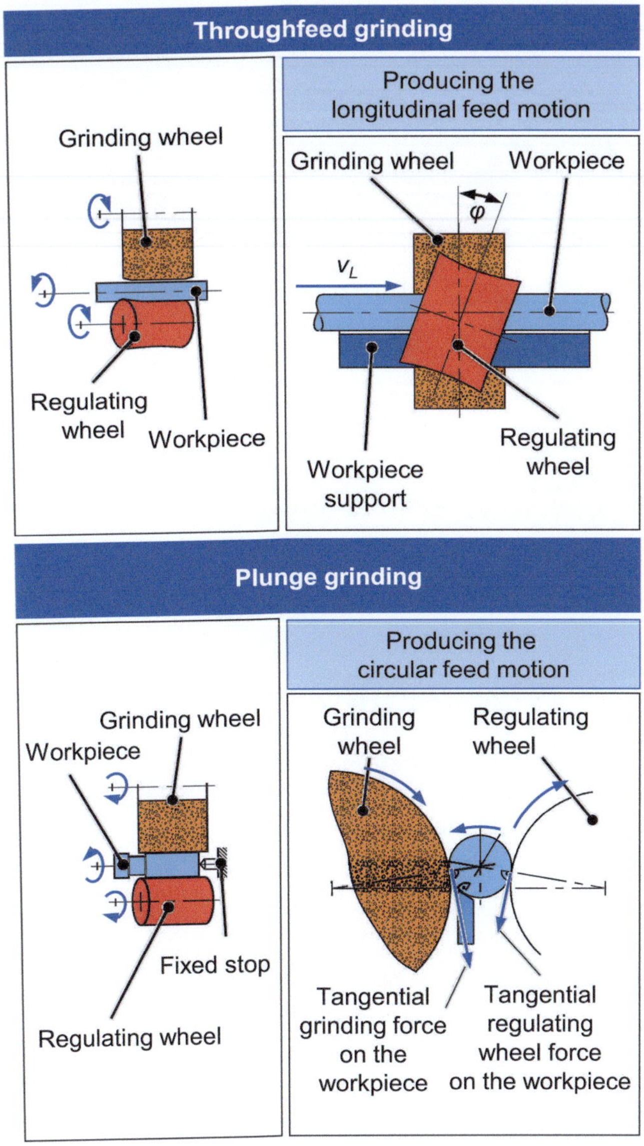

Fig. 5.9 Variants of centerless external cylindrical grinding

5.1.2 Face Grinding Machines

Flat surfaces are produced using face or flat grinding processes (Fig. 5.4). A distinction is made between circumferential and side grinding. Face or flat grinding machines can be fitted with reciprocating tables or rotary tables.

Figure 5.13 shows the structure of a typical face grinding machine with a horizontal spindle and reciprocating table. In this instance, the longitudinal and transverse feed motions are achieved by moving the workpiece along the X and Z-axes using a cross table. Moving the main spindle unit provides the vertical infeed. Linear rolling element guide systems and electromechanical feed drives based on ball screw drives are used in all three axes. In addition to the electrically driven ball screw drives used here, hydraulic cylinder-piston systems or direct electrical linear drives are used as an alternative for the longitudinal table feed. Magnetic clamping plates are often used to clamp the workpiece on universal flat grinding machines.

In the grinding machine shown in Fig. 5.14 all three linear axes are on the tool side. As when describing milling machines, in this instance, we also refer to this as a moving column design. The workpieces are clamped to the stationary table. If necessary, a workpiece swivel axis and/or a rotation axis can be clamped to the workpiece table, as shown in the figure. This allows complex geometries such as turbine blades to be ground in addition to pure face machining. The dressing unit with driven dressing rollers is located directly on the grinding spindle housing. Continuous infeed motion of the dressing rollers during the grinding process is possible due to NC-controlled swivel movement (continuous dressing or CD), see ▶ Sect. 5.1.4.

Figure 5.15 shows another face grinding machine in movable column design. In addition to the NC-con-

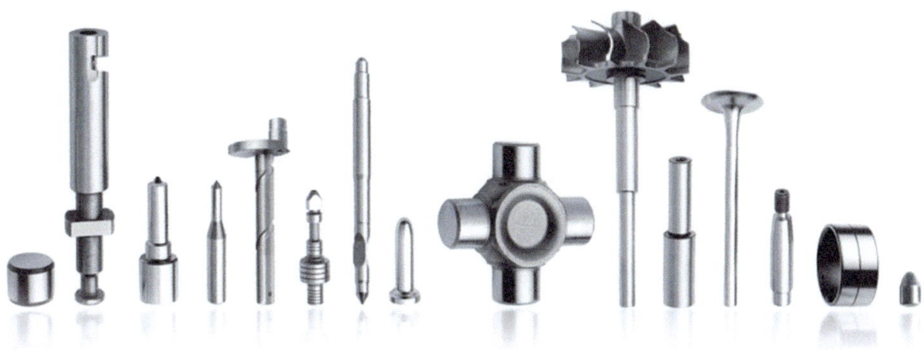

Fig. 5.10 Typical range of parts for centerless external cylindrical grinding. *Source* Junker

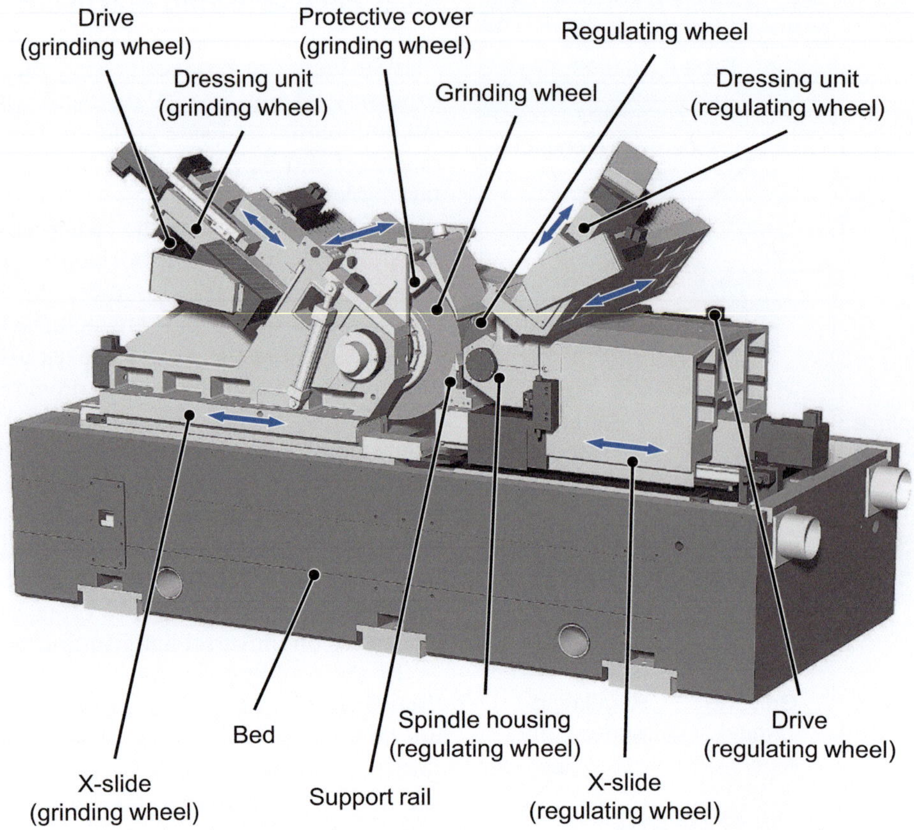

Fig. 5.11 Centerless external cylindrical grinding machine. *Source* Schleifring, Schaudt Mikrosa BWF

trolled linear feed axes, this machine has a 180° indexing table allowing machining of the workpieces during primary machining time. The machine can be fitted with either of the two dressing units shown. These are suitable for intermittent or continuous roller dressing.

Face grinding machines with a rotary table are particularly suitable for manufacturing rotationally symmetrical components. On these machines, translational positioning in two axes is usually achieved on the tool side while the machine table is solely responsible for producing the rotary feed motion. Figure 5.16 shows a grinding center with a rotary table suitable for the complete machining of several functional surfaces with the workpiece clamped only once. By using various grinding spindle units, each with their own vertical guide system in the movable column, in this instance, internal and external cylindrical grinding operations can be performed in addition to the face machining.

Figure 5.17 shows the structure of a double face grinding machine. This type of machine is typically used to manufacture workpieces with parallel surfaces in large-scale production. Appropriately, workpiece feed and removal operations are automated. In this machine, the feeder system moves the workpieces from a vibratory container into the perforated sheet workpiece carrier. By rotating the workpiece carrier, the workpieces are guided into the grinding gap. The workpieces are machined simultaneously on both sides by the spindles working in opposite directions. This machine can basically complete the machining as a throughfeed or plunge-cut process.

Guideway and surface grinding machines are face grinding machines in a column or portal design. They have to satisfy extremely high demands on the geometrical accuracy over large working area dimensions. The grinding machine shown in Fig. 5.18 is based on a static portal with the Y and Z-feed slides on the front of the machine. The longitudinal feed in the x-direction is produced by the hydraulically driven machine table. The twin table design also allows machining of the workpieces during primary machining time. To be able to machine horizontal, vertical and angled surfaces, e.g., prismatic guide systems, the grinding support has an additional swivel axis (A-axis).

5.1.3 Tool Grinding Machines

Tool grinding machines are used to manufacture and resharpen turning and planing tools, drill bits, milling

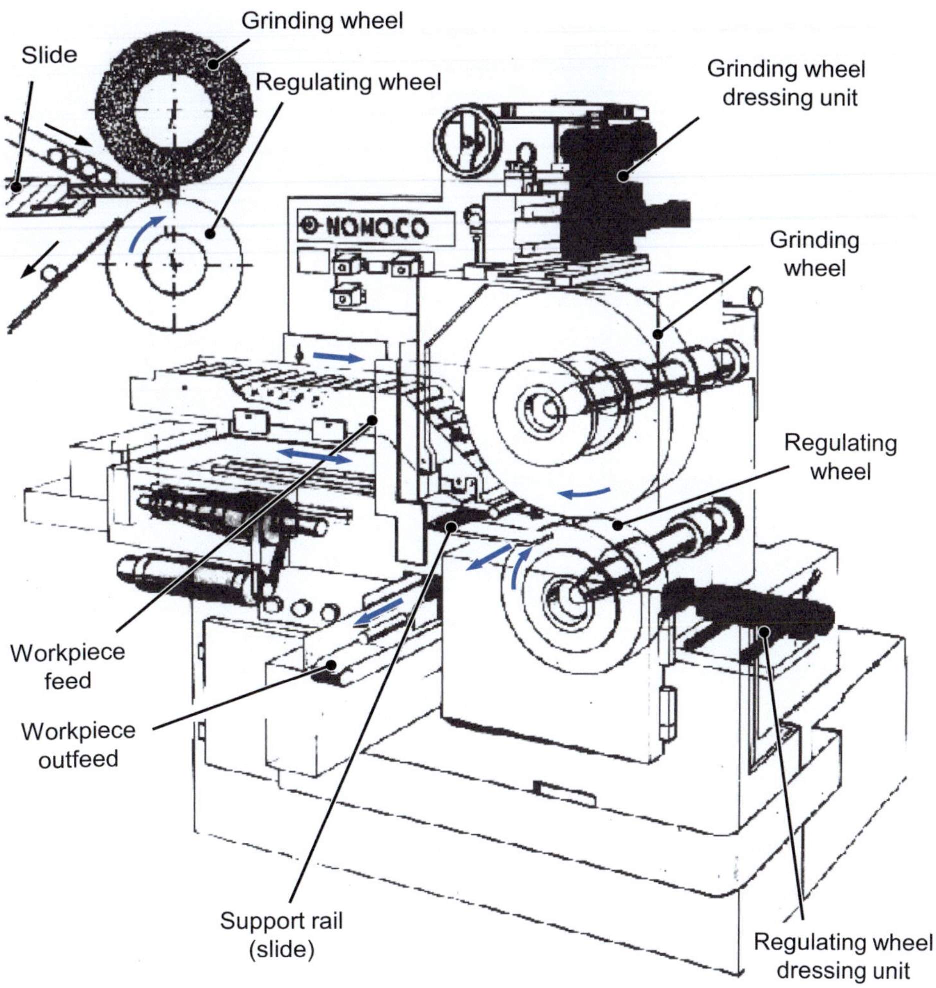

Fig. 5.12 Centerless external cylindrical grinding machine. *Source* Nomoco

tools and saws. Apart from simple chisel grinding machines where the chisel is guided against the grinding tool by hand, there are also universal and special tool grinding machines; these are used to grind tools with complex geometry.

The grinding machine shown in Fig. 5.19 is primarily used to produce milling tools. To produce the complex shape characteristics of this type of tool, such as spiral grooves, the machine has a total of five NC axes. The machine bed forms a single unit with the cooling lubricant tank, offering a balanced temperature distribution or low thermal deformation. The B-axis is at an angle of 30° to the vertical to improve stiffness. To permit the usual programming in orthogonal axes, coordinates are transformed in the control system. To provide the Z-feed, the complete grinding head on the column is moved in the horizontal direction. An additional compensation movement in the y-direction is required.

Figure 5.20 shows the arrangement of axes and the workspace for another tool grinding machine. The machine has a mineral cast bed and all the axes are driven by linear motors. The grinding spindle is supported centrally and can be fitted with several grinding wheels on both sides. This allows different geometrical elements to be machined without having to change the grinding wheel. Collet chucks, jaw chucks or hydro expansion chucks are generally used to clamp the workpiece.

5.1.4 Dressing Systems

Dressing provides the grinding wheel with a defined geometrical shape (macrogeometry) and also produces a defined grinding wheel topography. This topography is part of the microgeometry in addition to the position and structure of the grains. While the macrogeometry of the grinding wheel is the primary influence on the dimensional and geometrical accuracy of the workpiece, the surface roughness of the workpiece is largely determined by the microgeometry. During dressing, the

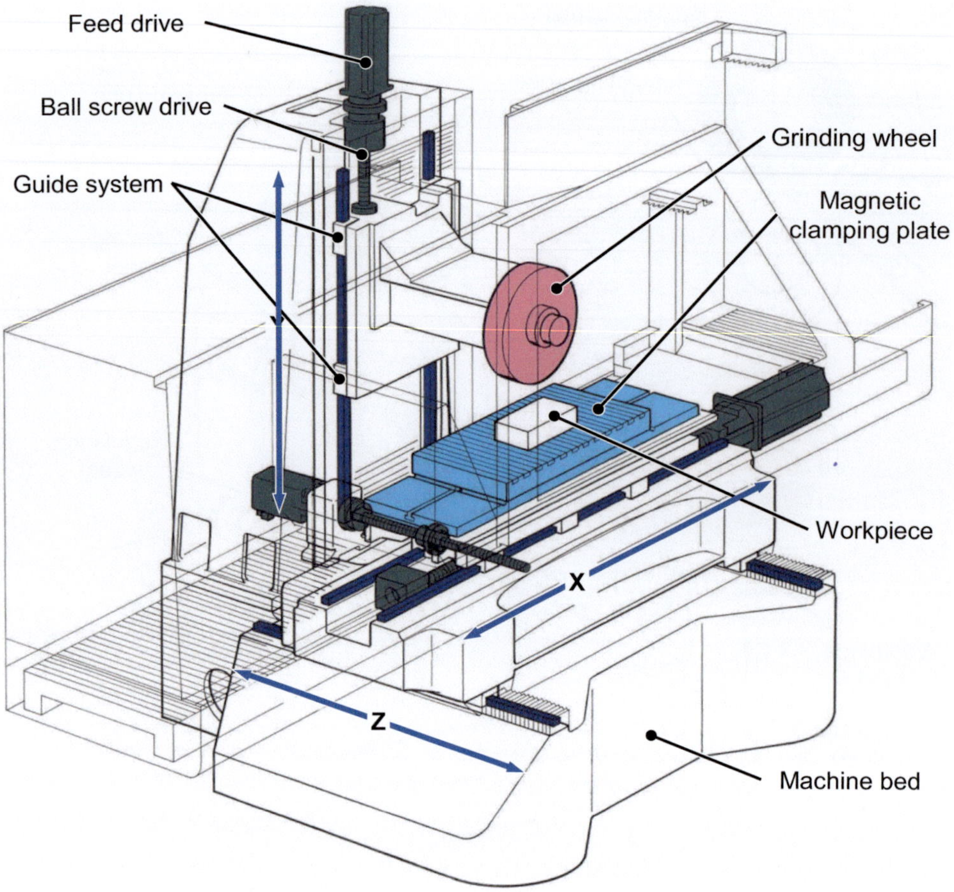

Fig. 5.13 Face grinding machine. *Source* Schleifring, Blohm

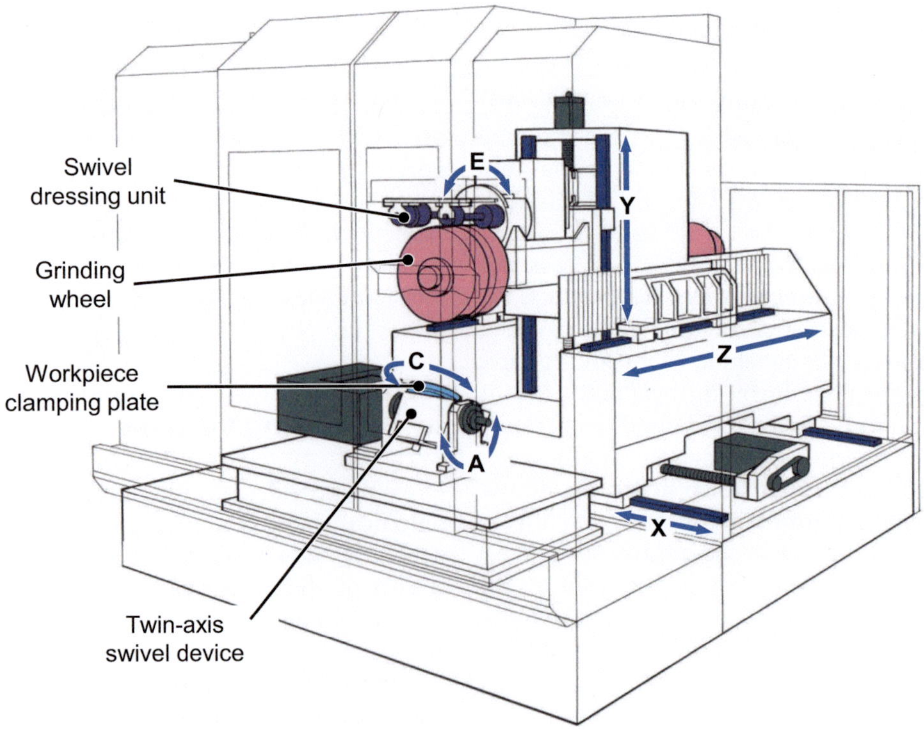

Fig. 5.14 Face grinding machine. *Source* Schleifring, Blohm

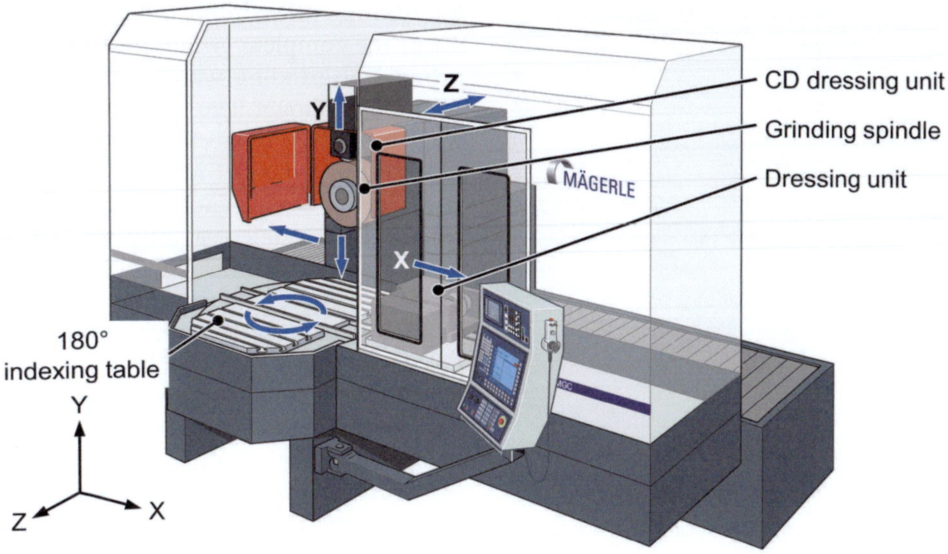

■ **Fig. 5.15** Face grinding machine. *Source* Schleifring, Mägerle

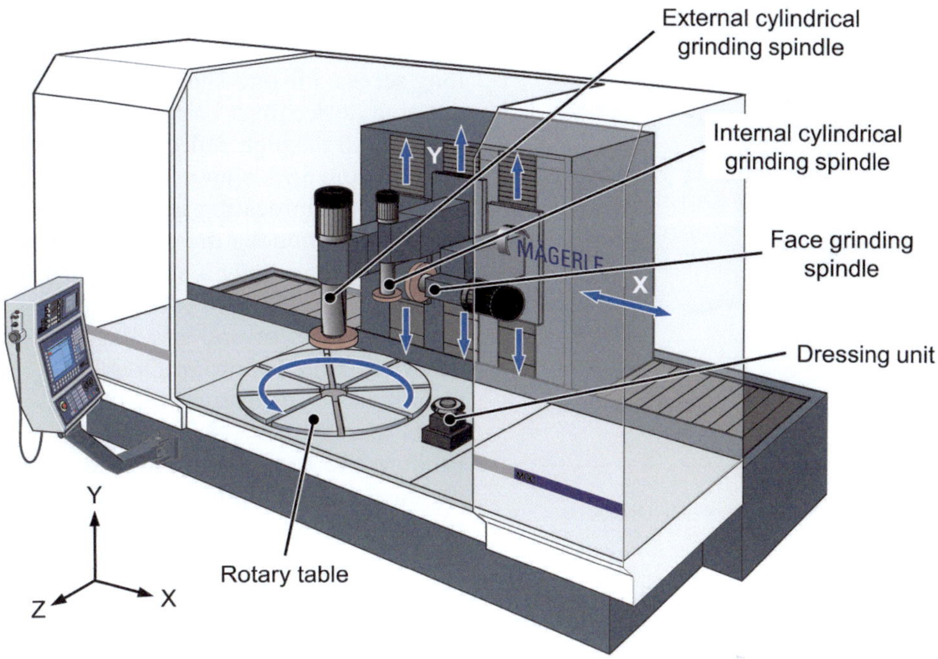

■ **Fig. 5.16** Grinding center. *Source* Schleifring, Mägerle

grinding grains are worn away by chipping or splintering, depending on the required geometry, using a dressing tool with diamond cutting edges.

The most significant processes for dressing grinding wheels are shown in ■ Fig. 5.21. A basic distinction is made between dressing with a stationary tool or a rotating tool. Dressing plates, single-point diamonds, and multi-point dressers are used in stationary tools. Dressing plates and single-point diamonds provide an almost point contact with the grinding wheel and are suitable for producing profiled surfaces such as radii. On the other hand, multi-point dressers are limited to generating straight profiles.

Two axes of motion are required to produce a cylindrical grinding wheel geometry, one for the radial infeed (x-direction) and one for the axial feed motion (z-direction). An additional swivel axis is often used to produce more complex profiles. This allows the dressing tool to be aligned perpendicular to the wheel surface.

Rotating dressing tools include shape rollers and profile rollers. Depending on their design, shape rollers

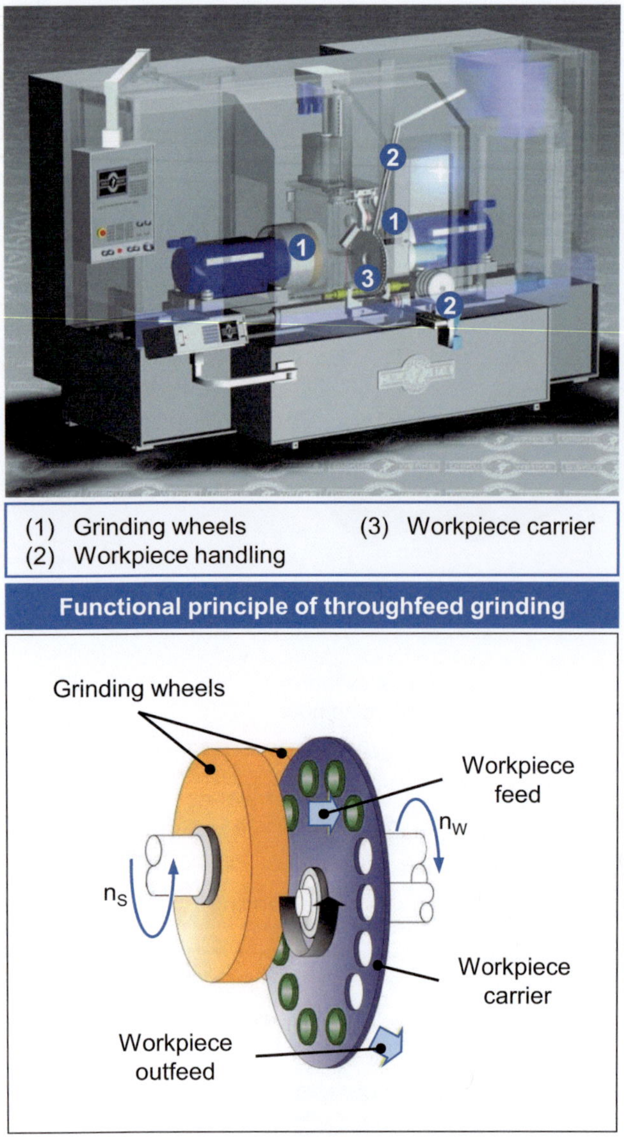

Fig. 5.17 Double face grinding machine. *Source* Diskus

have a point or linear contact at the point of interaction. The grinding wheel profile is produced in the same way as with stationary dressing tools by the movement of the dressing tool relative to the grinding wheel. Profile rollers have the negative profile of the grinding wheel contour to be produced. This profile is transferred directly to the grinding wheel by lowering the dressing roller radially.

The various dressing processes and the associated dressing equipment are described in greater detail below.

Profile rolling units with diamond profile rollers are used to profile grinding wheels as shown in Fig. 5.22. Having the dressing device above the wheel ensures it does not interfere with the contact between the grinding wheel and the workpiece. The profile roller has the negative form of the grinding wheel profile. Lowering the rotating roller into the rotating grinding wheel produces the complete grinding wheel profile. Producing a cylindrical grinding wheel, as in the almost rectangular profiles shown here, should be avoided as the flank wear cannot be compensated by repeated dressing. As can be seen from Fig. 5.22 lower right, either a precise vertical flank is not included in the design or the workpiece is angled slightly by suitable clamping so that the perpendicular end faces can be produced on one side by bevel edge grinding wheels.

One benefit of the position of the dressing device on the grinding headstock is that continuous dressing (CD) while machining is possible. Here, the reduction in the diameter of the grinding wheel caused by the wear on the grinding wheel during machining and the dressing process is continuously compensated by an appropriate infeed. Continuous dressing also avoids the pores becoming clogged with chips ensuring a permanently sharp grinding wheel. The diamond profile roller is driven by its own drive system to produce a favorable relative speed for dressing between the dressing tool and the grinding wheel. The benefits of diamond profile rollers compared to single-point diamonds are long service life and short dressing times. However, the dressing device must be designed with a very high stiffness due to the high deflection forces.

Figure 5.23 shows a comparison between conventional reciprocating grinding and creep-feed grinding with continuous dressing. In conventional reciprocating grinding, only a small profile depth is achieved with each grinding pass (Fig. 5.23, left). This generally means that several grinding passes are required to achieve the final contour. Dressing is intermittent after several passes depending on the wheel wear. By using a CD dressing unit (Fig. 5.23, right), it is possible to produce the final workpiece contour in a single grinding pass with a deep profile and the creep-feed grinding process. The tool wear occurring here is compensated by continuously adjusting the dressing roller and the grinding wheel. As a result, extremely accurate shapes can also be ground in a single pass in workpieces that are difficult to machine such as turbine blades.

If the dressing device is on the workpiece table (e.g., in longitudinal profile grinding to the height of the workpiece reference contour and when extending the longitudinal feed motion), the reduction in the grinding wheel diameter caused by dressing is automatically compensated as the relative position of the grinding wheel contact area to the workpiece remains constant. A dressing axis in addition to the existing machine axes is not required.

Shape rollers are being increasingly used as an alternative to profile rollers to produce profiled grinding wheel geometries. Figure 5.24 shows a dressing device with two diamond shape rollers fixed to the grinding spindle head. Both dressing rollers are held on a

Fig. 5.18 Guideway and surface grinding machine. *Source* Waldrich Coburg

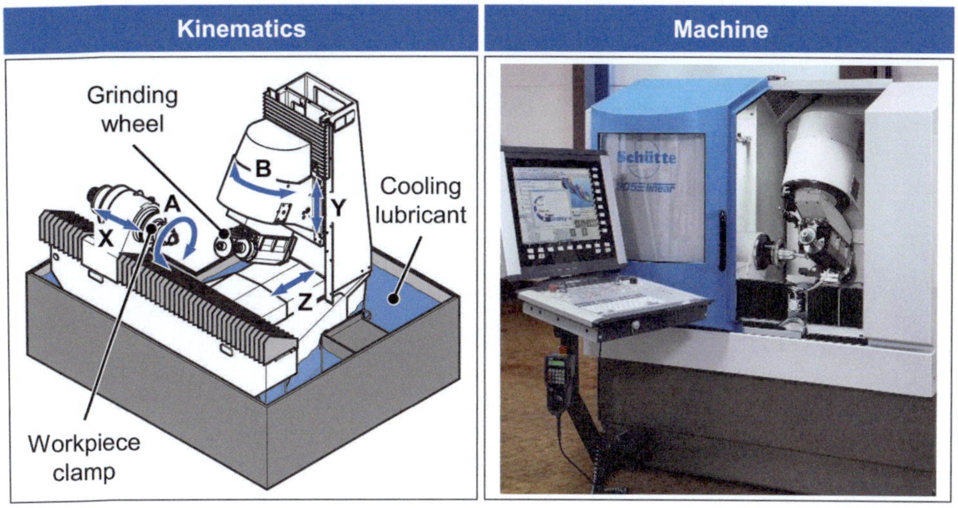

Fig. 5.19 Tool grinding machine. *Source* Schütte

common axis parallel to the grinding spindle axis and at a constant distance from each other. Each dressing roller profiles one grinding wheel flank. To achieve this, the NC-controlled x/y-slide passes by the dressing rollers on both grinding wheel flanks. NC dressing is a very convenient way of generating a profile. It is possible to make specific profile corrections which are all the more important as the grinding wheel profile to be dressed begins to deviate more and more from the theoretical negative form of the workpiece to be produced. This is the case when grinding cylindrical worms with extensive surface contact between the grinding wheel and the workpiece as shown in ◘ Fig. 5.24. Using dressing rollers that do not swivel makes it impossible to move the grinding wheel sideways.

The swivel axis B shown in ◘ Fig. 5.25 allows the single-point diamond to remain permanently perpendicular to the profile path in the xy-plane so that the contact conditions between the diamond and the grinding wheel remain almost constant along the path even when there are changes to the diamond cutting edge geometry caused by wear.

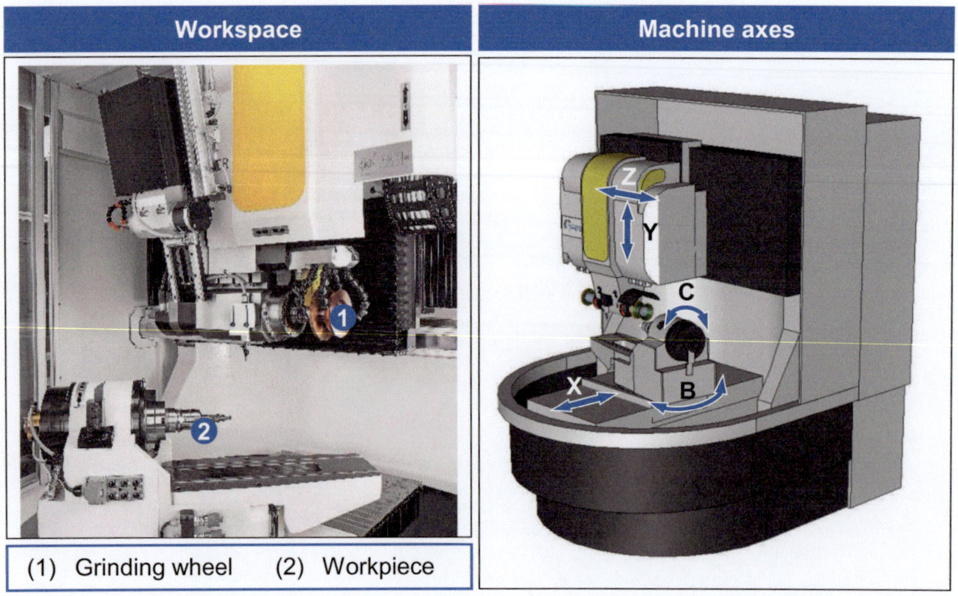

Fig. 5.20 Tool grinding machine. *Source* Walter

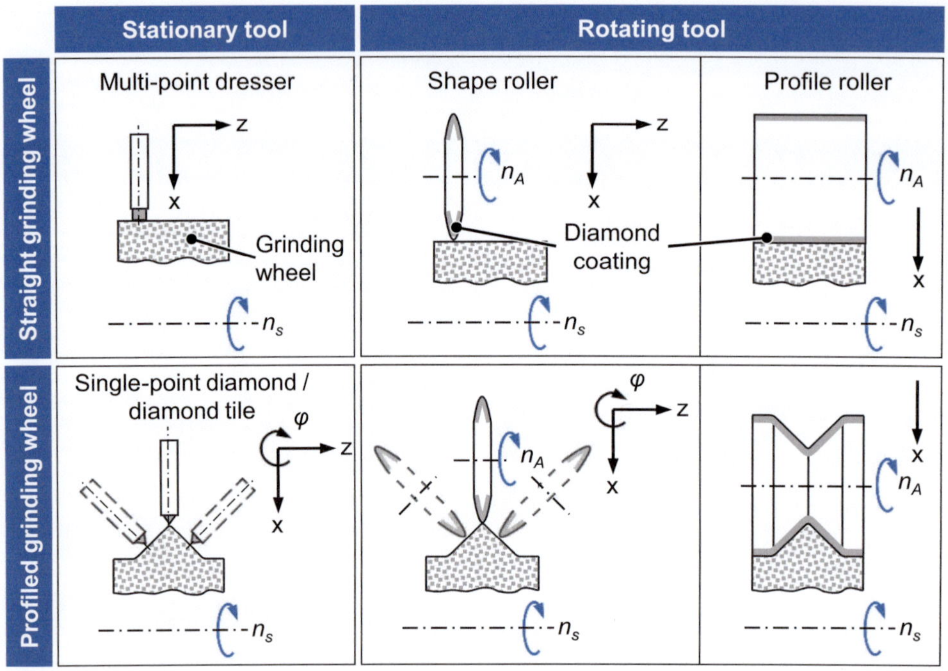

Fig. 5.21 Categorization of dressing processes from a kinematic perspective

If the dressing tool is guided along the grinding wheel reference contour during path operation, the dressing control system is required to maintain a specific, constant path velocity in order to achieve the required surface quality. When dressing path segments with small curvature radii, a relatively short NC line is run within the NC line sequence time and the next NC line must be made available. Depending on the path velocity and path curvatures, the NC line sequence times can be as short as one millisecond. The short downtime that occurs in conventional control systems when a short NC line has been run too quickly and the next NC line is not made available is not acceptable during dressing as this has an impact on surface quality.

Speed-controlled drives for dressing rollers are required if the roller speed is to be adjusted due to variations in the grinding wheel speed when grinding at a constant circumferential velocity (CD grinding). Furthermore, either down or up grinding is a benefit if spe-

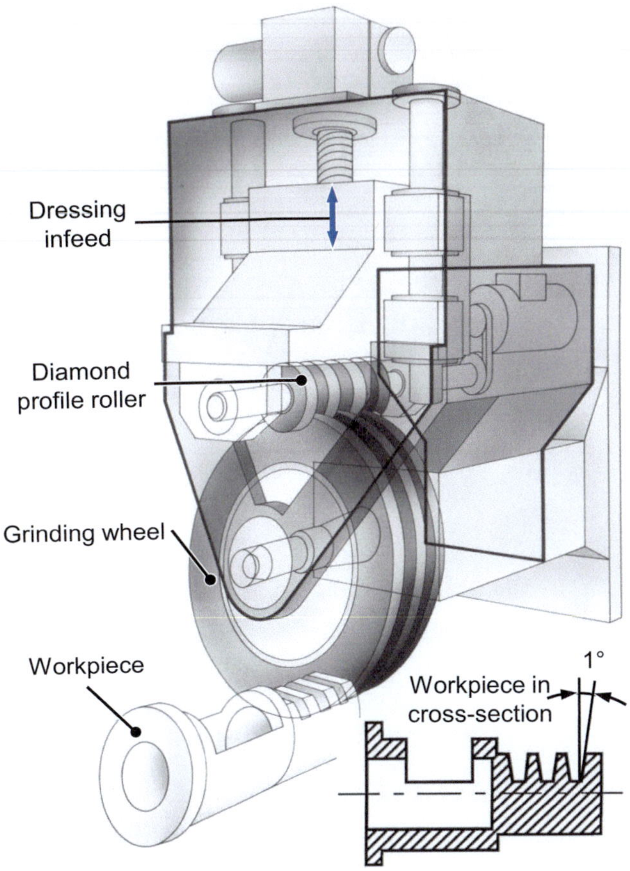

Fig. 5.22 Profile rolling unit. *Source* Elb

cific surface qualities are to be achieved. The speed setting of the dressing roller during continuous dressing is particularly important. Fine adjustments to the speed ratio between the roller and the grinding wheel are required to prevent axial and radial runout errors of the diamond roller being transferred to the workpiece.

In addition to the control system for the dressing device, external dressing monitoring systems are also used if changes in the dressing conditions have a detrimental effect on the overall machining process [WECK83]. To check the actual dressing infeed, the amplitude curve versus the dressing path is measured using a structure-borne sound sensor applied in a suitable manner. The structure-borne sound signal versus the dressing path and the measured dressing feed rate can be used to set up a monitoring system as shown in Fig. 5.26. This sensor system allows evaluations as to whether or not the dressing tool was profiling and sharpening effectively at every point on the grinding wheel. If the tolerance range for the structure-borne sound signal is exceeded, error messages are sent to the NC control system, which can trigger a repetition of the dressing process.

5.1.5 Balancing Units

Grinding wheels have to be balanced because the vibrations caused by imbalances have a detrimental effect on the surface quality and accuracy of the grind and lead to excess wear and an increased danger of the grinding wheel breaking. The imbalances are caused by geometrical deviations from the ideal shape due to clamping errors and inhomogeneities in the structure of the grinding wheel.

Geometrical imbalances can be rectified by dressing the circumference of the wheel and the parallel sides. When compensating the remaining imbalances, a distinction is made between static balancing (the radial plane of the center of gravity of the imbalance does not correspond to the plane of inertia) and dynamic balancing. Ideally, the balancing mass to be

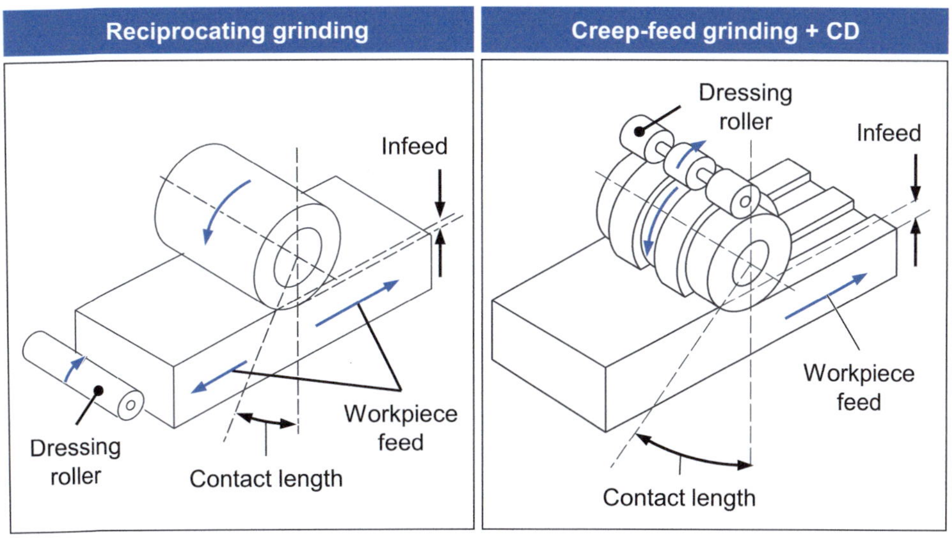

Fig. 5.23 Comparison between conventional reciprocating grinding and creep-feed grinding with continuous dressing. *Source* Elb

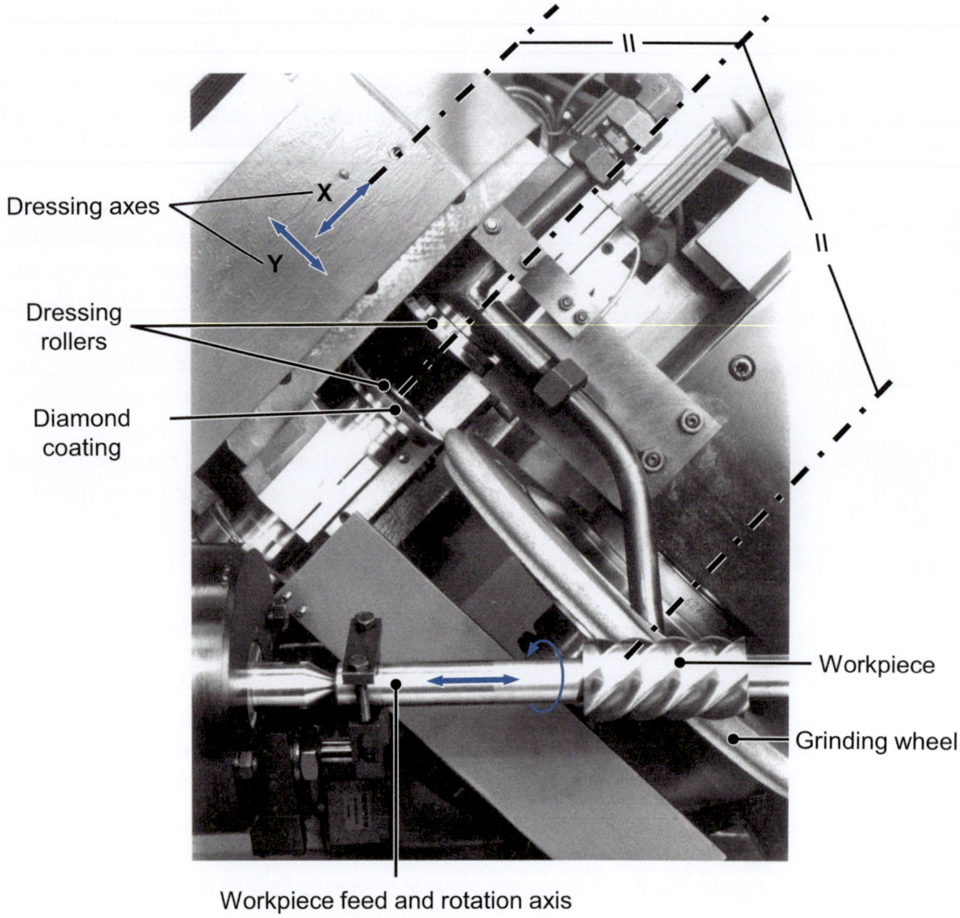

Fig. 5.24 Workspace of a gear grinding machine with two-roller NC dressing unit. *Source* Klingelnberg

added should lie in the rotational plane of the center of gravity of the imbalance. This is the only way to achieve dynamically perfect balancing. However, as this is not possible in practice (the balancing masses are frequently only attached to the end face of the grinding wheel), a dynamic moment perpendicular to the rotation axis is produced and this moment rotates at spindle speed. The moment is determined by multiplying the inertia or imbalance force by the distance between the rotational planes of inertia and the imbalance. This effect can be compensated with a second plane of inertia, e.g., on the back of the wheel. This is described as a dynamically balanced wheel. Grinding wheels are generally only statically balanced. Only very wide grinding wheels such as those used in centerless grinding may require dynamic balancing.

A static balancing process is frequently used to check balance errors on disassembled grinding wheels as this determines the stable position of equilibrium of the grinding wheel's center of gravity. To achieve this, the wheel is clamped on a mandrel with its support. This mandrel is then rolled on two horizontal cutting edges or a balancing stand. If the wheel has a static imbalance, it rolls back and forth until the center of gravity is vertically beneath the rotation axis. Two or three sliding blocks are moved in a groove around the circumference of the grinding wheel flange to balance the wheel. This simple but relatively inaccurate process is often used for pre-balancing.

By using balancing processes with rotating grinding wheels, a significant amount of time is saved due to the lack of assembling and disassembling times. In addition to this, balance errors due to possible inconsistencies between the clamping conditions of the removed wheel and the wheel installed in the machine are avoided.

A vibration transducer is attached to the spindle bearing block to capture the vibrations caused by the imbalances in a defined direction. The vibration measuring system is set such that only vibrations in the grinding wheel rotational frequency are recorded. This allows the magnitude of the imbalance force to be determined. The spatial position of the imbalance on the grinding wheel is located through phase measurement. To achieve this, the phase position is captured between a marking on the grinding wheel and

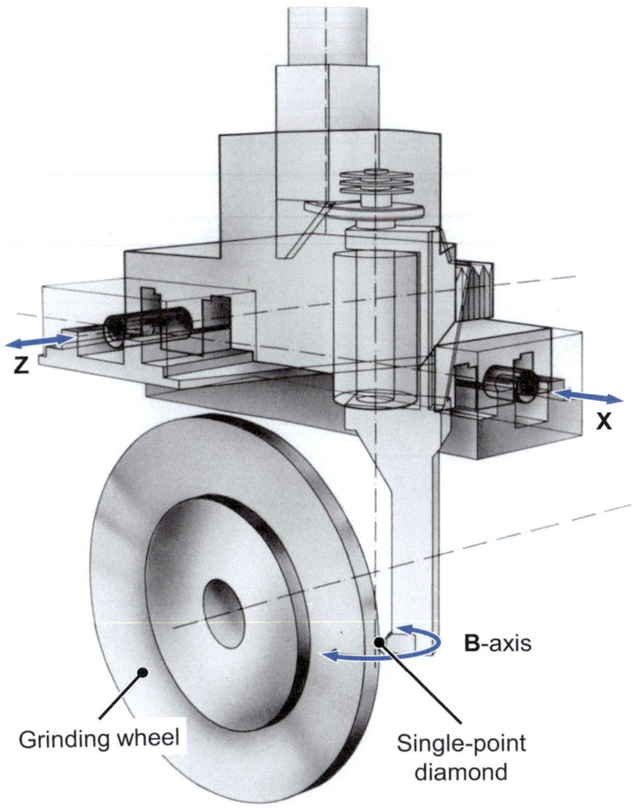

Fig. 5.25 Single-point diamond NC dressing unit. *Source* Elb

the maximum vibration occurring in a direction defined by the measurement. The phase relationship is revealed by a strobe light as shown in the stroboscopic balancer in Fig. 5.27. The flashes of light triggered by the signals from the vibration transducer produce a static image of the grinding wheel. The imbalance force points in the direction of the vibration measurement. The numbering allows the imbalance vector to be assigned to the position of the grinding wheel at the time of measurement. Other possibilities for determining the phase position are inductive or capacitive sensors mounted on the machine to evaluate special markings on the grinding wheel flange. The imbalance force is compensated using balancing masses; their position and effective value are altered in the area of the grinding wheel support. Sliding blocks provide a simple means of balancing; these are positioned in an annular groove on the grinding wheel support and can be moved around the groove and locked in place, see Fig. 5.27. The disadvantage is that the sliding blocks can only be adjusted when the grinding wheel is stationary.

Due to the significant time required by manual balancing systems, automatic balancing systems are used more frequently today; these are based on either a mechanical or a hydraulic balancing head. In addition to the time saved, these systems also offer the benefit of balancing during operation as the balancing condition varies.

A mechanical balancing head comprises two balancing masses whose angular position can be set independently of each other, as shown in Fig. 5.28. The spread α of the two masses allows the magnitude of the compensation force to be determined and the angle β provides the position of the compensation force with respect to the spindle. Both balancing masses can each be adjusted by an electric motor and an interconnected planetary gear. The balancing masses are automatically controlled by the evaluation and control unit based on the vibrations measured on the spindle housing. The transmission of energy from the control unit to the rotating spindle is via a contactless inductive transmission path.

A hydraulic balancing system consists of four chambers set at 90° to each other, Fig. 5.29. Depending on the size and position of the required compensation force, a liquid is injected into the individual chambers in a controlled manner by four solenoid valves. The machine's cooling lubricant is usually used as the liquid. Control is automatically based on the vibration signals captured on the spindle housing, similar to the system shown in Fig. 5.28. The disadvantage of these systems is that the grinding spindle has to be rebalanced each time it starts.

Figure 5.30 shows hydraulic balancing systems for balancing in one and two planes. Two-plane balancing allows dynamic and static imbalances to be compensated and is typically used for centerless grinding due to the large wheel widths.

5.1.6 Measuring Equipment

To achieve maximum quality on the workpiece during grinding, measurements are taken in-process. Program-controlled diameter measuring equipment that captures the workpiece diameters to be machined is suitable for this purpose. This makes it possible to compensate errors from temperature fluctuations, elastic deformation, contamination, and other effects from the workshop environment.

Figure 5.31 shows an example of an in-process measuring device. This measurement equipment measures the dimension of the workpiece during grinding and provides the machine control system and the operator with the necessary data to adjust the process parameters. Depending on the type of probe used, a wide range of diameters can be measured with this system. As the sensors are used during the process, they have to be resistant to the influence of the coolant used to ensure the measurements are not compromised.

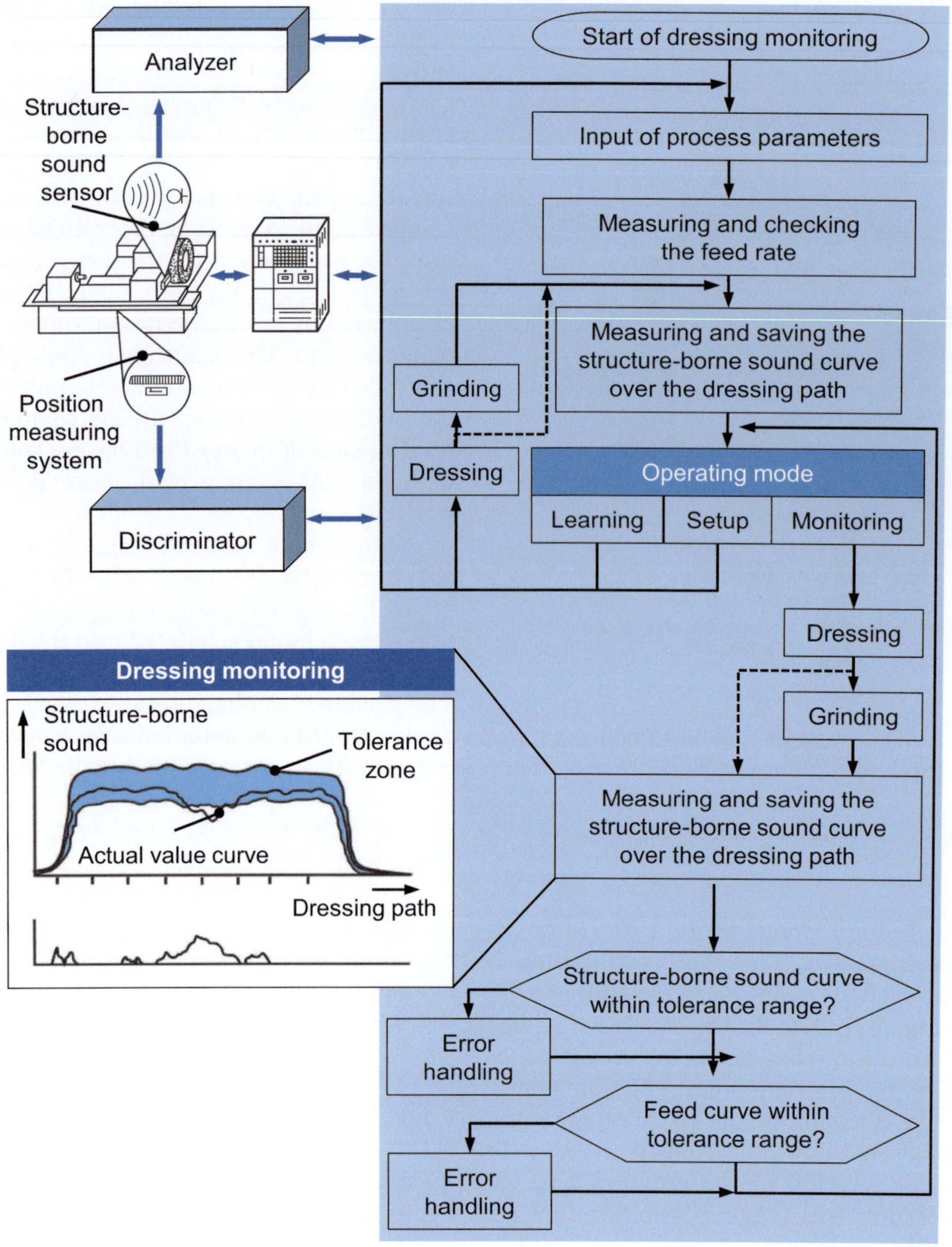

Fig. 5.26 Dressing monitoring operational sequence

5.2 Honing Machines

Wherever extremely tight tolerances for dimensional and geometrical accuracies cannot be cost-effectively achieved in precision drilling or grinding work, a honing stage is used. Unlike grinding, the cutting motion of the close-grained, ceramic or plastic-bonded tool (honing stone) is produced by the superimposition of two directions of motion, at least one of which is oscillating. This produces a high-quality surface and can be identified by the crossing machining marks. Honing machines are primarily used to achieve workpiece surfaces with a roughness of up to $R_a \leq 0.1$ µm. This reduces friction and wear in loaded tribological pairings (rolling bearing parts, plain bearing pins, guide columns, cylinder sleeves). Removing the ridged surfaces left over from previous machining stages increases the contact ratio by 15% to 40% to up to 95%. This produces components with a high load-carrying capacity (e.g., bearing seats, pins, sealing surfaces).

Increasing use is also being made of the ability of NC-controlled honing machines to compensate small shape and positional errors during machining by measuring actual values for the diameter, cylindricity, centricity and roundness or to influence the geometry of the workpiece by maintaining extremely tight toler-

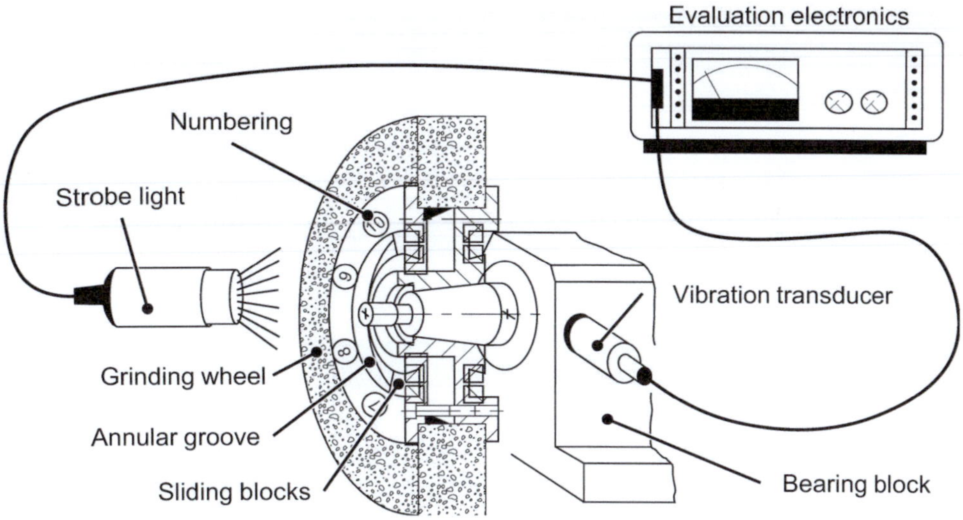

Fig. 5.27 Stroboscopic balancer

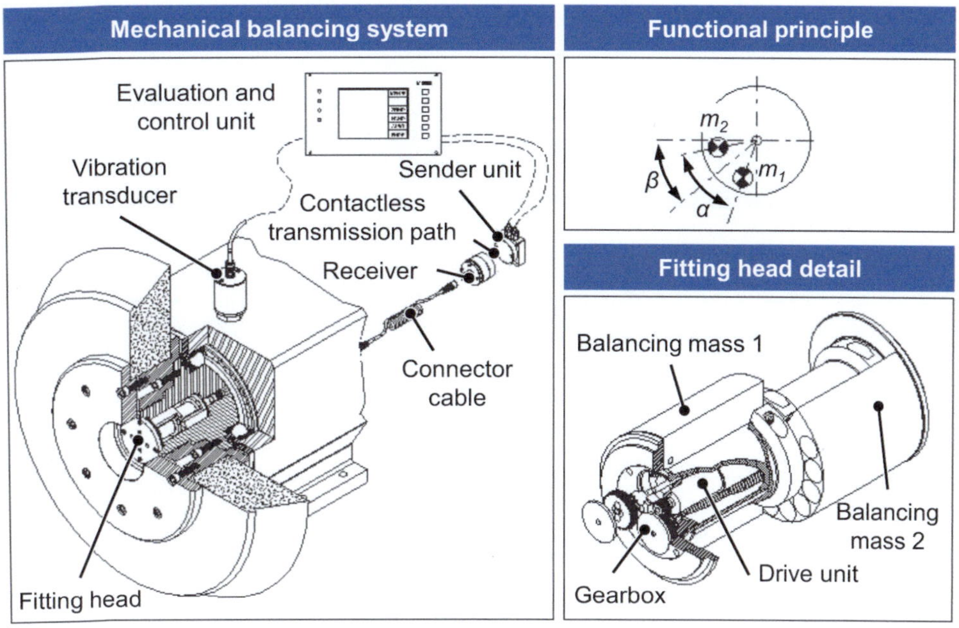

Fig. 5.28 Mechanical balancing system. *Source* Dittel

ances. More recent honing machines can produce conical holes and tapered or barrel-shaped rolling elements by varying the tool pressure over the length of the working distance. These types of special application areas are known as shape honing. Otherwise, the contour of the surface to be machined determines the manufacturing process. A decision also has to be made between external and internal honing depending on the position of the machining point on the workpiece, see Fig. 5.32.

In cylindrical honing, honing tools are generally used in which the honing stones are fed in with force locking by springs or a hydraulic system. The pressure on the honing stones is between 10 and 100 N/cm^2. The hardness and grain size of the honing stones is selected such that they sharpen themselves during use. A thin, filtered honing oil lubricates and cools the cutting point and rinses away the material removed from the workpiece and the material abraded from the honing stones.

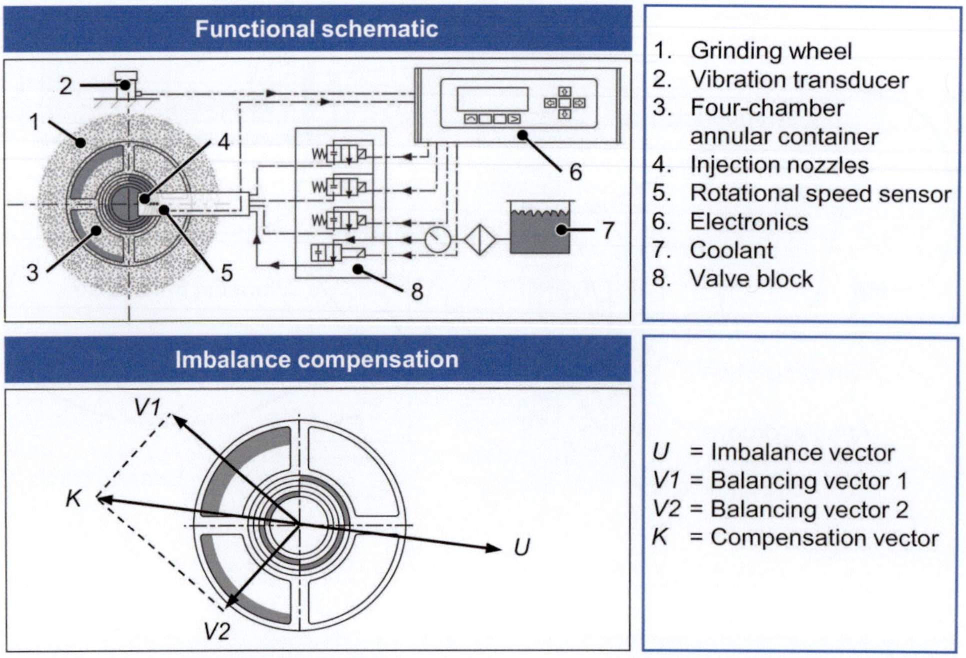

Fig. 5.29 Hydraulic balancing system. *Source* Hofmann

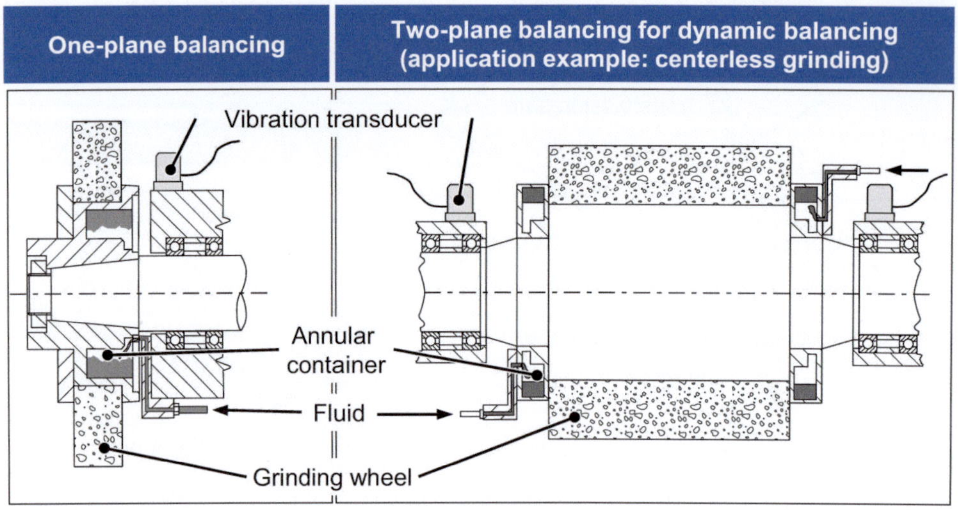

Fig. 5.30 Hydraulic balancing system. *Source* Hofmann

Long-stroke honing is used in cylindrical, flat and shape honing; here, the cutting motion comprises a rotational motion and a long-stroke oscillation motion. Figure 5.33 shows a cylindrical honing machine typically used to machine high-accuracy fuel injection pump components for the automotive industry. The machine has a total of three honing units for pre-processing, finishing and final processing. The workpieces are moved by an automated conveyor system from one station to the next. Measuring stations are located before and between the honing stations to detect the oversize as well as machining units for deburring. A special feature of the machine shown here is the stroke movement provided by linear motor drives. Otherwise, crank-rocker mechanisms or hydraulic cylinders are used to produce the stroke movement in the honing machines.

The internal cylindrical honing machine shown in Fig. 5.34 consists of a central column which can be fitted with up to four honing units and a maximum of five measuring stations for sequential machining operations with increasingly fine cutting grains. The honing units are fitted with low-maintenance direct drives for the stroke and the spindle rotation. The measuring stations allow the oversize and the honing stone wear to be measured and compensated. An up-

Fig. 5.31 Diameter measuring device for external cylindrical grinding. *Source:* Marposs

stream piston measuring station also allows match honing. This process is used, for example, to machine the diameter of the bore of high-pressure gasoline injection pumps so precisely that the piston alone seals the circumferential annular gap making a seal unnecessary. The NC rotary table arranged around the column ensures minimum transportation times (<1 s) even with large gimbal-mounted or floating fixtures for the workpieces.

Superfinishing

The microfinishing process of superfinishing, also known as short-stroke honing, and, according to DIN 8589–14, belongs to the family of machining with geometrically undefined cutting edge [DIN03a]. Surface qualities of less than $R_a = 0.05$ µm can be achieved.

As with other chip removal manufacturing processes, in the superfinishing process, a tool is brought into surface contact with a workpiece. This allows the creation of extremely accurate cylindrical and rotationally symmetric geometries. Figure 5.35 shows a short-stroke honing machine used to produce these geometries, for example, on roller-shaped rolling elements. The superfinishing tool is oscillated at a frequency of between 2 and 85 Hz and is pressed onto the rotating workpiece surface at a constant pressure of between 9 and 40 N/cm^2. The superposition of workpiece rotation and tool oscillation moves a single grain along a sinusoidal line which is typical for this process, Fig. 5.36, lower section. By superposing the individual sinusoidal lines, all the grinding grains in contact produce the machining marks which intersect at a defined angle. The tools used here are superfinishing stones (Fig. 5.36, right) or superfinishing belts

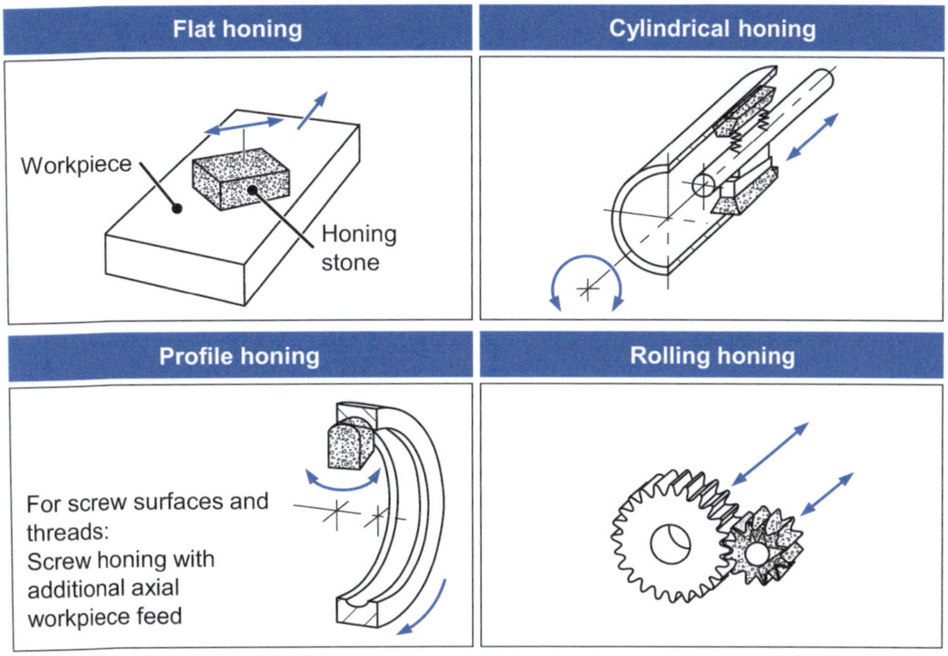

Fig. 5.32 Honing process, according to [DIN03a]

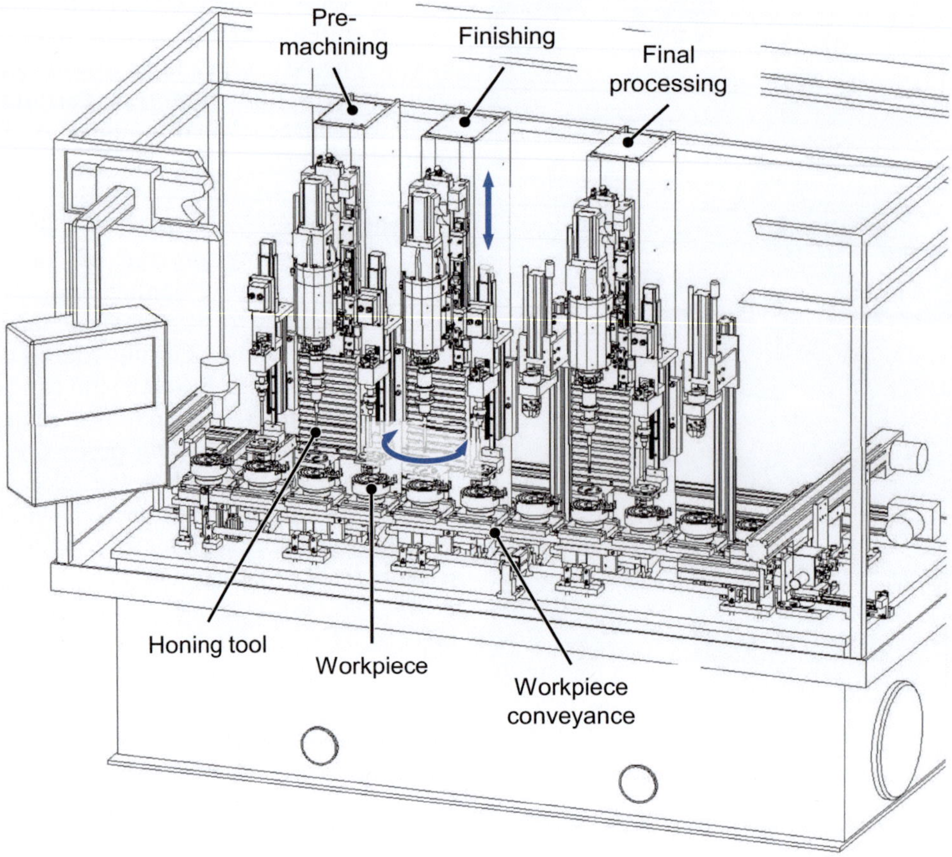

Fig. 5.33 Cylindrical honing machine. *Source* KADIA

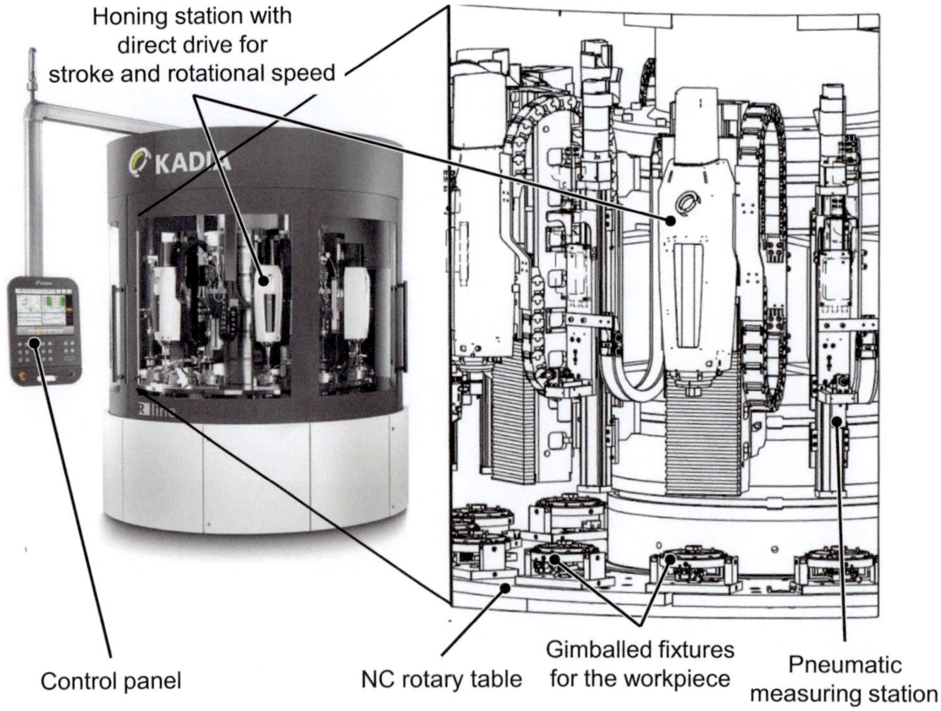

Fig. 5.34 Internal cylindrical honing machine. *Source* KADIA

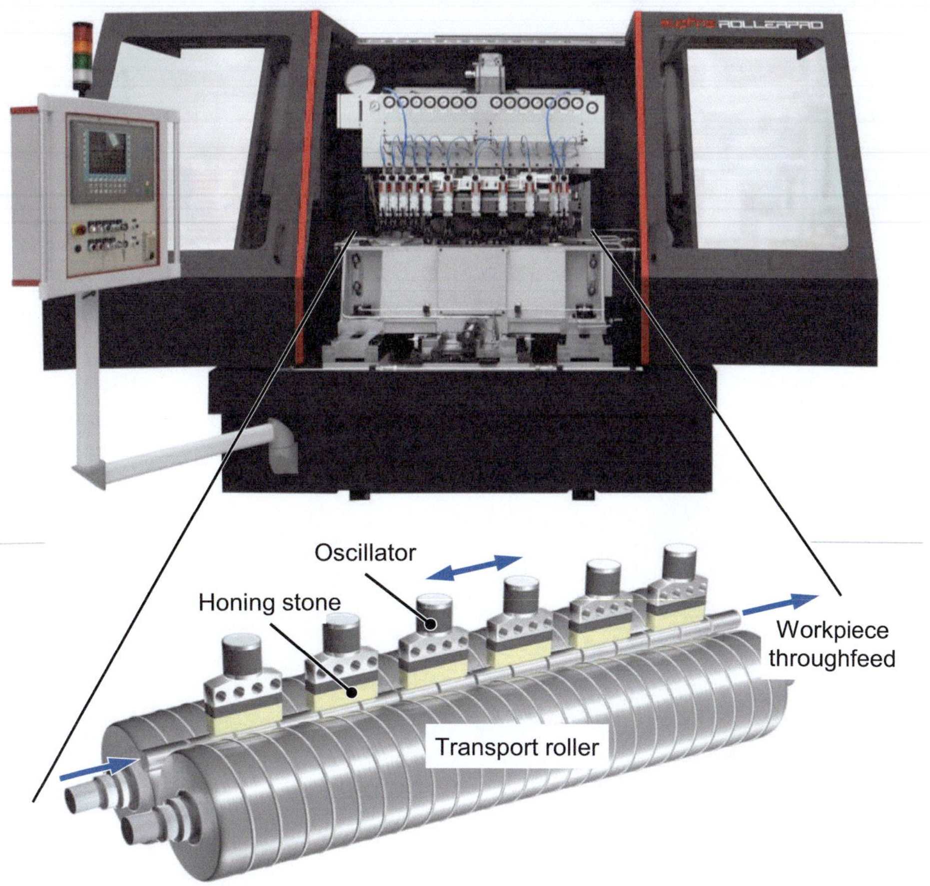

Fig. 5.35 Centerless throughfeed short-stroke honing machine. *Source* Supfina

(Fig. 5.36, left) where the abrasive is on a flexible backing with a mean grain size between 1 and 260 µm.

Flat Finishing
Cross-grinding superfinishing is a process primarily used to machine components with flat, concave, or convex surfaces, Fig. 5.37. The tool, rotated at a circumferential speed of between 1 and 20 m/s, is pressed with a defined pressure onto the workpiece, which is also rotating. Moving or tilting the tool axis creates a circular arc-shaped line if the movement of one cutting grain is followed. In this case, all the grinding grains in contact produce the crossing radial machining marks by superposing the individual arcs. Cross-grinding superfinishing uses bonded abrasives in cup or sleeve shape, or segmented grinding heads with a mean grain diameter between 15 and 70 µm.

Short-stroke honing devices are offered for mounting on lathes or grinding machines for one-off or short-run manufacturing. They essentially consist of just one tool and a pneumatic drive for the quick short-stroke movement. Equipment on the relevant carrier machine is used for the feeds and control.

5.3 Lapping Machines

5.3.1 Conventional Lapping Machines

Lapping is a machining process where the workpiece and the tool slide against each other with a contact pressure of approximately 10 N/cm^2. The added lapping medium comprises mineral oil and lapping grains with a grain size of between 3 and 45 µm depending on the application. The workpiece surface is cut by the grinding grains. The grains split due to mechanical stress always keeping their sharp edges. As the grain size continuously decreases, this method automatically improves the workpiece surface quality. As with honing, the aim of the process is to achieve extremely high dimensional and geometrical accuracy while improving the surface quality ($R_z \leq 0.1$ µm). The categorization of the manufacturing process is similar to honing (see ▸ Sect. 5.2). On a larger scale, lapping machines are only manufactured in a column or portal design to produce or improve the flatness (achievable evenness ≤ 0.3 µm) and parallelism of surfaces. When lap-

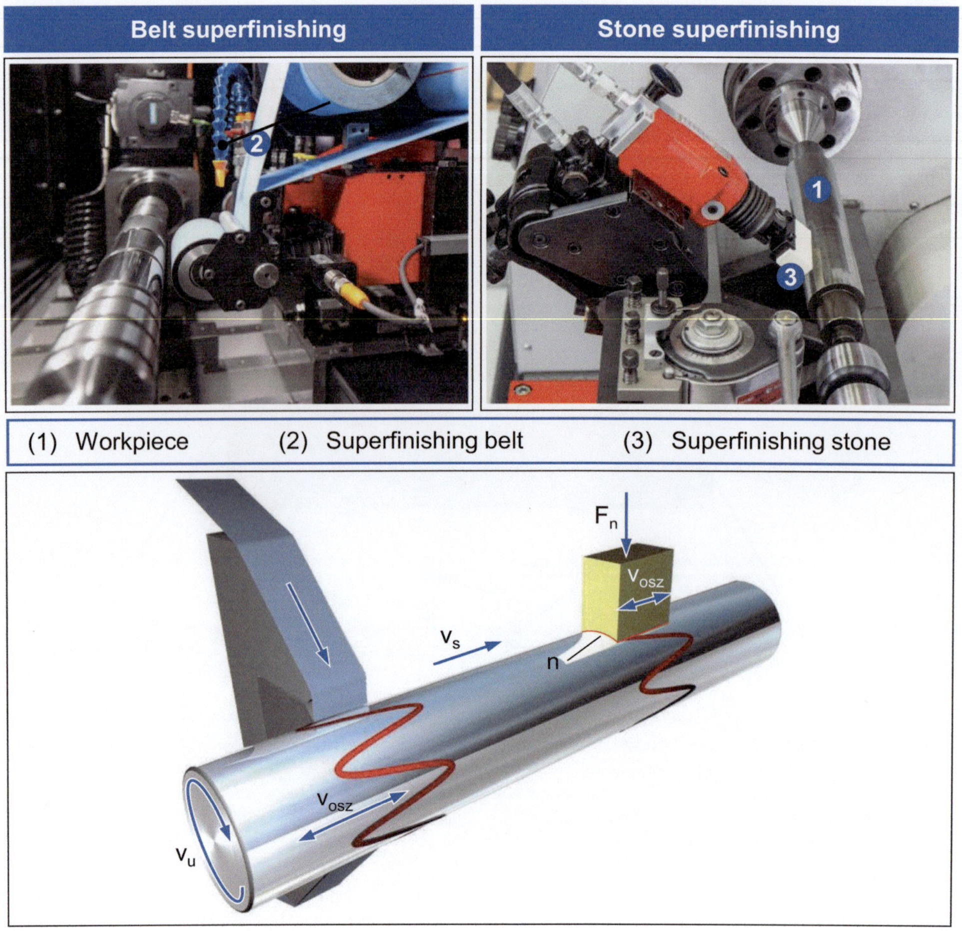

Fig. 5.36 Superfinishing devices. *Source* Supfina

ping a single workpiece plane (single-sided surface lapping), several workpieces are usually placed in a workpiece holder that matches the workpiece contour. Alternatively, for small parts, they are placed in lapping rings on the gray cast iron lapping wheel. The eccentric bearing of the holder is fixed to the machine or driven in the direction opposite to the driven lapping wheel. This produces a cycloidal relative path between the workpiece and the lapping wheel, Fig. 5.38. Light workpieces can be subjected to loading pneumatically on larger lapping machines to influence the chip removal, which is typically a few microns. Very thin, fragile components can also be safely machined due to the full-surface, uninterrupted support of the workpiece.

Twin-wheel lapping machines have two lapping wheels rotating either coaxially or eccentrically with respect to each other, Fig. 5.39. The workpieces are located between the wheels in a workpiece holder. The workpiece holders shown here allow the workpieces to be machined on both sides and have a planetary wheel arrangement. This produces the cycloidal relative motion of the workpieces with respect to the lapping wheels.

The structure of a twin-wheel lapping machine is shown in Fig. 5.40. The entire upper lapping unit can be adjusted vertically to accommodate a range of workpiece thicknesses. The fine feed or application of pressure is achieved via a balance arm with a pneumatic actuator. In order to ensure the contact pressure is uniform across the entire working area, the upper working wheel is supported by a self-aligning bearing.

The amount of material removed from the workpiece is matched by a corresponding amount of lapping wheel wear. This is compensated in machines with measurement control. To maintain the flatness of the lapping wheels, the workpieces are moved to the outer and inner edges of the lapping wheels by the workpiece holder so that the wheel wear remains consistent over the entire width. The lapping rings serve the same purpose on single-wheel lapping machines. The lapping wheel profiles are monitored with measuring rulers and probes and corrected if necessary, by turning with carbide metal turning tools as any uneven wear which may occur despite the above measures has a detrimental effect on the evenness.

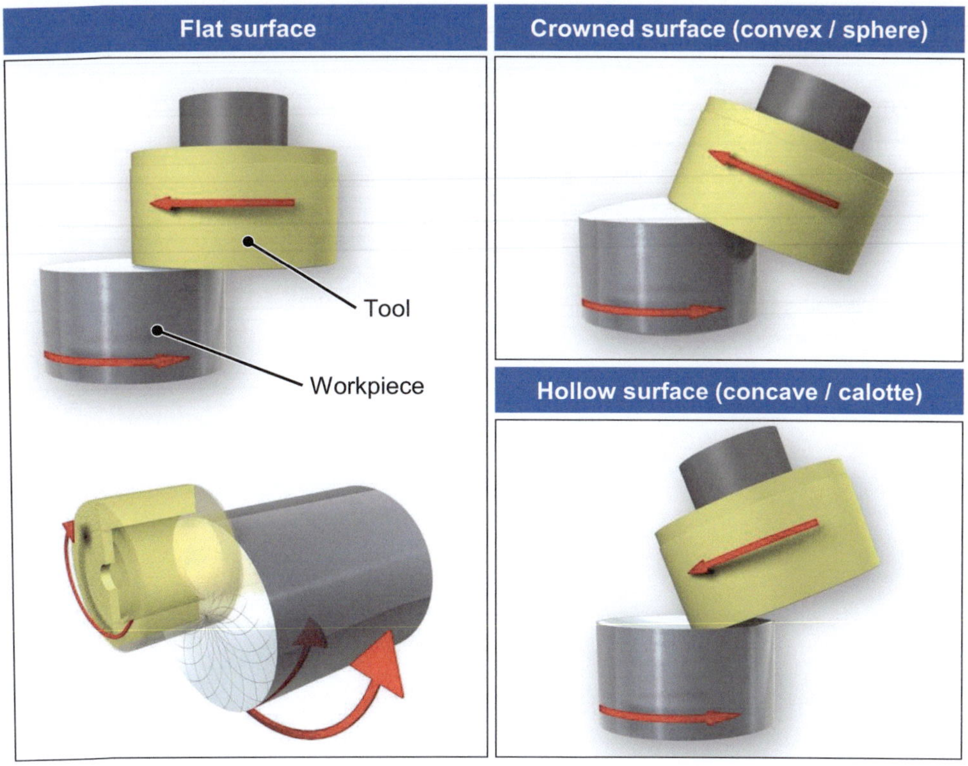

Fig. 5.37 Flat finishing. *Source* Supfina

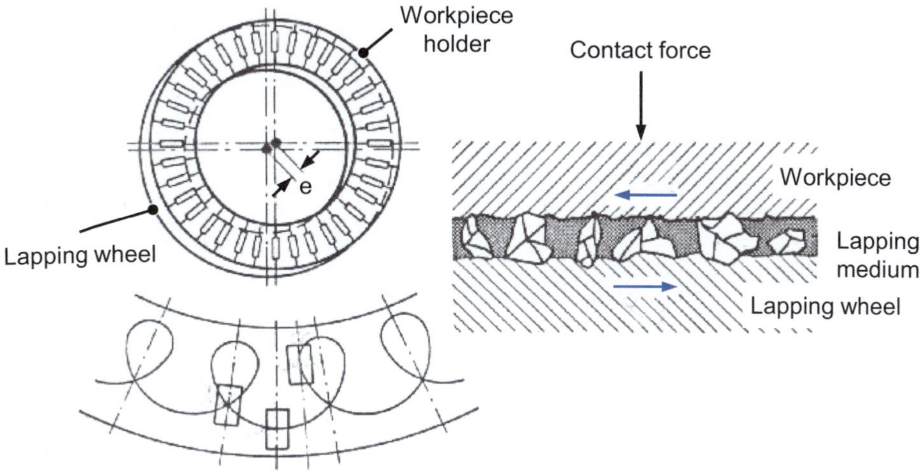

Fig. 5.38 Lapping process and relative motion of the workpieces in single-sided surface lapping

Even difficult to machine materials such as hardened steels, carbide metal, hard ceramics, glass, or silicon can be cost-effectively machined on lapping machines. On some machines, the lapping wheels are replaced with silicon carbide grinding wheels when removing larger amounts of soft materials.

5.3.2 Ultrasonic Lapping Machines

Ultrasonic machining systems are primarily used for sinking operations on electrically conductive and non-conductive brittle materials (glass, quartz, oxide ceramics, precious stones, carbides, germanium, silicon,

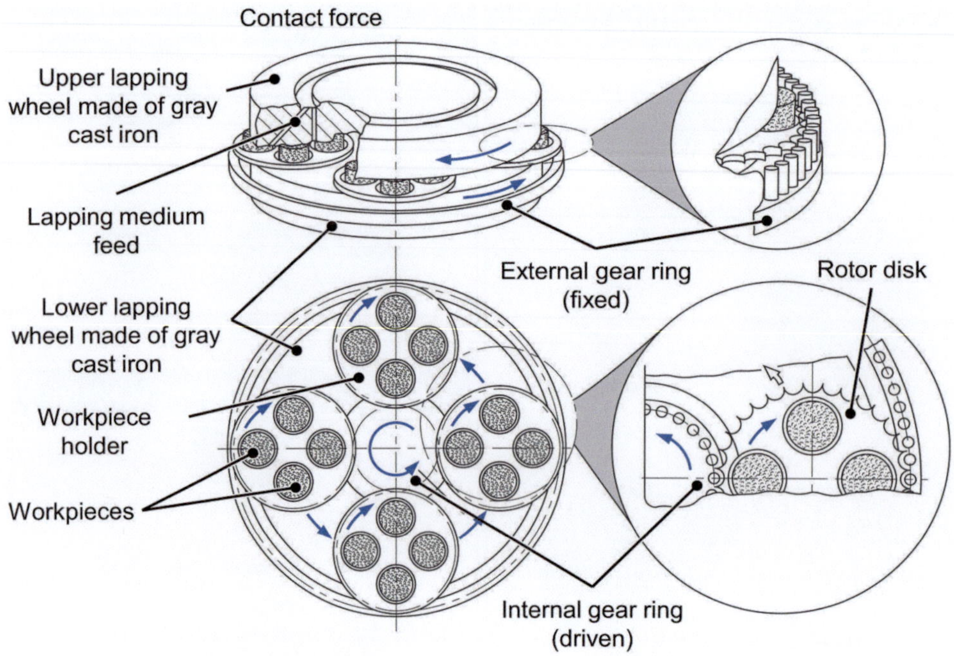

Fig. 5.39 Working principle of double-sided surface lapping

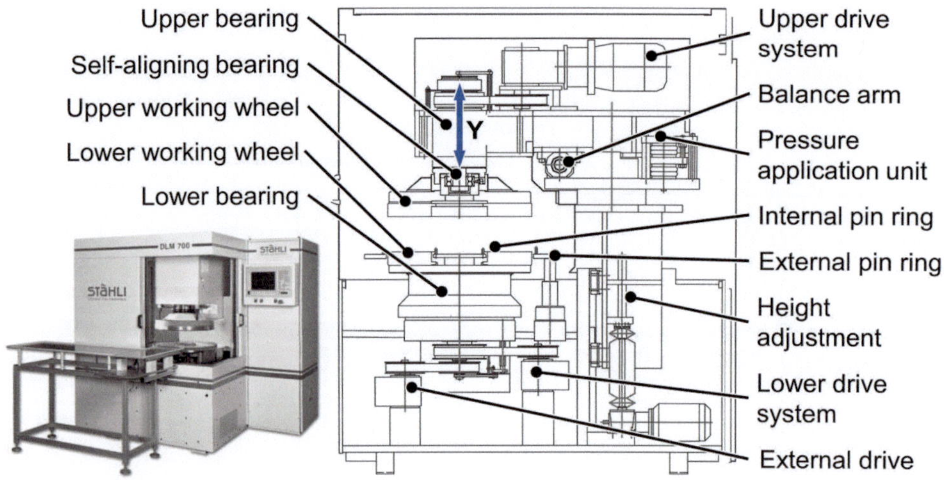

Fig. 5.40 Twin-wheel lapping machine. *Source* Stähli

graphite, or carbide metals) where electrical discharge machining (EDM) is not possible. The principle is illustrated in Fig. 5.41. A high-frequency generator excites the magnetostrictive oscillating head that sets the sonotrode with the soldered tool into high-frequency oscillations in the feed direction. The tool is only indirectly involved in the actual machining process. The material removal itself is effected by the small grains suspended in the abrasive suspension. These grains act as a number of cutting edges in simultaneous contact, similar to the lapping process. The abrasive suspension is fed into the machining point from the outside and is sucked out by the sonotrode.

Many older machines dispensed with a feed system altogether as, in a vertical configuration of the tool/workpiece system, the tool used its own weight force to engage with the workpiece according to the work progress. Contemporary machines however have a feed system that engages the sonotrode and apply a constant force, on average, to the workpiece. Such a system is shown in Fig. 5.42. The positioning system in the z-direction consists of a servo motor which drives a spindle nut system via a toothed belt. A force measuring plate is built into the working head to monitor the process forces. The feed is controlled in such a way that the measured force in the z-direction is constant. The

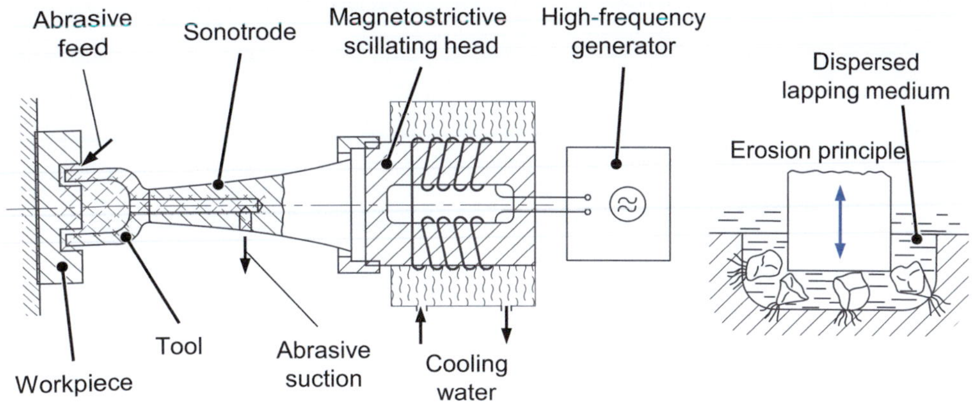

Fig. 5.41 Principle of ultrasonic machining

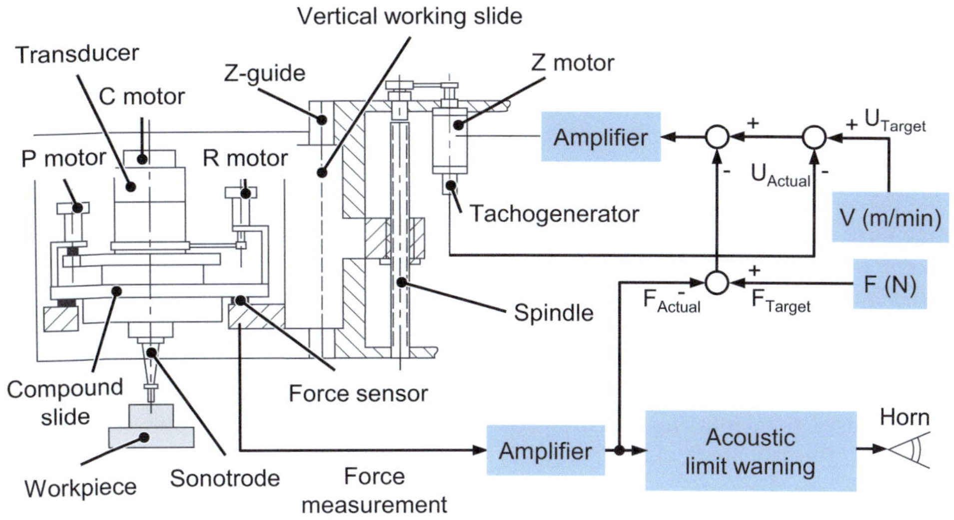

Fig. 5.42 Principle of the controlled feed axis for an ultrasonic machine. *Source* Dieter Hansen AG

Fig. 5.43 Mechanical deburring on turned parts. *Source* Rausch

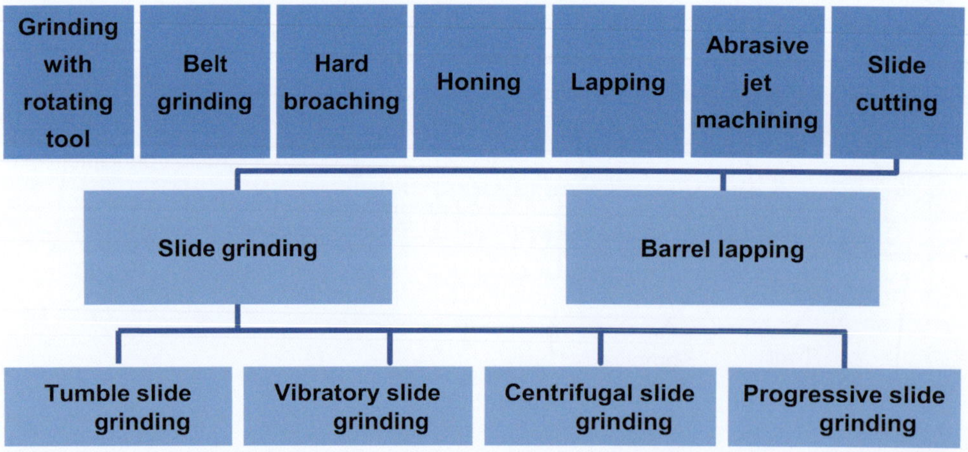

Fig. 5.44 The manufacturing process 'slide grinding' and its variants (as per DIN 8589)

drive head has two further drives to allow the planetary motions, as in EDM sinking.

5.4 Machines for Mechanical Deburring

When metal is machined by chip removal, the process causes burrs to occur at the edges which have a negative influence on the function and mountability of the workpieces. Burrs also represent a significant risk of injury. Various methods exist for removing burrs. Mechanical deburring is discussed in the following. Slide grinding, which is also used for deburring, is described in detail in ► Sect. 5.5. Electrochemical deburring is dealt with in ► Sect. 6.2 and thermo-chemical deburring in ► Sect. 6.5.

Mechanical deburring is not explicitly associated with a specific manufacturing process. Depending on the workpiece material, burr thickness, and workpiece geometry, deburring can be done by milling, grinding, or brushing. Mechanical deburring processes can be used if the burrs can be reached from the outside with a suitable deburring tool.

Figure 5.43 shows a simple solution for mechanical deburring on milled grooves and turned edges on turned parts The deburring process is automatic. Once the workpiece rotation is switched on, the rotating brush disk is moved towards the profile to be deburred by an actuator until it reaches a defined, adjustable contact force. The disk then follows the contour of the workpiece. After deburring, the workpiece and brush rotation are switched off and the brush is moved away from the workpiece. The degree of deburring, i.e., the chamfer width, is set by adjusting the relevant contact force between the brush and the workpiece. Grinding wheels, filing wheels, milling wheels, end mills, etc., are also suitable alternatives to steel wire brushes. A high degree of machine flexibility can be achieved for special deburring tasks by using different tool combinations.

5.5 Slide Grinding Machines

The term 'slide grinding' describes a process for rounding off edges, deburring, grinding, and smoothing which belongs to the manufacturing processes of machining with geometrically undefined cutting edges as shown in Fig. 5.44.

Material is removed by the irregular relative motion between the workpiece and a large number of loose abrasive chips. The machining efficacy can be increased even more by adding liquid compounds that generally consist of water with synthetic detergent raw materials, detergents, and corrosion inhibitors. Workpieces and the abrasive chips are kept clean or with maximum grip and machining residues are rinsed out of the system. Originally, irregularly shaped smelted and natural products were used. These were replaced by simply shaped ceramic chips. More recent developments use special abrasive shapes and other materials. For example, star and arrow-shaped ceramic chips have been developed as well as plastic-based shapes known as 'dragon-heads' and 'nipples'. The choice of shape, size, and composition is determined by the workpieces to be processed and the type and size of the slide grinding machine. Figure 5.45 shows a selection of possible abrasive shapes and compares the grinding efficiency and surface quality as a function of the abrasive chip materials used.

Centrifugal slide grinding is regarded as the most cost-effective of the four slide grinding processes shown in Fig. 5.44 due to the fact that semi- and full automation are relatively easy to achieve. In centrifugal slide grinding, the workpieces are placed in a fixed, cy-

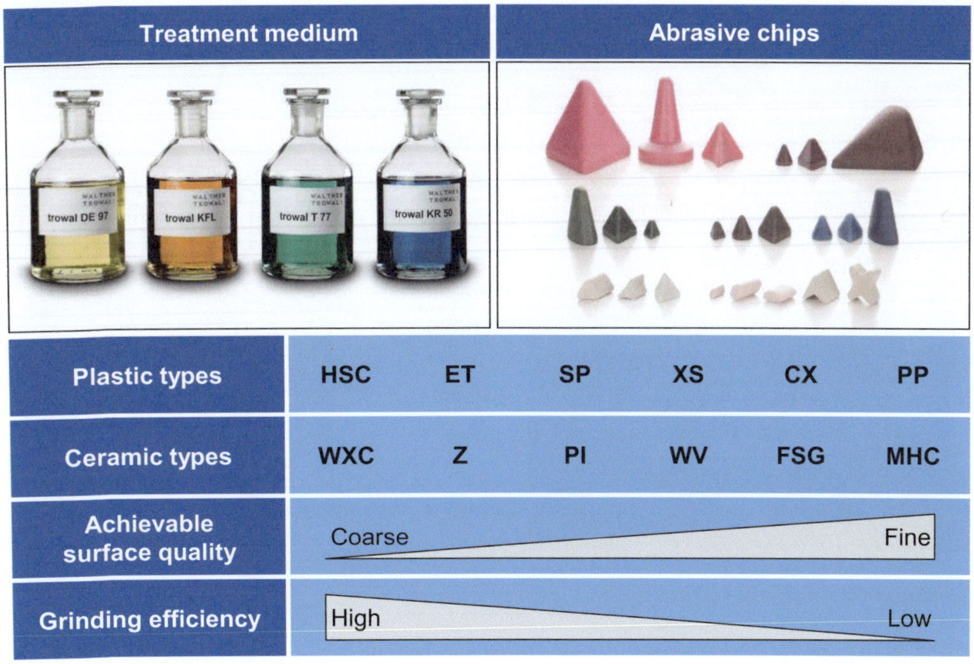

Fig. 5.45 Abrasive chips, compounds and comparison of grinding efficiency for various abrasive materials. *Source* Walther Trowal

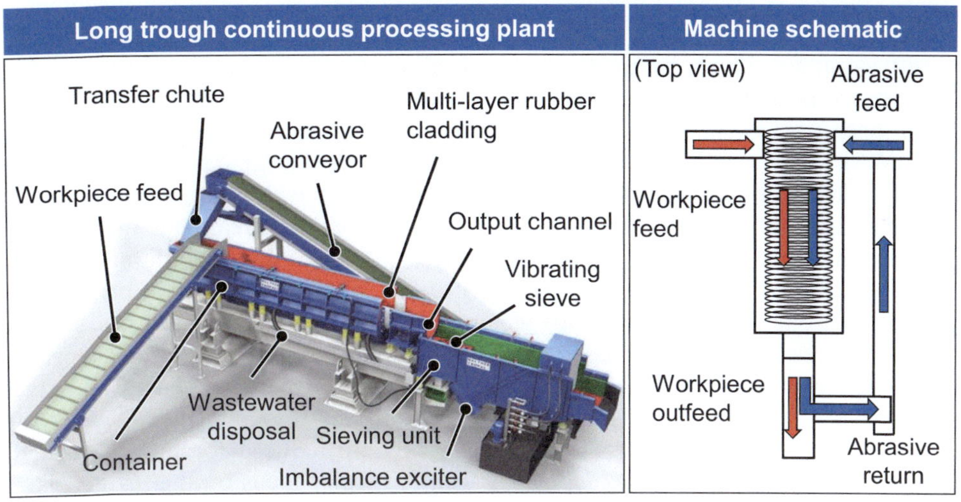

Fig. 5.46 Long trough continuous processing plant for vibratory slide grinding. *Source* Walther Trowal

lindrical container that is approximately 50–70% filled with abrasives and has a rotating base plate. Rotating the base plate (60 to 250 min^{-1}) accelerates the abrasive chip/workpiece composite up the inner wall of the container. Friction and gravity typically produce a spiral-shaped movement. The great grinding effect is primarily produced by the frictional contact of the abrasive chips which are forced against the workpieces with a high contact force. The circulation of these chips in the container produces an additional grinding effect. Compared to vibratory slide grinding, the grinding efficiency is up to 10 times greater and thus leads to significantly reduced machining times.

Figure 5.46 shows a long trough continuous processing plant which uses the vibratory principle for continuous and batch processing. It has a usable volume of 2080 L and can handle bulky workpieces such as light alloy engine blocks, crankshafts, and airplane spars.

The vibration for the container is generated by an imbalance system located underneath the container. The workpieces are fed into the rubber-clad container by a feed device. They take 5 to 25 min to pass through the container and are then separated from the abrasive chips by the sieving unit located at the end of the plant. The abrasive chips are returned to the top of the

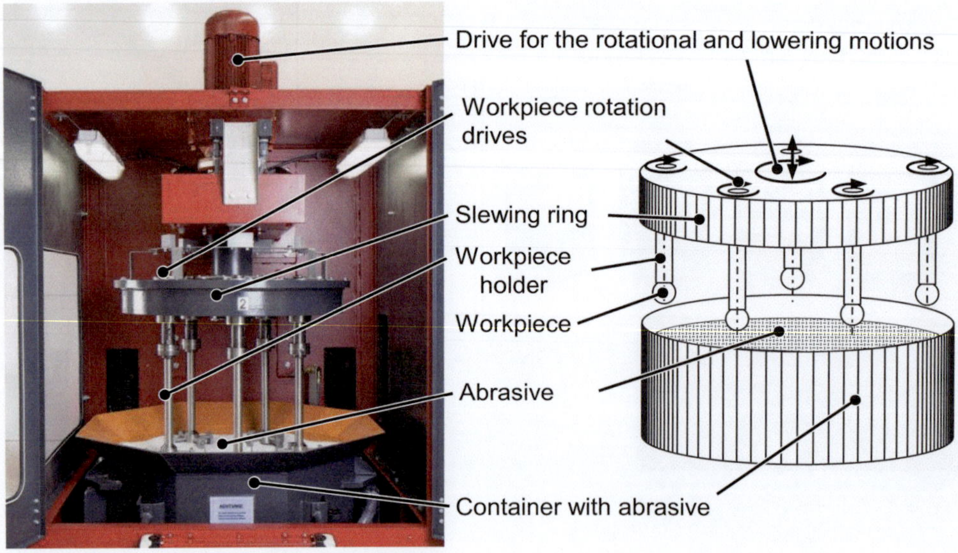

Fig. 5.47 Drag finishing machine. *Source* Walther Trowal

container via a conveyor belt. The output channel, located at the end of the container, has a flexible flanged connection and a separate drive. This channel controls the abrasive chip/workpiece mix and the throughput or dwell time per workpiece. It also lifts heavy workpieces and workpieces with complex shapes by means of a special transportation arrangement. The machine can be switched to batch operation by switching the output channel on and off. [WALT15]

Drag finishing is a further development of progressive slide grinding. The main difference from vibratory slide grinding is the significantly greater cutting performance. This is mainly achieved by moving the workpieces relatively quickly through the abrasive chips using a clamping device. Contact between and damage to the workpieces is avoided by using this positively driven method. Typical workpieces are pincers, gear components, or valve flaps.

Figure 5.47 shows a drag finishing machine. The cylindrical steel container with internal rubber coating is approximately 80% filled with abrasive chips. The workpieces to be processed are clamped in a slewing ring positioned above the container. The slewing ring is rotated and lowered, pulling the workpieces through the abrasive chip mixture. To ensure uniform results on all workpiece surfaces, the parts also rotate about their own axis during processing and are dragged through the mixture in the direction opposite that of the slewing ring at specific time intervals. By increasing the speed of the slewing ring, the freedom of movement of the chips decreases very quickly. The chips wedge up against each other. This supporting effect in the drag path, i.e., around the workpieces, produces a greater packing density of the abrasive chips increasing the contact force and the effective specific cutting action per surface unit. This improves the removal rate. The effectiveness is up to 20 times greater than in vibratory slide grinding machines and machining is significantly more intense than in centrifugal slide grinding machines. As the flow of the abrasive chips follows the contours of the workpiece, both convex and concave surfaces can be processed.

All of these processes have one thing in common: They produce wastewater containing abraded particles from the workpieces and the abrasive chips. There are also chemicals from the cleaning agents, oils, and greases from pre-treatment and the compounds, and the unavoidable treatment liquids for the slide grinding process. While machines are being continuously developed with optimized water and compound consumption, machines with wastewater recycling are now also an established part of system manufacturers' product portfolio. As the quantities of water and liquid compounds are reduced for process-related reasons, there is a particular focus on the further development of existing recycling technologies for processing wastewater such as centrifuges and flocculation technologies.

5.6 Water Jet Cutting Systems

In manufacturing technology, water jet cutting is used as a cold cutting process where lasers have reached their limits in terms of workpiece thickness or other processes either cannot complete the task or are too expensive. For example, dust-free water jet cutting has certainly proven to be cost-effective for contour machining of fiber-reinforced composites, felt-like insulating materials or components of rigid foam, and other non-metallic materials. It is defined in DIN 8580 as a cutting process and the subgroup of abrasive jet ma-

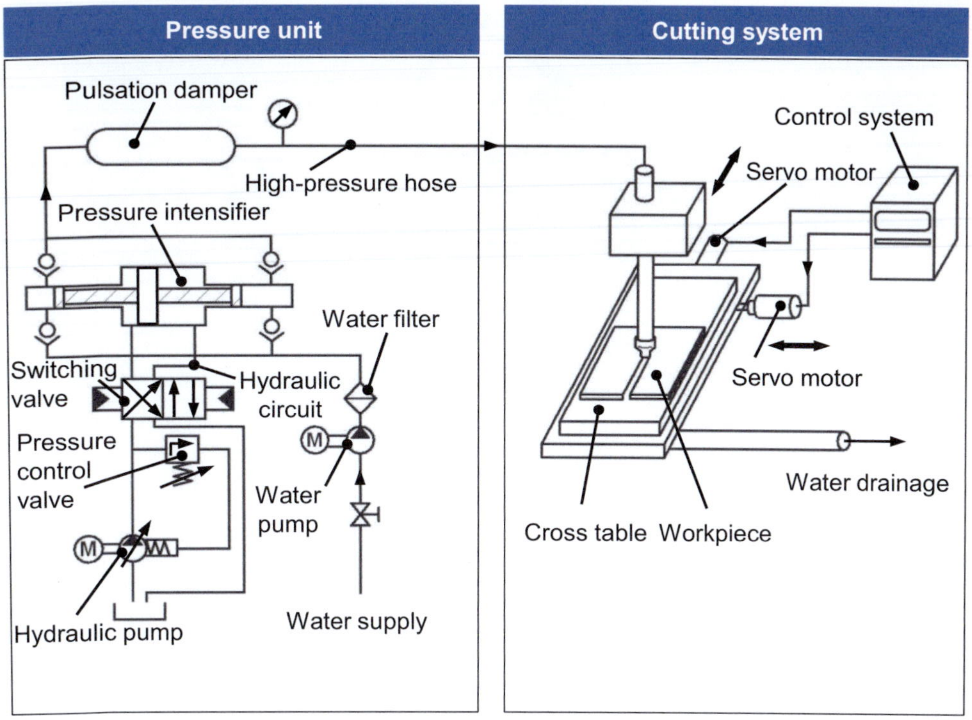

Fig. 5.48 Schematic of a water jet cutting system

chining is summarized in DIN 8200 [DIN82, DIN03b]. The principle of water jet cutting is based on the capability of material removal using an extremely high-speed jet of liquid. In principle, all pumpable liquids can be used as the jet. In most cases, however, water is used due to its availability, low cost and good physiological compatibility.

The design of a water jet cutting system is shown in Fig. 5.48. A distinction is made between the pure pressure unit, at left in the figure, and the cutting system, at right. The key element of the pressure unit is a piston-like pressure intensifier that is pressurized by a hydraulic pump using oil pressure. The oil pressure acts on the large piston surface so that the piston surface ratio produces a pressure intensification between the oil circuit and the working medium, i.e., the water. The pressure intensifier is alternately pressurized from both sides by oil pressure by a switching valve. It uses the stroke movement to produce the necessary high working pressure and volume flow in the water circuit. A high-volume compensation reservoir acts as a pulsation damper to compensate the pressure surges before the working medium (water) is forced to the cutting nozzle in the cutting system via a high-pressure hose. At the required working pressures of 500 to 5000 bar, the volume of the water decreases by up to 13%.

The cutting system basically consists of an NC-controlled cross table in which the workpiece is guided under the cutting nozzle in accordance with the contour to be machined. A collection tank is positioned underneath the cutting table to catch the residual water and the water is then filtered and drains into the sewer system. The narrow cutting nozzle opening accelerates the water jet, producing the required kinetic energy to remove material. The nozzles can only be made of high-strength materials such as carbide metal, sapphire, or diamond as the nozzle is subjected to high levels of wear from frictional forces and cavitation. The nozzle diameter for machining without abrasives must be extremely small, between 0.08 and 0.6 mm, to keep the volume flow rate, and therefore the pump output, low and to produce narrow cutting kerfs.

In special nozzle designs, abrasives such as silica sand, garnet, or corundum can be added to the fluid jet flowing at 800 to 900 m/s to increase the rate at which the water jet removes material. This can be achieved using either an injector or suspension principle. In the injector principle, the water jet sucks the abrasive medium automatically using negative pressure in the mixing chamber and accelerates it into a second nozzle known as the focus. If the abrasive medium is added to the water in the pressure circuit, this is known as the suspension principle. This requires a great deal of additional equipment as the solid material is held in a pressure vessel and has to be maintained as a suspension. Furthermore, these types of vessels can only be designed for relatively low pressures (max. 700 bar) so that a high suspension flow is necessary to achieve the required performance. However, the acceleration process for the suspension jet, comprising only water and

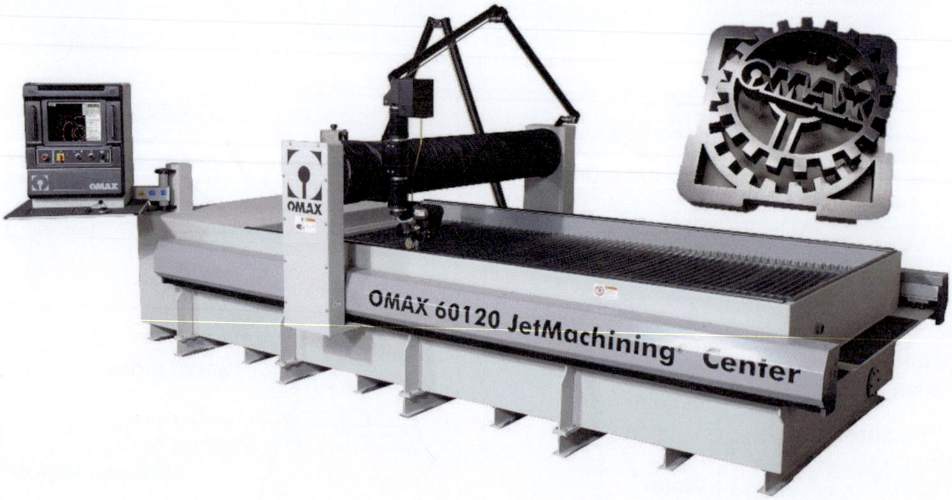

Fig. 5.49 Water cutting system and typical workpiece. *Source* INNOMAX AG

abrasive material, is more effective than that of the injector jet as this consists largely of air, thus reducing the accelerating effect of the water.

The diameter of these abrasive nozzles is 0.8 to 1.2 mm. Water jet cutting can cut materials with a hardness that is less than the abrasive material. Modern water jet systems can cut steels up to 100 mm thick and composite materials up to a thickness of 140 mm. The maximum cutting rates for water jet cutting with abrasives are 680 mm/min for tool steel and up to 8900 mm/min for ceramic tiles (5 mm thick). [FRIT12]

Figure 5.49 shows a water jet machining center and a typical workpiece. The water is routed to the cutting head via high-pressure hoses. The jet flows at approximately 800 m/s as it leaves the water nozzle. This jet now flows through a carbide metal nozzle. Using the Venturi effect, a sharp-edged abrasive material (garnet sand) is added. The water/sand/air mixture hits the workpiece at the speed of sound. The workpiece is fixed to a cutting table which controls the movement of the cutting head while removing the residual flow to the water tank. The abrasive material is automatically transported from a storage tank to the cutting head. Approximately 450 g are used per minute of cutting. The slurry from cutting is automatically cleansed from the tank by a desludging unit. The liquid slurry is passed through a filter to separate the water from the sludge. While the sludge can be recycled, the water is sent to the sewer system via the desludging unit. If necessary, the wastewater quality can be further improved by using stilling basins.

In terms of cost-effectiveness, water jet cutting can compete with laser cutting. Machining with lasers is more cost-effective than water-abrasive jets because laser cutting can achieve higher cutting rates at almost equal machine hourly rates. However, this only applies to the material thicknesses that can be cut by laser. This is approximately 20 mm for mild steel and 6–8 mm for aluminum or stainless steel. Water jet technology is primarily used for greater material thicknesses. Water jet technology provides similar benefits when cutting plastics, foams, or rubber because lasers have limited suitability here due to the toxic gases produced. However, the water mist created generates a greater potential for corrosion and a high noise level. The decision between laser or water jet machining is determined by the type of material to be cut and its thickness.

Water jet cutting is widely used in the medical field and the food industry as the jet medium is sterilized by the high pressures and there is no transfer of germs between the objects to be cut. It is also possible to cut out complex shapes from simple glass and bullet-proof glass as it does not have to be broken along straight edges using this method. Elastic materials can be cut without any significant distortion due to the high jet rates. Different material combinations such as composite materials can be cut without structural fraying.

One disadvantage of water jet cutting is the quality of the cutting surfaces. The workpiece surface tapers inward at high cutting rates and expands at low cutting rates. This geometrical error does not occur at specific intermediate rates and the cut surfaces run parallel. Other factors for a decreasing angular error ($\alpha \neq 90°$, see Fig. 5.50) are:

- decreasing distance between the nozzle and the workpiece
- increasing hardness of the workpiece
- increasing quality of the abrasive material
- decreasing material thickness
- increasing focusing capability of the nozzle.

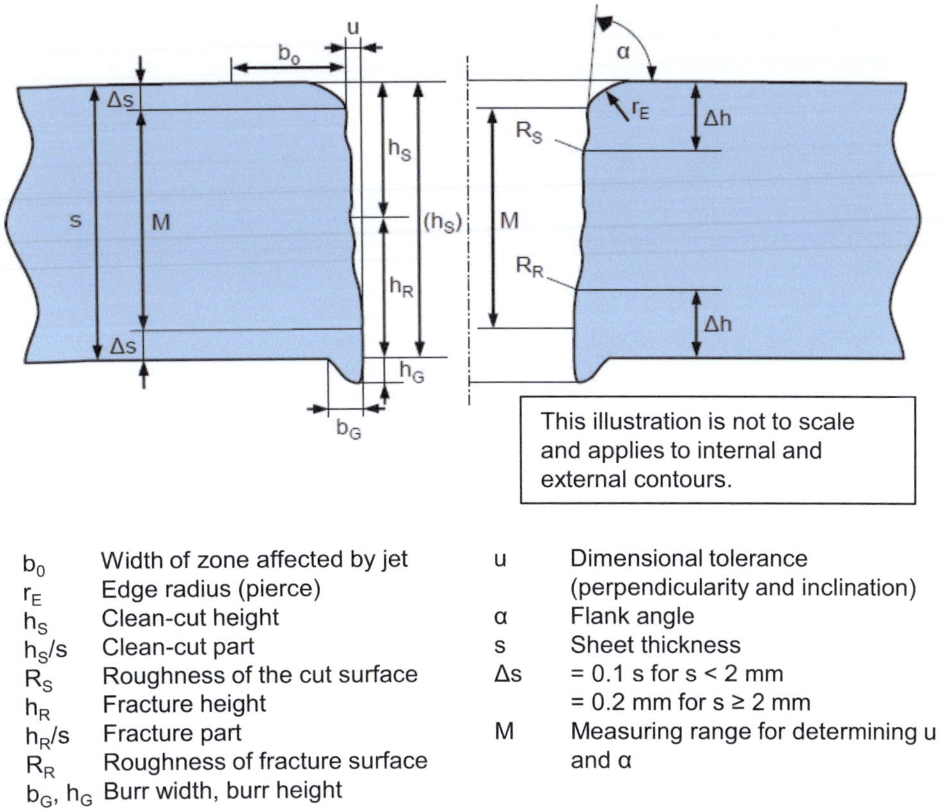

Fig. 5.50 Cut surface parameters for evaluating the cut surface quality of workpieces cut using abrasive water jets, according to [VDI94]

The dimensional tolerances and quality of the cut surfaces are dependent on the water pressure, the abrasive material and, particularly, the cutting rate. There are usually three quality stages for evaluating the cut surfaces and the necessary characteristics for this are determined in accordance with Fig. 5.50. It is not necessary to specify a tolerance for a rough cut or separation cut (Q1). For a normal or medium quality cut (Q2), the dimensional tolerances for the upper edge of the part are ±0.1 mm, the surface roughness is $R_z = 20$ to 50 µm and the angle deviation is 1° to 3°. For a high-quality cut (Q3), the dimensional tolerance is ±0.05 mm for the upper edge of the part, the surface roughness is $R_z = 10$ µm and the angle deviation is 0.5° to 1°. [FRIT12]

References

[DIN03a] Norm. DIN 8589 Teil 14 (September, 2003). Fertigungsverfahren Spanen—Honen; Ordnung, Unterteilung, Begriffe.

[DIN03b] Norm. DIN 8580 (September, 2003). Fertigungsverfahren—Begriffe, Einteilung.

[DIN03c] Norm. DIN 8589 Teil 11 (September, 2003). Fertigungsverfahren Spanen—Schleifen mit rotierendem Werkzeug; Einordnung, Unterteilung, Begriffe.

[DIN81] Norm. DIN 69651 Teil 1 bis 6 (Oktober, 1981). Werkzeugmaschinen; Werkzeugmaschinen für die Metallbearbeitung.

[DIN82] Norm. DIN 8200 (Oktober, 1982). Strahlverfahrenstechnik—Begriffe, Einordnung der Strahlverfahren.

[FRIT12] Fritz, A.H., Schulze, G.: Fertigungstechnik. Springer, Berlin (2012)

[KLOC05] Klocke, F., König, W.: Fertigungsverfahren 2. Schleifen, Honen, Läppen. 4. Aufl. Springer, Berlin (2005).

[VDI94] VDI-Richtlinie. VDI 2906-10. Schnittflächenqualität beim Schneiden, Beschneiden und Lochen von Werkstücken aus Metall—Abrasiv-Wasserstrahlschneiden (1994).

[WALT15] Walther Trowal: Durchlaufanlagen AV. Firmenschrift. Haan (2015).

[WECK83] Weck, M., Durst, H., Varlik, M.: Abrichtüberwachungssystem mit Bedienerführung. Industrie-Anzeiger. **105** (1983)

Material Removal Machines

I thank my staff member Mathias Dehn M.Sc. for his contribution in revising this chapter.

C. Brecher and M. Weck, *Machine Tools Production Systems 1*,
Lecture Notes in Production Engineering, https://doi.org/10.1007/978-3-662-68120-6_6

6

Contents

Material removal machines fall under the cutting manufacturing process (see Fig. 6.1). DIN 8590 [DIN03a] differentiates between chemical, thermal, and electrochemical removal. These three categories are subdivided into further manufacturing processes. Discussed below, however, are only the processes that can be performed with conventional processes and machine tools. Consequently, etching and electrochemical removal are discussed for electrochemical machining, and only thermo-chemical removal for chemical machining. For thermal machining, the numerous subcategories cover examples only of machines for commonly used processes having a very specific machine design. These include electrical discharge machining (EDM) and machining using beams, including electron beam machining. Theoretically, laser beam machining should also be covered here but its diverse applicability gives it its own chapter (see ► Chap. 7).

6.1 Chemical Etching Systems

In chemical etching, material is removed by the chemical reaction of a workpiece material with an active medium. Different bases and acids are used for the active medium. Machining occurs either in a bath (immersion etching) or by spraying (spray etching) the active medium onto the workpiece. The removal principle of chemical etching and the two processes mentioned are shown in Fig. 6.2. Chemical etching is used in the manufacturing of printed circuit boards, for example.

To generate defined workpiece geometries, the surfaces not machined are protected from the active medium by masks. The photoresist process or silkscreen process can be used to apply the masks. In the photoresist process, the whole workpiece is first coated with a light-sensitive varnish using rollers or immersion. The workpiece here consists of a copper-clad base material. Then a photographic image of the required geometry is projected onto the varnished component. Only those parts of the surface that are not to be machined later are exposed. On the exposed areas, the varnish is polymerized, making it acid-resistant. This creates the protective mask, also called resist [FEYE07]. In the silkscreen process, on the other hand, the protective mask is applied directly onto the workpiece.

Of particular importance in chemical etching is the prevention of uncontrolled etching undercuts, that, for PCB traces, for example, can cause deviations in the trace width. These faults can be due, for example, to accumulating etching agent on the component surface resulting in different etching rates on the top and underside of parts. Suction units are used in modern-day systems to remove the excess active medium [FEYE07]. After machining is finished, the strip process is used to remove the protective mask from the parts. To protect operating personnel, actual machining takes place in enclosed chambers with ventilation systems. Also, all of the parts of the system that come into contact with the active medium must be corrosion-resistant. In addition, harmful waste products such as metal salts are generated in the machining process. These salts are separated from the active medium with filters or centri-

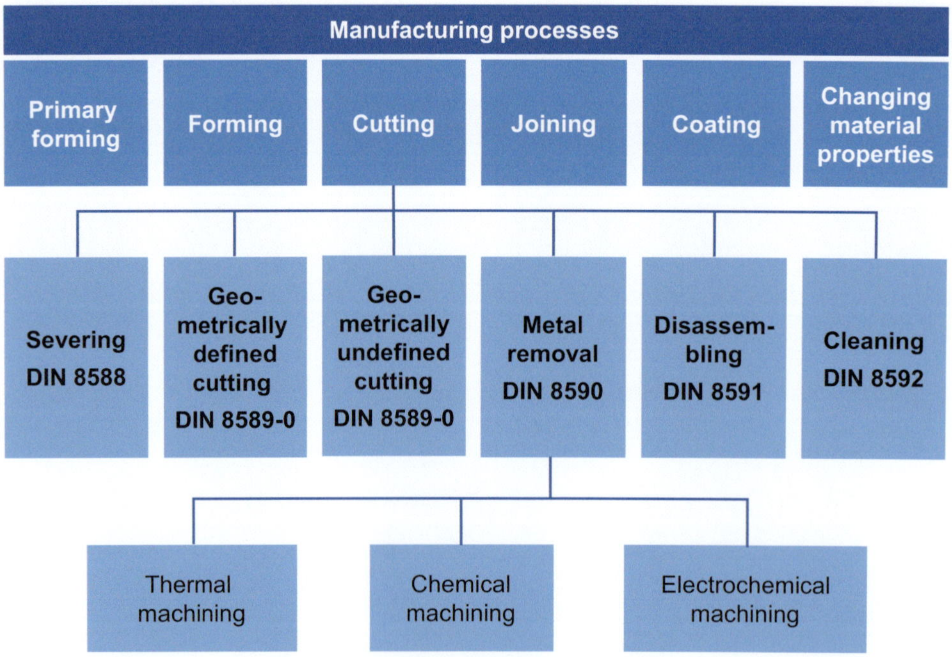

Fig. 6.1 Categorization of manufacturing processes in accordance with DIN 8580. *Source* [DIN03b]

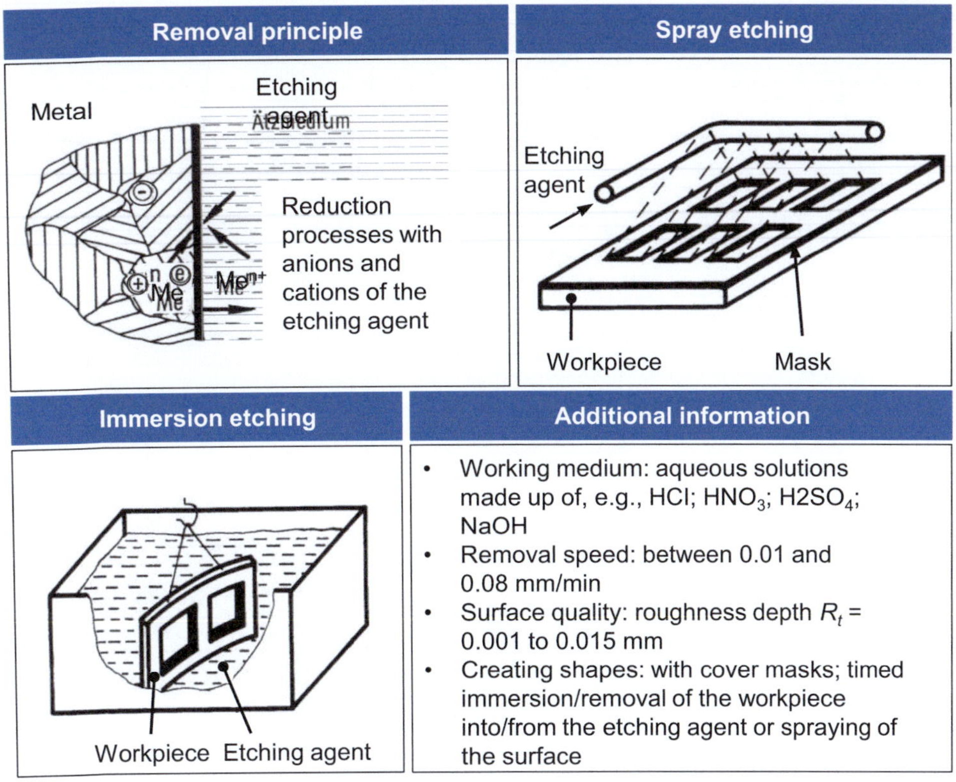

Fig. 6.2 Principle and processes of chemical machining (etching)

fuges to be disposed of separately. Figure 6.3 shows an example of a continuous line for spray etching. After being fed into the system, the workpieces enter the etching chamber, where the etching agent is applied and excess active medium is removed directly by suction units. Then the workpieces are flushed, dried and output from the machine via the outlet.

6.2 Electrochemical Machining Systems

In electrochemical machining (ECM), according to DIN 8590, a metallic material undergoes anodic dissolution by an electric current and electrolyte solution [DIN03b]. For this, a direct voltage source generates a voltage between the metallic workpiece (anode) and tool (cathode). Between the anode and cathode, an aqueous electrolyte solution provides a current flow. On the anode, i.e., the workpiece, metal ions that enter solution with the electrolyte are removed. The removal volume attainable can be calculated using Faraday's law.

$$V = V_{Sp} \cdot I \cdot t$$

where

$$V_{Sp} = a/(n \cdot \rho \cdot F) \tag{6.1}$$

I is the working current between the tool and the workpiece and t is the machining time. V_{Sp} represents a constant that is dependent upon the material used. It can be determined from the atomic weight a, solution valence n, density ρ, and Faraday constant F. The current here is a function of the voltage applied, the distance between tool and workpiece, machining surface, and the conductivity.

To prevent short circuits, a machining gap must be maintained between the workpiece and tool electrodes. This makes ECM a non-contact process. The electrolyte solution is pumped into the machining gap, and is used for the current flow and at the same time to convey the removed material and the heat energy away from the machining gap [DEGN84]. Accordingly, the workpiece is subjected to neither mechanical nor thermal loading. This is a particular benefit compared to other processes (such as chip creation processes). Used as the electrolyte, for example, are aqueous sodium nitrate ($NaNO_3$), sodium chloride solutions (NaCl) as well as acids (H_2SO_4, HNO_3, HCl, etc.) and bases. By using covers, with target flushing with the electrolyte solution and with a suitably shaped tool, a precisely determined amount of metal can be dissolved at any point. In the electrolyte, additional electrochemical reactions convert the removed material to metal hydroxides: solid waste products. No metal is removed from the cathode in this process [KLOC07].

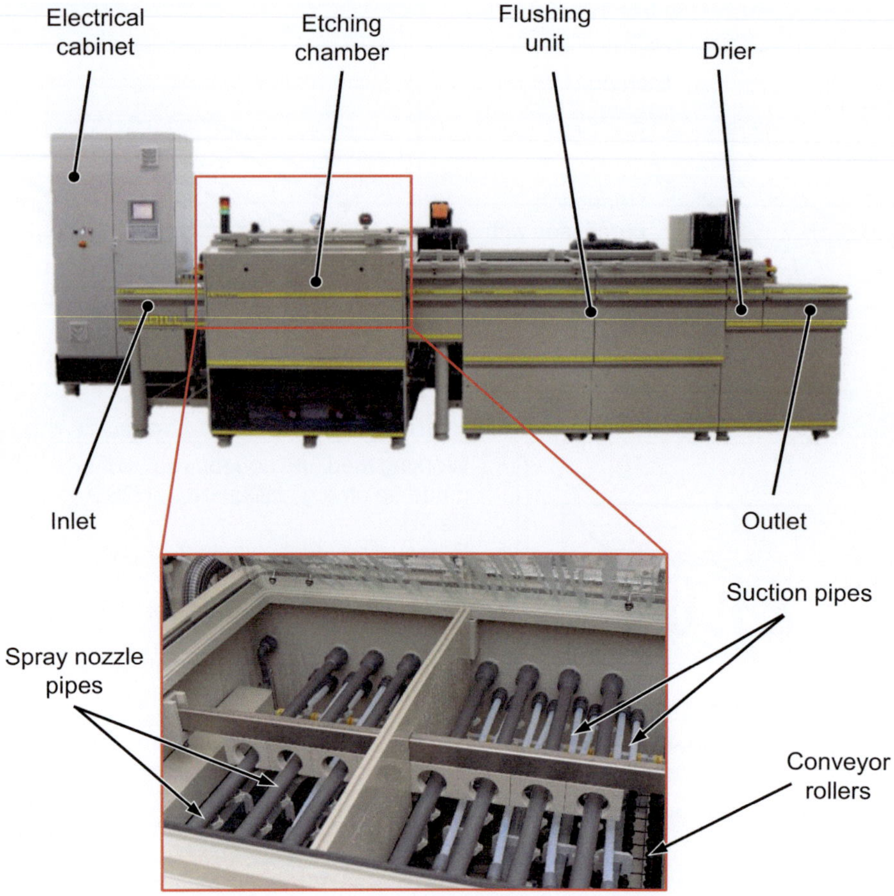

■ **Fig. 6.3** Spray etching machine, according to Pill

■ Figure 6.4 shows the schematic of an electrochemical machining system. It can essentially be divided into three main components: the generator, the machine tool, and the electrolyte management system [FRIT10]. The generator provides the voltage supply and comprises a transformer, reducing the mains voltage to a maximum gap voltage, and a rectifier. Since the gap voltage has a considerable bearing on the removal result, its setting is continuously variable. The direct voltage generator provides currents ranging from 150 to 3000 A for gap voltages between 5 and 65 V for deburring and currents of 20 kA for gap voltages between 8 and 24 V for sinking. The generator also features a short circuit cut-off that detects short circuits and prevents damage to the workpiece and tool by speedily cutting off the power. The electrolyte management system is responsible for supplying and processing the electrolyte solution [DEGN84]. Here, the supply pump conveys aqueous electrolyte solution from the tank to the machining point. During the process, this solution transports heat and removed material away from the machining gap. Given that the temperature and pH value of the electrolyte solution have an important bearing on conductivity and the removal result, the electrolyte supply must keep these parameters as constant as possible. As a result, a temperature controller and a device for measuring the pH value are integrated into the tank [BERG77]. The temperature control system consists of a cooling unit, a temperature controller, and heating elements. Also, the removal products generated in the form of metal hydroxides must be removed from the electrolyte solution before it can be returned to the machining point. Otherwise, the electrolyte solution will thicken over time leading to process disruptions and short circuits. Separators, centrifuges, and filter presses are used to remove the metal hydroxides in the electrolyte processing phase [KLOC07].

■ Figure 6.5 shows two examples of electrolyte management systems from EMAG.

The machine tool generally has clamping devices for the workpiece and tool as well as contacts for power and voltage transmission (see ■ Fig. 6.4) [FRIT10]. To keep the contact resistances low for the workpiece and tool, the contact surfaces must be of a particularly precise design for transmission. If possible, the feed lines for the electrolyte solution are also integrated in the clamping devices and tool electrodes. This means these lines can be routed directly to the

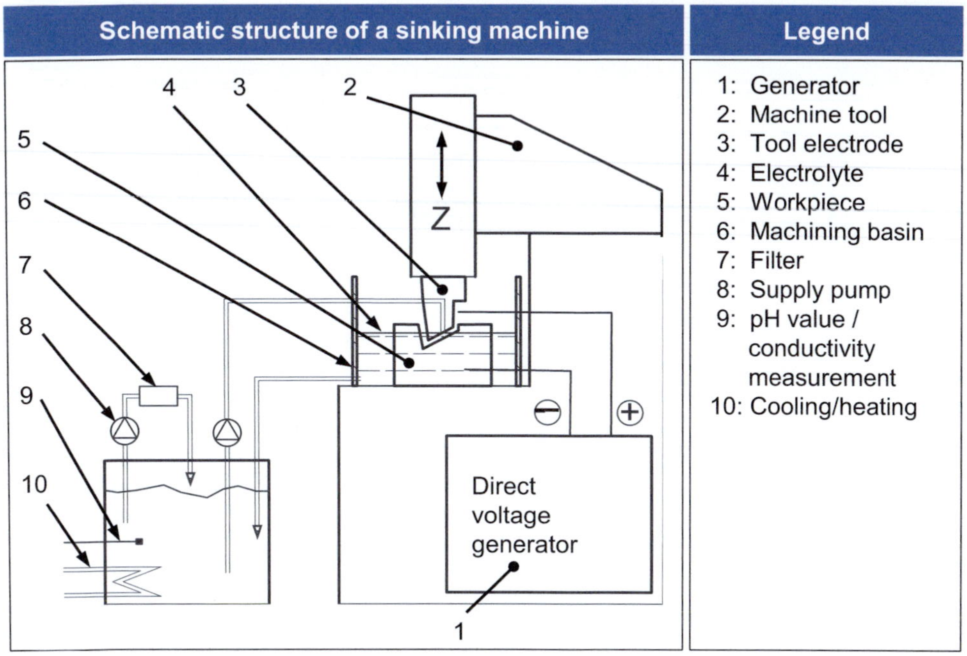

Fig. 6.4 Schematic structure of an electrochemical machining system

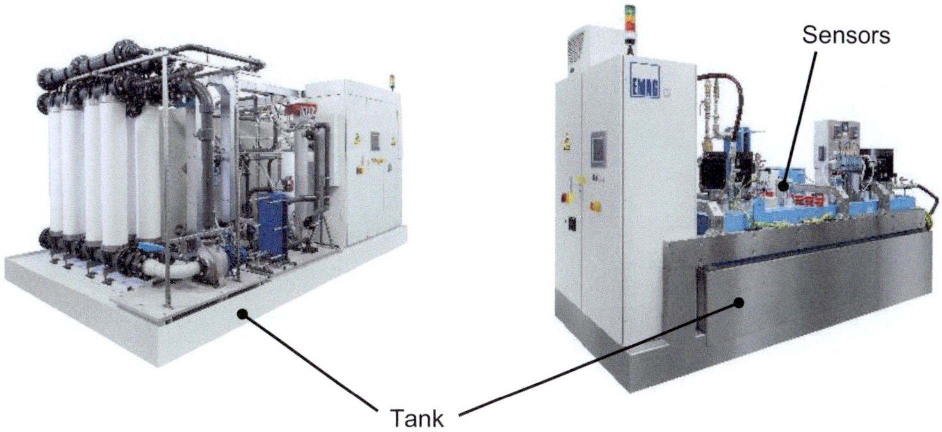

Fig. 6.5 Electrolyte management systems, according to EMAG

machining point. To prevent damage by the aggressive electrolyte solutions, all elements coming into contact with the solution must also be corrosion-resistant [DEGN84]. Materials such as copper, brass, graphite, stainless steels, and titanium are suitable here. The use of plastics to coat and isolate components is also common. The aim is also for a high thermal stability of the machine frame. This precluded changes of the relative position between the tool and the workpiece, and thus a reduction in quality of the removal result that otherwise may occur due to the heat energy generated in the machining gap [KLOC07].

In machines for ECM sinking that use a forming electrode, the machine tool must also have a feed axis for tracking the tool electrode [FRIT10]. To prevent short circuits, the axis must be capable of uniform and stick-slip free feed action. As relatively high electrolyte pressures (>20 bar) occur in the machining gap due to the small gap widths for the flushing process, the machine must also exhibit high stiffness. Similar to thermal stability, this prevents position changes between the workpiece and tool and the resultant accuracy loss. Accordingly, ECM sinking machines generally have a C-frame or portal design. In terms of the flushing pro-

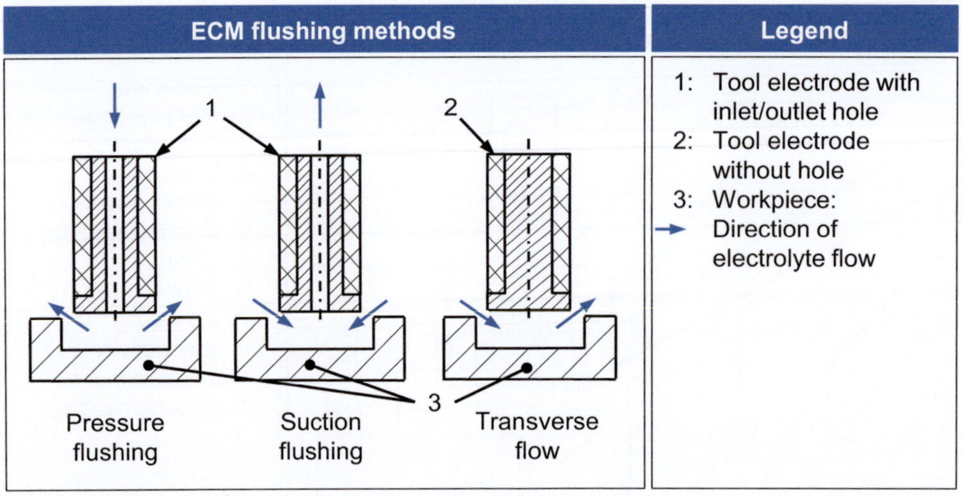

Fig. 6.6 Flushing methods for ECM sinking

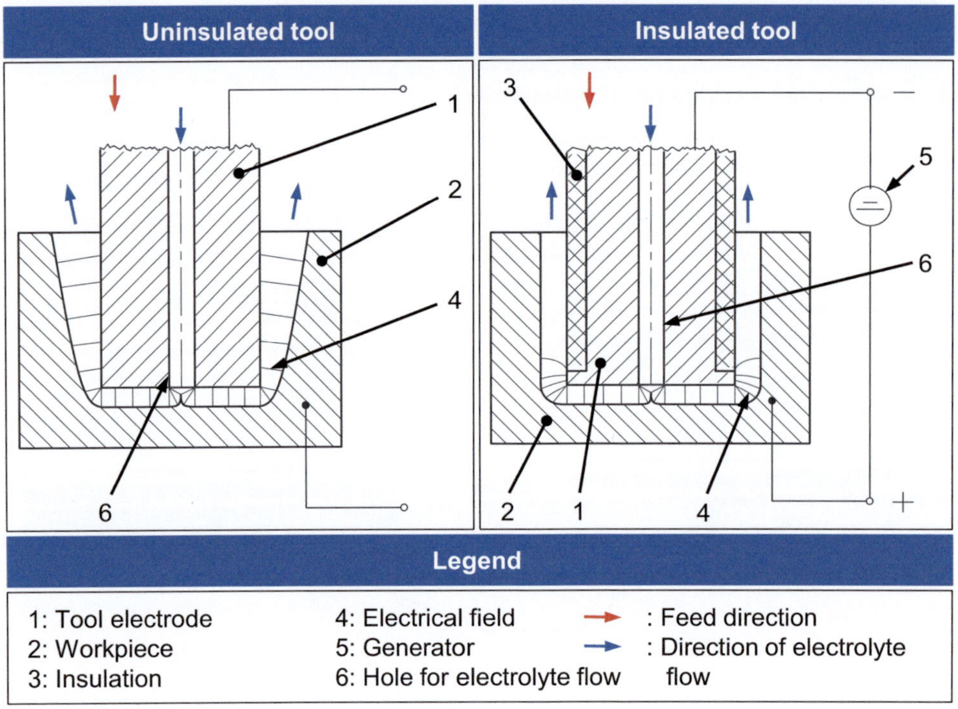

Fig. 6.7 Tools for the ECM sinking of cylindrical holes

cess, a distinction is made between pressure flushing and suction flushing. For pressure flushing, the electrolyte solution is routed through a hole in the tool electrode. The flushing action is in exactly the opposite direction used for suction flushing (refer to Fig. 6.6, left and center). If the tool electrode is too small for inlet and outlet holes, a transverse flow design is also possible (see Fig. 6.6, right) [KLOC07].

One challenge in electrochemical sinking is the dimensionally accurate representation of physical shapes. For example, no cylindrical tool can be used to make a cylindrical bore because there is a current flow at the front surface as well as at the shell surfaces. This can widen the gap the deeper the tool moves down into the workpiece. A tapered bore is created. To prevent this effect, the shell surface of the tool must be insulated. Figure 6.7 shows the different gap formations for insulated and uninsulated tools. [DEGN84].

The principle of electrochemical deburring corresponds to that of ECM sinking. Here, there is no feed motion of the tool electrode during the process, however. Instead, the machining gap is set before machining starts. Given that burrs and sharp edges generally jut out over the reference geometry of the workpiece, they are nearer to the tool than the rest of the workpiece. As a result, these items are removed mainly [KLOC07]. Also, the machining gaps in this process are greater than for sinking. The lower pressure in the machining gap means that the design of ECM deburring machines must not be as stiff as for sinking machines. ◻ Figure 6.8 shows the workspace of a machine that removes burrs electrochemically. The workpiece is clamped in a device, that also provides the power contacts. The tool electrode shapes and positions are adapted to the deburring task at hand. Before machining starts, the electrodes are moved up to the workpiece, and the machining gap is set.

In addition to classic electrochemical machining where direct current is used, there is also a variant where a pulsed voltage source is used. During the duty cycle, a higher current density can be used for working. This also means that the gap width can be reduced. The small machining gaps also means higher accuracies can be attained for complex shapes. Between the pulses, the electrolyte solution in the machining gap can be replaced by flushing. This dissipates the generated heat and transports removed material away. In this way, the electrolyte solution returns to its initial, optimal parameters during the next machining pulse. Short cir-

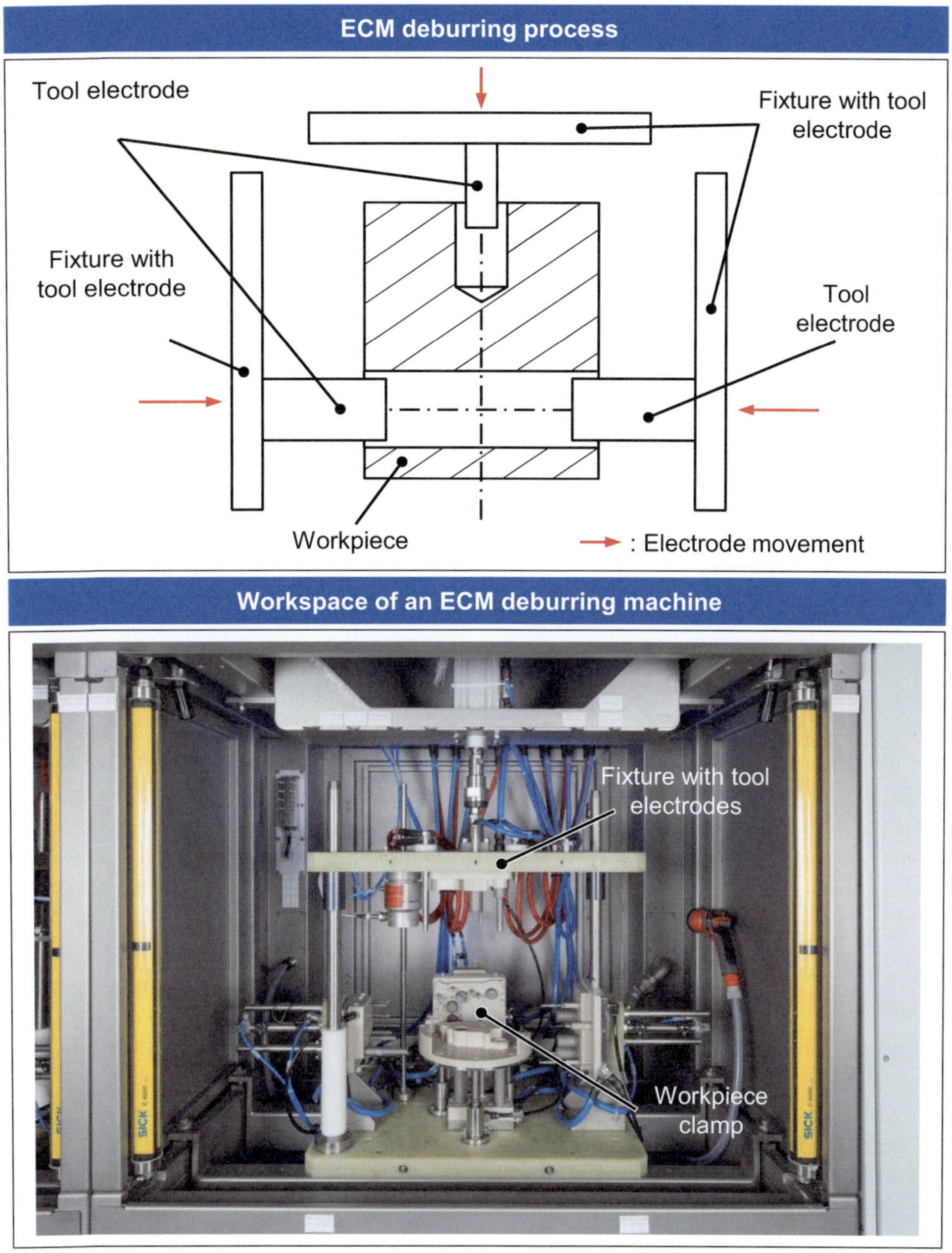

◻ Fig. 6.8 ECM deburring process, top: schematic, bottom: workspace of an ECM deburring machine, according to EMAG

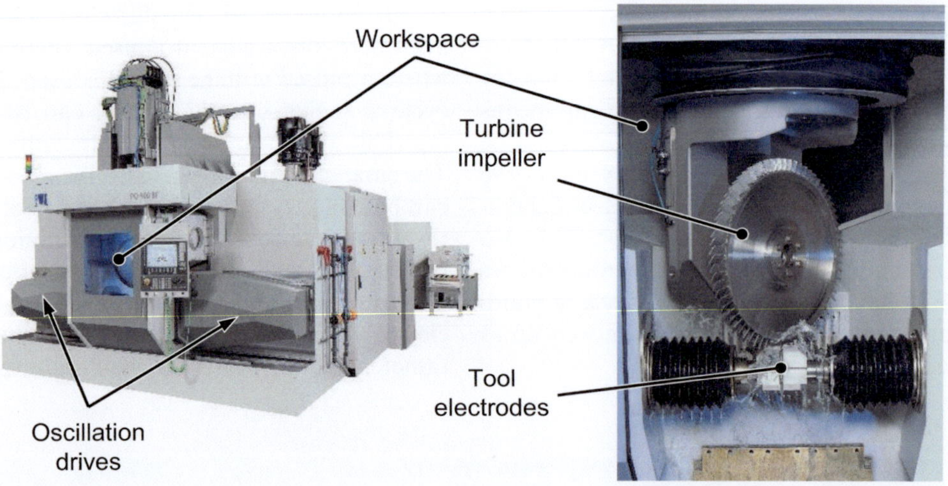

Fig. 6.9 ECM machine with pulsed voltage source and oscillating tool, according to EMAG

cuits due to impurities are prevented despite the narrow machining gap. It is possible to use an oscillating tool electrode as another variant in addition to the pulsed voltage source. During the switch-off time, the tool electrode is moved away from the workpiece momentarily increasing the machining gap. This has a beneficial effect on the flushing process and renders possible additional narrowing of the machining gap during the machining pulses. Accuracy can be increased further in this way. In addition, repositioning the tool electrode, and thus correction of the gap width, can be realized during the lifting movement. To generate the oscillation movement, suitable piezo drives can be integrated into the machine, for example [KLOC07].

Figure 6.9 shows an ECM machine capable of using a pulsed voltage source and oscillating tool electrode. It is used, for instance, to machine turbine blades for the aerospace industry (as also depicted in Fig. 6.9). The machine shown can generate a pulse frequency of 100 kHz. It can accommodate workpieces weighing a maximum of 500 kg and with a maximum diameter of 900 mm. The tool electrodes including the oscillation drive are positioned horizontally in such cases [EMAG11].

6.3 EDM Machines

A manufacturing process is denoted as electrical discharge machining (EDM) when material is removed by means of electrical discharges between the tool and workpiece. The basic operating principle of this process is shown in Fig. 6.10. Just like a parallel plate capacitor, a machining gap is set between the workpiece and tool, and the gap is filled with dielectric fluid [FRIT10]. Applying voltage pulses causes material to be removed. This process can be divided into three main phases for every pulse. In the pre-breakdown phase, applying the voltage causes a current increase delayed by the ignition delay time t_d (see Fig. 6.10). Current flow is generated once the dielectric strength of the working medium (dielectric fluid) is exceeded. In the process, an extremely high-energy plasma channel is generated between the workpiece and tool where high temperatures dominate. These temperatures in turn melt and vaporize the material. This is called the discharge phase (t_e). In the pause stage, the voltage is cut causing the plasma channel to collapse and the removed material to be ejected from the gap. This generates the typical irregular surface of a workpiece subjected to electrical discharge machining. The open circuit voltage u_i is between 70 and 250 V depending on the machining task. The frequencies of the machining pulses can be as high as several 100 kHz while the discharge current exceeds 200 A or near 1 A depending on the machining type (roughing or finishing).

EDM, like electrochemical machining, is a non-contact process. In contrast to electrochemical machining, however, there is also material removal at the tool electrode (anode). Attention must be paid to this in terms of the desired accuracy [KLOC07].

Electrical discharge machining can generally be divided into two main application areas: EDM sinking and EDM cutting. In EDM sinking, machining uses a suitably shaped tool electrode while a wire electrode is usually used for EDM cutting.

6.3.1 EDM Sinking Machines

In EDM sinking, the tool electrode is manufactured as the mirror image of the shape required. Then, the electrode is moved up to the workpiece and removes ma-

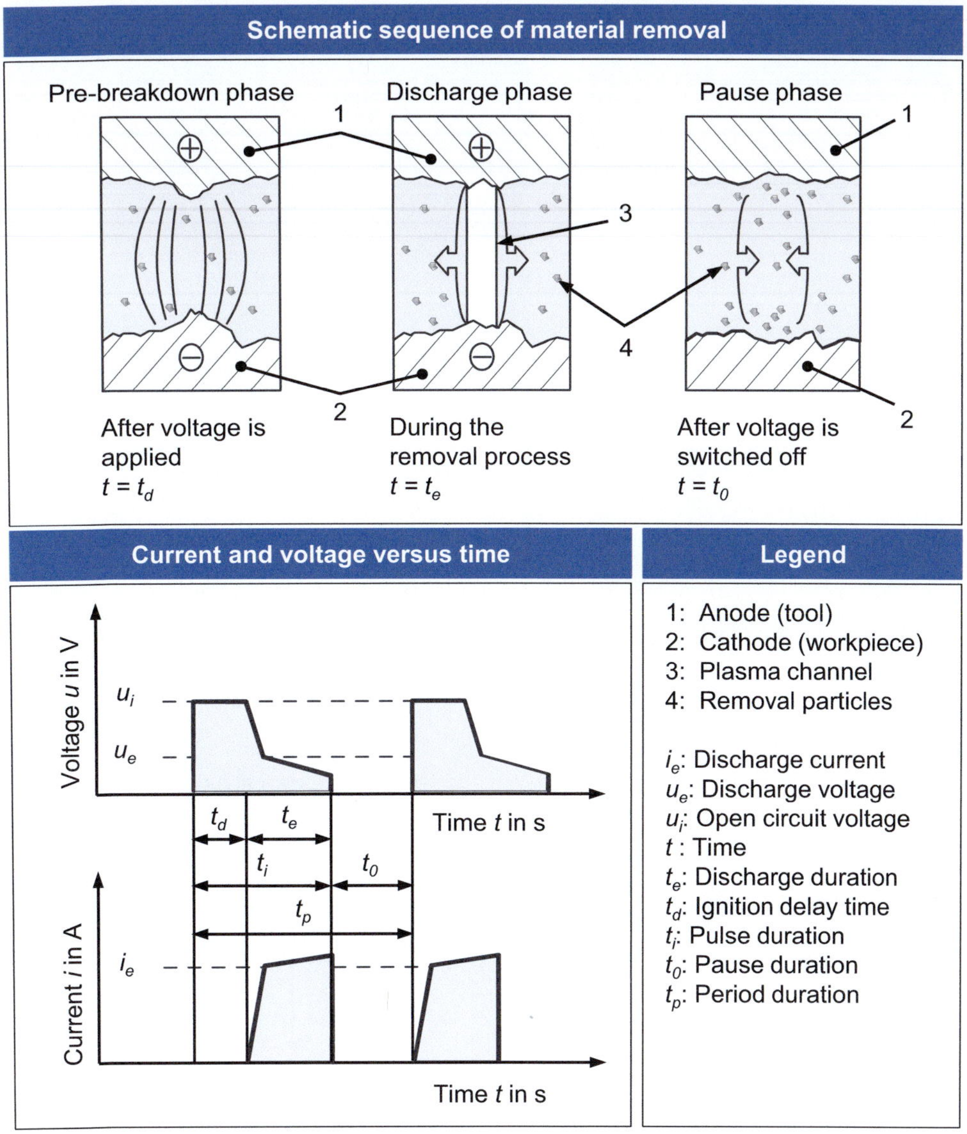

Fig. 6.10 Schematic of the EDM removal principle

terial from it using spark erosion until the contour required is complete [FRIT10]. Some machining examples with workpieces and tool electrodes are shown in Fig. 6.11.

Figure 6.12 shows the schematic structure of an EDM sinking machine. It comprises four main components: the generator, controller, actual machine tool, and unit for processing the dielectric fluid. The generator is responsible for applying the voltage pulses that cause material to be removed as described above. The controller sets the optimum gap width in line with the machining progress and the conditions inside the gap [BERG77] such that ignition takes place in the machining gap and material can be removed. If the gap is too wide, only the open circuit voltage is applied between the workpiece and the tool. As the dielectric strength of the working medium depends, among other things, on the gap width, no current can flow across the gap at this stage and thus no material can be removed (see Fig. 6.13). If the gap is too small, abnormal discharges and even short circuits can result. These can damage the tool and workpiece electrodes. However, as there is no method for determining the gap width during the machining process, equivalent variables need to be set instead. The gap voltage, or ignition delay time t_d between the voltage being applied and material being removed, can be used, for example [KLOC07].

Given that comparatively small forces occur in the machining gap for spark erosion, the stiffness requirements on the machine tool are not as strict as those for electrochemical machining. Most machines nowadays used for EDM have a C-frame design. Large machines, on which tools for manufacturing car body parts are machined, for example, have en-

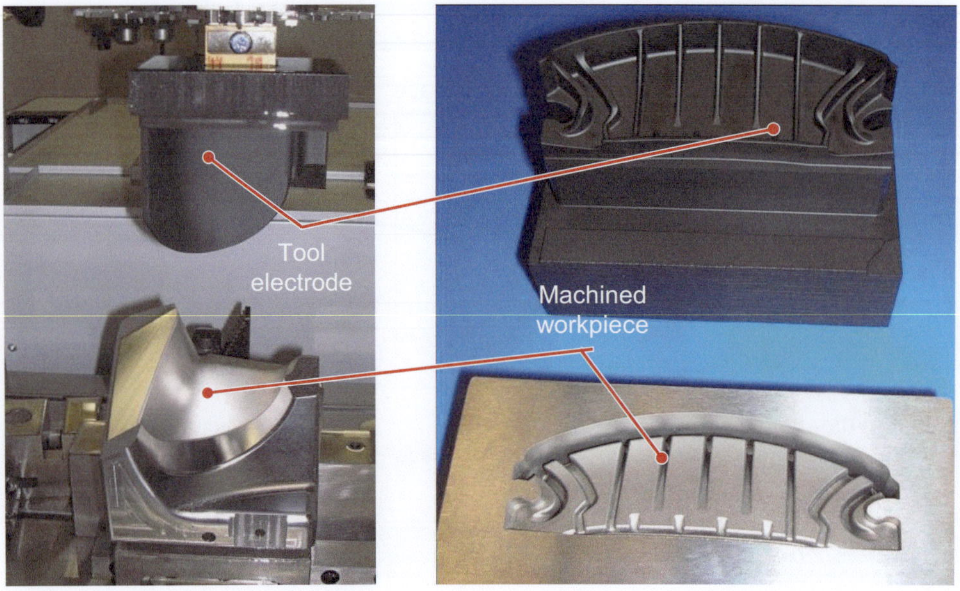

Fig. 6.11 EDM sinking examples, according to OPS-INGERSOLL

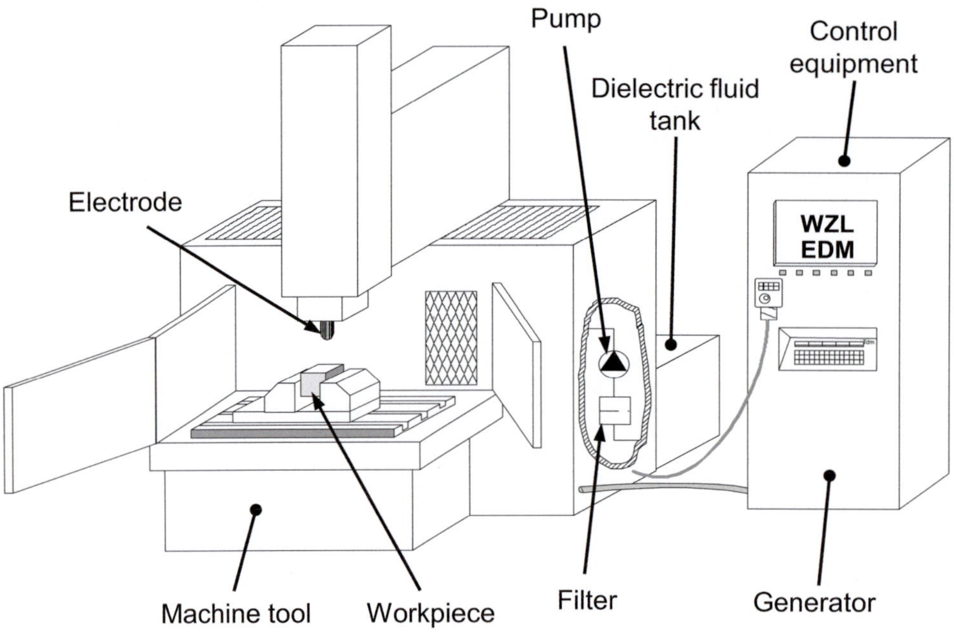

Fig. 6.12 Diagram of an EDM sinking machine, according to [KLOC07]

closed O-frames because the sizes of the tool electrodes used mean they are correspondingly heavy. The feed motion of the tool electrode is generally in the Z-direction of the machine coordinate system. For movement in the X and Y-directions, either the tool electrode or machine table (including clamped workpiece) can be traversed. Since machining must take place inside the dielectric fluid, a tank surrounding the workspace is also required. For small and mid-sized machines in particular, a lifting tank is used here. The sides of this tank can be raised and lowered. This allows access to the clamping plate. It also simplifies setup and removal of the workpiece. Using this kind of lifting tank, it is difficult to move the machine table. This means the machining head performs a movement in the X and Y directions in most cases. An automatic electrode changer can also be integrated into the machine to shorten setup times when many different tool electrodes are required.

Another important component of an EDM machine is the unit for the dielectric fluid. The dielectric

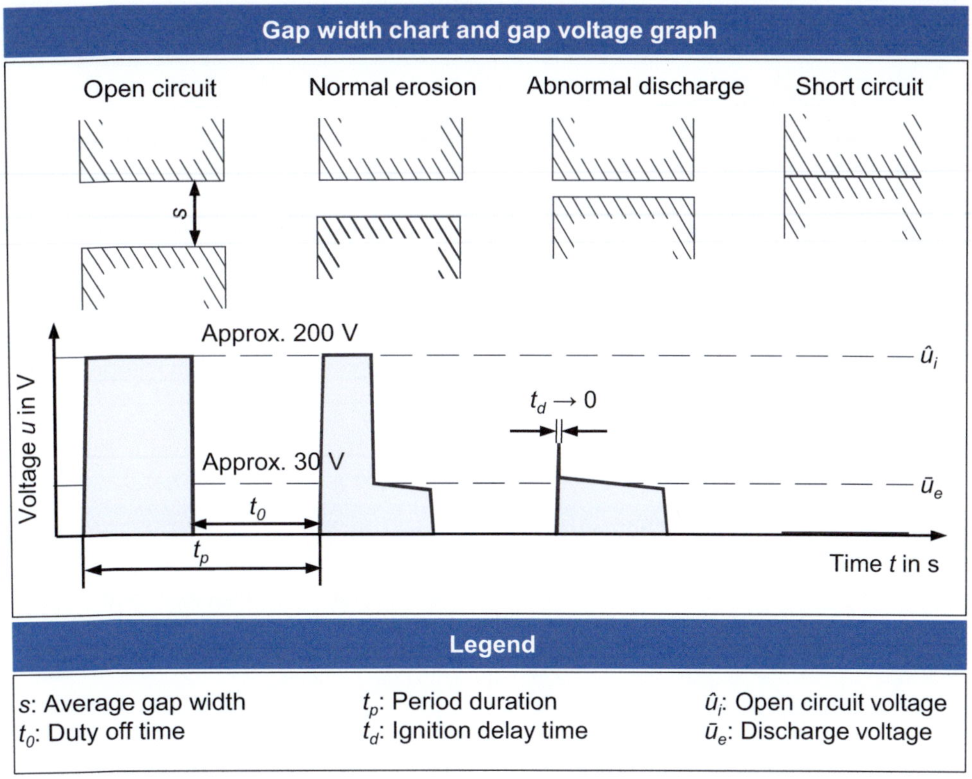

Fig. 6.13 Gap voltage and gap width for EDM sinking

fluid has several different functions here: One is to ensure a higher energy density by constricting the plasma channel. It also transports removed material out of the machining gap and cools the machining point. For the last two functions in particular, continuous replacement of dielectric fluid at the machining point is necessary [BERG77]. Three different flushing methods are used here for EDM sinking. These are shown in the diagrams of Fig. 6.14. For shallow sinking, lateral flushing can be easily implemented, for example, by increasing the pressure on one side of the gap. For deep sinking and narrow gap widths, this is not effective enough, however. Pressure or suction flushing can be used through the electrode if allowed by the electrode geometry. If not, flushing can also be realized by raising the tool electrode. It must be remembered, however, that this regularly interrupts the process and thus slows it. Combinations of different flushing methods are applied in many scenarios [KLOC07].

The flush supply (apart from flushing solely by the lifting movement of the tool) is realized by a pump that also fills the machining tank with dielectric fluid. From the machining tank, the dielectric fluid enters the processing tank. Paper filter cartridges, precoat filters, or gravel filters are used to remove the removed material. Also, the dielectric fluid must be cooled here because it is used to dissipate the process heat (as mentioned above). The cleaned dielectric fluid is finally sent back to the supply tank and can be reused.

Hydrocarbon dielectric fluids are usually used for EDM sinking processes because they do not cause corrosion and enable narrower machining gaps. They are however easily flammable and harmful to health and the environment, meaning compliance with various protection measures, such as extraction systems to remove toxic fumes, is a requirement [KLOC07].

Examples of EDM sinking machine designs found in industry are shown in Fig. 6.15. A classic EDM sinking machine with gantry design and exposed workspace is shown in the top left. The machine table here can carry 1000 kg and the lifting tank has dimensions of 770 × 670 × 440 mm (W × D × H). The axes have a maximum travel speed of 5000 mm/min [OPSI15a]. The close-up view of the workspace with the tool electrode fitted and workpiece clamped is shown at the top right in Fig. 6.15. Figure 6.15 is a machining center as a combination of a milling machine (left) and an EDM sinking machine (right). An additional module connects the two machines. This allows a fully automatic interchange of the tool electrodes be-

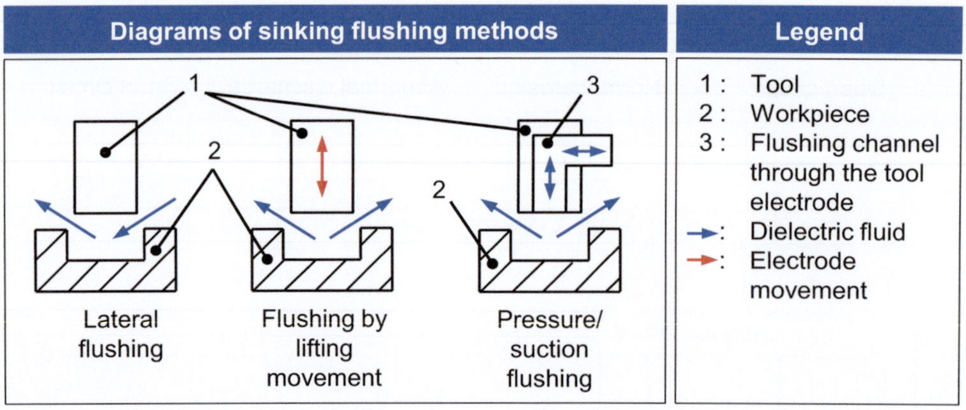

6

Fig. 6.14 Flushing methods for EDM sinking

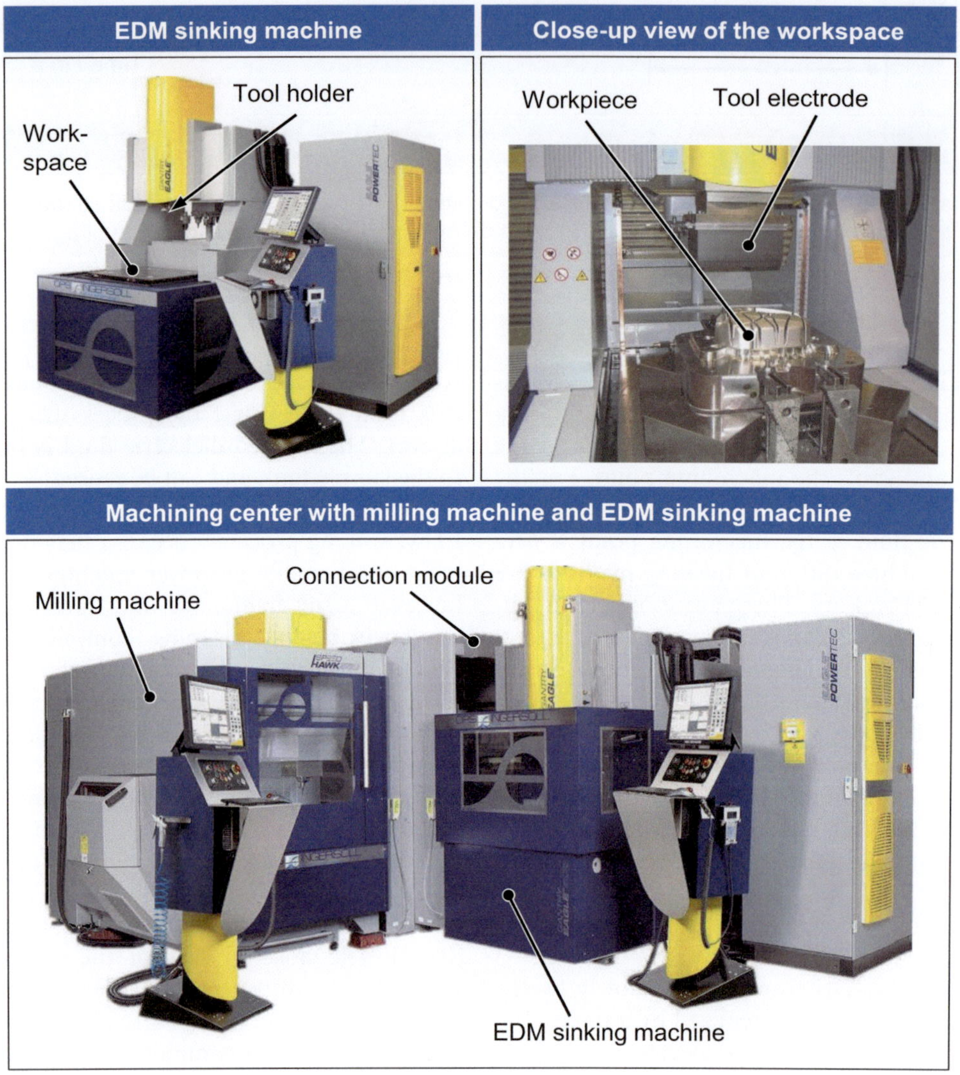

Fig. 6.15 EDM sinking machines, according to OPS-INGERSOLL

tween the milling machine and the EDM sinking machine. This way, in parallel to EDM, the next tool electrode required can be produced and transferred later to the EDM sinking machine. It is also possible, however, to measure previously used tool electrodes in the milling machine and to rework them as necessary, such that

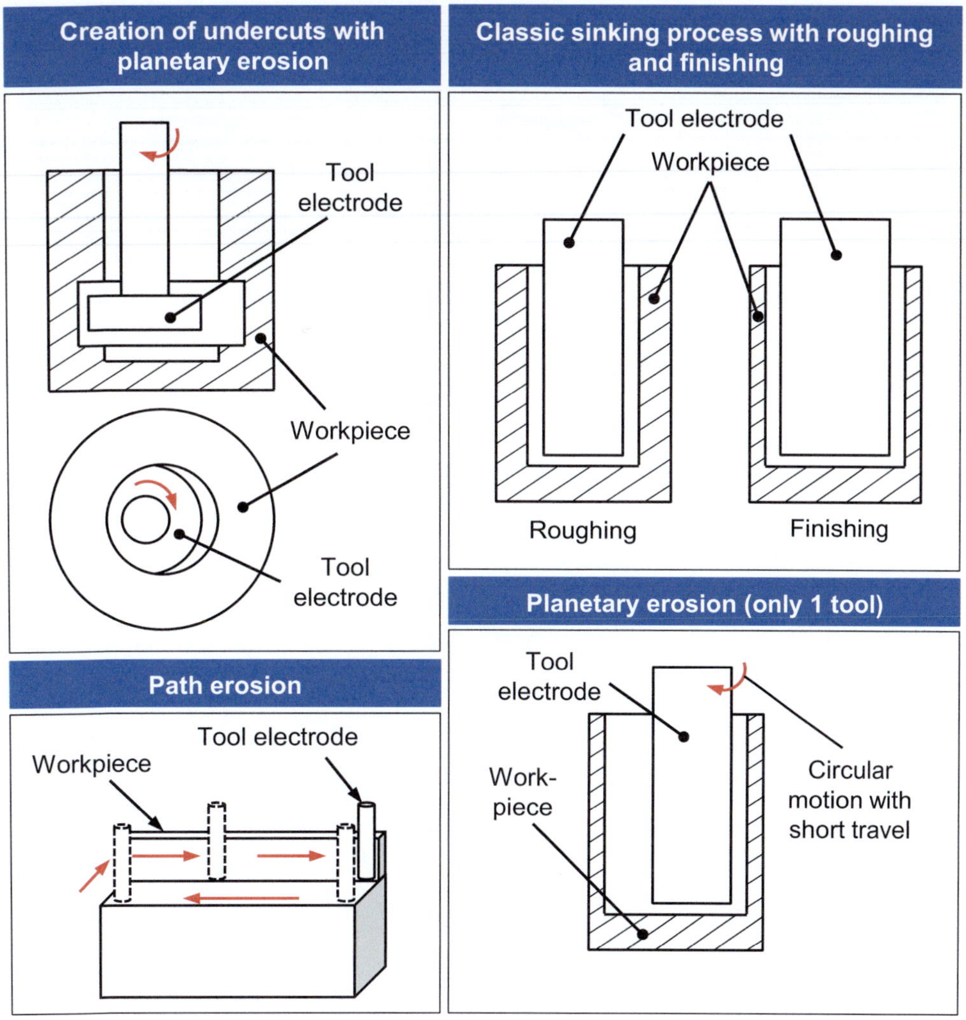

Fig. 6.16 Planetary and path erosion

the required product accuracy can be guaranteed [OPSI15b].

The classic sinking process has the downside that there can be no undercuts. In addition, the manufacture of electrodes can (depending on the form to realize) become very complicated and expensive, especially when multiple tool electrodes are required. This can be the case, for example, when after roughing, finishing is required and this requires a new electrode with the corresponding finishing allowance. The process modifications of planetary and path erosion introduced below remove these restrictions.

For planetary erosion, spatial, usually circular motion (e.g., on the XY-plane) is superimposed on the actual sinking movement (in the Z-direction). This means it is possible to generate undercuts and taper angle shapes with simple tool electrodes (see Fig. 6.16, top left). Also, the allowance of the finishing electrode required for the classic sinking process as compared to the roughing electrode can be compensated. As a result, roughing as well as finishing can be performed with just one electrode (see Fig. 6.16, right). As already mentioned, this can provide considerable savings in electrode manufacturing. Also, there is a beneficial effect on the flushing process because the gap width opposite the current machining point is slightly enlarged. In connection with the electrode movement in the plane, this creates an alternating pressure-suction flushing [KLOC07].

In path erosion, cylindrical electrodes are used. They are able to move on any programmed path within the space (see Fig. 6.16, bottom left). This means the realization of the electrodes is simpler than for planetary erosion. The path movement is realized here in a way similar to a milling process with an end mill cutter. The difference here is, that after every layer is removed, the effective electrode length needs to be determined for wear correction to be performed. Due to the simple electrode shape, standard electrodes can also be used here, meaning a special electrode is no longer re-

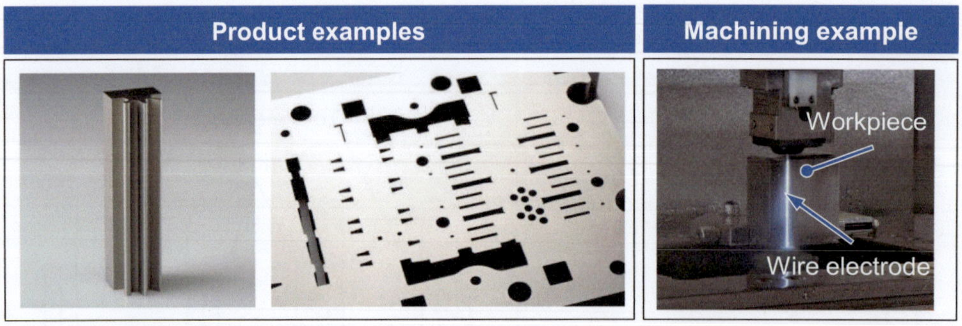

Fig. 6.17 Application examples for EDM wire cutting. *Source* GF Machining Solutions

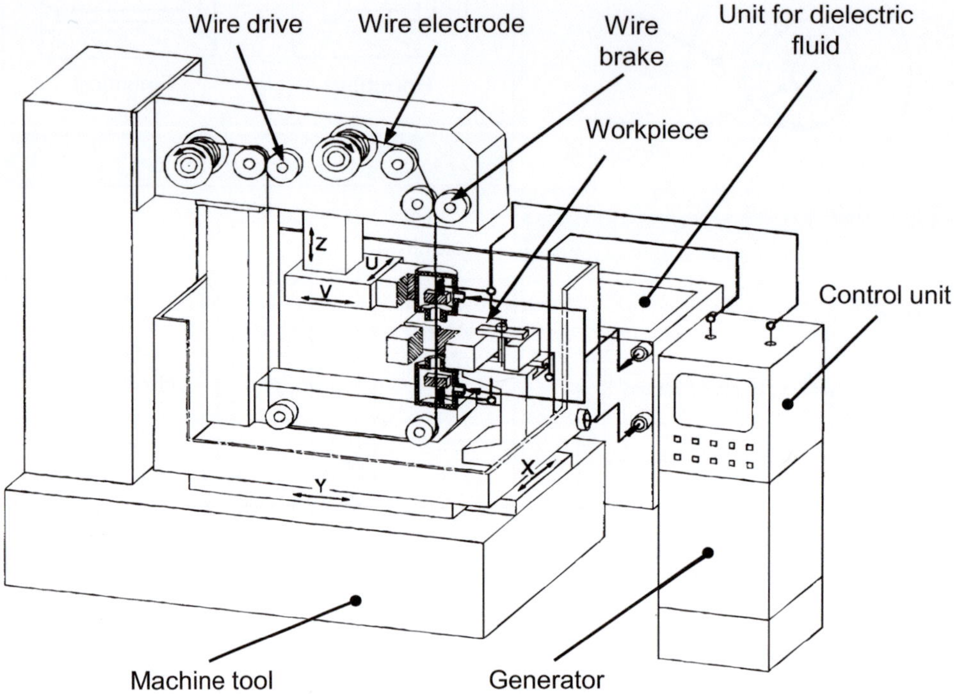

Fig. 6.18 Schematic structure of an EDM cutting machine, according to [KLOC07]

quired for every contour. The volume of material removed per layer is very low, however, and, just like milling, the creation of small-radius corners is restricted by the diameter of the tool electrode [KLOC07].

6.3.2 EDM Cutting Machines

In contrast to sinking, a wire electrode unwound from a coil and not a specially shaped electrode is used for EDM cutting. Micro machining, and the manufacturing of complex and fine structures, are just two of the applications using this process. Figure 6.17 shows a selection of machining examples for wire erosion.

The general design of an EDM cutting machine with a wire electrode can be seen in Fig. 6.18. The machine essentially comprises the same components as an EDM sinking machine: generator, controller, machine tool, and unit for the dielectric fluid. The main difference between cutting and sinking machines is the design of the machine tool that is inherent to the principle of operation. As mentioned, this process uses a wire as the tool electrode. The wire is unwound from the wire coil and routed by two guide arms on an arbitrary path relative to the workpiece. As the wire is subjected to wear by spark erosion, just like a forming electrode but with a smaller diameter, it is continuously fed through. In this way, there are no wire breaks or associated interruptions to the process. Furthermore, the voltage pulses applied are very short so the wire is not subjected to too high a thermal load. Worn wire is disposed of after a single use. The parameters of wire ten-

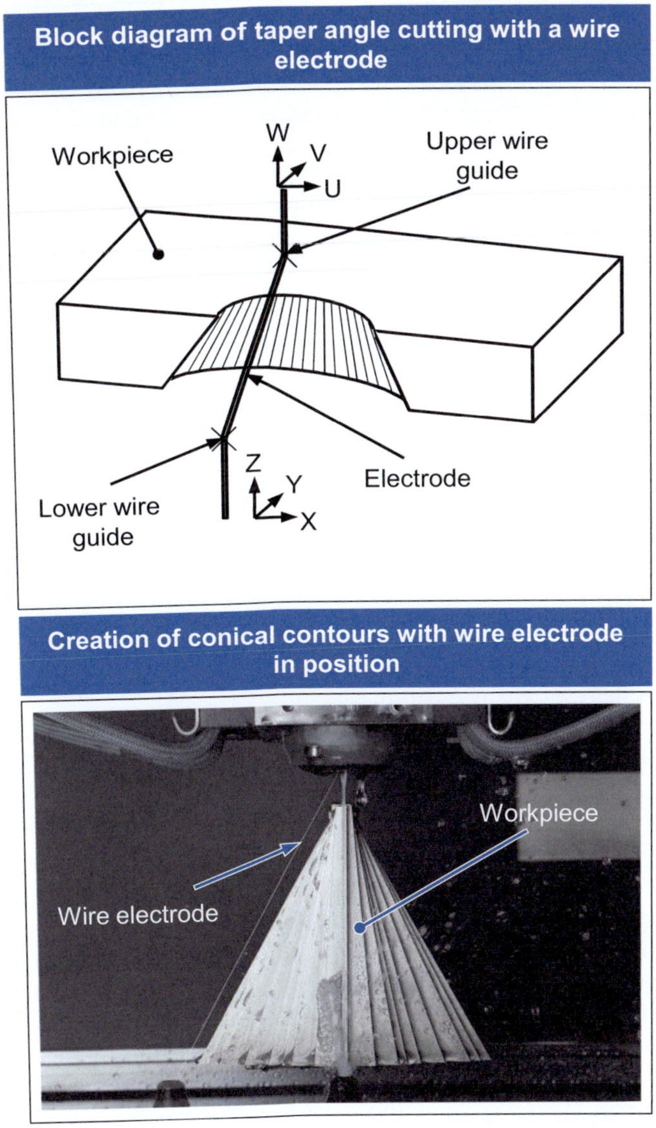

Fig. 6.19 Taper angle cutting, top: diagram according to [KLOC07], bottom: machining example. *Source* GF Machining Solutions

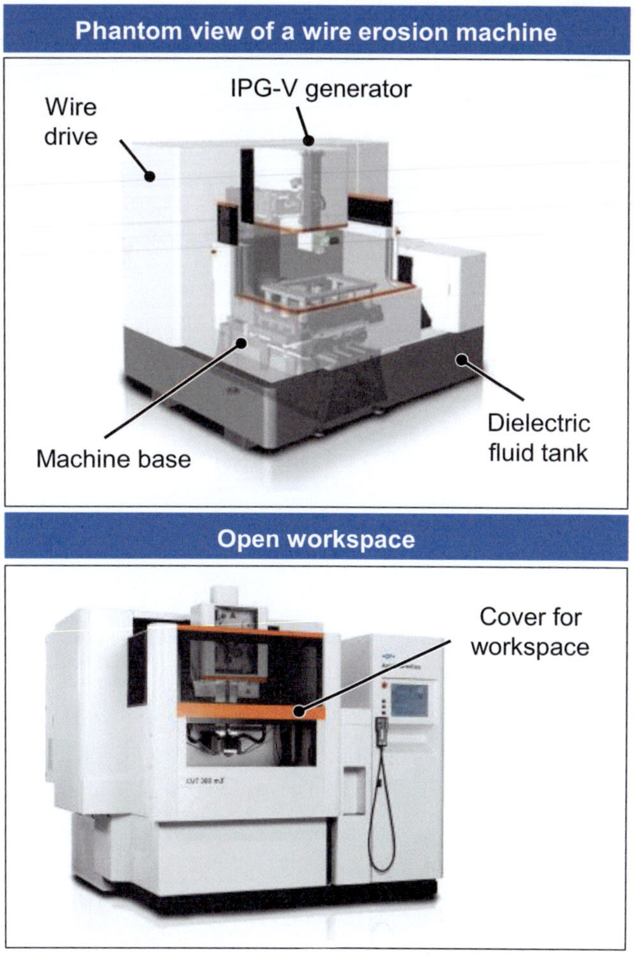

Fig. 6.20 EDM cutting machines found in industry. *Source* GF Machining Solutions

sile strength, wire run-off speed, and distance between guide arms are set in line with the machining task and workpiece dimensions [KLOC07].

The workpiece is clamped to the machine on the machine table. If the two guide arms are designed such that one of them (using auxiliary axes U and V) can be moved in relation to the X and Y-axes of the other arm (see Fig. 6.19), taper angle cuts with a constant or variable angle are also possible. This opens up a large spectrum of applications for EDM wire cutting [KLOC07].

Different wire erosion machine designs from GF Machining Solutions are shown in Fig. 6.20. The phantom view (Fig. 6.20, top) reveals the positions of individual machine components under the paneling. The picture at the bottom in Fig. 6.20 shows a machine with open workspace for loading and unloading workpieces. This machine is designed for a maximum workpiece weight of 1500 kg and allows tapering to 45° (for a workpiece height of 400 mm). Also, the X and Y-axes (or U and V-axes) can be moved at a maximum speed of 3 m/min [GF M12, GF M13].

Example wire guide arms are shown in Fig. 6.21. For process monitoring, measuring devices can also be integrated into the arms, such as here with an integrated CCD camera with backlight (see Fig. 6.21, bottom) [GF M15].

In addition, a device can be integrated for automatically threading the wire into the start hole at the beginning of the process. One such device is shown in Fig. 6.22. The taper angle shape means the wire can be positioned accurately in the start hole and be held by the lower wire guide arm. During the machining process, the device can be pivoted from the working area (see bottom of Fig. 6.21 and 6.22) [GF M15].

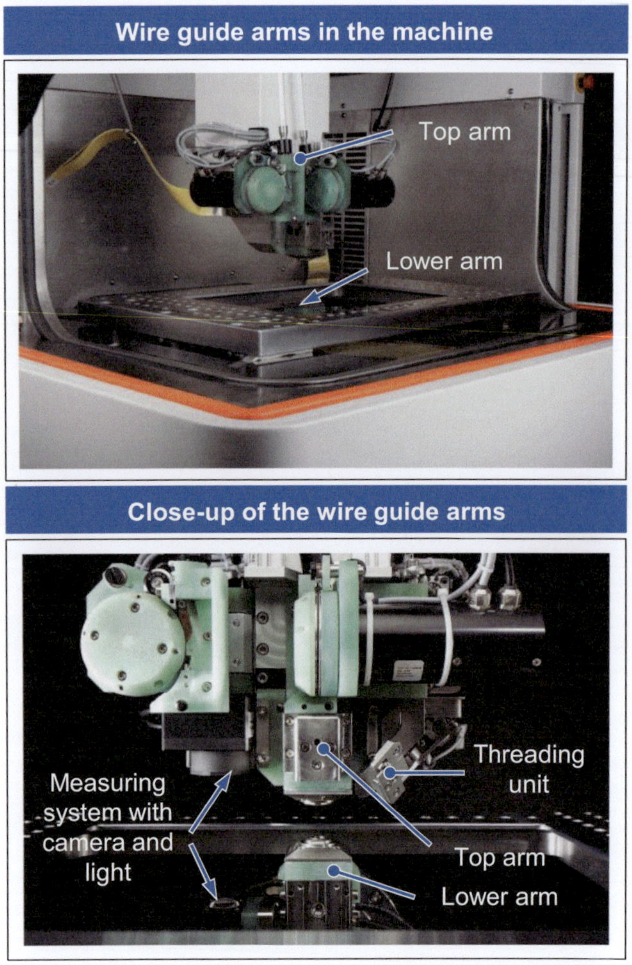

Fig. 6.21 Wire guide arms, according to GF Machining Solutions

In addition to the electrode shape, the flushing method is another difference between wire erosion and EDM sinking. Flushing inside the dielectric fluid bath coaxial to the wire is possible. Another option is a spray jet coaxial to the wire direction, without the work-space being completely flooded with dielectric fluid (see Fig. 6.23). In contrast to the sinking process, deionized water is used as the dielectric fluid instead of hydrocarbons because the water is not combustible or harmful to the environment and so no elaborate protection measures are necessary. The water must, however, be deionized first so it no longer conducts. The system must also be corrosion-resistant [KLOC07].

6.4 Systems for Thermal Removal by an Electron Beam

In contrast to spark erosion or ECM, electron beam systems can perform other processes, such as welding and surface treatment, in addition to material removal machining. The system design is generally independent of the process used. A distinction can be made between three main system components: the beam generation unit, the beam guidance/shaping unit, and the working chamber. Figure 6.24 shows the block diagram of a system for electron beam machining (EBM).

The beam generation unit comprises the acceleration cathode, the control electrode, and the anode. The cathode is heated electrically and emits electrons as a result. The electrons first form an electron cloud in front of the cathode and are accelerated to two thirds of the speed of light by the voltage between the anode (with drilled hole) and cathode [FRIT10]. The control electrode (also called Wehnelt cylinder) is used to control and focus these accelerated electrons. Between this electrode and the cathode, the block voltage is applied. This voltage ensures that only a certain number of electrons in front of the cathode can overcome the block. This regulates the power of the electron beam. The special shape of the control electrode, similar to a concave mirror, also ensures the beam is focused (see Fig. 6.24). [KLOC07]

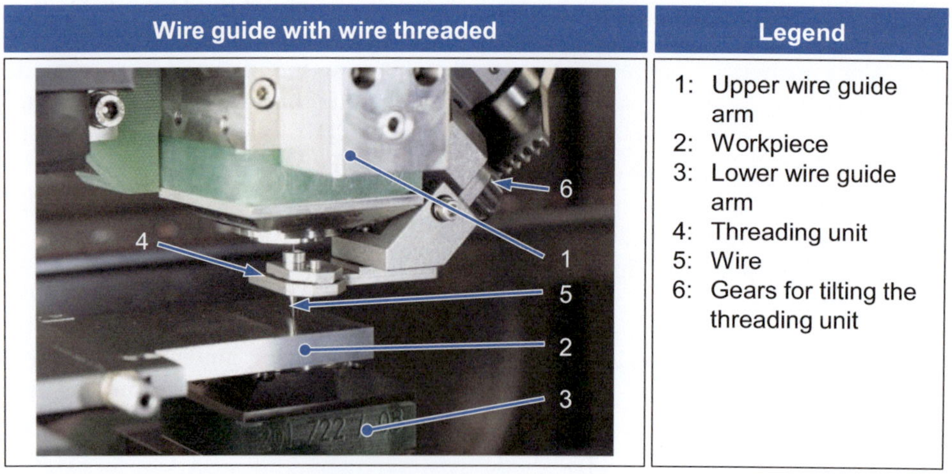

Fig. 6.22 Wire guiding, according to GF Machining Solutions

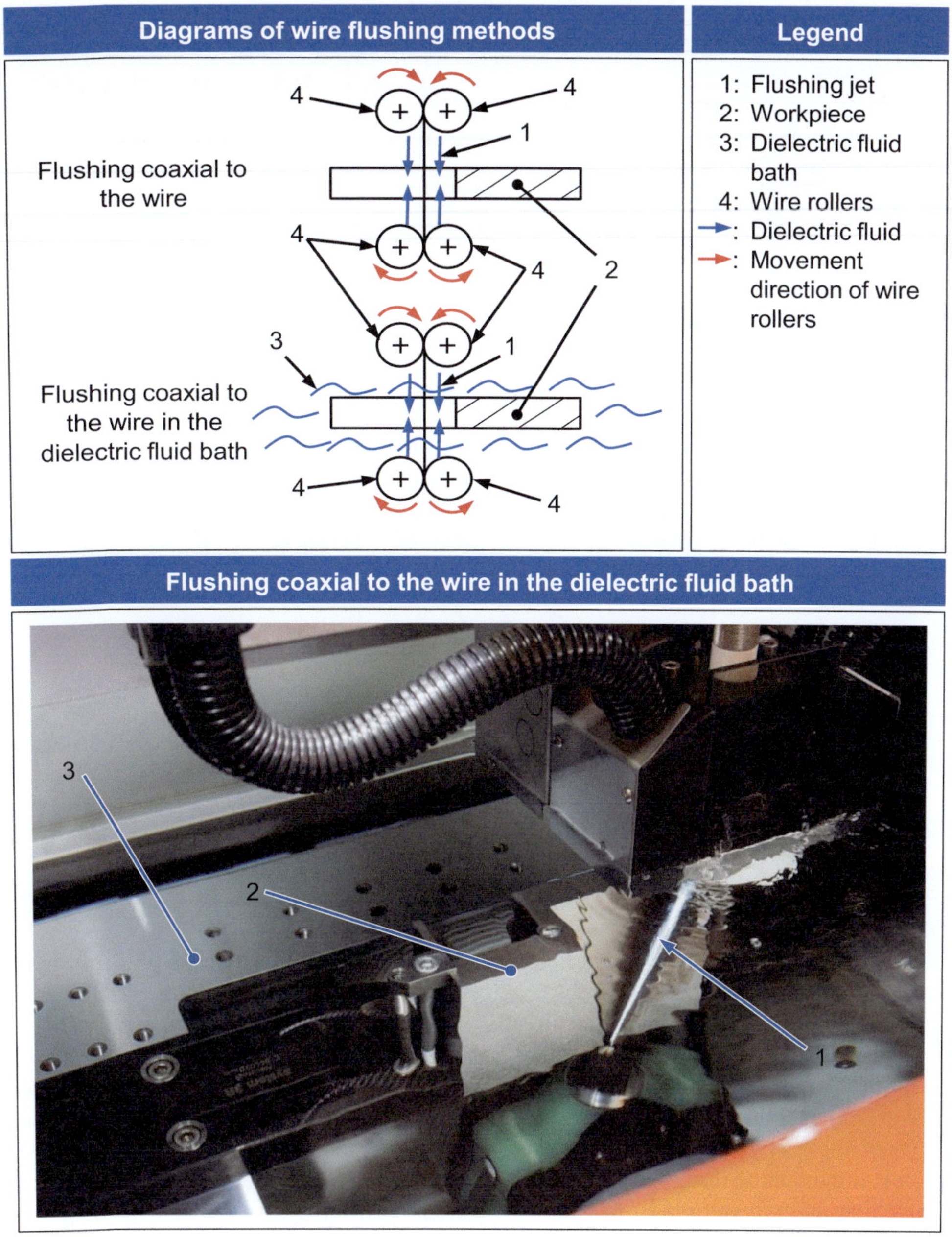

Fig. 6.23 Flushing methods for EDM cutting, top: schematic, bottom: actual implementation, according to GF Machining Solutions

After passing the anode, the beam diverges again, however, necessitating beam guidance/shaping. The adjuster coil ensures the beam is focused while the stigmator coil is used to correct any lens aberration. The electromagnetic focusing coil increases the power density of the beam by focusing it. An oscillating beam movement can be created with the deflector coil (see Fig. 6.25). Optics for beam positioning or video systems are usually integrated in the beam guidance to position the beam accurately above the machining position. Actual machining takes place in the working chamber.

A distinction is generally made in beam generation between high and low-voltage generators. They provide acceleration voltages of about 150 kV and 60 kV, respectively [STEI15a].

The beam must be generated in a high vacuum (e.g., $p < 10^{-4}$ mbar) to prevent oxidation of the cathode and reaction of the electrons with ambient air. This enables the beam to build up [BERG77]. Machining usually takes place in a vacuum because, even after the beam is generated and shaped, it diverges greatly when it meets atmospheric conditions due to scattering and reactions with the air. It is possible, however, depending on the requirements placed on machining quality, to perform the work process under higher ambient pressures or even at atmospheric pressure. Here a dis-

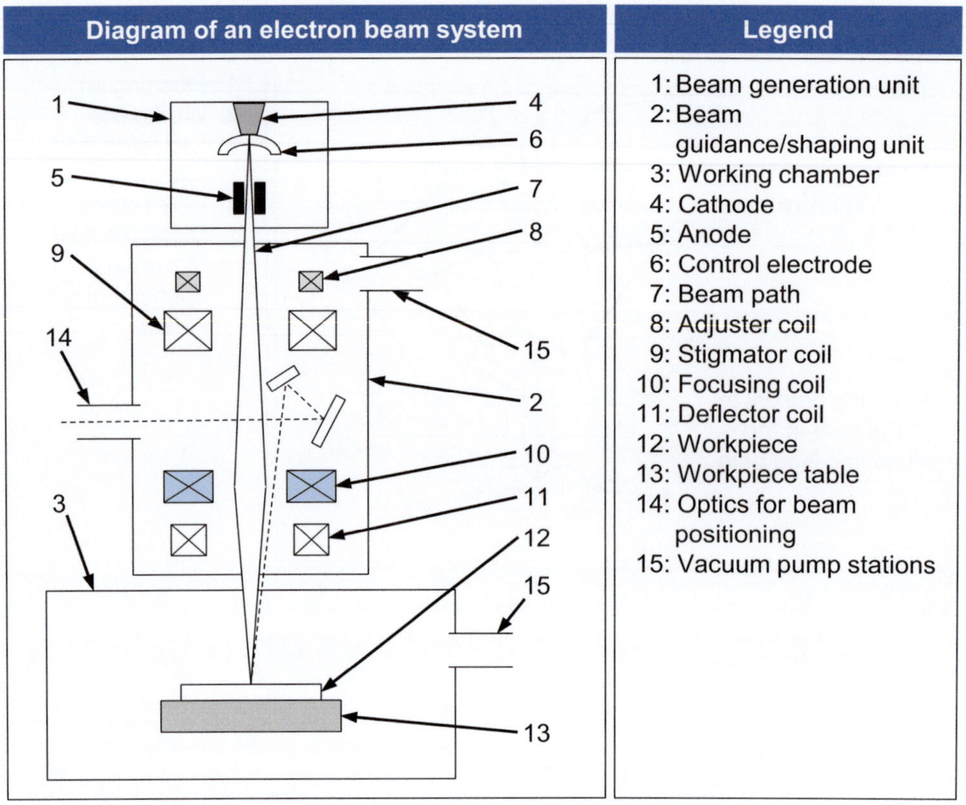

Fig. 6.24 Schematic structure of an electron beam system

tinction is made between systems with high vacuum (e.g., $p=10^{-4}$ mbar), medium vacuum (e.g., $p=10^{-2}$ mbar), and atmospheric pressure (also called non-vacuum electron beam machining) [STEI15b]. To generate the vacuum required, pumps, such as mechanical, oil diffusion, and cryo pumps are used.

In addition to the differences regarding acceleration voltage and working pressure, EBM machines can also differ in terms of the machine concept. The most common form here is the machine designed as a universal or chamber machine as the example in Fig. 6.26 shows. They are used mainly for the machining of single parts and have a working chamber that can be evacuated regardless of the beam generator. The size of the chamber is based on the workpieces to machine in each case. Given that movement of the beam generator is normally extremely complex given its size and complexity, machining movements are realized mostly on the workpiece side by movable, rotatable, and pivotable tables. One approach nowadays, however, is to have the beam generator move while reducing the size of the working chambers required using rotating disk drives. One to three eccentric disks with interleaving mounting, and one vertical drive, are used to move the generator (see Fig. 6.27). So far this has only been possible in combination with a low-voltage generator, however [STEI15c].

Phased machines have multiple working chambers that can be evacuated independently and are primarily suited to volume part production. The working chambers are aligned to the workpiece size. This allows the evacuation time to be shortened considerably. While one workpiece is being machined, the other chambers can be loaded and unloaded as well as evacuated and flooded. This shortens the cycle time and increases machine productivity (see Fig. 6.28, left). Continuous lines were designed in particular for the continuous machining of linear materials such as saw bands. Additional chambers with a partial vacuum are upstream or downstream of the actual working chamber for the material to be machined continuously and at the required pressure (see Fig. 6.28, right and bottom).

6.5 Thermo-Chemical Deburring Machines

The thermo-chemical deburring method (TEM, thermal deburring method) is a chemical removal process where burrs on metallic and non-metallic workpieces are burnt in an oxygen-rich atmosphere.

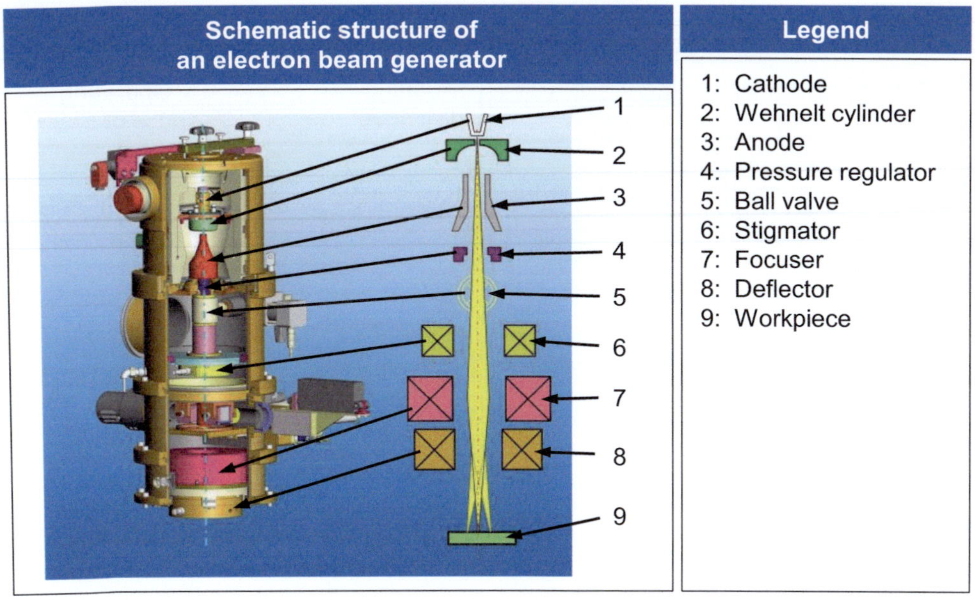

Fig. 6.25 Practical design of beam generators. *Source* Steigerwald Stahltechnik

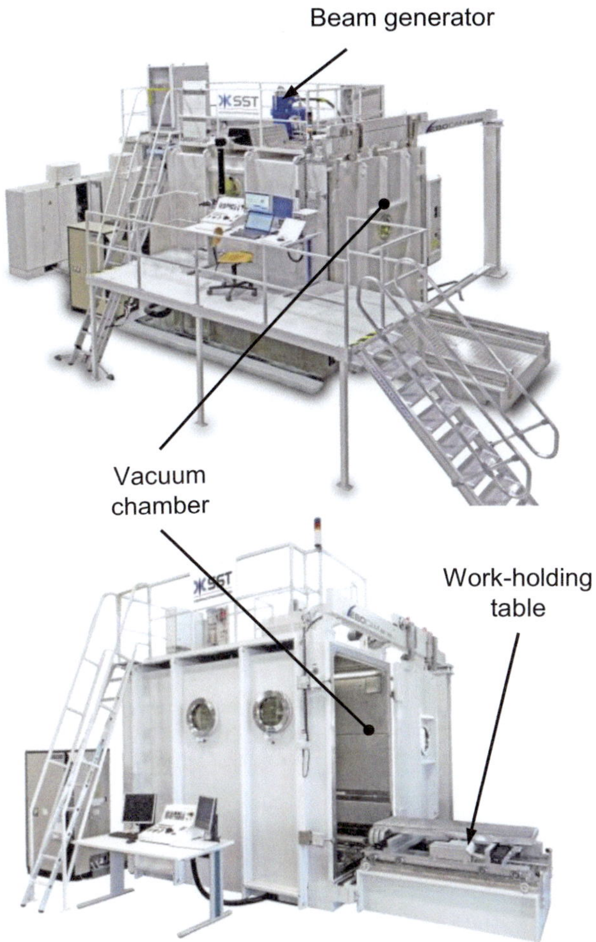

Fig. 6.26 Chamber machines, according to Steigerwald Stahltechnik

Hydrogen or a gas-methane mix is used as the combustion gas. Thermo-chemical deburring is a removal process without specific focus. All of the workpiece surface is exposed for about 20 ms to temperatures of 2500 to 3500 °C by the gas combustion with excess oxygen. The temperature effect is so short that only those places able to absorb a lot of energy due to their large surfaces yet low mass of material reach the oxidation temperatures. For every workpiece, these are the corners and edges and, in particular, detached burrs and loose chips. The workpiece only warms moderately. The uniform distribution of gas in the machining area means burrs in inaccessible cavities and bores can also be removed.

Figure 6.29 at the top shows the functional principle of a TEM machine. The deburring chamber hangs in a press frame. The chamber base, bearing the workpiece, is movable and is pressed against the chamber by a hydraulic closing system. The plastic seal inlaid into the base seals the chamber so as to be gastight. The plastic seal itself is protected from burning by a metallic pre-seal. The chamber is now filled to the pressure necessary for the machining task with an upstroke of the gas metering cylinders with combustion gas and oxygen, in the mixing ratio required. The settings for the upstroke of the metering cylinders and gas pressure in the cylinders are continuously adjustable before starting the process. The gas components mixed before entering the deburring chamber are ignited by a spark plug once the filling process is complete. The actual deburring process takes about 20 ms. The chamber closure plate acts as a workpiece carrier to shorten the machine cycle time. The example machine shown at the bottom of Fig. 6.29 has five of these closure plates,

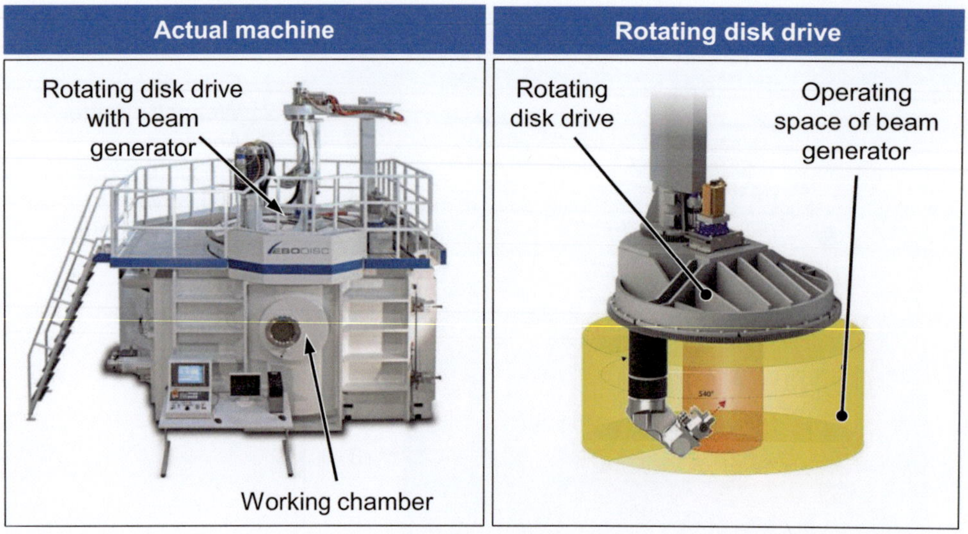

Fig. 6.27 Chamber machine with rotating disk drive, according to Steigerwald Stahltechnik

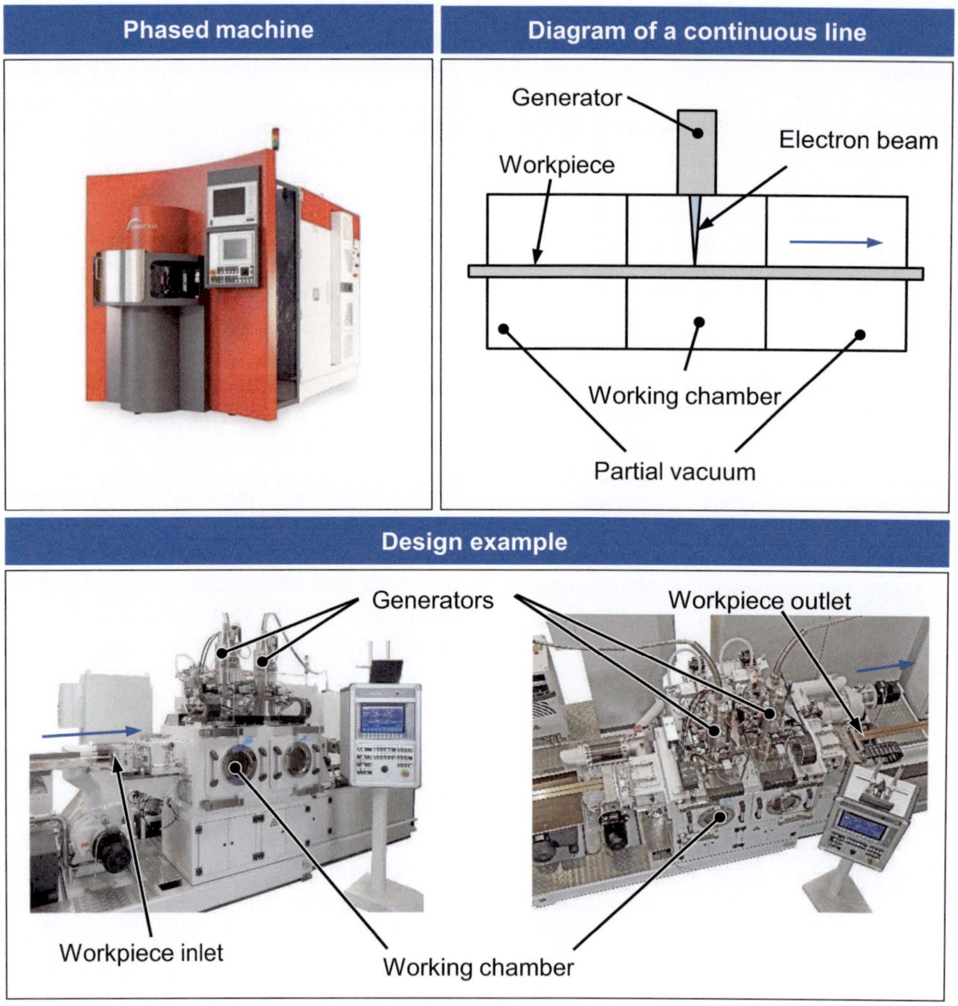

Fig. 6.28 left: Phased machine. Source: Steigerwald Stahltechnik, right: Block diagram of a continuous line, bottom: Example of a continuous line, according to Steigerwald Stahltechnik

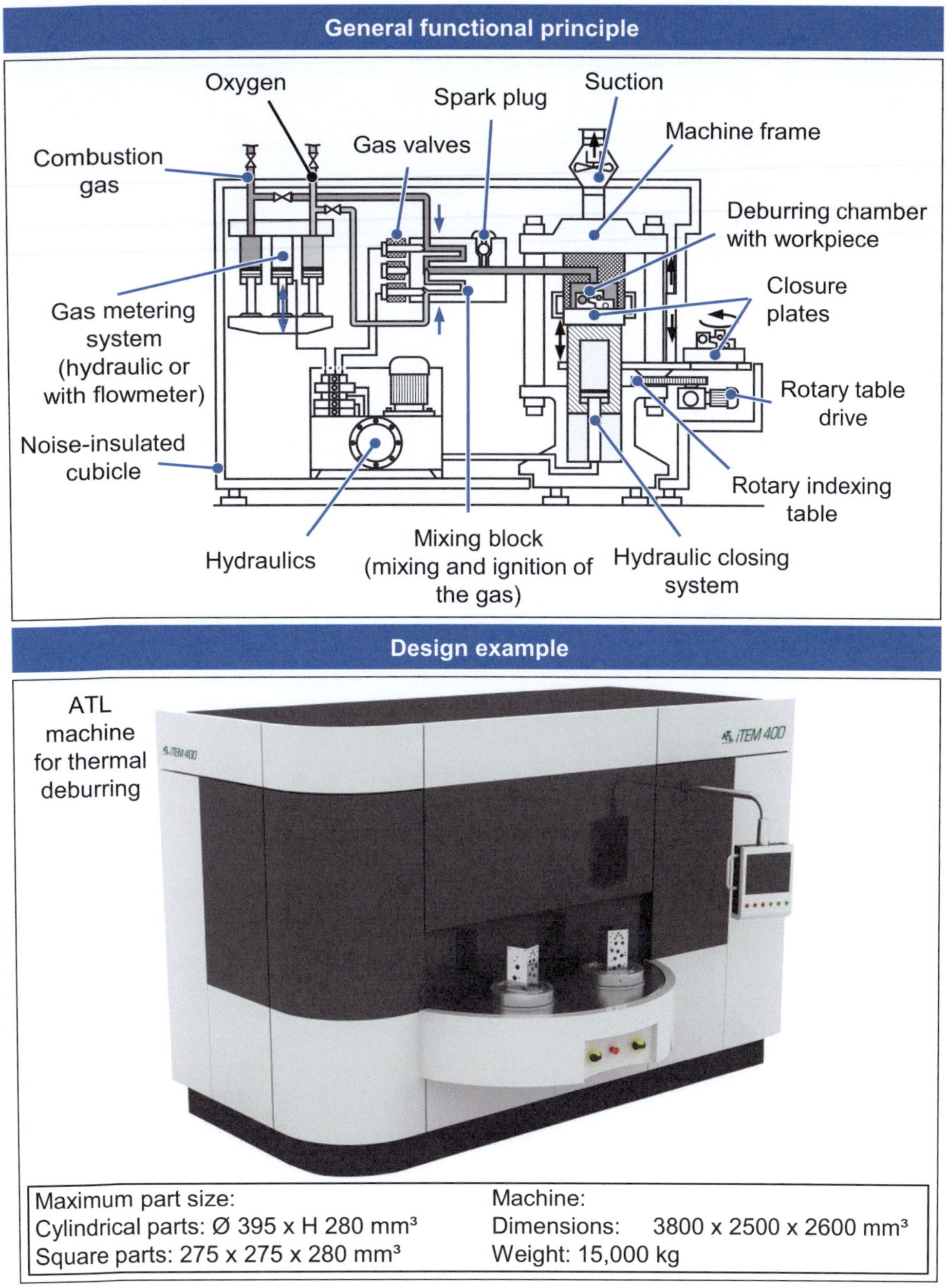

Fig. 6.29 Machine design for thermo-chemical deburring. *Source* Extrude Hone; ATL Thermisches Entgraten

that are positioned on a rotary indexing table. The machine is enclosed in a noise-insulated cubicle. This machine type can be used just as well for small batch sizes and large parts diversity as it can for large volume deburring. Small parts can be placed in cages on the closure plate.

Figure 6.30 shows different workpieces characteristic of thermo-chemical deburring. It is clear from the description of the process that all oxidizable materials can in theory be deburred using TEM.

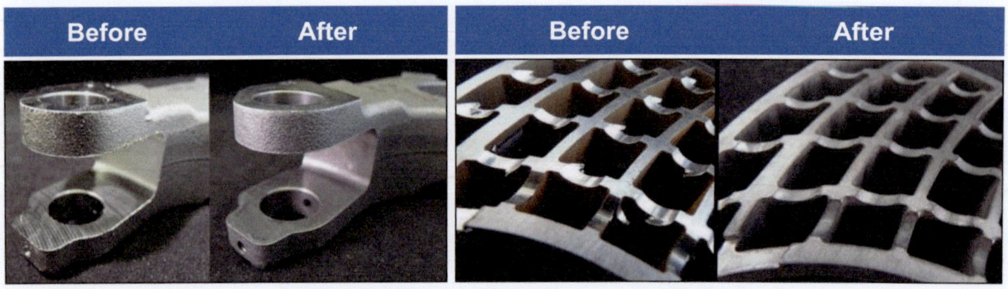

Fig. 6.30 Machining examples for thermo-chemical deburring. *Source* ATL Thermisches Entgraten

6

References

[BERG77] Berger, A.: Elektrisch abtragende Fertigungsverfahren. VDI-Verlag, Düsseldorf (1977)

[DEGN84] Degner, W., Nestler, K., Pirsig, W., Rößner, E., Wetzel, W., Wicht, H.: Elektrochemische Metallbearbeitung. VEB Verlag Technik, Berlin (1984)

[EMAG11] EMAG ECM: ECM/PECM Technologie. Gaildorf. Firmenschrift (2011)

[DIN03a] Norm DIN 8580 (September 2003). Fertigungsverfahren—Begriffe, Einteilung

[DIN03b] Norm DIN 8590 (September 2003). Fertigungsverfahren Abtragen—Einordnung, Unterteilung, Begriffe

[FEYE07] Feyerabend, Volker: Vakuum-Ätz-Technologie. In: PLUS—Produktion von Leiterplatten und Systemen—Fachzeitschrift für Aufbau- und Verbindungstechnik in der Elektronik, **8**, 1–3 (2007)

[FRIT10] Fritz, H., Schulze, G. (Hrsg.) Fertigungstechnik. 9. Aufl. Springer, Heidelberg (2010)

[GF M12] GF Machining Solutions: Kenndaten CUT 1000, CUT 1000 OilTech. Schorndorf. Firmenschrift (2012)

[GF M13] . GF Machining Solutions: Kenndaten CUT 200 mS, CUT 300 mS. Schorndorf. Firmenschrift (2013)

[GF M15] GF Machining Solutions: Bilder. Schorndorf. Firmenschrift (2015)

[KLOC07] Klocke, F., König, W.: Fertigungsverfahren 3. Abtragen, Generieren und Lasermaterialbearbeitung. (Reihe: Fertigungsverfahren. Bd. 3). 4. Aufl. Springer, Berlin (2007)

[OPSI15a] OPSIngersoll: Automation—Technologien intelligent verbinden. Burbach. Firmenschrift (2015)

[OPSI15b] OPSIngersoll: Funkenerosion auf höchstem Niveau. Burbach. Firmenschrift (2015)

[STEI15a] Steigerwald Strahltechnik: Aufbau eines Elektronenstrahl-Generators. Maisach. Firmenschrift (2015)

[STEI15b] Steigerwald Strahltechnik: EB-Atmosphärenschweißen (NVEBW). Maisach. Firmenschrift (2015)

[STEI15c] Steigerwald Strahltechnik: Kammermaschine: Baureihe EBOdisc. Maisach. Firmenschrift (2015)

Machines for Laser Machining

I thank my staff members Mathias Dehn M.Sc., Caroline Kiesewetter-Marko M.Sc. and Marcel Wittmann M.Sc. for their contribution in revising this chapter.

C. Brecher and M. Weck, *Machine Tools Production Systems 1*,
Lecture Notes in Production Engineering, https://doi.org/10.1007/978-3-662-68120-6_7

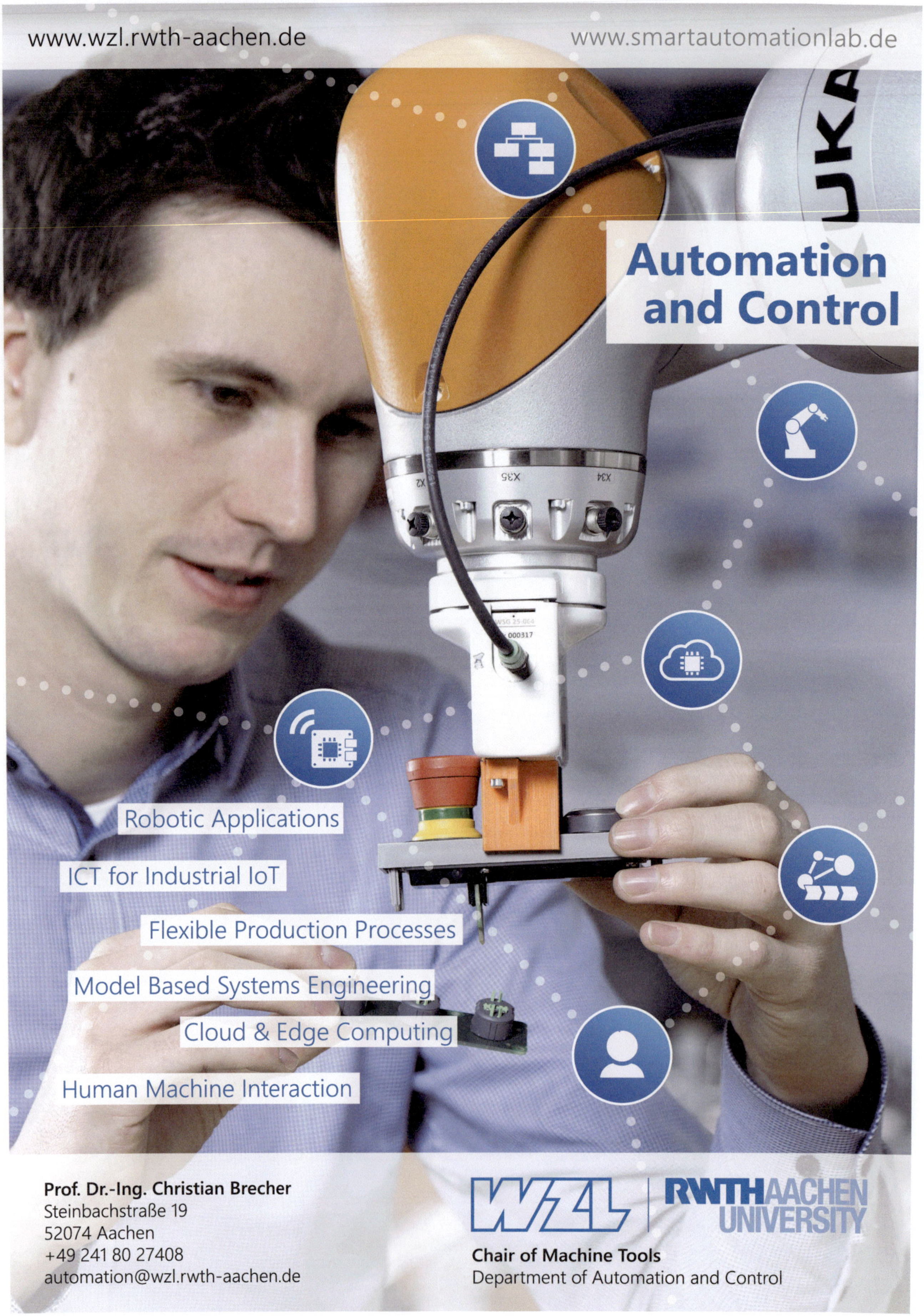
www.wzl.rwth-aachen.de
www.smartautomationlab.de
Automation and Control
Robotic Applications
ICT for Industrial IoT
Flexible Production Processes
Model Based Systems Engineering
Cloud & Edge Computing
Human Machine Interaction
Prof. Dr.-Ing. Christian Brecher
Steinbachstraße 19
52074 Aachen
+49 241 80 27408
automation@wzl.rwth-aachen.de
WZL
RWTH AACHEN UNIVERSITY
Chair of Machine Tools
Department of Automation and Control

Contents

The term LASER, which is widely used nowadays, is an abbreviation for a physical principle and stands for Light Amplification by Stimulated Emission of Radiation. The term LASER is now commonly used as a synonym for laser beam sources. A laser beam is created in a laser source by exciting an active laser medium to stimulate the emission of electromagnetic radiation with the following properties [POPR05]:

- Monochromatic: The wavelength is limited to a narrow band around the central laser wavelength. Depending on the active laser medium, the wavelength can be in the visible spectrum (between 380 and 780 nm) or the invisible range. A distinction is made here between ultraviolet (UV radiation: 1 nm to 380 nm) and infrared radiation (IR: 780 nm to 1000 µm). Laser beam sources for processing materials frequently emit radiation in the invisible spectrum. This requires particular precautions in terms of laser safety.
- Minimal divergence: The laser beam propagates in space through a narrow aperture angle (divergence angle). In an ideal laser source, the minimum angle of divergence is limited by the diffraction. The low divergence allows collimated laser beams to be guided over long distances and is a basic condition for focusing a laser beam to a very small diameter.
- Coherence: In an ideal laser beam, all waves have the same phase. This applies in both temporal and spatial terms: In other words, there is a fixed phase relationship for both different locations at the same time and the same location at different times. Coherence is therefore a condition for the production of extremely small focusing diameters—this requirement is only partially met in the laser beam sources used in material processing.

Laser beams used for processing material generally deviate from the ideal properties given above. The degree of deviation depends, among other factors, on the type and quality of the components used as well as the power output.

Laser beams are used in machine tools for a variety of different processes as flexible, almost non-wearing, and force-free photonic tools. For this reason, we have dedicated a chapter in this book to machines that utilize laser machining. First, the main components of a laser system are explained in this chapter. These are the laser beam source and components for beam guidance and dynamic beam shaping in a range of different designs. As with conventional machine tools, types of laser machines and hybrid machines for cutting, material removal, and additive manufacturing processes are explained below. Information on other laser processes can be found in the relevant specialist literature (e.g., [POPR05] or [POPR11]).

7.1 Laser Beam Sources

Various laser beam sources are available to generate laser beams for material processing. Although the various laser beam sources have a similar structure, they have a very broad range of different variants allowing them to match the demands of the particular application. The relevant features of a laser beam source which are pertinent to material processing are the operating mode, wavelength, maximum power output, beam quality, and electro-optical efficiency.

The operating mode depends on the laser process to be used: Laser sources with continuous emission can be used for thermal processes such as laser hardening or laser welding. Commonly used designations are CW laser sources (CW = continuous wave). Pulsed laser sources emit beam pulses with a defined pulse duration and repetition rate. The peak power of a laser pulse is significantly greater than the average power output of the laser beam source. Pulsed laser beam sources are classified in DIN EN 60 825 115 by their pulse duration into 'impulse' (pulse duration 0.25 s–1 µs), 'giant pulse' (pulse duration 1 µs–1 ns), and 'mode-locked' (pulse duration < 1 ns) [DIN15]. Due to their high peak power, pulsed laser sources with both low to medium output and, therefore, low thermal influence on the workpiece are typically used for marking processes and micro cutting and ablation processes.

Figure 7.1 provides an overview of typical process areas for various laser processes. It shows the basic relationship between irradiation time and laser beam intensity. The irradiation time ranges from continuous radiation to extremely short pulses of just a few picoseconds for pulsed laser beam sources. When using continuous laser beam sources, the intensities range from a few kW/cm^2 up to several GW/cm^2 depending on the laser power and the focal spot diameter. Significantly higher intensities are achieved with pulsed laser beam sources due to short-term peak powers. The product of the irradiation time and the laser beam intensity is a measure of the energy introduced into the workpiece: For micro processing, thermal influences on the workpiece can be minimized by extremely short, high-energy pulses. However, a macroscopic thermal influence occurs in processes such as cutting, welding, and hardening due to the long irradiation times at medium intensities. Laser processes generally have an upper and lower intensity limit for a given irradiation time. Below a threshold intensity, the process is not completed as required. If the intensity is too high, effects such as plasma formation increase in a non-linear manner. This can lead to the workpiece being shielded from the laser radiation and it is not possible to increase the power imparted to the workpiece any further [HÜGE09].

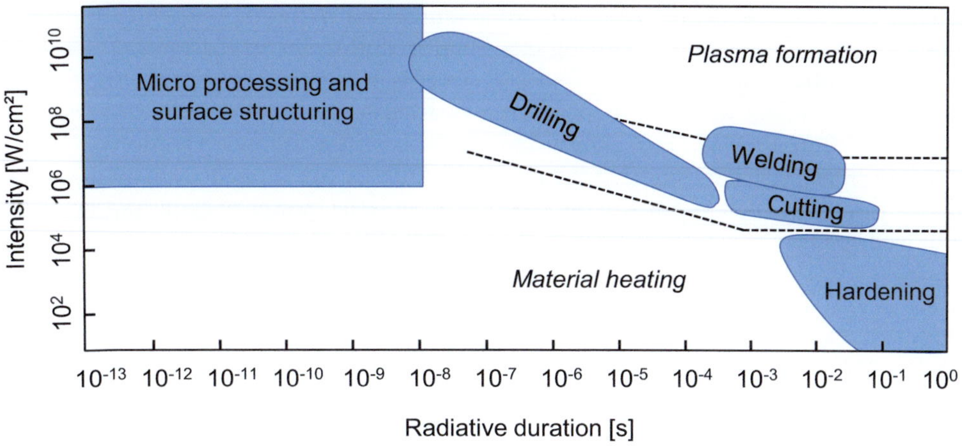

Fig. 7.1 Typical processing parameters for laser machining for various combinations of irradiation intensity and irradiation time, according to [HÜGE09]

The process areas shown in **Fig. 7.1** must be regarded as material-specific for subsequent process design as both the absorption characteristics of various materials (see Fig. 7.2) and the physical characteristics of the materials can have a decisive influence on the area thresholds or their existence for a specific material. In addition to this, environmental conditions such as ambient pressure or the flow of inert gas can influence the position of the process areas and their thresholds.

The wavelengths and the maximum power output of the laser system are selected to suit the requirements of the application. **Figure 7.2** shows absorption curves versus the wavelength for elementary metals and a typical steel alloy at room temperature. The introduction of energy into the workpiece can often be optimized by selecting a suitable laser wavelength. Discrete wavelength ranges of commercially available laser beam sources are available for this purpose. Solid-state lasers with a wavelength of approx. 1070 nm and diode lasers with a wavelength range of approx. 808 to 1060 nm have established themselves for standard applications. Shorter wavelengths are available by doubling the frequency (530 nm) or tripling the frequency (355 nm) of the laser radiation or by using alternative laser media. For example, a laser wavelength of 530 nm is frequently used for machining highly reflective materials such as copper alloys. Laser beam sources for wavelengths in the visible or ultraviolet wavelength range currently have comparably high procurement costs and comparably lower efficiencies. Other factors influencing the absorption of a material include the condition of the surface, the workpiece temperature, and the polarization of the laser radiation. When designing a process, it is recommended that the precise

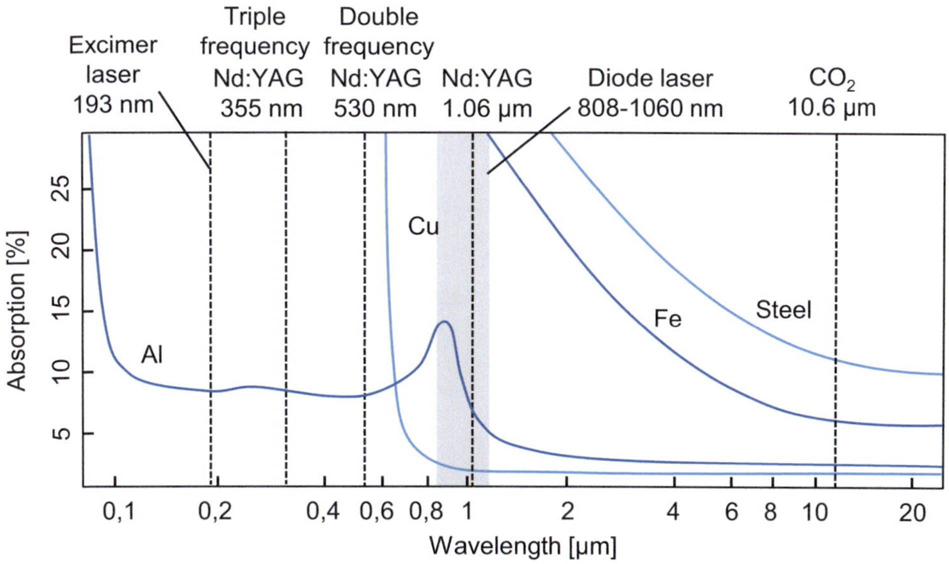

Fig. 7.2 Absorption spectrum for different metallic materials and common laser beam source wavelengths versus wavelength

absorption characteristics of the actual material are determined in advance as even individual alloying elements can affect the absorption behavior.

Laser systems available today are able to produce power outputs of more than 100 kW. This covers the majority of all current laser processes required for material processing. In principle, high beam quality and high power output lead to more expensive laser sources. It is therefore necessary to determine the required parameters of a laser beam source in advance for both technical and economic reasons.

The beam quality (M^2) and beam parameter product (BPP) are important parameters for characterizing laser sources and are laid down in DIN EN ISO 11145 [DIN08]. The beam quality factor allows the user to determine how precisely the laser beam can be focused to produce very small focal spot diameters creating greater power densities with the same laser power. If $M^2=1$, this is an ideal Gaussian laser beam (single mode) with optimum focusability. As M^2 increases, the beam quality decreases—the result is an increasing minimum focal spot diameter. Laser sources with very high power outputs generally emit multi-mode laser radiation: The laser beam consists of a superposition of various electromagnetic vibration modes and exhibits a correspondingly poorer beam quality ($M^2>10$). It is now common practice to use half the aperture angle in the far field for the beam parameter product (BPP) and the unit mm•mrad for the focal spot radius. The BPP allows the divergence of the laser beam to be determined for a given beam radius when designing optical systems and laser processes.

The electro-optical efficiency describes the conversion efficiency from the laser beam source's electrical power input into optical power output. The efficiency level includes the efficiency of the pumping source, the pumping process (quantum yield), the resonator, and losses in the optical elements. The efficiency level is heavily dependent on the inherent design. This is discussed in further detail for the various laser beam sources in the following sections.

Due to the limited availability of combinations of wavelengths, power output, and beam quality as well as economic considerations, selecting a laser source for a machining process usually involves compromise.

The excitation method is explained in the model shown on the left in ◘ Fig. 7.3: Electrons can migrate to a higher energy level within an atom by absorbing energy from an energy input. The energy levels within the atom are distributed discretely and their electron population decreases as energy increases in a thermal equilibrium. The higher energy levels are usually unstable meaning that the excited electron relaxes back to a stable energy level after a relatively short period of time of approximately 1 ns. In this process, which is known as spontaneous emission, energy is released either without radiation as heat or in the form of electromagnetic radiation (optical transition). A necessary condition for an active laser medium is the presence of a metastable state in which electrons can dwell for approximately 1 ms [EICH15]. In addition to the three-level system shown here, four-level systems and multi-level systems also exist to increase the efficiency of active laser media. Pumping electrons to the pump level and then quickly relaxing to the metastable state (laser level) without radiation allows a population inversion to take place between the laser level and the ground level. The interaction of a photon with an excited electron produces radiation with the same wavelength, phase position, and direction as the excited photon due to the electromagnetic interaction while the electron simultaneously relaxes to the ground state. This process

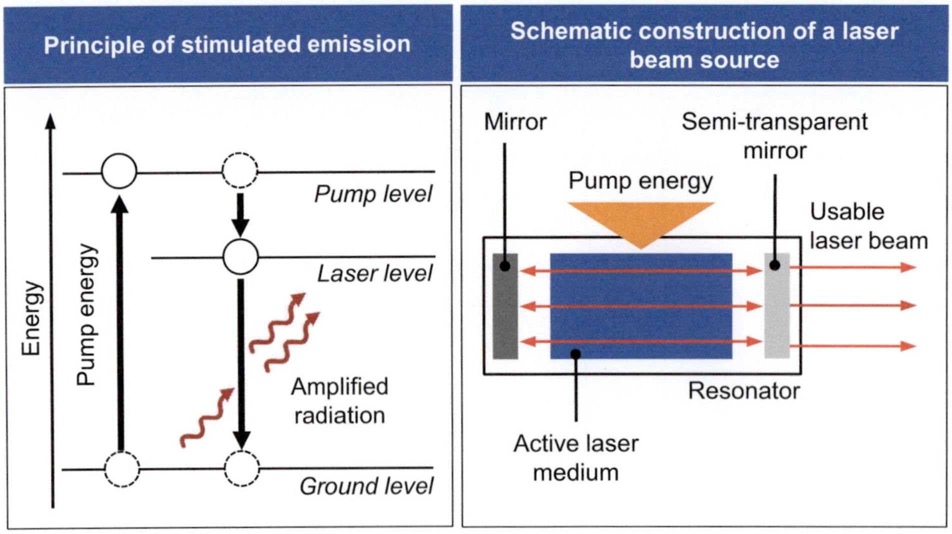

◘ **Fig. 7.3** Principle of stimulated emission (left). Schematic structure of a laser beam source (right) according to [FÖRS01]

is known as stimulated emission and allows the active laser medium to be used as an optical amplifier to produce high-energy laser beams.

In principle, the technical system of every laser beam source is constructed of three physical components: The active laser medium, the pumping source, and the resonator (see ◘ Fig. 7.3, right).

The energy fed from the pumping source is used in the active laser medium to produce and maintain the population inversion. This is the condition for the production of laser radiation through stimulated emission. The active laser medium can come in a solid, liquid or gaseous form. Gaseous and solid active laser media are relevant to the laser sources discussed here. Solid-state lasers have semiconductors or doped insulators as the active laser medium. A further distinction is made in doped active laser media between the substrate (generally in a crystal form, e.g., YAG (yttrium-aluminum-garnet)) and the doping element (typically multivalent metallic ions e.g., Nd^{3+} or Yb^{3+}). Laser beam sources are frequently designated by their active laser medium. A combination frequently used for fiber lasers is, for example, Yb:YAG with a central wavelength of 1070 nm [FÖRS01].

Spontaneous and also stimulated emission processes that are stochastically distributed take place in all spatial directions. The construction of the resonator amplifies electromagnetic radiation only along the axis of the resonator by repeated circulation of the laser radiation and repeated stimulated emission. One possible design has a highly reflective and a semi-transparent mirror. The mirrors amplify the laser radiation to the required power output while the mirrors are arranged such that a laser beam with the lowest possible spectral bandwidth and a high beam quality is produced by selecting specific mode shapes (transversal and longitudinal modes). (See ◘ Fig. 7.3, left). In this case, the usable laser beam is outcoupled from the resonator by a semi-transparent mirror [POPR05].

Energy is introduced into the active laser medium by the pumping source. The excitation can be electrical, optical, chemical, or by collisions between particles. To produce a high quantum yield, the pumping source is matched to the active laser medium so that most of the pumping energy is used to produce the population inversion. In solid-state lasers, for example, narrow-band excitation via laser diodes has prevailed over broadband excitation using flash lamps. This has led to significant increases in the efficiency of these laser sources.

In gas lasers (e.g., CO_2, CO or HeNe), the active laser medium is formed by gas molecules or atoms. The energy states in gas lasers primarily correspond to the various vibration modes of a molecule (e.g., stretching vibration or bending vibration). The transitions between the energy states that lead to radiation emission thus take place due to collisions between the gas molecules or atoms.

Semiconductor lasers (also known as diode lasers) are distinguished from doped insulators by the type of pumping process. Here, excitation is produced by applying an electric current at the p-n junction of the semiconductor. Recombination processes between electrons and the imperfections in the transition area cause laser radiation to be emitted.

A cooling mechanism is also required when operating a laser beam source to dissipate the heat produced in the power electronics and during excitation and generating the beam. Beam guidance systems, processing lenses, and fiber connectors, as well as the laser beam source, are actively cooled to ensure the temperature remains stable despite high power outputs.

7.1.1 Gas Lasers (CO_2 Lasers)

In gas lasers, a gaseous active laser medium is used to produce laser radiation. CO_2 lasers and other gas lasers are currently the most widely used in material processing. A CO_2 laser emits laser radiation with a wavelength of 10.6 µm and has an efficiency of up to 30%, depending on the design. These lasers use a gaseous mixture of helium (He), nitrogen (N_2), and carbon dioxide (CO_2) in a technical vacuum (100–250 hPa). The CO_2 is excited via the N_2 molecules. The electrical energy of the pumping source produces free electrons and is transferred to the N_2 molecule via electron collisions. The function of the helium atoms is to dissipate energy by means of collisions with the CO_2 molecule and to return the CO_2 molecule to its ground state [FÖRS01].

CO_2 lasers can be categorized in two ways by the pumping source: direct current excitation and high frequency excitation (HF) With direct current excitation, two electrodes are located within the gaseous mixture. The benefits of this method of excitation are greater overall efficiency and ease of technical implementation. However, burn-off occurs at the electrodes. This contaminates the gaseous mixture and causes the electrodes to wear. For this reason, this method of excitation is currently only rarely used. With high frequency excitation, the energy is introduced via two capacitor plates which produce a high-frequency electromagnetic alternating field. The capacitor plates are located outside the resonator, which avoids the technically pertinent wear on the electrodes. Also, the repetition rate of the laser pulses in pulsed operation and the maximum power output can be increased with HF excitation.

As a significant proportion of the pump output is converted into heat when generating a beam, the gaseous mixture must be constantly cooled. A distinction is made here between current common designs of diffusion-cooled and axial-flow CO_2 lasers.

The axial-flow CO_2 laser with high frequency excitation as shown in ◘ Fig. 7.4 typically achieves

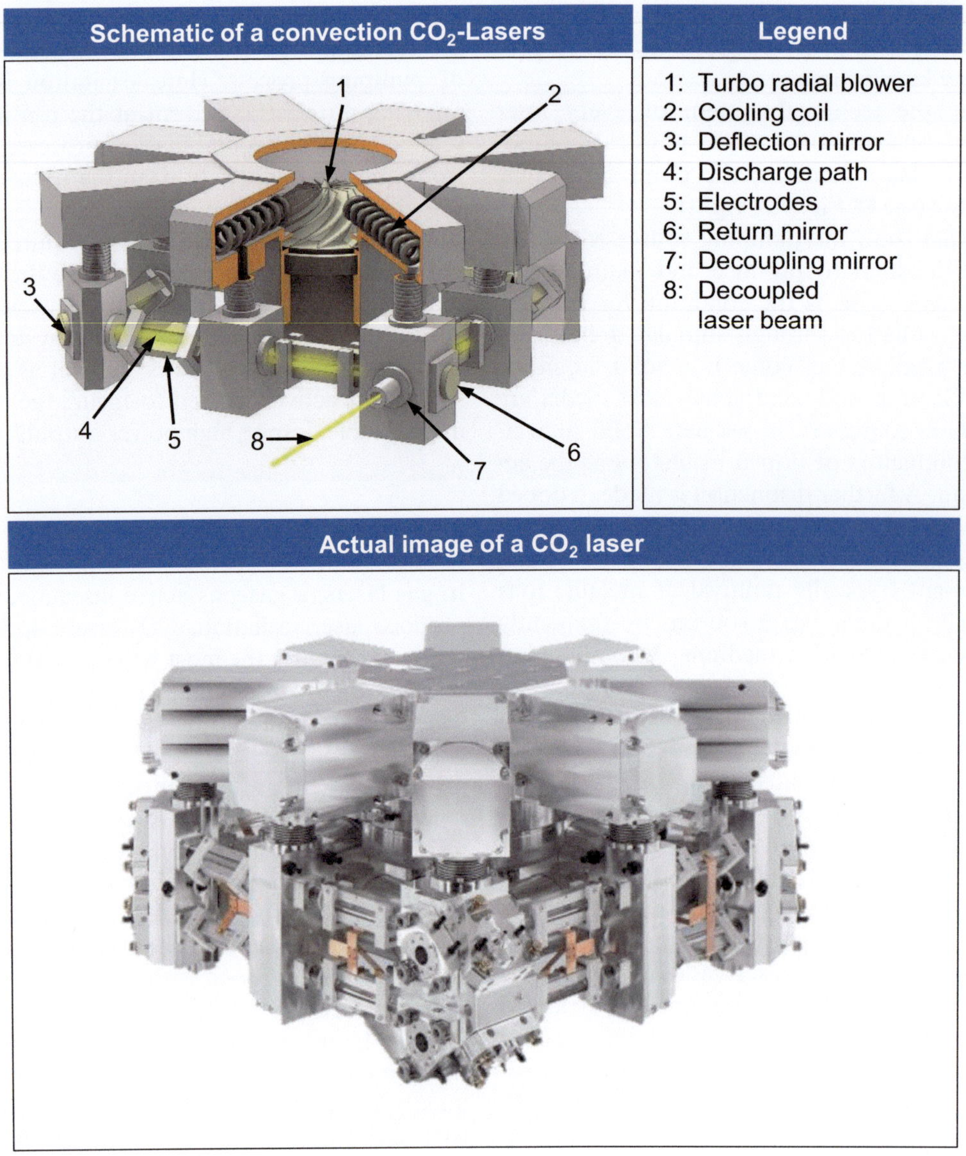

Fig. 7.4 HF-excited axial-flow CO_2 laser. *Source* TRUMPF Group

beam qualities of $M^2 = 1.8$ with power outputs of less than 10 kW. Even at very high power outputs of up to 20 kW, this design can produce laser beams with a comparably high beam quality of $M^2 = 5$ or a BPP of 17 mm•mrad. The resonator on the laser shown here is designed such that the laser beam is guided over a square beam path before outcoupling. Increasing the length of the resonator is known as convolution of the laser beam. This is approximately proportional to the achievable power output while maintaining relatively compact outer dimensions. This allows high-power laser beam sources to be designed sufficiently compact that they can be easily integrated into laser processing machines. In addition to this, the thermal stability of the laser source can also be improved due to symmetrical gas flow in the individual components of the resonator. The laser shown below can achieve power outputs up to 20 kW. The gas circulation is achieved with a magnetically suspended turbo radial blower in order to reduce maintenance requirements (see **Fig. 7.4**) [TRUM08].

There is no gas circulation in diffusion-cooled CO_2 lasers. Cooling uses convective heat transfer between the gaseous mixture and the resonator wall. Due to the closed system design, gas consumption is reduced and this lowers operating costs. A frequently used laser beam source with diffusion cooling is the slab laser in which the gaseous mixture is located between two water-cooled rectangular electrodes [FÖRS01]. **Figure 7.5** shows a typical diffusion-cooled CO_2 slab laser. The laser source achieves beam qualities of up to $M^2 = 1.05$ and power outputs up to 8 kW [ROFI14a].

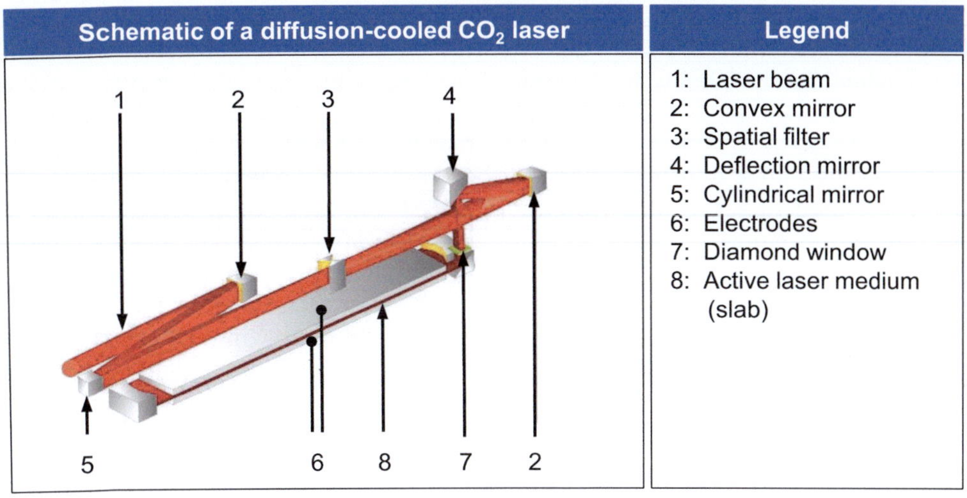

Fig. 7.5 HF-excited diffusion-cooled CO_2 laser. *Source* Coherent-ROFIN

7.1.2 Solid-State Lasers

In solid-state lasers, the active laser medium is solid and has a range of geometries. The excitation is generally optical, for example, using flash lamps, or with laser diodes as the pumping source. The latter method is currently regarded as the state of the art. Flash lamp-pumped lasers allow only relatively low overall efficiency to be achieved (approx. 5%) due to the wide wavelength spectrum that can lead to high operating costs and can make using them uneconomical. Diode-pumped solid-state lasers can currently achieve efficiencies of over 50%. The efficiency gain is due to matching the pumping wavelength of the diodes to the generally narrow absorption lines of the active laser medium (e.g., Nd^{3+} ions at 808 nm) and the improvement in the efficiency of the pump diodes, which allows greater efficiencies to be achieved. This excitation with lower losses produces lower thermal loads on the resonator. The lower thermal loads also lead to improved beam qualities at high outputs [POPR05].

Different designs of laser sources are also used to achieve a pulsed operating mode: In pulsed operation, discontinuous laser pulses are emitted from the laser beam source. The peak power can be a multiple of the average power output over the duration of the pulse. For example, peak powers of 10 kW can be achieved with a pulsed laser source having a pulse duration of 200 ns and an average output of 70 W. The following principles can be used to achieve pulsed operation in sold-state lasers:

The active laser medium can be excited to discontinuous emission by the pulsed operation of the pumping source. This type of excitation is usually used for pulses from several 100 ms down to 100 µs. Unlike other means of pulse excitation, this excitation method allows no or only a very low increase in power (10×) of the laser pulse to be achieved. However, there is the benefit of increased flexibility as the laser beam source can be used in both continuous wave operation with outputs of several kW as well as in pulsed operation with high pulse energies of up to 200 J [IPGL17].

Another type of pulsed laser source is the so-called Master Oscillator Power Amplifier (MOPA). In a MOPA laser, the radiation from a modulated laser diode (also known as a seed laser) is optically amplified to produce a laser pulse with a high power and short duration. By separating signal generation and signal amplification, laser diodes with relatively low power outputs can be used as seed lasers to produce a high beam quality, low spectral width, and much more precise control. The amplifiers consist of an active laser medium, e.g., in rod or fiber form, and are pumped to just below the laser threshold so that there is no stimulated emission as yet. When the preamplified signal enters the amplifier, stimulated emission takes place and high peak powers can be produced. This type has typical pulse durations of 10–100 ns and continuous wave operation is usually possible.

A Q-switch is an alternative to the MOPA structure. By using a Q-switch, the resonator can be influenced in such a way that no laser beam is produced due to high resonator losses, despite sufficient pump output. However, a high degree of population inversion can be maintained in the active laser medium so that when the Q-switch is switched (this leads to an increase in the resonator quality) a laser pulse of up to 10 ns with high pulse power can be produced. Once the laser pulse has been triggered, the Q-switch is opened again and the active laser medium is once again charged by pumping. The simplest type of Q-switch is rotating shutters. Q-switches with active (e.g., acousto-optic modulator) or passive (e.g., saturable absorber) non-linear, optical elements have an established market presence.

Mode locking is used to produce ultra-short pulses with a pulse duration of less than 50 ps. In this instance, several longitudinal modes vibrate in the resonator with a constant phase relationship. By introducing deliberate interference, a wave with a strong and very short increase in the amplitude is formed. Non-linear, optical elements are used to achieve the gain and outcoupling of the ultra-short pulses.

7.1.2.1 Rod Lasers

The central element in a rod laser is a cylindrical active laser medium with a diameter of up to 10 mm and a length of up to 150 mm. [EICH15] Flash lamps can be used for pumping. These lamps are housed in ellipsoidal resonators with the active laser rod material located in their common focal point. [POPR05] (see ◘ Fig. 7.6). In more up-to-date diode-pumped designs, pumping can take place either from the side or, as with most laser sources commonly in use today, coaxially to the laser rod by means of the high-reflectance mirror for laser wavelengths. End-pumped rod lasers allow the active laser medium to be more fully utilized and higher beam qualities to be produced.

The rod design of the active laser medium was primarily used in older solid-state laser types with low power outputs due to its straightforward construction. The cylindrical design produces thermal lens effects in the rod, primarily with high laser power. These effects can reduce the quality of the beam and the stability of the resonator such that stable, continuous operation of the laser source is no longer possible at high power outputs. In spite of this, diode-pumped rod lasers are primarily used with pulsed laser sources because the low average laser power allows thermal effects to be controlled and high beam qualities to be achieved. Fiber and disk lasers were developed to improve the cooling of the active medium. In both these designs, the problem of heat accumulation in the center of the rod was resolved by optimizing the geometry of the active laser medium.

7.1.2.2 Fiber Lasers

A fiber laser is a diode-pumped, solid-state laser in which the active laser medium takes the form of an optical fiber. The heat accumulation can be efficiently dissipated into the surrounding environment via the large surface of the very thin fibers. Fibers can typically be 60 m or longer with a small cross section to achieve correspondingly high power outputs.

The active fiber has a crystalline core structure, for example, yttrium-aluminum-garnet (YAG), which is doped with ytterbium- (Yb^{3+}). The pumping source radiation is connected either coaxially via a fiber end (see ◘ Fig. 7.7) or via pumping fibers spliced from the side to the active fiber. The resonator is created by mirror coating the fiber end surfaces with a fiber Bragg grat-

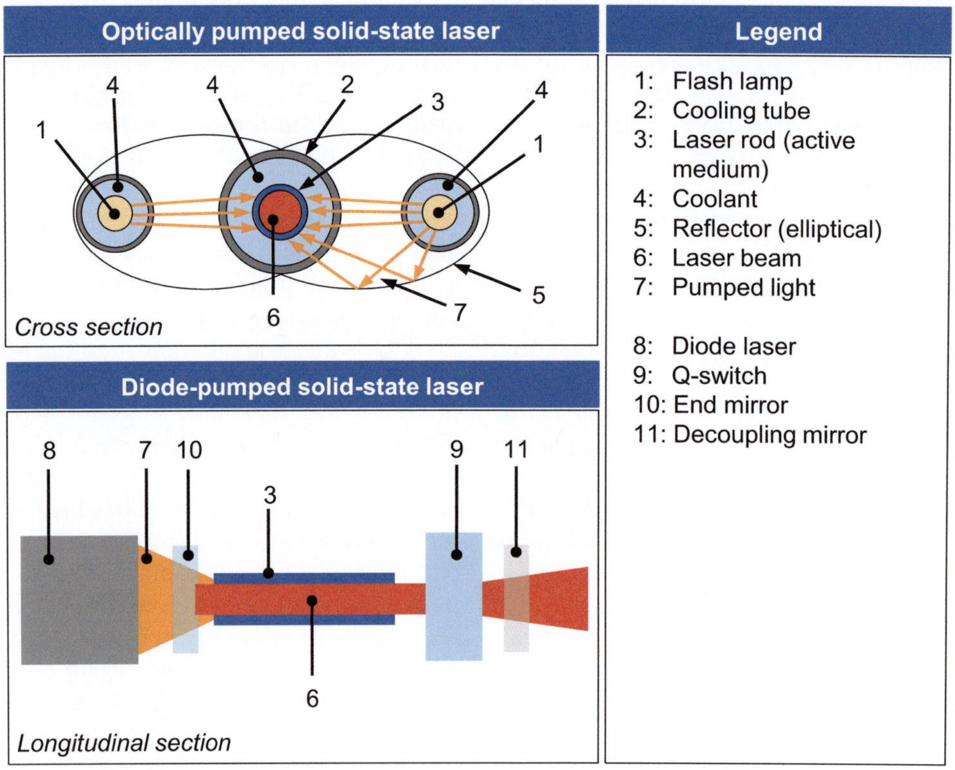

◘ **Fig. 7.6** Solid-state lasers with different pumping principles, according to [POPR05]

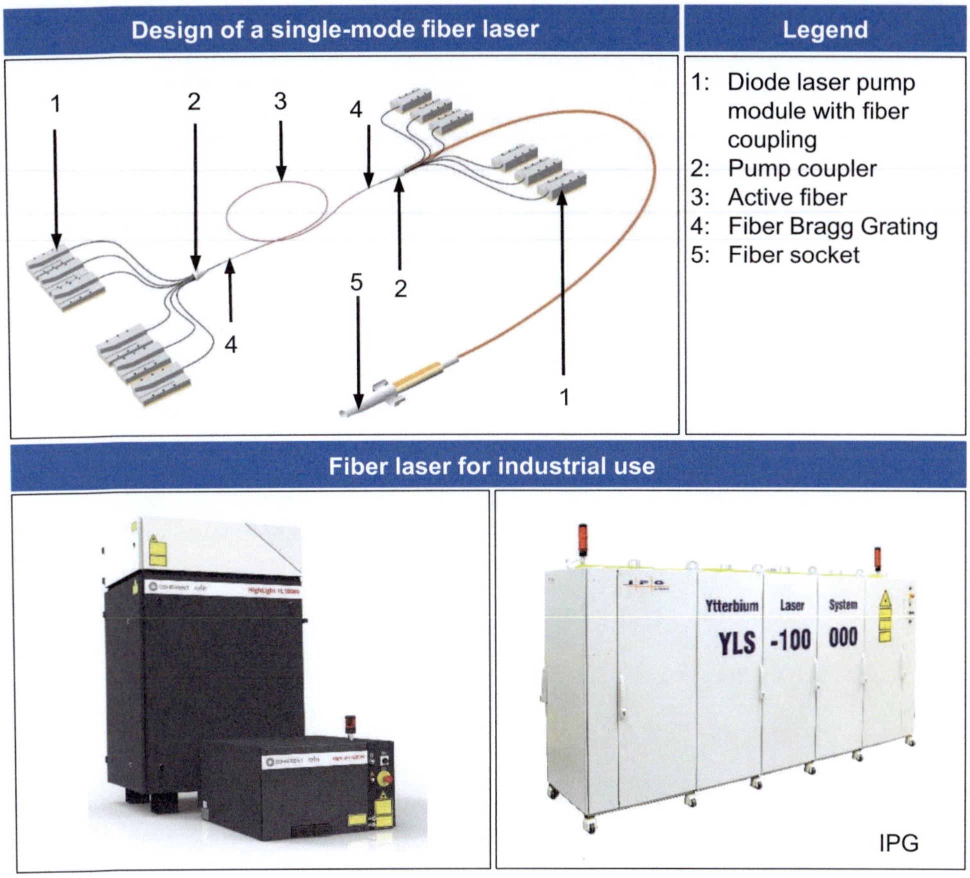

Fig. 7.7 Structure and view of an industrial fiber laser. *Source* Coherent-ROFIN, IPG Laser

ing (FBG) so that the pumping radiation remains in the fiber by means of total internal reflection until it is completely converted into laser radiation. Greater power outputs, very high beam quality, and almost maintenance-free operation can be achieved with fiber lasers. The fiber laser in Fig. 7.7 (lower left) achieves typical power outputs of up to 6 kW [ROFI14b].

To produce very high power outputs of more than 10 kW, a modular structure is used, for example, as shown in Fig. 7.7 (lower right), for better scalability. The laser power, comprising the individual fiber laser modules with power outputs up to 2 kW, is combined in a laser beam combiner to form a single fiber in order to produce high power outputs. This modular design also increases the robustness and the maintainability of the laser source as, in the event of damage, operations can continue with less power using individual modules or by activating a replacement module until the faulty module is replaced. It is possible to achieve laser sources with a power output of more than 100 kW and a wavelength of 1070 nm (see Fig. 7.7). Efficient excitation by pump diodes and loss-free optical-to-optical conversion allow electro-optical efficiency levels of up to 50% to be achieved with today's fiber lasers.

7.1.2.3 Disk Lasers

As with fiber lasers, disk lasers are diode-pumped, solid-state lasers. Here, the active laser medium consists of a thin Yb:YAG crystal disk, from 100 µm to 200 µm thick and with a diameter of approximately 15 mm, producing a laser beam with a wavelength of 1030 nm. The relatively thin crystal disk is mounted on a cooling element so that the rear surface of the disk can be cooled. This results in temperature gradients only occurring in the direction of the disk thickness. These do not have any negative influence on the beam quality. This means that particularly high-quality beams can be produced with disk lasers.

The contact surface with the disk is mirror coated and part of the resonator. Disk lasers also have a different pump beam guidance compared to other solid-state lasers. The radiation from the pumping source is redirected via a parabolic mirror and several tilted mirrors on the reverse of the cavity such that it passes through the disk many times. This can result in almost complete absorption of the pumping radiation (see Fig. 7.8). This allows a good optical-to-optical efficiency to be achieved. The output of this type of laser can also be scaled by placing several disk systems in a serial

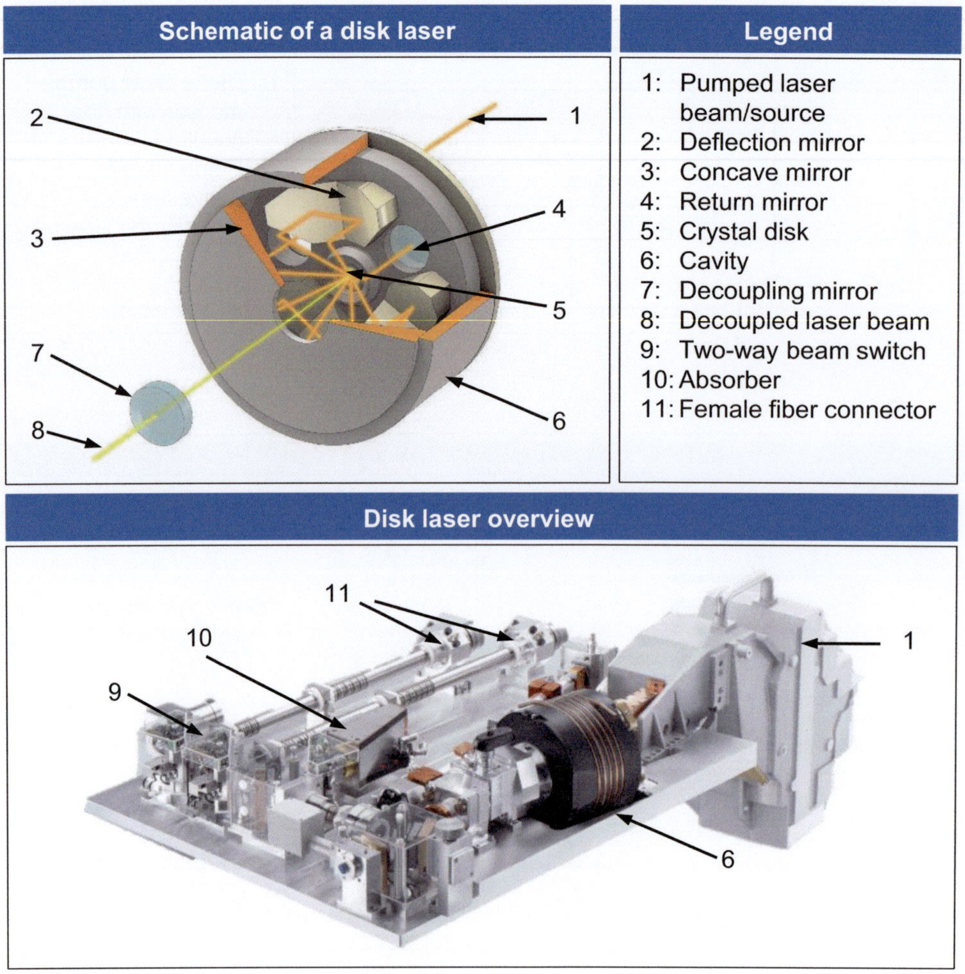

Fig. 7.8 Diagram and view of a disk laser. *Source* TRUMPF Group

configuration. The disk laser shown in Fig. 7.8 can produce a power output of up to 16 kW [TRUM08].

7.1.3 Semiconductor Lasers (Diode Lasers)

In semiconductor lasers, generally known as diode lasers, the active laser medium comprises doped semiconductor materials such as a gallium-aluminum-arsenide or gallium-arsenide system (GaAlAs or GaAs) [KLOC07]. A special feature of diode lasers is the direct excitation by an electrical current. This allows particularly high efficiencies of over 50% to be achieved as there is no need to convert electrical into optical pumping energy.

The active laser medium consists of a p-doped and an n-doped layer. The p-doped layer has fewer electrons than would be optimum while the n-doped layer has an excess of electrons [FÖRS01]. A voltage is applied to both of these layers (see Fig. 7.9). This results in an exchange of electrons from the conduction band to the valence band of the semiconductor at the transition between the two layers. This action produces radiation. As the wavelength of the laser is dependent on the composition of the semiconductor material, diode lasers can be produced with a wide range of different wavelengths from UV to IR [POPR05].

However, the active medium in a diode laser is restricted to very small dimensions for technical and physical reasons. This means that it is possible to achieve only very low power outputs with a single diode laser despite the high power density. To increase the power, approximately 20 to 30 laser diodes are arranged on a common semiconductor chip to form what are known as laser bars. The individual laser diodes can be powered in parallel electrically and will all emit radiation in the same direction. However, the laser radiation from the single diodes is coherent and there is no phase relationship between the laser beams from the various laser diodes. The superposition of several single beams causes no interference effects in the medium. Cooling is achieved by soldering the laser bars onto heat sinks.

When using diode lasers, it should be noted that the beam parameter product depends greatly on the

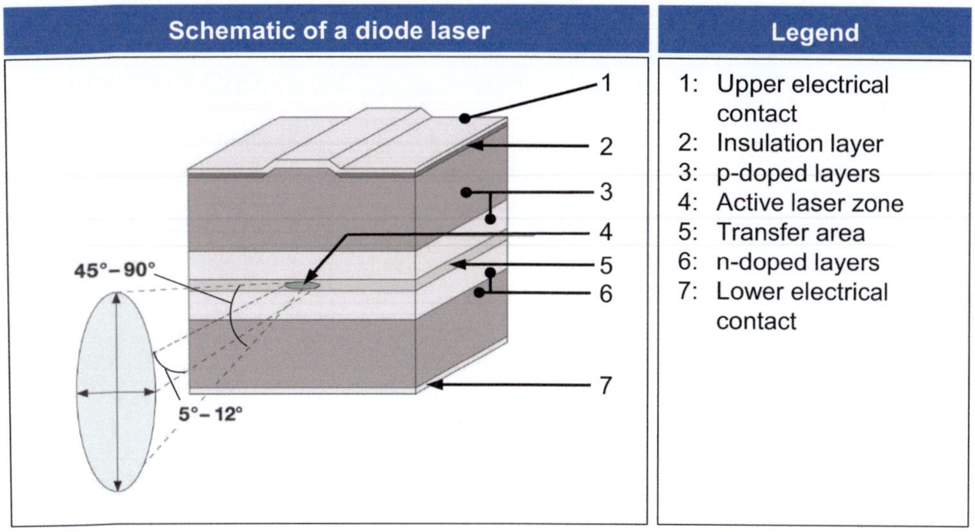

Fig. 7.9 Diode laser. *Source* TRUMPF Group

relevant direction with respect to the laser line. A relatively poor beam quality is produced along the laser line (in the lateral direction, also known as the slow axis) and the beam divergence angle is approximately 5°–12° (see Fig. 7.9). On the other hand, the beam quality is significantly better perpendicular to the laser line in the transverse direction (fast axis) where the beam divergence angle is substantially greater at between 45° and 90° (see Fig. 7.9). Cylindrical microlenses allow the angle in the transverse direction to be reduced such that the divergence is predominantly in the lateral direction. There are three different ways to increase the available power even further: stacking, polarization coupling, and wavelength coupling. In stacking, several laser bars, including the heat sinks and cylindrical microlenses, are brought together into what are known as laser stacks (see Fig. 7.10). The individual laser bars are arranged on top of each other. However, the beam quality deteriorates as the quantity in a stack increases. So, 6–15 laser bars per stack has proven to be a workable number.

On the other hand, polarization coupling exploits the fact that diode lasers generally emit linearly polarized radiation. Using polarization-selective optical elements, the radiation from two laser bars can be superimposed in such a way as to produce a beam that has twice the power density without impairing the beam quality. In wavelength coupling, the different wavelengths from the individual laser bars are superimposed using wavelength-selective mirrors. This allows the power density to be augmented without the combination process negatively affecting the beam quality. However, when focusing using optical elements, wavelength coupling leads to imaging errors (aberrations) and therefore to larger focal spot diameters. Regardless of the methods used for laser power scaling, the full beam can also eventually be parallelized further in the lateral direction using another cylindrical lens. However, it should be noted that, despite the above measures, higher power diode lasers achieve a significantly poorer beam quality when compared to diode-pumped solid-state lasers. But diode lasers do have some advantages—apart from their excellent levels of efficiency—as they are largely maintenance-free and have a particularly compact and lightweight design. The latter is particularly beneficial for use in moving axes or robot systems [POPR05].

A Laserline diode laser including a view of the wavelength coupler is shown in the lower section of Fig. 7.10. These laser sources can currently achieve power outputs of up to 20 kW.

7.2 Beam Guidance/Shaping System

Beam guidance systems are required to transport the laser beam from the laser source to the workpiece. These systems are to guide the laser beam to its point of use with a minimum of losses and with no deterioration in beam quality. Mirror systems or optical fibers are used to guide the beam. The choice of system to guide the beam depends on the application and the wavelength as well as the intensity of the laser radiation. The laser radiation of CO_2, UV and ultra-short pulsed laser sources can only be redirected using reflective optics. Beam guidance using optical fibers is not possible in the relevant laser power ranges because the absorption of CO_2, UV and ultra-short pulsed laser sources is too high and would lead to the core of the fiber being destroyed.

Beam guidance systems for CO_2 lasers generally consist of a series of water-cooled copper mirrors

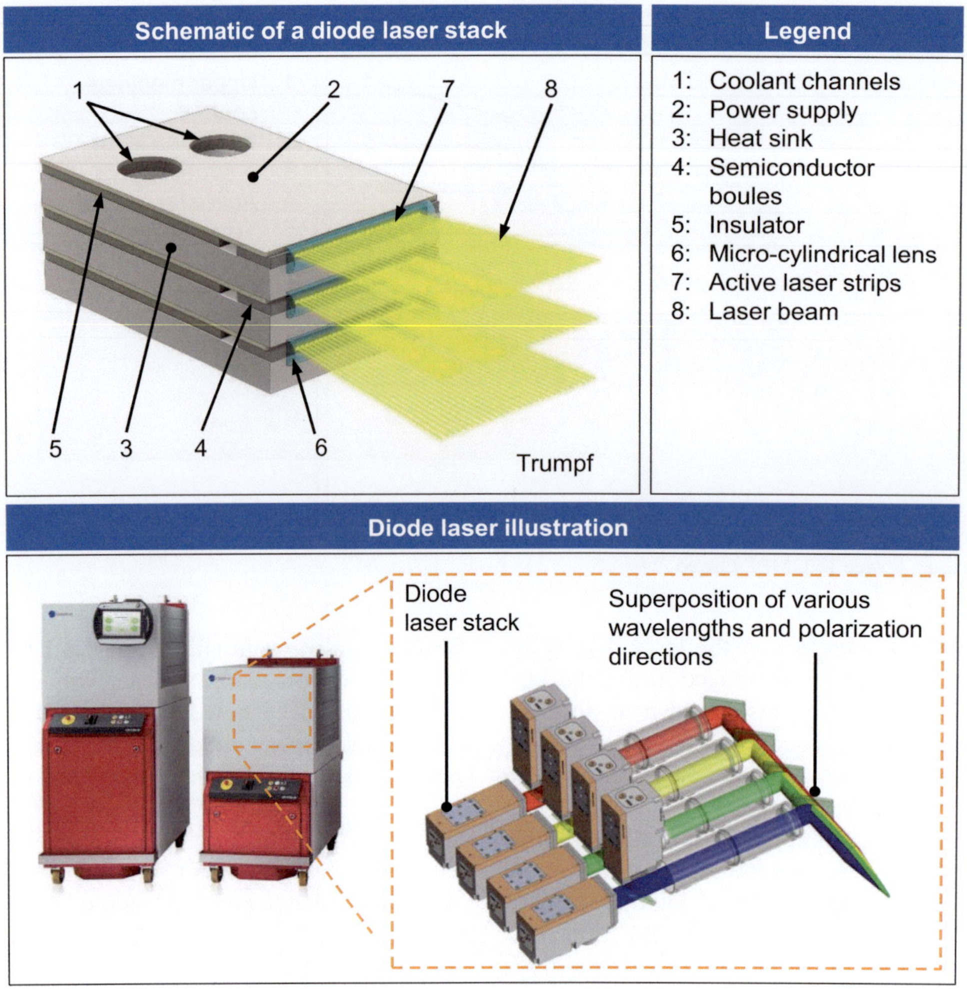

Fig. 7.10 Diode laser stacks and diode laser source. *Source* TRUMPF Group, Laserline

that guide the beam to the desired point. The complete beam path is encapsulated in a pipe system filled with gas to prevent particles in the surrounding air from contaminating the mirrors or scattering the laser beam. The encapsulation is also necessary for laser safety reasons to prevent employees from being harmed by scattered radiation. The number of deflection mirrors in beam guidance systems should be kept to a minimum as power losses of up to 1% to 2% of the laser beam power occur at every mirror. Spherical or parabolic mirrors or lenses are used to focus the arriving laser beam onto the workpiece. The lenses for CO_2 laser beam guidance systems are usually made from zinc selenide and are suitable for low to medium laser beam powers. Laser beams from UV and ultra-short pulsed laser sources are redirected using specially coated mirrors. Narrow or broad-band dielectric or metallic coatings on a glass substrate are used depending on the laser beam's emission spectrum and the power density of the beam on the mirror.

For continuous wave laser beam sources or short and ultra-short pulsed laser sources in the near infrared range used for material processing, beam guidance can be easily achieved with optical fibers (see Fig. 7.11). Optical fibers consist of a fused silica core and cladding. The total internal reflection can be used to guide the light with minimal loss caused by a sudden change in the refractive index between the cladding and the core (step-index fiber). The cladding must have a lower refractive index than the core. The acceptance angle of the optical fiber must be considered when coupling the laser beam into the fiber. This provides the maximum angle of incidence into the fiber at which the laser beam is still guided with total internal reflection. If laser radiation is coupled into the fiber at an angle greater than the acceptance angle, it hits the boundary layer between the fiber's core and cladding at an angle which is greater than the critical angle for total internal reflection. This is then not fully reflected but is outcoupled from the fiber and, in the worst-case scenario,

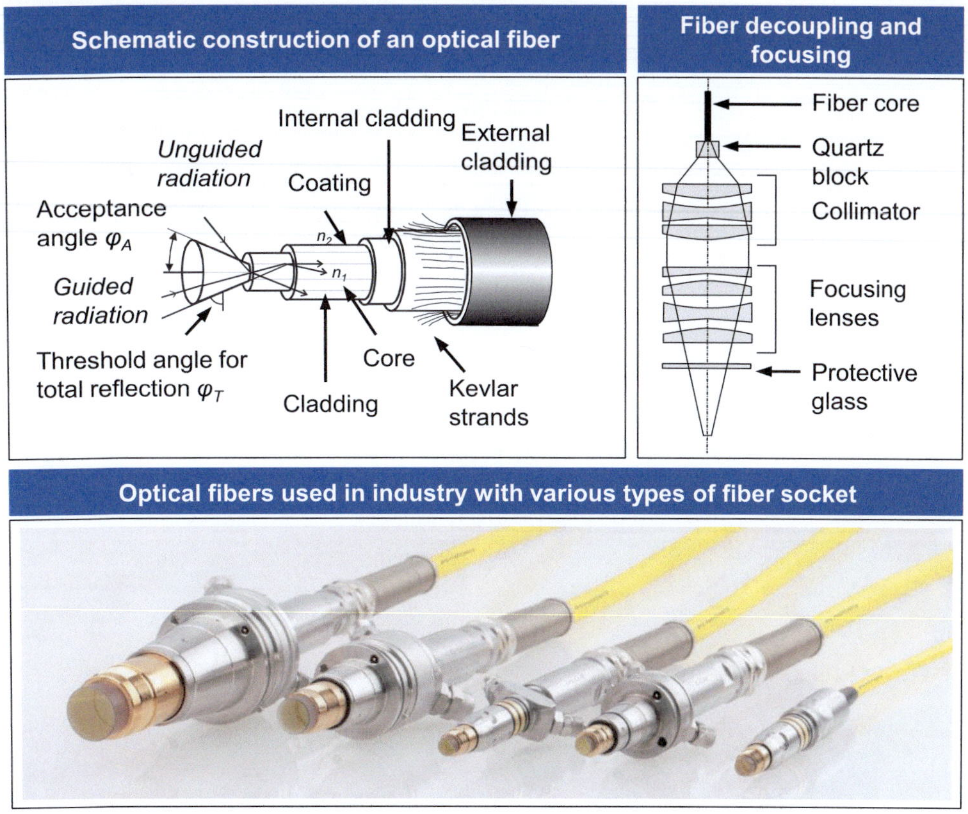

Fig. 7.11 Optical fiber for flexible beam guidance. *Source* [POPR05], IPG Laser

leads to the fiber being destroyed. The same happens in fibers that are not installed straight but bent. The angular relationships of the internal reflection are altered by bending the fibers. This means that a minimum (static) bending radius of approximately 100 mm usually has to be maintained to prevent outcoupling of the laser radiation. The core diameter for fibers used to process laser materials is between 30 µm and 2 mm. As the light emergence angle of the fiber is the same as its incidence angle, the divergent laser radiation after outcoupling is parallelized with a collimator.

Hollow core fibers are currently being developed to provide industry with access to the benefits of the most up-to-date optical fibers. These include laser radiation at wavelengths not suitable for beam guidance in the fused silica fibers currently in use (e.g., CO_2, UV or ultra-short pulse laser radiation). In addition to this, the transmissible power densities can be greater than the damage threshold of conventional fibers. This type of fiber is also made of fused silica but has a cylindrical cavity along the axis of the fiber used to guide the light. The optical boundary layer also has a mirror coated reflective layer (e.g., silver) to facilitate total internal reflection. Due to the relatively large diameters of the glass cladding, very large bending radii, compared to those for conventional glass fibers, must currently be maintained.

Laser beam deflection units, also known as laser scanners, are used for dynamic beam guidance. A laser scanner's task is to deflect and position the laser radiation within a two- or three-dimensional working area or space. In most of the laser scanners used for material processing, deflection in the xy-direction is achieved with two movable mirrors. These mirrors are rotated by galvanometer drives (see Fig. 7.12). Unlike machine tool axes, moving these optical axes involves moving only very small masses and significant translational displacement of the laser beam focus in the machining plane is achieved with small angular adjustments of the mirror. This means that high accelerations and high scanning speeds of up to 30 m/s can be achieved in the machining plane in typical system configurations. For three-dimensional machining, the laser beam's focal plane must be tracked according to the material removal or the component surface. The optical Z-axis of a 3D beam deflection unit can be achieved using a dynamically displaceable lens. However, two-dimensional deflection of the laser radiation in a working area is sufficient in many applications. In this case, the laser beam is focused by a processing lens once it has been deflected by the mirrors. Flat field scanning lenses (F-theta lenses) are mainly used for this purpose. They produce an almost constant focal spot diameter in the working plane by means of a multi-lens system. The

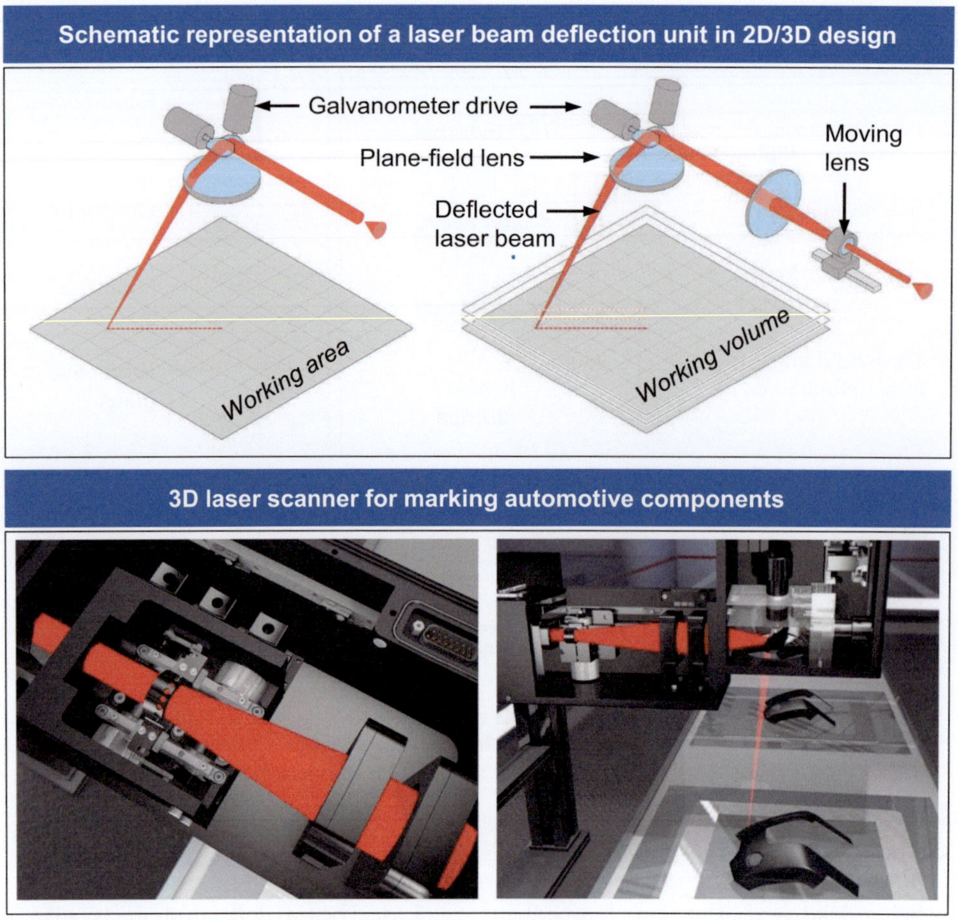

Fig. 7.12 Block diagram of a 2- or 3-axis scanning system. *Source* RAYLASE

deflection of the laser beam focus is thus proportional to the angle of incidence into the lens. In a few applications, e.g., high-accuracy laser texturing applications, the laser radiation must hit the surface to be machined perpendicular at every position of the machining area. As a result, the absorption conditions are identical at every point in the area. Telecentric lenses meet this requirement. The size of the working area can vary depending on the optical system being used. It is typically between 50 mm × 50 mm and 500 mm × 500 mm. Established application areas for laser scanners are remote welding and cutting, laser ablation, and laser hardening.

Polygon scanners were developed for even higher speeds and applications in which the laser beam only needs to be deflected dynamically in one dimension. Here, the laser radiation is deflected by a rotating faceted mirror and focused onto the workpiece by a one-dimensional flat field scanning lens. Scanning speeds in excess of 300 m/s can be achieved by rotating the polygon at extremely high speeds. Moving the component orthogonal to the scanning direction is generally achieved using a mechanical axis.

7.3 Laser Cutting Machines

Laser cutting machines utilize the thermal effect of laser beams and their high energy density versus the surface area. According to DIN 2310 part 6, laser beam cutting is classified within the process category of 'Thermal removal' [DIN03]. Compared to mechanical and other thermal cutting processes, laser cutting has many benefits:

- almost force-free machining
- high machining speeds
- small heat-affected zone
- low warping
- very narrow kerf widths
- virtually wear-free

Laser beams are predominantly used in industry to cut metallic materials. This typically includes mild steels, tool steels, and stainless steels as well as aluminum and its alloys. However, laser beams can also be used to cut non-metallic materials such as cardboard, wood, leather, and plastic. Thermal cutting using laser beams is subdivided into three processes:

- sublimation cutting
- fusion cutting
- laser-oxygen cutting

In sublimation cutting, the material is directly converted from solid to the gaseous state in the area of the kerf. As there is almost no melting, smooth cutting edges can be achieved. The heat-affected zone at the boundary of the cutting edge and the overall thermal load on the workpiece are both minimal. However, the high intensities required (10^8 W/cm^2), particularly with metals, are a disadvantage as these can only be achieved with powerful laser beam sources and high beam quality. Sublimation cutting also operates at significantly lower cutting speeds than fusion cutting [POPR11].

In the remote cutting process variant, thin metal sheets (<1 mm) are cut in a scanning process. The beam is passed over the cut contour multiple times in the remote cutting process. A specific ablation depth between 10 µm and 50 µm is achieved with each pass depending on the material and process parameters. The number of passes, and therefore the duration of the process, depends on the material thickness and the material removed per pass.

In fusion cutting, the material is melted by the laser beam in the area of the kerf and forced away with an inert gas jet. Striations and drag lines form at the edges due to the melt pool dynamics. A further process variant is high-pressure cutting where the cutting gas is introduced at a pressure exceeding 6 bar. This reduces the formation of burrs and slag at the cutting edge.

Laser-oxygen cutting is the most efficient cutting process and is primarily used to cut metals. Unlike fusion cutting, oxygen is used as the cutting gas. The resulting exothermic reaction during oxidation produces additional energy in the interaction area between the laser beam and the workpiece. The achievable cutting rates are greater than with laser fusion cutting by a factor of 2 to 6. The disadvantages are the oxide layer produced and the heavier striation and dragline formation.

In the most frequently used processes, fusion and laser-oxygen cutting are used at the laser machining heads combined with appropriate optics in order to focus the laser beam on the workpiece. The characteristic components of these types of working heads are the focusing optics and the nozzle. The latter serves to introduce process gases and protect the optics. The nozzle includes a distance sensor to compensate for workpiece unevenness during machining. This usually works capacitively for metallic materials while it can be by contact for other materials. The laser head tracks the unevenness and thus maintains a constant distance to the workpiece surface. As the distance to the laser head must be adjustable in a defined manner for machining, a relative movement between the focusing lens and the workpiece must be possible. This allows the allocation of the following kinematic axis movements to the appropriate handling systems:

- stationary lens with moving workpiece
- moving lens with stationary workpiece
- moving lens with moving workpiece

A further differentiating feature for laser cutting machines is the number of axes of motion. While 2D machines manage with three axes, up to five axes of motion are required for 3D cutting machines. The 2D cutting machines are most widely used to cut sheet metal (see ◘ Fig. 7.13). The laser beam source produces the laser beam required for machining. This is fed to the laser head by an encapsulated beam guidance system comprising an XY-traverse unit in a portal design. The cutting head moves in the Y-axis on the cross beam. The cross beam itself can travel in the X-direction. The laser beam is accurately positioned on the workpiece in terms of height in the Z-direction by means of a height-adjustable focusing unit. The metal sheets being machined are placed on a cutting grid. Emissions during cutting can be extracted directly from below. Pollutants from the extracted air are separated off by a filter system and the cleansed air can be released back into the environment. Depending on the application, the cutting nozzle in the working head can be supplied with a range of different gases (N_2, O_2, Ar, He) via gas feed lines. The differentiating features for a 5-axis laser machine compared to a flatbed laser machine are the greater space in the Z-direction and the two additional rotational axes. This allows the machine to perform both machining tasks on flat components and complex three-dimensional machining operations.

7.4 Laser Ablation Machines

Laser ablation is a subtractive process that is the opposite of the laser-based additive processes. The objective in laser ablation is to remove small volumes of material using a pulsed energy input and thus to produce the required reference geometry from a semifinished part. Due to the current low maximum achievable removal rates of a few cubic millimeters per minute, laser ablation is mainly only used in specific areas, e.g., to modify surface characteristics (dimples to improve the tribological characteristics, for example, in internal combustion engines), for chip grooves in CBN and PCD tools, laser marking, or for manufacturing small prototypes (micro injection molding tools, die punches).

Pulsed solid-state lasers with a pulse duration in the femtosecond to nanosecond range are used for laser ablation. In nanosecond laser processes (short-pulse laser

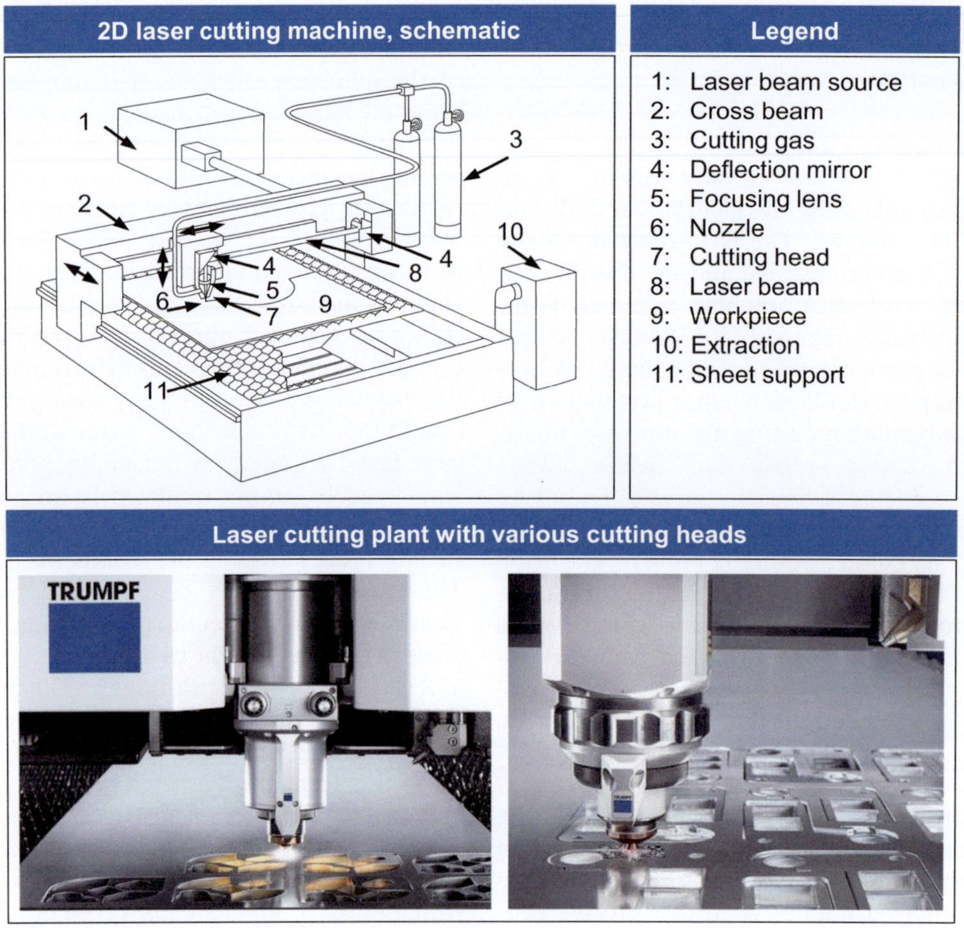

Fig. 7.13 2D laser cutting machine, according to TRUMPF Group

ablation), the material is heated, melted and vaporized. A macroscopic edge zone with melt adhesion is often unavoidable. In material removal using femtosecond or picosecond laser radiation (ultra-short pulse laser ablation), the material is largely directly sublimated by the high laser beam intensity and the extremely short pulse durations. Melt adhesion can be reduced to a minimum in this way. The peak power with short-pulse laser processes is in the kilowatt range and in the megawatt to gigawatt range with ultra-short pulse laser processes. However, the average power output is only approximately 10 to 1000 W. In addition to the laser parameters, the decisive factors for the removal rate and the surface quality of the structure are the speed and accuracy of the beam deflection. The beam deflection is usually produced by a laser scanner.

Laser ablation is characterized by, among other factors, its independence from material properties, the fact that it can operate within narrow local limits, the fact that it can be easily integrated into machine tools and the low environmental impact of the process. The process allows structures of 10 µm to be achieved with aspect ratios of 10–100 and surface qualities of 0.1 µm.

Machines for laser beam ablation have a different design to cutting machine tools. This is based on the requirements of the various machining processes. Particularly high static and dynamic stiffness is not a main point of focus for these machines as the processes completed by the laser are force-free. Instead, the key developmental points are a thermally stable design, high accuracy of the axis kinematics, the control-related combination of axis and scanning motions and measurement technology to calibrate the laser beam focus.

Figure 7.14 shows an axis system of a laser ablation machine (left) and another laser ablation machine (right). The axis system is mounted on a granite machine bed and has a portal design. Depending on the laser process, other or additional kinematics may be necessary (e.g., rotational axes for tool machining). The ablation machine shown here has laser beam sources with five wavelengths for machining a wide range of materials. They are deflected by a laser scanner. The machine has an axis system similar to the axis system shown on the left of Fig. 7.14 to position the component. The axis system and laser scanner together achieve an absolute accuracy of 10 µm. The machine

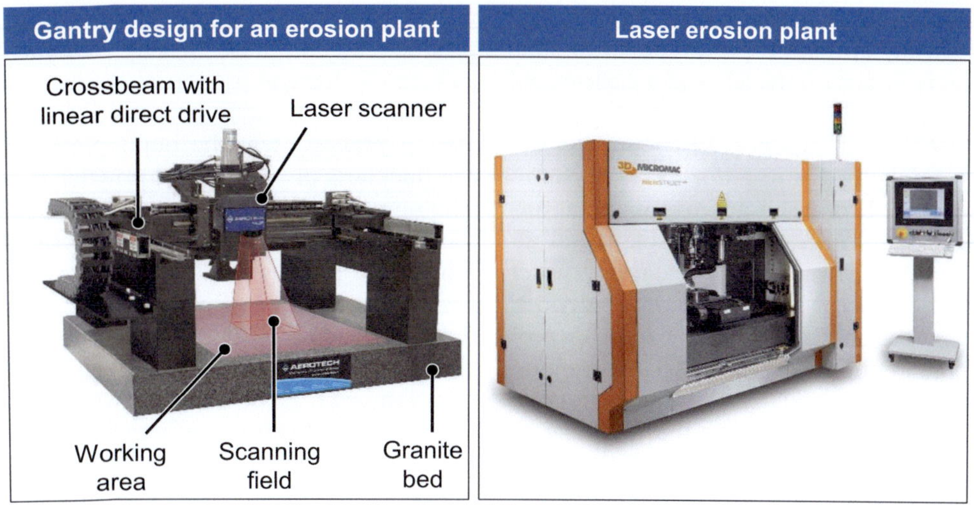

Fig. 7.14 Axis system and flexible laser texturing machine. *Source* AEROTECH, 3D-Micromac

also has measurement technology to calibrate workpieces and optical measuring devices to calibrate the scanner.

7.5 Machines for Additive Manufacturing

The ability to focus laser beams makes it possible to precisely and locally sinter or weld a range of materials. This can be used to produce three-dimensional objects from a powdered material using additive manufacturing. The workpieces are built in layers. In each layer, the material is fused only at those points required by the subsequent component. After this, other layers are added and combined with the previous layer until the component is completed. One benefit of this production process is the great geometric complexity that can be achieved compared to conventional manufacturing processes. An overview of the various additive manufacturing processes is given in ▶ Sect. 2.6. The laser powder bed fusion (LPBF) process and the corresponding laser system technology are discussed in greater detail below. Laser metal deposition (LMD) is dealt with in ▶ Sect. 7.6.

Additive manufacturing of metallic materials is currently limited to niche products. Such components are mainly produced for aerospace, medical technology, and occasionally for tool making. Metallic products are frequently produced using selective sintering or melting processes. Materials processed on commercial machines include stainless steels, aluminum alloys, nickel-based alloys, and titanium alloys. Figure 7.15

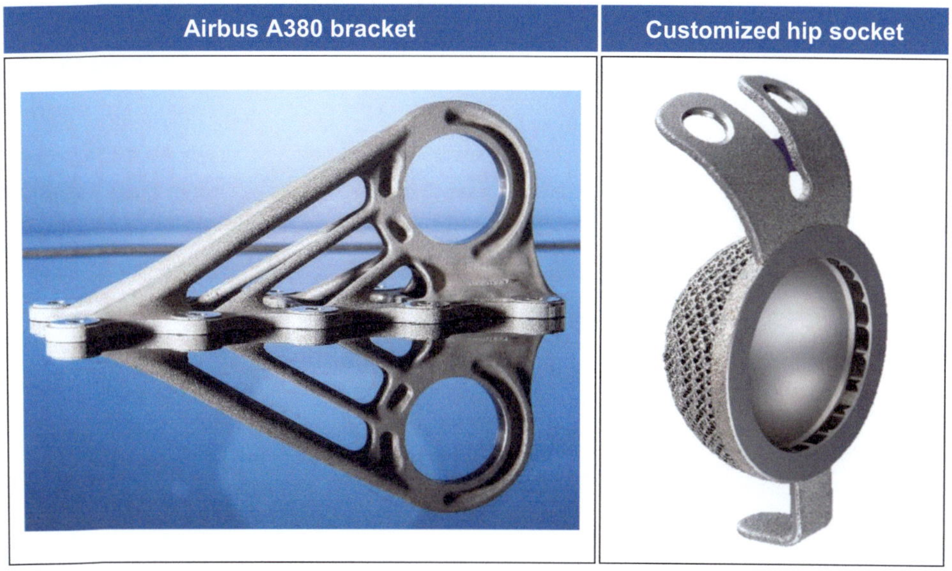

Fig. 7.15 Examples of metal products using additive manufacturing. *Source* Airbus, [POPR14]

shows some products typically made from metallic materials. A topologically optimized prototype for an Airbus A380 bracket is shown on the left and a customized hip socket is shown on the right.

The laser powder bed fusion (LPBF) process developed at the Fraunhofer Institute for Laser Technology (ILT) is a powder-bed-based additive manufacturing process [MEIN99]. Industrially available metal alloys are used as powder material. The laser powder bed fusion (LPBF) process sequence can be divided into three main steps: data preparation, additive process, and component finishing. In the data preparation step, a three-dimensional CAD model is transferred into an STL data format using the principle of triangulation, see Fig. 7.16. This format is then split into layers which correspond to the layer thickness to be produced. In the processing step, known as slicing, aberrations occur which result in aliasing. These errors increase as the layer thickness increases. The individual layers are then defined with scan vectors that specify the travel paths of the laser beam focus in the processing plane. In addition to this, further material in the form of supports is built up in areas of the component that are located in an overhang.

The process adds layer by layer using three repeating, successive process steps. In the first step, a powder dosing unit deposits the powder in a defined layer thickness of 20 µm to 200 µm. The powder is typically distributed uniformly using a brush, silicone blade, or roller. In the next step, the powder bed is selectively scanned by a laser. The powder is completely melted in the interaction area between the laser radiation and the material and, after solidification, forms a metallurgical bond with the adjacent and subjacent tracks. This produces a component density of almost 100% and results in mechanical properties corresponding to the values for conventionally processed materials. In the following process step, the build platform is lowered by the thickness of a layer and powder is deposited again. The number of times this cycle is repeated depends on the height of the component and the layer thickness. Unbonded powder can be reused after recycling. Finishing is performed after the additive process. For this, the components must first be removed from the build chamber and de-powdered. The support structures are then removed. Depending on where the component is to be used, individual surfaces can be subjected to specific heat treatment and reworking to achieve the required dimensional accuracy and surface quality.

Figure 7.17 shows the construction of an LBPF machine. This consists of a processing chamber with a build platform and the powder dosing system, a laser and optics unit, a control system, and a powder recycling system.

The LBPF process takes place in an inert gas atmosphere in a closed processing chamber, Fig. 7.18. The components are built up on the build platform, which can be lowered. The powder is conveyed to the build platform in front of a powder dosing unit and is uniformly deposited on the build platform by the linear movement of the powder dosing system. Excess powder is captured in the powder overflow containers, Fig. 7.18, right.

The optics consist of the laser beam source and a laser beam deflection unit. Fiber lasers are usually used to produce a laser beam with a wavelength of 1070 nm.

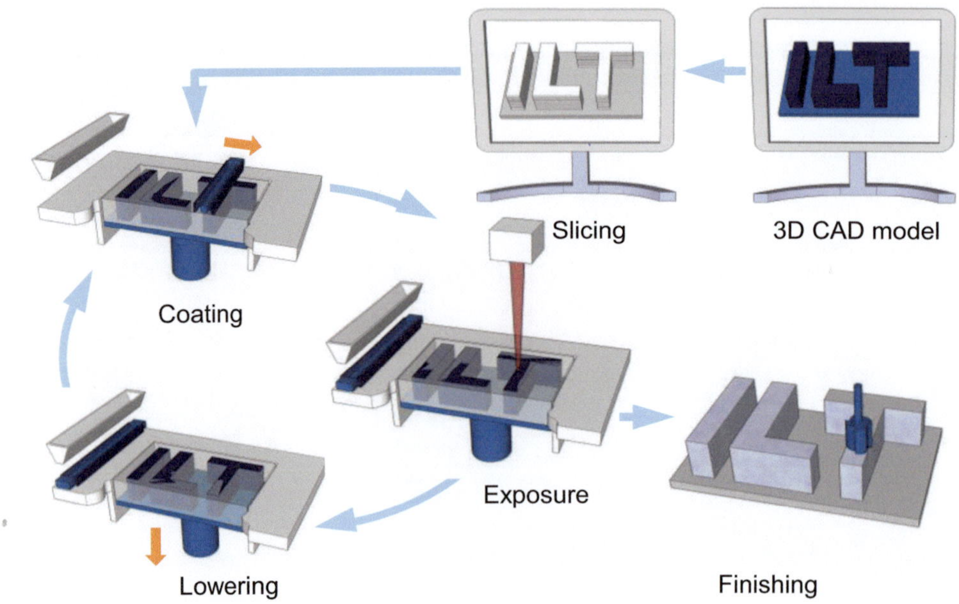

Fig. 7.16 Process schematic of the laser powder bed fusion (LPBF). *Source* [SCHL12]

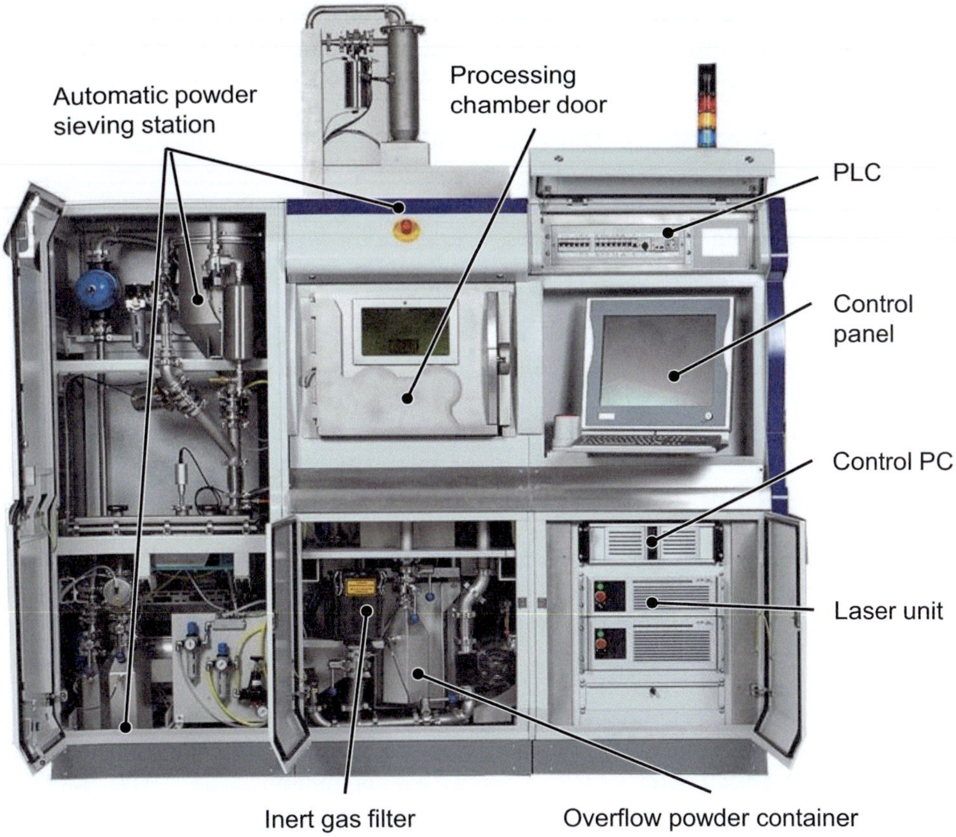

Fig. 7.17 Construction of an LBPF machine. *Source* SLM Solutions

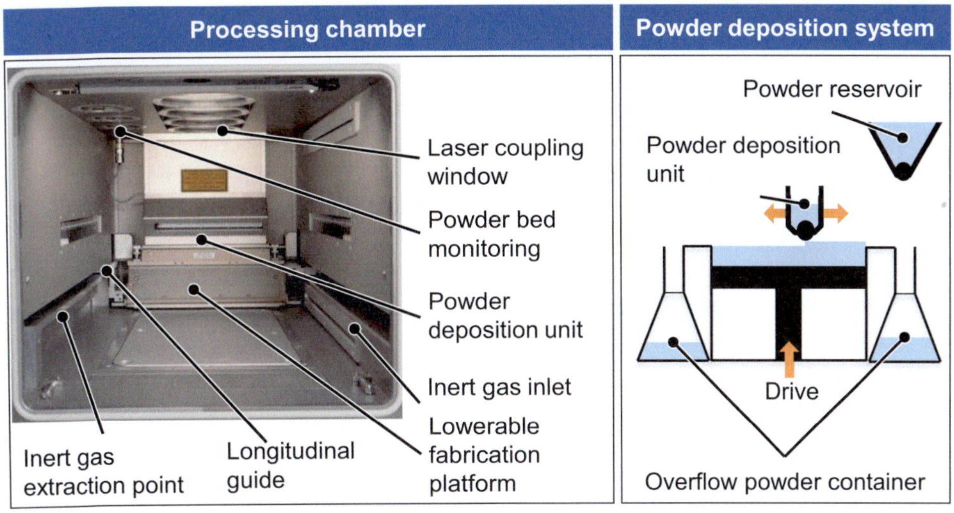

Fig. 7.18 LBPF machine processing chamber. *Source* SLM Solutions

The laser beam is conducted into the laser beam deflection unit via a fiber optic cable. The system is installed outside the processing chamber. The laser beam is focused by a system of lenses and is deflected onto the build platform by mirrors in the laser beam deflection unit. A coupling window protects the optics from being contaminated by the process gases in the processing chamber. The process gas atmosphere is cleansed by continuous recirculation and filtering to extract the flue gases and spatter created during the laser welding process. To achieve this, a flow of inert gas is introduced above the build platform and is extracted on the opposite side, see Fig. 7.18, left.

The market for commercial LBPF machines is characterized by a high degree of innovation and is extremely dynamic. Commercial machines have five key features: the deposition mechanism, the laser unit used, the available space, the inert gas flow, and the balance of the machine. Current trends in machine systems are multi-scanner systems that allow simultaneous scanning of the powder bed with several lasers. This does not just reduce the main process time. It also allows larger components to be produced economically by linking individual scan fields. Comprehensive process monitoring to check the quality of the deposited layer and the melted material as well as the temperature of the melt pool, the laser output parameters, and the inert gas atmosphere contribute to an optimized quality management system on LBPF machines.

7.6 Laser-Assisted Machining Systems

In addition to machines which process materials only using lasers, there are now hybrid machines that allow parts to be machined using both cutting and laser processes.

7.6.1 A 5-Axis Hybrid Machine for Milling and Additive Manufacturing of Components

Figure 7.19 shows such a hybrid machine. Laser metal deposition (LMD) and 5-axis milling are implemented on a single machine tool. The machine is based on a conventional 5-axis machining center with a movable column design and a supplemental laser processing unit for additive manufacturing. In the additive manufacturing process, a nozzle integrated into the laser machining head deposits a metallic powder. This additive manufacturing process is ten times quicker than additive manufacturing of components in a powder bed. The output of the laser source is matched to the specific application. Typically, a laser power of 2.5 kW with a wavelength of 1030 (±10) nm is used. The HSK-A63 tool holder on the milling and laser tool provides fully automatic tool changes.

During the milling process, the laser head remains in a secure tool magazine away from the workspace to protect it from coolant and chips. Difficult to reach areas, such as undercuts, can be milled between laser-added layers. This would not be possible on components manufactured using milling processes alone due to the geometry. As a result, components with higher surface qualities and complex 3D geometries can be achieved without having to add support geometries. Depending on the laser parameters and the geometry of the nozzle, wall thicknesses of 0.1 mm to 5 mm are feasible. Unlike laser melting in a powder bed, laser metal deposition (LMD) is characterized by a build rate of up to 1 kg/h. This allows large components to be produced relatively quickly. Rapid motion rates of up to 40 m/min and axis accelerations of up to 6 m/s^2 can be achieved in the milling process. The maximum rotational speed of the milling spindle is 20,000 rpm with a maximum torque of 100 Nm. Workpieces with a diameter of up to 500 mm or a height of 350 mm and a maximum mass of 600 kg can be machined on the NC swivel rotary table.

The machine shown in Fig. 7.20 is based on a two-axis lathe with a spindle and a counter spindle, which facilitates machining from two sides. Both spindles have their own cooling. In addition to the X-axis with a roller bearing guide and the Z-axis with a wear-free hydrostatic guide, a tool can be driven in the third C-axis, e.g., for eccentric axial bores. The slant bed of the machine is a welded design with the bed slanted at

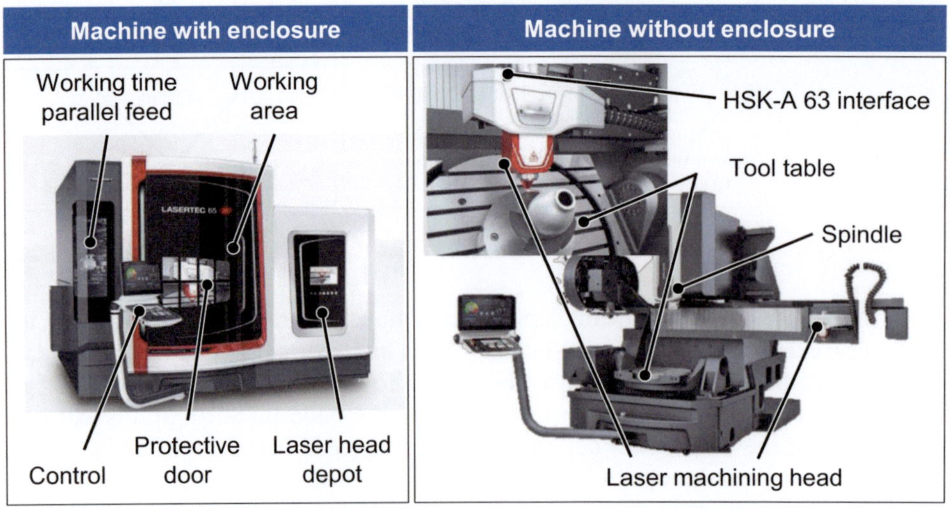

Fig. 7.19 Hybrid machine of 5-axis milling and laser metal deposition (LMD) on a Lasertec 65 Additive. *Source* DMG Sauer

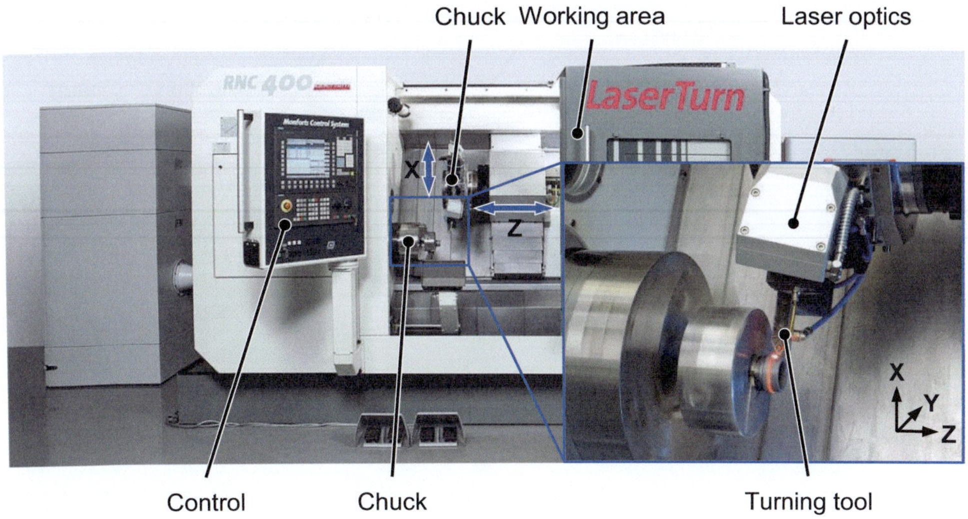

Fig. 7.20 LaserTurn machine for laser-assisted turning. *Source* Monforts

60° and resting on a concrete base. The cutting process is supported by a workpiece ejector in the counter spindle and workpiece removal in the door of the workspace. The maximum rotational speed of the tool spindle is 4000 rpm with a maximum torque of 240 Nm. The spindle unit has a runout accuracy of about 0.003 mm. All the axes are driven by ball screw drives with a preloaded double nut. The integral laser supports the actual turning process and heats the surface of the workpiece so that both soft and hard machining are possible. The laser beam required for the process is guided through the turret from behind by optical fibers and is targeted into the process zone by the laser optics. Potential laser machining processes for which the laser optics are designed include laser hardening and laser coating. Specifically, this allows the cost-optimized, laser-assisted cutting of sintered, ceramic materials. Ceramic workpieces are prevented from chipping by defined parting in the synchronous spindle operation. Machining ceramics produces nanometer-sized dust particles that are extracted directly at the clamping device. The benefits of hybrid machines lie in reduced cutting forces and cutting insert wear and increased material removal rate. The process chain is also shortened and the setup, transport, and idle times are reduced.

The machine shown in Fig. 7.21 is a multi-technology machine tool built as part of the first phase of the excellence initiative at the Laboratory for Machine Tools and Production Engineering (WZL) at RWTH Aachen University [BREC12]. The underlying machine is a five-axis milling center with a moving column design. It is characterized by the two identical workspaces, each with a rotary swivel table. Both workspaces can be approached by the moving column and the spindle. The workspaces are separated by a center console. This acts as the magazine for two laser machining heads. An articulated arm robot, which can cover the area in front of the machine, the console, and the workspaces, is integrated directly in front of the center console. The machine has an active laser safety enclosure for running tests. This ensures Class 1 laser safety.

The two laser machining heads are used for different applications: The texturing head has a pulsed laser source with an average power of 50 W and high accuracy guided by the spindle, which can be used for deburring and texturing component surfaces. The hybrid laser head has a power of 2.4 kW and is available for processes requiring high energy levels such as hardening or laser metal deposition (LMD) by the robot. For the latter process, the hybrid laser head is fitted with a hose assembly for the hot wire and inert gas supply. The supply lines to both heads are led away upwards together where they are each contained in a movable arm. As these media supply lines restrict the pitch angle of the laser beam to just a few degrees with respect to the Z-axis, both milling and laser processes can take place alternately on this multi-technology machine tool. However, the tool cannot handle hybrid processes using the laser beam and the cutting tool simultaneously. This machine does allow a workpiece to be finished without having to be reclamped [BREC12].

7.6.2 System Technology for Laser-Assisted Milling

High-performance materials such as titanium-based or nickel-based alloys and engineering ceramics such as silicon nitride (Si_3N_4) are becoming increasingly important in the industry due to their excellent thermo-mechanical properties. However, their high strength

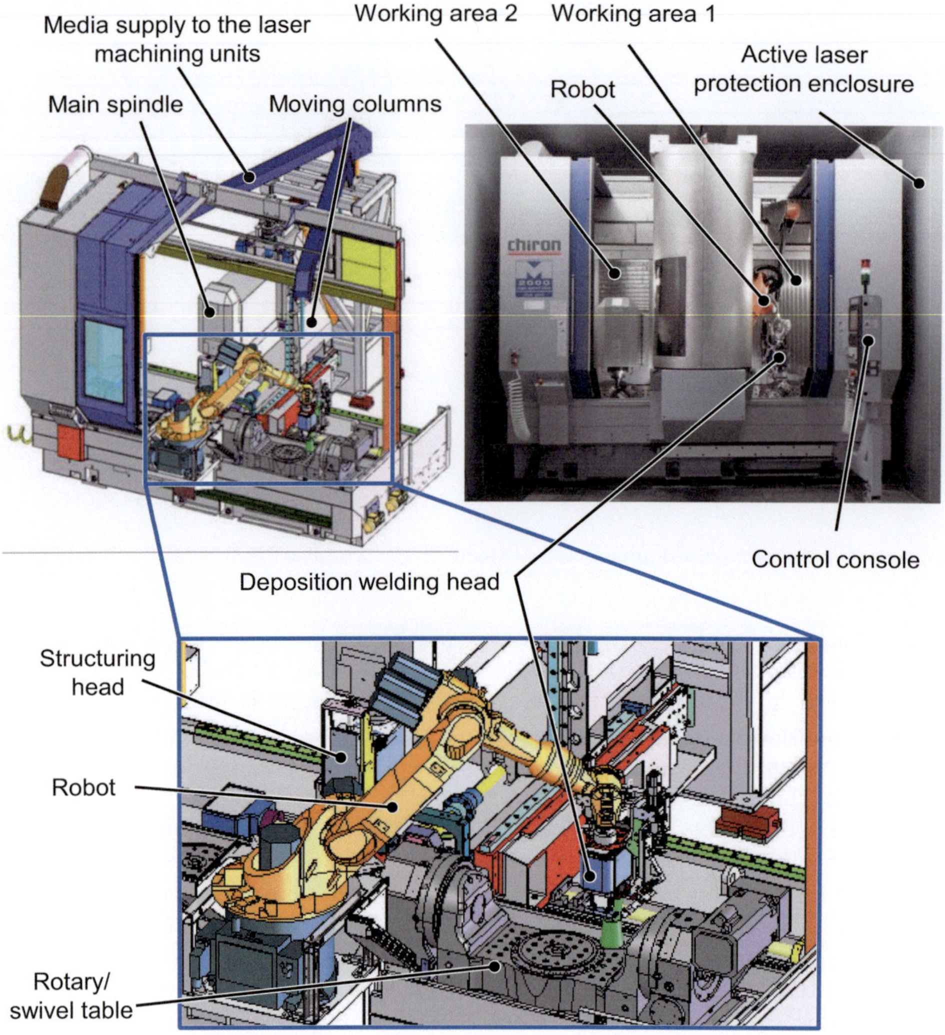

Fig. 7.21 Laboratory for Machine Tools and Production Engineering (WZL) multi-technology machine tool

represents a significant problem when it comes to processing them mechanically. Specifically, high tool wear often makes it difficult to use a cutting process with a defined cutting edge. By using a laser beam, is it possible to heat the workpiece immediately before tool contact significantly reducing the material's strength and allowing greater material removal rates to be achieved with lower tool wear [ROSE12].

To this end, the Fraunhofer Institute for Production Technology IPT has developed a compact and robust laser system technology for the integration of beam guidance systems into existing milling machines consisting of the following components:

- automatically adjustable collimator and adjustment unit with fiber connector socket
- spindle system with hollow spindle and rotating union
- double-edged milling tool with integral laser beam guidance
- real-time process control for multi-axis machining

The schematic structure of the laser system is shown in Fig. 7.22. The laser beam is guided through the hollow spindle via the rotating union after collimation and adjustment. The automatically adjustable collimation allows a range of beam diameters to be set. In this way, the laser beam focus can be adjusted to the contact conditions. Once the laser beam has been transferred into the specially developed milling tool with HSK63 interface, the laser beam is directed by combined lens-prism optics to immediately before the tool's indexable insert. By capturing the axis positions and the encoder signals from the milling spindle and coupling with a CAD-CAM processing model, the laser power is adjusted at a high frequency during tool contact to match the volume to be machined and the contact conditions. The system and process technology developed here allows high-performance materials to be machined with flexible geometries and simultaneously in several axes.

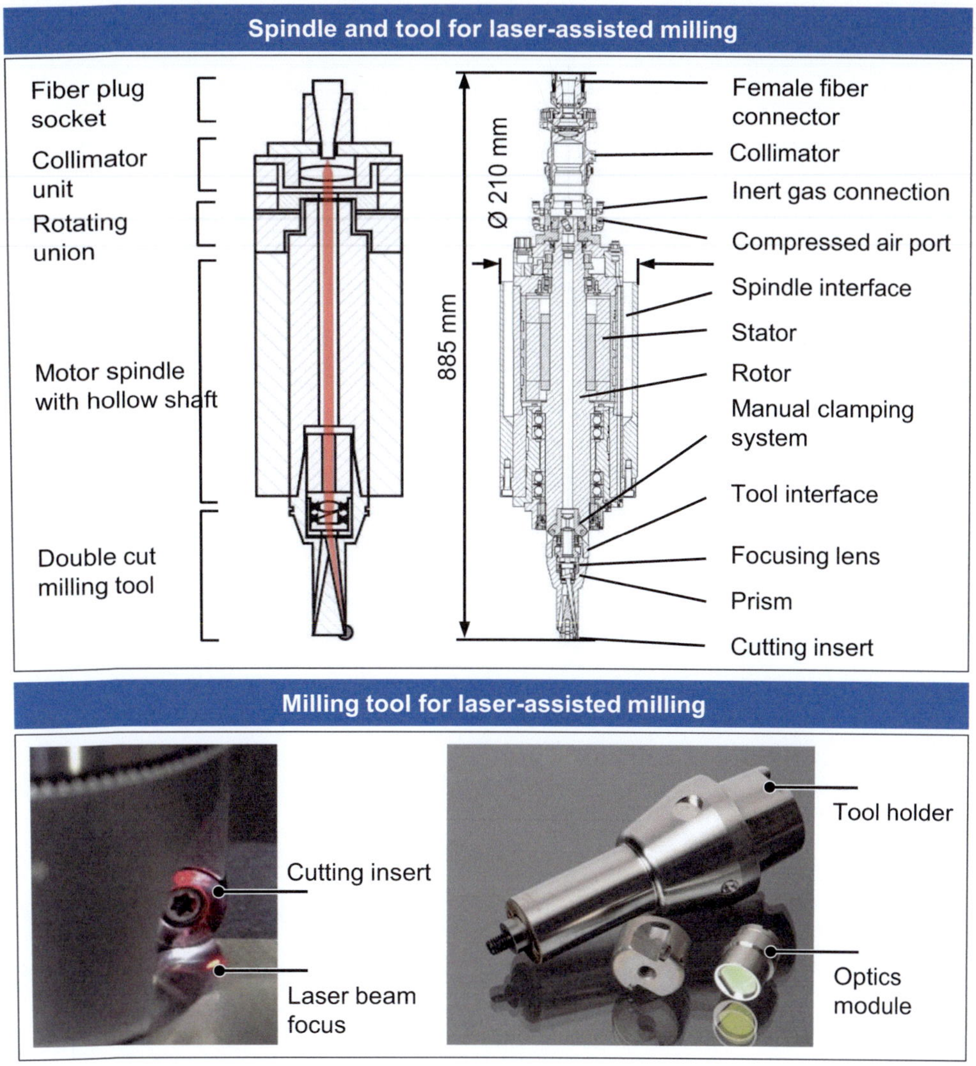

Fig. 7.22 Spindle and tool for laser-assisted milling. *Source* Fraunhofer IPT

References

[BREC12] Brecher, C. (ed.): Integrative production technology for high-wage countries. Springer, Berlin (2012)

[FÖRS01] Förster, D., Müller, W.: Laser in der Metallbearbeitung, 1st edn. Fachbuchverlag Leipzig im Carl Hanser Verlag, Leipzig (2001)

[KLOC07] Klocke, F., König, W.: Fertigungsverfahren 3. Abtragen, Generieren und Lasermaterialbearbeitung. (Reihe: Fertigungsverfahren. Bd. 3), 4th edn. Springer, Berlin (2007)

[MEIN99] Meiner, W.: Direktes selektives Lasersintern einkomponentiger metallischer Werkstoffe. Diss. RWTH Aachen (1999)

[POPR05] Poprawe, R.: Lasertechnik für die Fertigung. Springer, Berlin (2005)

[POPR14] Poprawe, R., Jauer, L., Meiners, W.: Selective Laser Melting für medizinische Implantate. Aachen, 11/2014

[ROFI14a] Rofin: Rofin DC-Serie. Hamburg, 2014. Firmenschrift

[ROFI14b] Rofin: Rofin FL-Serie. Hamburg, 2014. Firmenschrift

[SCHL12] Schleifenbaum, J.H.: Verfahren und Maschine zur individualisierten Produktion mit High Power Selective Laser Melting. Diss. RWTH Aachen (2012)

[DIN03] Norm DIN 2310 Teil 6 (Juni 2003). Thermisches Schneiden Teil 6 Einteilung, Prozesse

[TRUM08] Trumpf: Laser: Jede Aufgabe perfekt lösen. Ditzingen, 2008. Firmenschrift

[DIN08] Norm DIN EN ISO 11145 (November 2008). Laser und Laseranlagen – Begriffe und Formelzeichen

[DIN15] Norm DIN EN 60 825 1 – VDE 0837 1: Sicherheit von Lasereinrichtungen – Teil 1: Klassifizierung von Anlagen und Anforderungen, Deutsches Institut für Normung (eds.) Beuth Verlag, Berlin (2015)

[EICH15] Eichler, J., Eichler, H.J.: Laser: Bauformen, Strahlführung, Anwendungen. Springer, Berlin (2015)

[HÜGE09] Hügel, H., Graf, T.: Laser in der Fertigung, 2nd edn. Vieweg+Teubner, Wiesbaden (2009)

[IPGL17] IPG: IPG Laserquellen. Burbach, 2017. Firmenschrift

[POPR11] Poprawe, R.: Tailored light 2. Laser application technology. Springer, Berlin (2011)

[ROSE12] Rosen, C.J.: Laserunterstützte Fräsbearbeitung hochfester Werkstoffe. Dissertation, RWTH Aachen (2012)

Single-Purpose Machines

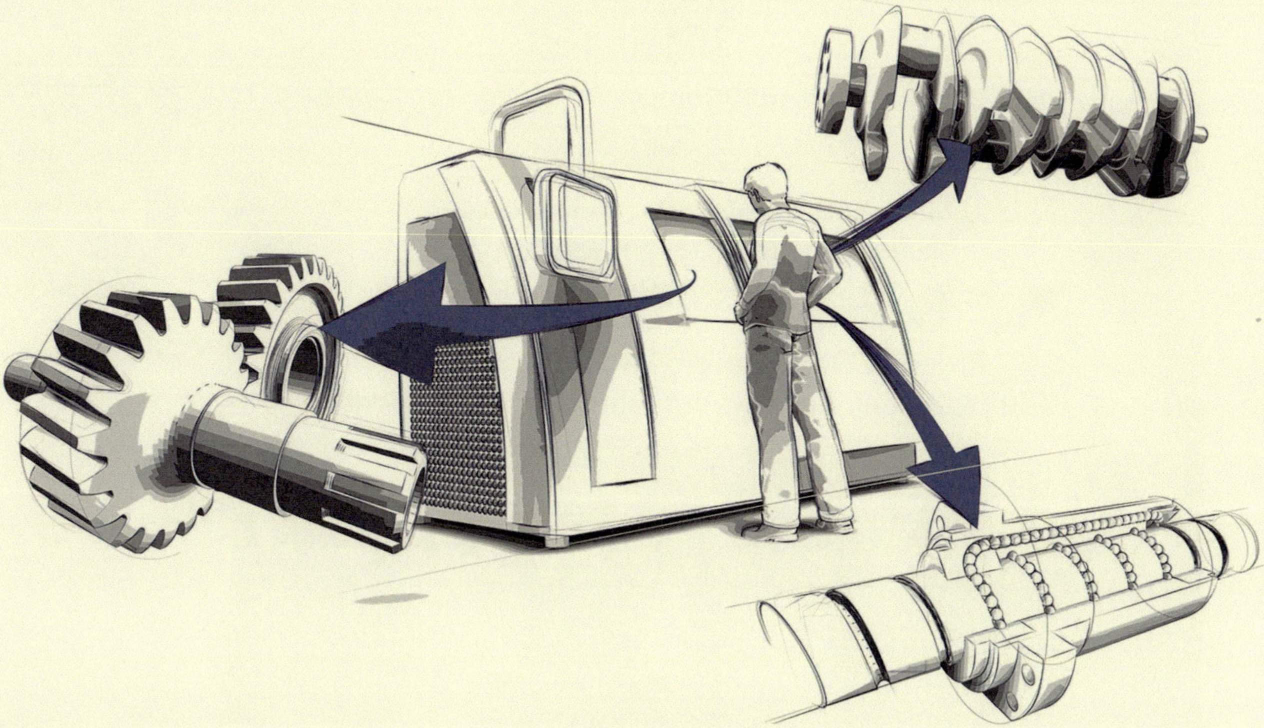

I thank my staff members Benedikt Biernat M.Sc., Mathias Dehn M.Sc. and Melina Kamratowski M.Sc. for their contribution in revising this chapter.

C. Brecher and M. Weck, *Machine Tools Production Systems 1*,
Lecture Notes in Production Engineering, https://doi.org/10.1007/978-3-662-68120-6_8

Contents

8

8.1 Gear Cutting Machines

To manufacture gears, high-precision machine tools with complex kinematic interdependencies of the machine axes movements are required. The geometrical diversity of workpieces depends on the numerous types of gears and gear transmissions in use. The different gear drive designs can be manufactured cost-effectively on very specific gear cutting machines using special gear manufacturing methods.

The different types of gear drives can be classified according to their axis orientation as in ◘ Fig. 8.1. The most common type of gears with parallel axes are involute gears, i.e. spur or helical gears. Special gears (such as cyclo gears and the circular arc gear shown in the figure) may also have parallel axes. Beveloid gears, also called conical gears, can be mounted parallel if they have the same cone angle.

For intersecting axes, bevel gears are used that can have straight or helical flank lines. For face gears, an involute cylindrical gear engages with a face gear. For both bevel and face gears, a shaft angle of 90° is commonplace but not an absolute necessity. For small shaft angles, beveloid gears are usually used.

For skew axis arrangement, hypoid gears can be used as a special application of a bevel gear. Worm drives consisting of a worm and a worm gear can also be used. Within certain technical limitations, crossing axes can also be realized with beveloid gears. The following sections focus on machines for manufacturing cylindrical, bevel and beveloid gears.

The process chain for manufacturing cylindrical gears is shown in ◘ Fig. 8.2. The system used for manufacturing bevel gears does not differ greatly from that for manufacturing cylindrical gears. A forged blank, or one shaped from a bar stock, is the starting point. The blank is machined to get the base body. To manufacture as cost-effectively as possible and to attain high gear accuracy, soft cutting is performed using high cutting speeds and high feeds. Depending on the dimensions, application and batch size of the gear, a selection can be made between the different rolling processes, such as gear hobbing, gear shaping, gear skiving and the now less common gear planing, and between the profiling processes, profile milling and broaching.To improve the gear quality and to save the time and cost for additional hard finishing, soft finishing (such as gear finish hobbing or gear shaving) can be performed after soft cutting. Subsequent to the soft finishing, heat treatment is carried out. [KLOC17]

Heat treatment is used to increase wear resistance on the edge while maintaining elasticity on the inside of the tooth. The gear teeth in contact are able to withstand the compressive forces applied on the tooth flanks and the high number of load cycles. For this reason, case hardening can be selected as the heat treatment process for gears with small and medium modules, and induction or flame hardening for large-module gears. However, due to distortions caused by the heat treatment, the gears may be hard finished. Grinding or other abrasive processes are mainly used here. Gear grinding, profile grinding and gear honing are the dominant processes for hard finishing. Hard finishing machines are quite similar to soft cutting machines in terms of kinematics. In line with the attributes of a grinding process, far lower cutting forces and higher circumferential speeds occur, meaning other machine concepts are essentially required for hard fine machining. A ready-to-install gear is at the end of the process chain. [FELT12]

The benefits of involutes mean cylindrical gears represent the most common application of gears with parallel axes. Insensitivity to changes to the center distance with a constant gear ratio is a standout feature

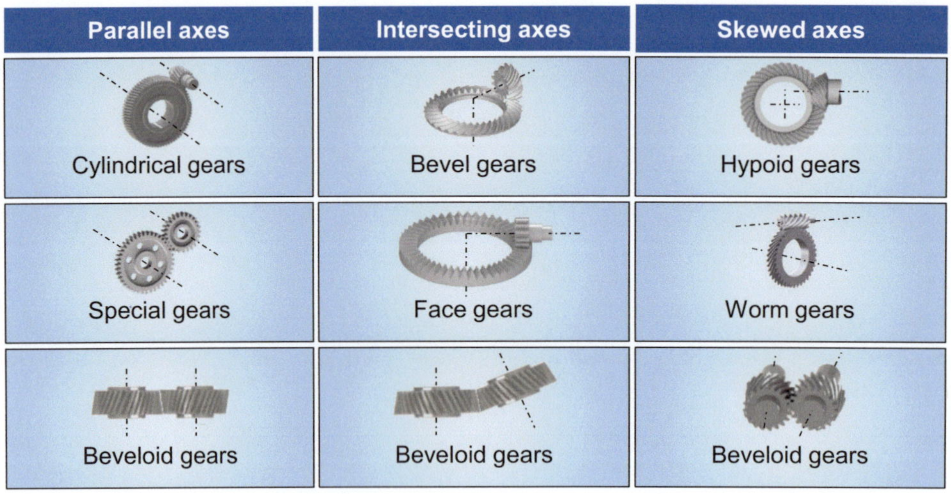

◘ **Fig. 8.1** Categorization of gears. *Source* [KLOC17]

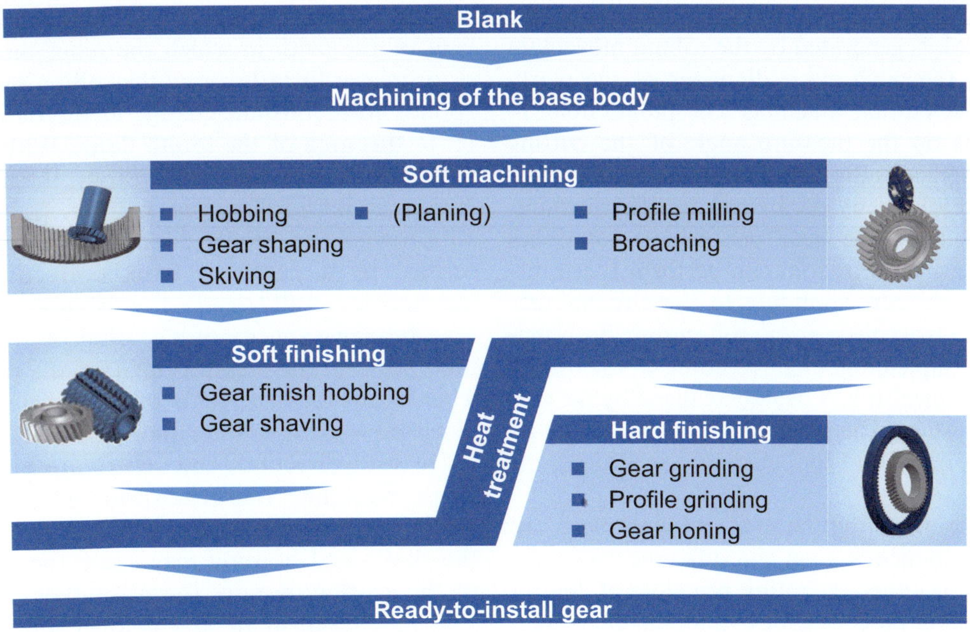

Fig. 8.2 Cutting processes for cylindrical gear manufacturing. *Source* [KLOC17]

of involutes realized as profile shapes of the tooth. The gears are characterized by low backlash and low sliding friction between the flanks. The constant direction of the tooth normal force results in comparatively low-excitation gears. Furthermore, an involute can be manufactured cost effectively using straight-flank tools. A downside, however, is the high flank pressure due to the rolling of two convex tooth flanks. Also, using an addendum modification represents a risk of undercut or pointed teeth—something to consider in the design phase. [MAAG85; SEHE13]

The relationship between the theoretical creation of involutes and the technical realization is depicted in Fig. 8.3. By unrolling a thread or rolling a tangent on a circle (base circle), the end of the thread or a point on the tangent forms the involute. Under the assumption of a slip-free and uniform gear ratio and the geometric relationships between the individual sections, the involute function can be derived as a function of the rolling angle ξ_y. [LITV04; SEHE13; ROTH98]

Arc segment ϑ_y can be determined by the involute function as in Formula (8.1) for any angle α_y.

$$\widehat{\vartheta_y} = inv(\alpha_y) = \tan(\alpha_y) - \hat{\alpha}_y = \widehat{\xi_y} - \widehat{\alpha_y} \tag{8.1}$$

By rolling a tangent on a base circle, the involute can be manufactured cost-effectively using straight-flank tools. One tangent each is required for the left and right flanks of a tooth. Both lines joined together result in the basic rack tooth profile having a geometry defined

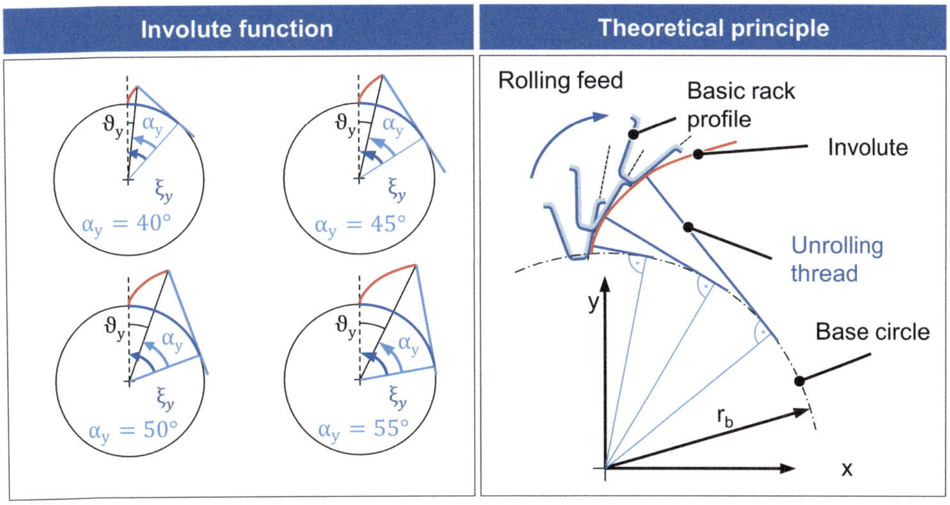

Fig. 8.3 The involute function and its theoretical implementation

according to DIN 867 [DIN05]. For rolling processes, the involute flank is generated by the rolling movement of the tool on the pitch circle diameter of the workpiece in the gear cutting machine. The pitch circle diameter depends on the pressure angle of the cutting tool. The theoretical rolling feed consists of two components of motion to be performed by the machine. The workbench with the gear performs a rotational motion. The workpiece or tool can perform the translatory feed motion corresponding to the circumferential speed of the workpiece at the roll radius. Both machine axes are coupled together according to the rolling condition. Nowadays, this is usually realized by the machine control system. The flank shape is created as an envelope of the individual tool cutting edges along the rolling feed.

The principle of rolling the tool and workpiece can be implemented depending on the different processes. ◘ Figure 8.4 shows the generating principle of the involute using rolling processes for cylindrical gear manufacturing. If the rolling feed is performed section by section for a tooth gap or group of tooth gaps, it is a partial rolling process (gear planing). Otherwise it is a continuous rolling process (gear hobbing, gear shaping, gear skiving). If no profile corrections are applied, the same tool can be used for all rolling processes as long as the gear module is constant, i.e. different diameters. [KLOC17]

In addition to generating the involute profile, manufacturing has the task of creating defined tooth flank modifications. To guarantee a certain dynamic and acoustic behavior or to adjust a particular contact pattern to a load case, modern high-performance gearboxes require modifications in profile line and lead direction as shown in ◘ Fig. 8.5.

Tip and root reliefs are modifications at the tip or root of the teeth in which the relief amount increases or decreases linearly along the rolling path. This corresponds to a constant change in the base circle diameter in this area of the toolth flank. With this modification to the tooth flank topography, the meshing impact is reduced for a rolling gear pairing. [KLOC17]

Crowning in profile direction is a modification described by a parabolic flank correction. This modification generally leads to an improvement in the load-bearing capacity of a gearing. With a profile angle modification, the pressure angle of the gear is changed. This means displacements of the gearing can be compensated when operating under load. [KLOC17]

In lead direction, end reliefs are applied to modify the flank line. End reliefs can be linear or parabolic. When operating under load, the elastic deformation of the shafts and housing may cause the contact pattern of the gearing to move towards either end of the gear. This may result in an edge contact. End reliefs are applied to prevent this. [KLOC17]

As in profile direction, a parabolic modification in width direction or a lead line modification can be applied in lead direction. Due to the described modifications, the behavior of a gear under load can be influenced positively. However, with an increasing parabolic modification, the contact surface of the two tooth flanks decreases. Thus, a compromise between the robustness attained and the contact surface is required when choosing the modification. Moreover, topological corrections can be applied to the tooth flank. The magnitude of the correction can be specified as required using a raster. The functions available for the modification profile extend beyond linear and parabolic modifications. [KLOC17]

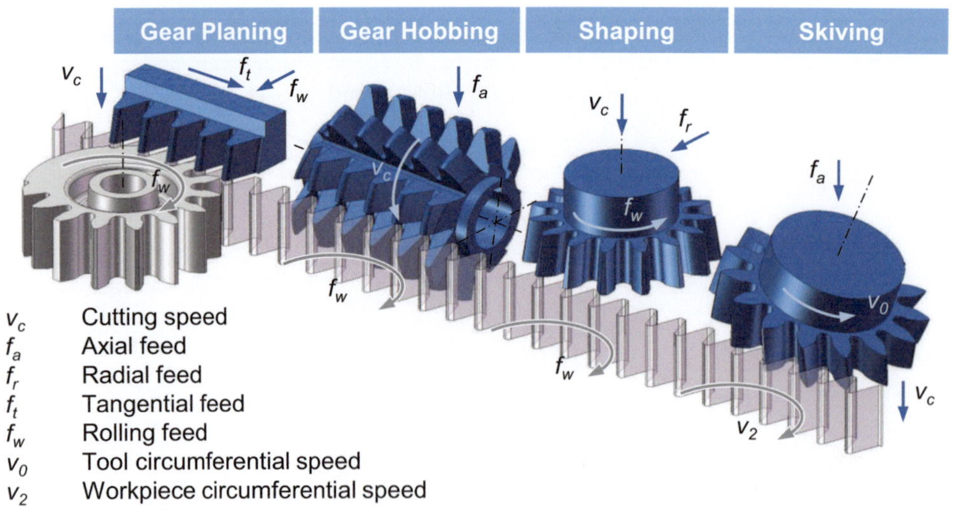

◘ **Fig. 8.4** Principle for generating an involute using rolling processes

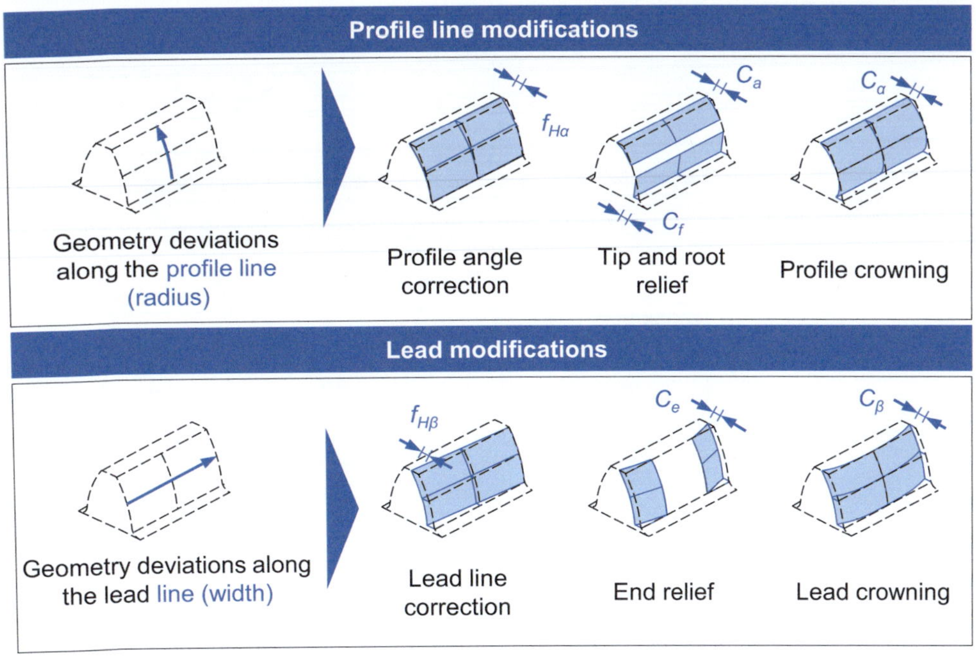

Fig. 8.5 Modifications to the tooth flank topography in profile and lead direction. *Source* [KLOC17]

Depending on the hard machining process used, the gear modifications mentioned are realized by modifying the tool geometry or by changing the machine control data, i.e. the machine movements. The ability to realize the tooth flank modifications mentioned essentially depends on the machining process used. [KLOC17]

8.1.1 Rolling Gear Manufacturing Machines

For rolling processes, the tooth geometry can be represented by the envelope of the tool geometry for all tooth profiles in terms of the number of teeth and the addendum modification. The continuous manufacturing process realizes a uniform pitch on the workpiece, which results in high requirements on the accuracy of the individual movements of the machine axes as well as their synchronization.

8.1.1.1 Gear Hobbing Machines

Gear Hobbing represents the manufacturing process with the highest productivity and cost-efficiency for manufacturing external cylindrical gears [BAUS11]. Given its high flexibility, gear hobbing is a very widely used process. The application is limited mainly to external gears with adequate axial clearance [FELT12]. Increasing requirements on the gear quality result in increasing requirements on the accuracy of the machine axes.

The principle of the gear hobbing process is depicted in Fig. 8.6. The involute is created by rolling the basic rack profile with the workpiece. For the theoretical creation, the rolling movement is split into a translational main movement of the tool and a rotational movement of the workpiece. The basic rack profile is wrapped around the hob in one or more threads for the technical realization of the creation [MAAG85]. The basic rack profile is interrupted by gashes with relieved cutters. The gashes are used to remove the chips. The relieved cutters guarantee the clearance angle necessary for the cutting process.

During the machining process, the tool and workpiece mesh comparable to a helical rolling type gear transmission. The rotation of the hob generates the cutting speed v_c for the cutting movement and the translational rolling component by the tangential rotation of the cutters. The workpiece rotation generates the rolling feed f_w. The swivel angle η is given by the direction and magnitude of the helix angle of the gear to manufacutre β_2 and the thread angle γ_0 as in Formula (8.2).

$$\eta = \gamma_0 \pm \beta_2 \tag{8.2}$$

The feed motion f_a in the hobbing process is generated by moving the tool in tooth width direction (axial milling) [KLOC17].

To distribute the load evenly on all cutters and prevent premature wearing of individual teeth of the hob, the tool can be moved tangentially to the workpiece continually or incrementally at certain time intervals. This tangential displacement to the workpiece is realized by the hob being moved axially parallel to its rotation axis. This causes the teeth of the hob not yet in-

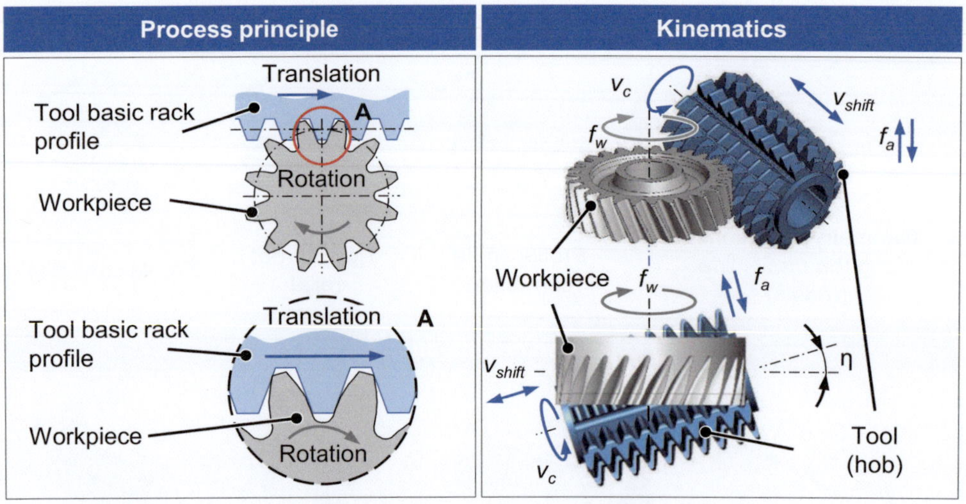

Fig. 8.6 Principle of the gear hobbing process. *Source* [KLOC17]

volved in the cutting process to get involved in cutting the workpiece. This process is called "shifting" and takes place when the workpiece and tool are not in contact. Diagonal milling is the term used when the tangential feed for the shifting occurs continuously next to the axial feed. [FELT12; PFAU76; MAAG85]

By varying the machine settings, any number of teeth, all addendum modifications and any inclination angles can be manufactured with one hob for its inherent module and pressure angle. Only the working area of the machine and the modifications applied are limiting factors.

Figure 8.7 shows the general view of a 6-axis CNC gear hobbing machine with direct drive (decoupled from the tool axis) of the workpiece table and additional counter holder. Implementing the roller coupling with a mechanically coupled differential gear is less important in current designs. All machine axes required for gear hobbing are shown. The plunge depth of the hob is specified by the radial feed axis X_1. The tool is inclined by the swivel axis A_1. Tangential feed axis V_1 enables the shift movement, i.e., displacement of the milling spindle on the milling slide. The actual rolling movement is created by coupling the axes of miller rotation B, table rotation C_1 and axial feed Z_1. The gear hobbing machine shown is suitable for machining with or without cutting fluid. The machine table in an upright and fixed position is a prerequisite for the small space requirement of the machines. [PFAU76; OV14; KLOC17]

The drive for the table rotation C_1 may use one of two different designs, shown on the right in Fig. 8.8. A split, two-speed spur-gear gearbox with a clearance-free preload by the axial displacement of an intermediate shaft can be used. Alternatively, workpiece tables with direct drive can be used. For both designs, a counter holder (Z_2 axis) can also be provided to sup-

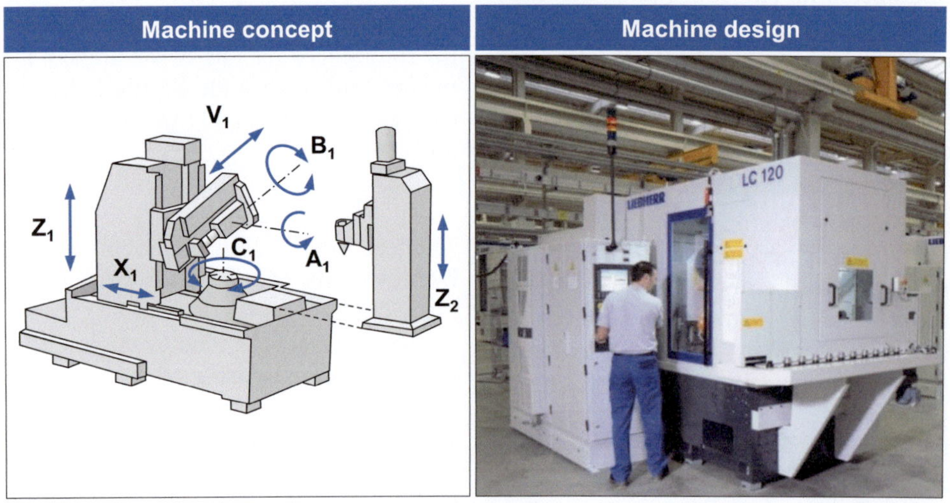

Fig. 8.7 Structure of a gear hobbing machine. Example from Liebherr

Fig. 8.8 Workspace and table drive for a gear hobbing machine. Example from Liebherr

port shafts that, at the same time, enable automated workpiece handling using a workpiece loader that rotates around the column of the counter holder (C_1 axis) [PFAU76; OV14; KLOC17].

Shown on the left in Fig. 8.8 is the working area of a gear hobbing machine allowing the manufacturing of gears with or without cutting fluid. The tool spindle (B axis) and the workpiece spindle (C axis) have direct driven, maintenance-free drives for the Liebherr LC120 selected as an example. The vertical workpiece spindle arrangement means the chips fall unobstructed onto the chip conveyor. Short workpieces can be held in an overhung position, i.e., without counter holder, and long workpieces or shafts between tips with the Z_2 axis being used as an aid [OV14; KLOC17]

Different hob designs are used as gear cutting tools, depending on the conditions, that are implemented as block, roughing or indexable insert hobs. Block hobs, as shown in the bottom right of Fig. 8.8 are made entirely of the cutting material used. As the tooth tip area of the tool in the hobbing process is subjected to greater stress than the flank area, increased wear results at the tip of the cutters of block hobs. Roughing hobs have, in addition to indexable inserts, tip cutters to make the machining process more productive for the same load on the hob. For large cutting volumes, roughing hobs are cost-effective and thus used for gears with large modules. Indexable insert hobs comprise a steel base body and indexable inserts bolted on. The gap profile of the tool is split across several indexable inserts. Worn tools are reconditioned by replacing individual indexable inserts, meaning cost-effective use of the tools for large tool diameters is guaranteed without having to use carbide for the hob base body. [KLOC17].

In series production in the automobile sector, dry cutting with hard-material-coated high-speed steels on a powder metal base or coated carbide hobs are used. Their sensitivity to fracture has to be considered when carbide tools are handled. However, higher cutting speeds, compared to hobs made of high-speed steel can be achieved, enabling high productivity. As a large part of the energy from the cutting process is dissipated as heat into the chips, the workspace is lined with sheet steel and monitored to ensure no chips can build up. This prevents thermal displacement of the machine axes due to chip temperatures. [KLOC17]

8.1.1.2 Gear Shaping Machines

Gear shaping machines are continually working gear cutting machines. In these machines, the roller kinematics during shaping are emulated by the workpiece and tool in the form of a gearbox with parallel axes. The main application of shaping is soft machining of internal or external spur and helical gears [BAUS11; MAAG85; ROTH98]. In contrast to gear hobbing, gear shaping can also be used to manufacture gears with interfering contours [BAUS11; MAAG85]. Another application for gear shaping is soft machining of beveloids and face gears [BAUS11; ROTH98].

Figure 8.9 shows the principle of the gear shaping process. The tool corresponds to a gear with an involute profile. The base circle pitch of this gear is the same as for the workpiece [PFAU76; ROTH98]. The involute on the tooth flank is created by the tool and workpiece rolling on each other with parallel axes. The roller coupling depends on the number of teeth on

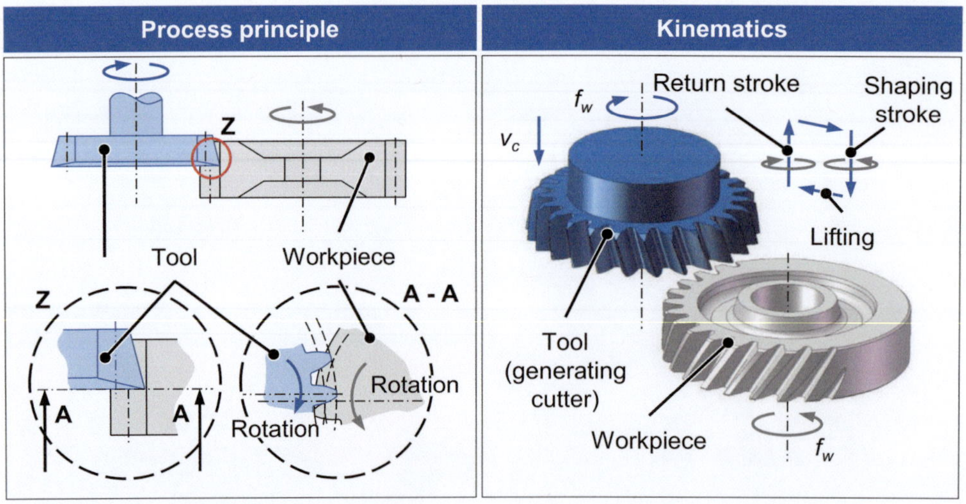

Fig. 8.9 Principle of the gear shaping process. *Source* [KLOC17]

the tool and workpiece, meaning no additional partial movement is necessary [MAAG85]. If a helical gear is to be manufactured, also a helical tool is used. The rolling movement of the workpiece is superimposed on the rotation to create the helix angle [BAUS11].

The tool teeth are relieved, so they generate the clearance angle necessary for cutting at the tip and the flank of the workpiece [PFAU76]. The rolling feed f_w between workpiece and tool realizes the rolling of workpiece and tool [BAUS11]. The shaping stroke, also called the working stroke, generates the cutting speed necessary for cutting and thus causes chip removal [BAUS11; FELT12]. If the translational tool axis is moved by a crank drive, there is no constant cutting speed. Instead, the movement has a sinusoidal waveform with the cutting speed reaching its maximum at half the shaping stroke. The subsequent lift-off is radial to protect the tool from wear and to guarantee good surface quality. The return stroke, also called the idle stroke, is used to move the tool back to the initial position without interrupting the rolling movement [ROTH98; KLOC17].

Helical generating cutters are required to manufacture helical gears. In addition, the generating cutter must be screwed through the workpiece during the stroke. To realize a certain helix angle β in an inclined guide brush with a constant thread height of the screw guide, the diameter of the generating cutter must assume a certain value. The diameter of the generating cutter d_s is calculated according to Formula (8.3).

$$d_s = \frac{p_z}{\pi} \cdot \tan \beta \tag{8.3}$$

On older machines, this rotational movement was realized by a hydrostatic inclined guiding of the shaping spindle, which needed to be adapted for every helix angle. Changing the inclined guide represented major effort and had to be considered in workpiece design. Nowadays, the screw movement necessary can be realized by the computerized numerical controller for the individual axes.

Due to the short tool runout paths and parallel axis arrangement of the tool and workpiece, the main applications for gear shaping are creating internal gears and gearings with limited axial clearance, such as double helical gears and gears on stepped shafts, clutch gears and similar. For clutch gears, special fixtures are required to incline the table for rear shaping of the chamfers.

The kinematics of gear shaping are realized using the machine axes of a gear shaping machine. Fig. 8.10 shows this using a Gleason-Pfauter P600 ES as an example [15]. The tool and workpiece each have their own rotational axis, axes C_1 and C_2. Roller coupling synchronization can be guaranteed using direct drives, which influences the resulting gear quality significantly [BAUS11]. The radial feed of the X-axis is used to adjust the plunging depth. The plunging depth is set before or during the rolling movement depending on the process modification [ROTH98]. Axis Z_1 is used to adjust the stroke position necessary to machine different tooth widths. The actual lift movement of the shaping tool is performed by axis Z_2. The separation into stroke position and stroke movement of the Z-axis is especially useful for applications with higher flexibility requirements [BAUS11]. The radial infeed axis X is used to provide a specific lift-off angle [BAUS11]. This prevents collisions on the return stroke [BAUS11] and allows special profiles to be machined. The tool can be inclined by the swivel axis A. Many machine types can have a counter holder for the workpiece (not shown in the picture) parallel to the C_2 axis [KLOC17].

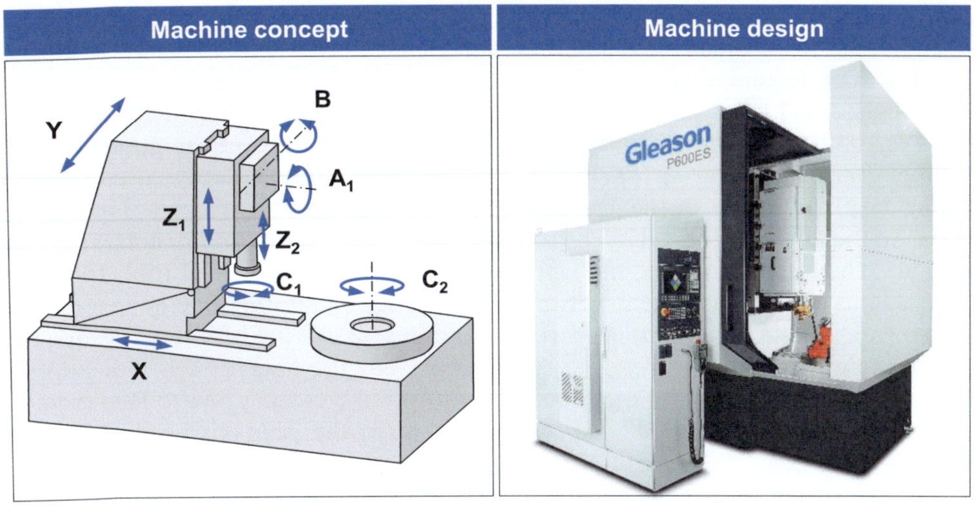

Fig. 8.10 Structure of a gear shaping machine. Example from Gleason

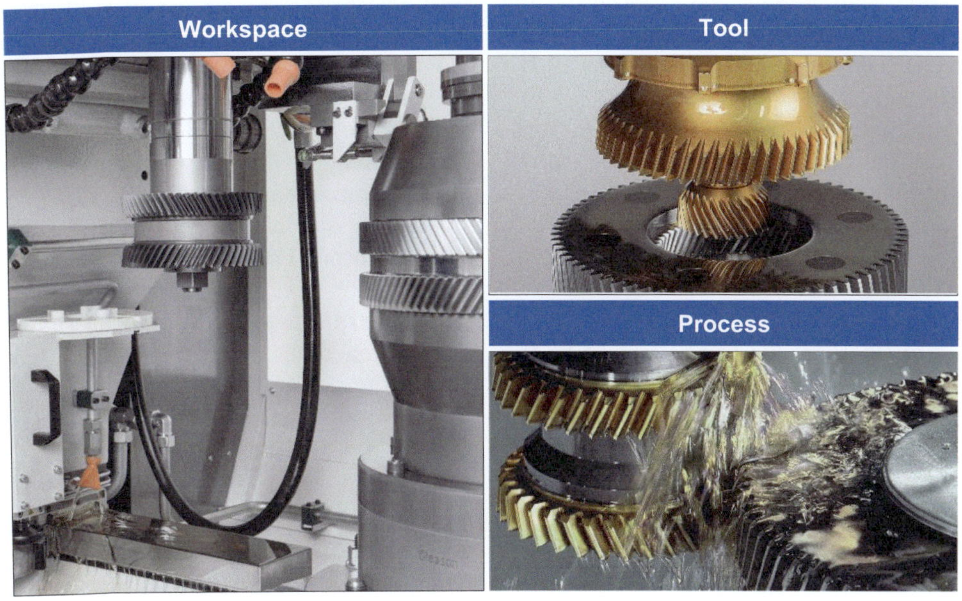

Fig. 8.11 Workspace and tools in a gear shaping machine. Example from Gleason

Figure 8.11 shows the workspace of a gear shaping machine with a tool for manufacturing double helical gears. The nozzles above the tool are used to supply cutting fluid. In the foreground on the right, the workpiece is chucked into the workpiece clamp. Workpieces with small runout are a main application for shaping [KLOC17]

Different designs can be used for gear shaping tools depending on the application. A distinction is made between disk, bell, shaft and cavity-type shaping tools. Tool designs are conical in all cases and mainly hard-material-coated high-speed steels are used. Grinding the rake face of a worn tool reconditions it. Separation errors from the grinding process influence the gear quality in all shaping tool designs directly. As an alternative to conventional tools, single-use wafer tools can be used. They are thin, profiled gears that are clamped in a serrated seating and are replaced after their service lives. [KLOC17]

8.1.1.3 Gear Skiving Machines

Gear skiving is a gear cutting process derived from gear shaping (▶ Sect. 8.1.1.2). The rotational axes of the workpiece and the tool are inclined to each other by the axis crossing angle. If the axis crossing angle is increased starting from 0°, which is the case for gear shaping, the rolling movement becomes a screw rolling movement with increasing screw component and decreasing rolling component. The screw component results in a cutting speed, which leads to chip removal if the axis crossing angle is sufficiently large, without needing any additional lift movement of the tool. Gear skiving is used to manufacture internal and external spur and helical cylindrical gears [BAUS11]. Gear skiving can be used for both, soft and hard machining in the process chain of gear manufacturing [ROTH98].

The principle for creating gears with the gear skiving process is shown on the left in ◘ Fig. 8.12. The tool has an involute profile. The rotation of tool and workpiece causes a continuous rolling movement between tool and workpiece. As a result, no additional partial movement is necessary to generate the profile.

In gear skiving, the workpiece and tool mesh comparable to a crossed helical gear pair [ROTH98]. The tool can have a cylindrical or conical outer contour and be straight or helical. During the manufacturing of helical gears, the axis crossing angle Σ results from the inclination angles of the tool and workpiece, as shown in the detail on the right in ◘ Fig. 8.12. The cutting speed v_c results from the different rotational speeds of tool and workpiece. The axial feed f_a is parallel to the workpiece axis [ROTH98].

Using the tool rotational speed v_0 and the workpiece inclination angle β_2, the cutting speed v_c is derived according to Formula (8.4).

$$v_c \approx \frac{v_0 \cdot \sin \Sigma}{\cos \beta_2} \tag{8.4}$$

The axis crossing angle Σ results from the sum of the tool inclination angle β_0 and the workpiece inclination angle β_2. During the machining process, there is an offset of the tool parallel to the workpiece axis at feed velocity v_a in addition to the rotational movements of the tool and workpiece. At the same time, the workpiece performs an additional rotation relative to the tool. The superposition of offset and additional workpiece rotation results in a screw movement [KLOC17].

The machine axes of a gear skiving machine, as depicted in ◘ Fig. 8.13 using a Gleason P600 PS as an example, provide the kinematic movements necessary. The Z_2 axis performs the axial feed. The Z_1 axis is used to securely adjust the height of the tool. Different feed strategies are realized using a combination of X and Y-axes movements. The axis crossing angle is realized by rotating the A axis. If a cylindrical tool is used instead of a conical tool, the tip clearance angle can be set by rotating the tool around the B-axis. Also, during gear skiving of internal gears, the B-axis is used to prevent the back of the tool interfering with teeth opposite the cutting point. This is particularly important when the gear skiving tool has a large diameter [SCHM03; KLOC17].

Tool and workpiece each have their own rotational axis, axes C_1 and C_2, that are coupled using direct drives. The process kinematics described require high tool and workpiece speeds to achieve sufficiently high cutting speeds. Accordingly, strict requirements are placed on the precision of clamping devices for the tool and workpiece to prevent imbalance. Shown as examples on the right in ◘ Fig. 8.14 are modular workpiece clamps that provide the clamping described for different workpiece diameters [KLOC17].

The classic design of gear skiving tools are made of solid material. These single-piece tools can have a plane rake face or be sharpened on the rake face. Tools with step cut are shown in the bottom right of ◘ Fig. 8.14. Special machines are necessary for sharpening helical, step-cut tools [KLOC17].

8.1.1.4 Gear Shaving Machines

Gear shaving is used as a soft finishing process in mid and large-volume production [FELT12; ROTH98] to attain improvements in gear quality prior to heat treatment [BAUS11]. The primary functions are to eliminate

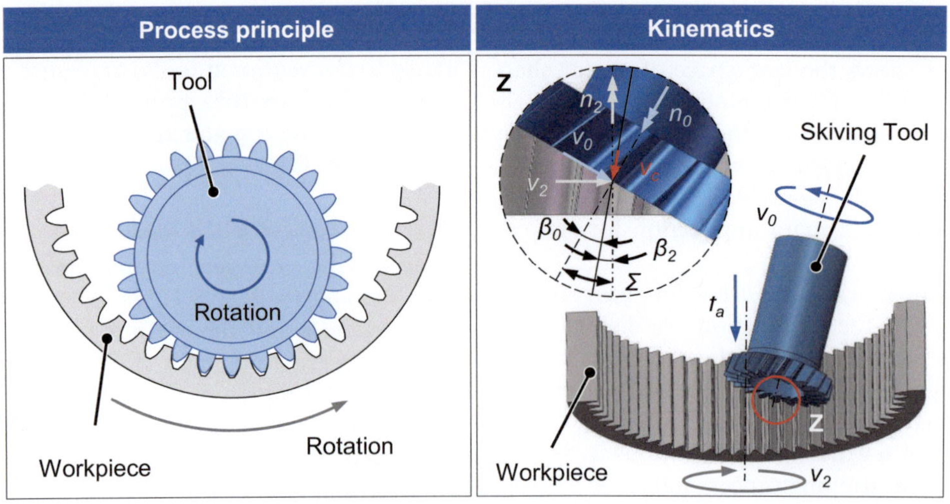

◘ **Fig. 8.12** The principle of the gear skiving process. *Source* [KLOC17]

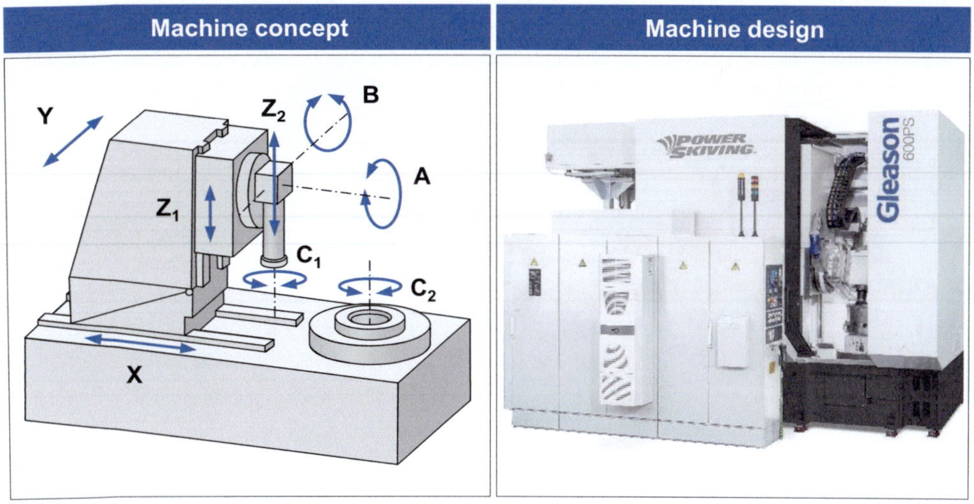

Fig. 8.13 Structure of a gear skiving machine. Example from Gleason

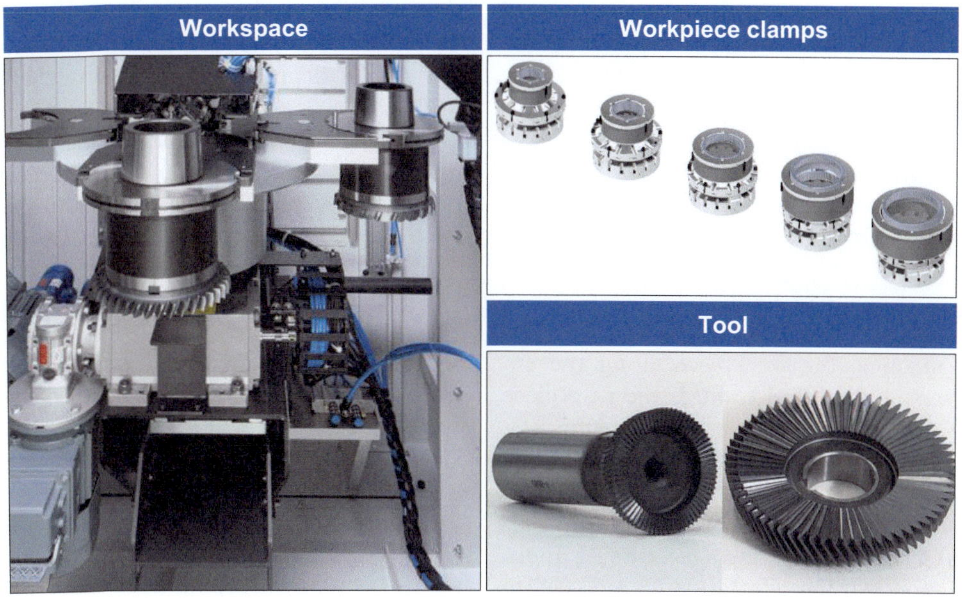

Fig. 8.14 Workspace and workpiece clamp for a gear skiving machine. Example from Gleason

deviations from the involute and also to incorporate any other modifications into the tooth profile. The additional process step can be cost-effective because quality increases of two to three tolerance classes are possible [BAUS11; ROTH98]. Gear shaving exhibits similar kinematics as gear skiving [BAUS11] and can be used for straight and helical as well as for external and internal gears [ROTH98].

The principle of the gear shaving process is shown in Fig. 8.15 using a workpiece with external teeth as an example. The tool also has an involute tooth profile and makes contact with both flanks of the workpiece when rolling [BAUS11]. Accordingly, both gear flanks are manufactured simultaneously. Theoretically, there is only one point of contact between the flanks of the workpiece and the flanks of the tool [FELT12].

Due to the kinematics of gear shaving the workpiece profile is generated from a mixture of cutting, skiving and forming components [BAUS11]. The high cutting performance is achieved by several individual cutting edges positioned behind each other on the tool [BAUS11]. The cutting movement results from a translatory main movement, consisting of the rotations of the workpiece and tool. Because of the axis crossing angle Σ the workpiece and tool mesh comparable to a hyperbolic helical type gear transmission. This causes a longitudinal sliding movement between the tool and workpiece flanks. This longitudinal sliding movement

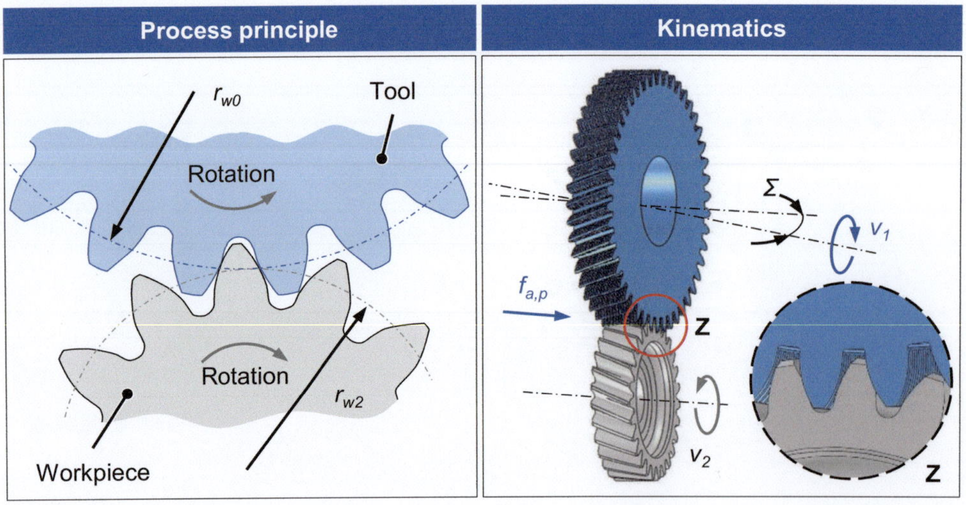

Fig. 8.15 The principle of the gear shaving process. *Source* [KLOC17]

causes machining marks in line with the magnitude of the slide. These usually regular marks are typical for gear shaving. For parallel gear shaving, the feed $f_{a,p}$ is parallel to the workpiece axis. In addition to parallel gear shaving, there are more process variants such as diagonal, traverse and plunge gear shaving [BAUS11].

The contact pressure between the gear and tool flanks necessary for chip removal is applied by the machine using the radial convergence of the two wheels. There is in theory a point of contact for screw drives that expands into a contact zone due to the contact forces. To machine the workpiece along the entire width, a corresponding feed towards the workpiece axis is required. The structure of a gear shaving machine is shown in Fig. 8.16 using a Gleason 130SV as an example. With the exception of the workpiece rotation, all movements are in relation to the tool axis [BAUS11]. Axis A is used to set the axis crossing angle. Tool rotation C_1 and workpiece rotation C_2 are used to set the relative speed between tool and workpiece on the pitch circle. The feed direction is set with the X, Y and Z-axes depending on the process variant [o.V.15; KLOC17].

Figure 8.17 shows the workspace of a gear shaving machine. In addition to the axes described that perform the movements necessary for the process, the machine can have other traverse axes for automatic workpiece loading. Using a disk cam-driven changer with two gripper arms, the workpiece can be replaced by a blank from the magazine at the end of the process. Coated, powder metal-based high-speed steel has proven to be the most effective tool material for gear shaving [SCHR06; KLOC17].

8.1.1.5 Gear Grinding Machines

A general distinction is made in gear grinding between discontinuous and continuous processes [KLOC05]. In discontinuous gear grinding, also known as discontinu-

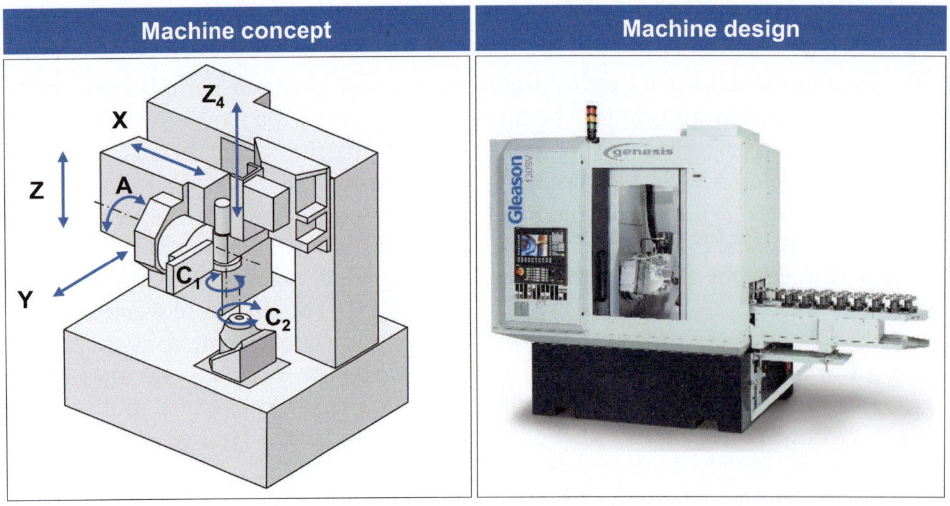

Fig. 8.16 Structure of a gear shaving machine. Example from Gleason

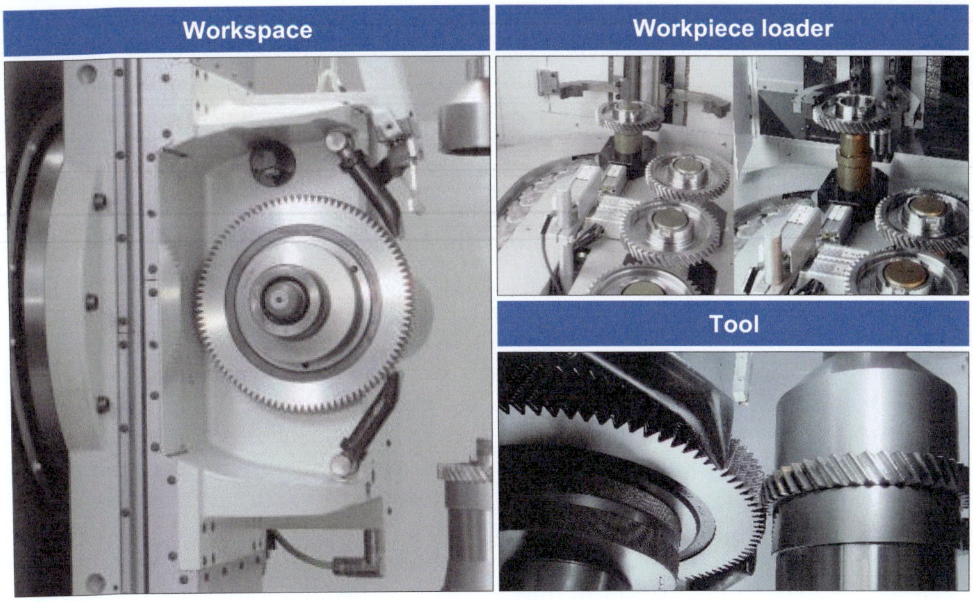

Fig. 8.17 Workspace and tools in a gear shaving machine. Example from Gleason

ous generating gear grinding, the tooth gaps are ground in sequence and indexed after one tooth gap is ground grinding wheels with a geometry according to the basic rack profile are used as tools. Both flanks of the gear can be machined simultaneously as well as consecutively [BAUS11; ROTH98].

Due to the discontinuous indexing typical for discontinuous processes, these processes are limited in terms of productivity and only used in special applications in the industry. The focus below is thus on the more productive continuous gear grinding. In series production of gears with a normal module of $m_n = 0.5$ mm to $m_n = 10$ mm, generating gear grinding is one of the dominant precision machining processes. The process is used to manufacture external spur and helical gears with a cylindrical or a conical shape. In generating gear grinding, cylindrical grinding worms having the basic rack profile of the gears in the normal cut of the grinding worm are used [KLOC05]. During generating gear grinding, multiple tooth gaps are machined at the same time. The pitch is realized by the rotation of the workpiece, which results from the ratio between the number of the teeth of the workpiece and the number of starts of the grinding worm. This synchronization by the electronic gearboxes is also called roller coupling. Similar to gear hobbing, this coupling generates the involute. In generating gear grinding, the translatory movement component of the basic rack profile is realized by the rotation of the grinding worm and the spiral of the starts of the grinding worm, as shown on the left in Fig. 8.18. The rotation of the gear creates the rotational movement com-

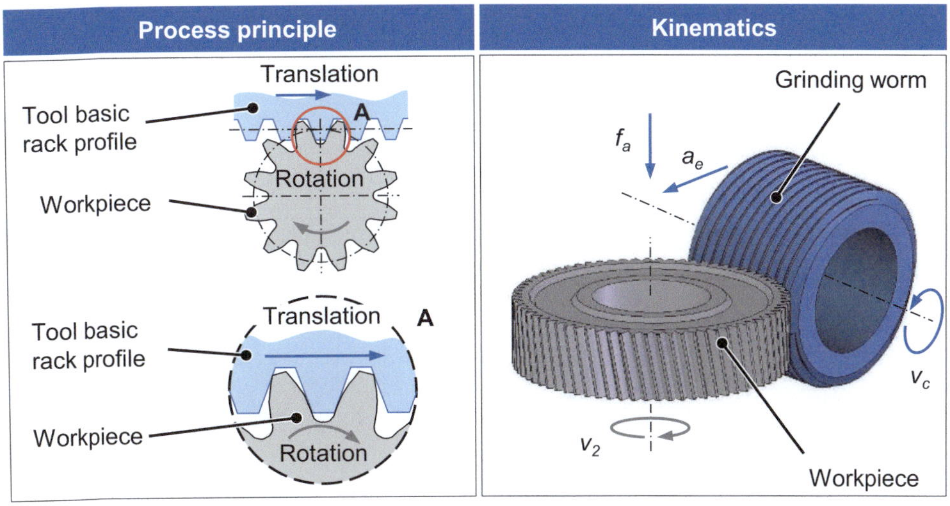

Fig. 8.18 The principle of the continuous gear grinding process. *Source* [KLOC17]

ponent. In addition to the roller coupling, the grinding worm is also pivoted to compensate for the incline of the starts. This necessitates an increased runout length of the tool. Hence, the application of the process is limited in regard to interfering contours, e.g., for shaft machining [BAUS11]. Deviations inherent to the process, such as twists occurring during manufacturing of flank modifications for helical gears, must also be taken into account. Continuous gear grinding enables an array of profile and lead modifications such as tip and root relief, profile and lead crowning, profile angle or lead line modifications and end and corner reliefs. Also, twists precisely applied as modifications are possible to compensate the natural twist mentioned before [BAUS11; KLOC17].

Dressable and non-dressable tools are used depending on the process. Non-dressable tools include galvanically bonded CBN worms or disks. These tools already have the basic rack profile required and can only be used for a special gear process. Dressable tools made of ceramically bonded corundum, for example, can be dressed to the basic rack profile required using the gear grinding machine. Usually, disk-shaped dressing tools are used for this [BAUS11; KLOC17].

As described above, the grinding worm and the gear rotate with a defined ratio for continuous gear grinding. Other movements besides rotation, shown on the right in Fig. 8.18, are part of process kinematics. To machine a tooth flank along the tooth width, the grinding worm is guided along the gear with axial feed f_a. For material to be cut at the gear, the grinding worm is fed into the gear with the radial infeed a_e. The stock allowance can be calculated using the pressure angle and radial infeed. By means of defined interaction between axial feed, radial infeed and rotation of the gear, the tooth flank modifications mentioned above can be generated in lead direction (see Fig. 8.5). A shift movement is performed to move unused areas of the grinding worm into contact and thus to attain uniform wear along the grinding worm. The shift movement takes place along the rotation axis of the grinding worm, the same as for shifting in the hobbing process. Different shift strategies are possible. Shifting can take place after a gear is machined, which is known as discontinuous shifting. Alternatively, continuous shifting can be used. For continuous shifting the grinding worm is shifted during the machining process of a gear. Continuous shifting is commonly used for roughing. After roughing, the grinding worm is shifted to an unused position by means of discontinuous shifting and the gear is finished using the axial cut without a shift movement. The cutting speed vector, which is tangential along the tooth flank, is a result of the tool rotation. [SCHR08; KLOC17]

The machine data, necessary for realizing the process, are derived from the process kinematics. Figure 8.19 shows the schematic structure of a gear grinding machine using a KX 100 Dynamic from KAPP NILES as an example (shown with axis names) [OV15b]. The B_1-axis rotates the tool and the axes C_1 and C_2 rotate the workpiece. The tool spindle can be traversed along axis Z_1 on the vertical axis and axis Y along the tool rotation axis. Axis Z_1 represents axial feed f_a of the slide and axis Y is the shift movement of shift carriage v_{Shift}. Radial infeed a_e is set by axis X_1. The whole grinding tool can be rotated by axis A_1 meaning the swivel angle of the grinding worm can be set. The swivel angle corresponds to the sum of the helix angle of the gear and the thread angle of the grinding worm. The entire tailstock column can be rotated about the CTT axis such that the dressing unit can be moved to the dressing position, or the workpiece rotation axes C_1 and C_2 can be moved into the machining position. A two-spindle system is used to perform load-

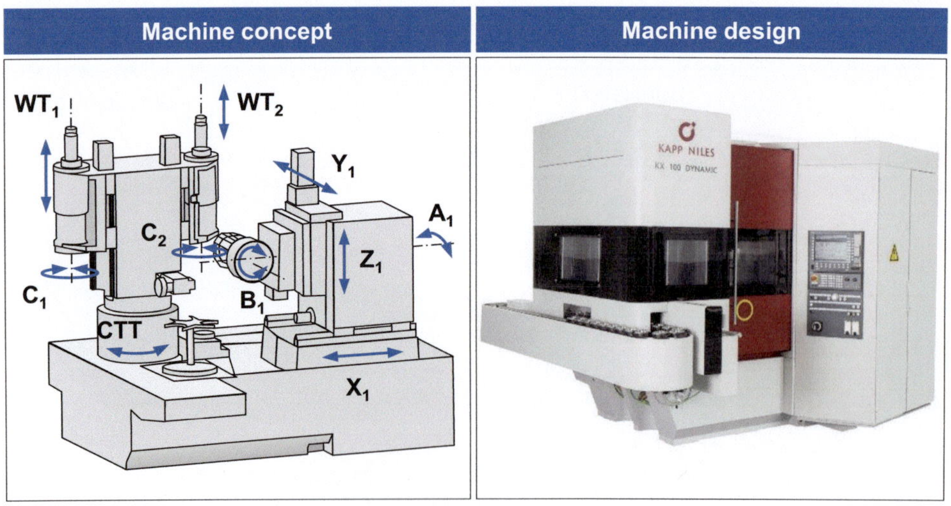

Fig. 8.19 Structure of a gear grinding machine. Example from KAPP NILES

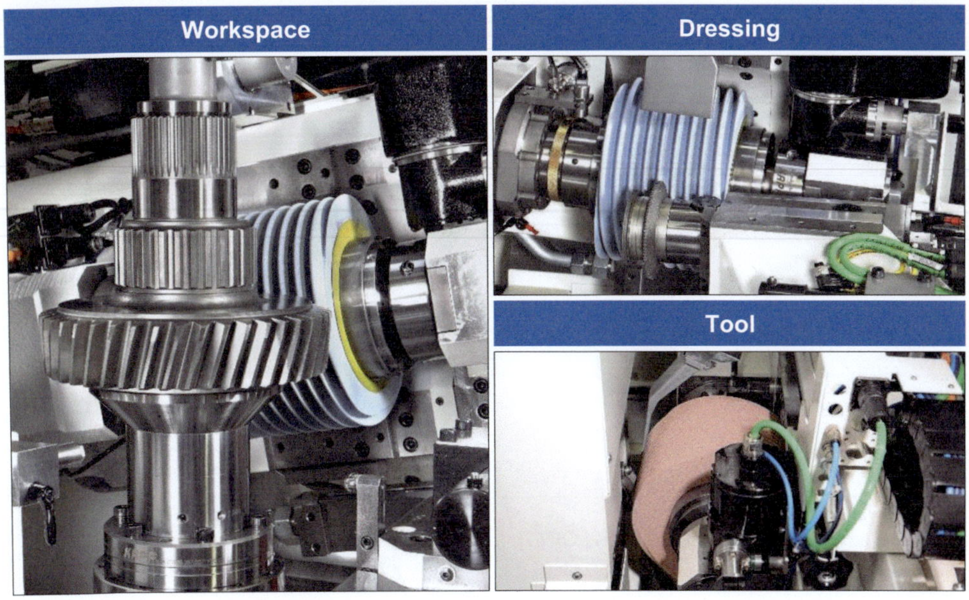

■ Fig. 8.20 Workspace and tool in a gear grinding machine. Example from KAPP NILES

ing and unloading as well as alignment processes in production during manufacturing. [BAUS11; SCHR08; KLOC17]

The gear has, due to the grinding machine C-axis described, a rotational degree of freedom about the workpiece axis. The workpiece can also be clamped by a tailstock, which can be moved by the WST_2 or WST_1-axis [SCHR08]. To manufacture a beveloid, the tool must be moved along a conical base body. This is generally achieved by a coupling of the axial feed along the Z_1-axis and radial infeed in X_1-direction. If the requirement is to apply lead crowning to the gear, the radial infeed is also adapted along the gear width. Hence, the grinding worm no longer moves in vertical direction along a straight line but along a parabolic trajectory in X_1-direction. To realize end and corner reliefs, the movement of the X_1-axis is also adapted. A lead line modification is realized by adapting the coupling between the axes C_2, Z_1 and Y_1. All profile modifications (crowning, relief and profile angle modifications) are achieved by modifying the tool profile during the dressing process. Topological gear grinding represents a special application. For topological grinding, the tool can be dressed such that the tool profile continually changes along one thread of the grinding worm. If such a grinding worm with modified kinematics is used, it is possible to compensate for the natural twist that results from the manufacturing of center-distance based flank modification of helical gears during generating gear grinding, or to apply specific twists. It is also possible to implement free tooth flank modifications by manipulating the axis movements in combination with a varying tool profile. [BAUS11; SCHR08; KLOC17]

■ Figure 8.20 shows the workspace of a gear grinding machine with the workpiece and grinding tool in contact. When used over extended periods, grinding tools exhibit macrogeometric wear, such as out-of-roundness, waviness and profile loss, that can result in disruptions in the grinding process and/or the loss of accuracy of workpieces. The combination of profiling to create a defined tool profile and sharpening to create a defined tool microtopography is called dressing. Dressing is possible directly on the machine so that the tool does not need to be removed.

8.1.1.6 Gear Honing Machines

Gear honing is used for hard finishing internal and external spur and helical gears. It can be used as a standalone process or to complement the aforementioned hard finishing process [FELT12]. Given the low tool runout, honing is very well suited for hard finishing of gears with interfering contours [ROTH98].

The principle of the gear honing process is depicted in ■ Fig. 8.21 using an external workpiece as an example. The tool has, for this workpiece, the shape of an internal gear and is called a honing ring [BAUS11]. Hence, large overlapping between tool and workpiece is possible. This has a positive effect on the resulting gear quality for the gear honing process. Tool and workpiece mesh by both rotating in the same direction. It is possible to machine both flanks at the same time or one flank after another. By meshing the gear with the involute internal honing ring, the involute is created in profile direction of the gear.

The tool and workpiece form a hyperbolic helical rolling type gear transmission [BAUS11]. The

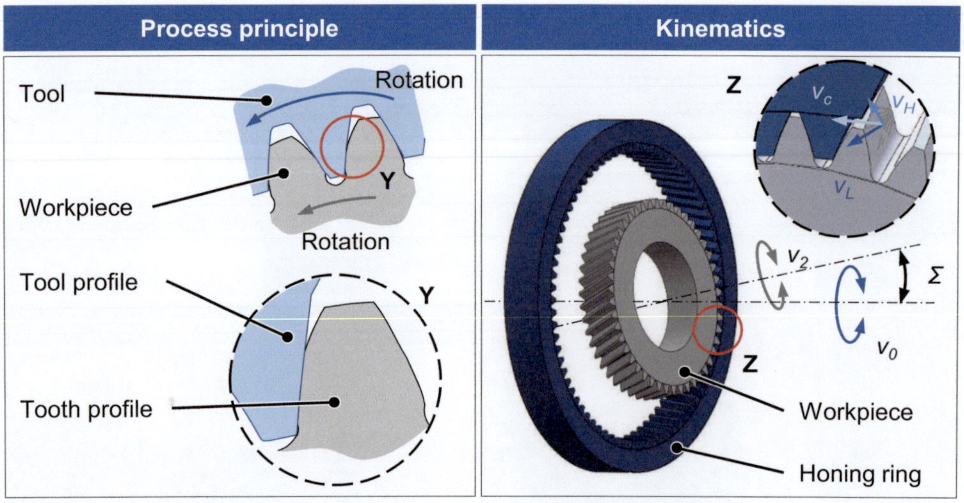

Fig. 8.21 Principle of the gear honing process. *Source* [KLOC17]

movement sequence corresponds to the kinematics of gear shaving in the plunge shave method (see ▶ Sect. 8.1.1.4) and gear skiving (see "Triangle", 8.1.1.3). The axis crossing angle Σ is given by the sum of the helix angles of the workpiece and the tool. The constant rolling is superimposed by the sliding of the partial bodies [BAUS11] comprising vertical slide speed v_H and longitudinal slide speed v_L. The machining marks characteristic of gear honing are created due to the changing cutting speed distribution along the workpiece flank [FELT12].

Figure 8.23 shows the design of a gear honing machine using a Fässler HMX 400 from Daetwyler as an example. A direct drive creates the tool rotation via the C_1-axis. The workpiece rotation is realized by the two axes C_2 and C_3, also using direct drives. The dual workpiece rotation axes mean the workpiece loading process can be performed in short non-productive times with rotation of the B_2-axis and a longitudinal movement of the headstock in the Z_2-direction [BAUS11; OV15c]. Radial infeed of the honing ring is realized by the X-axis [OV15c]. The longitudinal feed of the Z_1-axis performs an oscillating motion, meaning the tool is used uniformly over its width [BAUS11; OV15c; KLOC17] (Fig. 8.22).

Rotation of the honing ring to generate lead crowning is provided by rotation about the B_1-axis. Another way to create lead crowning is to use the dressing tool to apply it directly onto the honing ring. As a result, an additional rotation of the tool axis is not necessary while the workpiece is being machined. [OV15c; KLOC17]

On the right, Fig. 8.23 shows the unconventional variant of the dressing process of a honing ring with a ceramically bonded CBN profiled grinding wheel. The ceramically bonded corundum honing ring is usually

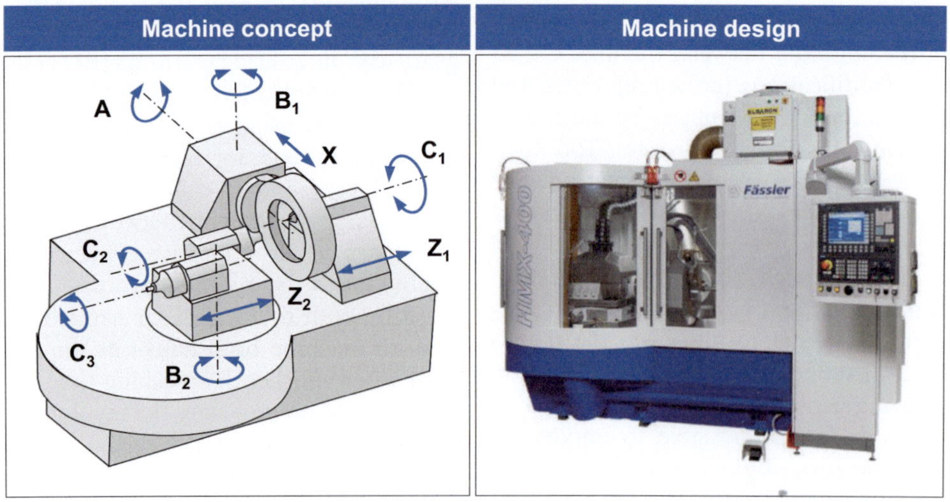

Fig. 8.22 Structure of a gear honing machine. Example from Daetwyler

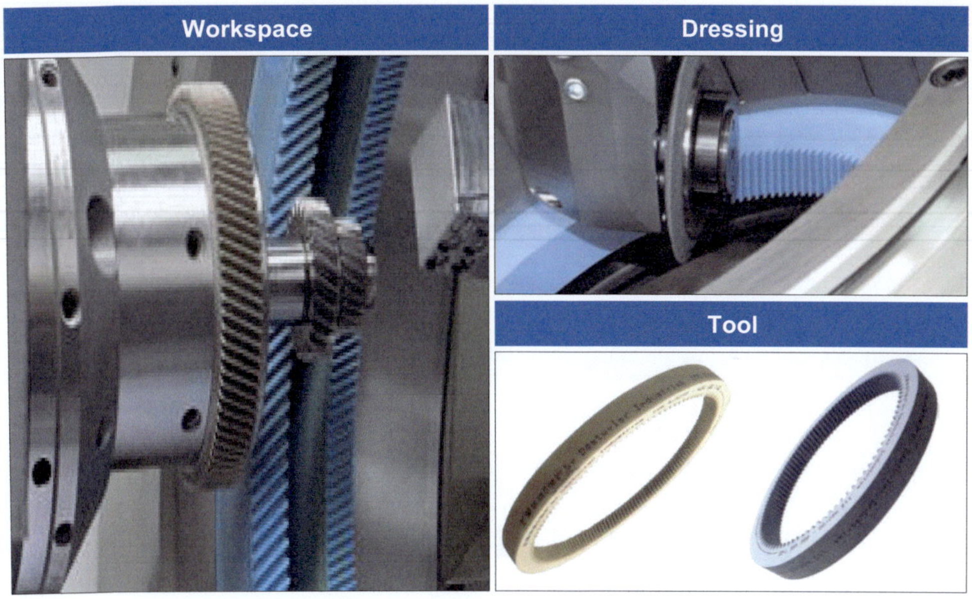

Fig. 8.23 Workspace and tools in a gear honing machine. Example from Daetwyler

profiled and sharpened by a diamond-studded dressing gear. In addition to the ceramic bond most common for industrial applications, synthetic resin-bonded corundum honing rings can be used. As with gear grinding, dressing of the tool and hard finishing of the workpiece can take place on the same machine. [KLOC17]

8.1.2 Profile Gear Cutting Machines

For manufacturing using profiling processes, the tool (profile cutter, grinding wheel, forming tool) has the contour of the tooth gap to be manufactured. For most of the forming processes, every tooth gap is cut individually and then the workpiece is indexed by one pitch for machining of the next tooth gap (discontinuous indexing). The tool profile must correspond exactly to the tooth contour. This method is thus used mainly in single-part manufacturing of gears with a large module and series production of gears with small modules.

The benefits of Profile machining of gears are in the contouring and thus the exact representation of involutes and their modifications in profile direction. High cutting performance can also be attained. A downside is that a different tool needs to be used for every tooth profile. Due to the indexing movement necessary, strict requirements are made on accuracy and synchronization of the machine axes for machine tools used for Profile gear cutting machines.

8.1.2.1 Profile Milling Machines

Profile milling is a versatile process that is used not only to manufacture involute internal and external gears but also gear profiles that cannot be manufactured with rolling processes, such as special gears [BAUS11; ROTH00]. It is also used for worm gears with a small number of teeth and small pitch [BAUS11]. In all cases, the profile of the gear to be manufactured is applied to the milling tool and the milling tool is traversed axially with respect to the workpiece. The workpiece does not rotate during the manufacturing of spur gears. For helical gears, the workpiece rotates during the axial feed such that the tool moves on a helical line relative to the workpiece in accordance with the helix angle of the gear.

The principle of the profile milling process is shown in Fig. 8.24 using a circular milling cutter with indexable inserts as an example. The tool can also be an end mill [BAUS11; ROTH98]. In the axial section, the tool has an involute profile that is mirrored in the tooth gap [BAUS11]. In profile milling, every tooth gap is manufactured individually. Then the workpiece is rotated by one pitch and the next tooth gap is manufactured until all of the teeth have been cut.

As no rolling is necessary to generate the involute profile, the process kinematics of profile milling are determined by cutting speed v_c, radial feed f_r and axial feed f_a. The cutting speed v_c is generated by the rotation of the circular milling cutter about its own axis and thus causes the formation of chips. The workpiece is machined along the gear width by axial feed f_a. The radial feed f_r defines the infeed of the tool into the workpiece. [KLOC17]

Figure 8.25 depicts the structure of a profile milling machine. As a general rule, machines having an axis arrangement similar to that for gear hobbing are used for profile milling. This means that profile milling

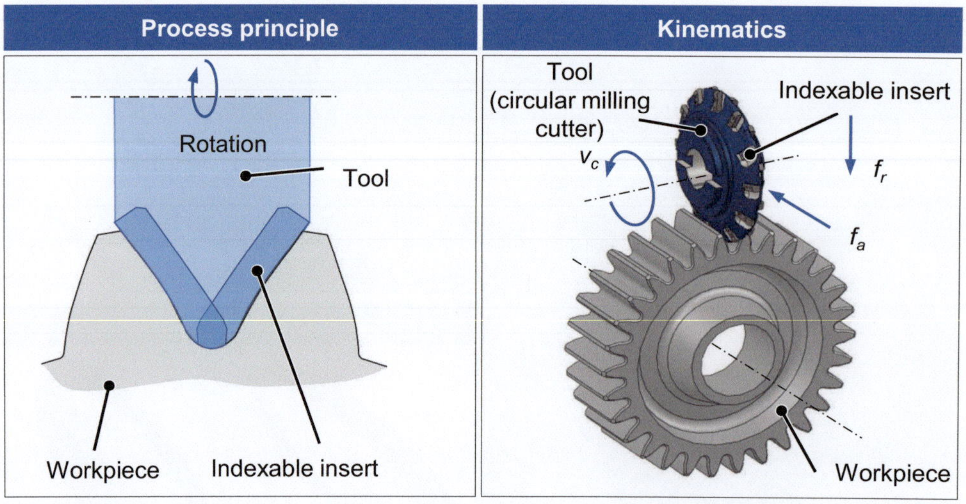

Fig. 8.24 The principle of the profile milling process. *Source* [KLOC17]

and gear hobbing operations for gears can often be realized on the same machine [DIET14]. The kinematics for discontinuous profile milling differs from continuous gear hobbing in that the A-axis used to realize the helix angle is clamped. When a helical gear is milled, the B-axis rotates with the gear clamped on the machine table. The rotation is coupled to the axial feed of the milling tool in the direction of the axes B and Z such that the defined helix angle β is milled. The radial infeed towards the X-axis is usually constant. Profile milling can also be realized on conventional machining centers and milling machines under certain conditions.

An example workspace of a profile milling machine is shown in Fig. 8.26. The tool holders are often designed so that they can hold both gear hobbing and profile milling tools. In addition to the tool shown with indexable inserts made of carbide, end mills can also be used. However, end mills require a different tool drive concept. [KLOC17]

8.1.2.2 Profile Grinding Machines

Different process variants are available for profile grinding [BAUS11; MAAG85]. They can be categorized into continuous processes such as grinding with a globoid worm, and discontinuous processes such as grinding with grinding wheels or sanding tips [BAUS11]. The focus of this chapter, in line with industrial relevance, is on discontinuous profile grinding with a grinding wheel. In terms of its path of motion, profile grinding corresponds to profile milling described in ► Sect. 8.1.2.1. The profile of the grinding wheel corresponds to the counter profile of the tooth gap required. The grinding wheels must be profiled accordingly for every gear geometry.

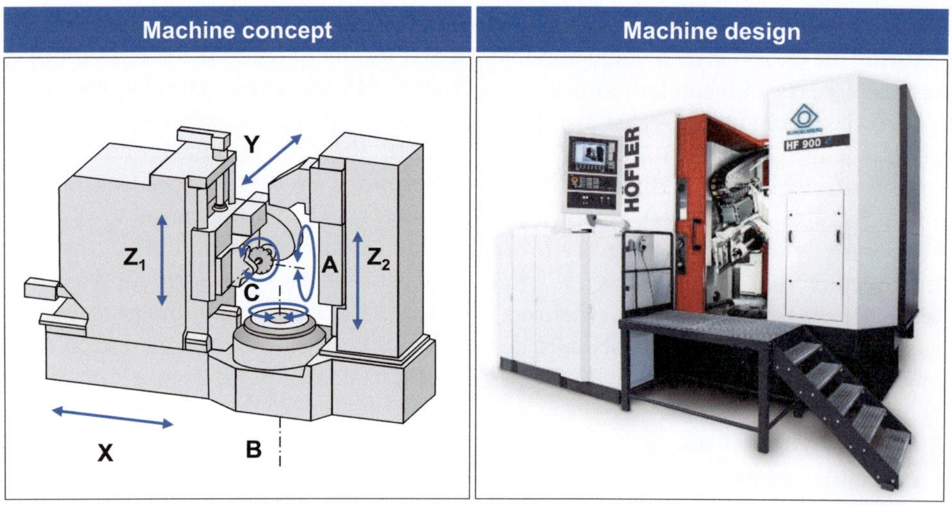

Fig. 8.25 Structure of a profile milling machine. Example from Klingelnberg Höfler

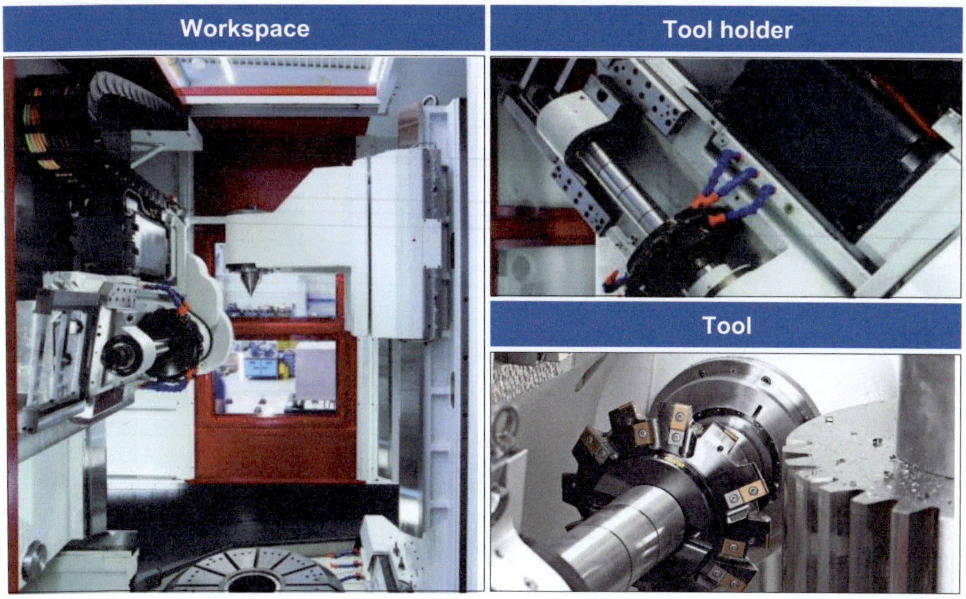

Fig. 8.26 Workspace and tools in a profile milling machine. Example from Klingelnberg Höfler

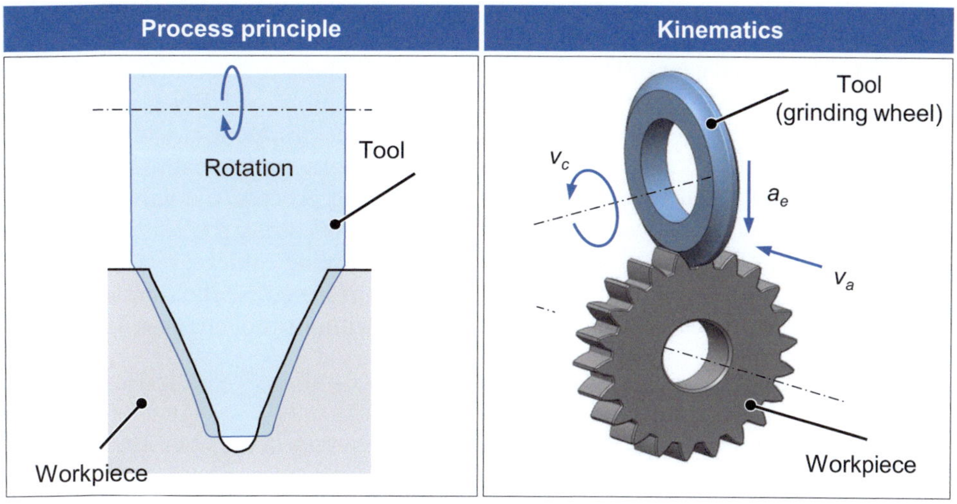

Fig. 8.27 Principle of the profile grinding process. *Source* [KLOC17]

The principle of the profile grinding process is depicted in Fig. 8.27. As with profile milling, the tool exhibits an involute profile in the axial section. Also, the profile modifications required are integrated into the tool profile. The tool profile is mirrored in the tooth gap of the workpiece. Also, in the same way as profile milling, one tooth gap is machined and the workpiece is then indexed by on pitch for machining the following tooth gap. [KLOC17]

The cutting speed v_c necessary for the cutting process is provided by the rotation of the grinding wheel. The axial feed speed v_a enables machining of the tooth gap along the gear width. Radial infeed a_e determines the stock allowance to be cut. Modifications in lead direction can be applied by a change in the axial distance over the gear width. The process control can be synchronized or reversed. Similarly, the stock allowance can be ground in a single stroke or with one or more forward and return strokes.

Figure 8.28 shows the structure of a profile grinding machine using the VX 55 from KAPP NILES as an example. The A-axis rotates the workpiece for the necessary separation movement and as an additional movement for the profile grinding of helical gears. The cutting speed of the tool necessary for the cutting process is generated by the C-axis rotation. Linear axis Y is the radial infeed axis. In combination with the axial feed axis X, corrections to the workpiece in lead direction can be applied. As in generating gear grinding (see ▸ Sect. 8.1.1.5), profile

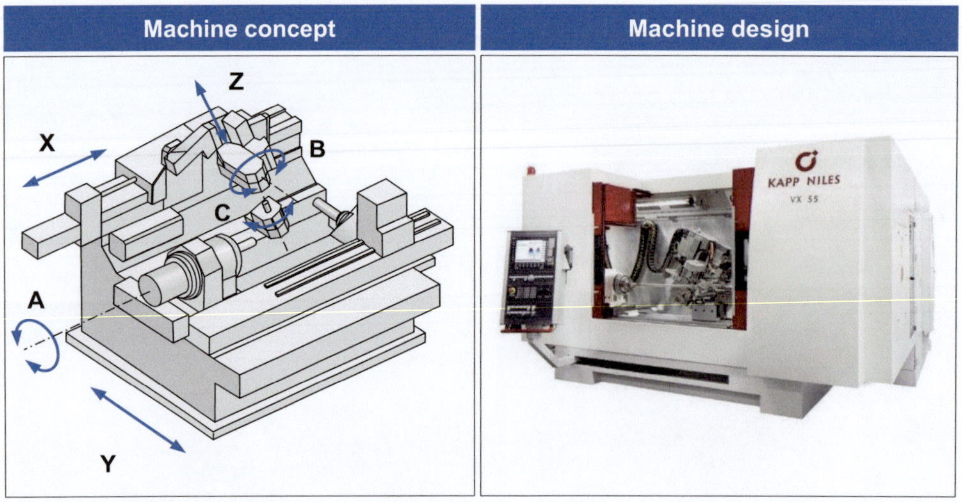

Fig. 8.28 Structure of a profile grinding machine. Example from KAPP NILES

grinding produces a natural twist in the case that the modification of helical gears causes a change in the center distance between the grinding wheel and the workpiece along the gear width. Swivel axis B is used to adjust the grinding wheel to the required helix angle for profile grinding of helical gears. The Z-axis is used to move the grinding wheel, to apply corrections to the tooth flank or grind each flank individually. The movement is also necessary to position the grinding wheel towards the integrated dresser. In addition, a specific topology can be applied to the tooth flank by coupling several axes. For this, additional degrees of freedom in the dressing movement are necessary to modify the tool profile. [KLOC17]

Figure 8.28 shows the workspace of a profile grinding machine. The herringbone gear with a spline is clamped on the C-axis in an overhung position. Tools used can be grinding wheels clamped in an overhung position. Either dressable corundum, sol-gel or CBN grinding wheels are used for prototyping and low-volume production, or CBN grinding wheels are used for high-volume production, with a steel body and a single, non-dressable, galvanically bonded CBN coating. In the dressing process, the galvanically bonded diamond dresser moves along the contour of the grinding wheel. As with honing and gear grinding machines, dressing can be performed on the same machine as manufacturing and without tool changes. [KLOC17] (Fig. 8.29)

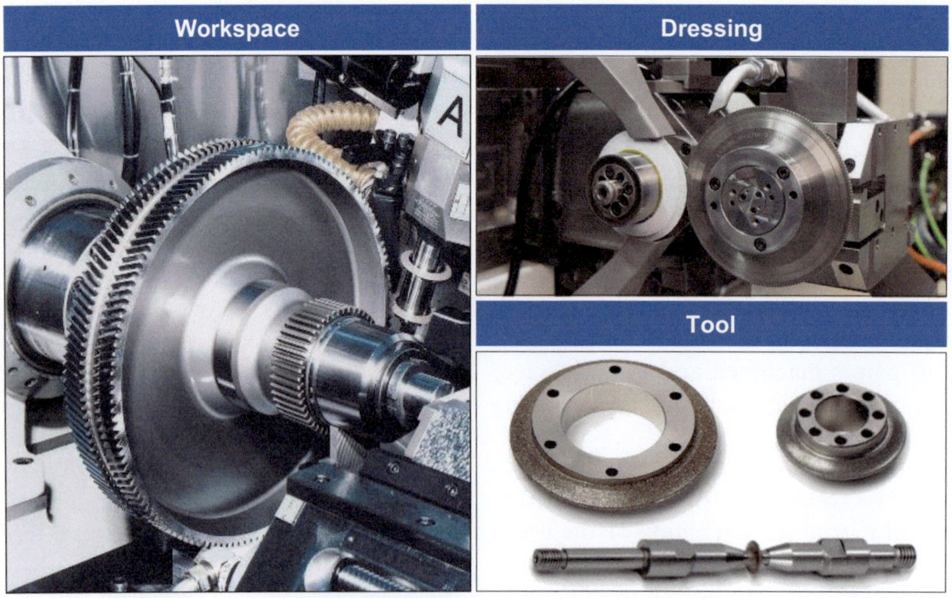

Fig. 8.29 Workspace and tools in a profile grinding machine. Example from KAPP NILES

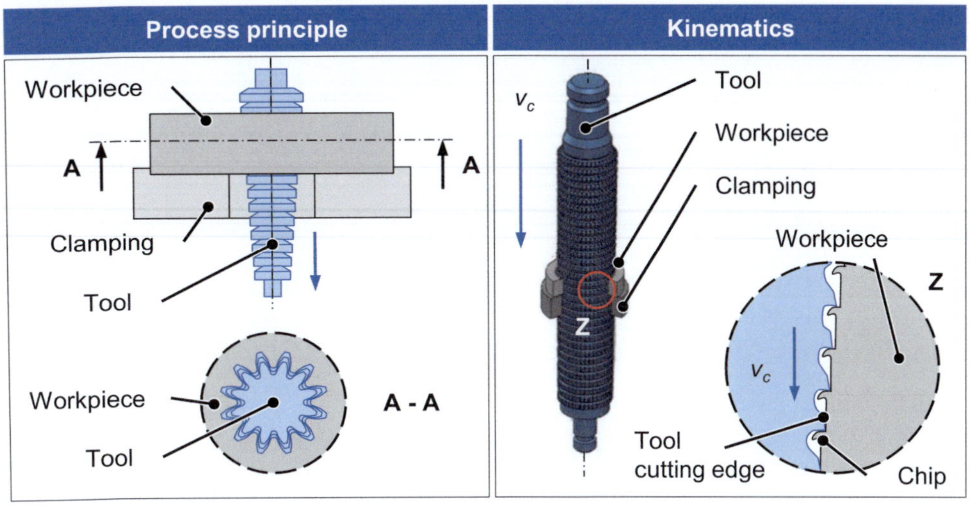

Fig. 8.30 Principle of the gear broaching process. *Source* [KLOC17]

8.1.2.3 Gear Broaching Machines

Gear broaching machines use a shape cutting process without a rolling movement. In this machine, the contour of the tooth is applied to the tool, i.e. the gear negative. Broaching of gears can be used for soft as well as hard machining of external and internal spur and helical cylindrical gears [BAUS11]. Furthermore, face gears can be soft machined by broaching [ROTH98]. The great cutting capacity, with high tool cost, means that gear broaching is used in high volume production. The machine structure and kinematics broadly correspond to those of conventional broaching machines. Thus, it is only the additional movement to generate helical gears and the design of the broaching tools that are gear-specific.

The principle of the gear broaching process is shown in Fig. 8.30 using a cylindrical internal gear with as an example. The broaching needle has a conical design. With a small outer diameter, the broaching needle first enters the pre-drilled workpiece and thus performs a roughing process [FELT12; ROTH98]. When the broaching needle exits the workpiece, the large outer diameter performs the finishing of the gear [BAUS11; FELT12; ROTH98]. In the finishing area, the tool has the negative geometry of the gear geometry required.

In gear broaching, the cutting speed v_c is realized by the tool being pulled through the workpiece [BAUS11]. Gears are usually manufactured in a single stroke [FELT12]. This means every cutter only gets into contact once per workpiece [BAUS11]. The infeed is provided by the multi-cutter tool [BAUS11] and thus requires no additional tool movement. Undercuts generate the necessary rake angle for chip removal. The space between individual tool cutters in radial direction has an influence on the chip thickness [FELT12]. The chips collect in the chip space between tool cutters. The chips remain in the chip space until the tool is removed from the workpiece. Then, the chips fall into the chip container of the gear broaching machine [BAUS11] and can be removed.

Gear broaching machines can be classified into vertical and horizontal broaching machines depending on the feed direction of the broaching needle. Figure 8.31 shows the structure of a broaching machine with horizontal design for broaching spur gears, using a machine from Forst as an example. Helical broaching machines are used for helical gears. The feed axis drive that moves the broaching needle in feed direction is positioned above the actual machine frame. The high process forces arising in the broaching process are absorbed by the double-column design. [BAUS11; FELT12; KLOC17]

8.1.3 Special-Purpose Gear Cutting Machines

In addition to the machine tools introduced in Sects. 8.1.1 and 8.1.2 for manufacturing cylindrical gears, there are more special-purpose gear cutting machines. They are used, for example, to soft machine bevel gears such as on mechanical bevel gear milling machines and 6-axis universal milling machines. Alternative processes and machines are also introduced, as well as soft machining on existing 5-axis milling machines (see ▶ Sect. 4.3).

The particular feature in the design and manufacture of bevel gears is that both processes are interdependent: precise knowledge of the machine settings is required to design and calculate the flank geometry. However, the machine settings cannot be calculated without knowledge on the contact conditions of

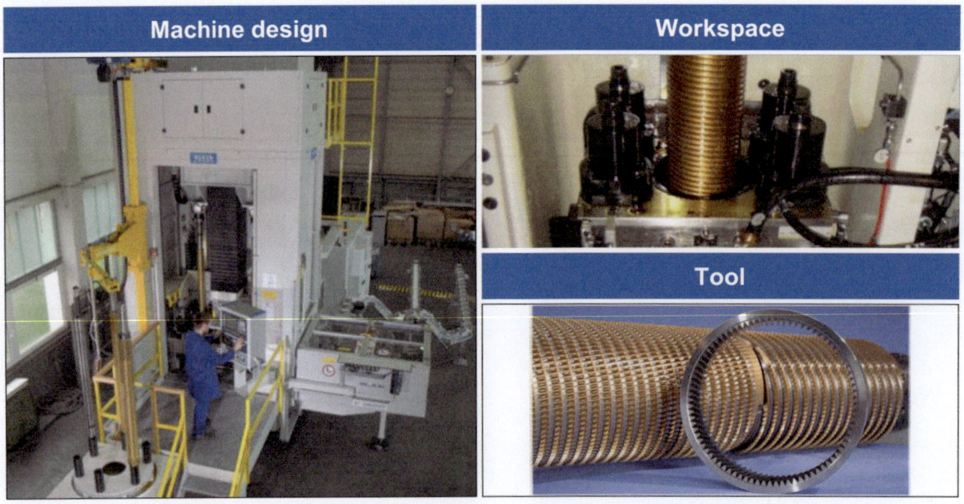

Fig. 8.31 Structure of a broaching machine. Example from Forst Technologie

the bevel gear flanks. Nowadays, this circular logic is addressed with modern calculation programs. Using these, a user can determine the machine settings and flank geometry depending on a basic design. These calculation approaches are based on the knowledge of the machine kinematics of the virtual machine. The machine kinematics and the virtual machine are described below with the help of a mechanical bevel gear milling machine. [KLIN08; KRUM67; MÜLL13; STAD13]

The principle of generation of a bevel gear is depicted in Fig. 8.32. The theoretical generation of bevel gear flanks can be derived in analogy to the principle of a rolling rack with cylindrical gears explained at the beginning of ► Sect. 8.1. A cylindrical gear that rolls on a trapezoidal basic rack profile can roll uniformly with a cylindrical gear created with the same theoretical basic rack profile. Accordingly, bevel gears also exhibit a common basic rack profile that can roll with both ring gear and pinion. The joint profile is the generating gear. In general, the teeth of the generating gear can be straight, curved or tapered. As the principle of the generating gear is a theoretical construct, the number of teeth can be non-integer values. [KLIN08; KRUM67; STAD13]

During manufacturing of bevel gears, the rolling motion of the generating gear with either the gear or the pinion is represented by the motion of the tool. For bevel gear cutting a multi-part tool system consisting of a cutterhead and removable stick blades is used. One tooth of the generating gear is represented by the entire tool or a blade group. Rolling of the generating gear with the workpiece is represented by the movement of the cutterhead in space. [KLIN08; STAD13; KRUM67]

8.1.3.1 Mechanical Bevel Gear Milling Machines

Based on the setup of mechanical machines used for bevel gear cutting, up to twelve translatory and rotational axes describe the positioning of the tool in accordance with the principle of the generating gear. The movements of individual axes are coupled together with roller belts and cam disks, as shown at the bottom of Fig. 8.32 using Gleason No. 116 as an example. Used as reference values for production are the cradle angle α, representing the rotation for the rolling movement between generating gear and workpiece, and tool rotational angle ω, realizing the necessary cutting speed with a cutterhead rotation. For mechanical machines, there are a number of variants with fewer axes. As a result, these machines are limited in realizing the manufacturing processes used, the tooth flank modifications and profiled shapes of bevel gears. An overview of more possible adjustable axes on mechanical bevel gear machines is in Table 8.1. [KLIN08; MÜLL13; STAD13]

8.1.3.2 6-Axis Universal Milling Machines

While the axes of mechanical bevel gear milling machines mentioned in ► Sect. 8.1.3 illustrate the positioning of tool, generating gear and workpiece, they exhibit mathematical redundancies in the positioning of tool and workpiece. Due to advances in the development of drives and control technology of electric motors, the necessary axis movements for manufacturing bevel gears can be represented by six CNC axes. These represent all possible degrees of freedom in space through three independent linear axes and three independent rotational axes. By reducing the number

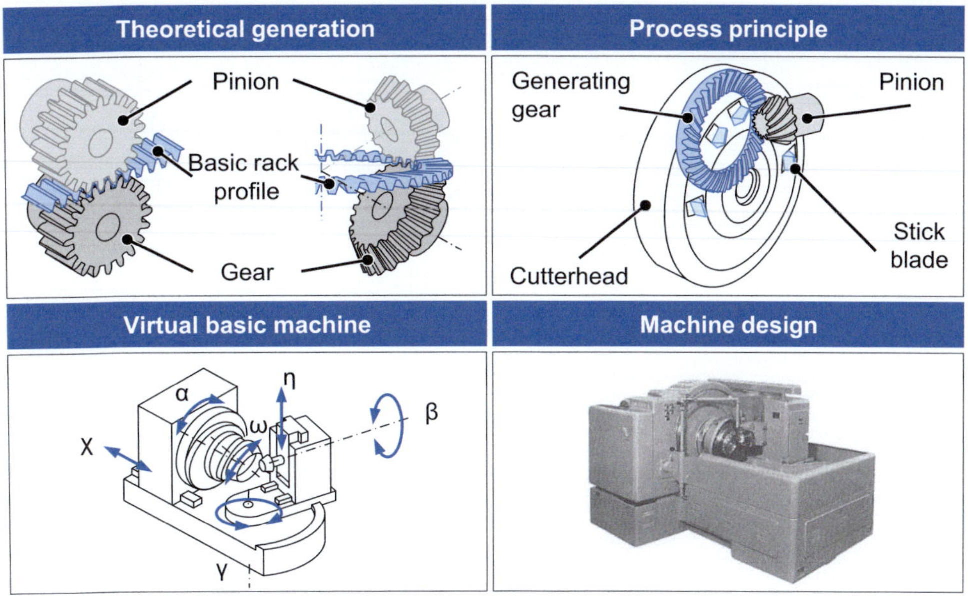

Fig. 8.32 Principle of the bevel gear milling process. Example from Gleason, according to: [KLOC17]

of machine axes, a higher accuracy was achieved with higher stiffness of the machine tools, since CNC machines have fewer joints. Furthermore, the development of 6-axis universal milling machines also reduced the costs and complexity of bevel gear milling machines. [KLIN08; MÜLL13; STAD13]

The manufacturing processes for spiral bevel gears can generally be divided into the categories of continuous and discontinuous indexing as well as plunging and rolling processes. Depending on the indexing method, different tool systems are used. Figure 8.33 shows an overview of the process kinematics and the tool concepts used for manufacturing bevel gears on a 6-axis CNC milling machines. [KLOC17]

The manufacturing processes depicted are the processes most commonly used. For special applications, there are other processes that will not be discussed in detail here. For the discontinuous process, which is also referred to as face milling, the tooth gaps are manufactured consecutively. After a tooth gap is cut, a reverse stroke of the cutter head takes place and the workpiece is rotated by at least one pitch before the next tooth gap is machined. For the continuous process, which is also known as face hobbing, both the cutterhead and the workpiece rotate during manufacturing. Hence, the trajectory of the cutterhead is an epicycloid due to the process kinematics. The pitch is realized by the rotation of the workpiece, which results from the ration between the number of teeth of the workpiece and the number of blade groups of the cutterhead. [KLIN08; STAD13]

Regardless of the manufacturing process, a cutterhead with removable stick blades is the most common tool concept for cutting spiral bevel gears. Currently, there are two cutterhead concepts applied. Cutterheads

Table 8.1 Overview of adjustable machine axes on mechanical bevel gear milling machines according to [KLIN08]

Axis	Name	Movement
β	Workpiece rotation	$\beta = \beta(\alpha,\omega)$
γ	Root angle	$\gamma = \gamma(\alpha)$
ε	Horizontal motion	$\varepsilon = \varepsilon(\alpha)$
η	Machine offset	$\eta = \eta(\alpha)$
σ	Swivel angle	$\sigma = \sigma(\alpha)$
τ	Tilt angle	$\tau = \tau(\alpha)$
φ	Radial motion or tool eccentricity	φ = const.
χ	Sliding base	$\chi = \chi(\alpha)$
mccp	Machine center to crossing point	mccp = const.

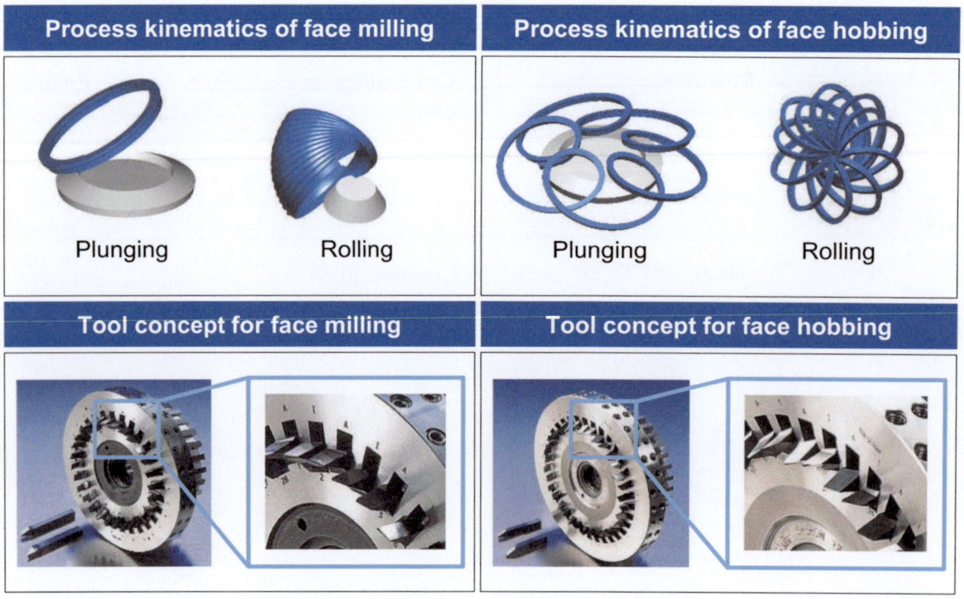

Fig. 8.33 Process kinematics and tool concepts for CNC bevel gear milling machines. Example from [KLOC17]

either comprise of an inner and outer ring welded together or are made from one base body. The cutterhead concept depicted consists of one base body, in which the slots for stick blades are manufactured using electrical discharge machining (EDM). The positioning of the slots and thus the stick blades depend on the indexing process. For the discontinuous process, the slots are arranged in an arc along the nominal cutter radius. In comparison, for the continuous process the slots are tangential to a virtual rolling circle. To influence the radial position of the stick blade, a shim can be inserted into the slot. The stick blades are fixated in the slots on the cutterhead using one or two bolts. [KLIN08; STAD13; KLOC17]

In principle, a distinction can be made between two design concepts of 6-axis universal milling machines, which are compared in Fig. 8.34. In the horizontal design, the workpiece and tool axes are arranged horizontally. This offers the benefit of easy tool changes. The arrangement of CNC axes on one column results in short force flux paths and symmetric positioning of workpiece and tool in relation to the spindle axes. Both have a positive effect on the deformation behavior of the axes from the process forces occurring. Due to the axis arrangement, chips can fall into the chip container of the machine and be removed without needing any other special equipment. [STAD13]

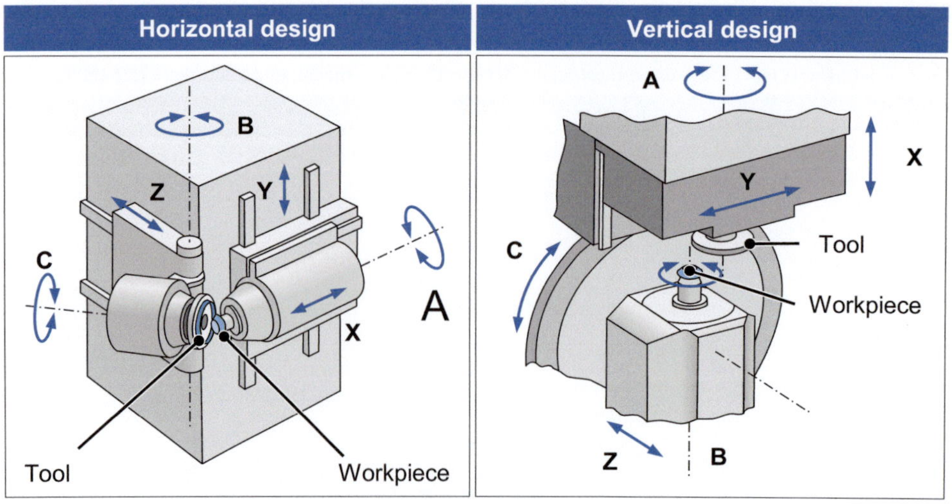

Fig. 8.34 Types of numerically controlled 6-axis universal milling machines

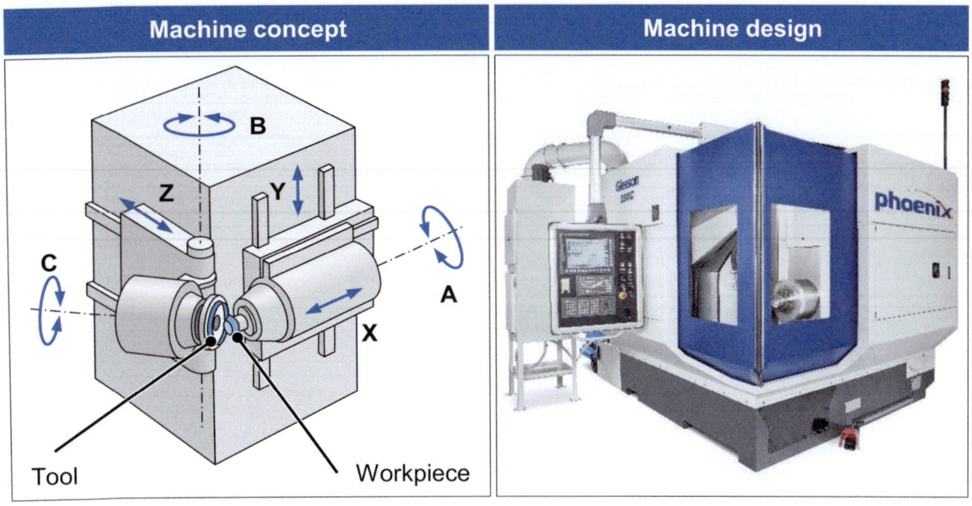

Fig. 8.35 6-axis universal milling machine. Example from Gleason

The benefit of the vertical design of 6-axis universal milling machines is low deformation of the tool and workpiece axes due to gravity. Since the manufacturing kinematics do not allow the use of counter holders for the workpiece and tool, this deformation can have an influence on the gear quality, especially for large and thus heavy components. In the vertical design, the workpiece is easily accessible, which simplifies changes and reduces susceptibility to errors. Because of the large mass of its machine bed universal milling machines with vertical design have a positive damping behavior. [KLIN08; MÜLL13]

Figure 8.35 shows the horizontal design of a 6-axis universal milling machine. Instead of tilting the workpiece for machining, as is the case for the mechanical bevel gear milling machines described in ► Sect. 8.1.3.1, the tool is tilted by the rotation of the B-axis. This tilting movement is realized by a directly driven ball screw drive. The angular position of the B-axis is determined by an incremental angle encoder connected directly to the top pivot point of the axis. Due to the large linear movements required for production and the resulting high linear accelerations, this design avoids losses in positional accuracy. The linear axes X, Y and Z are moved by directly driven ball screws. The temperatures of the individual direct drives are regulated by active cooling to avoid misalignments due to thermal expansion. [STAD13; OV15a]

Figure 8.36 shows the vertical design of a 6-axis universal milling machine. In contrast to grinding, higher torques and lower speeds are used for milling. For this reason, the spindle bearing arrangement of tool spindle A can be realized using rolling elements. The milling spindle axis is positioned as close as possible to the guide systems of vertical axis X to provide

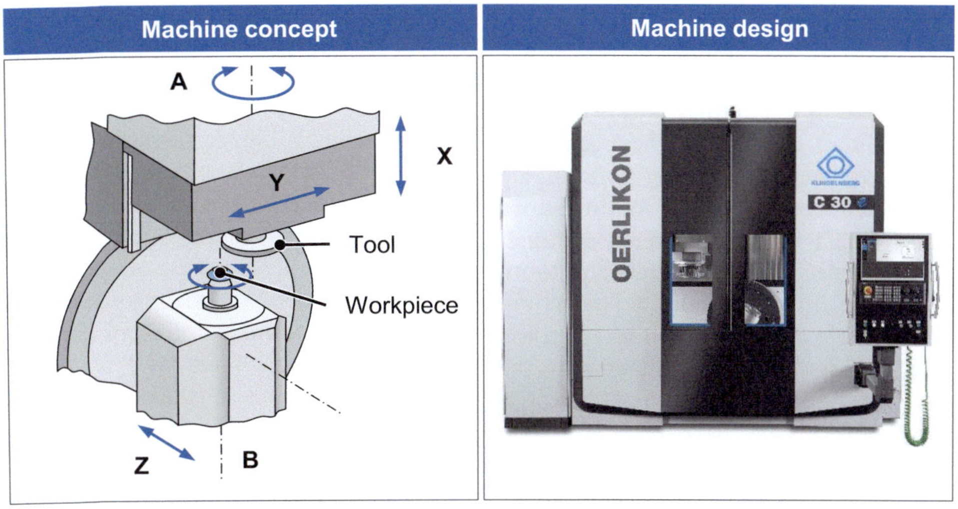

Fig. 8.36 6-axis universal milling machine. Example from Klingelnberg

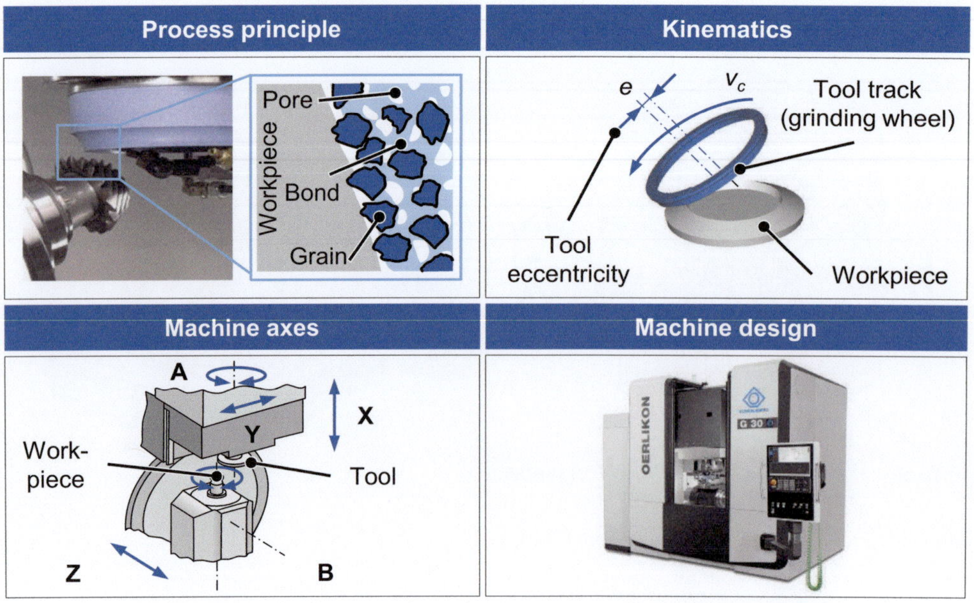

Fig. 8.37 Principle and machine structure for bevel gear grinding. Example from Klingelnberg Oerlikon

8

high stiffness. The milling spindle has a fixed/floating bearing arrangement. The double-row preloaded spindle bearing of the fixed bearing is mounted at the lower end of the spindle. Hence, the forces occurring in the milling process are transmitted into the machine bed over a short distance. The radial bearing is on the upper end of the spindle using a double-row spindle bearing in an X arrangement. This results in a small axial length increase of the spindle when heated. [MÜLL13]

8.1.3.3 Bevel Gear Grinding Machines

As in the manufacturing of cylindrical gears, the grinding of bevel gears is mainly used for hard finishing and can also be used for soft machining of small batch sizes and for prototyping. Figure 8.37 provides an overview of the principle and design of the latest CNC bevel gear grinding machines.

For bevel gear grinding, grinding wheels in the shape of a hollow cylinder are used. The grinding wheels are usually made of corundum grains in a ceramic bond. The profile of a grinding wheel generates the bevel gear tooth profile in the rolling or plunging process and contains the required profile modifications of the tooth flank. Due to the rotationally symmetric, circular grinding wheel the manufactured tooth flank has the shape of a circular arc in lead direction. Hence, grinding is suitable for discontinuous indexing, also referred to as face milling. The diameter of the grinding wheel is adapted to the curvature of the tooth flank in lead direction. [KLIN08; MÜLL13; STAD13; KLOC17]

The cutting speed v_c results from the rotation of the grinding wheel. The control speed $v_{w,0}$ is generated from the machine settings as with CNC milling machines for soft machining of bevel gears. The control speed also includes the workpiece rotation v_{w2} necessary for the rolling generation of the tooth profile. Since the grinding wheel and flank in lead direction of the workpiece have similar curvature radii, there is surface contact between the grinding wheel and workpiece. This introduces a high level of energy into the workpiece and increases the risk of thermal influence on the peripheral zone. For this reason, eccentricity e, in range of a few hundredths of a millimeter to one millimeter, is applied to the control movement. This eccentricity causes the grinding wheel to perform an oscillatory movement. As a result, the surface contact at the flanks is interrupted for a short time. This additional movement is called the Waguri motion. [KLIN08; MÜLL13; STAD13]

In order to realize an optimal grinding pressure between workpiece and tool while preventing the tool from being displaced, high stiffness of the machine axes is necessary for grinding bevel gears. The speeds used are much higher for grinding than for milling. Hence, the dynamic behavior of tool spindle A has a major influence on the surface quality to be achieved. Due to the high energy input into the material described above, a suitable volume flow of cooling lubricant must be directed into the contact between the grinding wheel and workpiece. For this reason, bevel gear grinding machines have a system of adjustable nozzles and fixtures for the supply of cooling lubricant and for extinguishing systems. The previously described eccentric movement of the grinding wheel is realized by an eccentric spindle within the tool spindle. Since kinematics similar to soft machining are necessary for gen-

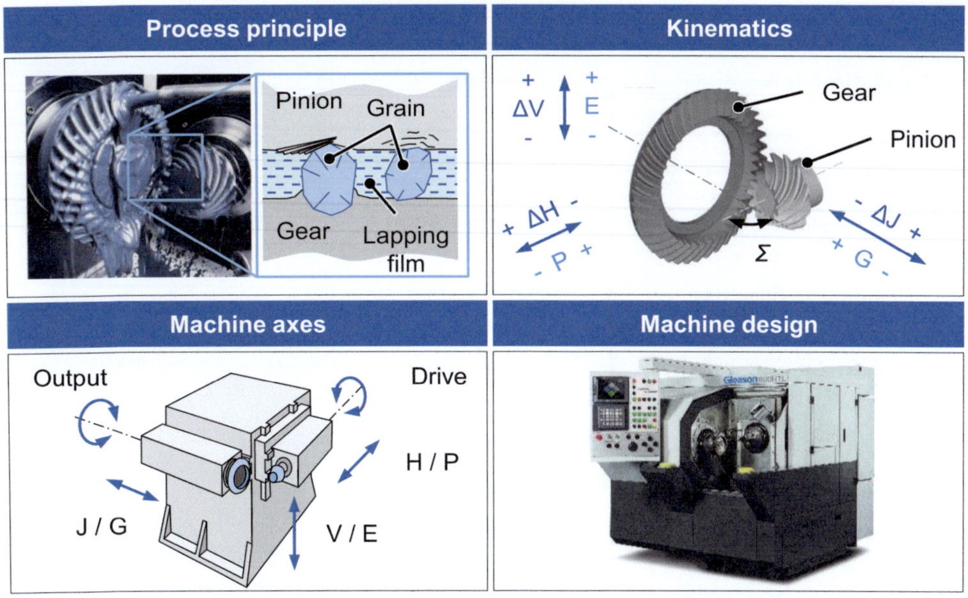

Fig. 8.38 Principle and machine structure for bevel gear lapping. Example from Gleason

eration of the flank geometry, CNC machines for bevel gear grinding are also designed with six independent axes. Bevel gear grinding machine drives and bearings differ due to higher spindle and cutting speeds. [KLIN08; MÜLL13; STAD13]

8.1.3.4 Bevel Gear Lapping Machines

Lapping can be used as an individual hard finishing process for bevel gears or to complement the grinding process to improve the surface quality and geometry of tooth flanks [KRUM67]. Figure 8.38 provides an overview of the process and the design of the latest CNC bevel gear lapping machines.

During lapping, fine grains for chip removal are in an oil-containing lapping film. Gear and pinion mesh, which results in machining due to the relative movement of the flanks. In this way, the gear and pinion flanks in contact are machined simultaneously. By rolling both gear and pinion, a surface is machined that corresponds to the contact pattern of the respective flanks. The amount of the material machined depends on the relative speed between the tooth flanks. Different lapping positions can be set via the axes, which are referred to as V, H and J positions or E, P and G positions depending on the machine manufacturer. This shifts the contact pattern and increases the machined area on the flanks. Since the gear and pinion are machined simultaneously and each bevel gear pairing has an individual contact pattern, the pairing of gear and pinion must be maintained in application. In case of damage, the whole bevel gear pairing needs to be replaced [KLIN08; STAD13; KLOC17]

A lapping machine has three linear and two rotational axes. An example design is shown in Fig. 8.38. Axis J or G adjusts the gear position for lapping. Axis H or P positions the pinion. Axis V or E adjusts the shaft offset between the gear and pinion. The drive is provided via the pinion, while the output is provided via the gear. The torques applied are comparatively low [KRUM67]. The column design shown for a lapping machine results in short force transfer distances and a compact workspace. This simplifies loading and unloading of workpieces [KLIN08; STAD13; KLOC17].

8.1.3.5 5-Axis Milling Machines

Due to increasing demands on flexibility for manufacturing systems, especially in low volume production and prototyping of cylindrical and bevel gears, milling gears on five-axis milling machines is attracting more attention [JAST12]. Five-axis milling is characterized by a high degree of flexibility in terms of machine and tool selection. Due to the high machining time and associated low productivity rate, five-axis milling complements conventional gear cutting processes but is not a replacement [KLOC13].

The main applications for 5-axis machining of gears are the manufacturing of prototypes and individual parts as well as pre-development. Particularly in the production of individual parts and the repair and supply of spare parts, five-axis milling is used for large gearboxes such as marine gearboxes or cement mills [KLOC13].

The process characteristics of 5-axis milling are defined by the tool selection, the input data generation and the machining strategy. This results in process-specific workpiece properties [POTT13; SCHU15]. Figure 8.39 provides an overview of

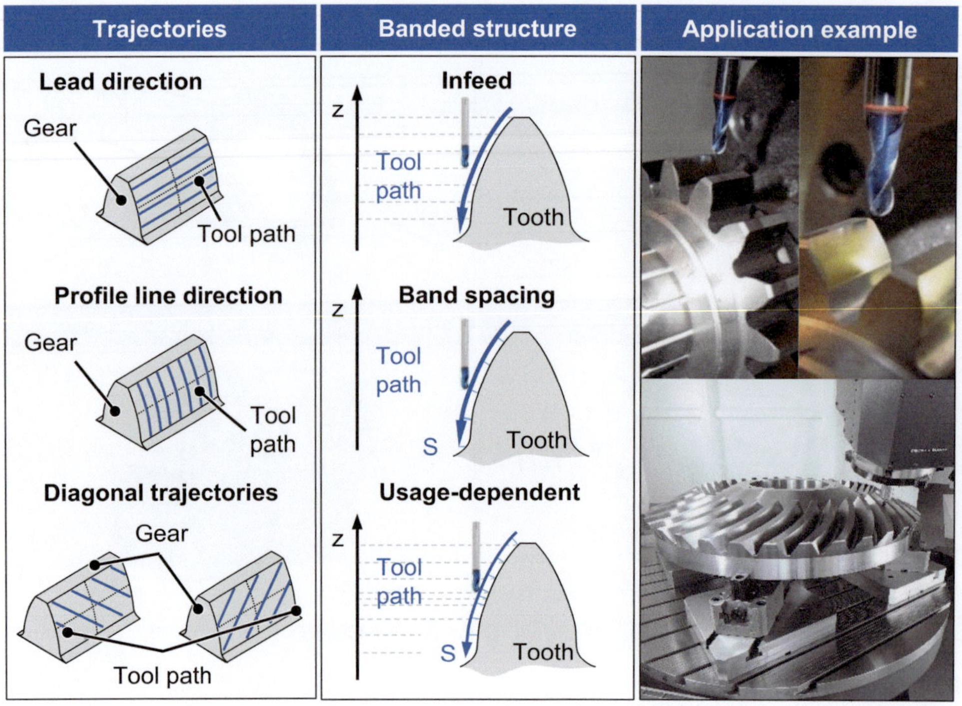

Fig. 8.39 Machining strategies for the 5-axis milling of gear teeth. Example from [KLOC17]

possible machining strategies for the 5-axis milling of gears. The trajectory of the milling tool can be selected as desired to imitate structures or define new structures. The trajectory of the cutter can be in the lead or profile direction of the gear. Diagonal trajectories exhibit more complex movements of the machine axes, which is why the benefits of the surface manufactured must be weighed against the loss in cost efficiency [KLOC13].

The banded structure defines the number of tool overruns necessary to machine a tooth flank to completion and the pattern showing how the bands are distributed on the tooth flank. The milling head can be fed at a constant rate. However, the infeed can also be selected such that the band spacing remains constant. If the aim, for example, is low roughness on the pitch circle, the banded structure near the pitch circle diameter can be adapted to the application such that the spacing of the bands is reduced. [KLOC13]

8.1.4 Forming Gear Manufacturing Machines

In gear manufacturing, the benefits of forming and shaping processes can also be utilized. These include a short manufacturing time for a given workpiece quality and almost complete utilization of the material. However, given the limited flexibility and the high time and cost involved in product and process design, the use of these processes is only economical for high volume production. [BAUS11]

Forming machines are used for gears to produce blanks [FELT12]. Sintering and powder forging of gears take place on machines that are explained in ► Sect. 2.7. Die-casting machines described in ► Sect. 2.4 are mainly used for special applications with non-ferrous materials, such as synthetics and zinc, tin and lead alloys [FELT12].

8.1.4.1 Rolling Machines

Rolling represents a step-by-step forming process under compressive conditions with one or more tools [BAUS11]. Spur and helical gears, as well as special gears such as Wildhaber-Novikov gears, can be created [ROTH98]. A distinction can be made between different rolling processes. Longitudinal rolling with profile rollers splits the forming work into several individual steps [BAUS11; FELT12]. Transverse rolling is performed using racks or roller wheels. Transverse rolling with roller wheels, also called transverse rolling of shapes [ROTH98], is discussed in more detail below [BAUS11; FELT12].

Figure 8.40 provides an overview of the principle and structure of a machine for transverse rolling with roller wheels. The change of the center distance of the axis-parallel, external tools initiates the forming process [FELT12; ROTH98]. The tooth form is generated in a locally limited forming zone one after the other by rolling the workpiece between two tools.

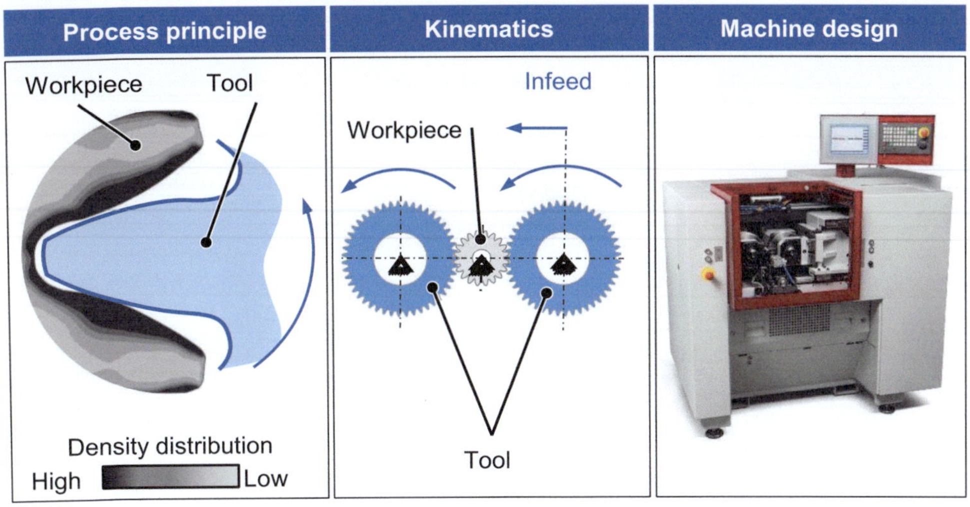

Fig. 8.40 Principle and machine structure for gear rolling. Example from Profiroll. *Source* [KLOC17]

Then, the rolling direction is reversed. Since there are different stresses on the leading and trailing flank, the reversing process ensures a consistent quality on both flanks. The material flow on the tooth flank causes the tip diameter in the rolling process to increase. The blank diameter required must be determined theoretically or empirically in preliminary tests [FELT12; ROTH98].

The machine design is shown on the right in Fig. 8.40. One tool is fixed to machine the blank while the other one is fed into the blank linearly. The tool spindles are path-controlled electronically. Given that the workpiece quality of rolled gears depends directly on the control of the individual workpiece axes, strict requirements are made on axis positioning and control of the tools. [BAUS11; KLOC17]

8.1.4.2 Orbital Forming Presses

Orbital pressing is a forming process used for rotationally symmetric parts and components with significant diameter differences. It allows good forming even when the workpiece cross-sections are critical [BAUS11]. It is used mainly for forming of blanks into face gears. Figure 8.41 provides an overview of the principle and machine design for orbital forming presses.

In orbital pressing, the upper die executes an orbital motion around the rotation axis. The orbital motion can be generated by a circular or spiral movement, by a straight-lined eccentric method or by the multi-blade cam movement. The multi-blade cam movement radiates out in a star shape from the center position of the upper die [FELT12]. The orbital motion means the tangential force only acts upon one partial

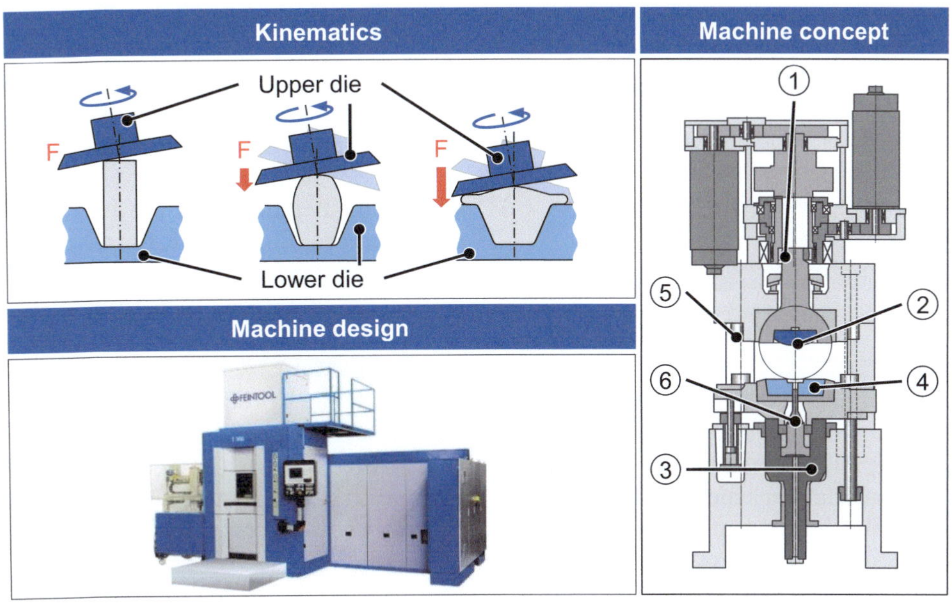

Fig. 8.41 The principle and machine structure for orbital pressing of gear teeth. Example from Feintool

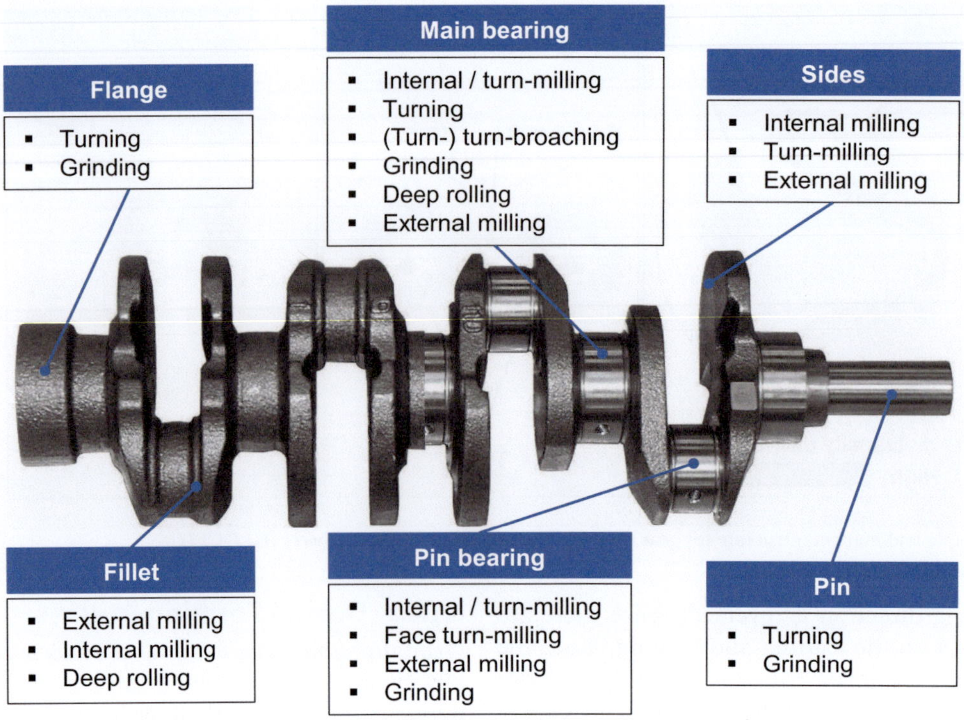

Fig. 8.42 Machining processes for the sections of a crankshaft. *Source* MAG

8

surface of the blank. As infeed continues, the forming force acting upon the blank also increases. Great deformations can be achieved by a combined pressing and rolling process using a comparatively low forming force [BAUS11]. This results in lowered friction forces between the blank and die, which leads to a positive effect on the surface quality of the part [BAUS11; KLOC17].

Example machine axes and machine design are shown in Fig. 8.41. The orbital head (1) generates an orbital motion of the upper die (2) mechanically using two eccentric sleeves driven independently. The pressing force is generated hydraulically in the lower part of the machine and transferred by the press ram (3). As a result, the lower die (4) performs a linear motion that is directed upwards. The stop (5) is used to restrict the movement. After the forming movement, the workpiece is released from the lower die by the ejector (6). [FELT12; KLOC17]

8.2 Machines for Crankshaft Production

A number of different manufacturing processes are used to make crankshafts. This is attributable firstly to the complex geometry of the workpiece, resulting primarily from the eccentric arrangement of the pin bearings and secondly from the tight manufacturing tolerances. Modern-day crankshaft production lines use milling and turning operations as well as broaching, rolling and grinding processes. Different processes are combined in manufacturing depending on the section of the crankshaft. Figure 8.42 assigns the machines and machining processes introduced in the following chapters to the individual sections of a crankshaft. The greatest challenge in production is machining the eccentrically positioned pin bearings.

8.2.1 Crankshaft Milling

The main and pin bearings of crankshafts can be machined with internal as well as external milling operations and with the sides being machined at the same time. In these main processes, a distinction is made with regard to the movement of the workpiece in internal milling with a fixed or rotating workpiece and with regard to the arrangement of the tool relative to the workpiece in peripheral milling with parallel axes and external milling. The interrupted cut and high attainable cutting capacities are characteristic for milling processes. In contrast to turning processes, the crankshaft either rotates at a low circumferential speed or does not rotate at all. Imbalances, as they occur at high workpiece speeds (for turning and turn-broaching operations for example) are prevented as a result.

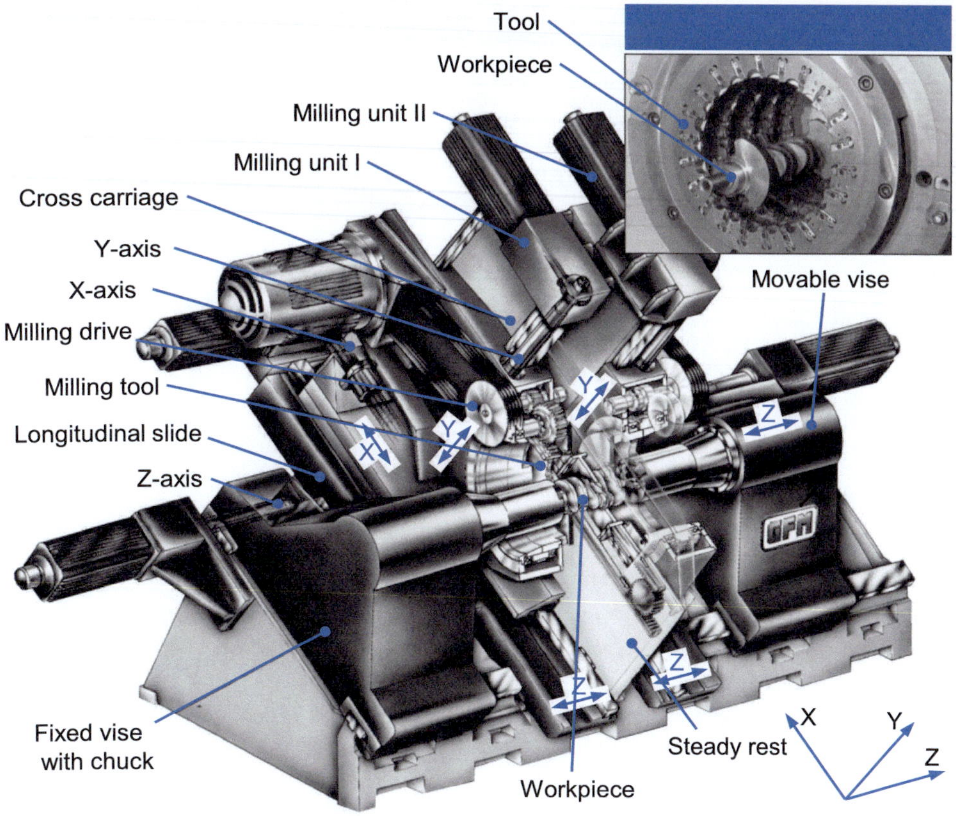

Fig. 8.43 Internal milling with fixed workpiece. *Source* GFM

8.2.1.1 Internal Milling with a Fixed Workpiece

In the past, main and pin bearings were often machined using internal milling with a fixed workpiece down to the grinding allowance. Nowadays, this process is no longer used because internal milling with a rotating workpiece, introduced in the next chapter, is much more efficient. Given its importance over many years, the process of interior milling with a fixed workpiece for crankshaft production is explained at this juncture.

In the process, ring-shaped internal milling tools with indexable inserts are used. These tools are numerically controlled to move on concentric paths around the bearing axis relative to the workpiece axis. The tool cutters are arranged on the inner circumference of the ring-shaped inner milling head. In the center of these, the workpiece is moved axially and positioned. The rotating tool is positioned axially at the main/pin bearing of the crankshaft, and then fed in radially by two CNC axes perpendicular to the workpiece axis. It is then guided once around the bearing points to be machined along a concentric path (circular feed). After one circulation of the milling unit, the bearing point is pre-machined to the grinding allowance. The machine shown in Fig. 8.43 was used for the mass production of 3-stroke to 6-stroke automobile crankshafts. It has two independent milling slides that can also be equipped with multi-cutter heads to shorten production times.

Refer to Fig. 8.44 for the basic milling sequence of the main and pin bearings of a 4-stroke crankshaft using the process described above. In this machining process, the machine is equipped with a cross carriage carrying three separate tool holders that can be traversed in the vertical direction.

The workpiece is chucked at both ends as well as clamped by two steady rests between the milling tools. Five cutter heads are loaded on the tool holders for milling the main bearing pins. With one joint plunge-cut, and within one circulation of the cutter heads moved by the cross table, the main bearing pins are machined to the grinding allowance.

For milling the pin bearings, the middle tool holder carries two cutter heads, and the two outer holders carry one cutter head each. After the axial positioning of the crankshaft, the middle tool holders are traversed up by the stroke and the two external holders are traversed down by the same amount. Then, the four opposing pairs of pin bearings are milled simultaneously.

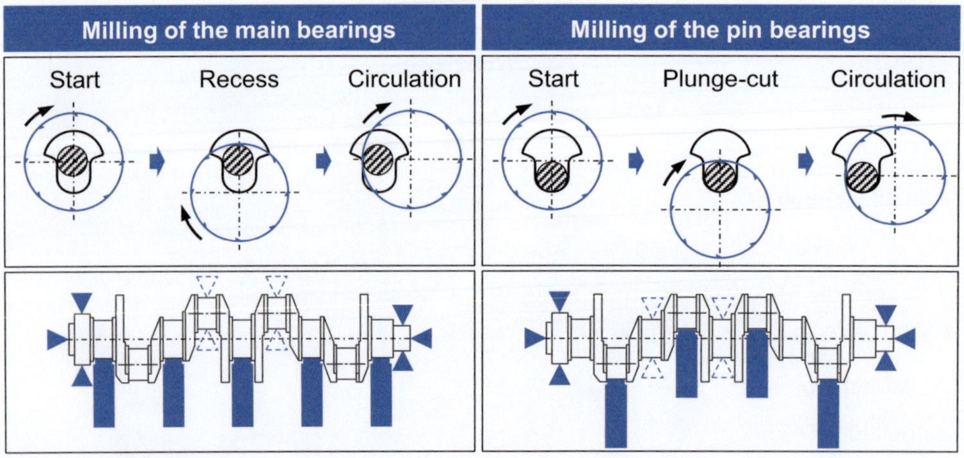

Fig. 8.44 Machining process for CNC crankshaft milling. *Source* GFM

8.2.1.2 Internal Milling with a Rotating Workpiece

Crankshaft internal milling differs from internal milling with a fixed workpiece in terms of how the feed is generated. Circular feed is provided by the rotation of the crankshaft. For machining the bearing pins, the rotating cutter head is positioned axially at the location to be machined, with the workpiece fixed. It keeps being advanced in the radial direction with a CNC-controlled X-axis until the required diameter amount is reached. Afterwards, the crankshaft is rotated. Milling is complete after exactly one workpiece revolution.

To machine the eccentrically positioned pin bearings, the milling unit moves in a pendulum motion. During this process, rotation of the centrally clamped crankshaft is controlled to achieve uniform chip removal. The process of pendulum stroke milling is shown in Fig. 8.45 for the machining of a pin bearing.

A typical machine for internal milling with a rotating workpiece is shown in Fig. 8.46. Unlike internal milling with a fixed workpiece, the milling unit of this machine has only one X-axis because circular feed is provided by the C-axis. To increase the levels of stiffness and process stability, a crankshaft is supported on one side by simple steady rests and on the other by a stabilization bezel. The stabilization bezel has three chucks that clamp the adjacent pin and thus increase system stability.

8.2.1.3 External Milling with Parallel Axes (Turn-Milling)

The process of turn-milling or external milling with parallel axes for crankshaft bearings was developed in the 1980s. This process has widely replaced internal milling. In turn-milling, the tool and workpiece axes are parallel to each other as shown in Fig. 8.47. A crankshaft rotates slowly in a central chuck about its

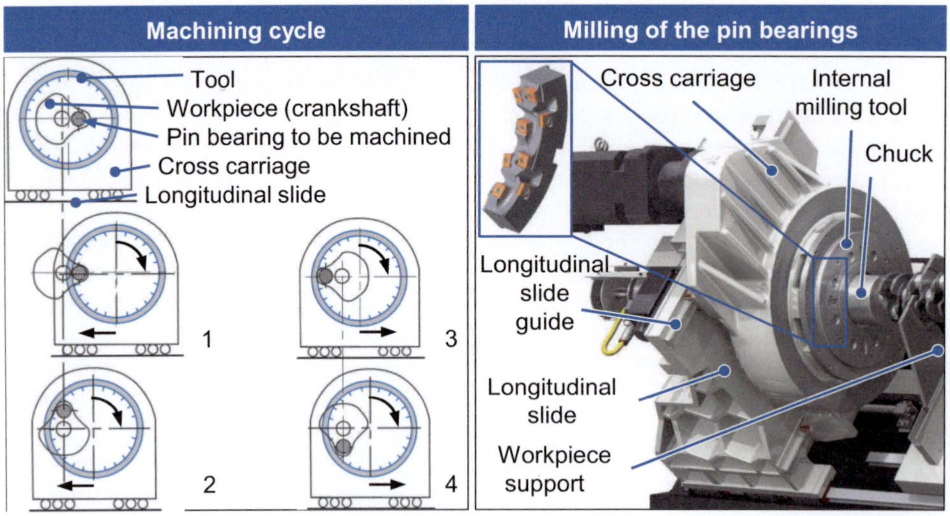

Fig. 8.45 Internal milling with a rotating workpiece. *Sources* Heller, MAG, Sandvik

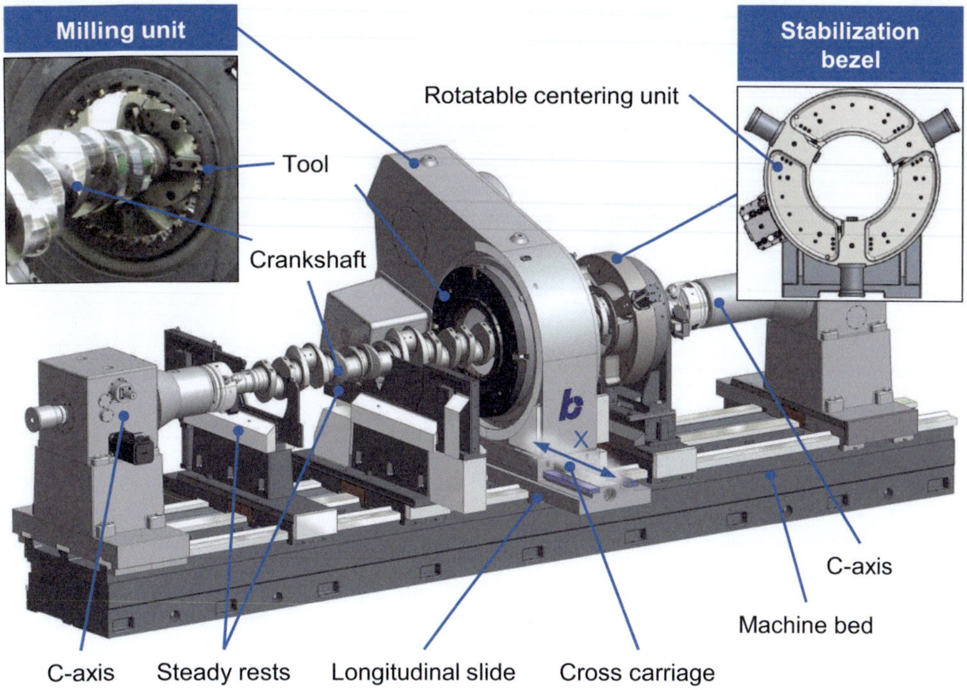

Fig. 8.46 Internal milling machine. *Source* Burgsmüller

main bearing axis (circular feed) while the side milling cutter is fed in radially. The rotation of the side milling cutter generates the cutting speed necessary. The movement of the entire milling unit is generated by CNC-controlled axes, allowing for the machining of pin bearings using a radial pendulum stroke motion (pendulum stroke milling) while the crankshaft is chucked centrally. This means that all the main and pin bearings, as well as the sides of a crankshaft, can be machined in one clamping operation. Side milling cutters with diameters up to 700 mm, comprising individual segments fitted with cutting inserts, are used as tools. For example, this results in 240 cutting inserts for a milling wheel with a profile layout of four cutting rows, which need to be fitted with tight tolerances to meet the axial runout requirements of the crankshafts. No cooling lubricant is required for the milling process. The cutting speed conventionally attainable for external milling is 250 m/min. Nowadays, with improved tool materials in hyperspeed milling, cutting speeds up to 600 m/min are possible when finishing with carbide. Even faster cutting speeds are possible with ceramic cutting tools. Since the middle of the 1990s, turn-milling with parallel axes has been deployed on modern production lines, usually for pin bearing machining, due to its higher cost efficiency because there is no clamping in eccentric chucks (necessary for turn-broaching). The process is also flexible with regard to any depressed centers and changeable bearing widths.

Low cutting forces (compared to internal milling) are inherent in this process due to the smaller overlap between the tool and workpiece, the resulting high workpiece accuracy and the ease of machining eccentrically positioned pin bearings. The independently operating tools oscillate along the respective X-axis, depending on their angular position. Workpiece loading can be easier and faster compared to internal milling because there is no positioning of the crankshafts into hollow tools. This means the cycle times for external milling are marginally shorter. Using turn-milling processes, low surface roughnesses can be attained, reducing to a minimum the necessary allowances for subsequent grinding.

8.2.1.4 External Milling (Face Turn-Milling)

In the face turn-milling process, the milling axis is positioned orthogonally to the workpiece axis as shown in Fig. 8.48. The process differs from turn-milling with parallel axes mainly due to the additional vertical Y-axis. This, together with the X-axis, enables any eccentric and polygon profiles to be created making this machining process particularly versatile. Crankshafts of all types can be made cost-effectively, even in small batch sizes.

Figure 8.48 shows the creation of cylindrical surfaces in the face turn-milling process at the bottom [SAND13]. In the first cut, the milling tool generates a cylindrical surface. In this process, the milling axis is offset by E_{W1} with respect to the pin axis to be machined and the face cutting surface runs par-

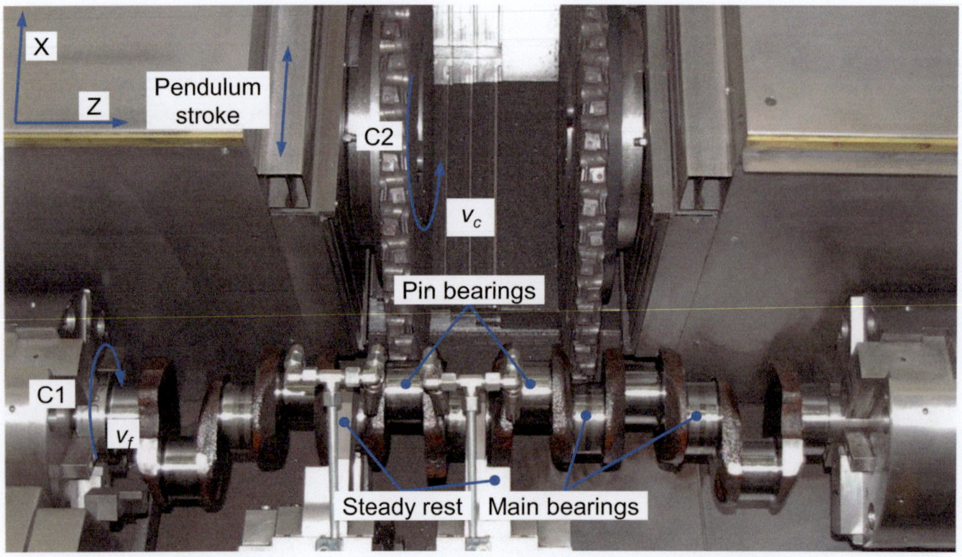

8

Fig. 8.47 External milling with parallel axes. *Source* MAG

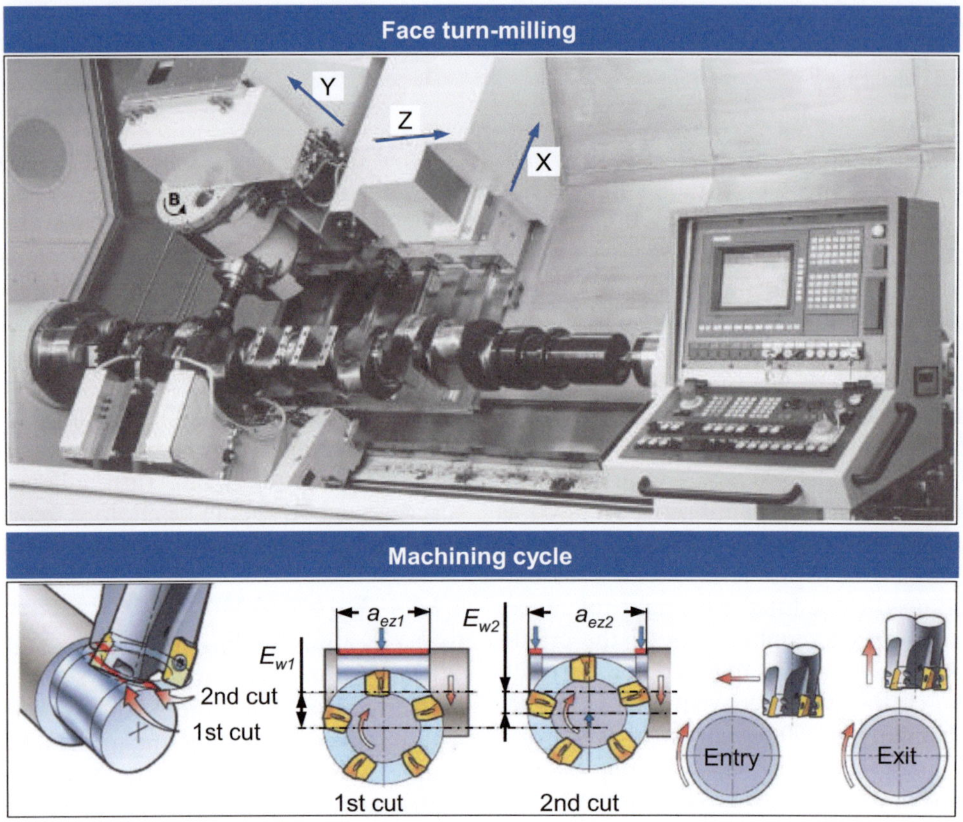

Fig. 8.48 Face turn-milling of crankshaft pin bearings. *Sources* Niles-Simmons, Sandvik

allel to the workpiece axis. Here, the eccentricity depends on the possible face cutting length of the tool, formed by the external tool diameter (corner radius) and the radial minor cutting edge length. This means the center of the pin is machined cylindrically, and distorted areas remain at the edges on the left and right. Another cut with reduced eccentricity E_{W2} is required to eliminate these distortions. In the borderline case, the tool axis is positioned centrally above the pin axis.

8.2.2 Crankshaft Turning and (Turn-) Turn-Broaching

Turning and turn-broaching are other processes of crankshaft production. The cutting speed is generated by the rotating workpiece in both operations. To prevent a crankshaft with low torsional rigidity from twisting, it is usually driven synchronously at both ends. The high rotational speeds make the effects of imbalance forces on accuracy unavoidable. In the turning process, pin bearing machining uses eccentric chucks that clamp a crankshaft in a way that the center of the pin bearing aligns centrally with the rotational axis of the main spindle. Undesired vibration and deformation of the component often occur due to imbalance. In addition, the workpiece spindles are subjected to high stresses by the imbalance forces. In such case, the rotational speed, and thus the cutting speed, must be reduced. As a result, the turning process is not cost-effective. Furthermore, the use of eccentric chucks is not retooling-friendly and there are no correction options for stroke and angle deviations. Nowadays therefore, external milling is used mostly for the machining of pin bearings. Other sections (see ▶ Sect. 8.2), however, are often machined by turning.

8.2.2.1 Turning

Turning plays a key role in the volume production of crankshafts. This process is applied mainly for machining flange and pin, external counterbalance diameter and side end faces and in the finish turning process for main and thrust bearing end faces and encoder ring contours. ◘ Figure 8.49 shows an example of machining main bearings and the associated processes and tools for roughing (a–c) and finishing (d–f).

CNC lathes, configured based on the processing sequence and specific requirements, are used. The machines usually have two disk-type turrets with up to 45 tool slots. The narrow disks enable collision-free plunging between the crank webs, with a short cantilever length, and thus maximum stability of the tool holders. The tool holders are loaded in such a way that it results in an economically optimal processing sequence with minimal tool change times. To increase the overall usage duration of the loaded turret heads, they are equipped with several redundant tool sets. The process is flexible with regard to changeable component dimensions because these changes can be made by adapting the CNC program, meaning no reloading of a tool holder is necessary. If radial recesses and axial undercuts are required in crankshaft machining, the turning process cannot be replaced by any other cutting process. Multi-tooth tools, or chasing tools, are used to reduce machining times for finishing bearing diameters (see ◘ Fig. 8.49).

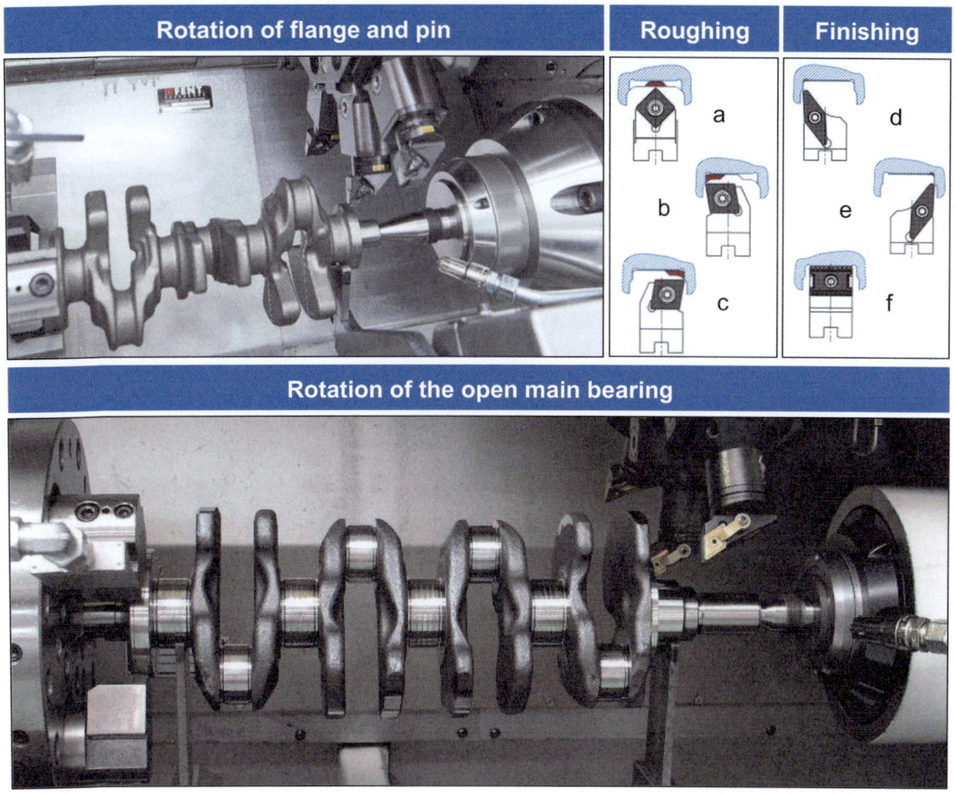

◘ **Fig. 8.49** Turning. *Sources* MAG, Sandvik

8.2.2.2 Turn-Broaching

Turn-broaching was commonly used in the past for machining main and pin bearings but has widely been superseded due to the development of alternative, more efficient processes. Given its historical importance, the process is introduced here, nevertheless. Based on its kinematics, turn-broaching is assigned to the turning and broaching group as per DIN 8589-5 [DIN03b]. In the turn-broaching process, the rotating workpiece is machined by a circular broaching tool moved rotationally. The broaching tool unwinds a straight broach on the circumference of a cylindrical disk. As every tool cutter is only in contact once in this machining process, there is low mechanical and thermal tool loading. In the turn-broaching process for the main bearing, the crankshaft is centrally chucked on both sides into the two synchronously driven main spindles. The workpiece speed set determines the cutting speed. The tool support shown in ◘ Fig. 8.50 has a CNC-controlled transverse axis X and a rotational axis C2. These axes determine the radial position and movement of the multi-cutter broaching tool. The broaching tool is comprised of several segments. These are loaded with cutting inserts and adjusted using an external fixture. As is the case for the conventional broaching process, infeed is by the tool cutters arriving in succession at the contact. A unique, fixed position is assigned to each cutter. Thus, the feed is provided by the continuous, slow rotation movement of the broaching tool.

A broaching tool is divided into a roughing zone and a finishing zone. After the last cutter of the tool makes contact, the machining process for the main bearing is complete. To machine multiple bearings, such as the simultaneous machining of all main bearings, sets of tools from identical tool disks can be used. But this also means a new tool set needs to be set for every dimensional change to the crankshaft. Thin workpieces with low stiffness can be supported by a steady rest. In such cases, the bearing seat to be supported is machined first. After the steady rest is applied, the remaining places are machined. Besides machining the sides and center bearings, it is also possible to machine the pin bearings. For this purpose, special chucks with eccentric workpiece clamps are used (see ◘ Fig. 8.50, top right). Turn-broaching enables achieving high dimensional accuracies. It is highly suitable for finishing prior to subsequent grinding. The low flexibility level of the process with the fixed tools means turn-broaching has lost importance over recent years and has now been almost entirely replaced by turn-turn broaching.

8.2.2.3 Turn-Turn Broaching

The turn-turn broaching process has been widely adopted in industry since the start of the 1990s. It combines the advantages of turning (high cutting capacity) for roughing and of turn-broaching (high accuracy and surface quality) for finishing. As a result, turn-turn broaching has established itself as a particularly cost-effective and flexible machining process, replacing the individual processes of turning and turn-broaching as the standard. In general, the entire crankshaft can be machined using eccentric chucks. However, on modern production lines, typically only the centric main bearings are machined with turn-turn broaching, while the pin bearings are machined on turn-milling machines. Similar to turn-broaching, the crankshaft is clamped centrically and driven on both ends to largely prevent the

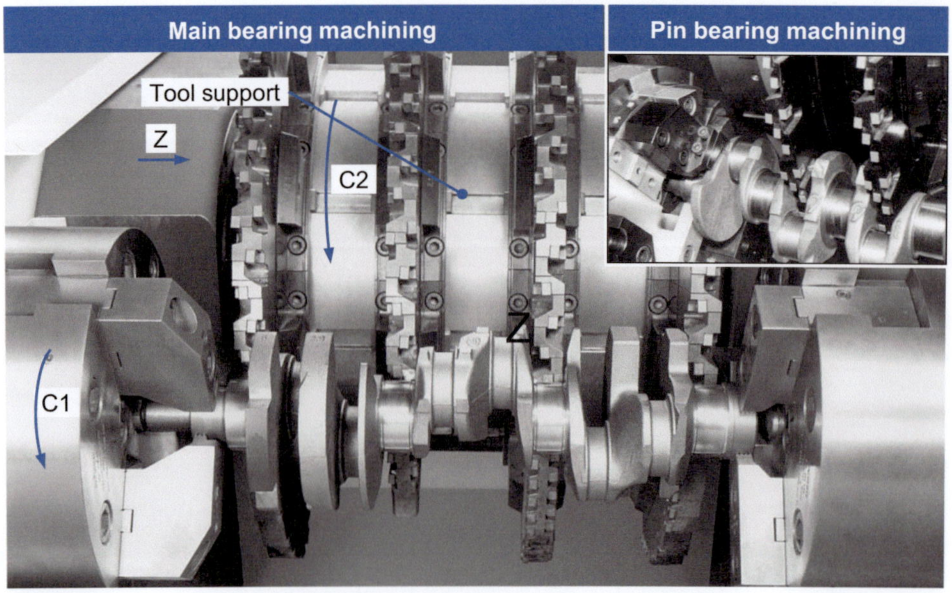

◘ **Fig. 8.50** Turn-broaching. *Source* MAG

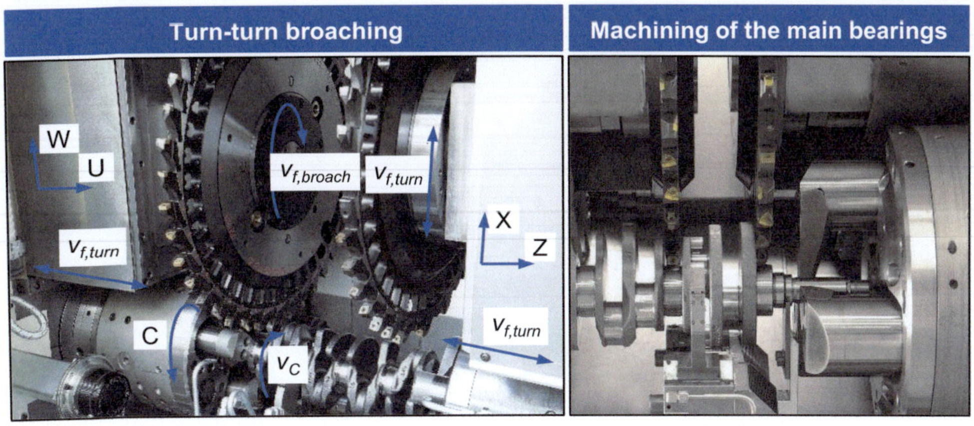

Fig. 8.51 Turn-turn broaching. *Source* MAG

workpiece twisting during the machining process. The tool supports, equipped with CNC-controlled transverse axis X/U and rotational axes C1/C2, also feature an additional CNC-controlled longitudinal axis Z/W for axial tool positioning, providing flexibility for dimensional changes. Two independent supports are commonly used, as shown in the example in Fig. 8.51.The tools are mounted on large disks, usually also loaded with redundant tools to maximize service life. Following several cutting inserts for turning, smaller broaching segments set to the final contour are used for finishing, taking up less space compared to conventional turn-broaching. This extra space on the tool disk can be utilized for broaching segments with different geometries.

A high level of flexibility is a feature of turn-turn broaching. The bearing widths can be changed by the axial positioning while the number of mountable tool segments enables several crankshaft types to be made on the same machine without tool exchange. Tool changes are made by continued turning of the disk. During the turning process, the rotational axis of the tool disk is clamped hydraulically. This provides support for the torque generated by the machining forces. For subsequent broaching, the clamp is undone, and the corresponding broaching segment is moved at a constant circular feed. Firstly, all turning operations like roughing of the contour and thrust collars, side machining and any recesses or undercuts are performed. In the last step, the bearing diameters are broached. The steady rests prevent bending deformation of the crankshaft during the machining process.

8.2.3 Crankshaft Grinding

Grinding is necessary for reworking both the main and pin bearings. While in the normal plunge grinding process the main bearings are ground between tips or with centered tension, machining the pin bearings can only be done using an eccentric clamp on conventional machines. In this case, the crankshaft must be positioned in such a way that the originally eccentrically running pin bearing aligns centrally with the axis of rotation of the workpiece drive. Figure 8.52 illustrates this using external cylindrical grinding of 1-stroke and 4-stroke crankshafts as an example. For the 4-stroke shaft, the two each pin bearings, sides and fillets on one stroke level are ground with two grinding wheels in one plunge-cut. To compensate the large imbalance forces and torques occurring, eccentric chucks often have mass balances (see Fig. 8.52, left).

On crankshafts for multi-cylinder engines, the pin bearings are often positioned at different angles. For machining the pin bearings, the crankshaft must be moved into the central position for several directions. The displacement process in the chuck must be very exact to maintain the tolerances stipulated. Furthermore, the large imbalances cause high forces acting upon the chuck heads, which must be very stiff. In addition, due to the large imbalance torques, high load fluctuations result for the drives. These can cause manufacturing inaccuracies. Conventional grinding becomes very complex. In particular, this is due to the eccentric chucking of the crankshafts. Moreover, the flexibility of conventional machines is severely limited, as new eccentric chucks with new displacement directions are required for every workpiece change.

The pendulum stroke concept in Fig. 8.53 avoids this complex mechanical structure of a conventional grinding machine. In pendulum stroke grinding, the workpiece is held centrically between tips. The reciprocating motion, necessary to balance the eccentricity of the pin bearings, comes from the infeed slide on the machine (X-axis). The machine kinematics are comparable with the machine for external milling with parallel axes (see ► Sect. 8.2.1.3). To ensure that the cylindrical deviations remain within the specified tight tolerances, the maximum permitted workpiece speed in this pro-

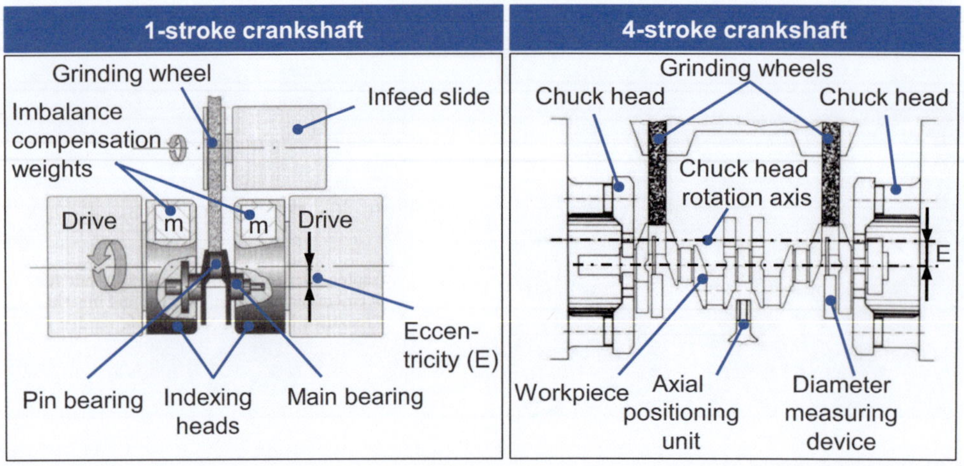

Fig. 8.52 Conventional grinding. *Source* EMAG

cess depends directly on the stroke of the crankshaft and on the drive dynamics of the feed axis X. As increasing the workpiece speed leads to reduced grinding times, high feed dynamics, i.e., feed generation with low subsequent error, are essential for this grinding process. The kinematics of the contact point shown on the right in Fig. 8.53 illustrate the machining sequence during a reciprocation cycle. The plunge-cut is made at the first point (P1) at a position of 0°. While the crankshaft rotates, the grinding wheel is tracked along the X-axis. In this way, contact points P2, P3 and P4 occur between the surface to be machined and the grinding wheel in the extreme positions (90°, 180° and 270°).

A modern, flexible machine concept with two independent tool carriages is depicted in Fig. 8.54. With this machine concept, over 30 different machine variants can be created by combining individual assemblies. The 2-carriage variant shown can be loaded with up to four grinding wheels. All grinding units have a dressing unit and contactless, automatic balancing systems on the grinding spindles.

The machine table is inclined at an angle of 20° to the base surface and, like the machine bed, is perfectly matched to the force flux of the grinding process. The stop rails, one positioned on each side, guarantee during retooling the best repeat accuracies for the positions of the table superstructures. The stiffness of the overall system is attained by the dovetail guide system of the table superstructures and the stiff connection to the machine base frame. A different guide system can be deployed depending on the machining task. The options here range from low-cost linear rolling element guide systems, slide guides for circular grinding to hydrostatic guides with hydrostatic threaded spindle for cam profile grinding. The use of a linear motor for crankshaft machining is also taken into consideration from a design perspective. For the 2-carriage version, the design of the guideways enables both carriages to

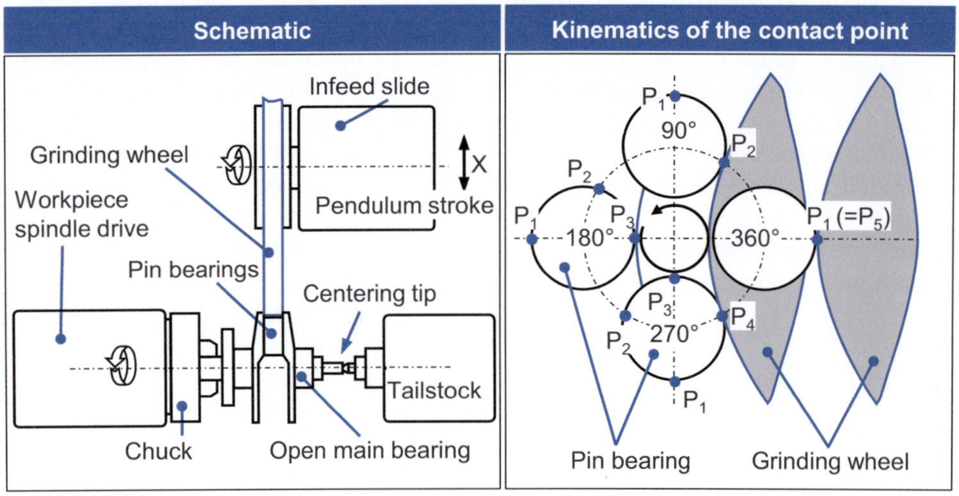

Fig. 8.53 Machine concept for the pendulum stroke grinding of crankshafts. *Source* EMAG

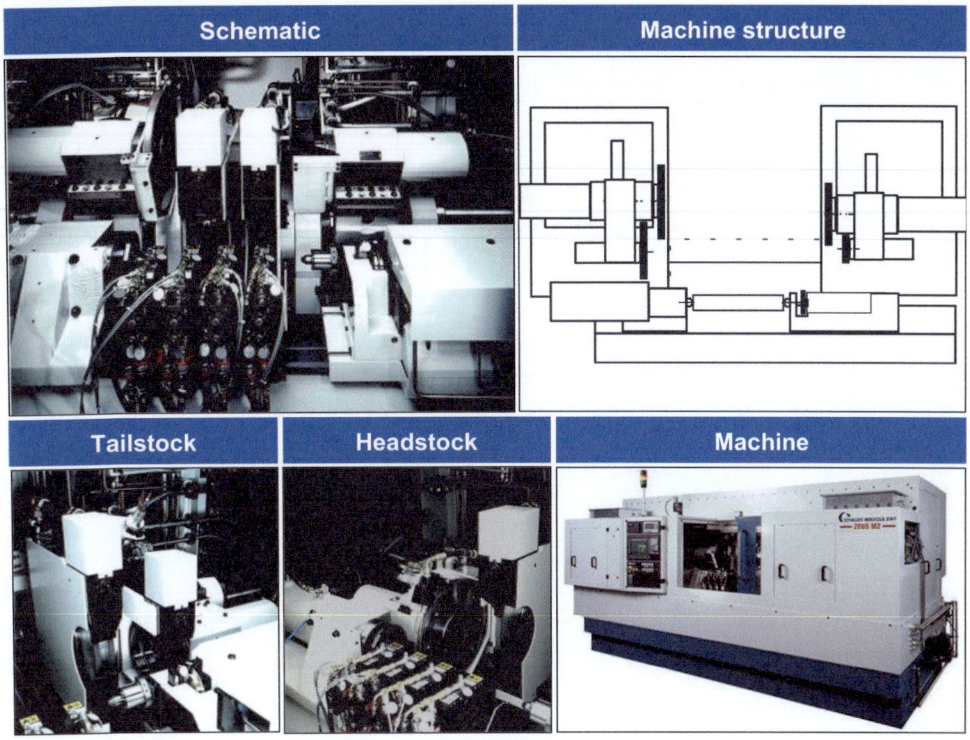

Fig. 8.54 2-carriage version of a crankshaft grinding machine. *Source* Schaudt

move toward each other up to a very small safety distance for the protective covers. This means even cams or diameters that are close together can be ground simultaneously. During the machining process, a CNC controller handles dynamic collision monitoring. The machine is also equipped with an automatic high-speed balancing system, that provides a contactless encoder and pickup directly on the grinding spindle. The adequately dimensioned travel distances of the axes enable only one dressing system to be used. This can be implemented as a fixed diamond tile or a rotating dressing device, such as a shape and profile roller or cup wheel.

8.2.4 Crankshaft Rolling

Crankshafts surfaces are typically finished using the rolling process. A distinction is made among three rolling processes depending on the machining result required: deep rolling, roller burnishing and roll straightening.

The purpose of deep rolling is to prolong the life of a crankshaft. Its fatigue strength is increased significantly with targeted application of internal compressive stresses. Plastic deformation of the surface layers results in material solidification, which increases both the surface hardness and the dynamic load capacity of the entire crankshaft. Deep rolling is typically performed after hardening the bearing contact surfaces, especially in the fillet area.

Roll straightening is commonly combined with the deep rolling process and is used to minimize the crankshaft radial deflection or runout error with no reduction in strength. For angle-dependent deep rolling of the fillets, the axial runout deviations of the oil composites are reduced by adapting the rolling force to the different compliances of the oil composites.

Roller burnishing of crankshafts is used to create functional surfaces, such as the thrust collars of thrust bearings, shoulders and flanges and the bearing contact surfaces. Roller burnishing represents a forming precision machining process that attains low roughness depths with a high profile contact area ratio and high surface hardness.

A deep rolling roller, the principle of deep rolling and the principle of roller burnishing are shown on the left, center and right side of Fig. 8.55, respectively. The fillets are machined with the deep rolling modules shown at the bottom. These comprise a deep rolling head and a support roller head. The bearing contact surfaces are machined with a rolling head that applies a narrow roller burnishing needle over the entire bearing width with the rolling force. In the rolling process, a hydrostatic tracker system ensures a rolling force that is proportional to the fluid pressure set and automatically compensates manufacturing tolerances.

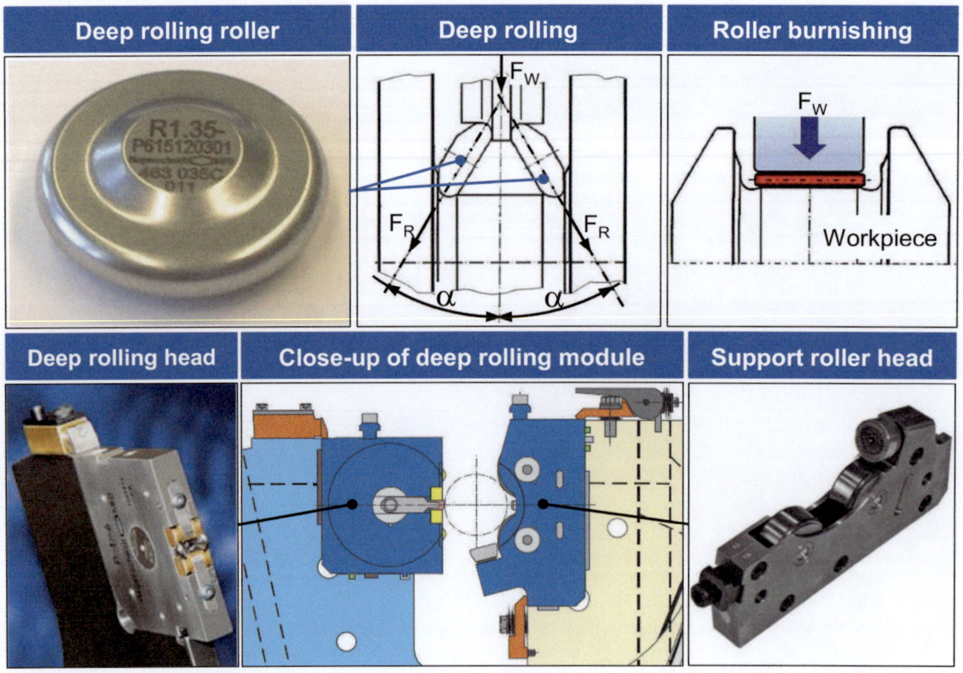

Fig. 8.55 Deep rolling and roller burnishing of crankshafts. *Source* Hegenscheidt-MFD

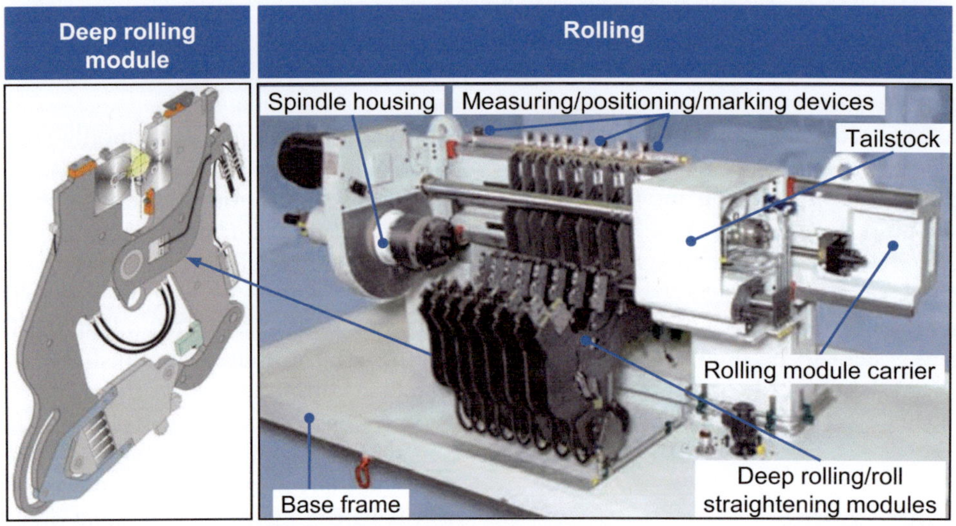

Fig. 8.56 Machine concept for rolling crankshafts. *Source* Hegenscheidt-MFD

The mechanical structure of a deep rolling and roll straightening machine for crankshafts is shown in Fig. 8.56. All machine modules are housed on a base frame. Specifically, these are the spindle housing with chuck and 3-phase servo motor, the tailstock with quill, the deep rolling and roll straightening modules with their tools and the associated measuring, positioning and marking devices used to position the rolling modules axially on the rolling module carrier. While on earlier machines master crankshafts controlled the movement of the rolling devices, nowadays, the functions of all deep rolling and roll straightening devices can be actuated individually. The machine shown can accommodate up to 13 deep rolling devices.

The individual rolling modules are designed for deep rolling and roll straightening. All deep rolling modules used for the machining of bearings are suspended by pneumatic cylinders to relieve the weight. The deep rolling modules on the open main bearings (on pin or flange side) are supported axially from the side. The deep rolling modules can be rotated into their working/park positions as required. The connec-

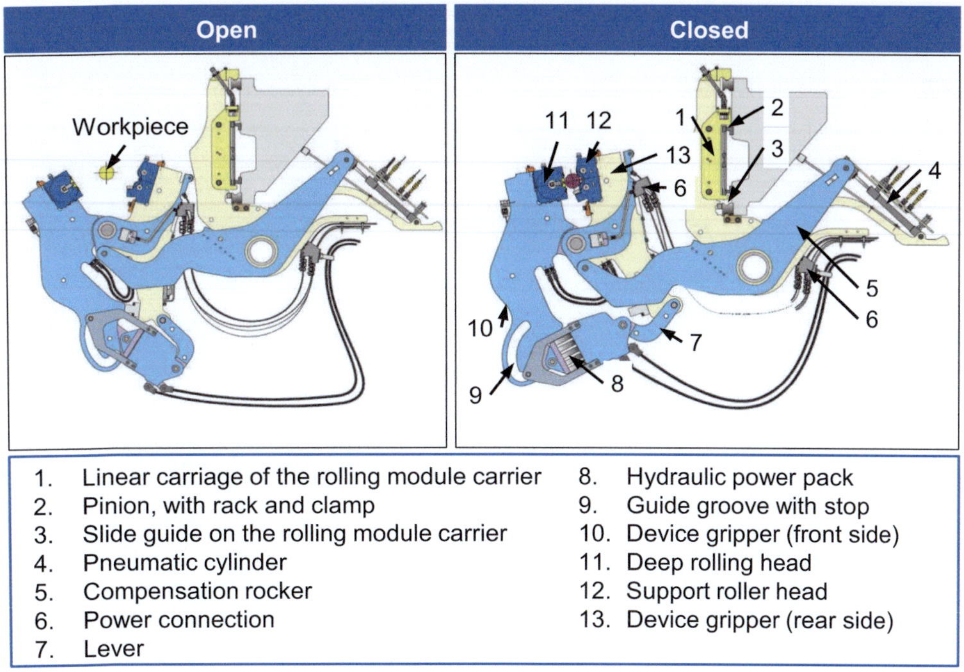

Fig. 8.57 Structure of a deep rolling module. *Source* Hegenscheidt-MFD

tors for hydraulics, rolling oil, pneumatics, force sensors and position encoders are integrated into the deep rolling modules. Figure 8.57 shows a deep rolling module in the working and park positions and the associated components.

The rolling module carrier of the machine is designed with slide guides (3) and roller guide systems for accommodating the deep rolling modules. Each deep rolling module has a central power connection at the rear of the machine. A rack and pinion (2) and clamp are used to manually adjust the linear carriages (1). The compensation rocker (5) holds the device gripper with the front (10) and rear sides (13). The device gripper can be held up, down, or in the unloaded weight state using a pneumatic cylinder (4) with an integrated clamp. The cylinder clamp can hold the device gripper in any position. The front and rear sides joined by a bearing, together with the closing mechanism and cylinder, form the collet chuck of the deep rolling module. The hydraulic power pack (8) between the front and rear sides generates the rolling force necessary. Individual force control for each module is realized by polling the actual force values on force sensors in conjunction with a change in hydraulic pressure at the unit. The hydraulic power pack opens and closes the device grippers using a lever (7). The wide open sides form the space for loading and maintenance work. The deep rolling head (11) and support roller head (12), adapted to the appropriate crankshaft bearing type, are chucked on the sides with clamping holders [HEGE11b].

The workspace of a modern-day crankshaft rolling machine for deep rolling and roll straightening is shown in Fig. 8.58.

In order to accomplish the final machining in as few clamping operations as possible, the trend is to combine several machining types in hybrid machining centers. Figure 8.59 shows a hybrid roller burnishing tool and precision lathe used for cost-effectively finishing crankshaft thrust bearings [HEGE11a]. In the first machining step, the width of the thrust bearings is turned to the final dimensions in a precision turning process. The parallelism of the bearing plane surfaces is very high due to the machining accuracy of the precision turning process. Subsequent roller burnishing significantly increases the contact area ratio of the plane surfaces. Additional CNC-driven cross slides can machine other sections of the crankshaft in a clamp (such as flange faces and encoder ring contours), something that has a positive effect on the axial manufacturing tolerances.

8.3 Machines for Camshaft Production

Camshafts are round bar-shaped machine elements and in most cases have cam contours running parallel to the shaft axis. Camshafts are mostly used in piston engines as part of the valve operating mechanism to control the inlet and outlet valves. Up until a few years ago, camshafts were only cast, or in exceptional cases forged and then machined by cutting. Recent years have seen

Fig. 8.58 Workspace of a crankshaft rolling machine. *Source* Hegenscheidt-MFD

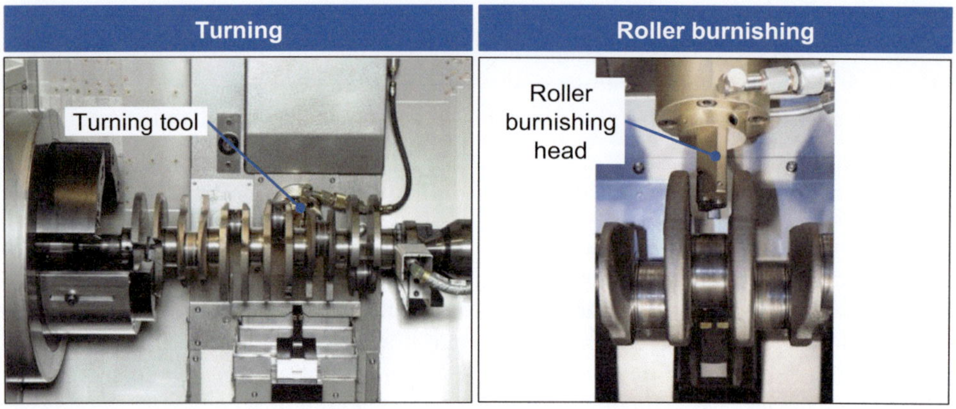

Fig. 8.59 Combined roller burnishing tool and precision lathe. *Source* Hegenscheidt-MFD

increased numbers of assembled camshafts, where the cams and shaft are made separately and then joined, coming onto the market. The benefits here are lower manufacturing costs, reduced mass, higher flexibility in production and the ability to create new camshaft geometries. Despite the benefits of constructed camshafts, the majority of camshafts used are made of cast or forged single-piece blanks.

The general categorization of manufacturing processes as in DIN 8580 is shown in Fig. 1.10. To manufacture single-piece camshafts, cutting manufacturing processes with geometrically defined cutting edge (milling) and geometrically undefined cutting edge (grinding with rotating tool) are used. The machines used for machining single-piece camshafts are introduced in more detail below. Previously, turning and post-forming turning were the classic methods for pre-machining cast and forged camshafts for subsequent hardening and grinding. The complicated mechanical systems were neither powerful nor easy to retool. For this reason, the mechanical machining of single-piece camshafts almost exclusively uses CNC pendulum stroke milling with subsequent circular or non-circular grinding (see Fig. 8.60).

8.3.1 Camshaft Milling Machines

To be able to mill all cam profiles, even those with straight-line or concave contour profiles, side milling cutters with external teeth are used. Using profile cutters, even the bevels on the cams of the blanks can be

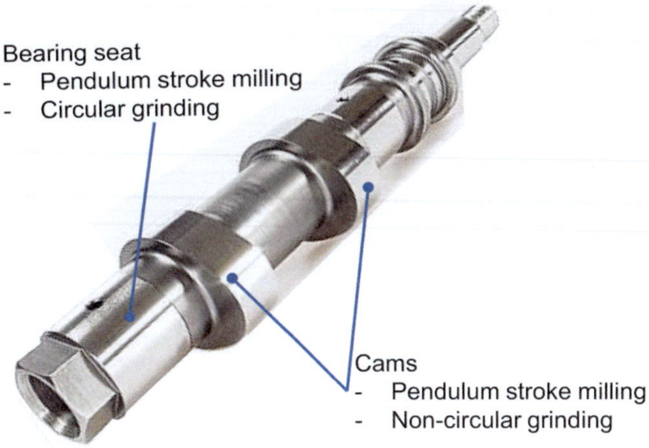

Fig. 8.60 Machining processes for sections of a cast camshaft. *Source* Windschiegl Maschinenbau

machined in one operation. Figure 8.61 illustrates the machining principle and shows the basic structure of a camshaft milling machine. For the machining process, the blank (pre-machined at the ends and bearing points) is inserted automatically into the machine, clamped in chucks at both ends and supported at the bearing points with steady rests to ensure clearance-free machining. Milling is performed while the camshaft is turning. To attain uniform chip removal per tooth, the circular feed and pendulum stroke of the milling unit are adapted to the cam profile by the machine control. After one workpiece revolution, one cam is milled to completion and the process repeats until all cams have been machined.

8.3.2 Camshaft Grinding Machines

The cam profile grinding process is similar to cam milling in terms of configuration and movement of workpiece and tool. Here too, the cam contour is created using superposition of the cam rotation and the lift movement of the grinding carriage. To generate a precise cam profile, the controller calculates, at small angle intervals, the feed path necessary depending on the current grinding wheel diameter. The turning feed of the camshaft is based on the material volume to be removed per angle unit. This is greatly dependent on the cam profile. The goal is to keep the material removal rate as constant as possible. Figure 8.62 shows the workspace of a single-slide universal grinding machine with maximum two grinding wheels for workpieces having maximum lengths of 650 mm. Machines of this type are capable of completing the machining of camshafts in a single clamping operation. This is possible because the bearings, thrust bearings, shoulders and the ends of the camshaft can be machined without reclamping. The machine is suitable for machining camshafts as well as other shaft-shaped workpieces. The Z-axis of the grinding stock spindle is a sliding guide in V-form and the X-axis is a guideway fitted with rolling elements. The machine is designed for cooling lubrications with emulsions and oil. CBN grinding wheels, and galvanically coated or metal-bonded tools can be used. The maximum wheel diameter is 480 mm. Modern-day machines for camshaft machining allow in-process measurement of the tool geometry. Measurements are taken during the grinding process. This renders manual adjustments by the machine operator redun-

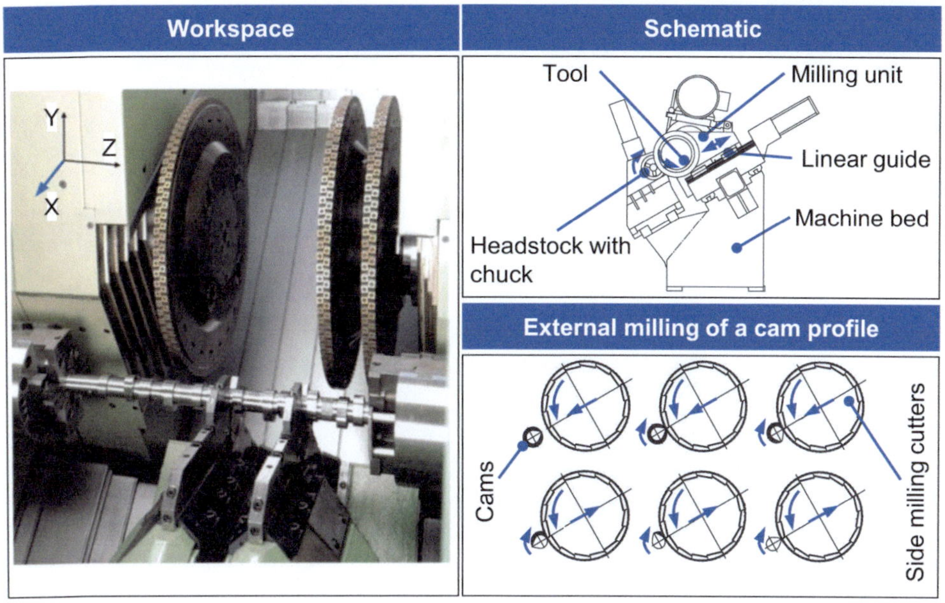

Fig. 8.61 Milling of camshafts. *Source* Heller

dant and unwanted downtimes are shortened considerably.

8.4 Machines for Producing Threaded Spindles

To machine workpieces, process forces and relative movements between the tool and workpiece are required. Both are provided by feed axes in machine tools. A feed axis can generally be implemented with a direct drive, for example a linear motor. However, the most common feed drives consist of a servo motor and associated mechanical components that convert rotational motion into translational movement. The mechanical feed components used most in machine tools include ball screw drives. The basic structure of a ball screw drive (BSD) is shown on the left in ▪ Fig. 8.63. Ball screw drives comprise a threaded spindle and nut. Balls in the thread profiling minimize friction in the thread contact. In this way, sliding friction is replaced by rolling friction with lower loss. To keep the balls in the gap between spindle and nut profiling, the ball return system feeds them back all the way through the nut either after one thread revolution or after the full cycle.

In addition to ball screw drives, machine tools also use trapezoidal spindles, hydraulic threaded spindles, and roller screw drives (RSD), which are also classified as spindle or screw drives. In these systems as well, a rotating drive motion is converted to a linear feed motion by a threaded profile. While for trapezoidal spindles the contact between nut and spindle has no intermediate elements, the friction in hydraulic screw drives is reduced by oil pockets and in roller screw drives by planetary rollers. The basic structure of a roller screw drive is shown on the right in ▪ Fig. 8.63. Ball screws and roller screw drives are discussed in detail in Volume 2, Chap. 6 of this book series.

In the production of the spindles for the spindle drives above, a general distinction can be made between two manufacturing tasks: production of the thread profiling and production of the spindle ends. This chapter covers manufacturing processes and machines for producing threads. Processes from the following groups are used mainly for producing the thread geometry: forming under compressive conditions, cutting with geometrically defined cutting edges and cutting with geometrically undefined cutting edges (▪ Fig. 8.64). The most important processes for thread manufacturing are briefly described below.

The process of thread rolling is widely used in manufacturing spindles. The process belongs to the group forming under compressive conditions (▪ Fig. 8.65). In the thread rolling process, a thread profile is created by cold forming the shell surface of circular blanks using one or two rollers under the effect of very high forces. The processes of skew rolling of solid sections and transverse rolling of solid sections are both known as thread rolling. In the process of skew rolling, the axes of the rollers are positioned at an angle to the workpiece axis. The rollers are groove-profiled and rotate about their own axes. A thread geometry is pressed into the workpiece surface by the superimposed feed and ro-

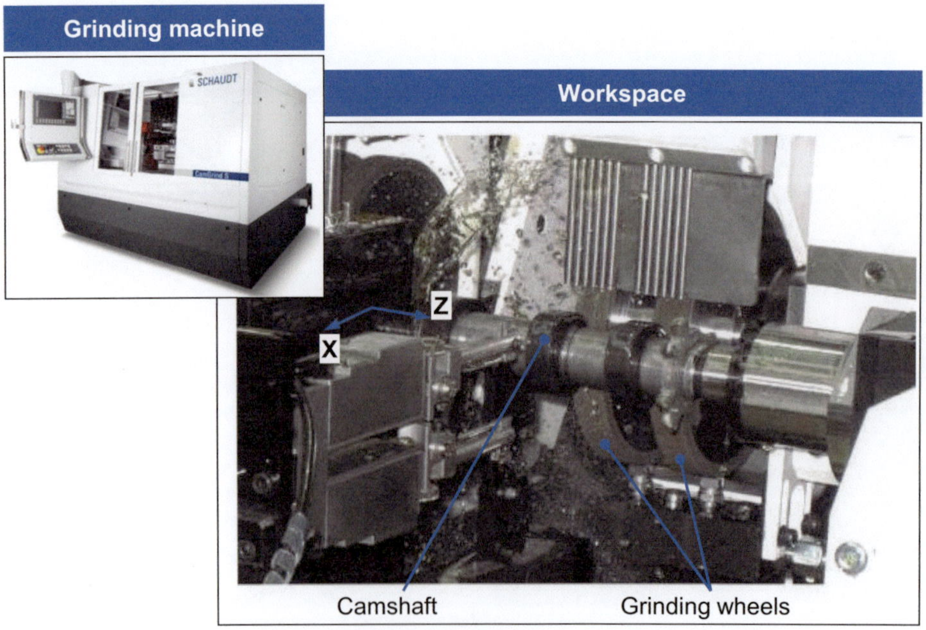

▪ **Fig. 8.62** View and workspace of a camshaft grinding machine. *Source* Schaudt

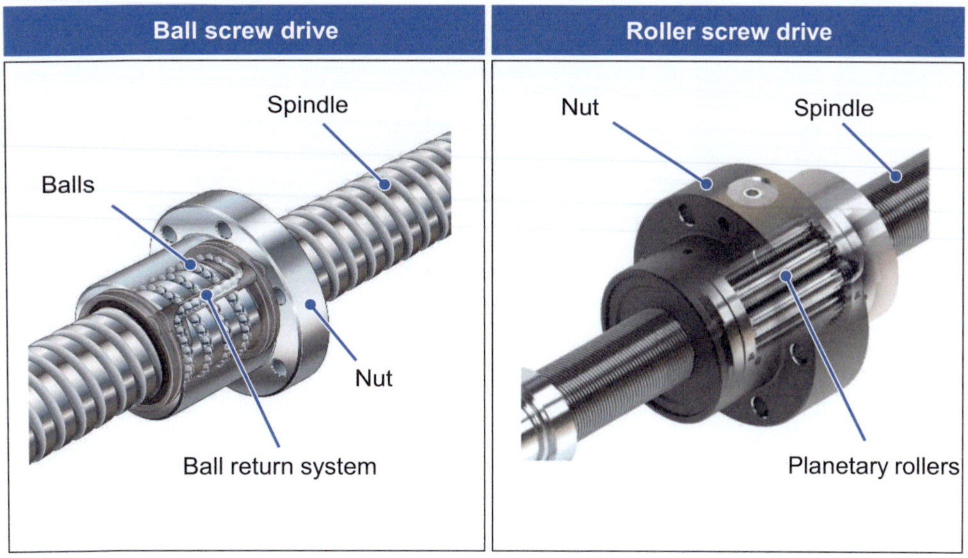

Fig. 8.63 Example of a spindle drive structure. Left: ball screw drive, right: roller screw drive. *Source* Bosch Rexroth

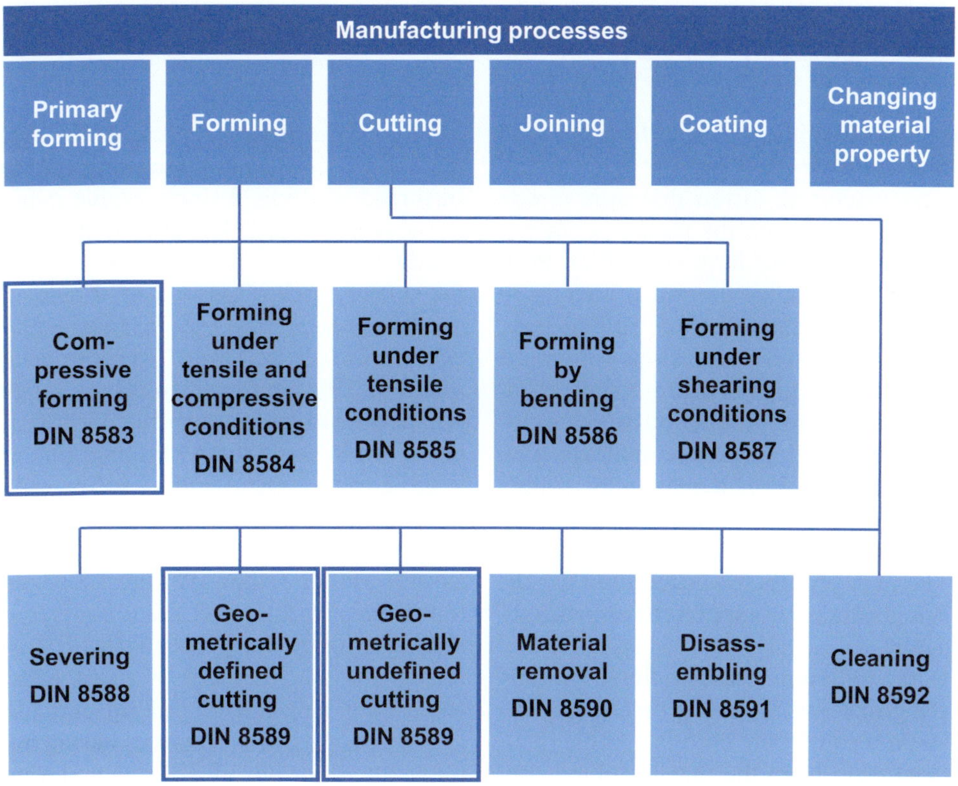

Fig. 8.64 Categorization of manufacturing processes according to DIN 8580 [DIN03a] (only an excerpt)

tation motions of the workpiece. The process is particular suited for producing long threaded rods and screw drive spindles (Fig. 8.65, right). In the process of transverse rolling of shapes, the rollers have a thread profile and are positioned parallel to the workpiece. During the machining process, the rollers and workpiece rotate about their own axes. There is no axial motion of the workpiece, meaning the thread length possible is limited by the corresponding roller length (Fig. 8.65, left).

Spindles made with thread rolling exhibit a number of positive properties. They are inherent to the process

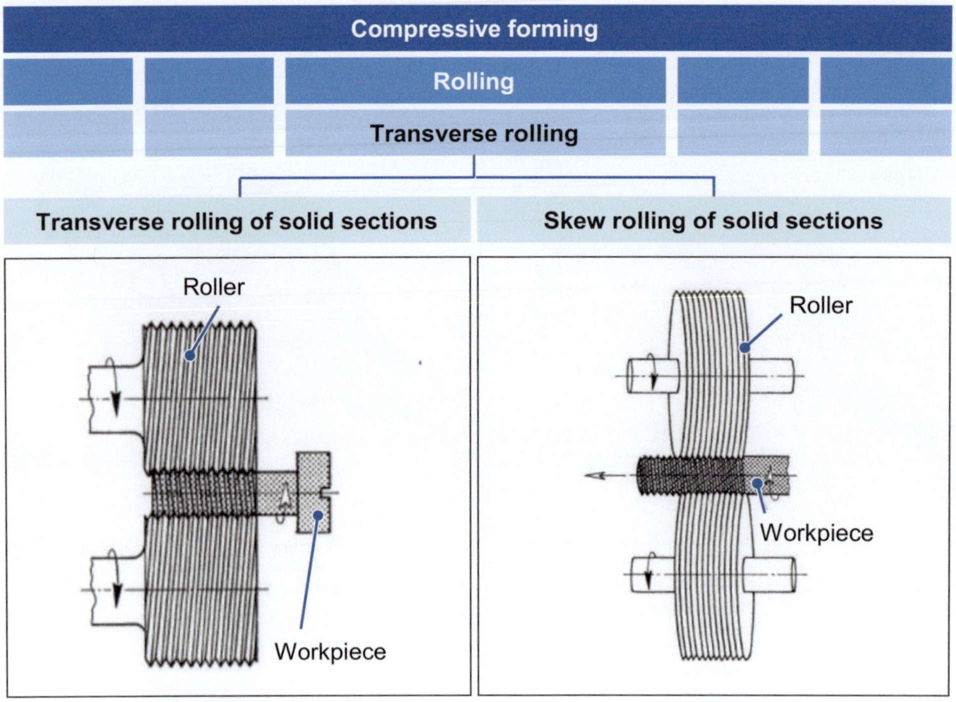

Fig. 8.65 Thread rolling according to DIN 8583-2 [DIN03e]

and include increased strength from cold forming with reduction of notch sensitivity and good roughness values on the surfaces machined. Given that appropriate rollers are required for every specific thread geometry, thread rolling is used mainly for large quantities. Rolling machines are used to manufacture rolled threads. In addition to machines specialized for one rolling type, there are also universal machines that allow transverse rolling as well as skew rolling. It is also possible to integrate rolling machines into production lines. Figure 8.66 shows a thread rolling machine that can handle both rolling processes. With a maximum rolling force of 300 kN, workpieces with a maximum diameter of 75 mm can be manufactured. The maximum thread pitch is 7 mm. The high process forces required for the rolling process can generally be applied either hydraulically or mechanically.

In addition to the rolling processes, different milling processes from the group cutting with geometrically defined cutting edges are commonly used in spindle manufacturing. The processes from the thread milling subgroup are of particular importance here. One of these processes is long-thread milling. One variation of long-thread milling is thread whirling (also called thread peeling). This process represents the most cost-effective method of thread manufacturing (Fig. 8.67, center). In thread whirling, a single-profile milling tool with internal teeth is used. The axis of this tool is inclined towards the thread pitch and its feed corresponds to the thread pitch. Classic long-thread milling has similar kinematics but uses a tool with external teeth (Fig. 8.67, left). Thread whirling brings time savings of up to 75% compared to classic long-thread milling. As the thread length possible corresponds to the milling width in short-thread milling, this process (also belonging to the thread milling subgroup) is less suited to producing spindle profiles.

Special machines are used for long-thread milling, as they are for rolling. Figure 8.68 shows a machine for thread whirling and the manufacture of a demonstration workpiece. Depending on customer requirements, the machine can be fitted with a milling tool for classic long-thread milling and for thread whirling. Production of workpieces having a thread profile, a maximum diameter of 200 mm and a maximum length of 9000 mm is possible depending on installation size. Single-start and multi-start profiled spindles can be manufactured. Measurement of the finished geometry is offered as an option. This feature is integrated into

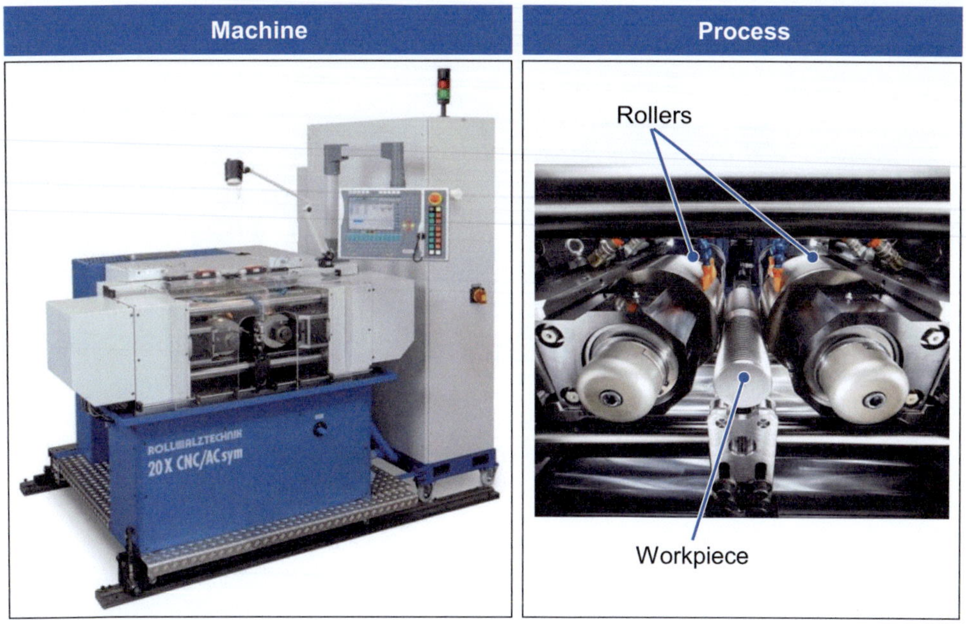

Fig. 8.66 Thread rolling. Left: machine, right: process. *Source* Rollwalztechnik Abele + Höltich

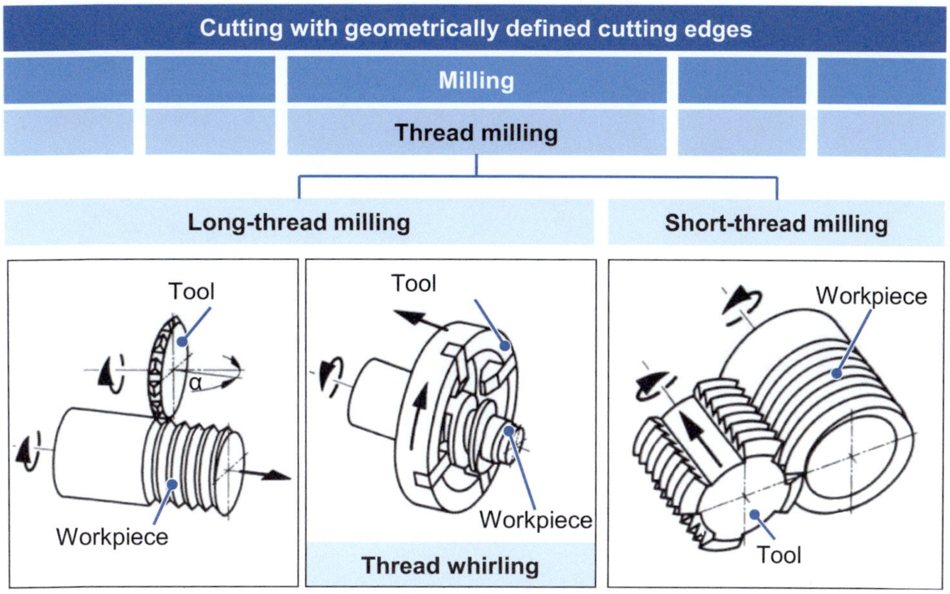

Fig. 8.67 Milling processes for manufacturing spindle profiles according to DIN 8589-3 [DIN03c]

the process and uses a measuring probe with wireless transmission.

After creation of the geometry of the ball screw spindle by one of the processes above, the surfaces are hardened and, as required, ground. Grinding is used to attain higher spindle pitch accuracies and to improve the surface roughness after milling. Longitudinal and transverse external thread grinding are often used. In longitudinal external thread grinding, a single-rib grinding wheel is used (Fig. 8.69, left). The grinding wheel and workpiece rotate. The grinding wheel is first advanced along the workpiece axis. The thread geometry is created by the axial feed motion of the workpiece (spindle) superimposing the

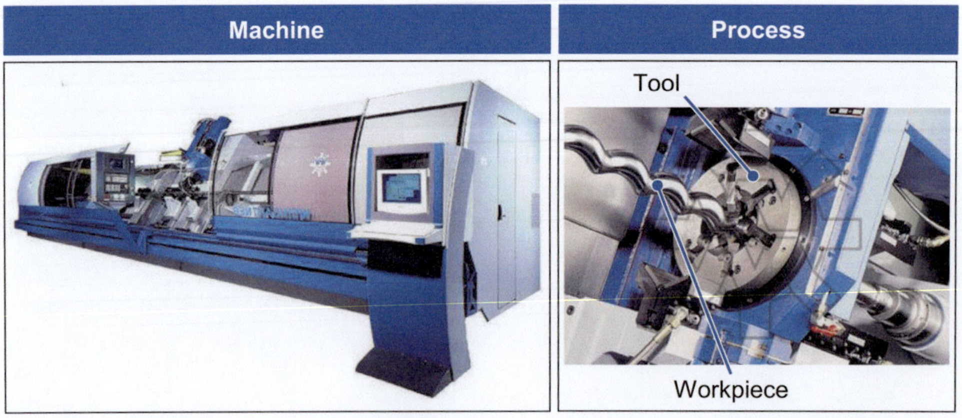

Fig. 8.68 Thread whirling. Left: machine, right: thread whirling of a demonstration workpiece. *Source* Weingärtner Maschinenbau

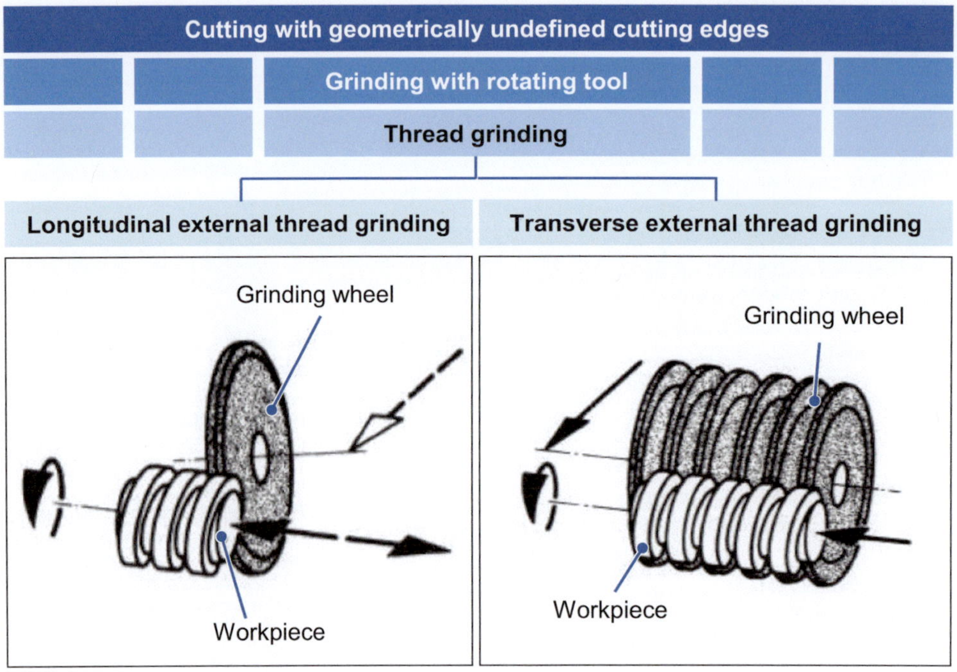

Fig. 8.69 Grinding processes for manufacturing spindle profiles according to DIN 8589-11 [DIN03d]

rotation. After the required thread length has been traversed, the workpiece is moved back to the initial position. The grinding wheel is advanced again towards the spindle axis and the workpiece is moved. These steps are repeated until the required thread depth is reached. For transverse external thread grinding, a multi-profile roller is used. This roller rotates together with the workpiece (Fig. 8.69 on right). At the same time, the workpiece is moved axially and the roller is advanced continually along the workpiece axis. Since the grinding wheel must be at least as long as the screw surface to be ground, this process is only suitable for shorter spindles and for machining spindle ends.

Figure 8.70 shows a thread grinding machine for longitudinal external thread grinding. With a workpiece length up to 1000 mm, maximum workpiece diameter of 300 mm, thread pitch angles to 40° for right-

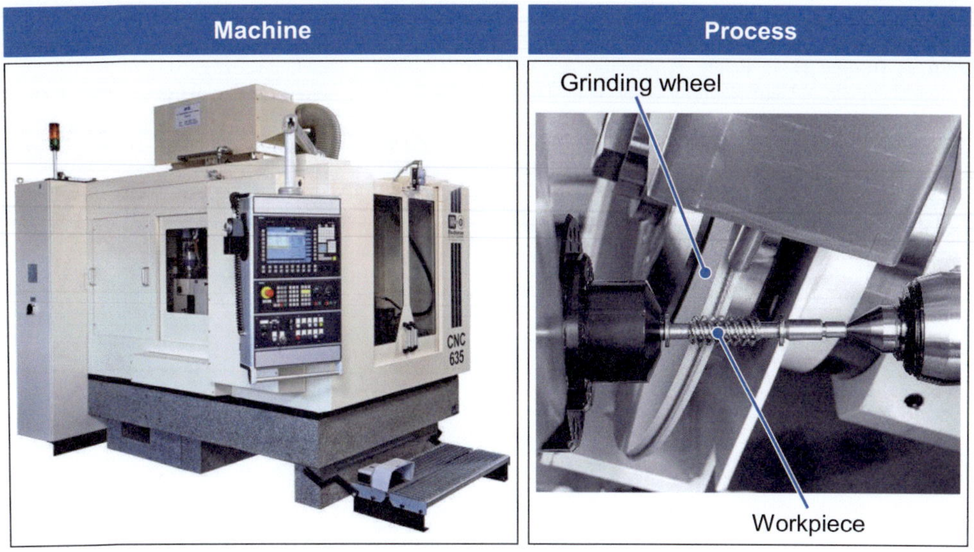

Fig. 8.70 Thread grinding. Left: CNC 635 thread grinding machine, right: longitudinal external screw grinding. *Source* Buderus Schleiftechnik

handed and 35° for left-handed threads, and a broad selection of achievable profile shapes (such as the Gothic pointed arch profile and the radius profile), the machine can be used to produce all common spindle profiles.

References

[BAUS11] Bausch, T.: Innovative Zahnradfertigung. 4. neu bearb. und aktualisierte. Aufl. Renningen. Expert-Verlag (2011)
[DIET14] Dietsche, F.: Zur IMTS mit vier Stars. Freiburg im Breisgau, 28.08.2014
[DIN03a] Norm DIN 8580 (September 2003). Fertigungsverfahren – Begriffe, Einteilung
[DIN03b] Norm DIN 8589-6 (September 2003). Fertigungsverfahren Spanen – Teil 6: Sägen – Einordnung, Unterteilung, Begriffe
[DIN03c] Norm DIN 8589-3 (September 2003). Fertigungsverfahren Spanen – Teil 3: Fräsen – Einordnung, Unterteilung, Begriffe
[DIN03d] Norm DIN 8589-11 (September 2003). Fertigungsverfahren Spanen – Teil 11: Schleifen mit rotierendem Werkzeug – Einordnung, Unterteilung, Begriffe
[DIN03e] Norm DIN 8583-2 (September 2003). Fertigungsverfahren Druckformen – Teil 2: Walzen – Einordnung, Unterteilung, Begriffe
[DIN05] Norm 867 (03/2005). Bezugsprofile
[FELT12] Felten, K.: Verzahntechnik, 3rd edn. Expert Verlag (2012)
[GLEA15a] Gear Shaving Solutions. Ludwigsburg (2015)
[GLEA15b] Gear Shaping Machines. Ludwigsburg (2015)
[HEGE11a] Hegenscheidt-MFD GmbH & Co.KG: Universale Feindreh- und Glattwalzmaschine zur Bearbeitung von Kurbelwellen TYP 7794. Erkelenz, 2011. Firmenschrift
[HEGE11b] Hegenscheidt-MFD GmbH & Co.KG: Universale Fest- und Richtwalzmaschine Typ 7893R. Erkelenz, 2011. Firmenschrift
[JAST12] Jaster, M.: Multifunctional advancements. The latest in big gear machining with DMG/Mori Seiki. Gear Technology (1), 42–44 (2012)
[KLIN08] Klingelnberg, J.: Kegelräder. Springer Berlin Heidelberg, Berlin (2008)
[KLOC05] Klocke, F., König, W.: Fertigungsverfahren 2. Schleifen, Honen, Läppen. (Reihe: Fertigungsverfahren. Bd. 2). 4th edn. Springer, Berlin (2005)
[KLOC13] Klocke, F., Brumm, M., Staudt, J.: Potentiale des Freiformfräsens von Verzahnungen. Tagungsband Zahnrad und Getriebetechnik. Tagungsband zu Zahnrad- und Getriebetechnik. (Bd. 54.). Aachen, 15.–16.05.2013: WZL der RWTH Aachen (Veranst.), S. 15
[KLOC17] Klocke, F., Brecher, C.: Zahnrad- und Getriebetechnik. Hanser, München (2017)
[KRUM67] Krumme, W.: Klingelnberg Spiralkegelräder, 3rd edn. Springer, Berlin (1967)
[LITV04] Litvin, F.L., Fuentes, A.: Gear geometry and applied theory, 2nd edn. Cambridge University Press, Cambridge (2004)
[MAAG85] Maag-Zahnräder AG (eds.): MAAG Taschenbuch. Zürich (1985)
[MÜLL13] Müller, H.: Kegelradmaschinen. (Bd. 360). Moderne Industrie, Landsberg (2013)
[OV14] o. V.: Die CNC-Wälzfräsmaschinen LC 80 – LC 180. Kempten, 07/2014
[OV15a] o. V.: 275HC Bevel Gear Cutting Machine. Ludwigsburg (2015)
[OV15b] o. V.: RZ 550. Wallisellen (2015)
[OV15c] o. V.: HMX 400. Bleienbach (2015)
[PFAU76] Pfauter, H.: Pfauter-Wälzfräsen. Springer, Berlin (1976)
[POTT13] Potts, M.: Steigerung der Tragfähigkeit freiformgefräster Kegelräder im Großmodulbereich. DMK 2013. Tagungsband zu: DMK; Dresdner Maschinenelemente Kolloquium; DMK 2013 (Veranst.)
[ROTH00] Roth, K.: Zahnradtechnik. (Bd. 1). 2nd edn. Springer, Berlin (2000)
[ROTH98] Roth, K.: Zahnradtechnik – Evolventen-Sonderverzahnungen zur Getriebeverbesserung. Springer, Berlin (1998)

[SAND13] Sandvik AG: Drehfräsen. Düsseldorf, Wien, Luzern, 2013. Firmenschrift

[SCHM03] Schmidt, J., Tröndle, H.-P., Felten, K., Bechle, A.: Anforderungen an die Neuentwicklung einer Wälzschälmaschine – mechatronische Analyse des Maschinenkonzeptes. wt Werkstattstechnik online **93**, 549 (2003)

[SCHR06] Schröder, T.: Analyse der Werkzeugbelastungen beim Zahnradschaben. Diss. RWTH Aachen (2006)

[SCHR08] Schriefer, H.: Kontinuierliches Wälzschleifen von Verzahnungen, 1st edn. Reishauer, Wallisellen (2008)

[SCHU15] Schug, P.: Modellierung, Bewertung, Analyse und Optimierung von CAx-Prozessketten. Diss. RWTH Aachen (2015)

[SEHE13] Seherr-Thoss, H.C.v.: Die Entwicklung der Zahnrad-Technik. o. O.: Springer (2013)

[STAD13] Stadtfeld, H.J.: Gleason Kegelradtechnologie. Renningen: expert (2013)

[o.V.15] 130SV(C) Shaving Machine. Ludwigsburg (2015)

8

Multi-Machine Systems

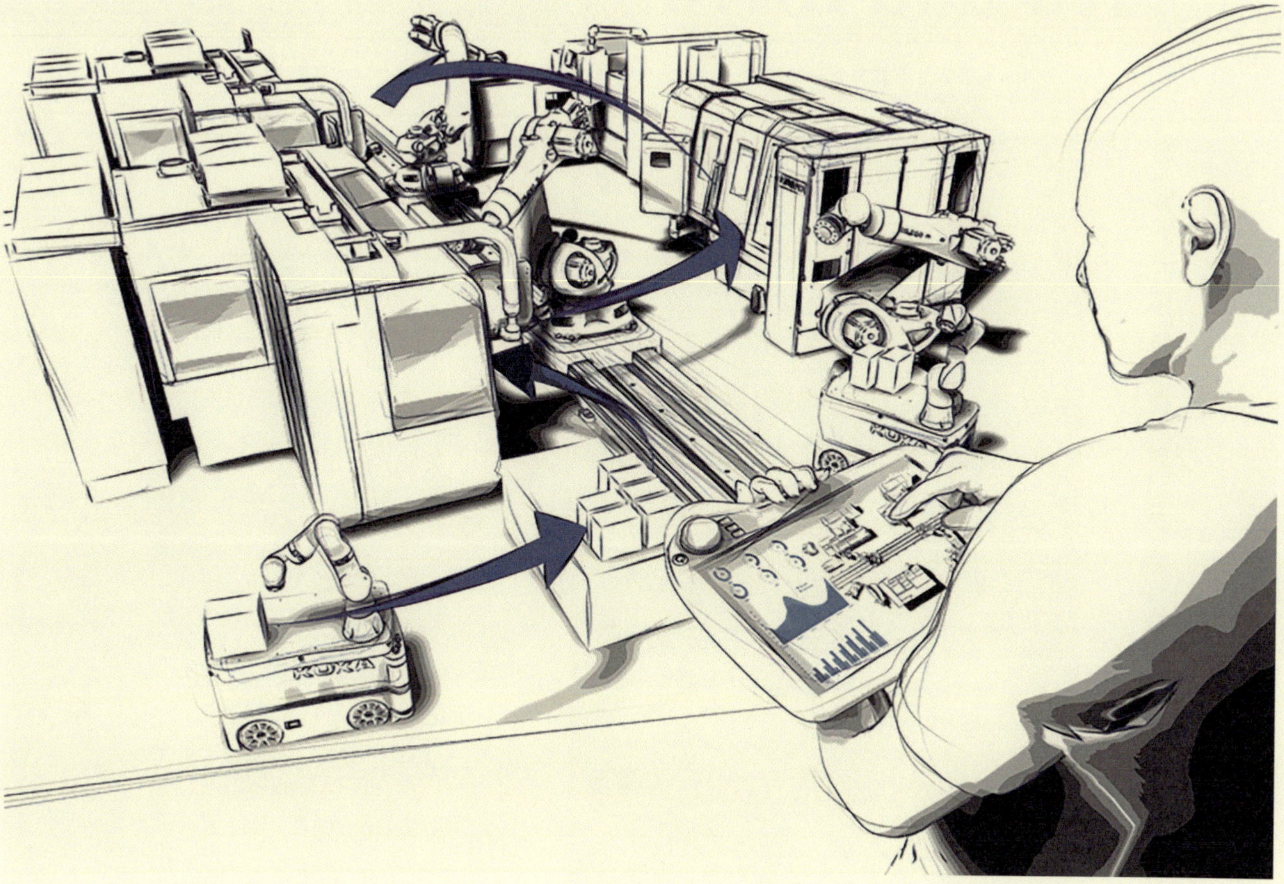

I thank my staff members Manuel Belke M.Sc. and Mathias Dehn M.Sc. for their contribution in revising this chapter.

C. Brecher and M. Weck, *Machine Tools Production Systems 1*,
Lecture Notes in Production Engineering, https://doi.org/10.1007/978-3-662-68120-6_9

Contents

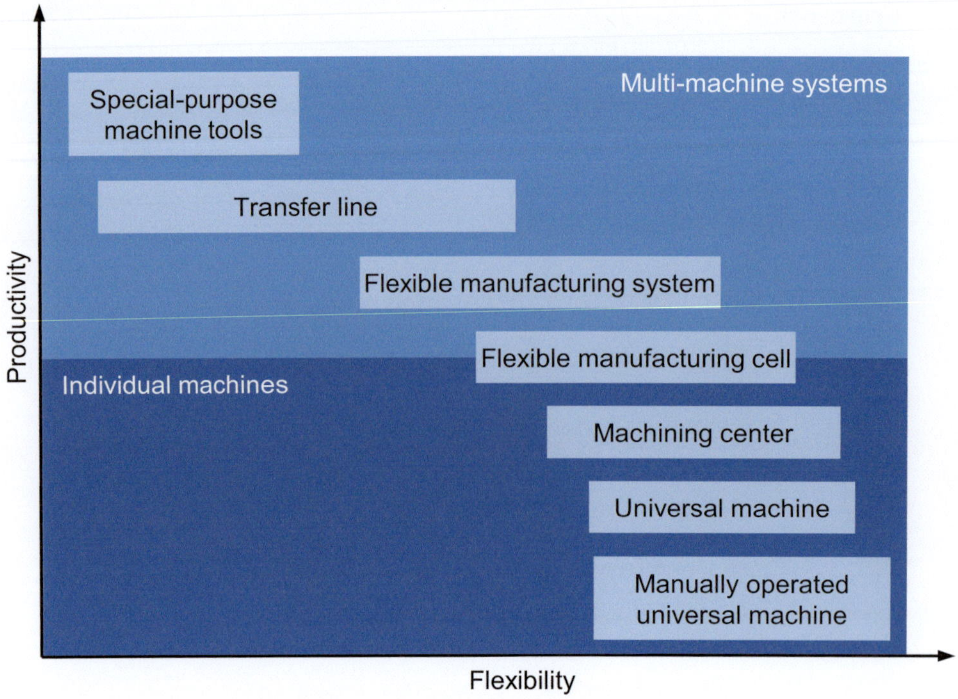

Fig. 9.1 Productivity and flexibility of different manufacturing systems

There has been a demand for continually increasing flexibility and productivity of industrial production since machine tools have been in existence. Figure 9.1 shows how the multi-machine systems described in this chapter can be classified with regard to the fundamentally conflicting demands for the maximum flexibility and productivity. In addition to the group of multi-machine systems, individual machines are also listed. The range extends from hand-operated universal machines exhibiting high flexibility yet low productivity, to special-purpose machine tools that are highly productive but highly inflexible.

Special-purpose machine tools and transfer lines are workpiece-specific multi-machine systems, i.e., the individual machining stations of a system tailored to the special machining tasks of a workpiece, and possibly also a limited range of workpiece variants (single-purpose machines tools). In light of product life cycles generally becoming shorter, broader product variety, and increasingly smaller quantities and batch sizes, volume production required increasing the product and quantity flexibility of fixed transfer lines. This led to the development of flexible transfer lines. These are characterized by an adaptability to different machining tasks by interlinking numerically controlled machining units.

The demand for greater flexibility and extensive automation, both in terms of machine usage and freedom in floor space configuration, as well as with regard to system extensions and reuse, has resulted in the use of flexible manufacturing systems in low, medium and large volume production. A distinguishing feature of these systems is that multiple universal NC machines of similar or different types are connected using a shared, automatic workpiece supply and possibly a tool supply with an integrated computer control to form a closed system.

Flexible manufacturing cells are mainly used in production sectors where smaller batch sizes must be made cost-effectively. A flexible manufacturing cell is a system with great flexibility using a universal machine (machining center) as the central component of the system. Coupled with this are a workpiece magazine, equipment for workpiece handling and tool handling, if necessary. However, the productivity in this case is much lower than for transfer lines.

The following chapters each introduce examples from applications of special-purpose machines, fixed and flexible transfer lines, flexible manufacturing cells and flexible manufacturing systems.

9.1 Fundamentals of Machine Interlinking

Multi-machine systems are individual systems tailored to the process being used. They are not designed to any fixed assembly plan. Modular systems can aid in assembling the system as early as the process planning stage. The following chapter first introduces different manufacturing forms that are used in practice. The selected manufacturing form determines how individual machines are arranged or grouped within multi-machine systems. Further, machine tools typically used in multi-machine systems are described. Finally, machines and systems for interlinking into multi-machine systems are discussed.

9.1.1 Manufacturing Forms

If different process steps are necessary to produce a component, it is necessary to machine the workpiece on different machine tools. The possible machine arrangements are known as manufacturing forms and can be categorized into the process and object-oriented manufacturing forms shown in ◘ Fig. 9.2 in terms of their typical floor space configurations and key features.

In classic workshop manufacturing, machines having the same or similar manufacturing processes are grouped. All machines are stationary while the workpieces are moved between the machines without interlinking in line with manufacturing requirements. The benefit of workshop manufacturing is flexibility in terms of changing production tasks. The high transport cost is a disadvantage [GRUN09]. Workshop manufacturing is one manufacturing form that is process-oriented.

Another manufacturing form that allows some process-oriented manufacturing but also object-oriented manufacturing is nest production. In nest production, individual machining stations are moved together, usually arranged in a ring shape. The selection of the

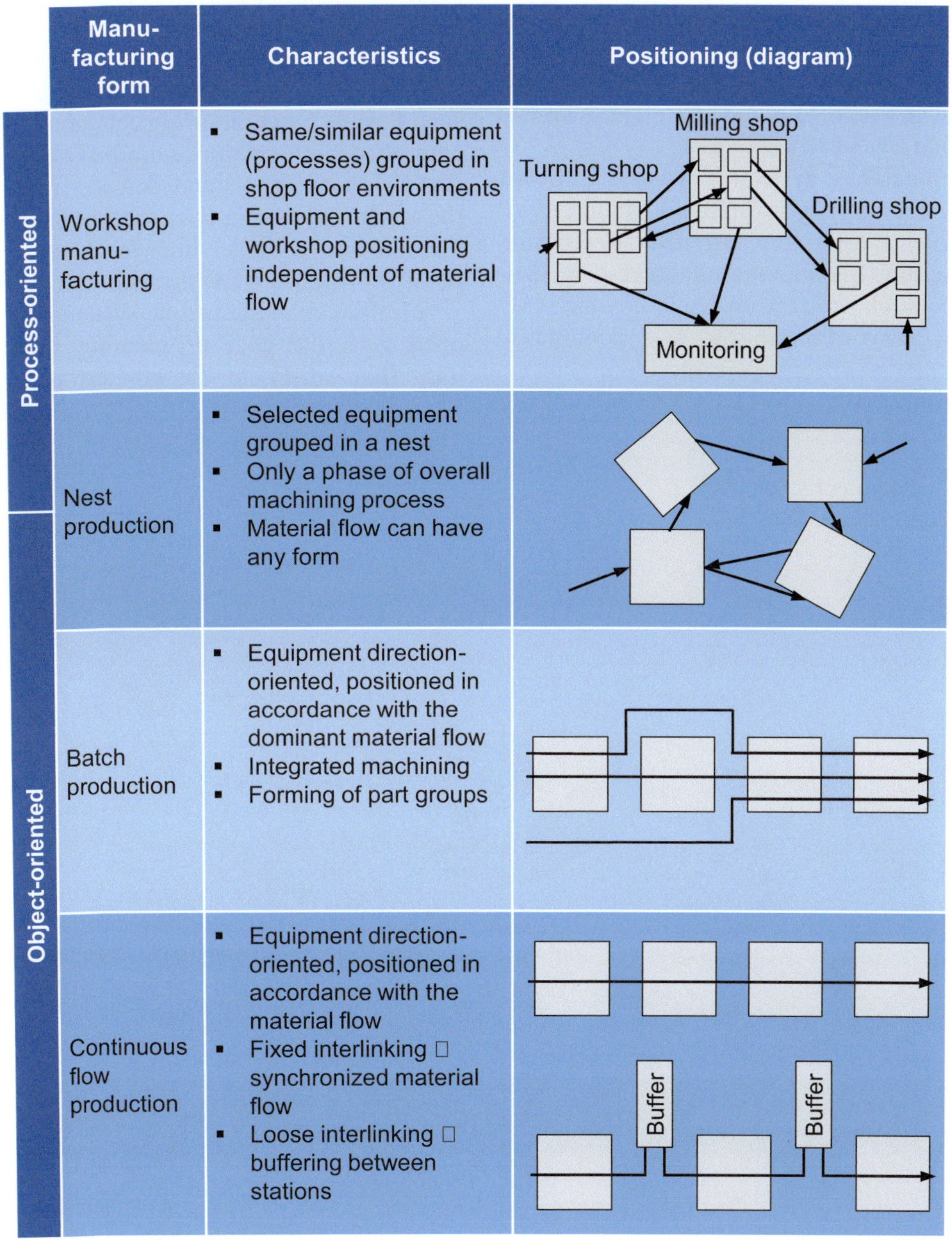

◘ **Fig. 9.2** Manufacturing forms, according to [GRUN09]

machines depends on the work steps to be performed on the workpiece. A nest usually only covers one phase of the overall machining process. Material flow within the nest can take any form.

Batch production represents a completely object-oriented manufacturing form. For the end-to-end machining of workpieces, the machine tools necessary are positioned in line with the dominant material flow. It is possible to stop machining workpieces at defined stations or supply the production line separately. This is not possible in continuous flow production. The difference between this and batch production is fixed or loose interlinking. A multi-machine system with continuous flow production and fixed interlinking follows a fixed cycle. This presupposes that the process times in the individual stations are synchronized to each other and are not subject to major fluctuations. If this can not be guaranteed, a loose, flexible interlinking can be used. In loose interlinking, there are buffers between the stations that are able to compensate (to a limited extent) for fluctuations in process times.

The multi-machine systems discussed below fall predominantly in object-oriented manufacturing forms. Transfer lines are assigned to continuous flow production. Flexible manufacturing systems are allocated to batch production while flexible manufacturing cells very often follow nest production [GRUN09].

9.1.2 Interlinked Machines in Machine Systems

For greater understanding, strict delineation of the terms "machining center", "flexible manufacturing cell" and "flexible manufacturing system" is key. For this reason, systems having these designs are delineated from each other below in the detailed consideration of the material flow (workpiece and tool).

An NC production machine with automatic tool change and a tool magazine is called a machining center. While workpieces are usually changed and clamped individually by hand here, on the next-highest automation level (the flexible manufacturing cell), the machining center is normally connected to a workpiece magazine by a workpiece change system (see ◘ Fig. 9.3). Automatic workpiece transport is undertaken, for example, by handling systems or roller conveyor systems that clamp workpieces on pallets. One such flexible manufacturing cell, controlled by a master computer, can work autonomously over a long time period. Monitoring systems detect errors and faults with the machinery and in the process cycle. In this way, production over a longer time (such as an entire shift) is possible with just a few operators or even none at all.

Flexible manufacturing systems have been developed with the goal of creating manufacturing systems that satisfy, to the greatest possible extent, the

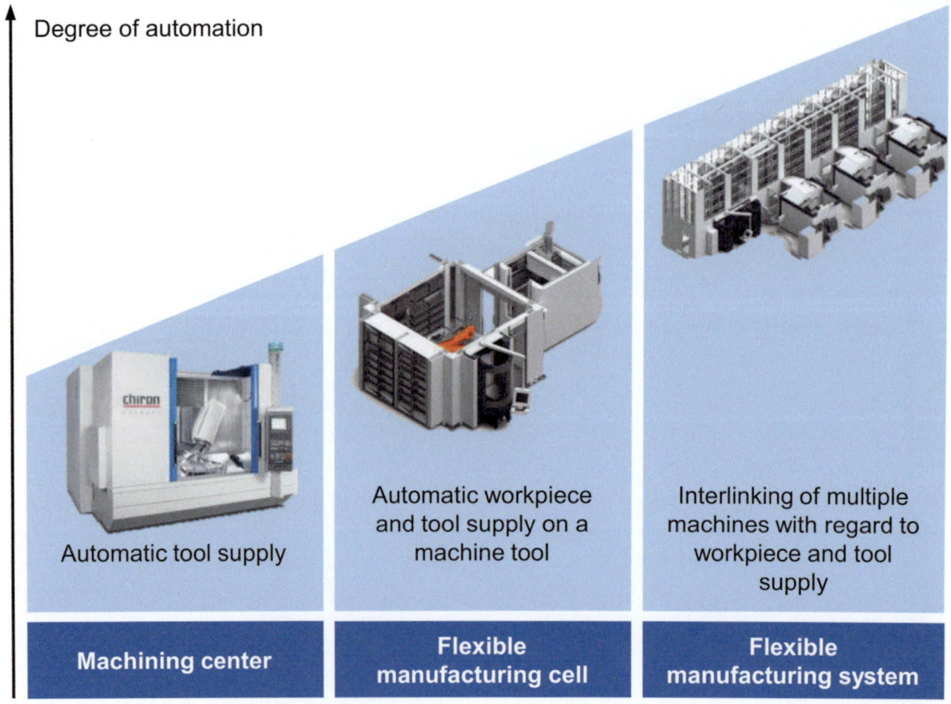

◘ **Fig. 9.3** Degrees of automation for different manufacturing systems, according to CHIRON Werke and PROMOT Automation

conflicting requirements for high productivity yet high flexibility. Several NC machines, such as machining centers, measuring machines and cleaning machines, are connected by a computer-aided workpiece transport system. In conjunction with flexible manufacturing systems, interchangeable machine tools are understood to be machines of the same design that have the same machining functions and can be replaced by each other. In such cases, the term used is single-stage manufacturing systems, i.e., a workpiece moves to only one machine tool at a time for its finishing process. If machines in a manufacturing system are designed to complement each other with regard to their machining tasks, the term used is complementary machines. In such cases, a workpiece must move to multiple machine tools for its finishing process. Accordingly, such a system is called multi-stage manufacturing system. In addition to the NC machines, some of the automatic workpiece magazines for blanks and finished parts are coupled to the conveyor systems. In order to extend the scope of applications, i.e., flexibility, automatic provisioning and changing of tools are possible using automatic conveyor and handling systems. The design of flexible manufacturing systems is tailored to the requirements of individual part and small volume production, in particular the production of spare parts and families of parts. Large volume production trials using flexible manufacturing systems as an alternative to customary transfer lines are underway.

In the 1980s, the aim was to increase the technical availability of manufacturing systems to above 90% using flexible manufacturing systems. Another goal was to save costs by reducing the deployment of personnel. The purpose of automating tool management was to fully utilize the tool life and thus reduce tool costs. With these goals in mind, manufacturing facilities comprising up to 30 interlinked machine tools were implemented and were able to make a range of parts covering several hundred different workpieces [VIEH92].

These objectives for the performance of flexible manufacturing systems from the 1980s and 1990s could usually not be achieved. The great complexity of such systems, including their control solutions and in part inadequate process reliability, frequently resulted in unjustifiably long downtimes, meaning the stipulated system availability and operational reliability could not be attained [LAY01]. Also, operating and service personnel were frequently overtaxed [HOFM00]. Systems could usually only be operated by experts who were able to quickly locate and rectify faults. In hindsight, the large investment for these fully automatic, interlinked multi-machine systems were not proportional to the objectives achieved such as increased productivity, higher quality and greater machine utilization. The level of complexity in view of the probability of system failure, the level of employee training of those responsible for the system and the utilization level, must always be evaluated as accurately as possible in terms of cost effectiveness.

Owing to experience gained and technical progress over recent years, flexible manufacturing systems have overall become more process-reliable, more mature and smaller, generating increased interest from users in such systems. This trend is also reflected in the market development of single-station unit construction machines and multi-station transfer machines made in Germany. Sales of these rose sharply from € 466 million to € 902 million (+93%) between 2010 and 2014. Within the same time period, German production of cutting machine tools rose by about 55% (from € 5.092 million in 2010 to € 7.912 million in 2014) [VERE15].

9.1.2.1 Machine Tools

As an example of a machining station for multi-spindle machining in a multi-machine system, ◘ Fig. 9.4 shows a modular machining unit with a spindle head changer and integrated multi-spindle head magazine. The machining unit can hold up to six single and multi-spindle heads with task-specific spindle diameters and bearings. When single-spindle heads are used, a tool magazine with changer can also be provided. It is also possible to load more spindle heads from an additional magazine at the back. The modular system features downward and upward compatibility. The base module comprises a 3-axis NC unit with exchangeable single and multi-spindle heads. The simplest variant is a 1-axis NC module (Z-axis) with fixed spindle head. The unit has a spindle head change time of seven seconds and a chip-to-chip time of ten seconds.

Another multi-spindle unit with drill head change system is shown in ◘ Fig. 9.5. A drill head change uses translational and rotational movement of the tool holder. After the tool holder (orbiter) moves back, the new drill head is taken (see ◘ Fig. 9.5b). Then the orbiter is rotated by 180° (see ◘ Fig. 9.5c). The old drill head is put down and the new one is brought into the working position by a translational movement of the tool holder. Alternatively, four drill heads can be positioned around the orbiter. These are then brought into the working position with rotation of the carrier. The drill head change times of such systems is nine seconds.

In multi-machine systems, multi-spindle machining centers (such as dual-spindle machines) are seeing increased use to either increase throughput compared to standard machining centers or to reduce system space requirements and investment costs. There are essentially two basic machine concepts:

- Multi-spindle machining on the same machine column
- Dual-spindle machining with a separate column for every spindle unit

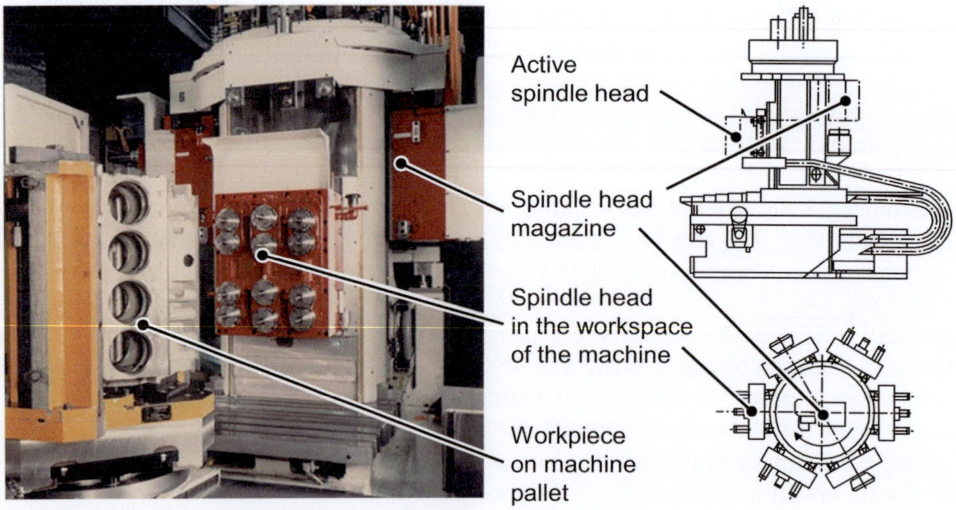

Fig. 9.4 Drill head change unit with integrated drill head magazine, according to Honsberg

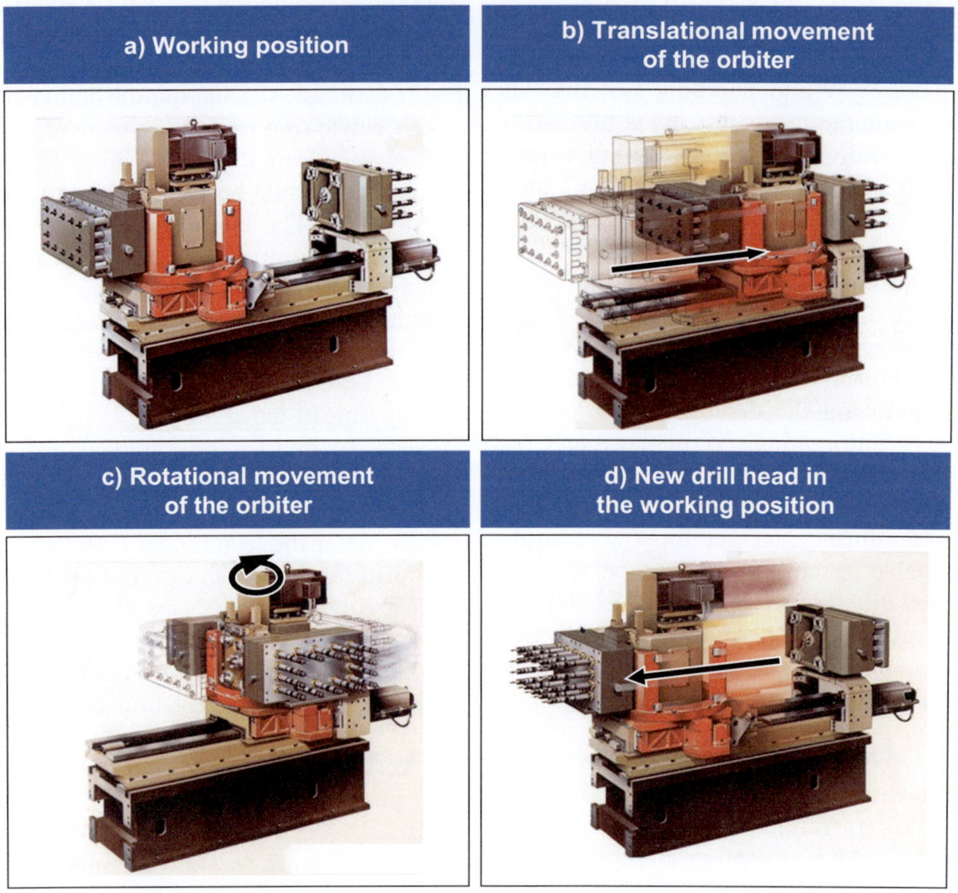

Fig. 9.5 Automatic drill head changer in flexible transfer line, according to MAG IAS

If the spindles machine several identical parts at the same time throughput per machine hour increases. In this case, the machines are equipped with up to eight work-holding tables. These tables are positioned next to each other and identical workpieces can be machined by the parallel spindles. One such machining center with four spindles is shown in Fig. 9.6. The four spindles use the same traversing units. Consequently, four workpieces having identical contours can be made. The comparatively limited dynamic properties

■ Fig. 9.6 Multi-spindle machining center. *Source* CHIRON Werke

represent one downside of this concept. The cutting capacities, in a stable process, are lower compared to two individual mill centers. As a result, the productivity of a dual-spindle machine only corresponds to 1.7 to 1.8 times that of a similar mill center (depending on the design). In addition to the design with four spindle units on the same machine column, there are also mill centers with separate columns for every spindle unit. The dual drive and guidance systems and the greater space required are disadvantages. These mill centers can however work on one workpiece with different NC programs simultaneously. Each of these spindle units can be fitted with one or more spindles [JUNG99].

Simultaneously machining two workpieces in separate clamping operations with two identical tool spindles is a feature of the dual-spindle machining center shown in ■ Fig. 9.7. It is possible to machine all workpieces on five sides. Simultaneously clamping two workpieces reduces single part production times through lower non-productive times and unit cost savings by up to 40%.

Increasing the number of working spindles in action at the same time results in parallel machining and to a proportional increase of production output. Also, the machine uses a tool magazine with automatic tool change system for every spindle.

Another feature is the ability to safely load and unload the machine tool during primary machining time through a center partition. This workpiece changer enables, in one enclosed machining area, the finished workpieces to be removed and the machine to be loaded with new parts. In the other enclosed machining area, dual-spindle machining of workpieces is happening at the same time. Parts are then transported simultaneously from one enclosed machining area to the other by a swivel system. Important to remember in this machine concept, however, is that the work processes taking place simultaneously excite each other dynamically. This can lead to process instabilities. Such situations must be prevented by selecting the process parameters accordingly.

9.1.2.2 Cleaning Machines

The demand for high throughput while adhering to defined cleanliness specifications is a challenge for companies in the metalworking industry, in particular for parts destined for the automotive industry, precision engineering and hydraulic applications [SCHU12]. Compliance with the strict requirements on workpiece cleaning is mandatory here. For this reason, machining stations in multi-machine systems are often equipped with special cleaning machines to remove oil, emulsion and chips on workpieces.

A cleaning result is essentially measured on the basis of two evaluation criteria. First, there must be no particles or residue on the component exceeding a

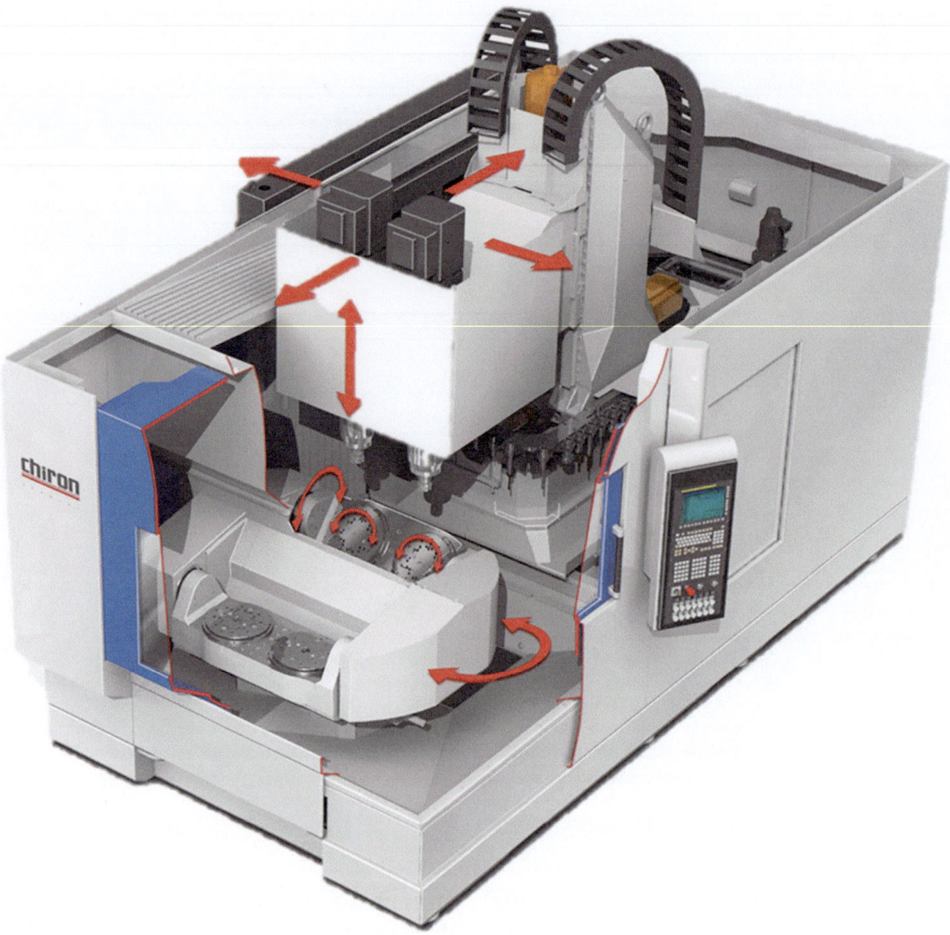

Fig. 9.7 Dual-spindle machining center. *Source* CHIRON Werke

defined size threshold, and second, the total residue not removed is measured. A typical residual dirt requirement for a drive shaft stipulates that no particles or residue greater than 200 µm and a maximum residual dirt amount of 0.5 mg per component may remain [HAIM13].

Cleaning processes are workpiece-specific combinations of the spray method, injection flooding cleaning, washing using the spray method and washing using the injection flooding method [ISEN11]. In spray cleaning, a jet of cleaning agent with high kinetic energy impacts the component. The water jet pulse removes particle contamination in particular, such as chips. This cleans machining residues from blind holes and undercuts with cavities. In the injection flooding processes, the entire component is dipped into a cleaning medium and washed using a high flow rate. Drying using the air circulation method or a vacuum comes next [ISEN11]. In vacuum drying, the boiling temperature of the water residue is reduced by evacuating a vacuum chamber. The result is that the residue evaporates at ambient temperature and the workpiece dries.

Recent decades have seen centralized cleaning concepts give way to decentralized, task-specific solutions. The often modular solutions offer shorter logistical paths and maybe even redundancy to avert bottlenecks [KOBL14]. One such modular cleaning system is shown in Fig. 9.8.

The cleaning system shown is divided into four modules: interlinking, airlock, main system and vacuum dryer. Workpieces are fed to the airlock via interlinked conveyor. Here, components are gripped either by a 6-axis industrial robot (see ▸ Sect. 9.2.1.2) or a SCARA manipulator, designed specifically for handling workpieces in cleaning systems. The corresponding robot is located in the process area of the cleaning system and manipulates the workpieces in line with the cleaning process at the various stations (spray or injection flooding method). The workpieces can either be fixed on a workpiece carrier or be gripped directly by the robot.

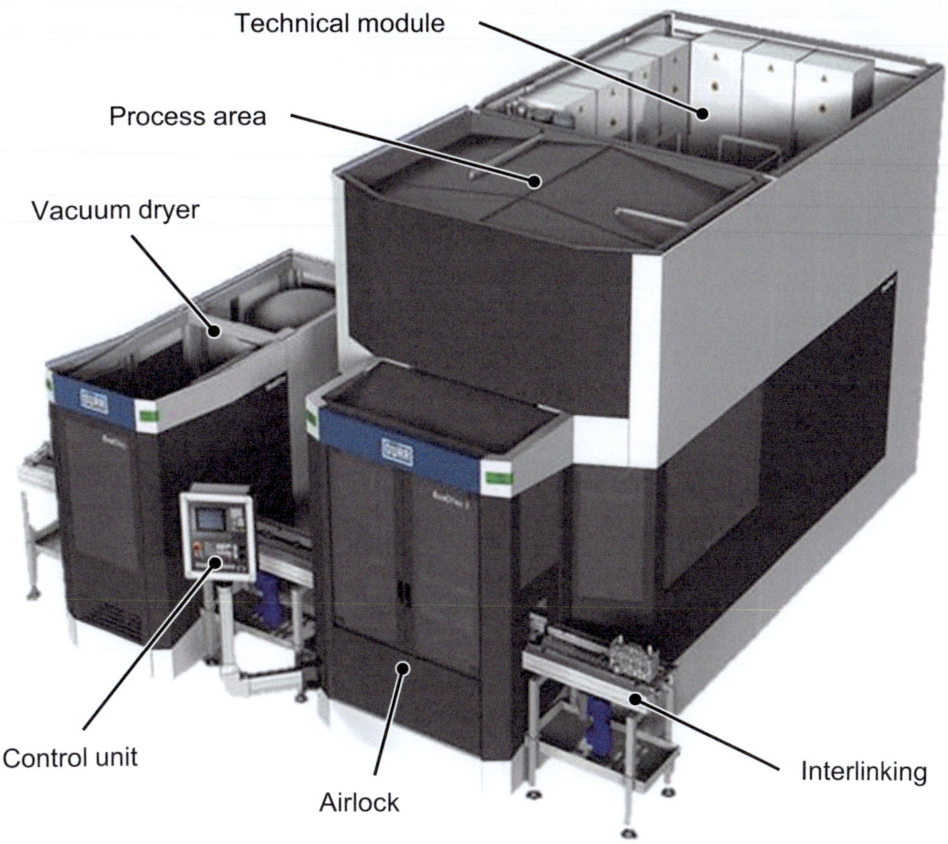

Fig. 9.8 Cleaning system, according to Ecoclean

The back part of the cleaning system houses the technical module. This contains all the systems required for machine operation. These systems include standalone devices, i.e., for high-pressure preparation, filtering and heating of the cleaning media, system and robot control as well as electrical cabinets for cabling/wiring. After the cleaning cycle, the workpiece is placed back into the airlock by the manipulator. The cleaned component is fed to the dryer via the interlinked conveyor, while the cleaning process is run again for the next component. In the system shown, the dryer is a vacuum dryer.

9.1.3 Interlinking Systems

Figure 9.9 shows an overview of typical interlinking systems in multi-machine systems. In the variant shown in Fig. 9.9a, a conventional articulated arm robot is used to load the machining centers. Workpieces are fed and removed by roller conveyors (cell interlinking). On the one hand, this concept is mid-range regarding the space requirement and costs as compared to other variants, but it has downsides regarding accessibility, volume flexibility and extendability. Beneficial on the other hand is the straightforward integration of buffer places in the cell (storage function of the cell) because the robot is able to reach any position within its radius of action. The variant shown in Fig. 9.9b represents a simple and low-cost solution. The cells are also interlinked with roller conveyors here, where feed devices reroute workpiece carriers into the individual machines. This solution is particularly used for linear material flows with just a few machines working in parallel. If more flexibility regarding the capacity is required, individual cells are fitted with multiple machining centers. The supply of workpieces to and removal from machines are realized by gantry loaders that take workpieces from an intermediate storage. This can be realized with a workpiece circulation conveyor, for example, as shown in Fig. 9.9c. Workpiece flow across cells uses other conveyors of a portal design. These carry the workpieces between the workpiece circulation conveyors of the individual cells. Of benefit here in particular are volume flexibility and extendability, as well as very good accessibility to the individual machining stations. However, the gantry loaders increase investment costs and reduce availability. For this rea-

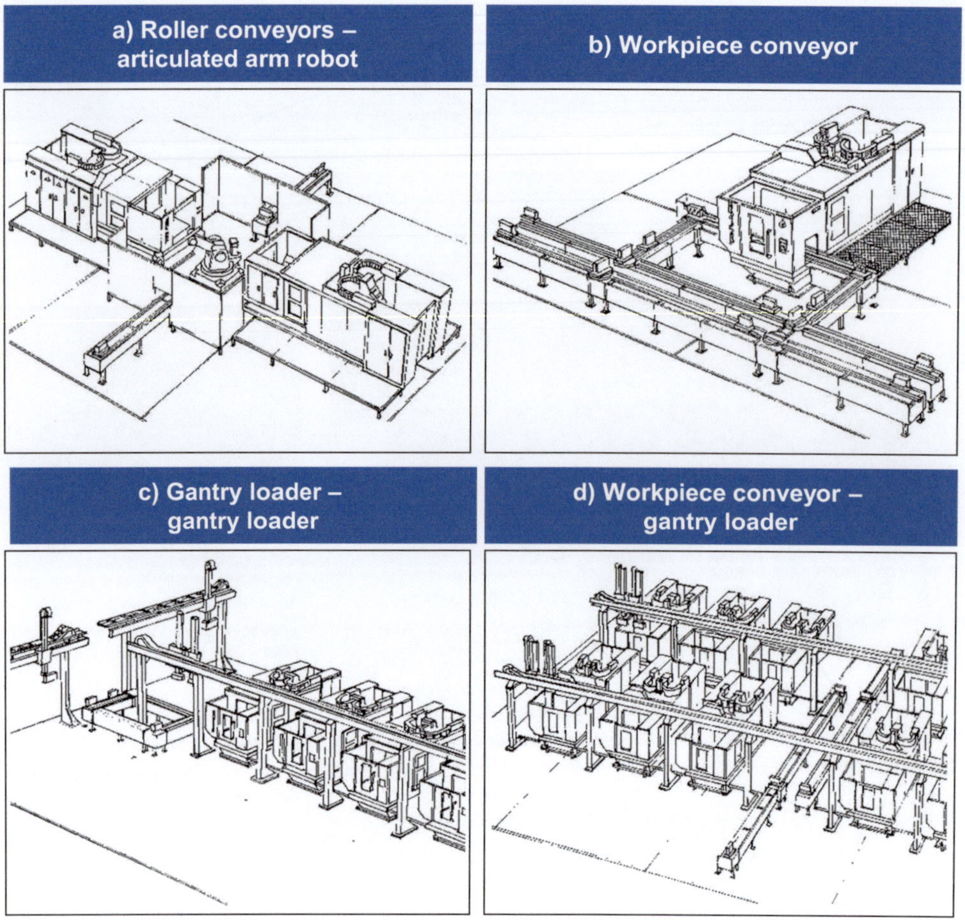

Fig. 9.9 Interlinking types in multi-machine systems, according to MAG IAS

son, large transfer lines are often designed like the variant shown in Fig. 9.9d. Gantry loaders are also used here to load the machines while roller conveyors are used for cell interlinking. Here, as shown, a roller conveyor (left) can be used to deliver blanks or semi-finished parts. The finish machined parts are placed on a second roller conveyor (right) for further transport. The roller conveyors are used at the same time as intermediate storage. This also achieves the highest availability of the interlinking concepts shown.

The lift/lower conveyor system shown in Fig. 9.10 is used in transfer lines for moving workpieces within a machining station. This conveyor system is usually used in the automotive industry to transport and position chassis and car body parts. The movements use two regulated servo drives. One of the drives enables the up and down motion via a cylinder crank while the second drive provides the longitudinal motion via a form-fit belt system. The up and down motion has five steps. One cycle starts with the upward movement of the workpiece carrier. A component is grabbed and then lifted to the conveyor position. Then the system lowers. Once the component is placed on the positioning points, the workpiece carrier moves to its end position. One such cycle takes less than eight seconds. These types of transport systems are normally used for transfer lines with multiple, consecutive machining stations. The individual transport systems must be synchronized so that a finished workpiece is taken from a machining station by the downstream transport system before the transport system upsteam of the machining station passes on a new workpiece. The benefits of the transport system shown are the small space requirements, jerk and impact-free lifting and lowering, variable step lengths and lifting heights as well as short cycle times.

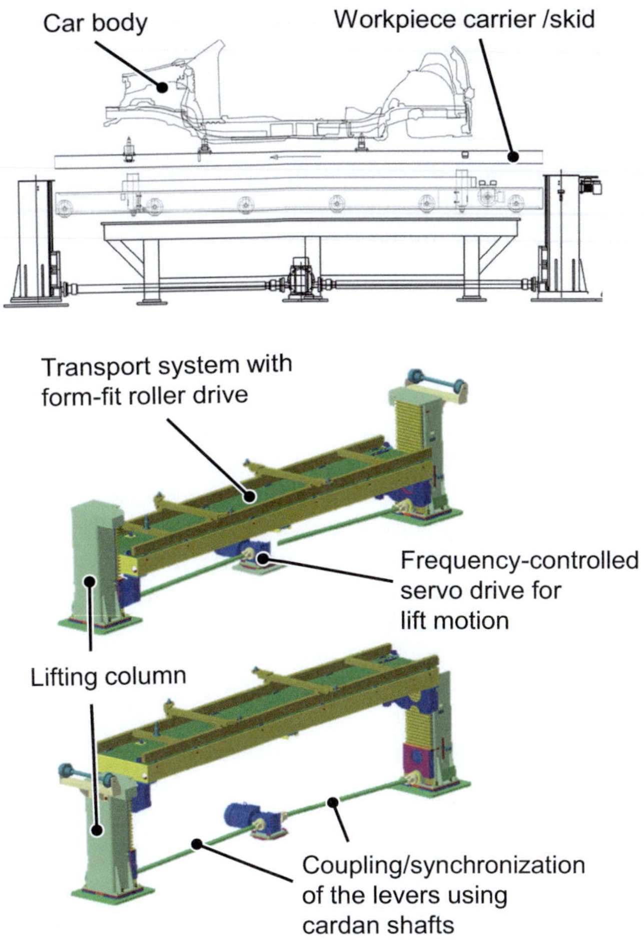

Fig. 9.10 Lift/lower conveyor system for transfer lines, according to Expert-Tünkers

9.2 Handling Systems

For multi-machine systems in the form of transfer lines and flexible manufacturing systems, it is necessary to manipulate workpieces with handling equipment. For flexible manufacturing cells, handling equipment can also be used for unmanned shift work. In addition to feeding workpieces, it is possible to extend automated handling to tools. An existing tool change system can be operated here or the tool changed on the spindle by an external handling equipment.

9.2.1 Placement Devices and Industrial Robots

Placement devices are used when simple, recurrent motions are required for machine feeding. Having kinematics aligned for the task at hand means they often have very short cycle times. They are used in great numbers for machine tool feeding in series and volume production (in particular for presses, lathes, cylindrical grinding machines and pressure die casting/injection molding machines), on automatic assembly lines and for packaging tasks. Acceleration and velocity curves are generally not programmable for mechanical solutions. Path information can be specified by stops and switching cams in conjunction with switches and valves. The mechanical positions of these stops and switching cams must be adjusted when the placement task changes. Nowadays a rigid controlled movement performed this way is almost always replaced by a programmable motion controller. Here, the movement is performed by axes that have displacement transducers and are moved into defined positions using a programmable logic controller (PLC) or motion control (see Volume 3, Chap. 10). Programming includes not only the target positions but also parameters defining speed and acceleration.

9.2.1.1 Placement Systems

Integrated Placement Systems

Figure 9.11 shows the integrated workpiece loading system of a turning center. The arm, reaching into the workspace of the machine using a lateral feed shaft, first takes the workpiece to be machined from the spindle using a pivoting double gripper with three-jaw chuck and then switches to the blank carried in the second chuck. The sideways position of the loader allows easy access to the integrated workpiece magazine.

Modular Handling Systems

To reduce costs when the requirements made of a handling system change, devices comprising individual modules can also be used for assembly and packaging tasks. The benefits of the modular design are that the components can be combined with each other in any way with the result that the kinematics can be tailored to the new application (see Fig. 9.12). Also, the modules can be reused when requirements change. Most modules are driven pneumatically but electrically operated components (motor and ball screw drive) are also available on the market. Drive modules with integrated distance measuring system have the benefit that they can be positioned anywhere within the possible range of movement and they permit, to a limited extent, modular, interlinked kinematics to be created.

9.2.1.2 Industrial Robots

Industrial robots are being used increasingly for flexibly automated manufacturing tasks. Their spectrum of tasks ranges from simple workpiece handling to complex, sensor-controlled tool handling, such as for arc welding, deburring and polishing. The majority of robots installed in Germany are used for handling, assembly, and path and spot welding [HAUN13]. Also refer to the application areas of industrial robots in Volume

Fig. 9.11 Integrated placement system, according to Promot Automation

3, Chap. 13, and position measuring systems for NC machines in ▸ Sect. 2.3. In multi-machine systems, robots are normally used as handling systems. Their functions are restricted to the changing of workpieces and tools. At the start and end of a multi-machine system, palleting, depalleting and packaging tasks can also be handled by industrial robots.

The key differentiating factor of robot designs is their kinematics, configurable primarily using rotational and linear axes. From the multitude of possible designs, the common variants listed in Fig. 9.13 have asserted themselves.

Two designs of industrial robots are widely used in multi-machine systems. First, robots with six rotational axes of motion (articulated arm robot), see Fig. 9.14a, are universally deployable and position their end effector in space with six degrees of freedom. Second, those with Cartesian structure in portal designs (gantry robots), see Fig. 9.14c, can, in most cases, move workpieces translationally freely within space. For handling heavy loads, industrial robots with a parallelogram axis structure have proven themselves, as shown in Fig. 9.14b, because their stable mechanical design allows a more precise sequence of motions with greater loads than articulated arm robots.

The design-specific mechanical structure specifies the static, dynamic and thermal properties and the performance limits of a robot. These parameters have a direct bearing on the application area and its limits (Fig. 9.15). The positioning accuracy, load capacity, action dynamics and workspace characteristics are compared in the figure. As can be seen, physical laws mean that the characteristics sometimes behave in a contradictory manner. A high load capacity, for example, given adequate stiffness results in a heavy robot. These features cannot be combined with a strict requirement for absolute positioning accuracy with high action dynamics. The lower part of Fig. 9.15 clearly shows that the robot type (kinematics) is determined solely by the workspace required.

Vertical Articulated Arm Robot

In terms of its axis configuration, the vertical articulated arm robot shown in Fig. 9.16 corresponds to the standard kinematics of vertical articulated arm robots. It comprises six rotational axes driven by 3-phase servo motors with integrated stopping brake. Refer to Volume 3, Sect. 13.2.1 for more detailed information

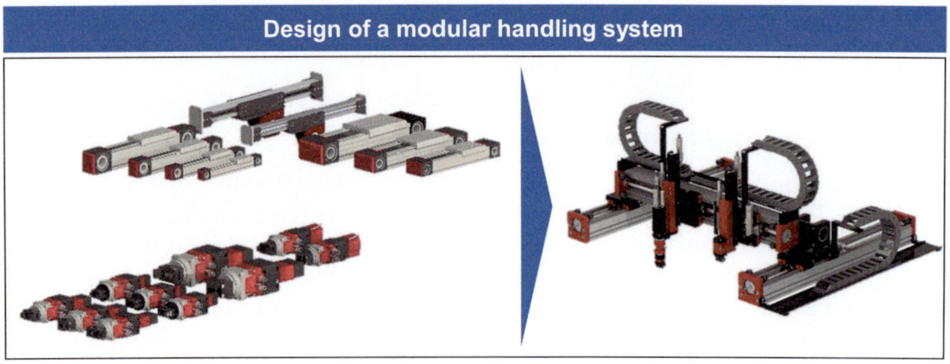

Fig. 9.12 Modular handling system, according to Afag

	Name	Principle	Axes	Load capacity	Workspace
Main axes	Vertical articulated arm	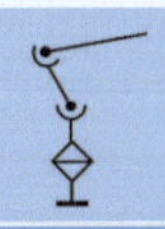	3	≤ 2300 kg	Hemisphere r ≤ 4.6 m
Main axes	Horizontal articulated arm (SCARA)	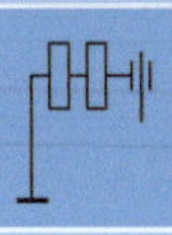	3	≤ 20 kg	Sickle r ≤ 1.5 m h ≤ 0.5 m
Main axes	Cylindrical coordinates	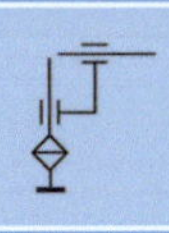	3	≤ 80 kg	Cylinder r ≤ 2 m h ≤ 1 m
Main axes	Spherical coordinates	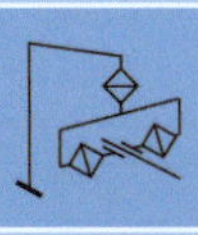	3	≤ 200 kg	Sphere r ≤ 2 m
Main axes	Line gantry	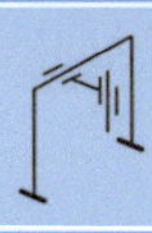	2	≤ 2400 kg	Surface l ≤ 50 m h ≤ 2 m
Main axes	Area gantry	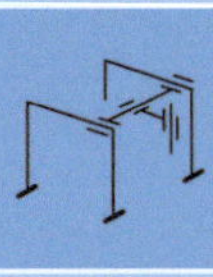	3	≤ 1300 kg	Cuboid l ≤ 60 m b ≤ 12 m h ≤ 2.4 m
Secondary axes	Double-angle hand	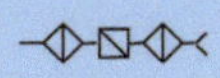	3	≤ 150 kg	Hemispherical surface
Secondary axes	Central hand	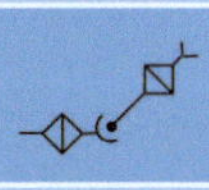	3	≤ 200 kg	Hemispherical surface
Secondary axes	Rotate-swivel joint	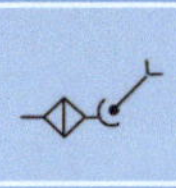	2	≤ 200 kg	Hemispherical surface
Secondary axes	Rotary joint	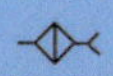	1	≤ 500 kg	Circle (point)

Fig. 9.13 Types and parameters of industrial robots, specified by main and secondary axes, according to ABB, EPSON DEUTSCHLAND GmbH, Fanuc, KUKA, Liebherr-Verzahnungstechnik

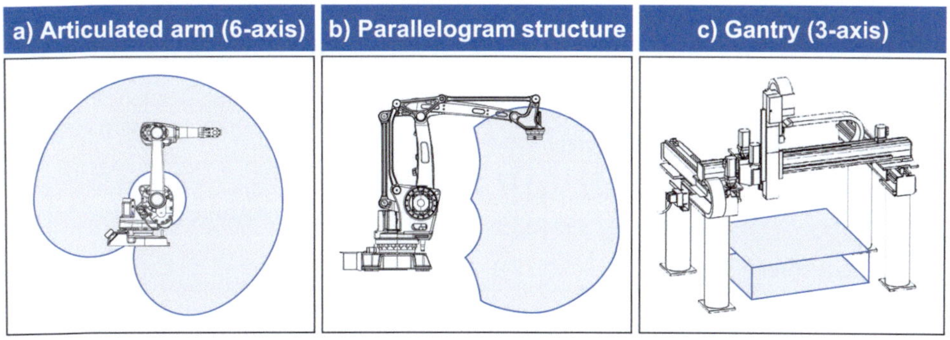

Fig. 9.14 Robot types and their work zones, according to ABB, KUKA

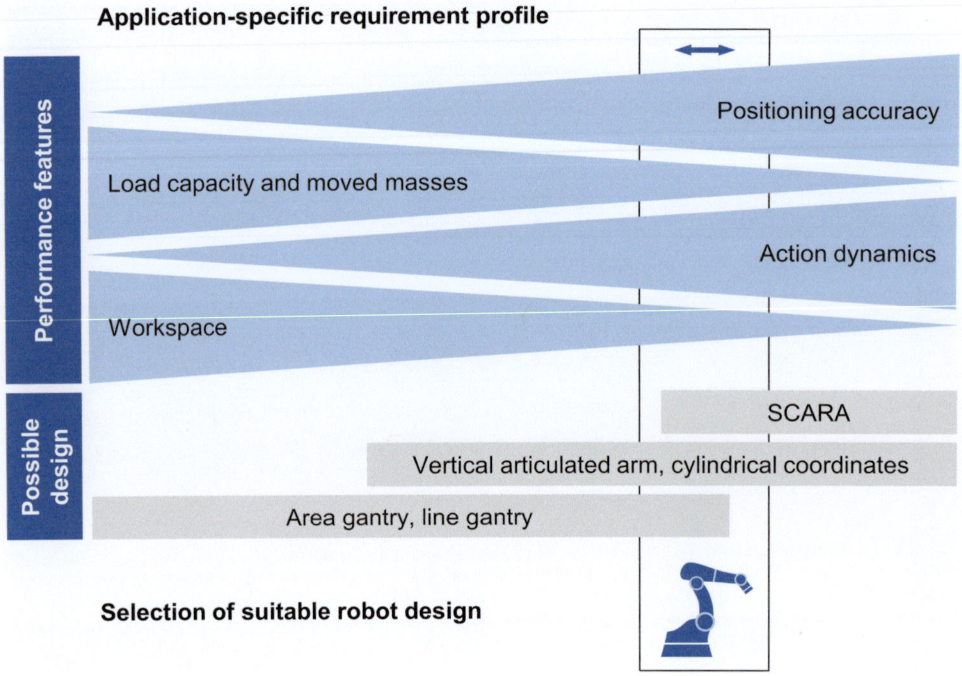

Fig. 9.15 Performance limits of robot designs

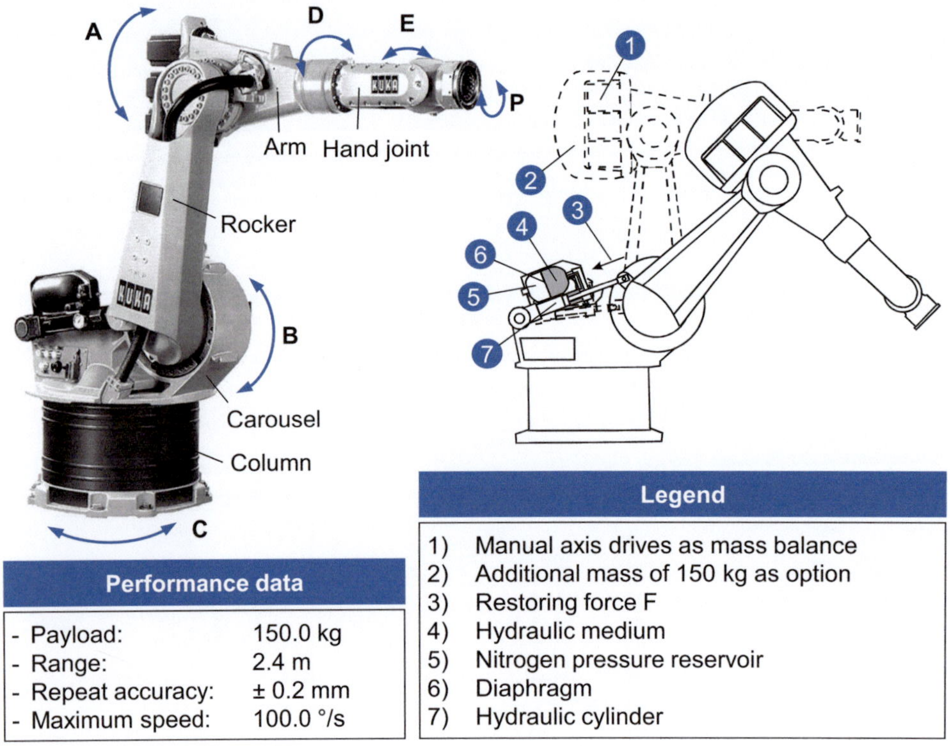

Fig. 9.16 Vertical articulated arm robot with hydraulic mechanical balance, according to KUKA

on the design of individual drive axes and measurement-based determination of the current position.

Horizontal Articulated Arm Robot

The horizontal articulated arm robot is similar to a human arm in terms of structure and is thus called a SCARA (Selective Compliance Assembly Robot Arm). Its speedy movement and repeat accuracy mean it is suitable for assembly and joining tasks, such as populating PCBs in the electronics industry (refer also to Volume 3, Sect. 13.2.2). High stiffness in the vertical direction is a main contributor because

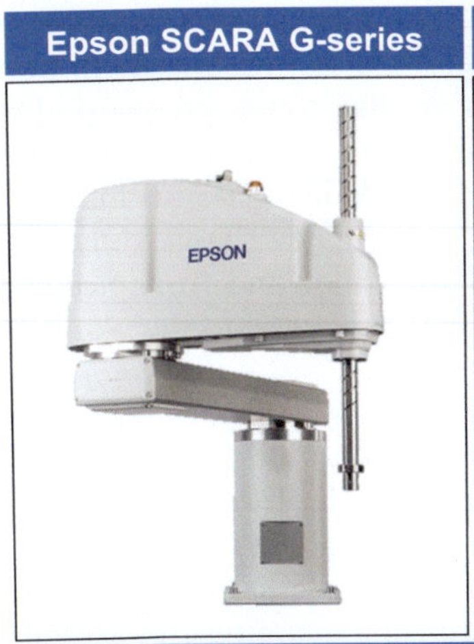

Performance data

Type:	Inner-oriented horizontal joint arm
Load capacity:	Max. 0.5 - 20 kg
Range:	175 - 1,000 mm (horizontal)
	100 - 420 mm (vertical)
Repeat accuracy:	± 0.01 m

Fig. 9.17 4-axis horizontal SCARA robot in a production line, according to EPSON Deutschland GmbH

all forces occurring and the robot's own weight act as a torque perpendicular to the bearing rotation axes (Fig. 9.17). The 4-axis handling equipment shown has three rotation axes and one translation axis.

Parallel Kinematics

In contrast to serial kinematics comprising a concatenation of arm parts, on parallel robots that found their way into the machine tool industry in the mid-1990s, several prismatic joints act directly upon the moving tool holder. Figure 9.18 shows a 4-axis robot comprising 3-axis parallel kinematics with an additional rotation axis in the robot hand. It is suited particularly well to pick & place applications, such as in the packaging industry. The platform retains its orientation constantly during movement within space, i.e., gripper axis P remains in its perpendicular position. This is achieved by the special design. Coupled with cardan joints on

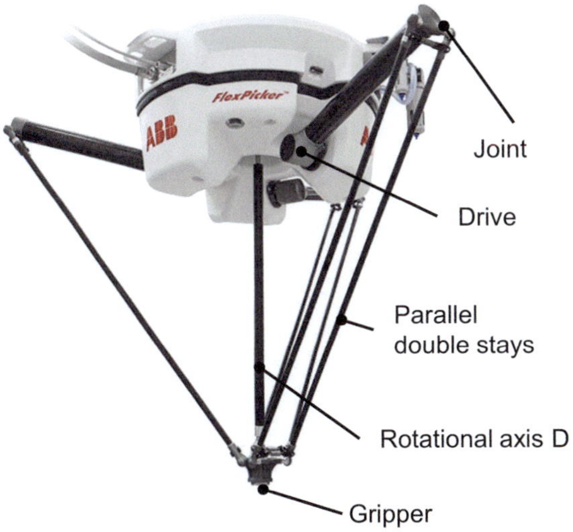

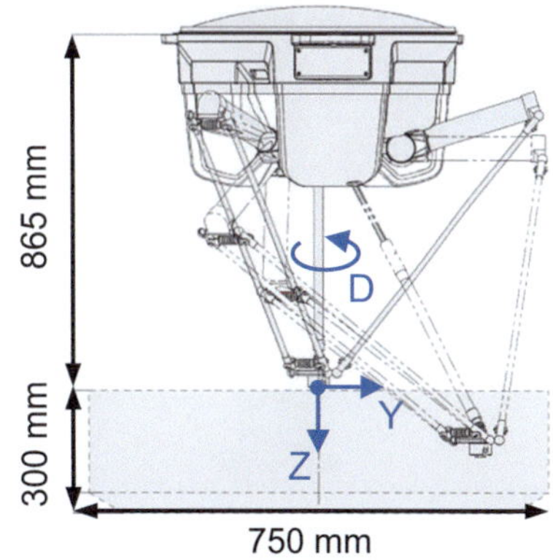

Performance data

Axes:	4
Nominal load:	1 kg
Workspace diameter:	750 mm
Repeat accuracy:	0.1 mm
Travel speed:	10 m/s
Acceleration:	15 m/s^2

Fig. 9.18 4-axis parallel kinematics for pick & place operations, according to ABB

each of the three rotatory motion arms on the base frame, there are two parallel double stays joined on the other side to the moving platform with ball joints.

Hybrid Kinematics

To marry the properties of different designs, hybrid kinematics combine two mechanical structural concepts. ◘ Figure 9.19 are 6-axis hybrid kinematics combining the benefits of the linear concept with those of an articulated arm robot. The very low overall height, resulting from the compact structure of the robot and of the combined linear articulated arm kinematics, is also suited to low factory buildings, meaning the application areas are very flexible.

Coordinate Systems and Reference Points

The specification of coordinate systems in the workspace and of reference points on the industrial robot are required to describe the motion operations and form the basis for the unique definition of movement instructions in the robot program. Different coordinate systems can be selected depending on the handling tasks to address. A distinction is made, for example, between base coordinates (base frame) joined rigidly to the base of the robot and workpiece coordinates (object frame) that are joined to the workpiece.

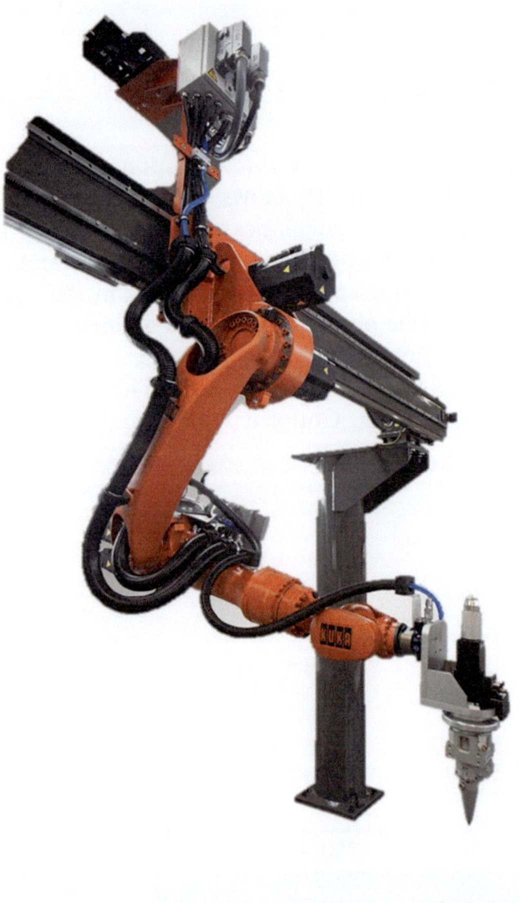

Performance data	
Axes:	5 rotation axes / 1 linear axis
Maximum load:	16 kg
Workspace (linear):	To 30 m
Repeat accuracy:	± 0.05 mm

◘ **Fig. 9.19** 6-axis hybrid kinematics. *Source* KUKA

Operating and Programming Robots

The objective of robot programming is to generate control commands in the form of programs to realize an automation task set. The spectrum of tasks can extend from simple workpiece handling functions to complex, sensor-controlled machining and assembly tasks.

Different programming methods exist, some of which are computer-aided. The suitability of these methods varies depending on the application area. The most important procedures are the online, offline and hybrid methods. They are essentially differentiated by the place of program entry. For more detailed information, refer to Volume 3, Sect. 13.5.

Using Industrial Robots in Multi-Machine Systems

◘ Figure 9.20 shows an example of a 6-axis articulated arm robot, attached to a linear axis to increase its radius of action. In contrast, the 6-axis articulated arm robot is usually used where the handling of workpieces and tools is restricted to a small space. In most applications, an articulated arm robot is used to remove workpieces from a magazine and feed them to the machining process. Workpieces and tools can both be manipulated with an articulated arm robot. One key benefit of such a configuration compared to gantry robots is that there is no need for the high components of the portal kinematics.

9.2.2 End Effectors

Grippers and gripper systems form the interface between the handling system and workpiece. They play a special role in the transportation of workpieces and tools. For this reason, the requirements and variants of different gripping techniques are discussed below. Tool change systems represent another peripheral component of machine tools. In addition to the classic systems with tool magazine, tool holders with turret design and multiple spindles are commonly used in multi-machine systems.

9.2.2.1 Grippers

Grippers are differentiated into categories by the working principle selected and the respective design of the technical system. For many applications, special designs are being developed that are not transferable to

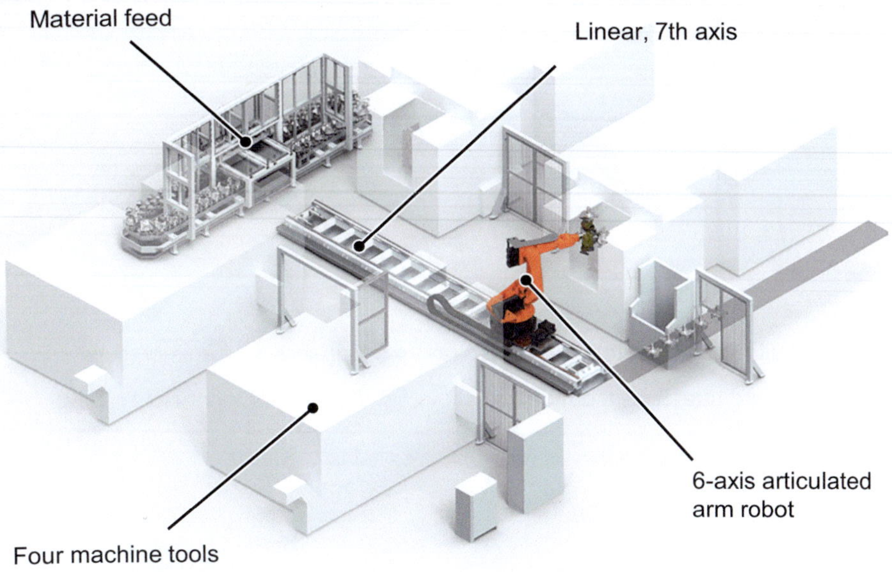

Fig. 9.20 Flexible manufacturing system, according to PROMOT Automation

other applications. Due to this, only standardized mechanical and pneumatic gripper types are discussed here. These can also be found in the machine tools environment in particular. Other gripper types are in Volume 3, Sect. 13.8.3.

Mechanical grippers
Mechanical grippers are commonly used in robot systems. They use the principle of form fit or stable equilibrium of forces to hold the gripped object (Fig. 9.21). This equilibrium is achieved by the gripper applying forces to the gripped object. So, for the design of mechanical grippers, the limits to the degrees of freedom of the gripped object and whether the (frictional) forces necessary for securing can be attained, are crucial.

The gripper used most in industrial robot systems is the parallel jaw gripper, introduced below as an exemplary gripper type. It has two parallel gripper jaws that are moved towards or away from each other during the gripping process. The usually symmetrical opening and closing processes mean both gripper jaws are actively involved in the movement. For this, the force generated by the drive is split by a gearbox (kinematics) and routed to the individual gripper fingers. Wedge hook kinematics (Fig. 9.22) are characterized by the high gripping forces generated, which can be applied to both actions, i.e., on opening and closing. These kinematics are an example of this type of gearbox.

Wedge hook kinematics can also be used in 3-finger grippers that have a 3-part gripping device (see Fig. 9.23). Here too, all of the gripper fingers are

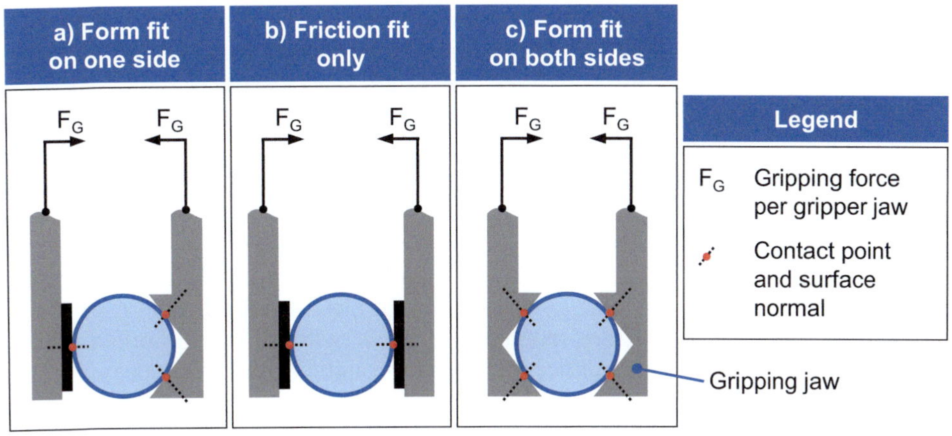

Fig. 9.21 Options for exerting gripping forces

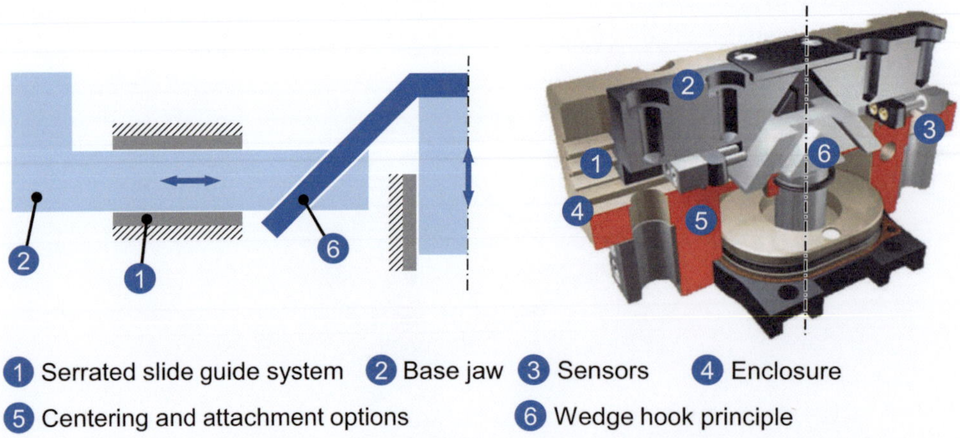

Fig. 9.22 Wedge hook kinematics for mechanical parallel-jaw grippers, according to Schunk

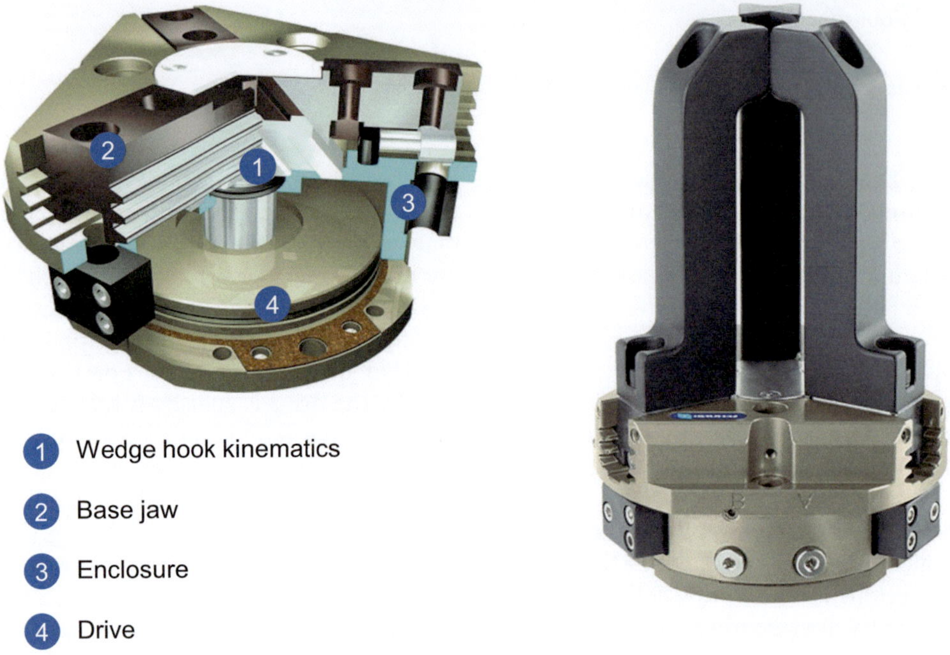

Fig. 9.23 3-finger self-centering gripper, according to Schunk

moved at the same time and follow the same path back during the closing process. The object gripped is kept moving until its movement is hindered by the other gripper fingers, and thus centered. Centering makes the 3-finger gripper particularly suitable for the joining processes of cylindrical components. This process imposes strict requirements of a defined component position and the gripping force that can be exerted.

Pneumatic grippers

In contrast to the grippers introduced before, pneumatic grippers do not have a gearbox or kinematics for force transformation. Instead, the grippers make contact with the gripped object using a pressure difference [HESS04]. Pneumatic grippers can be categorized into pressure and vacuum grippers. Pressure grippers are characterized by the fact that the forces applied are pressure forces and point in the direction of the gripped object. In the other case, a pressure reduction is used to generate gripping forces that point away from the object gripped towards the gripper. Colloquially, this principle is also known as a vacuum, leading to the name of the group of vacuum grippers.

An air-jet gripper (Fig. 9.24) represents the physically simplest structure of a pressure gripper. When the gripper is switched on, compressed air escapes as an air flow through one or more openings. The shape of these openings is changed using nozzles forming a defined air jet. If the jet is directed at the workpiece to grip, back pressure forms that lifts, and with the aid of a slide

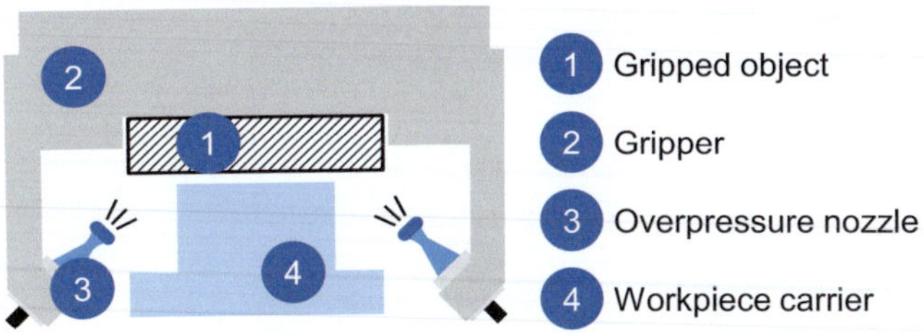

Fig. 9.24 Schematic working principle of an air-jet gripper, according to [HESS04]

guide, grips the workpiece. Closing the gripper valves releases the workpiece.

There are also other variants of pressure grippers where the compressed air is routed into enclosed cavities. These cavities deform due to the inner pressure increase. This change in shape means objects are clamped temporarily and movement forces are applied. To release objects gripped, the cavity pressure is reduced and the gripper returns to its original form. The deformable cavities are usually formed by thin diaphragms. They deform unhindered in the gripping direction and can adapt to the form of the object gripped in this way. These types of grippers are known as diaphragm grippers (see Fig. 9.25).

More detailed explanations on the working principle of vacuum grippers can be found in Volume 3, Sect. 13.8.3.2.

Application example of a gripper system on an industrial robot

A typical arrangement of industrial robot and gripper system is shown in Fig. 9.26.

An industrial robot equipped with a gripper system loaded with two clamping devices in one machine tool. The gripper system has a dual-gripper design. With such a configuration, the gripper can either transport two finished parts or blanks with one travel cycle or remove one finished part from the clamping device and place a blank that is in the second gripper. If part size and load capacity permit, retooling times can be saved by using multi grippers. Refer to Volume 3, Chap. 13 for a detailed description of gripper systems and comprehensive information on gripper technology.

9.2.2.2 Gripper Systems

A large part of gripper technologies used today is covered by the different gripping principles and concepts described. Furthermore, more types of jigs and fixtures exist at the interface between gripper and handling equipment used to aid the handling and assembly processes. The combination of fixtures and grippers individually attached to the handling equipment is called a gripper system. This comprises the mechanically active

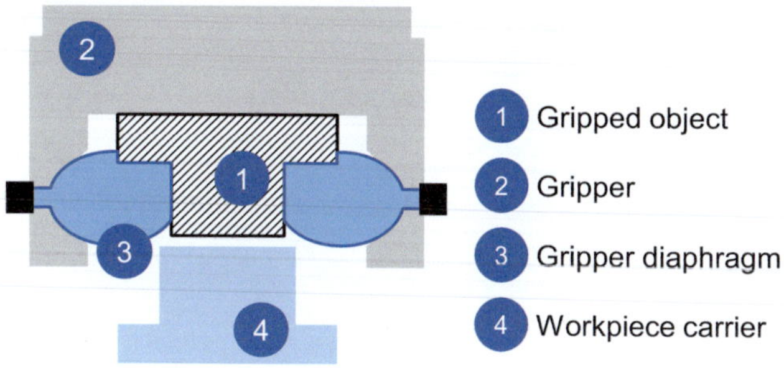

Fig. 9.25 Principle of a diaphragm gripper, according to [HESS04]

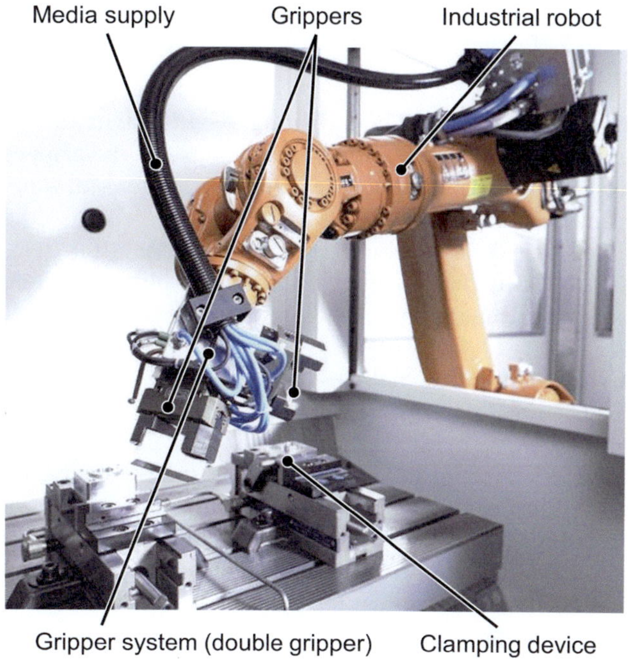

Fig. 9.26 Gripper system on a robot system, according to DMG MORI

actuators, guide elements and controller units that monitor the information flow of the gripping process.

Changing devices

Grippers for automated handling devices are usually product-specific. As a result, it is frequently the case that they can not be reused when the product or its properties change. To nevertheless cover a wide variety of variants with one set of handling equipment, the handling system must be capable of changing its gripping technique at runtime. Two-part changing systems are mounted for this purpose to the handling equipment and gripper to establish and end its connection in a controlled manner (Fig. 9.27).

In addition to the mechanical connection, changing devices also provide an interface for different working

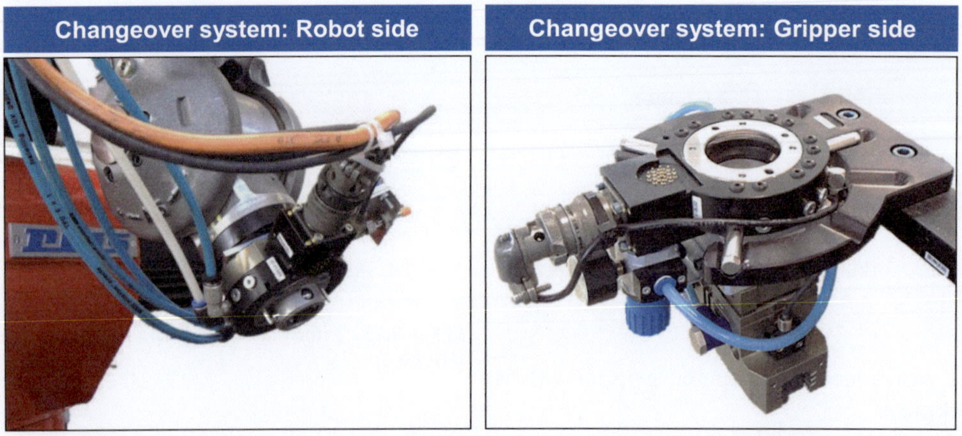

Fig. 9.27 Gripper change fixture with signal and media feedthroughs

media and signals. The individual feedthroughs can be assigned as required in the startup phase of the gripper system. When the individual modules are reused, however, the relevant feedthrough must be matched to the controller code. Malfunctions often occur after grippers are switched to different handling equipment using compatible changing systems.

Gripper controllers and sensors

The drives of a gripper are controlled using different actuation methods depending on the energy source used. Electrically driven grippers often have their own controllers and a number of alternative hardware and software interfaces for sending signals. The interfaces are used to set the different controller parameters as well as for integration in the field for actuating the gripper drive units.

Pneumatic grippers require valves that are actuated individually via the outputs of the internal robot controller or an external PLC. These valves are often positioned near the place of use in the form of valve terminals to lower pressure losses.

The increasing level of sensor monitoring for gripping processes and continuously feeding readings back to the control process enable a control structure to be created. This allows handling operations to be performed in a more tolerant way with regard to incorrect positioning.

Additional information on gripper systems is in Volume 3, Sect. 13.8.4.

9.2.2.3 Tailored Tool Change Systems in Multi-Machine Systems

One of the factors for machining stations attaining high productivity levels in multi-machine systems is very short downtimes. Tool holders with turret design and multiple spindles are used, for example. These allow speedy tool changes by advancing from one working spindle to the next.

Figure 9.28 shows a variant of a crown-type turret with a total of eight slots for driven tools. The number of tools available for machining can be increased even more by using multi-spindle heads in place of single tools. The tool axis forms an angle of 45° with the turret swivel axis. This means that the non-active tool paths are as collision-free with the workpiece as possible.

A particular feature of this crown-type turret is the use of one single AC servo motor to rotate the turret as well as the tool drive. The switching principle is illustrated in Fig. 9.29.

While the tool is operating, the motor is connected directly to the tool by a shaft-hub connection. The individual working spindles with tool holders can be tailored specifically to customer requirements so that either driven tools from the turret manufacturer or tools with standardized interfaces can be used.

For a tool change, the motor is stopped and converted to swivel operation via hydraulic line A. For this, pressure is applied to line A so that the tool drive shaft is pressed to the right and the locking mechanism of the Hirth serration unlocks from the tool holder disk. The shaft-hub connection for the tool is released and the cylinder pinion engages with the crown gear for swivel operation. As soon as the switch from tool to swivel operation is complete, the tool holder disk is rotated to the next working position. Then, pressure is applied to hydraulic line B. The tool holder disk is locked in this position, the drive shaft is connected back to the tool and machining of the workpiece can continue.

The 3-part Hirth serration guarantees the positioning accuracy of the tool holder disk. As locking and unlocking do not require lifting of the tool holder disk, the entire turret mechanism is protected from external factors as best as possible. The fact that locking and unlocking are performed parallel to the carriage path is one factor for the short tool change times.

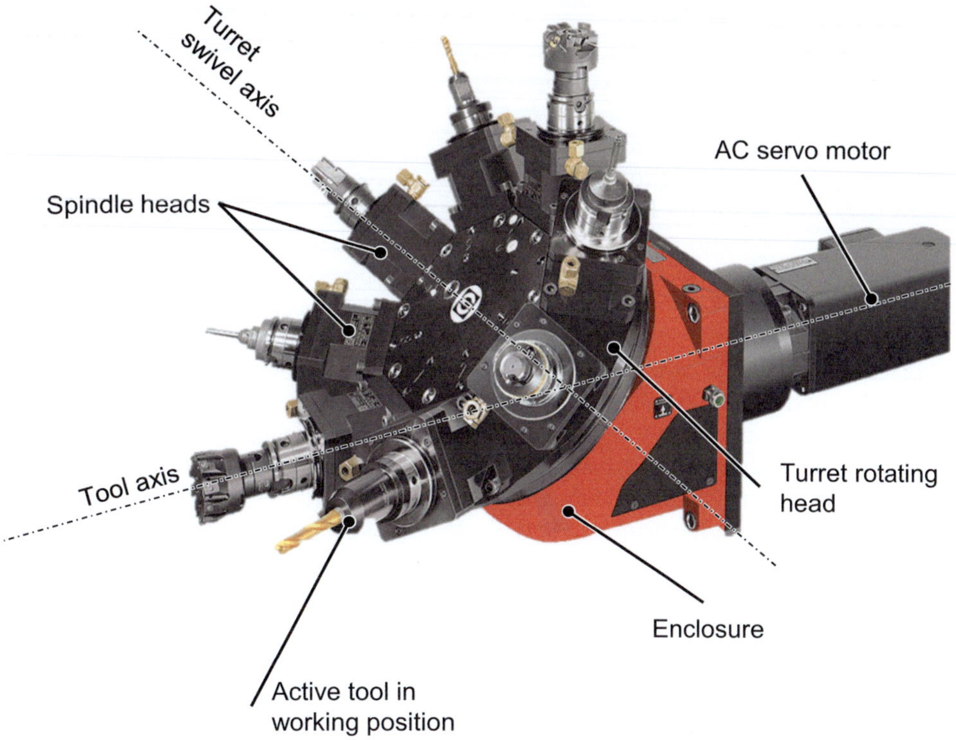

Fig. 9.28 Tool crown-type turret, according to Sauter

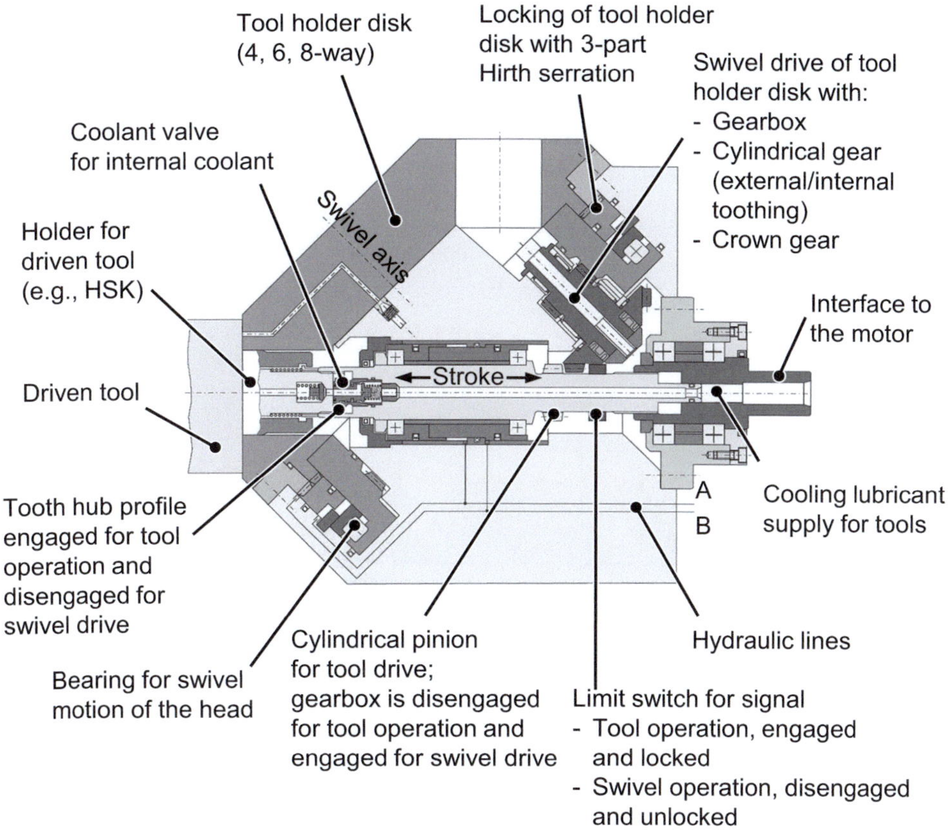

Fig. 9.29 Switching principle of a tool crown-type turret, according to Sauter

9.3 Special-Purpose Machine Tools

The term special-purpose machine tools comprises manufacturing systems that are tailored to special machining tasks. They usually only have one workpiece type (possibly with minor geometry variations) that can be machined. Given their low flexibility and high productivity levels, these kinds of systems are only used in series and volume production. If the machining task is so comprehensive that it cannot be completed with a single machining station, it is split across multiple stations. Here, workpieces are forwarded from station to station in a linear and synchronized manner (transfer line)—refer to ► Sect. 9.4. The machining stations can also be arranged in a circle. The term used here is rotary index machine (see ► Sect. 9.3.1). The increasing variant diversity and correspondingly shrinking batch sizes are also leading, in series and volume production, to stricter flexibility requirements being made on the means of production without the demands on productivity being withdrawn. Special-purpose machines and transfer lines are designed so that retooling a part family at short notice, and the general conversion of a machine to completely different production tasks, are possible quickly and cost-effectively [BULL03]. To meet these requirements as well as possible, an attempt is being made to broadly replace workpiece operation-specific machining stations with modular, multi-axis NC units with tool magazine and automatic tool change units. By using NC technology and components from a modular system, flexibility of the systems can be greatly increased. The modular structure of transfer lines with standardized machining stations also results in the reduction and standardization of the interfaces, an increase in process reliability and a reduction in the investment costs.

9.3.1 Rotary Index Machines

Rotary index machines belong to the group of special-purpose machine tools. They are used for single or multi-sided machining of complex workpieces in medium-sized or quantities that can require drilling, countersinking, threading and milling operations. Compared to transfer lines with linear workpiece flow, the concept of the rotary index machine offers, in addition to minimum cycle times and small space requirement, the benefit that clamping devices automatically return to the workpiece change position without additional overhead after one cycle. This needs to be resolved more extensively on production lines. The limited number of stations and the dimensions of workpieces are downsides, however [LOTT12].

◘ Figure 9.30 shows a rotary index machine that can be fitted with up to 16 machining stations for different manufacturing steps, such as drilling, milling, turning and grinding. The workpieces at the individual workstations are moved along synchronized with the manufacturing tasks. The workpieces are machined at the same time at all machining stations. Up to three simultaneously operating single or multi-spindle machining units can be installed in every machining station. This means the workpieces can be machined from five sides. If a turning station is used in place of a machining station, the sixth workpiece side can also be

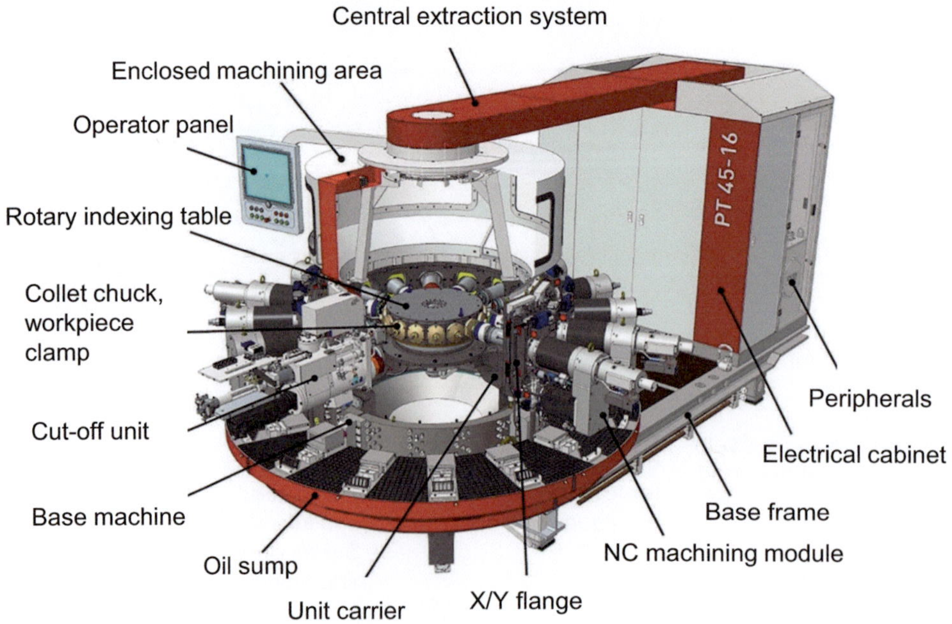

◘ **Fig. 9.30** Rotary index machine, according to PFIFFNER

machined. The workpiece clamps in the rotary index machine can take the form of spindles. This makes it possible to machine workpieces when rotating.

The individual axes of a rotary index machine can be implemented mechanically using an upright shaft or be NC/PLC/CNC-controlled (refer to the electronic line shaft in Volume 3, Sect. 8.4). In the case of PLC-controlled systems, the machining units are moved to the workpieces at the feed rate preselected. After the units reach the end position of their feed path, determined by fixed stops or limit switches, they move back in rapid motion. Since the motion paths of the tools are programmable, the low flexibility level of PLC-controlled rotary index machines can be increased considerably by using CNC machining units. As soon as all of the machining units have completed their operation, the table moves one step further. In addition to the machining stations, measuring stations can also be set up to check, using measuring probes or diameter mandrels for example, particular quality characteristics on every workpiece. One workpiece is complete after every work cycle. This workpiece is then removed from the machine manually or by a loading gantry and replaced with an unmachined workpiece. The cycle times of such rotary index machines are between 3 s and 1 min, but in most cases between 10 and 30 s. When three shifts are operating, annual volumes of 200,000 to 1.5 million are possible. Tool lives, measured with counters, are in the five-figure range for the machining of aluminum (10,000 to 50,000 workpieces) and between 500 and 2000 for cast iron and steel. This means tool changes lasting about two minutes are necessary once or twice a shift. To prevent damage and defective workpieces due to tool fractures, the machines also have electronic or mechanical systems for monitoring tool fractures.

9.3.2 Linear Transfer Machines

The inherent design of rotary index machines limits the number of machining stations. If it is not possible to combine more than one rotary index machine, linear transfer machines can be used. The workpiece flow is linear with the start and end being physically separated. Compared to rotary index machines, linear transfer machines make better use of the production area and are easier to access for maintenance work [LOTT12]. In production, linear transfer machines are often used for assembly automation. In terms of the material flow and layout of transfer lines, the properties overlap those of linear transfer machines. However, as transfer lines represent a combination of several machines, the term linear transfer machine is not used in the context of multi-machine systems.

9.4 Transfer Lines

Transfer lines comprise a number of machining stations positioned in sequence (turning, drilling, grinding, honing, measuring stations, etc.). The aligned, linear material flow is characteristic for a transfer line, i.e., the workpieces pass through the system with a set station sequence. The stations are linked to each other spatially, and often chronologically, by an automatic workpiece transport mechanism. In this arrangement, the workspaces of the machining units are directly in the workpiece flow. The work sequence is synchronized, or not synchronized using intermediate storage. As soon as all of the machines have performed their machining tasks (on synchronized systems), the workpieces are advanced by one increment to the next machining station. Given their low level of flexibility (i.e., the ability to be adapted to different machining tasks) and high investment costs, fixed transfer lines can only be used cost-effectively in high-volume production.

9.4.1 Fixed Transfer Lines

◘ Figure 9.31 shows a transfer line for machining automobile engine blocks. The blanks are routed to the system via interlinked conveyors and moved onto the transfer line by a gantry loader. The workpiece transport system, laid out as a vertical step conveyor, is in the lower part of the figure. It operates mechanically and is driven by a servo motor-driven carriage unit with ball screw drive. Z-units (1-axis), cross carriage units (2-axis) or 3-axis units are used as machining stations. The stations can be placed on one or both sides of the transport system and have modular multi-spindle heads. The chip-to-chip time attainable is 1 s. If a machining operation takes longer than the preceding operation, it is possible to machine two workpieces simultaneously in identical machining stations. To guarantee the highest machining quality, measuring devices are in the most important machining stations. These adjust the infeed of the axes in the event of dimensional deviations.

9.4.2 Flexible Transfer Lines

As the demand for greater flexibility is also increasing in series and mass production, modern-day transfer lines often no longer consist of single-purpose machine tools but modular NC machining units instead. These types of manufacturing facilities for continuous flow production are now called "flexible" rather than "fixed" transfer lines.

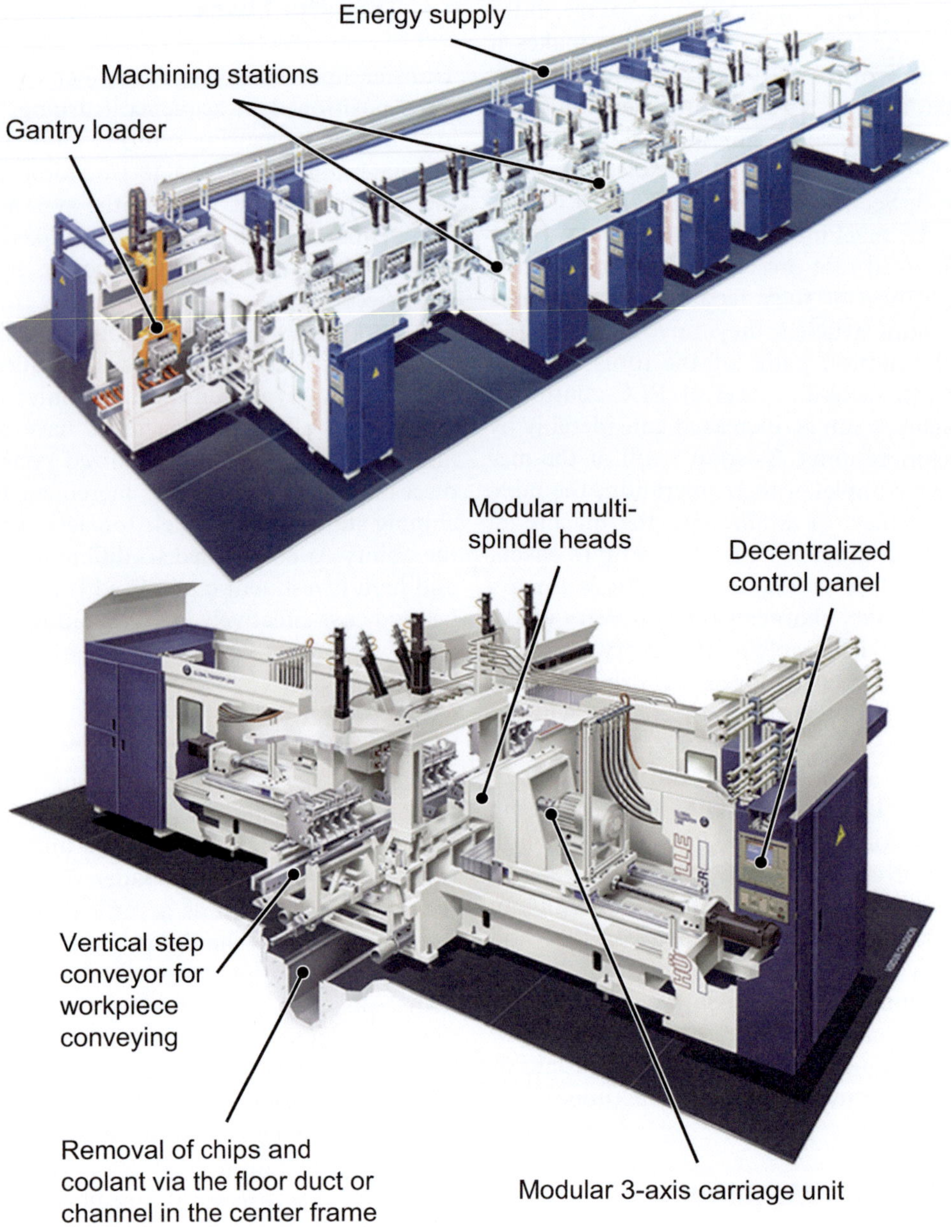

Fig. 9.31 Fixed transfer line, according to MAG IAS

A flexible transfer line is used to manufacture a range of parts where the same machining stations, in the same sequence, are involved in the production. Machining stations are, unlike fixed transfer lines, multi-axis NC machine tools with tool magazines and tool change equipment, not single-purpose machine tools with rigidly defined tools. This makes universal use of the individual station possible. The productivity downside compared to a fixed transfer line where multiple tools are usually in contact for every unit is that an attempt is made to reduce the time lost to tool changes (2 to 5 s) for NC units by means of high-speed cutting (HSC). In this process, the machine is moved between machining operations at high accelerations (1–2 g) and feed rates of up to 100 m/min. This requires highly dynamic drives and that the moved structural components have low mass. Material flows between the machine stations arranged in a row are, just like with fixed systems, either synchronized or realized in a not synchronized form using roller conveyors or workpiece conveyor belts with intermediate storage [NEUG12]. For the user, flexible transfer lines offer flexibility in particular with regard to future workpiece changes and potential system extensions depending on market demand for the finished products [JUNG99].

Figure 9.32 shows a flexible transfer line for making constant-velocity joints. The blanks arrive on pallets, are transferred to a conveyor and then transported

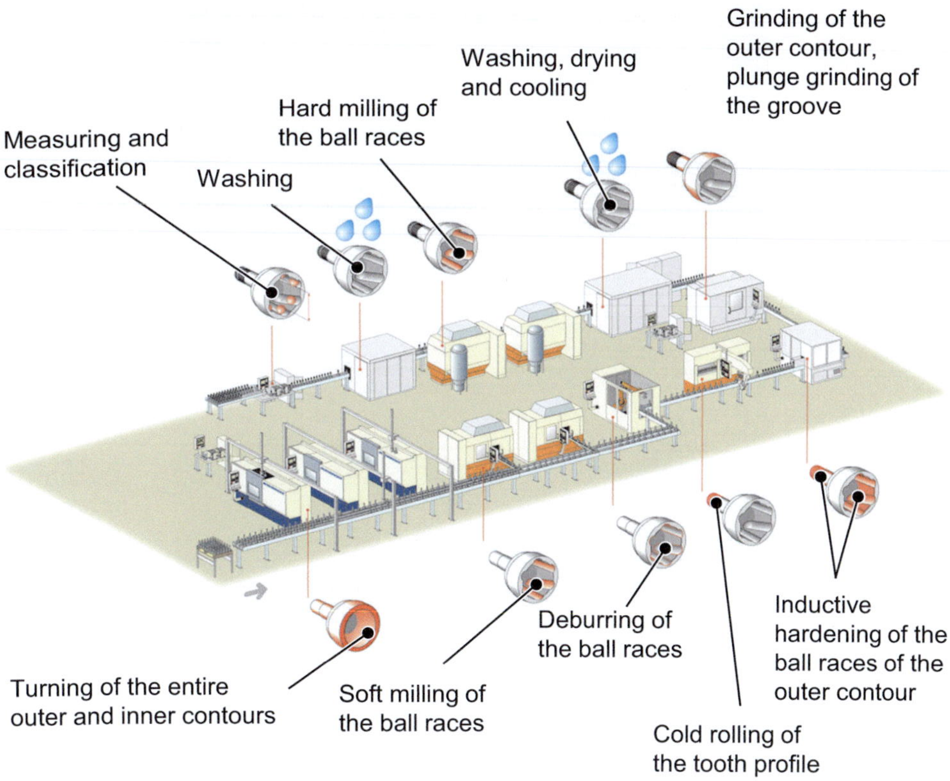

Fig. 9.32 Flexible transfer line, according to MAG IAS

to one of three turning centers. Here, the complete outer and inner contours are turned simultaneously in two operations. A loading gantry loads the lathes. The finished turned parts are placed back on the conveyor and the transported to one of two high-speed mill centers. This is where the ball races are milled. Next, the parts are deburred in a robot-aided cell. Following cold rolling of the tooth profile comes the inductive hardening process of the ball races and the outer contour that is ground in the subsequent machining step. The workpieces are cleaned in a cleaning machine before the ball races are hard milled in one of two mill centers. Then another wash cycle is performed, and the finished parts are measured and classified. The system shown is made up entirely of standard NC machining centers that can be used at any time for other workpiece types. This gives the user higher security of investment than with special-purpose machines.

The flexibility increase for flexible transfer lines coming from using numerically controlled machining stations with single tool machining has the downside of lower productivity compared to fixed transfer lines with adapted multi-spindle drill heads. Changeable multi-spindle heads are also used in flexible transfer lines to benefit from the high productivity of multi-spindle units and provide sufficiently high flexibility. When a new workpiece with a changed geometry is added to the machining program for the line, new and suitably adapted multi-spindle heads must be integrated into the system. The machining area of a special machining center for flexible transfer lines is shown in Fig. 9.33. It depicts a boring mill for boring holes in gearboxes. The machine also has an additional multi-spindle turret head for other drilling and milling operations. The gripper of the gantry loader protrudes into the machining area from the top and removes workpieces from the hydraulic clamping device.

The flexible transfer line in Fig. 9.34 is used to make automobile crankcases of one type (see [BRAN00]). The pre-machined blanks are delivered from the foundry. The machine shown is used to machine them to completion ready, ready for assembly. The workpiece spectrum is divided into two workpiece types with different numbers of cylinders. The system is also laid out such that it can be expanded to manufacture another four types of crankcases. The overall system is a hybrid system comprising 18 high-producing universal machining centers and five numerically controlled special-purpose machines. Machining processes not suitable for the machining centers from a technical or economic viewpoint can be combined into special-purpose machines. Given the poor flexibility levels of such special-purpose machines, the machining tasks are primarily tasks that cannot be changed at all, or only slightly, such as the pre-machining and precision machining of crankshaft bearings [JUNG99].

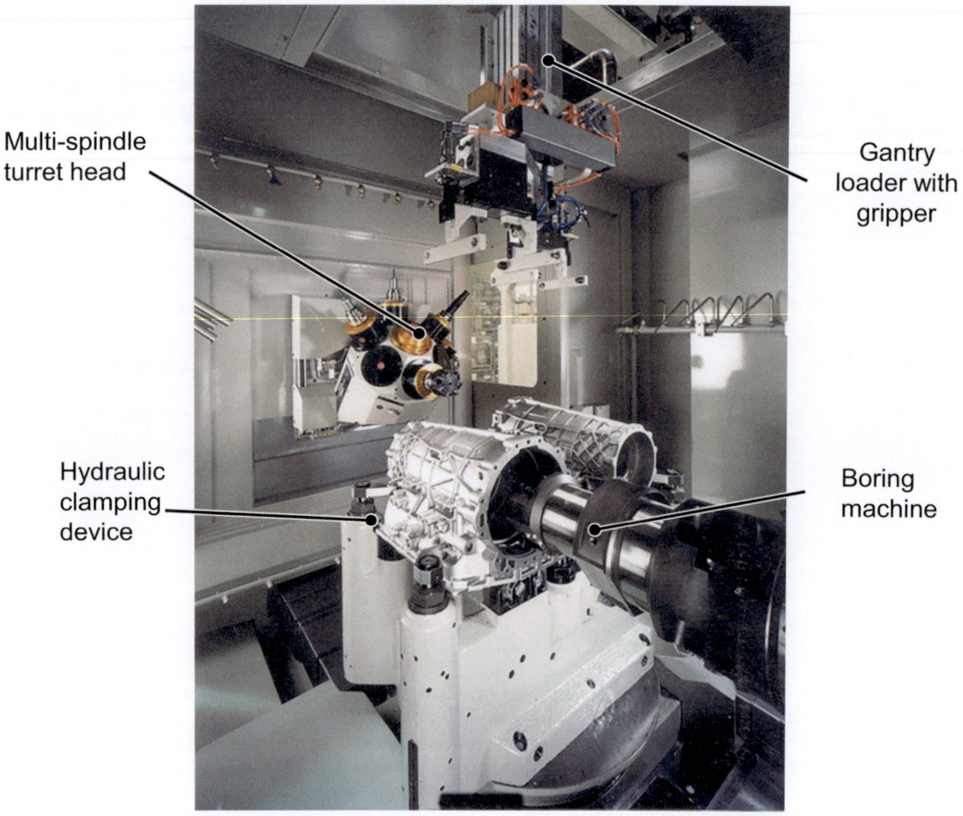

Fig. 9.33 NC machining station in a flexible transfer line, according to HELLER

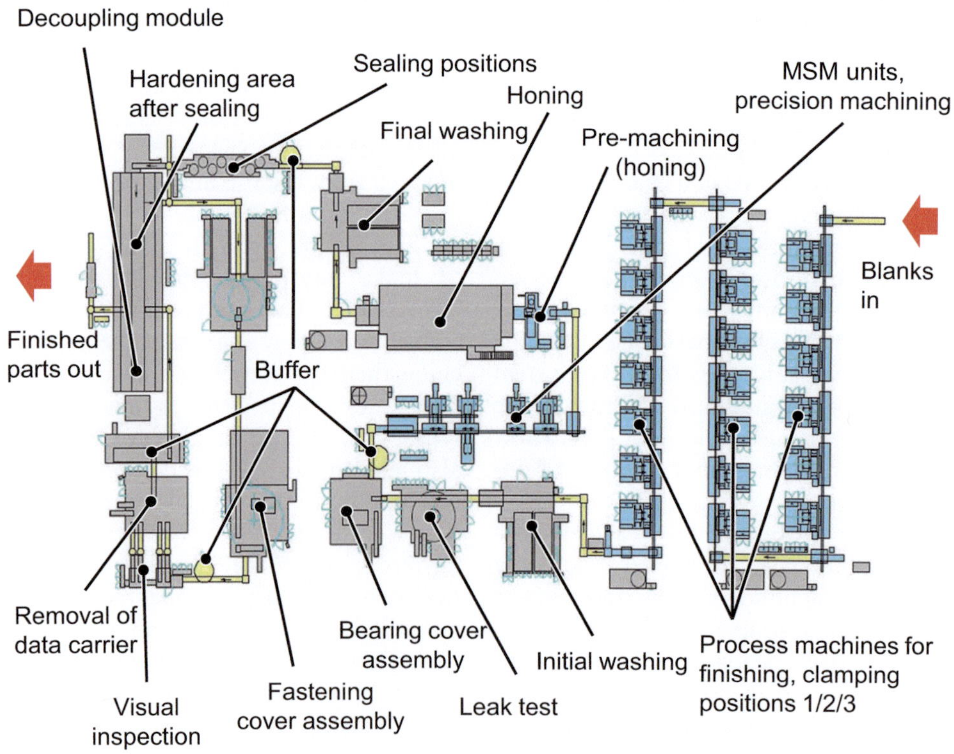

Fig. 9.34 Flexible transfer line for automobile crankcases, according to GROB-WERKE GmbH & Co.KG

Every crankcase has a data carrier including information about the workpiece, such as the ID number, workpiece type, date of manufacture and details pertaining to process and quality. These data carriers are used to identify workpieces in the machines and to start the required workpiece type-specific machining programs in the numerical controller.

◘ Figure 9.35 shows an example of the inside of a machining center as used in flexible transfer lines. A workpiece is fitted on an adapter plate, that in turn is secured by the collet chucks of the hydraulically actuated clamping device (see ▶ Sect. 10.4).

◘ Figure 9.36 shows another design of flexible transfer lines for 5-sided machining of complex cubic workpieces having a maximum edge length of 200 mm. The compact linear transfer system comprises multiple, directly connected, dual-spindle NC machining modules. The tool supply in every module uses two disk-type magazines each with 18 tools. For the machining process, workpieces are routed alternately to the

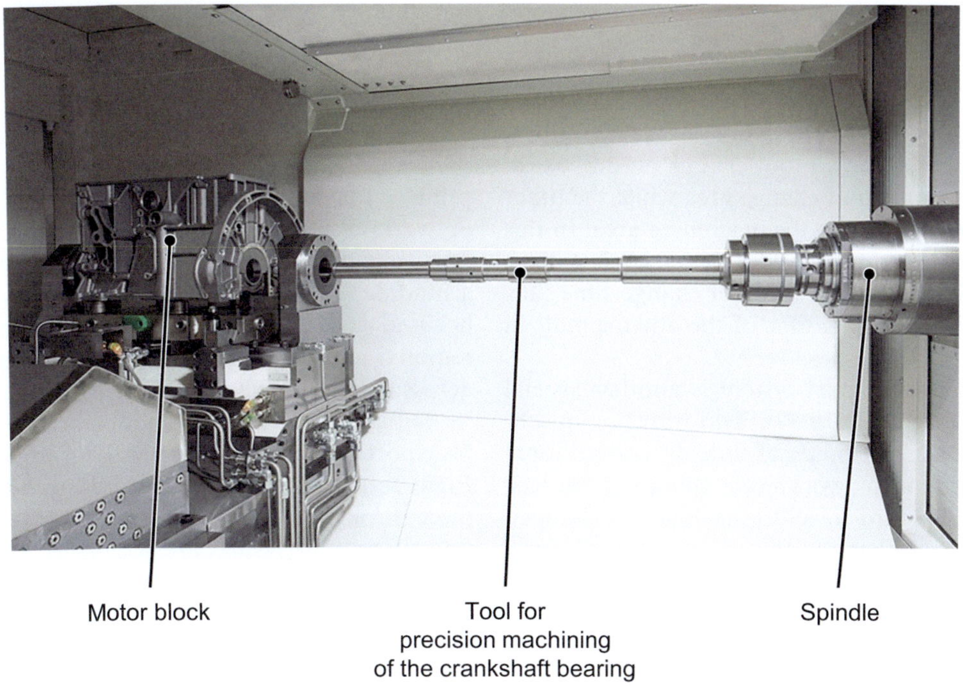

◘ **Fig. 9.35** Precision machining of a crankshaft bearing, according to MAG IAS

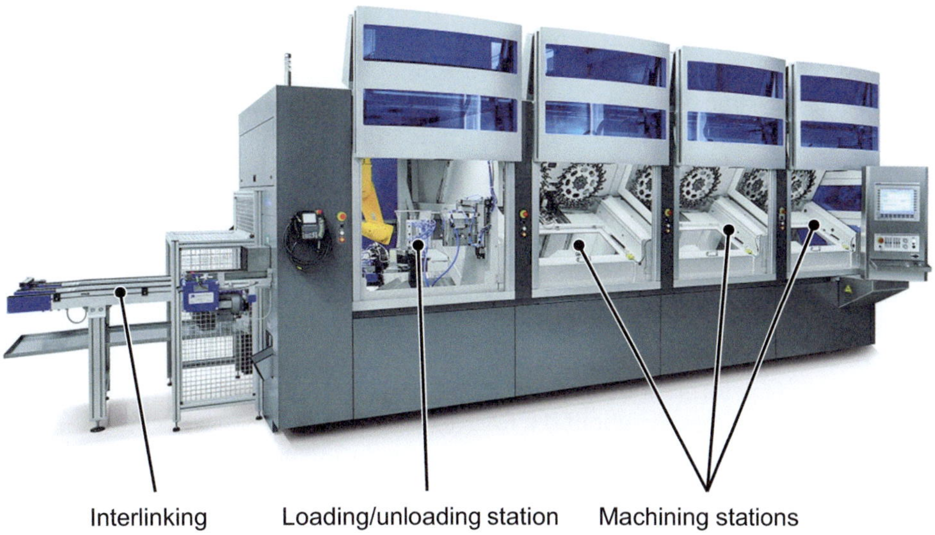

◘ **Fig. 9.36** Linear transfer system, according to Mikron SA Agno

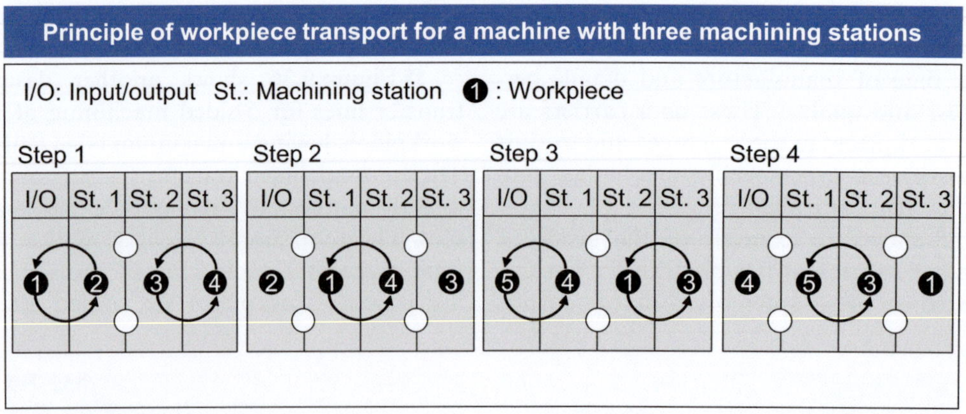

Fig. 9.37 Example workpiece transport

two working spindles. For a tool change on one working spindle, it is in the tool change area while the other spindle performs its work in the machining area. In this way, extremely short chip-to-chip times of less than 1 s can be achieved because the tool change time falls into the primary machining time of the other spindle in each case.

The workpiece transport system is another special system feature (see Fig. 9.37). Two transport mechanisms are required for material flow on conventional transfer systems with workpiece clamping pallets: One is used for forward travel to machine the clamped workpieces and another for the return of the empty clamping devices. These expensive aids are secured during return transport, something that has a negative effect in terms of cost. These downsides are averted in the transfer system shown in Fig. 9.36 with reorganization of the workpiece transport. Using double grippers, the workpieces including workpiece carriers move both forward and backward as one continuous flow with alternate transfer in pairs between two adjacent stations each time. The travel of each workpiece from the first station to the last takes half the machining cycle. After this, the workpieces are transferred back in the reverse direction using the same conveyors and machined to completion. This means a workpiece is machined twice in each station, apart from the end station.

The continuous flow comes from the fact that only every second double gripper is used on an alternating basis so either one or the other group of converters is working. The transfer movement between two adjacent stations takes 1 s. Locking and unlocking the workpiece carriers and exchanging information take about 3 s. This results in a chip-to-chip time of about 4 s for workpiece changes. This process is slow compared to the short time for changing the working spindles (about 1 s). For this system, the very short downtimes result from the aforementioned short non-productive times of the machining module tool change running during primary machining time. So that the longer downtimes of workpiece changes do not accumulate counterproductively, the minimum time for workpieces to stay on a module should be at least 20 s. The system can then be used cost-effectively when many tool changes are required and work is able to be distributed uniformly across the individual machining stations.

Figure 9.37 shows the principle of workpiece transport by way of example for four transport steps. Eight transport steps are required to advance a workpiece through the entire system. Automatic loading and unloading of a system are performed by an industrial robot, that takes blanks from a DIN pallet and places them into the automatic clamping device. Finished parts are transported on the same route back to the open places on the pallet.

9.5 Flexible Manufacturing Cells

Given the high investment costs for NC machining centers, the aim is to shorten the period of amortization by operating machines over multiple shifts, and to reduce down-, retooling, idle and transport times to enable more cost-effective production. To operate a machine over multiple shifts in combination with reduced personnel, a necessary condition is connecting the machine to a workpiece magazine with automatic workpiece change. The workpiece magazines of the flexible manufacturing cell (workpieces on chucking pallet or workpiece carrier in conjunction with handling systems) are loaded with a sufficient number of unmachined parts during the day shift. The throughput time can be reduced considerably by clamping the workpieces during primary machining time.

The machining of usually different workpieces is sequential in line with the order specified by the clamping or loading process, or in line with a working sequence defined by the master computer. The assignment of machining programs to workpieces is also available from the cell computer, or the relevant information stored on a coded information carrier (such as a RFID transponder) can be read. The corresponding NC machining program is enabled automatically after the workpieces are exchanged. In this way, the machine is able to autonomously work through the machining tasks specified in second and third shifts (unmanned or with reduced personnel).

Large tool magazines, such as chain or tower-type magazines, can also be used to broaden the areas of application by making a number of different parts, and by providing an adequate number of replacement tools when failures occur or wear limits are exceeded. The capital outlay for the tools provided and the sizes of the tool magazines are just two of the limits here.

Further machine properties are required to realize shifts with reduced personnel. Tool life monitoring on its own can be deemed state of the art. In addition, tool fracture detection and tool wear monitoring systems are necessary. However, nowadays these systems often do not yet attain the level of confidence demanded (see Volume 3, Chap. 6). For flexible manufacturing cells, therefore, the cutting values (cutting speed and feed) are commonly reduced in the unsupervised time to ensure that the tools survive this time undamaged. The use of diagnostic systems for determining the machine condition is another prerequisite for preventing damage to the machine and the associated, time-consuming and costly system downtimes. Furthermore, appropriate manufacturing cells of today are usually equipped with remote maintenance systems to resolve major problems in a speedy, uncomplicated and low-cost manner in conjunction with service technicians working for the machine tool manufacturer (see Volume 3, Chap. 6).

A flexible manufacturing cell for turning, drilling and milling is shown in ◘ Fig. 9.38. The machining center depicted is a universal automatic lathe that is loaded by a gantry loader. For workpiece handling, a linear gantry with a pivoting double gripper is used. The function of the gantry robot is to exchange a blank and a finished workpiece in the machine. The gripper can hold two workpieces. This allows speedy workpiece exchanges in the machine and short chip-to-chip times.

The gantry loader shown in ◘ Fig. 9.39 is a H-loader (dual gantry loader) with two gripper arms. On this H-loader, two independent arms with grippers are positioned parallel in the direction of motion on the overhead gantry crane. One arm is used to hold the blank, the other arm carries a finished part. The distance between the arms is fixed. This arrangement is

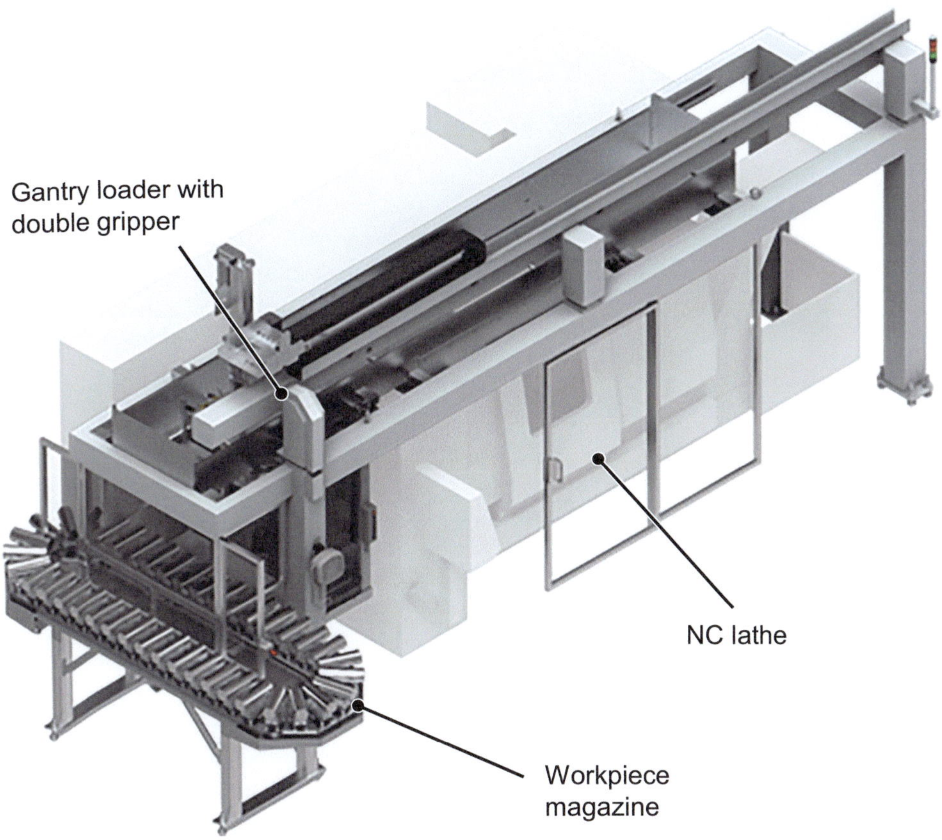

◘ **Fig. 9.38** NC lathe with gantry loader and workpiece magazine, according to PROMOT Automation

used for long shaft parts, for example, where a pivoting double gripper can not be used. For the provision of blanks or the storage of finished parts, different part magazines can be used. These include pallet belts, bar magazines for automatic bar lathes, synchronized palletizers and stacking cells. To retool to another workpiece type, the grippers or gripper fingers of the gantry robot might need to be exchanged and the robot program adapted. Usual times for retooling processes are between 5 and 10 min.

Additional machining centers and stations can be added to the flexible manufacturing cells depicted in ◘ Figs. 9.39 and 9.40. These centers and stations are responsible for turning and ejecting parts for the collection of statistical test data, or for marking and deburring tasks.

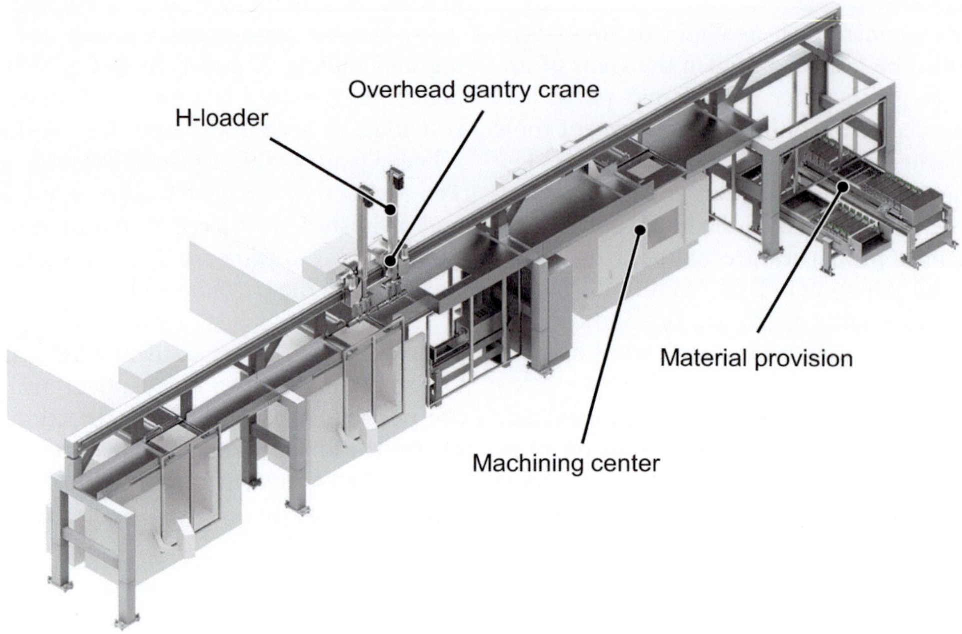

◘ **Fig. 9.39** NC manufacturing cell with H-loader, according to PROMOT Automation

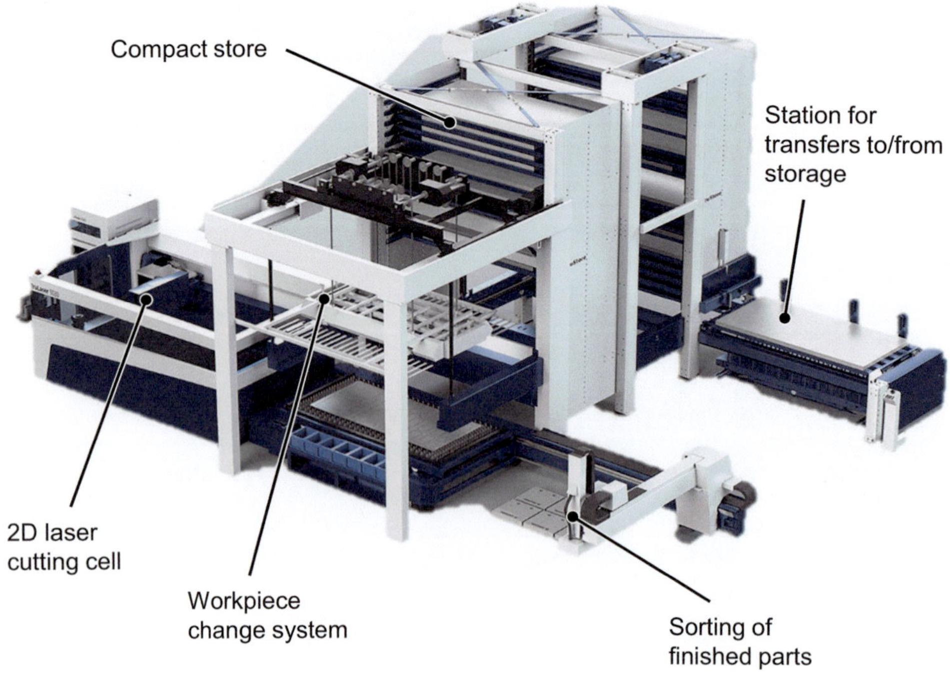

◘ **Fig. 9.40** Flexible, automated laser cutting cell, according to TRUMPF Group

Flexible manufacturing cells are also used in sheet metal processing. Here, numerically controlled cutting machines are used in the automatic production of smaller and medium-sized series (see ◘ Fig. 9.40). The aim is to make sheet metal parts as specified in orders with as few waste cuttings, and with as short a throughput time, as possible while at the same time minimizing the material stock. Achieving this objective is aided in the planning phase by CAM systems with nesting optimization and warehouse management systems. Parts from the same material and with the same thickness are combined from different production jobs within a planning period. Metal sheets with known dimensions are assigned such that the waste part is minimized.

◘ Figure 9.40 shows a sample flexible manufacturing cell for 2D laser cutting of sheet metal. The blank metal sheets can be retrieved from a compact store. For this, the required material is moved out by an automatic rack feeder and removed by a linear workpiece change system. This is able to take several parts and place them at the location required. This is necessary to transport back into storage parts cut to completion, or residual sheets, while the cutting process is taking place. The loading table in front of the machine is usually fitted with a centering or positioning device. The metal sheets are then transferred to the machine via roller conveyors or line gantries with suction grippers. After the cutting operation, the remaining lattices and finished parts are conveyed out of the machine. The linear workpiece change system transports the remaining lattices to a separate storage station. In the meantime, a linear destacking system places the finished parts onto DIN pallets or into boxes depending on the part sizes. Alternatively, finished parts can be transported back to storage by the rack feeder. Storage has separate stations for transfers from and to stock, from where the manufacturing cell can be supplied with new raw material, and finished parts can be removed.

9.6 Flexible Manufacturing Systems

The basic modules of every flexible manufacturing system (FMS) are multiple, numerically controlled individual machines, normally machining centers that can be extended with suitable peripherals for supplying workpieces and tools. The interconnection of the individual system components is controlled by a central computer for the workpiece and tool transport systems as well as NC data and production data handling. The flexible transport systems used in flexible manufacturing systems represent the key difference compared to flexible transfer lines. A flexible transport system is designed such that the machining units can be actuated in any order. For flexible transfer lines, however, the machining sequence is specified by the floor space configuration of the machines.

While, at the start of the 1980s, flexible manufacturing systems were developed with the aim of realizing completely automated and effective series and single part production (i.e., change of job without time loss in conjunction with system operation with a reduced number of personnel), the 1990s saw a recession-caused change of attitude towards the development of flexible manufacturing systems. In the past, the demand for end-to-end automation has resulted in systems with extremely high complexity and very high investment costs. Flexible manufacturing systems with 70 or more individual machines were not uncommon in the automotive industry. The flexibility of such large-scale systems was lower than originally assumed, however. Also, the complexity of systems resulted in considerable restrictions with regard to system availability and operational reliability as well as enormous problems with acceptance by the operating personnel [HAMM94].

Now, flexible manufacturing systems are often designed as lean "low-cost" solutions, frequently with two to five individual machines. The systems comprise standardized assemblies. This considerably reduces the initial investment costs for conversion work and upgrading of the systems. In the development of modern, flexible manufacturing systems, there is a clear trend towards user-specific, flexible automation, i.e., systems are customized to a manageable range of parts with a limited number of variants. An attempt is made here to prevent over-engineering as was practiced in the past [LAY01]. What makes the most economic sense is paramount—not what is technically feasible.

The control and operation of flexible manufacturing systems are subjected to strict requirements. Proactive resource scheduling for planning the production processes, usually lasting a longer time, is a prerequisite for efficient operation. Unlike the situation for individual machines, a controller system for flexible manufacturing systems has the task of distributing workpieces of different variants, so those with strongly divergent machining times, across different machine tools. The aims here are to optimize machine utilization and to minimize throughput times. Warehousing of blanks and finished parts as well as the provision of tools and clamping devices for the machine tools, are additional tasks of the master controller. Refer to Volume 3, Chap. 14 for additional information on production and process control.

◘ Figure 9.41 shows a flexible manufacturing system. Here, the workpiece is transported by a rack feeder that moves machine pallets between the storage locations of the centrally located high bay racking and the machining stations. The manufacturing system shown has three set-up stations. The system operator clamps blanks onto empty pallets, performs re-clamping operations on semi-finished parts and removes finished parts at these stations. Three standard

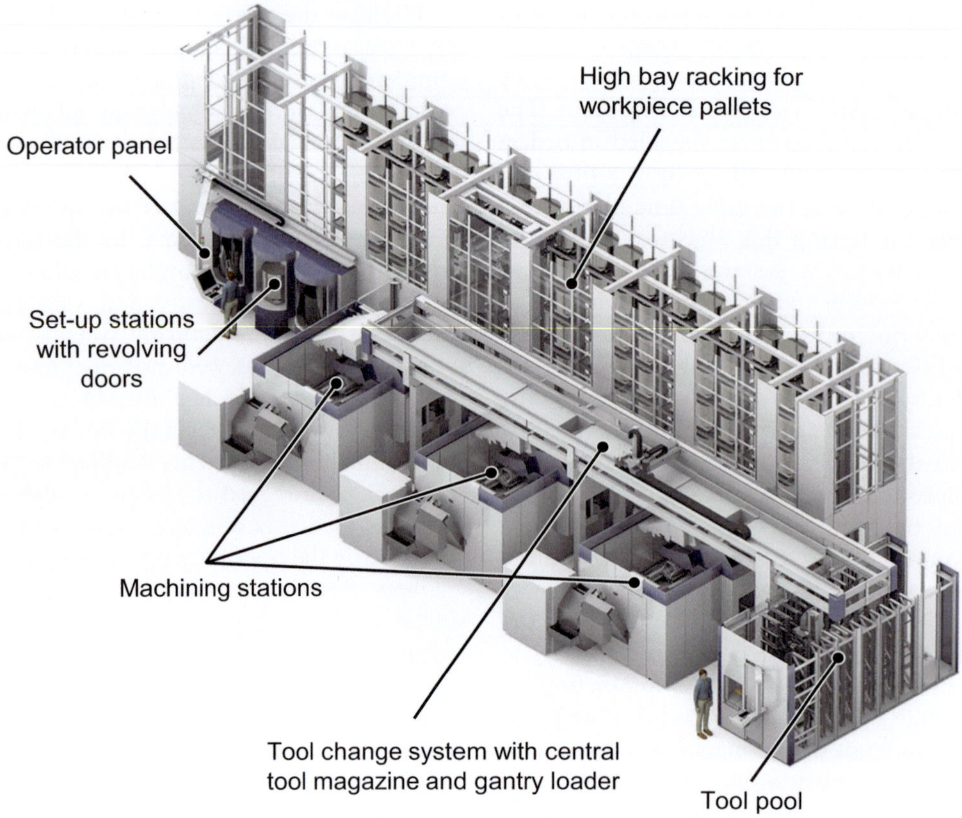

Fig. 9.41 FMS with automated workpiece and tool flow, according to PROMOT Automation

NC machines, loaded by a rack feeder using a pallet changer, are used as machining stations. A gantry loader, equipped with a pivoting double gripper, is used for computer-controlled tool exchange between the machine-bound tool magazines and the shared tool pool in a rack-type magazine. This system is used for the real-time supply to machines of the tools necessary to machine the workpiece spectrum scheduled. Worn tools are replaced automatically and during primary machining time, while tools are measured by operating personnel using an (optional) adjuster unit connected directly to the main production control computer. The system is operated from an operator panel that coordinates the overall process cycle for job processing, workpiece handling, tool planning and checking, and production data acquisition. The storage system as well as the machining capacity can easily be extended in line with the structure shown by connecting additional machines.

The use of gantry machining centers (see ▸ Sect. 4.3.1) is required to machine large workpieces. The flexible manufacturing system shown in Fig. 9.42 comprises three interchangeable gantry machining centers. Each of these vertical machines has a tool change system for up to 182 tool slots. Fixed angle heads, horizontal spindle units and other custom spindle units can also be interchanged. The machines can be upgraded to 5-axis machines using supplementary 2-axis numerically controlled heads that also permit tool swivel movements. In this way, simultaneous 5-axis machining can be performed in addition to 5-side machining. This machining is required, for example, in the production of turbine blades, tool shapes and marine propellers,

The pallets are transported with a pallet transport carriage between the workpiece tooling station, machining centers and workpiece storage locations. Accommodating two workpiece carriers on the pallet transport carriage provides short change times and temporary storage of pallets on the pallet carriage.

A tooling station is available for the manual clamping and unclamping of workpieces.

Figure 9.43 shows another flexible manufacturing system with machining centers for large aerospace components made of aluminum. The parallel kinematic machining head is a feature of the machine tools depicted. This head enables highly dynamic 5-axis milling of integral airplane components. The high installed spindle power of 120 kW at 30,000 min^{-1} means material removal rates of 10 l/min can be attained from fully high-strength aluminum. The machines each have a Siemens controller (Sinumerik 840 D), a rack-type tool magazine with 125 slots, vacuum clamping

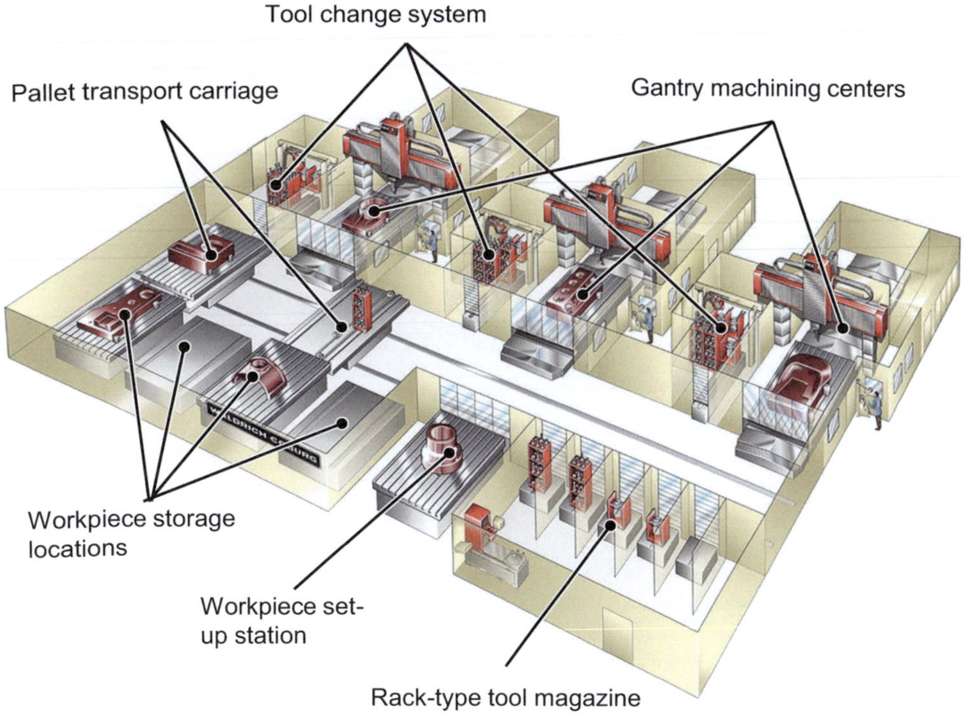

Fig. 9.42 FMS with gantry machining centers, according to WALDRICH COBURG

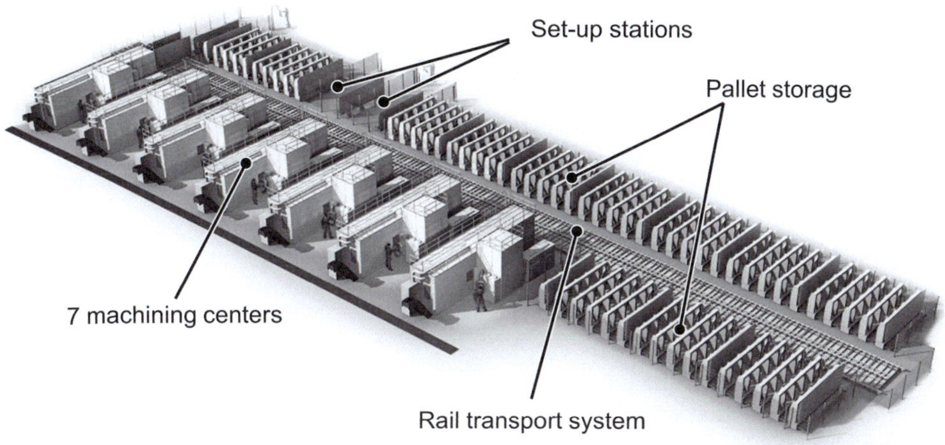

Fig. 9.43 FMS for large aerospace components, production line Ecospeed, according to Starrag Technology GmbH

systems for workpieces, and video monitoring of the working area. A rail transport system with 63 stations for loading and unloading as well as storage connects the seven machines into a flexible production environment. Here, 60 pallets (sized 2000 mm × 6000 mm) ensure high-producing manufacturing is realized in the form of dynamic switching between set-up and machining. The flexible manufacturing system is controlled by a master computer. It regulates all of the automated functions such as pallet changes, tool changes and starting of machining programs. This master computer controls the system such that the cutting machines are utilized to the full for production. The storage stations for pallets are used as buffers to prevent bottlenecks due to individual system modules. The trend towards flexible manufacturing centers can be seen generally for large parts. This is due to the requirement for higher productivity and reduced operator costs in this sector, too.

Flexible manufacturing systems are being increasingly used in sheet metal processing, too. The manufacturing system for sheet metal processing shown in Fig. 9.44 represents the build level of a flexible manufacturing cell as in Fig. 9.40. Here, the material

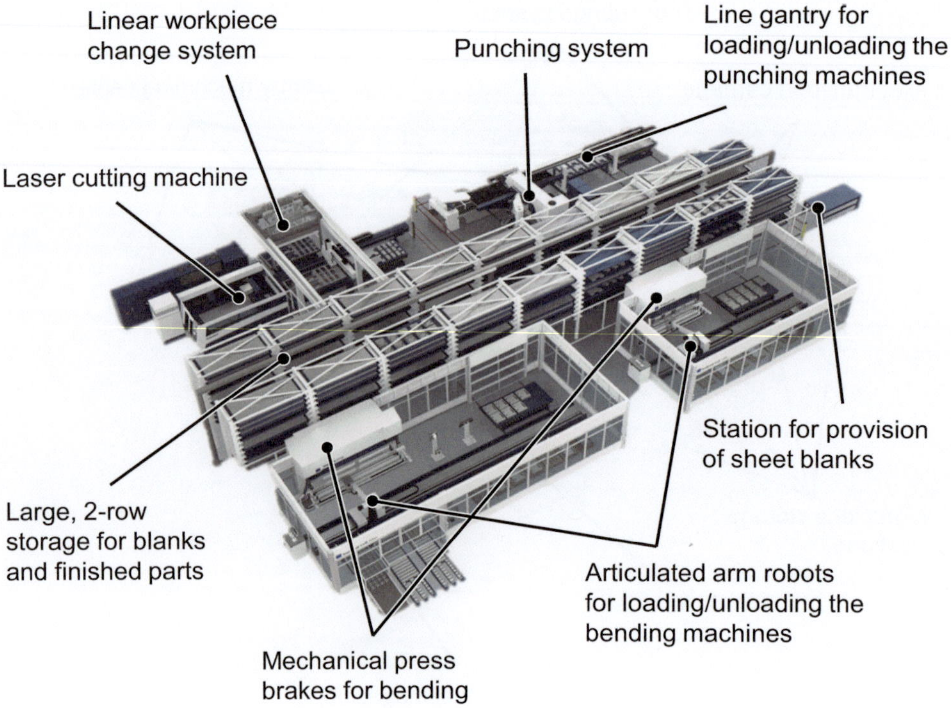

Fig. 9.44 FMS for sheet metal processing, according to TRUMPF Group

store comprises a large, 2-row storage that is loaded and unloaded by a rack feeder positioned in the middle. A punching/cutting system and 2D laser cutting center are available for machining.

The two mechanical press brakes are also connected to the automated material flow. They are loaded by articulated arm robots that then stack the finished parts on DIN pallets. The punching system shown is loaded and unloaded by a line gantry. The blank metal sheets come in on a rail transport carriage from storage. Finished parts on the other hand, are destacked onto pallets, provided by a chain conveyor. Small parts can be diverted into boxes by sorting devices. The machine has a separate, external tool magazine that expands the tool capacity of the machine by an additional 40 or 70 slots. The tool magazine has its own cantilever with swivel arm. It uses two grippers to interchange tools between the machine and magazine. The tool magazine can be set up during primary machining time.

References

[BRAN00] Brandner, H., Jung, P.: Flexible Fertigung. Kombinierter Einsatz von Bearbeitungszentren und flexiblen Transferstraßen. WB Werkstatt und Betrieb **133**(4), 55–58 (2000)

[BULL03] Bullinger, H.-J., Warnecke, H.-J., Westerkämper, E.: Neue Organisationsformen im Unternehmen. (Reihe: Engineering online library). 2nd edn. Springer, Berlin (2003)

[GRUN09] Grundig, C.-G.: Fabrikplanung, 3rd edn. Hanser, München (2009)

[HAIM13] Haimerl, C.: Reinigen im Rundtakt. JOT – Journal für Oberflächentechnik **53**(3), 54–55 (2013)

[HAMM94] Hammer, H.: Flexible Fertigungssysteme im Lean-Trend. Werkstattstechnik – Produktion und Management **84**(3), 85–90 (1994)

[HAUN13] Haun, M.: Handbuch Robotik. Springer, Berlin (2013)

[HESS04] Hesse, S.: Robotergreifer. Hanser, op. München (2004)

[HOFM00] Hofmann, G.: Flexible Fertigungssysteme von heute. WB Werkstatt und Betrieb **133**(3), 46–49 (2000)

[ISEN11] Isenburg, T.: Zylinderköpfe energieeffizient reinigen. JOT – Journal für Oberflächentechnik **51**(6), 68–70 (2011)

[JUNG99] Jung, P.: Bearbeitungszentren für Groß- und Mittelserien. Bearbeitungszentren contra Transferstraßen. WB Werkstatt und Betrieb **132**(5), 48–51 (1999)

[KOBL14] Koblenzer, G.: Maßgeschneidert modular. JOT – Journal für Oberflächentechnik **54**(3), 62–64 (2014)

[LAY01] Lay, G., Schirrmeister, E.: Sackgasse Hochautomatisierung? Mitteilungen aus der Produktionsinnovationserhebung. Karlsruhe (2001)

[LOTT12] Lotter, B., Wiendahl, H.-P.: Montage in der industriellen Produktion. Springer, Berlin (2012)

[NEUG12] Neugebauer, R.: Werkzeugmaschinen. (Reihe: SpringerLink: Bücher). Springer Vieweg, Berlin (2012)

[SCHU12] Schulz, D.: Eine perfekte Anlagentechnik allein reicht nicht. JOT – Journal für Oberflächentechnik **52**(12), 50–51 (2012)

[VERE15] Verein Deutscher Werkzeugmaschinenfabriken e. V. (VDW): Marktbericht 2014. Frankfurt (2015)

[VIEH92] Viehweger, B.: FFS als wesentlicher Bestandteil von Fertigungsarchitekturen. CIM **8**(2), 10–32 (1992)

Equipment Components for Machine Tools

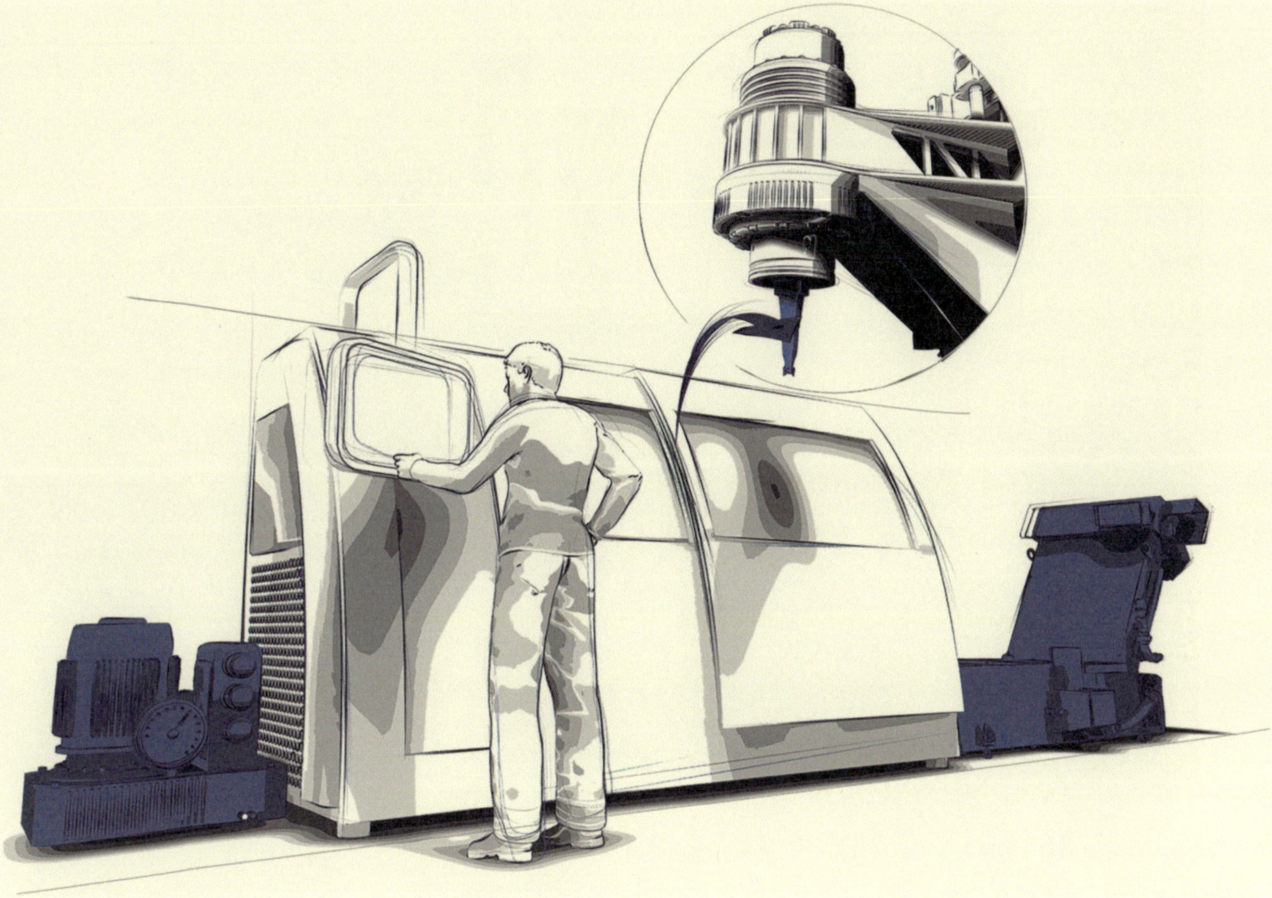

I thank my staff members Mathias Dehn M.Sc., Niclas Klumpen M.Sc., Marcus Gärtner, M.Sc. and Guifeng Zhao M.Sc. for their contribution in revising this chapter.

C. Brecher and M. Weck, *Machine Tools Production Systems 1*,
Lecture Notes in Production Engineering, https://doi.org/10.1007/978-3-662-68120-6_10

Contents

A comprehensive explanation of standard and additional equipment for all machine tool types is not possible within the scope of this book. The focus of the discussion here is equipment for drilling and milling machines as well as lathes. After describing the tasks and relevance of equipment components, their functional principles are explained. The following discussion of individual components using examples ranges from production resources to change and magazine equipment for tools and workpieces.

10.1 Overview

Production resources (tools, fixtures, measuring and test equipment) are an absolute necessity to operate a machine tool. In addition, different change and magazine equipment for tools and workpieces is required for the current high degree of automation.

Explained generally below are the nature and function of equipment accessories using a lathe as an example. The possible accessory components for a lathe are illustrated in ◘ Fig. 10.1. Tools, clamping devices, handling and magazine equipment, as components of a tool and workpiece system, can take on different forms depending on the machining task, machine design and degree of automation. Important equipment components are introduced in the following chapters, especially with regard to automated NC production.

Shown in ◘ Fig. 10.2 are the functional relationships between the machine tool with its part components, the workpiece and tool holders, the clamping elements and the resulting interfaces. Production resources enable machine tools to perform the various machining tasks. Workpieces and tools are held or clamped on the appropriate carriers. In terms of exchangeability and flexible adaptation of machine tools to different machining tasks, special importance is attached to the design of the mechanical interfaces between the various production resource components and the machine. This chapter describes the functions of part components as well as the interfaces.

In addition to machine-related interfaces, machine tools also have interfaces for the automatic loading and unloading of tools, clamping devices and workpieces. Today, the machine-tool interfaces have been broadly standardized. On the workpiece side, however, there is the difficult task of flexible adaptation to changing workpiece geometries. Given their different designs and sizes, workpieces require flexible clamping devices.

Tools and workpieces must be positioned and clamped at a certain orientation relative to the machine coordinate system in line with the machining task stipulated. Precise position determination, i.e., the defined spatial configuration of tools and workpieces in relation to each other, is with geometrically well-defined securing. The requirements made of repeated, automatic changing demand uniform interfaces that allow speedy and unambiguous position determination of tools, workpieces and their fixtures. In addition to determining the position and orientation, clamping is a key function within the scope of workpiece, tool and machine tool. On the workpiece side, the functions mentioned are implemented by fixtures, carrier pallets and indexings on the machining tables or workpiece spindles. The same constructive requirements are made on the tool side. These are easier to realize, however, due to the design similarity. These relationships are shown in ◘ Fig. 10.3 [VDI90a].

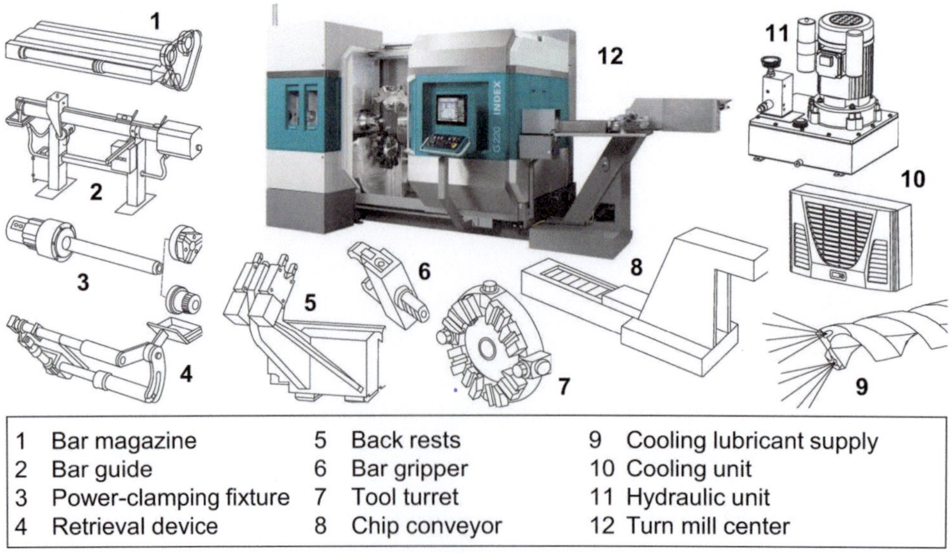

◘ **Fig. 10.1** Equipment accessories of an NC machining center for task-specific automation. *Sources* Gildemeister, INDEX, LUKAS Hydraulik, Rittal

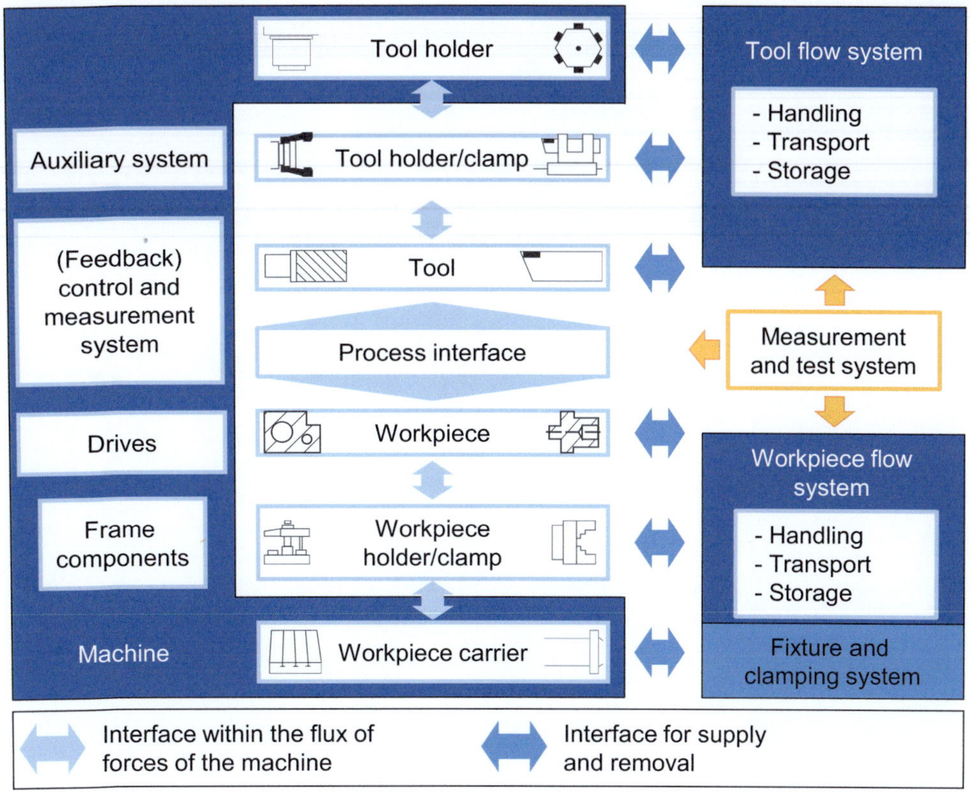

Fig. 10.2 Interfaces of the equipment components to the machine tool and surrounding environment

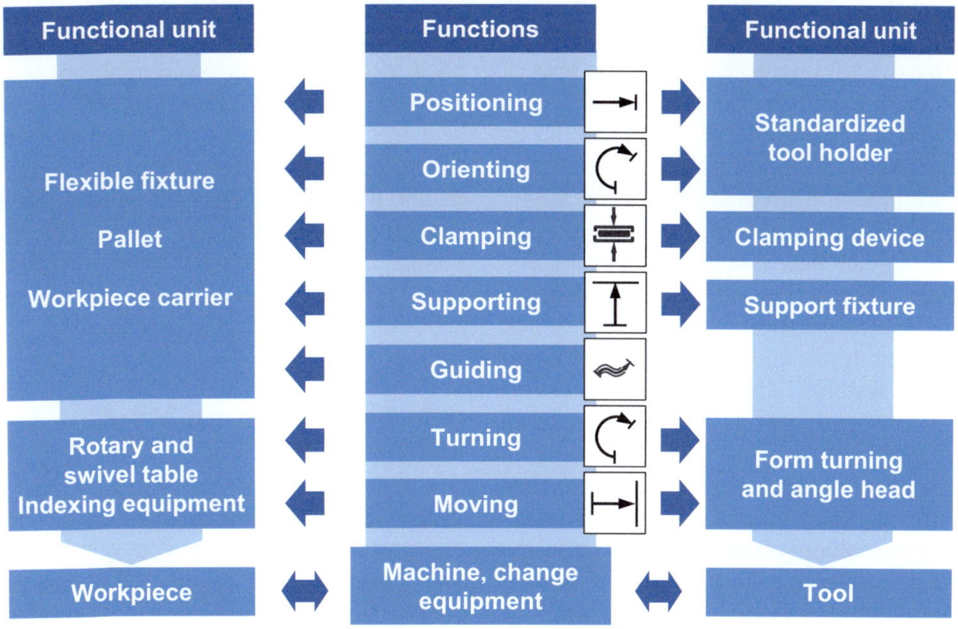

Fig. 10.3 Functions of carriers, holders and clamping devices

In addition to the orientation determination and clamping functions mentioned, additional tasks of the production resources in interacting with machine tools and handling equipment must be performed. Elements with a supporting function are used to reduce the compliance of thin-walled workpieces and cantilevered tools, for example. If required, the exact position of the tools in relation to the workpiece is guaranteed by an

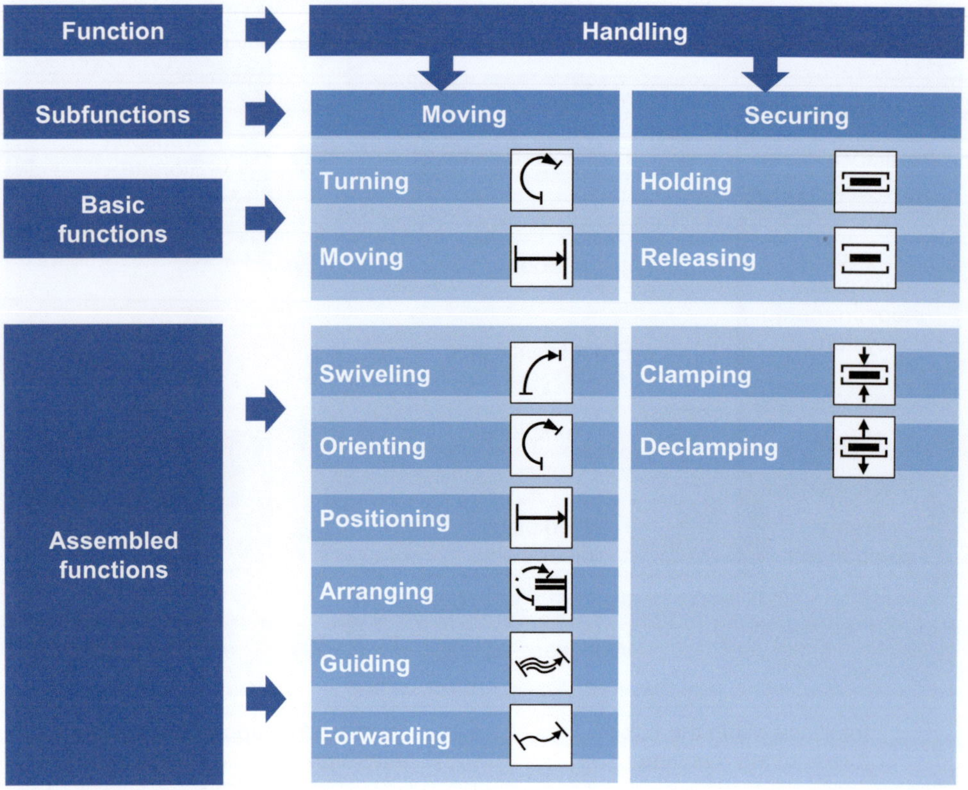

Fig. 10.4 Functional organization of the handling concept

additional guide system. The functional units of modular devices and tool systems must be connected to realize the respective overall function. Other functions of devices, such as turning, shifting or swiveling, are used to enhance the machine functions, e.g., for machining on all sides and defined, repeated position determination (linear longitudinal and circular parts) [LOCK84].

High importance is attached to automatically operating tool and workpiece flow systems on current machine tools. Different equipment components, depending on the degree of automation sought, are required for flexibly supplying tools and workpieces to single and multi-machine systems. The handling, transport and storage functions must generally be implemented (Fig. 10.4).

In addition to the equipment described above, various peripherals are necessary to operate a machine tool. The function of the **cooling systems** is to maintain the temperature level as constant as possible inside the machine tool. In metal-cutting machine tools, for example, the main spindles, linear direct drive motors and feed drives, hydraulic systems, electrical cabinet and cooling lubricant are cooled. Another peripheral is the **cooling lubricant supply unit**. This dissipates process heat from the tool-workpiece contact point, uses lubrication to reduce the heat generated and removes chips from the machining point. The chips are then transported on or processed by **handling systems** such as chip conveyors, chip centrifuges and briquet machines. Lubricant is supplied to the most varied lubrication points of machine tools by **central lubrication systems**. The functionality of bearings, guide systems and gearboxes can be ensured in this way using metered quantities. With the help of **hydraulic units**, functions such as tool changes, workpiece clamping, pallet changes, and weight compensation of vertical axes are implemented in machine tools. The designs of these peripherals, together with how they work, are described in the following chapters.

10.2 Peripherals for Machine Tools

10.2.1 Cooling Systems

The manufacturing quality of a machine tool suffers greatly from thermo-elastic deformations of the machine components from various heat sources. Internal process-dependent factors (such as power loss of motors, gearboxes and main spindle bearings or process heat in the tool and chips) and external factors (such as fluctuating shop floor temperatures) are time-variant and generate an unsteady temperature field over the machine structure. So, when there are strict accuracy requirements, temperature control units must be used to maintain constant temperatures. These control units are capable of both cooling and heating.

Cooling systems are used in metal-cutting machine tools to cool the main spindles, the linear direct drive motors and feed drives, the hydraulic systems, the electrical cabinets and the cooling lubricant.

The basic operation of these devices and their components are shown in ◘ Fig. 10.5. In an external circuit (cooling circuit), cooling water circulates in heat exchangers to keep the temperatures of the hydraulic oil, the cooling lubricant, the electrical cabinet (e.g., air-water heat exchanger) and the frame components of the machine tool as closely as possible to a temperature level and to track the level of room temperature. The cooling water temperature is set by an internal cooling circuit or external heating unit.

The most important design criterion for machine cooling is the accuracy of the cooling water temperature control in the tank. For standard machining centers, a larger hysteresis (variance from the reference temperature ±1.5 K, for example) than for high-precision machining processes (variance ±0.5 K, for example) can usually be selected [ROTH10; ABEL13]. Depending on the temperature control requirements, differentiation is made between the conventional methods, clocked compressor and hot gas bypass, for providing cooling capacity based on need. The frequency-controlled compressor and digital scroll compressor methods show clear energy-saving potential [ROTH10].

A **clocked compressor** runs under full load until the tank temperature has reached the lower threshold. When the minimum temperature is reached, the compressor drive and fan switch off. The heat exchanger transitions to a base state in which only the cooling water pump runs. In comparison to systems where the compressor runs continuously, switching the compressor off increases energy efficiency. The number of times a compressor cooler switches per hour is limited to 10 to 12. This comes from the fact that frequent switching has a negative impact on compressor life [ROTH10]. Furthermore, no narrow temperature control accuracies (hysteresis less than ±0.5 K) can be achieved by continuously switching the compressor on and off. It is important that the compressor not be overengineered for the application.

For high-precision machining processes where hysteresis less than ±0.5 K is required, temperature control cannot use a clocked compressor but rather the **hot gas bypass** method [RITT10b]. The compressor runs continuously in this case. The cooling capacity is regulated by changing the quantity of coolant and is infinitely variable in the range of 0 to 100%. Downstream of the compressor, the excess quantity of coolant

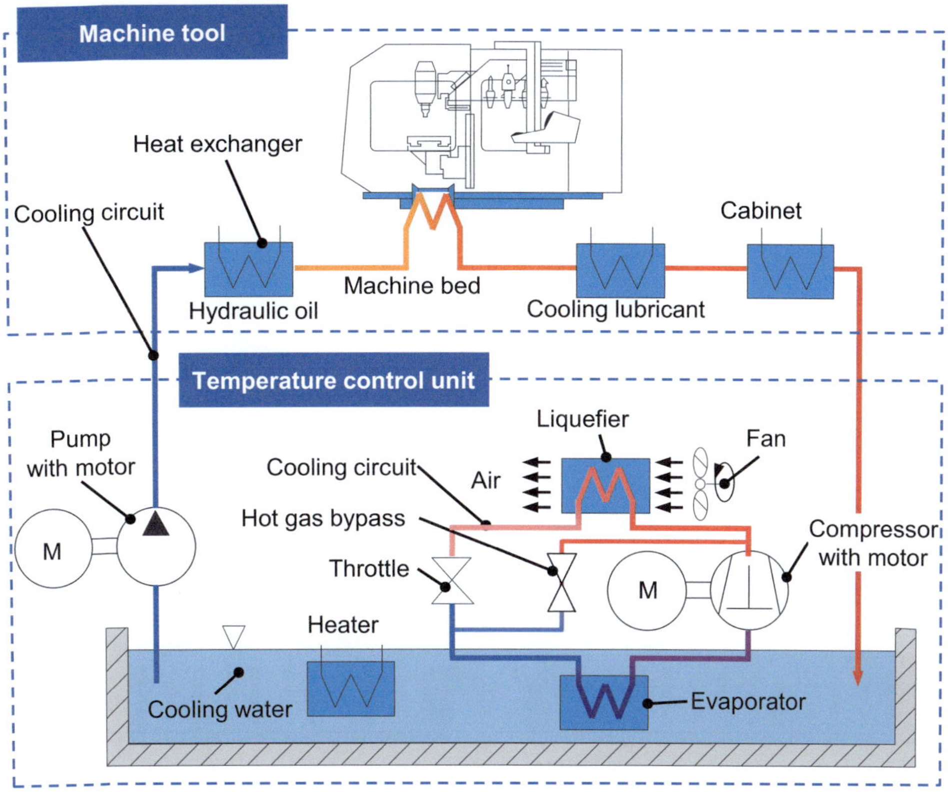

◘ **Fig. 10.5** Circuits of a temperature control unit, according to Behr

compressed is injected by a bypass directly upstream of the evaporator. This bypasses the liquefier and throttle (see ◘ Fig. 10.5). In this way, the hysteresis required can be achieved but the drive energy applied proportionally is dissipated in full. Although the power consumption of the compressor drops by 15 to 20% in part-load mode compared to full-load mode, cooling systems using the hot gas bypass method have low energy efficiency levels (in the part-load range in particular) because excess cooling capacity is dissipated by artificial loading in the bypass [DENK11; RITT10b; ROTH10; SIEM15].

To address the poor energy efficiency of the hot gas bypass method and at the same time to provide a compact design for cooling, two new principles have been developed:

- the frequency-controlled compressor
- the digital scroll compressor

These have not been on the market long or will be soon ready for series production. Both systems have the downside of higher investment costs, however.

Frequency-controlled compressors and fans represent a new approach for uniform temperature control and attain an increase in energy efficiency compared to the hot gas bypass method. Energy efficiency is increased primarily in the part-load range because cooling capacity need not be dissipated in the bypass. According to the manufacturers, savings of up to 70% of the energy costs for machine cooling are possible with this method [RITT10a]. Frequency control uses inverter technology. A rectifier converts AC to DC. From the DC, the inverter generates an artificial 3-phase AC with a certain frequency and amplitude. The artificial 3-phase AC is routed to the compressor and fan drive. The inverter controls the speed of the drives by changing the frequency and it can change the capacity of the cooling system from 0 to 100% [TOSH].

The cooling capacity of the **digital scroll compressor** is infinitely variable from 10 to 100% when the compressor speed is constant. Two Archimedes' spirals, called scrolls, with a constant winding gap, form the core of the digital scroll compressor (see ◘ Fig. 10.6, top right). The structure and the compression process are shown in the top left of ◘ Fig. 10.6: The spiral joined to the base surface is stationary (stator), while the cover surface spiral, rotated by 180°, is offset from the first one by an eccentric. After one revolution, a sickle-shaped surface that is maintained in its shape by the outer pressure is generated. This surface is reduced further on continued rotation. If the gas is compressed to its maximum, it flows out through the center of the stator. Pressure control is achieved by controlled separation of the two spirals by about 1 mm (bottom of ◘ Fig. 10.6). For this,

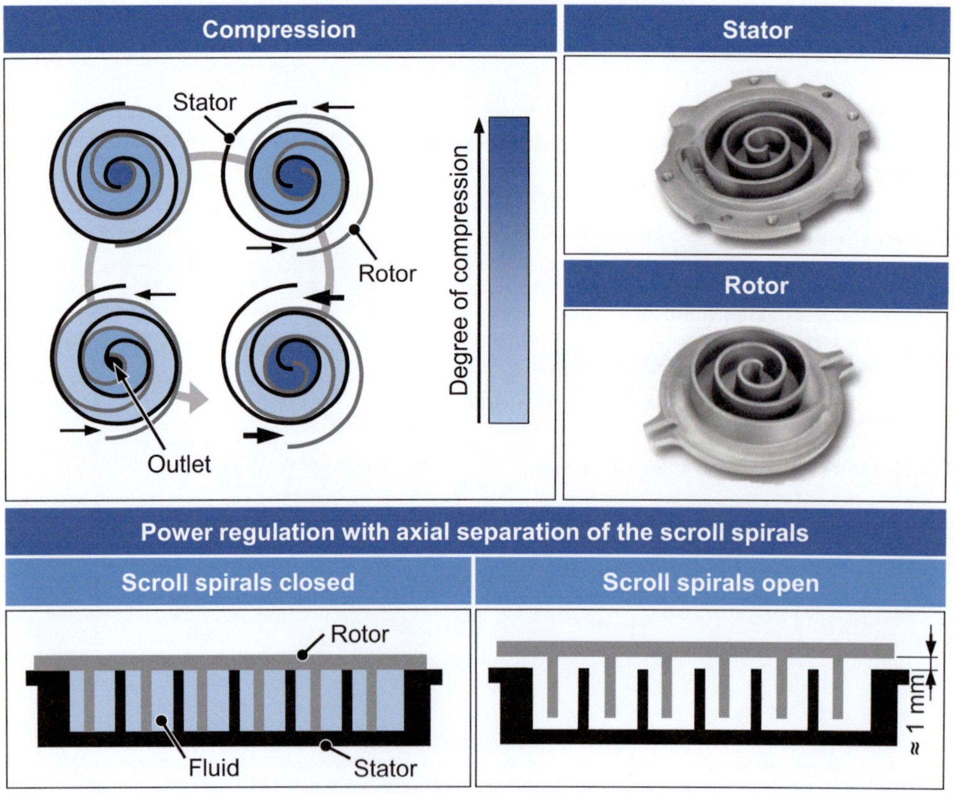

◘ **Fig. 10.6** Design and functional principle of a digital scroll compressor. *Source* Emerson Climate Technologies

the connection between the pressure and suction sides is opened and closed by a solenoid valve. When the valve is open, a control piston separates the stator from the moving spiral. This pressure equalization means that compressed gas is not generated between the pressure and suction sides. When the valve is closed, the spirals are pushed back together, and the compressor operates normally. The ratio of lock time of the valve to overall cycle time determines the cooling capacity control of the compressor [ENER13].

The digital scroll compressor exhibits improved energy efficiency, in the part-load range in particular. The downsides of this recooling unit are the high investment costs in the form of additional control equipment and a complex controller configuration.

Heat exchangers are important components of the cooling circuit in addition to the compressor. Heat exchangers transfer the cooling capacity of the compressor cooling circuit to the coolant (evaporator) and ensure that the heat of the peripherals is transferred to the coolant. These heat exchangers are described below.

Figure 10.7 shows the functions and structures of different heat exchangers. The plate heat exchanger on the right in the figure is used for heat exchange between two liquid media. In the air cooler on the left in the figure, the heat of an electrical cabinet, for example, is transferred to the ambient air (air/air heat exchanger). The temperature in the electrical cabinet is thus always above room temperature. Using an air/liquid heat exchanger, shown in the center of Fig. 10.7, the heat from an electrical cabinet, for example, can be transferred to a cooling circuit. The intake area for warm air is normally at the top. The cold air is routed downwards.

10.2.2 Cooling Lubricant Supply

Cooling lubricants represent an important resource for machining metallic materials. The oils and water-mixed emulsions used here usually reduce the process temperatures, increase tool service lives and possible cutting capacity as well as improve the attainable surface qualities and accuracies. Cooling lubricants also help in removing metal chips generated during the machining process.

After the cutting process, cooling lubricants stick to the rough chip surfaces. They are discharged from the machine tool together with the chips. The transport and recycling of metal chips contaminated by cooling lubricant are costly and are becoming increasingly restricted by environmental regulations. On the other hand, costs accrue for replacing lost, valuable cooling lubricant, meaning that from economic and ecological viewpoints, greater significance is being attached to separating out and recovering cooling lubricant in production facilities.

In the following chapter, the basics of cooling lubricant supply are explained by way of introduction. Then different systems for minimum quantity cooling lubrication and separation processes for cleaning cooling lubricants are introduced. When combustible, non-water-miscible cooling lubricants (cutting oils) are used, precautions must be in place for oil fires and weak oil mist explosions (deflagrations). Finally, fire precautions are explained.

10.2.2.1 Basics of Cooling Lubricant Supply

To reach high performance levels, modern cooling lubricant systems must perform the following main tasks [KLOC08]:

- Dissipation of the process heat from the tool-workpiece contact point by cooling
- Reduction of heat generation using lubrication
- Removal of chips

As shown in Fig. 10.8, cooling lubricants in DIN 51385 are divided into three categories, viz., water-immiscible, water-miscible and water-mixed cooling lubricants [DIN13d].

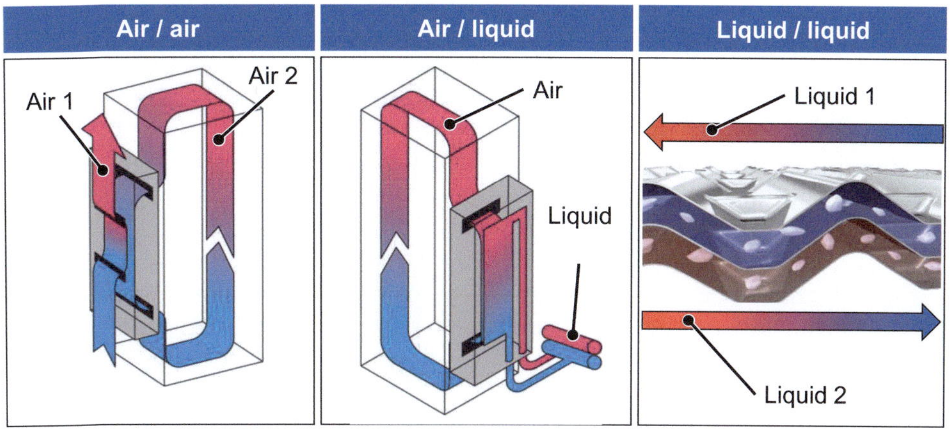

Fig. 10.7 Types and structures of different heat exchangers. *Sources* Kelvion PHE, Rittal

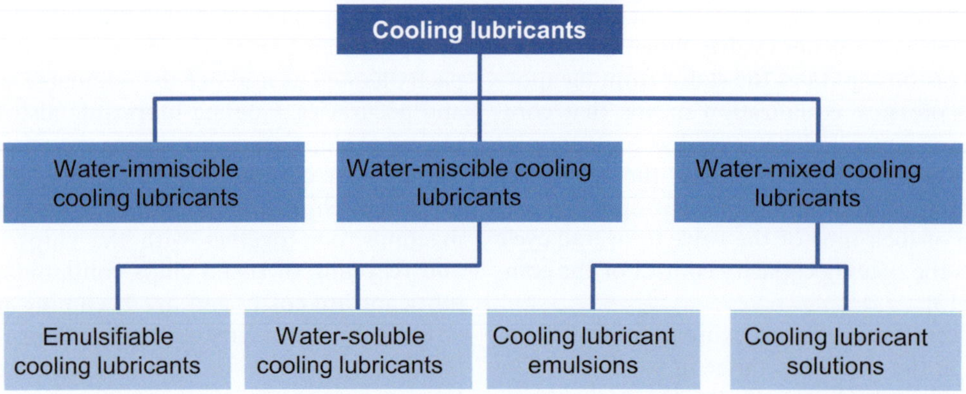

Fig. 10.8 Division of the most important cooling lubricants for metal processing, in accordance with DIN 51385 [DIN13d]

Water-immiscible cooling lubricants usually comprise mineral oils that also contain additional active ingredients to improve lubricity, wear protection, corrosion protection, aging resistance and foam behavior. Additives that improve lubrication (antiwear additives) are used to reduce friction at the cutting point. For this, natural grease oils (palm oil, rapeseed oil) or synthetic greases (ester) are added. Low-viscosity ester oils are used in addition to mineral oils. Mineral oils are currently being used the most (for about 90% of cases). Water-miscible cooling lubricants are provided as concentrates and diluted with water by the user before use. The high water content (up to 99%) is the reason for the good cooling effect but also for the properties that speed up corrosion. Rust protection is guaranteed by the slight alkaline character of the liquid (pH value: 8–9). An emulsifiable cooling lubricant can form a discontinuous phase of an oil-in-water emulsion, whereas water-soluble cooling lubricants form a solution when mixed with water. The most important additives to cooling lubricant emulsions are emulsifiers. Their function is to disperse the oil in the water so that a stable oil-in-water emulsion results after water is mixed in. Water-mixed cooling lubricants describe ready-to-use mixtures and so need not be diluted with water by the user. They correspond to ready-mixed water-miscible cooling lubricants [DIN13d; KLOC08].

Figure 10.9 shows the typical structure of a cooling lubricant supply system. A cooling lubricant is pumped from the clean tank to the machine tool by the pumps. In addition to the supply of cooling lubricant for the machining point (inner and outer supply, refer to ► Sect. 10.2.2.2), pumps for workspace flushing can be installed. Dirty cooling lubricant flows into a chip conveyor with the chips. A continuous scraper belt, comprising two chain sections joined by conveyor belts (see ► Sect. 10.2.3.1), transports the chips continuously on the container base into the chip container via the discharge point. The cooling lubricant is pumped into the dirt tank via a coarse filter with a pump controlled by a level switch. In the dirt tank, a filter pump, connected to the inner chamber of the filter drum, sucks the cooling lubricant into the inside of the drum. The dirt particles on the surface of the filter drum are stopped by very fine fabric. Other mechanical separation processes are used besides this one (see ► Sect. 10.2.2.3). The low pressure increases due to the resistance of the filter cake building up. When an adjustable low-pressure threshold is reached, the drum is washed automatically by a slush pump (from the inside outwards). The dirt particles settle and are moved by a continuous scraper belt (working in fixed cycles) on the container base to the dirt container. The cleaned cooling lubricant is pumped from the inside of the drum to the clean tank by the filter pump.

In addition to the technological benefits that using cooling lubricants offers, there is also a certain risk for the people working with them and for the environment. If water-immiscible cooling lubricants are used, safety procedures must be in place for machine tools that guarantee fire protection (see ► Sect. 10.2.2.4). Cooling lubricant components such as bactericides and fungicides, as well as reaction products generated and impurities introduced, can also trigger illnesses. Leakage and drag-out losses, emissions, wash water and the disposal of used cooling lubricants are harmful to soil, water and air. Working with and disposing of cooling lubricants are strictly regulated by law and trade associations. Information on the requirements of cooling lubricants and associated equipment, on handling and disposing of cooling lubricants, and on laws, provisions and trade association rules and regulations to observe, can be found at the following places (not exhaustive):

- German Trade Association Regulations BGR 143: Working with cooling lubricants [HAUP06],
- VDI Standard 3397: Cooling lubricants for metal-cutting and forming manufacturing processes [VDI07]
- VDMA Guide: Cooling lubricants—Fresh air at the workplace [VDMA02].

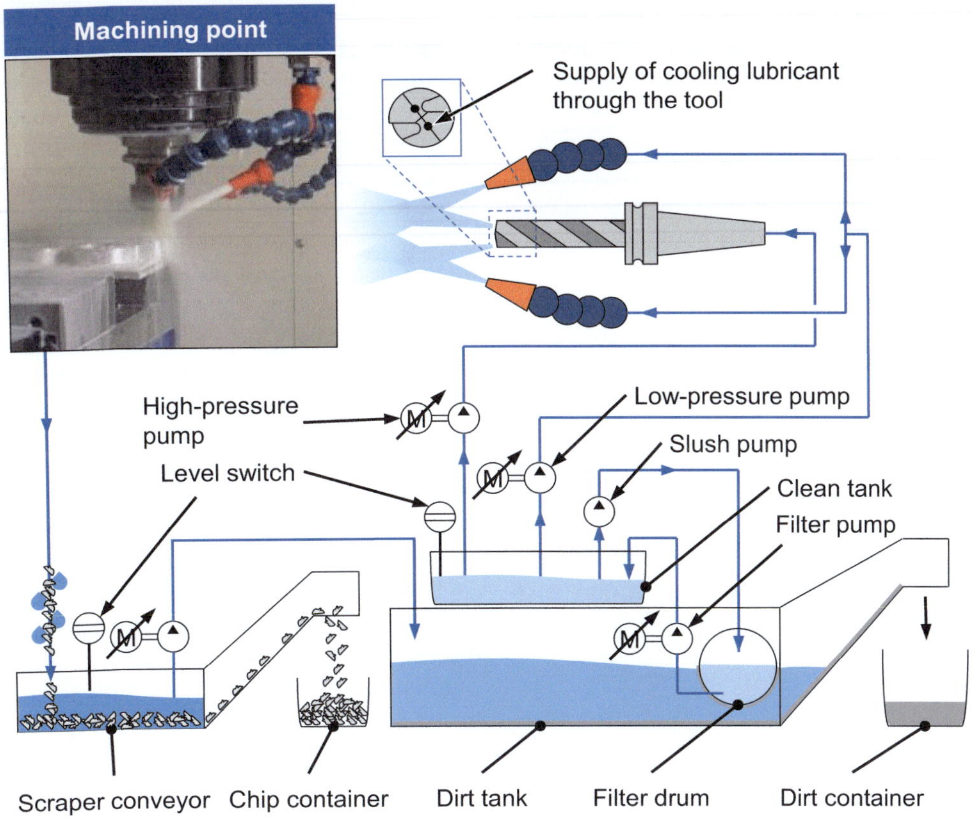

Fig. 10.9 Structure of a cooling lubricant supply system

For companies, these stipulations and regulations primarily mean increased financial costs—in addition to the considerable responsibility towards their employees. Production is therefore increasingly challenged to meet environmental requirements without compromising cost-effectiveness. One approach is to reduce the use of cooling lubricants or cut them out altogether. The aim of lowering cooling lubricant usage is to supply the process with only the amount required technologically. The secondary functions of a cooling lubricant, such as transportation of chips inside the machine or temperature control of the chips, must be resolved by taking different actions. These actions pertain to the tool as well as the machine tool [KLOC08].

10.2.2.2 Systems for Minimum Quantity Cooling Lubrication

In contrast to wet machining, dry machining does not use cooling lubricants, usually the media for the lubrication, cooling and chip transport functions. The motivation behind the trend towards dry machining is mainly the saving potential in investments for operating, keeping clean and disposing of cooling lubricants. Other important reasons are occupational hygiene and keeping the working environment clean.

For the full dry machining process, however, certain combinations of machining methods and materials exhibit strong adhesive and abrasive wear, meaning lubrication cannot be omitted completely in these cases. Boring, reaming and thread cutting are often critical. It has been shown that even the smallest amounts of lubricant (10 to 50 ml per process hour) can greatly decrease cutting temperatures as well as tool wear. Minimum quantity cooling lubrication (MQCL) is understood to be the targeted supply of lubrication medium to the points of use for machining or cutting and has a maximum volume flow of less than 50 ml per process hour [WEIN99].

Lubrication between the tool and workpiece is realized by an air flow with finely dispersed oil droplets, or aerosol.

Hydrocarbons, fatty alcohols, polyglycols and fatty acid esters are used as lubrication media. These substances are transported to the cutting points with a jet of compressed air. Generally, the transport effect of compressed air for the chips, and its cooling effect, are inferior to the liquids in wet machining.

The selection of suitable lubricant components is subject to a number of requirements and should, in addition to technological applicability, take into account primarily toxicological and dermatological compatibility with regard to the machine operator and compatibility with the design materials of the machine tool (such as for seals) [KLOC98]. In the design and operation of a machine, it is important to ensure that occupational safety regulations such as technical guideline concentrations (TGC) and maximum workplace concentrations (MWC) with regard to the auxiliary machining materials used are observed [PFLA98; BUND06].

The methods shown in ◘ Fig. 10.10 enable more or less comprehensive conclusions to be drawn about the spray pattern of a nozzle and allow a functional test. The procedures described can be transferred to the inner minimum quantity cooling lubrication. The supply of lubricant medium can be split into the inner and outer supply through the spindle. These supply variants are described below. The aim of current developments is the sensor-based functional test of the quantity at the cutter, for the purposes of reliable machining process capable of being automated.

Outer Supply

Great importance is attached to the nozzle for the outer supply because it must generate a uniform spray pattern with no mist or dripping immediately after switch-off. As they are exposed directly to flying chips, they are wearing parts and should be tested regularly for correct operation.

The minimum quantity lubrication system shown in ◘ Fig. 10.11 comprises a lubricant reservoir (4), a mixture regulator and lubricant lines with spray nozzles (9). The compressed air supplied via a solenoid valve (1) applies pressure to the lubricant reservoir. This compressed air transports the lubricant separately via a duct and line system to the spray nozzles. Regulation of the required lubricant (5, 6) and atomizer (7, 8) air quantities, and setting of the internal pressure (2, 3) for the lubricant reservoir, are set manually using the control valves on the lubricant reservoir. The lubricant lines are coaxial lines to transport lubricant and atomizer air separately to the spray nozzle. The spray nozzle uses a 2-component nozzle because two different substances are mixed. The actual aerosol generation takes place at the nozzle outlet. The compressed air flowing past atomizes the lubricant into very fine particles and sprays them onto the cutting point.

Straightforward adaption when retooling conventional workshop machines and bridging larger spray distances (to 300 mm) are among the benefits of the outer supply. Chips flying around, vibrations and interfering contours can have a negative impact on the accurate nozzles to cutting point alignment required. Effective use presupposes workpieces and machining processes with good accessibility for the spray jet to the point of the cutter contact.

Method	Overhead	Sprays? (Yes/No)	Size of the spray cone?	Continuous spraying?	Drop size
Spray on finger nail	Low	◕	◔	◔	◔
Spray on blotting paper / mirror	Medium	◕	●	◔	○
Spray on moving blotting paper	High	◕	◕	●	○
Laser spectral analysis	Very high	◕	○	○	●

● Very good conclusion ◕ Good conclusion
◔ Possible conclusion with restrictions ○ No conclusion possible

◘ **Fig. 10.10** Functional test of MQCL systems. *Source* Steidle

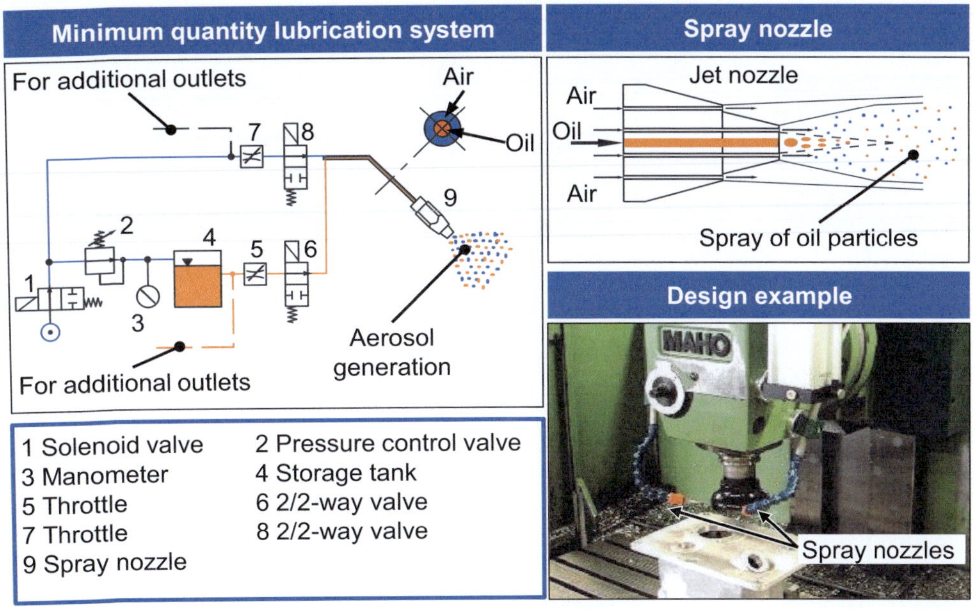

Fig. 10.11 System for outer MQCL supply. *Source* SKF Lubrication Systems

Inner Supply

The inner supply enables direct transport of the lubricant to the point of use. The lubricant is transported by the rotating union, the spindle and the inner cooling channels of the tool. In contrast to the outer supply, no adjustment of supply nozzles is necessary and only minor scatter losses occur. Inner supply of the medium is selected when outer supply is no longer adequate (e.g., for drilling with large l/d ratios) or when there is a requirement to increase machine productivity to attain shorter machining times. Two working principles have established themselves that depend on where the mist is generated. On the 1-channel system, the aerosol is generated in the container, i.e., before entry into the spindle. On the 2-channel system, both media are routed separately through the spindle, and are only mixed into an aerosol when exiting the spindle directly in front of the tool (Fig. 10.12) [DGUV10].

The 1-channel system shown in Fig. 10.13 is based on an aerosol generator creating an average oil droplet size of about 0.5 µm. This small size means the droplets barely have inertia or fall velocity. So longer transport distances, high deflection levels and the centrifugal forces of rotating tool spindles hardly have a demixing effect and the lubrication particles are available continuously at the tool cutter. Individual aerosol quantities and compositions required for each tool and machining task can be set by enabling valves. This allows the use of a number of different tools. This is customary for modern-day machining centers with an automatic tool changer.

On 2-channel systems with inner MQCL supply, air and lubricant are routed through the spindle with separate lines, and only brought into contact directly in front of the tool. A 2-channel rotating union suitable for the maximum spindle speed and capable of unlubricated operation is required.

The system shown in Fig. 10.14 shows the design of the 2-channel, inner minimum quantity cooling lubricant supply.

A ring line carries the oil from the unit to a quick-acting valve. A gear pump provides adjustable and constant pressures in the ring line. Oil metering and air flow are regulated by the quick-acting valve and proportional valve, respectively. The cycle and opening time of the quick-acting valve, and the air quantity, are specified by the machine control system/machine operator. A lance mounted in the center of the spindle ensures the separate supply of oil in the inner channel and compressed air in the outer channel (ring line). The front end of the lance projects into the mixing chamber of the pipe nozzle. From here, the oil-air mix generated is delivered to the tool. The pipe nozzle pressed into the standard coolant pipe comprises a mixing chamber and pipe piece, that is cut to length depending on the particular assembly situation [DGUV10].

Oil and air are available near the process at the tool interface. Routing the lubricant through the tool directly to the tool cutter is a benefit of the inner supply. This ensures supply is reliable and robust. Other benefits are the oil supply that can be calibrated and is independent of the rotational speed and the short response times of the system.

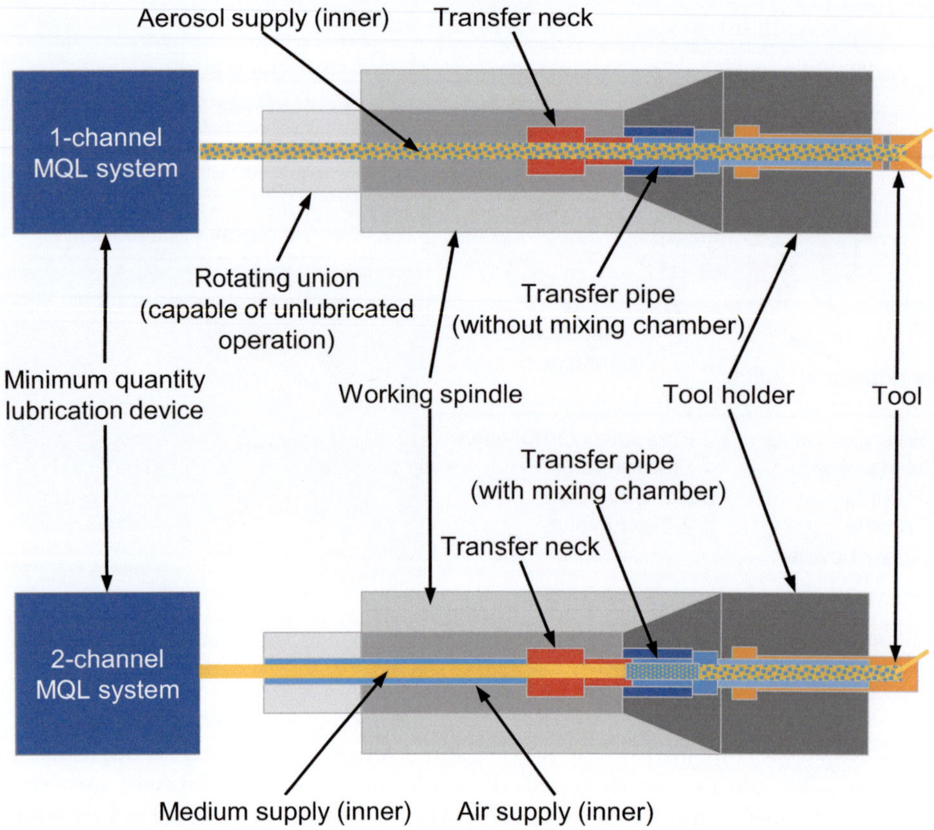

Fig. 10.12 Comparison of 1-channel and 2-channel MQCL systems [DGUV10]

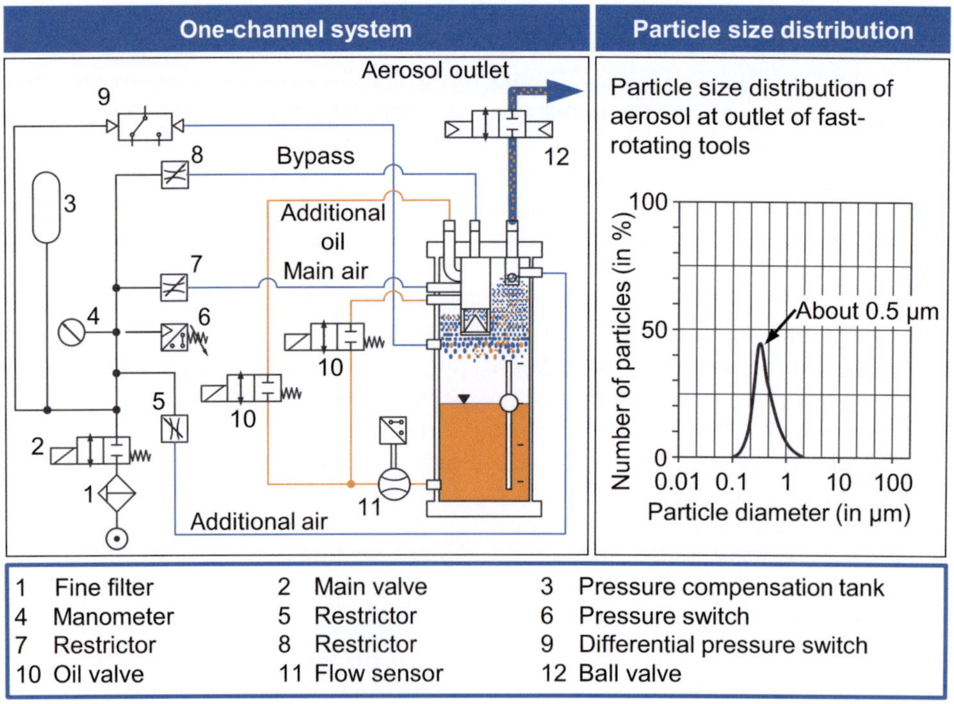

Fig. 10.13 Aerosol preparation for inner, 1-channel MQCL supply. *Source* SKF Lubrication Systems

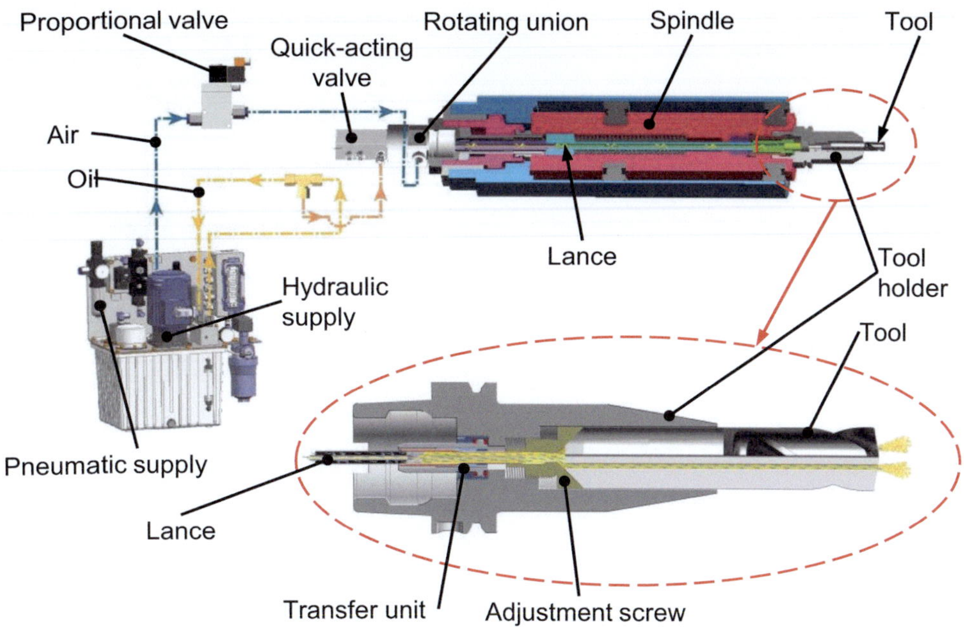

Fig. 10.14 Structure of inner, 2-channel MQCL systems. *Source* BIELOMATIK

10.2.2.3 Separating Processes for Cleaning Cooling Lubricants

Among the factors required to maintain constant conditions when machining using cooling lubricants is continuous cleaning of the cooling lubricant that has been dirtied by chips and dirt in the machining process. Sedimentation, filtration and magnetic separation in particular are familiar processes suitable for mechanical separation processes. The associated equipment with limitations on use and application areas are summarized in Fig. 10.15 and described below.

The term sedimentation describes the separation of solids from a suspension using inertia forces. In practice, gravity or centrifugal forces are used for this. For filtration, different physical filters are used depending on the contamination of the filter and the cleanliness level required after filtering. Generally, a high filter fineness causes a high flow resistance and restricts the realizable volume flow rate. A magnetic separator is a system for separating ferromagnetic chips from cooling lubricants. These procedures are contrasted below.

Sedimentation

Figure 10.16 shows the principle of a simple sedimentation basin. The intake near the bottom is free of turbulence to stop small particles already separated from being swirled up. The drain near the fluid surface must be as far from the intake as possible. Taking the simplified assumption of spherical solid particles in a Newtonian fluid, Stokes' law gives the specified stationary settling velocity v_T for the particles. It must be greater than the mean settling velocity v_{Fl} of the fluid, given by the vertical component of the volume flow in the basin. The minimum effective tank surface $b \times l$ required can be derived from it. The settled substances are transported to a dirt tank by a belt conveyor.

Centrifugal forces in a hydrocyclone can also be used to separate solids and fluid. The dirty liquid enters the cyclone cell shown in Fig. 10.17 on the left and is set into rotation by the helical channel in the inlet piece (primary vortex). Any solid particles also in the liquid settle on the side of the nozzle due to their greater densities and are discharged at the bottom together with a small amount of liquid. Most of the liquid flows back up in an inner vortex (secondary vortex) rotating in the same direction. Cyclone systems are suitable for low-viscous media (such as emulsions) with a lot of contamination.

Filtration

Belt filters are normally used to clean the cooling lubricant of chips from turning, milling and grinding processes. Figure 10.18 shows the structure and principle of a gravity belt filter. The medium to filter passes through the inlet box into the trough-shaped filter basin, which uses a chain screen to move a filter belt made of paper or fleece fabric. While the liquid trickles into the container, impurities with a grain size of about 10 µm accumulate on the filter material.

The filter cake forming causes ever-finer filtration but also inhibits the flow (changing the ratio of incoming to outgoing volume flow rates). A floater on the now rising wastewater level activates a switch that trig-

Method	Separation threshold [µm]	Particle contamination of medium	Application
Sedimentation			
Sedimentation basin	30	High	Grinding, drilling, turning (low-viscosity fluid with discontinuous chips)
Hydrocyclone	10	Very high	Grinding (low-viscosity fluid with crumbly or powdery chips)
Filtration			
Gravity belt filter	10	High	Turning, milling, drilling, grinding (chips of all kinds)
Suction belt filter	5	High	Turning, milling, drilling, grinding (chips of all kinds)
Rotation belt filter	15	Medium	Turning, milling, drilling, grinding (short chips, max. size 20 mm)
Backflush filter	20	Medium	Grinding, drilling, turning (low and high-viscosity fluids with discontinuous chips)
Settling filter	5	Low	Superfine filtering (no emulsion)
Magnetic separator			
Magnetic rotor separator	5	Medium	Grinding, drilling, turning (low and high-viscosity fluids with ferromagnetic, discontinuous chips)

Fig. 10.15 Mechanical separating processes for cleaning cooling lubricants

gers removal of the used filter belt into the dirt tank as well as the supply of fresh filter material.

The filter capacities of belt filters can be increased by creating low pressure on the clean side by pumping. Figure 10.19 shows a diagram of a suction belt filter system. The dirtied cooling lubricant flows from the machine tool into the dirt tank of the low-pressure filter (1). The cooling lubricant is sucked through the filter medium into the low-pressure chamber by a pump (2). At the same time, the pump pumps cleaned cooling lubricant from the low-pressure chamber back to the machine tool (3). In the process, a partial flow fills the compensation chamber (4). The differential pressure between the dirty and clean sides is recorded as a measure of how dirty the fleece is. When a threshold is exceeded, regeneration of the filter is triggered (Fig. 10.19, right). Here, opening a cut-off valve (5) routes cleaner cooling lubricant from the compensation chamber to the low-pressure chamber. The dirty fleece is ejected from the filter by the scraper bars all around the filter (6). The used fleece is rolled up on the fleece winding device (7).

Figure 10.20 shows a rotation belt filter that works without filtering aids. The cooling lubricant to

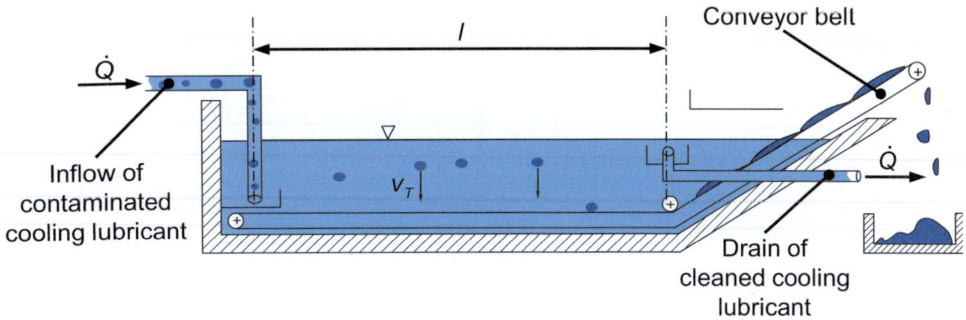

Settling velocity of particle: $v_T = \frac{d^2 * \Delta\rho * g}{18\eta}$

Mean rise speed of fluid: $v_{Fl} = \frac{\dot{Q}}{b * l}$

Separation conditions: $v_T > v_{Fl} \Rightarrow b * l > \frac{18 * \eta * \dot{Q}}{\Delta\rho * g * d^2}$

$\Delta\rho$ = Difference between particle and fluid densities
η = Fluid viscosity
$\dot{Q}$ = Volume flow of cooling lubricant
d = Particle diameter
g = Gravitational acceleration
l = Effective tank length

Fig. 10.16 Functional principle of a sedimentation basin

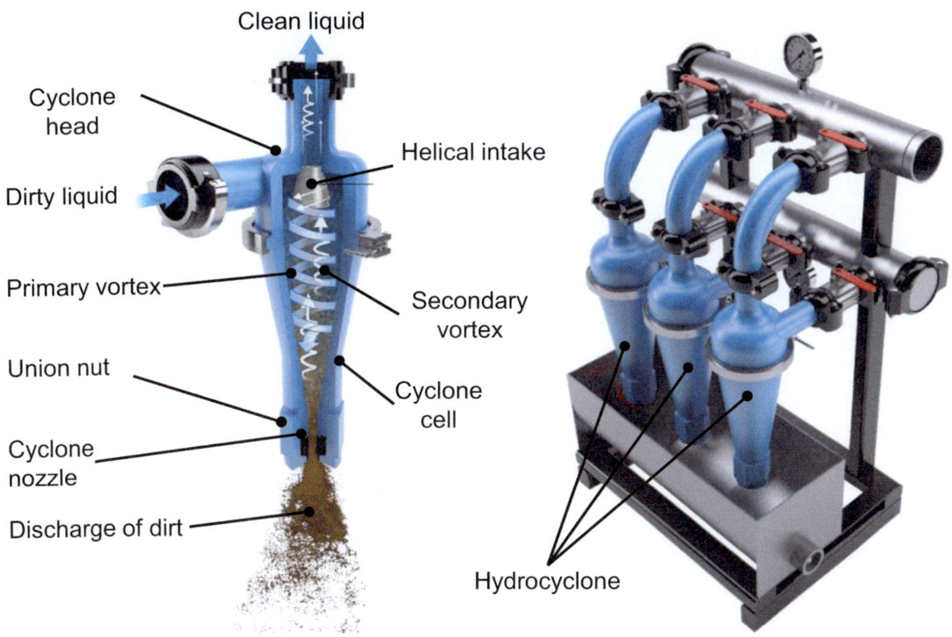

Fig. 10.17 Functional principle of a hydrocyclone, according to [COMP17]

filter is pumped from the dirt tank to the filter housing and filtered by a rotating stainless steel drum. The filter cake created is removed by a rotating brush and collected in a dirt container. Coarse particles that settle in the filter housing are transported along by the strips affixed on the drum. These particles are also removed by the brush. The stainless steel filter mesh is cleaned with flushing, with pure cooling lubricant being pumped from inside the drum to the outside. This filter principle for cleaning cooling lubricants is suited in particular to filtering oils and emulsions used in grinding processes. The maximum chip size is limited to

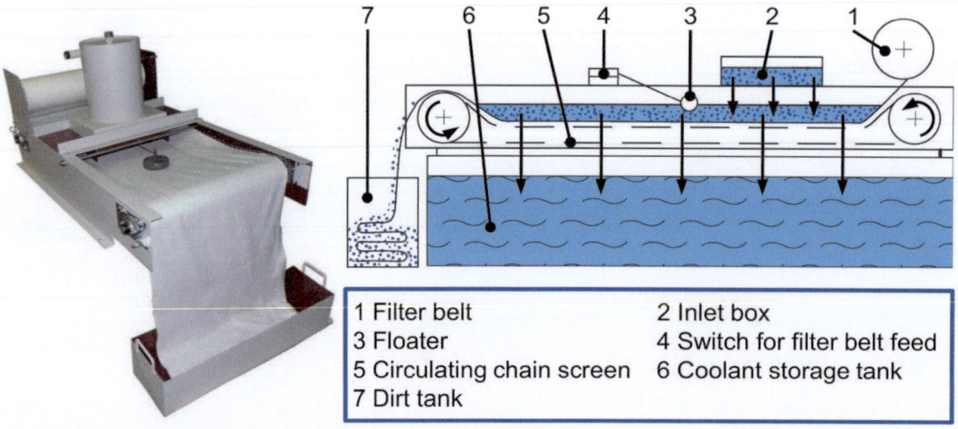

Fig. 10.18 Structure and principle of a gravity belt filter. *Source* FAUDI

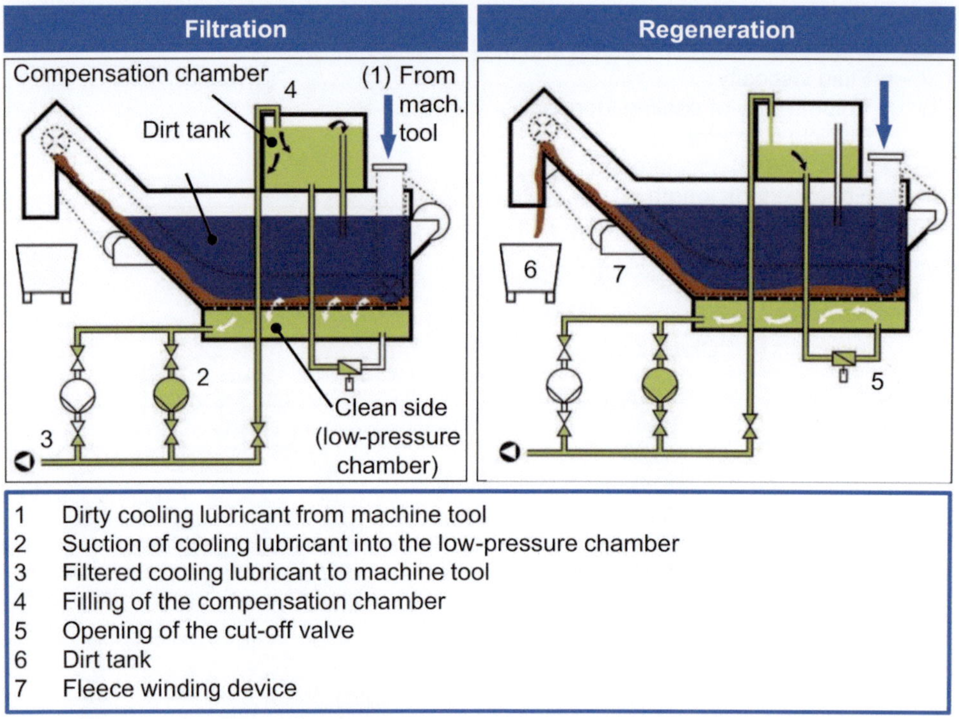

Fig. 10.19 Structure and principle of a suction belt filter system. *Source* FAUDI

20 mm. Depending on the model, the system achieves a throughput of 50 to 1500 l/min cooling lubricant with a degree of cleaning of 15–40 µm.

Backflush filters are used to clean machining fluids of higher viscosity. In the variant shown in Fig. 10.21, there is flow (from the outside in) through filter candles positioned in a cylindrical container. Increasing particle contamination of the filter candle raises the differential pressure between the dirty and clean sides. When an adjustable threshold value is reached, the intake is blocked, and the backflushing process initiated. For this, compressed air or clean cooling lubricant flows through the filter candles from the inside outwards to remove the layer of contamination from the outsides of the filter candles. The residues are passed to a belt filter, for example.

The filter fineness of such a system can be increased by, before the actual filtering process, applying an additive (usually diatomaceous earth) on the fabric fibers of the filter to reduce its permeability.

Magnetic Separation

Magnetic separators are suitable for separating cooling lubricants and ferritic chips. The principle shown in Fig. 10.22 is used. The contaminated working liquid flows, calmed by flow breakers, free of turbulence

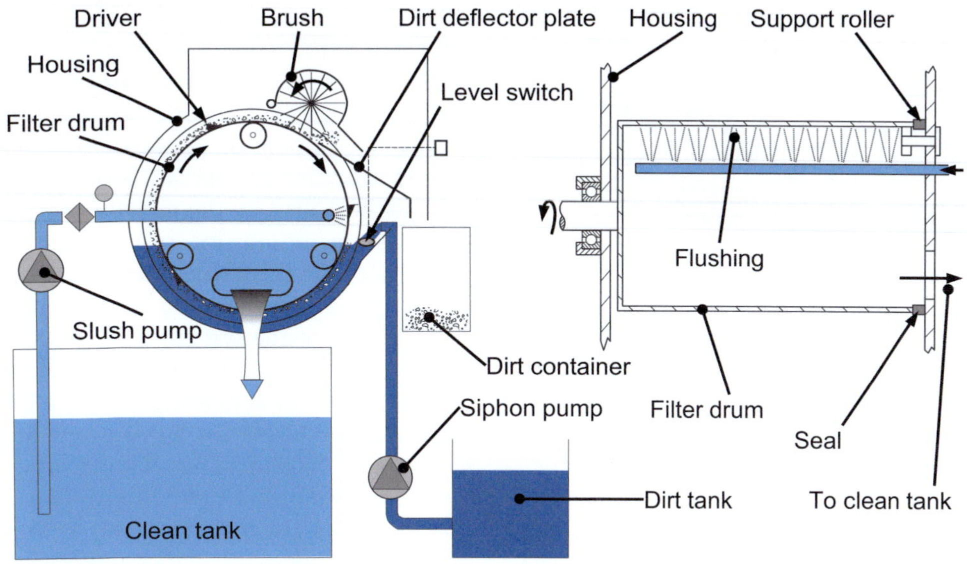

Fig. 10.20 Structure and principle of a rotation belt filter. *Source* Mayfran

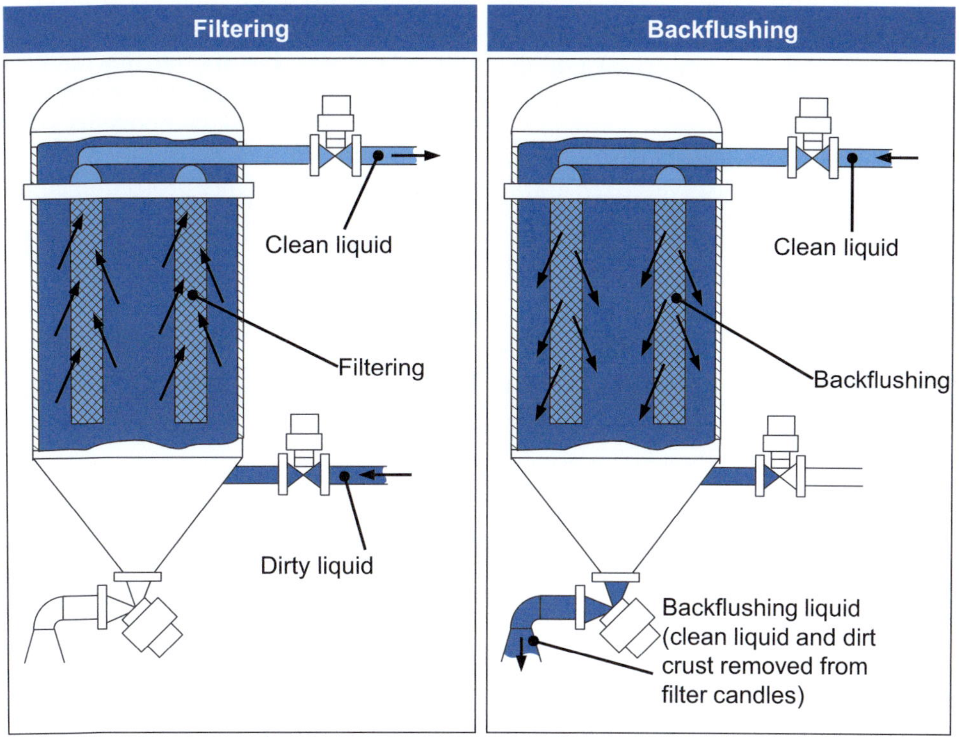

Fig. 10.21 Function of a backflush filter. *Source* Mann & Hummel

into a container in which a cylindrical permanent magnet rotates slowly. The rotational motion is continuous or is controlled by mirror height sensors. Ferritic impurities are attracted by the magnetic forces and transported to the wiper positioned at an angle. The length of the wiper and the slight angle enable liquid residues to flow back. The sediment removed by the rotor is pushed to the dirt tank.

10.2.2.4 Fire Protection Using Water-Immiscible Cooling Lubricants

When combustible, water-immiscible cooling lubricants (e.g., cutting oils) are used, precautions must be in place for oil fires and low-pressure oil mist explosions (deflagrations) in the form of a safety concept for the machine tool. These concepts are often developed

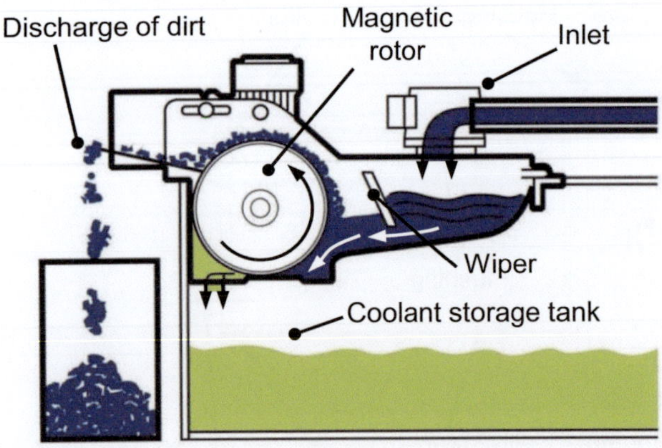

Fig. 10.22 Magnetic separator. *Source* FAUDI

together with the later company operating the machine tool. In this way, they can exhibit alternative actions and have the same protection goal [WECK].

High tool speeds and cooling lubricant supply pressures (to 80 bar) atomize the cooling lubricant when it enters the workspace and form a finely atomized aerosol. If the aerosol concentration exceeds a concentration known as the lower explosion limit (LEL) of the air-cooling lubricant mix, an ignition source can cause an explosive burn-off.

Low-vaporizing cooling lubricants with a flash point >140°C and a minimum viscosity of 10–16 mm^2/s at 40°C should be used as preventive measures. A powerful extraction system must be installed for continuous discharge of very fine cooling lubricant aerosols and vapors from the workspace of the machine tool. Proper functioning of the extraction system represents a basic condition for operation of the machine tool and so must be polled by the machine control system (by a low-pressure monitoring system, for example). In the event of a fire, the extraction system with central separator requires decoupling of the individual machine from the rest of the system.

The complex and varied machining processes in the workspace of a machine tool make difficult a general statement about the hazard potential of a machining process. Extraordinary events often result in ignition sources. For example, too low a cooling lubricant supply, a tool fracture, incorrect workpiece clamping, or collisions can cause sparks and hot surfaces. Ignition cannot be ruled out completely when combustible cooling lubricants are used [HIRS97].

An extinguisher system must be provided for oil fires. Automatic extinguisher systems to protect machine tools usually work with argon or carbon dioxide quenching gas. For the oil machining of magnesium, argon should be used consistently. Figure 10.23 shows the important components of an extinguisher system and provides an overview of the workspace of a machine tool with fire protection equipment installed.

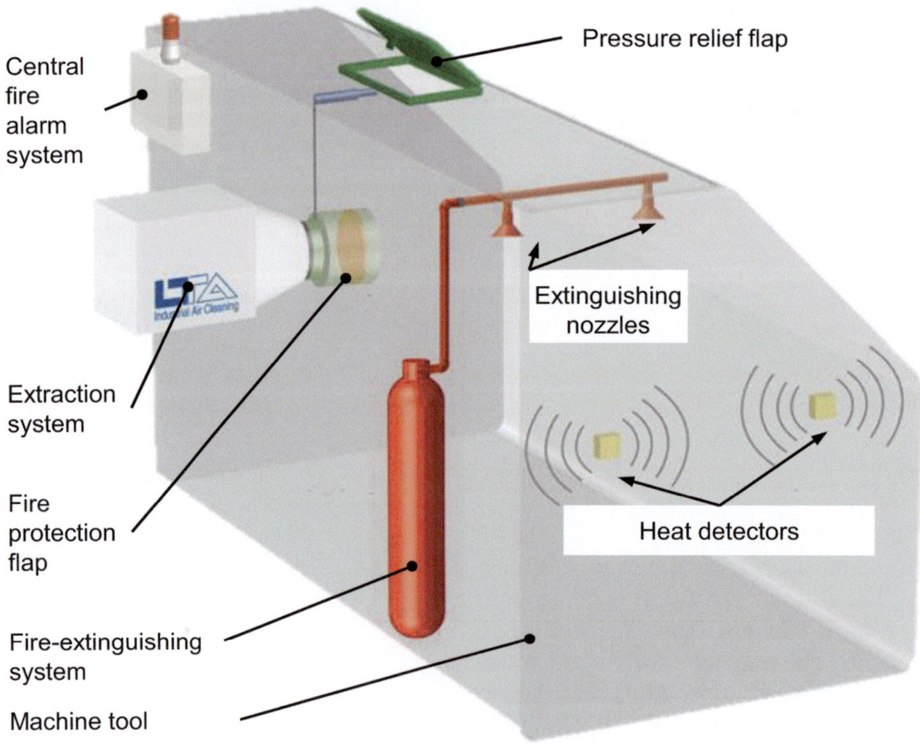

Fig. 10.23 Safety components for oil wet machining. *Source* LTA Lufttechnik

The components of an extinguisher system must have a power supply that is separate from the machine tool. Accumulators keep the system ready for use in the event of a power failure. Two thermo maximal detectors connected together as a logical OR detect a fire in the workspace. The response temperatures of both detectors are set to about 60°C. Speedy detection can be achieved with the alternative use of infrared flame detectors. In addition, a button for manual triggering is located at a fixed, clearly visible and accessible place near the operator panel. Detector signals are relayed directly to the extinguisher system controller by cables that are monitored for breaks. A weigher monitors the fill levels of the quenching gas cylinders to detect a malfunction (due to slow escape for example).

If the doors for the workspace are closed, the quenching gas is introduced, and the machine tool transitioned to the "Emergency off" status when an alarm is triggered. For reasons of personal safety, quenching gas must not be introduced while people may still be in machine areas where quenching gas concentrations posing a suffocation risk can be reached.

The consequence of these considerations is that the opening state of doors permitting access to areas where quenching gas is released must be monitored. The door opening states must be transmitted to the extinguisher system controller in a way that is independent of the machine voltage and safe from manipulation. The consequences of an alarm are summarized below:

- Machine transitions to "Emergency off" status, doors lock in the event of a fire
- The extinguisher system starts operating as soon as the doors are closed
- Visible and audible warnings are issued by the horn and strobe lamp on the machine
- Interruption of the suction air volume flow by the fire protection flap
- Transmission of fire signal information from the extinguisher system controller to the central fire alarm system (if available).

As deflagrations are possible in the workspace, hot combustion gases or flames must be stopped from entering the area of the loading station for manual loading during primary machining time. To protect the machine operator, who might be there with the loading door open, other possible automatic safety protection flaps are necessary in addition to the partition to seal and separate the loading station from the workspace. Unavoidable gaps at access doors to the workspace must resist flame penetration using labyrinth seals overlapping on both sides. Correct seal designs can improve penetration safety [VDW98].

When there is a deflagration in the workspace, the machine operator must be protected from the escape of flames and hot combustion gases. To limit the effects of a deflagration, the machine tool must be equipped with an effective pressure relief device. Lamella valves as shown in Fig. 10.24 have established themselves over bursting disks and foils.

When a deflagration occurs, they open a defined relief area that limits the pressure rise in the housing to a noncritical value. It is possible to design relief areas using the nomograms specified in [VDMA95] (Fig. 10.25). The experience that, for oil mist deflagrations, the expected explosion overpressures that are

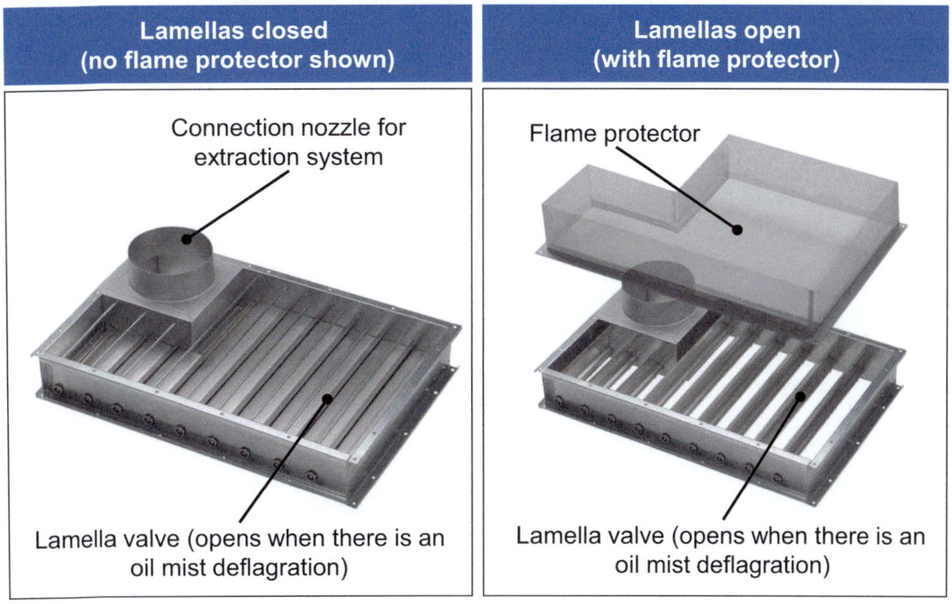

Fig. 10.24 Pressure relief lamella valve. *Source* RERUCHA

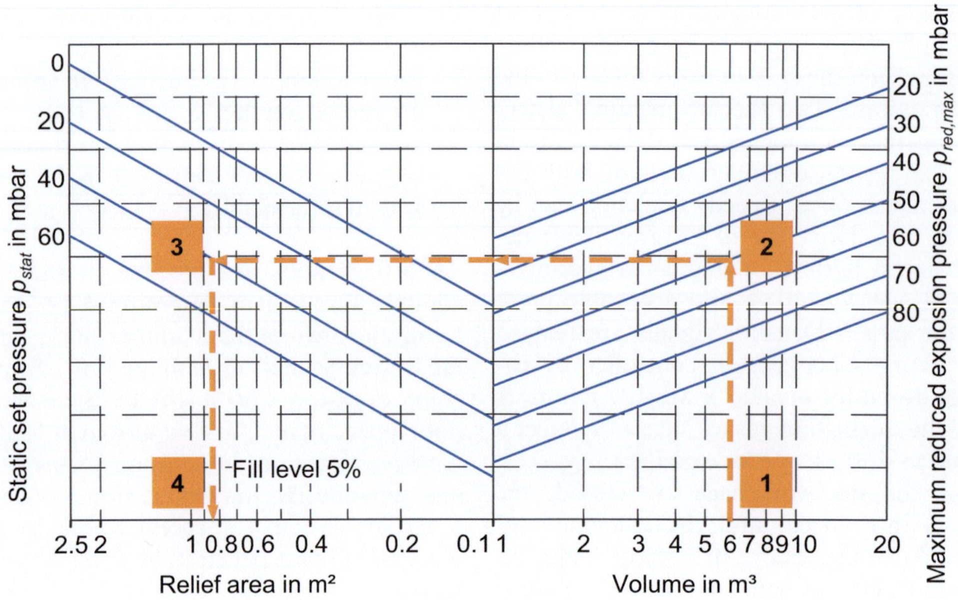

Fig. 10.25 Nomogram for the design of pressure relief areas using bursting foils, according to [VDW96; VDMA95]

used as the basis for calculations generally do not exceed 0.1 bar, and the containment strengths of the machine housing are not greater than 0.1 bar [VDW96].

Necessary in addition to the housing volume (1) are the explosion strength of the machine housing and the reduced explosion pressure required (2). The manufacturer specifications for the pressure relief device show its static set pressure (3), enabling the relief area (4) to be provided structurally to be read. The fill level (5% in the example) provides the expected share of the potentially explosive atmosphere as a percentage of the total workspace volume. In the design of lamella valves, their low pressure relief capability (compared to bursting foils) must be taken into account. In this way, the lamella valve's pressure relief area equivalent to the relief effect of a bursting foil is greater by a construction/manufacturer-specific factor.

Reference must be made in the machine documentation that flames can escape near the pressure relief areas if they do not have reliable flame barriers. These places must be marked as hazard areas as required and cable trays, for example, must not be installed there without appropriate protection.

10.2.3 Systems for Handling Chips

10.2.3.1 Chip Conveyors

The required high cutting speeds and the growing performance of numerically controlled machine tools are leading to large cutting volumes per unit of time. Different conveyors are used depending on the chips generated over time, the bulk weight of the chips, the chip shape and the quantity of cooling lubricant used in addition.

Slat conveyors are the most common design in the machine tool industry (Fig. 10.26). A continuous slat-band with steel slats rotates in an enclosed housing and carries the chips discarded on the upper run to a container standing ready. The slats are

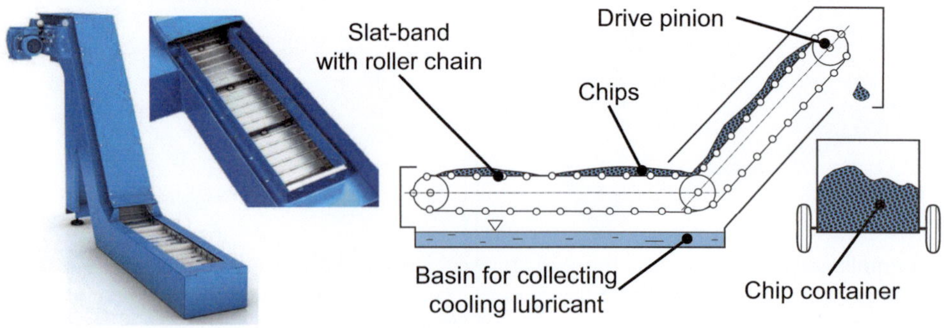

Fig. 10.26 Slat conveyor with coolant container. *Source* KABELSCHLEPP

pulled and guided by roller chains affixed on the sides (▫ Fig. 10.27). Perforated hinged slats enable coolant separation (▫ Fig. 10.27, middle). The base of the conveyor is formed as a tray. From here, the coolant collected is returned to the coolant container after being cleaned. To transport small punching waste and nibbling chips, a belt made of synthetic-coated fabric is used instead of the slat-band. To convey "sticky" parts, slats with beads are used (▫ Fig. 10.27, right).

Discontinuous chips can be transported by a scraper conveyor of a simpler design (▫ Fig. 10.28, left). There is no conveyor belt here. The chips falls onto the base of the trough-shaped housing and are pushed on the lower run side to a container by scraper plates secured between two roller chains. Larger material pieces and chip clusters can easily block the system. A distinction is made between the following designs: straight (a), straight/inclined (b) and straight/inclined/straight (c). See ▫ Fig. 10.28, right.

Conveyor belt speeds can vary for all three designs. These range from 1.3 to 3 m/min for chips moistened with coolant and 1.3 to 12 m/min for dry chips. In addition to the conveyor belt speed, the width of the belt has a linear effect on the possible transport capacity of the conveyor. Besides the structural constraints inherent to the machine tool design, the following have a bearing on the belt width: bulk density of the chips (in kg/m^3), mass flow rate of the chips (in kg/h) and the amount of coolant adhering to the chips. For a standard slat conveyor, this means with a belt width of 600 mm and conveying speed of 8 m/min for example, a possible chip conveying capacity of 10 m^3/h for a horizontal conveying section. A reduction of 30% (60%) in conveying capacity must be factored in for an angle of inclination of 30° (60°).

Magnetic conveyors can be used to transport steel chips. The chips transported fall onto a wear-resistant, austenitic steel strip. Permanent magnets are moving below this strip. The magnetic fields of these magnets act upon the chips through the strip and move them along.

Given their small sizes, the screw conveyors shown in ▫ Fig. 10.29 are used primarily in confined spaces. Chips are moved by rotation of the helical screw conveyor. The comparatively high energy requirement and susceptibility to wear of screw conveyors are consequences of the high friction level and tendency to jam between the screw and trough sides. Screw conveyors are susceptible to coarse parts and should only be used for small chips.

In addition to the mechanical conveyance of chips, greater importance is being attached to pneumatic conveyance as part of the dry machining process. Moist chips from wet machining can also be collected using an extraction system. ▫ Figure 10.30 is a stationary, central extraction system with multiple collection points. After being scraped mechanically from the machine bed, the chips fall into a collection tank. They are extracted from there. In contrast, two variants of collection near to the tool, directly where the chips are generated, are shown on the right in ▫ Fig. 10.30. The designs of chip collection devices vary from relatively simple and universally usable nozzles (▫ Fig. 10.30, top right) to complex designs adapted specifically to the tool design (▫ Fig. 10.30, bottom right).

In the design of stationary and mobile extraction systems, it is important to consider whether they are being used for substances that pose a risk of dust explosion, such as light metals. In such cases, special safety regulations must be observed and corresponding safety precautions be put into place [BERU90; HAUP05].

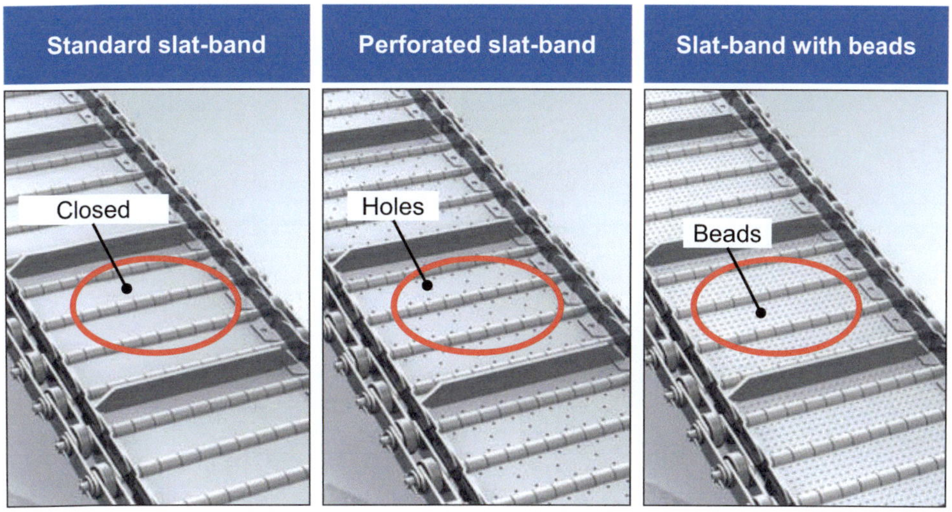

▫ **Fig. 10.27** Design variants of slat conveyors. *Source* KABELSCHLEPP

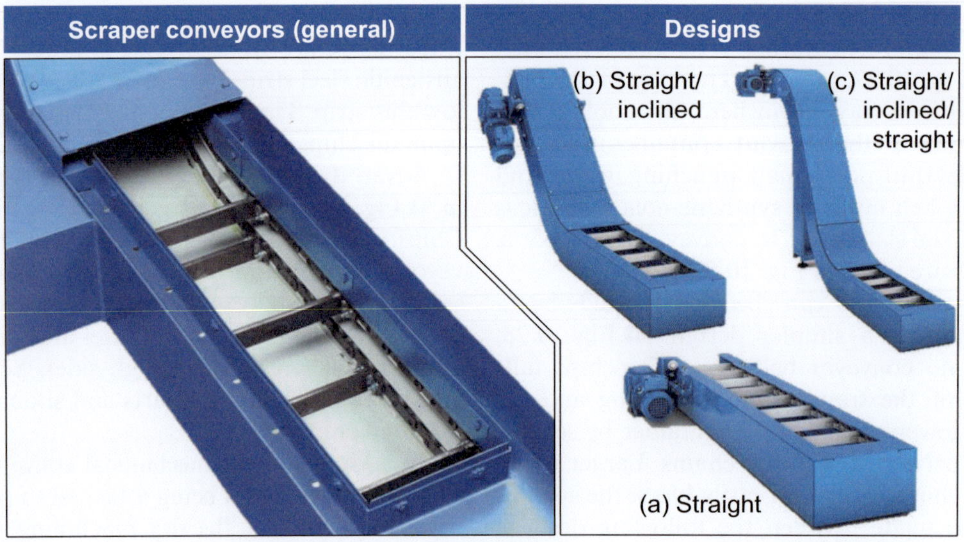

Fig. 10.28 Scraper conveyors. *Source* KABELSCHLEPP

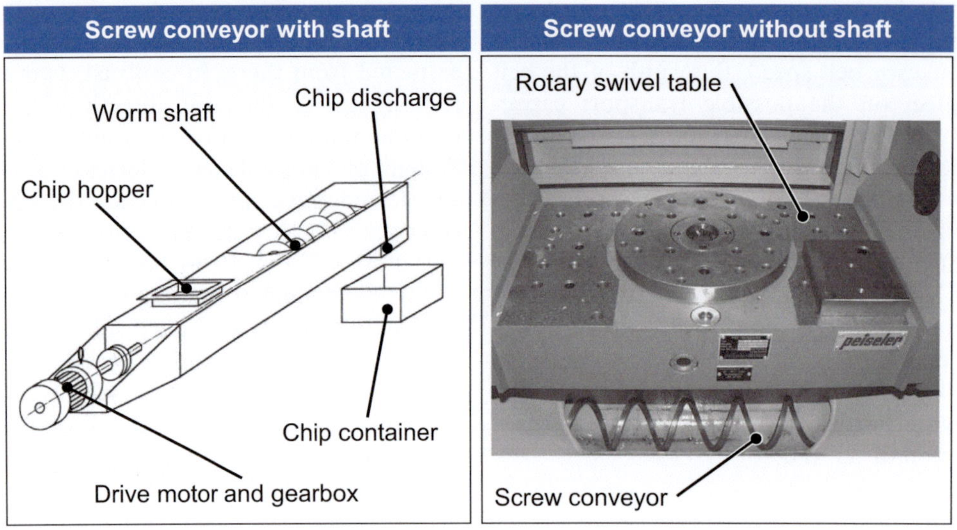

Fig. 10.29 Structure of a screw conveyor. *Source* Kabelschlepp

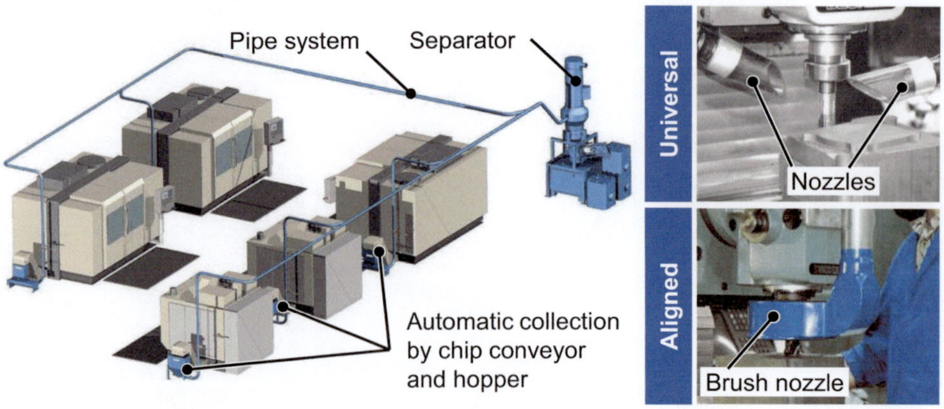

Fig. 10.30 Central extraction system, examples of particle collection near the tool. *Source* Ringler/Kärcher Group

10.2.3.2 Chip Centrifuges

Chip centrifuges separate the liquid from the metal chips using the centrifugal forces occurring at high speeds. If such a centrifuge is located directly at the machine tool, filling using the chip conveyor of the machine, and direct returning of the centrifuged cooling lubricant into the cooling lubricant circuit of the machine tool are possible. The maximum attainable drying level for the chips is about 3% residual liquid depending on the material cut, chip shape and cooling lubricant. In contrast, about 50% of the adhesive cooling lubricant can be recovered with simple dripping under the influence of gravity (onto a chip conveyor for example). Standard throughput rates of commercially available chip centrifuges for use near machines are generally 0.2 to 4 t/h (steel chips), given drive power ratings of 1 to 7 kW and drum speeds from 750 to 1500 min^{-1}.

Differentiation is essentially made between:

- continuously working chip centrifuges and
- discontinuously working chip centrifuges
- depending on their function.

For continuously working centrifuges, conical shaping of the internal drum surfaces, for example, generates an axial force component that acts on the chips and is parallel to the axis of rotation of the centrifugal drum from the centrifugal force (◘ Fig. 10.31). This causes the chips to move along the axis through the drum towards the biggest drum diameter. Here, the cooling lubricant is separated from the chips by a recessed slotted screen and moved out of the drum. The dried chips are ejected from the drum over the top edge and fall into a chip container or onto a chip conveyor.

This principle means continuous operation is possible without additional loading and unloading processes. How long the chips stay in the centrifuge, and the level of drying attainable as a result, are largely determined by the inflow of chips and the operating speed of the drum. For discontinuous feeding of the centrifuge, an additional metering unit is usually required for uniform filling of the drum. For discontinuously working centrifuges with simple, cylindrical drum geometry, the loading and unloading processes and the actual centrifugation process are normally time-controlled. The loading or unloading process requires the drum rotation speed to be slowed down, possibly to a complete standstill, depending on whether the drum needs to be emptied or the drum body (realized in this case as a chip container) needs to be completely changed. The acceleration and deceleration processes necessary here require considerably higher power consumption and more complex control, than for continuous centrifuges. Considerably simpler and more variable adaptations to different operating conditions are possible, however.

For the variant of a discontinuously working centrifuge shown in ◘ Fig. 10.32, the chips introduced into

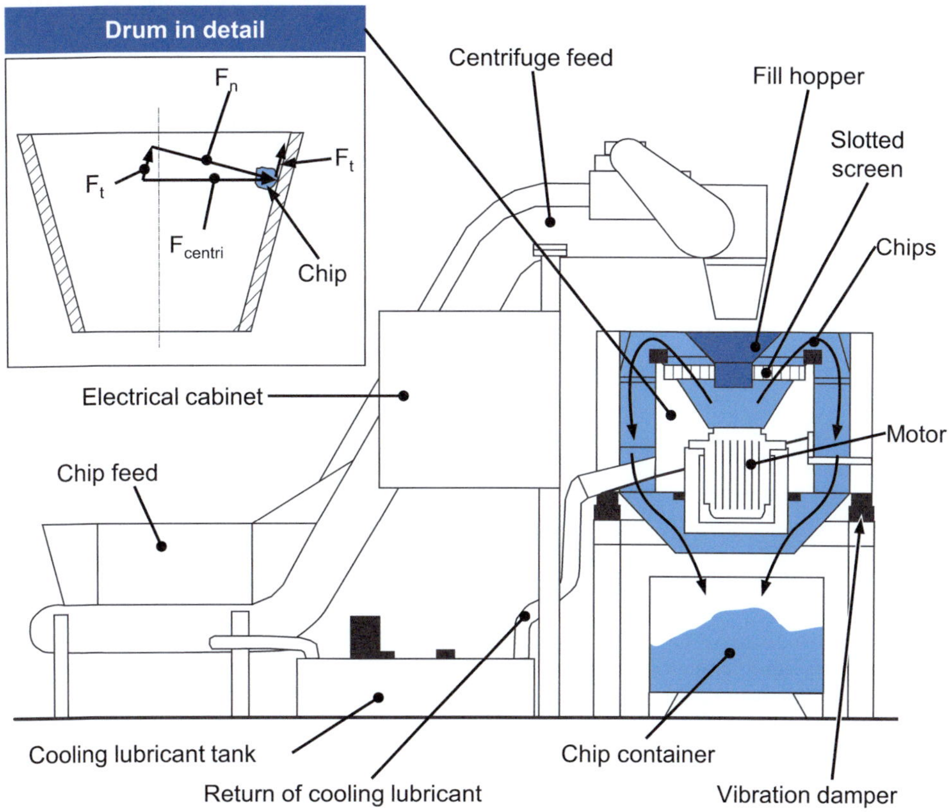

◘ **Fig. 10.31** Functional principle of a continuous chip centrifuge. *Source* CEPA

the drum via a filling hopper are accelerated by a rotating disk and sprayed onto the drum wall, also rotating. The liquid is thrust outward by the centrifugal force and flows through the perforated drum wall. Once the capacity volume of the drum is reached, the supply is stopped, and the time-controlled centrifugation process starts. For unloading, the drum speed is reduced, and the chips are removed from the drum wall, aided by a jogging motor and with compressed air. They fall downward out of the centrifuge.

To increase the efficiency of the chip centrifuge used, or if there are special requirements, a hot air blower can also be used. Here, additional drying or a reduction in the viscosity of the cooling lubricants used, and therefore improved separability of chips, can be attained, but this increases energy use. Unfavorable chip shapes, as is the case with thread and ribbon chips, that can clog the centrifuge and so result in poor efficiency, may necessitate the additional use of an upstream chip breaker.

10.2.3.3 Briquetting

Since the new Recycling Management and Waste Law came into force in Germany, waste recycling has

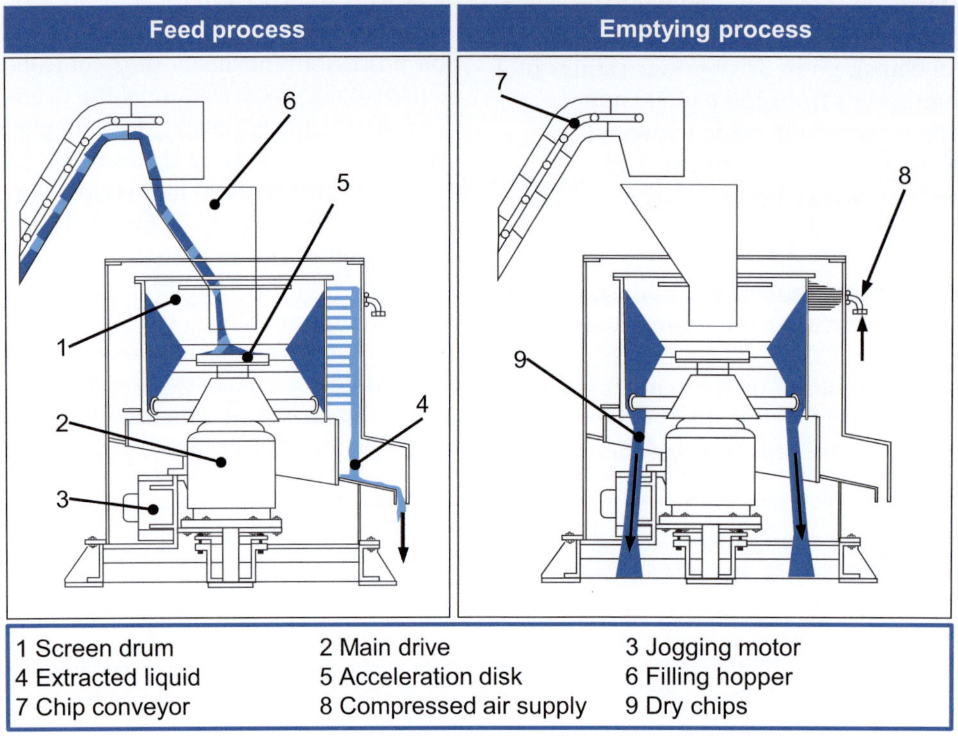

Fig. 10.32 Working principle of a discontinuous chip centrifuge. *Source* Steimel

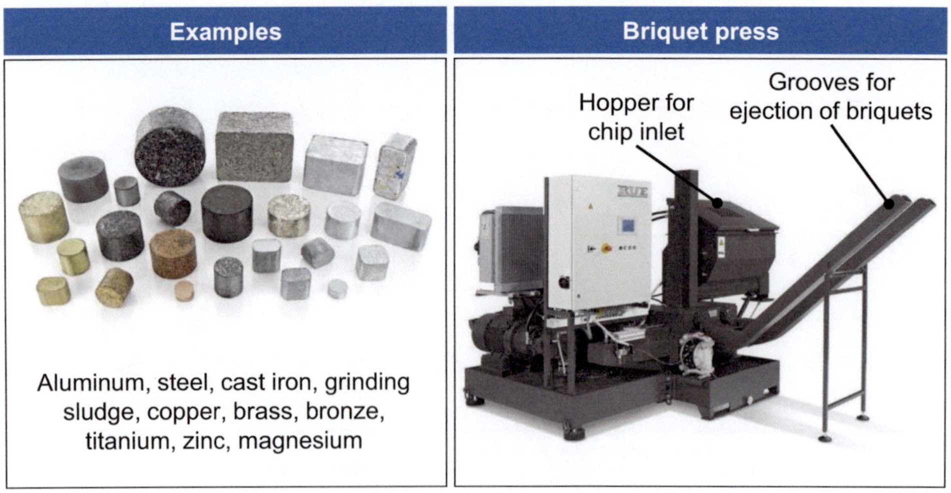

Fig. 10.33 Briquet press and briquets made of different materials. *Source* Ruf Maschinenbau

been the clear preference over waste disposal. The storage, disposal and reuse of chips can be simplified by compressing the loose material into cylindrical briquets using a hydraulic stamping press. Shown on the left in ◘ Fig. 10.33 are examples of chip briquets made of different materials. On the right is a briquet press.

Generally, materials that can be made into briquets range from metal chips and grinding sludge and powdered ironworks waste as generated in steel and cast iron production, to paper shavings and plastics. Besides the reduction in volume, an extensive separation of clinging cooling lubricants is achieved for wet materials.

Increasing machining work using magnesium materials results in contact with highly flammable magnesium chips more frequently. Immediate briquetting prior to storage or continued handling of these chips provides more safety for the production facility. This comes from greatly reducing the risk of ignition with subsequent rapid spreading of fire. ◘ Figure 10.34 shows the operation of a briquet press.

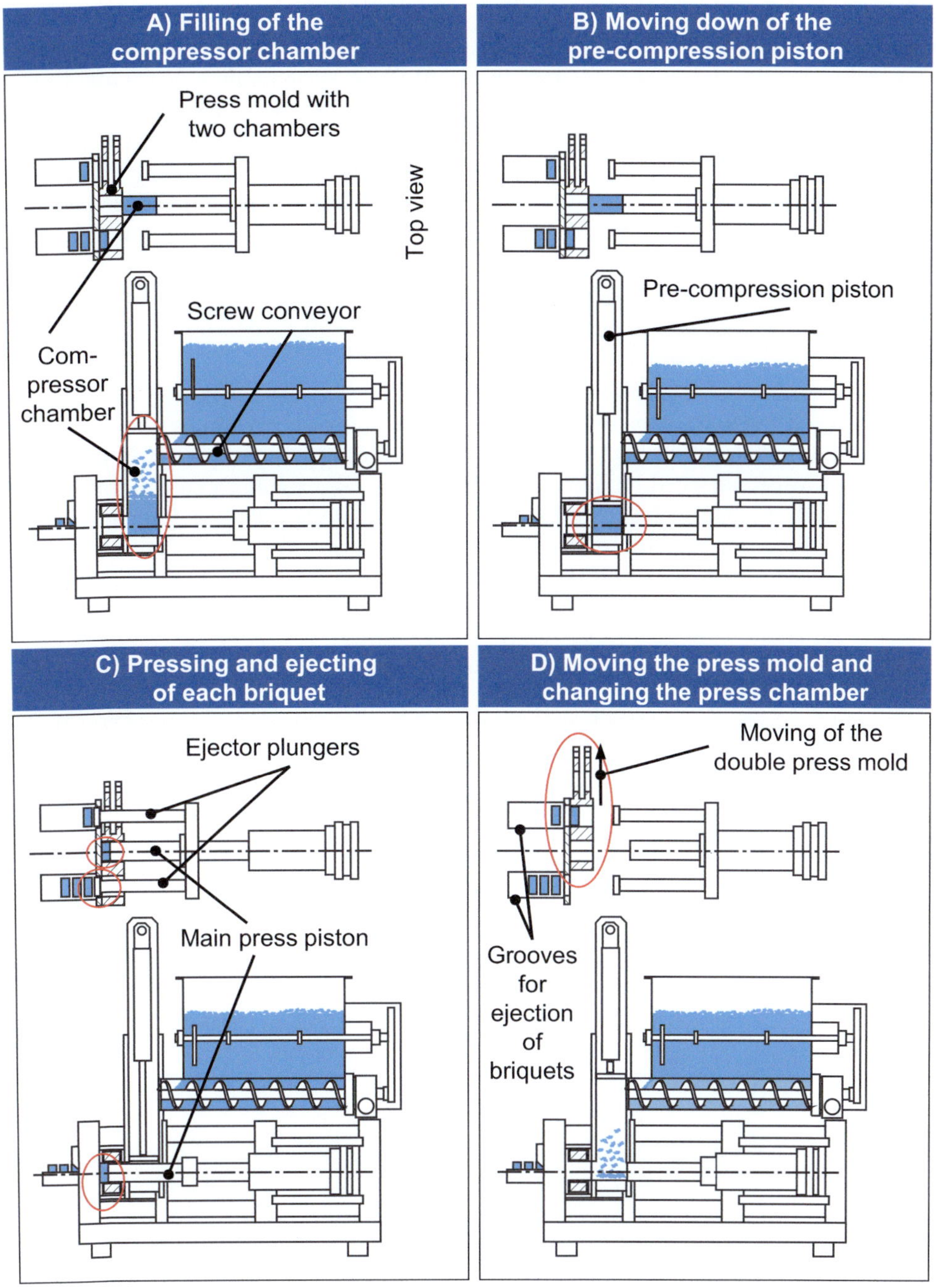

◘ **Fig. 10.34** Operation of a briquet press. *Source* Ruf Maschinenbau

The screw conveyor transports chips from the collection tank to the pre-compressor (A). The runtime of the screw is calculated by the controller. After the appropriate amount of chips has been moved into the pre-compression chamber, the compression piston moving down (B) causes the material to be pre-compressed perpendicular to the horizontal main pressing direction. Then the main press piston presses the material into one of the two press molds and compresses it to a solid block (C). In parallel, an ejector plunger, coupled to the movement of the main press piston, moves into the second press mold and ejects an already finished briquet out of the ejection opening. After the pre-compression and main press pistons retract, the press mold changes, and a new cycle starts (D).

Pressing such a cylindrical briquet with a 60 mm diameter and 60 mm length takes about 7 s per briquet.

10.2.4 Central Lubrication Systems

Friction and wear occur at all material contact points with relative movement. To keep losses caused within limits, machine tools are fitted with central lubrication systems that reliably supply the bearings, guides and gear units with metered amounts of lubricant. A distinction is made [VOGE03] between consumption lubrication systems for oil, oil mist, oil-air and grease lubrication where the lubricant is supplied to lubrication points where it is used and circulation lubrication systems for oil lubrication where oil (possibly after filtering and recooling) is returned to the storage tank.

The basic structure of a consumption lubrication system comprises a lubricant reservoir, a pump for lubricant transport, a distributor and flow regulator, the main lubricant feed line between the pump and distributor and the lubricant line between the distributor and lubrication points. After use the lubricant is not being reused. It mixes with the cooling lubricant of the production process or is collected in tanks. Circulation systems, on the other hand, return the lubricant to the lubrication point after use. Their only essential difference from consumption systems is that they have an oil collection system and additional return line. Both types are discussed together below.

◘ Table 10.1 provides an overview of the most common system types. Differentiation is made between single-line systems, progressive systems, dual-line systems, multi-line systems and restrictor systems described in more detail in Sects. 10.2.4.1 to 10.2.4.5. In the attempt to supply the rolling element bearings of fast-turning main spindles with minimal amounts of lubricant to keep friction as low as possible, systems for oil-air lubrication are increasingly being coupled to the central lubrication system. These devices, grouped under the term "compressed air lubrication systems", are discussed in ▶ Sect. 10.2.4.6.

10.2.4.1 Single-Line Systems

The single-line system is the most reasonably priced and widespread system type in the machine tool industry. It is used for both consumption and circulation lubrication with oils and fluid greases. They are suitable for supplying from 10 to several hundred lubrication points. The lubricant is fed by a pump to a piston

◘ **Table 10.1** Central lubrication systems

System type	Metering volume in cm³/pulse	Lubricant	Lubrication points	Area of application
Single-line systems	0.01–1.5	Oil, fluid grease	10–100	Consumption lubrication on small and mid-sized machines and systems
Progressive systems	0.05–3.2	Oil, grease, fluid grease	10–100	Consumption and circulation lubrication on small and mid-sized machines and systems
Dual-line systems	0.03–15	Oil, grease, semi-fluid grease	50–50	Consumption and circulation lubrication on mid-sized and large machines and systems
Multi-line systems	0.5–200	Oil, grease, fluid grease	2–36	Consumption and circulation lubrication on small machines
Restrictor systems	10–10,000	Oil	Over 100	Consumption and circulation lubrication on large machines and systems

distributor system and, from there, fed uniformly to all lubrication points. Electrically, hydraulically or pneumatically operated piston or gear pumps are used.

The piston distributors ensure that, in every lubrication cycle, a constant amount of lubricant, independent of the pump runtime, is provided to the lubrication points. The pump runtime must be set such that reliable filling of all distributor chambers is ensured. The pump is usually switched on by a time signal. It is switched off via a limit switch on the pressure in the lubricant line farthest away after an additional safety after-run time of about 15 s. The pump displacement per lubrication cycle only depends on the geometric dimensions of the distributor. Values attainable with commercially available distributors are between 0.01 cm^3 and 1.5 cm^3 per lubrication pulse.

Figure 10.35 shows the cross-section of a typical piston distributor. In the idle position, the oil to pump is in the spring chamber. Applying pump pressure to the main line presses the piston up against the spring force. This pushes the delivery volume upwards to the lubrication points. Once the pump pressure has dropped, the spring pushes the piston back. Lubricant flows into the spring chamber through a fixed overflow pipe in the middle of the cylinder. The distributor is now ready for the next lubrication cycle. The independent distributor cylinders make expanding or reducing the size of an existing system comparatively easy by flange-mounting additional distributor blocks or sealing individual outputs (but monitoring of the system is difficult to the same degree). Generally, only the supply to individual, important lubrication points and the pressure in the main line are monitored.

10.2.4.2 Progressive Systems

A progressive system feeds lubricant progressively to the different lubrication points, i.e., gradually and successively. In this case, all lubrication points are allocated the same amount regardless of the flow resistance. The working principle of a progressive distributor is illustrated in Fig. 10.36.

In the start position, the top front surface of the right-hand piston is under pump pressure. The piston moves down and pushes the corresponding cylinder content through connector 1a. This piston movement frees up the connection of the main line to the top front surface of the middle piston. The piston moves, pushes the lubricant through connector 2a and initiates movement of the third piston. This cycle continues until the delivery of the pump is switched off. So, the fineness of metering depends greatly on the pump

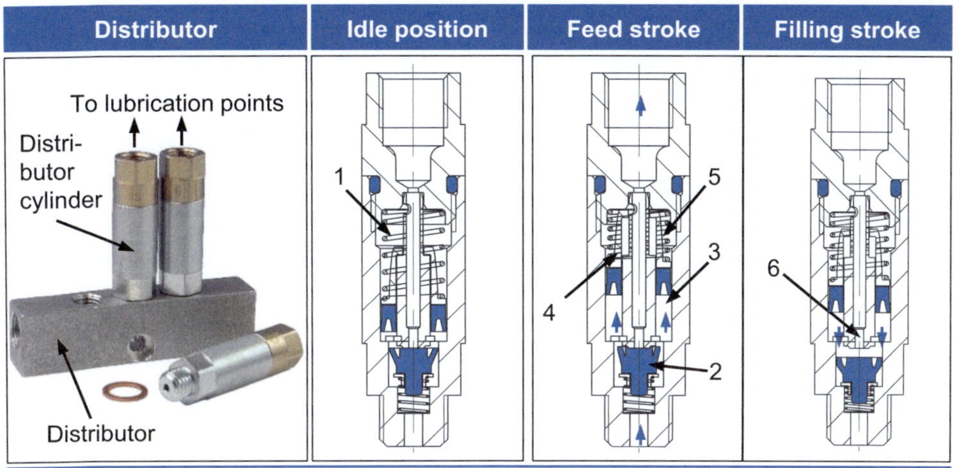

Idle position:	The lubricant is in spring chamber 1
Feed stroke:	The pump starts and pumps lubricant past control collar 2 into piston chamber 3. Piston 4 moves upwards against the force of spring 5 and pushes the lubricant from the spring chamber to the lubrication point.
Filling stroke:	The pump is switched off, The piston is pressed back into the starting position by the spring. The control collar is pressed downward against the spring—this releases overflow hole 6 and the lubricant is moved into the spring chamber.

Fig. 10.35 Piston distributor for a single-line system. *Source* SFK

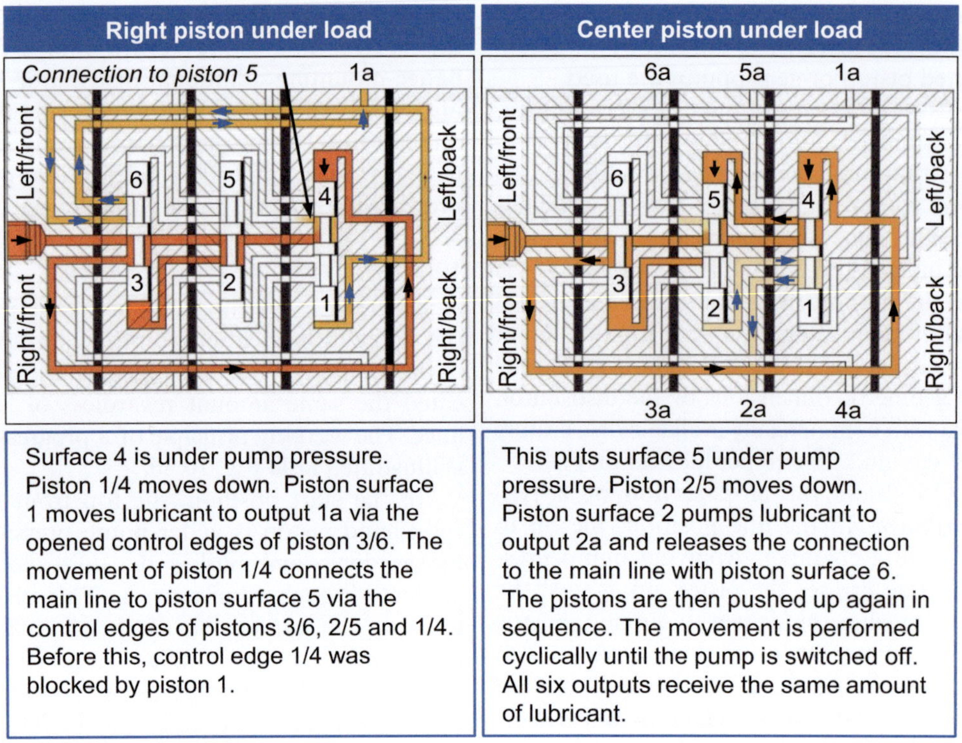

10

Fig. 10.36 Piston distributor of a progressive system. *Source* SKF Lubrication Systems

performance data. The volume of a metering chamber is between 0.05 cm^3 and 3.2 cm^3.

Progressive systems represent an alternative to single-line systems with medium metering amounts. They are more robust and thus also suitable for supplying grease. The forced operation of the individual distributor pistons means monitoring of one lubrication point is sufficient to assess the functionality of the overall system. Changing the number of lubrication points is laborious however as spare connectors may not be sealed (and instead must be returned to the storage container to maintain the motion cycle of the pistons).

10.2.4.3 Dual-Line Systems

Unlike the two types described above, a dual-line system comprises two main lines that supply alternately the two chambers of the distributor cylinder (Fig. 10.37) via a changeover valve. The delivery volume is set by the lubrication pulse-to-interval ratio. After the interval time ends, the pump starts and delivers lubricant into the left-hand metering chamber through the channel released by the control piston. The metering piston moves right and pumps lubricant from the right-hand metering chamber to the lubrication point connected via the right-hand outlet and line 13 (on the right). For the next pulse, the changeover valve is actuated, and pump pressure is applied to the right-hand metering chamber. Lubricant flows to the lubrication point on the left. This process repeats until the timer switches off the pump. The pump displacement per stroke is between 0.03 cm^3 and 15 cm^3.

Dual-line systems are exceedingly sturdy and thus well suited to greases and semi-fluid greases. Given the increased overhead for the mechanical setup (two main lines and the changeover valve), dual-line systems are only cost-effective for 100 or more lubrication points.

10.2.4.4 Multi-Line Systems

The major component of a multi-line system is a multi-piston pump with up to 36 delivery cylinders. The outlets are connected directly to the lubrication points, so a main line and distributor are not necessary. Multi-line systems are used primarily for lubricating small machines using grease.

10.2.4.5 Restrictor Systems

Restrictor systems represent the simplest system type structure. However, they are only suitable for machines and systems requiring larger oil quantities and with a requirement for uninterrupted supply. A pump delivers the lubricant into a network of pipes connected to all lubrication points by branches and distributor blocks. In the individual branches of the pipe system, the volume flow rate is set in line with the restrictor losses by the flow resistance of the pipe and possibly by adjustable threaded chokes also provided.

In the design of a system with two lubrication points, the total oil volume flow required for all

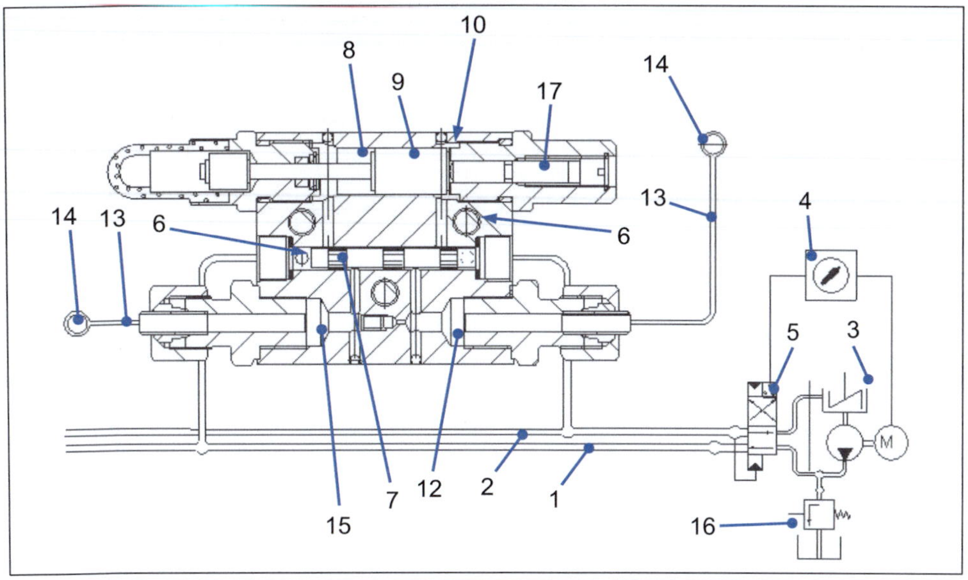

1. Main line 1
2. Main line 2
3. Pump with lubricant tank
4. Control unit
5. Changeover valve
6. Left control piston chamber
7. Control piston
8. Left metering chamber
9. Metering piston
10. Right metering chamber
11. Right control piston chamber
12. Right outlet
13. Lubricant line
14. Lubrication point
15. Left outlet
16. Pressure relief valve
17. Metering screw

Fig. 10.37 Piston distributor for a dual-line system. *Source* SKF Lubrication Systems

lubrication points, or the connection value, is first determined as in Fig. 10.38. It must match a delivery value adjustable on the pump. Splitting the volume flow is proportional to the added pressure losses caused on the way to the lubrication point by the pipework, threaded chokes, T-pieces and distributor blocks. The internal diameters and lengths of the pipes must be matched to it. The pump power rating required is calculated as the product of this pressure loss and the overall oil volume flow.

The tendency for narrow gaps and holes to become dirty must be factored into the selection of restrictors. Where adjustability is not necessarily required, priority should be given to restrictor pipes. Careful filtering of the oil is essential for smooth operation, especially for circulation lubrication.

10.2.4.6 Compressed Air Lubrication Systems

The three systems in Table 10.2 are used for the oiling of compressed air in pneumatic systems (cylinder-piston systems) or for the oil-air lubrication of fast-turning rolling element bearing spindles. For large machines with many lubrication points, a central lubrication system (usually a single-line system) is fitted with a mixing valve where compressed air and oil are mixed using the Venturi principle (Fig. 10.39). To prevent sedimentation of oil in the lubricant line, mixing takes place as near to the lubrication point as possible.

For lubrication with minimum amounts (oil-air minimum quantity lubrication), injection oilers are used (see Fig. 10.40 for their functional principle). Oil falls under the influence of gravity from the storage container into the metering chamber through a down-pipe. The lubrication cycle is triggered by a signal transducer after an interval time set by the user. First, the mixing chamber is flooded with compressed air. This moves the control valve up and frees the actuation hole for the metering piston.

This applies air pressure to the piston of the oiler. The piston moves against the spring force to the right and pushes the oil in the metering chamber through the injection nozzle. During the push-out phase, the oil supply hole of the metering chamber is closed by the piston so that only the cylinder volume is supplied regardless of the travel speed of the piston. In the interval time, uncontrolled oil leakage is prevented by a check valve.

The oil-air flow is routed past an injection oiler in the bypass and mixed with the continuous air flow. The small amount of oil covers the side of the tube and is spread lengthwise by the air flowing past to make a streaky-shaped drop. Provided the tube is long enough,

Procedure

1. Define the volume flow of all lubrication points:

$$Q_{pump} \geq \sum_{i=1}^{2} Q_i \quad \text{, here n = 2}$$

2. Set the line length and cross section such that they give the flow split required:

$$\frac{Q_1}{Q_2} = \left(\frac{d_1}{d_2}\right)^4 \cdot \frac{l_2}{l_1}$$

3. Check that the pressure loss the pump must sustain is covered by the maximum pump pressure:

$$\Delta p = \frac{128 \cdot \eta \cdot l_1 \cdot Q_1}{\pi \cdot d_1^4} + \sum \Delta p_{chok} + \Delta p_{cons} \leq \Delta p_{pump}$$

(proof is only required for one lubricant line because the pressure loss is identical in all lines)

Diagram of a restrictor system

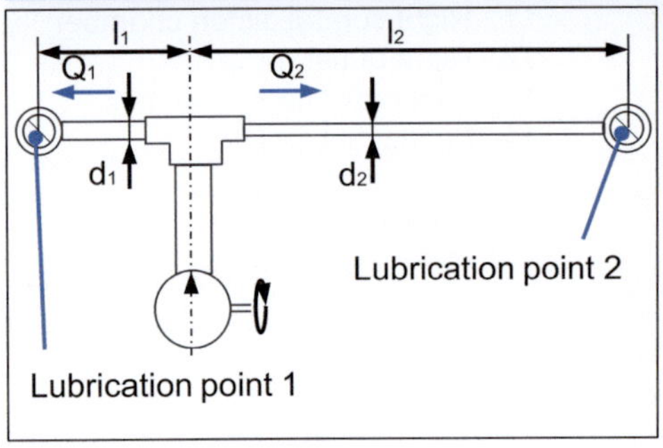

d - Pipe diameter
l - Pipe length
η - Dynamic viscosity of the lubricant
Q - Volume flow
Δp_{chok} - Pressure loss in threaded chokes, T-pieces and distributor blocks
Δp_{cons} - Pressure loss of consumer

Fig. 10.38 Design of a restrictor system

Table 10.2 Compressed air lubrication systems

System type	Metering volume in cm³/pulse	Lubricant	Lubrication points	Area of application
Central lubrication systems, e.g., single-line systems	0.01–1	Oil, lubricating grease	10–100	Lubrication of compressed air cylinders
Injection oiler	0.003–0.03	Oil, lubricating grease	1–10	Oil-air lubrication of rolling element bearings, chain lubrication, wetting of molds and dies
Cyclic oiler	0.03–0.3	Oil	1–100	Lubrication of compressed air tools

there is a constant lubricant volume flow at the lubrication point. With increasing tube length however, the system inertia also increases. This means that in practice an ideal line length from the oiler to the lubrication point is 1 m to 3 m.

Compressed air tools operating with a high but irregular volume flow requirement need a sturdier oiler whose metering of lubricant quantity is decoupled from the air extraction. The cyclic oiler suitable here is similar to an injection oiler as regards design and functional principle. To prevent over-oiling of the air, the metering chamber here is not necessarily filled after every pulse of compressed air but by a pump that is actuated in a time-dependent manner.

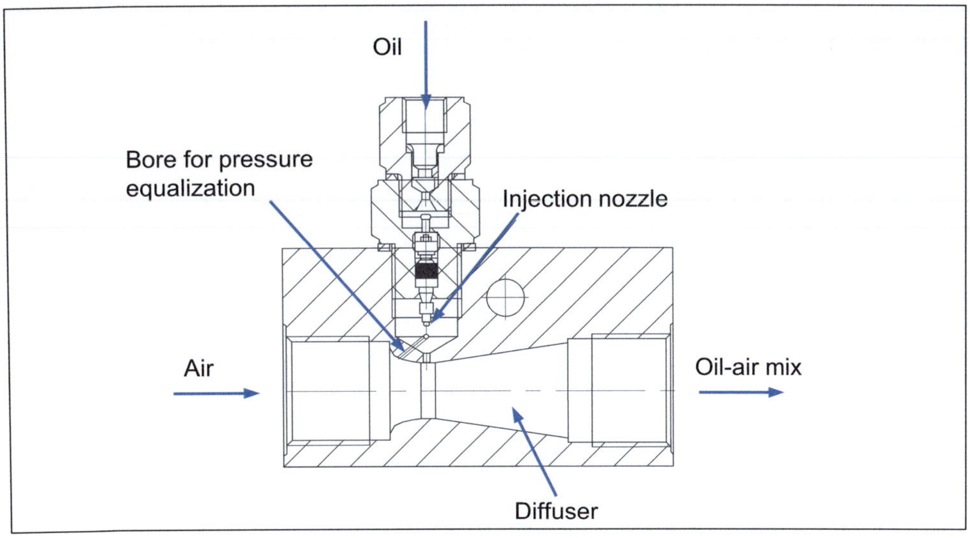

Fig. 10.39 Mixing valve for oil-air lubrication with a central lubrication system. *Source* SKF Lubrication Systems

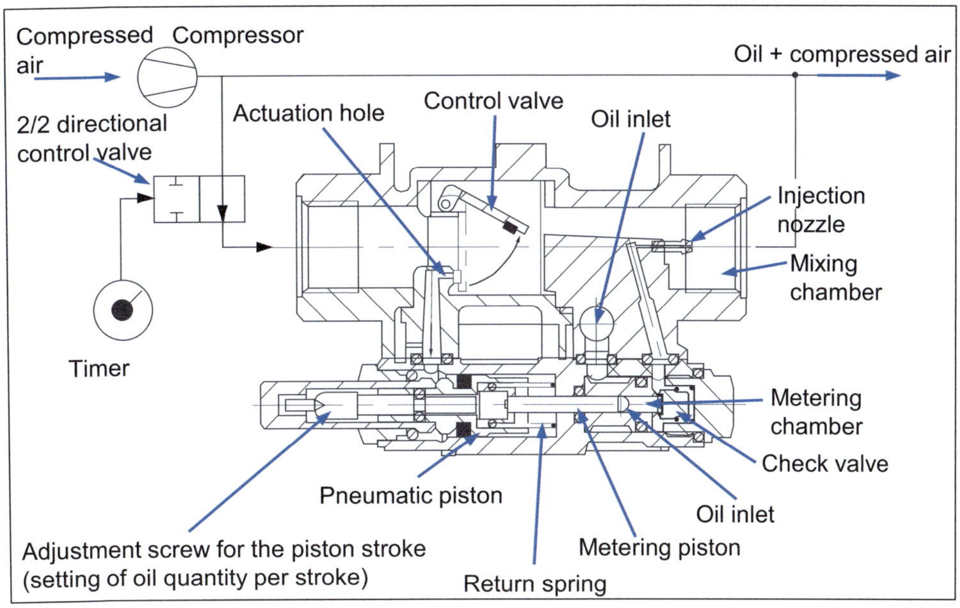

Fig. 10.40 Function of an injection oiler. *Source* SKF Lubrication Systems

10.2.5 Hydraulic Units

10.2.5.1 Functions and Importance of Hydraulic Systems

Hydraulics is about the transfer of energy and signals through fluids. Hydraulic fluids based on mineral oils and natural or synthetic esters are used. For specific requirements such as flame resistance and operating brakes, flame resistant and special fluids are used. The field of hydraulics is divided into hydrostatics and hydrodynamics.

In hydrostatics, energy is transferred using static pressure. Hydrostatic machines normally operate with high pressures and low flow velocities. In hydrodynamic power transfer, however, the kinetic energy of the flowing liquid is used. High flow velocities occur here at low pressures [MURR12].

In systems with hydrostatic energy transfer, the kinetic energy is negligibly small compared to the pressure energy. Conversely for hydrodynamic energy transfer, the pressure energy in the delivery is usually negligible. In mechanical engineering, hydrostatic drives play a considerably greater role than hydrodynamic drives. Consequently, hydrostatics are discussed in the following section. The term hydraulic system is used as a synonym for hydraulic drive [GROL10].

The primary function of a hydraulic unit is power transfer, combined with the transfer of pressure and motion. The hydraulic functions of a machine tool are, for example, tool changing, workpiece clamping, pallet changing and the weight compensation of vertical axes. Furthermore, the transfer of signals can be realized for (feedback) control purposes. With a share of 74%, mineral oil is the most widely used hydraulic medium [HERS11].

Figure 10.41 summarizes the benefits and downsides of hydraulic drives and controllers as seen today.

10.2.5.2 Basic Structure of a Hydraulic System

A hydraulic system comprises different components depending on the functionality required. The most significant components and their function are described below using a simple hydraulic system as an example. The left side of Fig. 10.42 shows a section drawing of the individual components of the hydraulic system. The schematic on the right side contains the standard DIN ISO 1219-1 symbols used for these components [ISO12].

The hydraulic pump (2) (in this instance, an external gear pump) sucks oil from the tank (1) and converts the mechanical energy supplied by the drive shaft into hydraulic energy. This now exists primarily as pressure energy in the oil flow (delivery) at the pressure connection of the pump. Depending on the mechanical/hydraulic efficiency of the pump, losses occur during the energy conversion process in the form of heat.

In order to retract the piston rod, the directional control valve (7) is switched (solenoid (b)) so that the oil flow delivered by the hydraulic pump flows into the piston side of the hydraulic cylinder (9) via the check valve (3), the directional control valve (7) and the check valve of (8*). The oil in the hydraulic cylinder on the piston rod side is fed to the tank via the throttle valve of (8), the directional control valve (7) and the filter (4).

To retract the piston rod, the directional control valve is activated. The oil flow delivered by the hydraulic pump is now routed to the piston rod side of the hydraulic cylinder via the check valve of (8): The piston rod retracts and the oil on the piston side in the hydraulic cylinder is routed to the tank via the throttle valve of (8*). The pressure in the system is monitored using manometers (5).

The function of the directional control valve is to control the direction of the pump flow to allow the piston rod on the hydraulic cylinder to be extended or retracted as required. The oil pressure generated by the pump depends on the load on the piston rod. This is designated the load resistance of the pressure oil end user.

The pressure relief valve (6) is set to the maximum permissible pressure for the hydraulic system. When the pressure set is reached, the valve opens and some of the

Benefits	Downsides
⊕ Simple realization of linear movements	⊖ Sensitivity to dirt and wear of the components
⊕ High power density	⊖ Power consumption (losses from friction and internal leaks)
⊕ Good control characteristics	⊖ Noise generation
⊕ Good lubrication and dissipation of heat loss by the fluid	⊖ Risk of fire
⊕ Simple overload protection by pressure relief valves	⊖ Viscosity and compressibility depending on pressure and temperature affects operational behavior
⊕ Good behavior over time through low inertia resistances of the hydraulic motors and working cylinders	⊖ Relatively high tendency to vibrate due to high power density and low damping
⊕ Infinitely variable change of the output motions	⊖ Environmental pollution from hydraulic fluid
⊕ Simple to reverse the direction of motion	
⊕ Startup possible under full load from standstill	
⊕ Simple display of load by pressure gauges	

Fig. 10.41 Benefits and downsides of hydraulic drives and controllers, according to [HERS11; MURR12]

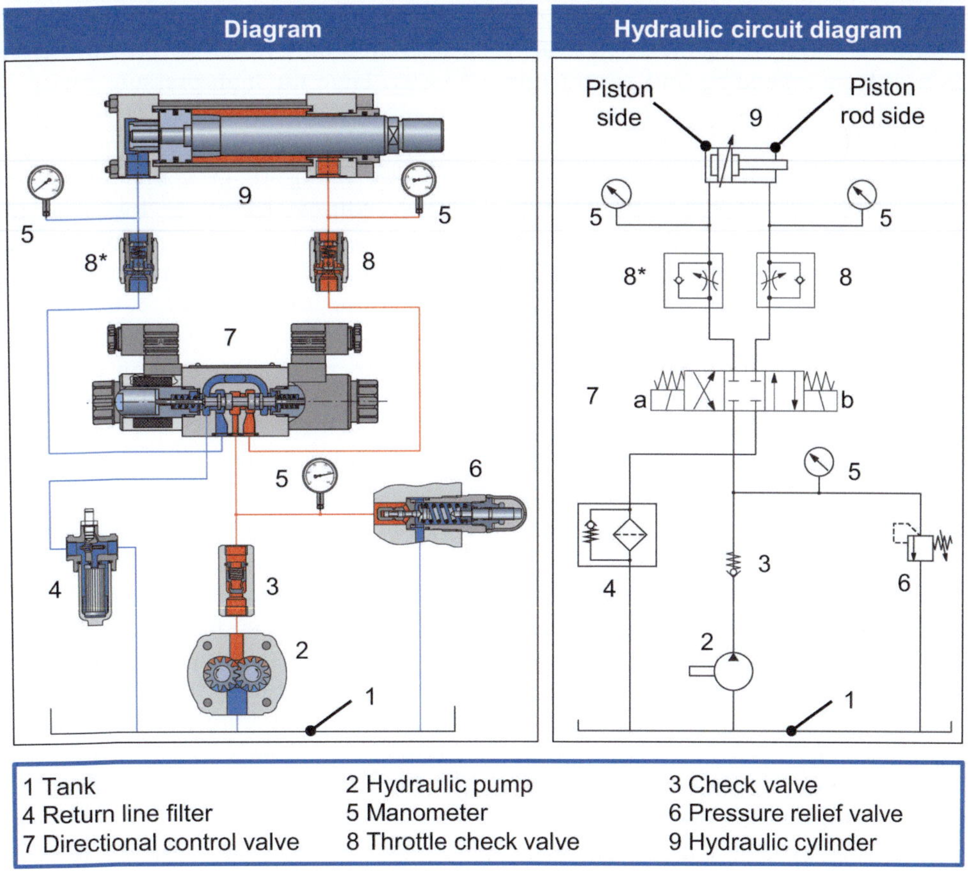

Fig. 10.42 Basic structure of a hydraulic system. *Source* Bosch Rexroth

oil returns to the tank. This prevents a further increase in the pressure applied to the pressure connection on the pump. The pressure relief valve acts as the hydraulic system's safety valve.

The oil flow fed to the hydraulic cylinder, and thus the extending speed of the piston rod, can be changed by the throttle valves of (8) and (8*) in combination with the pressure relief valve. If the throttle cross-section is reduced using the adjustment screw on the throttle valve, the pressure between the pressure connection of the hydraulic pump and the throttle valve increases. If this pressure reaches the maximum permissible pressure for the hydraulic system, as set at the pressure relief valve (by a further reduction in the throttle cross-section), the valve opens and only some of the total oil flow fed by the hydraulic pump reaches the hydraulic cylinder while the rest flows to the tank via the pressure relief valve. The result is a reduction of the extending/retracting speed of the piston rod. This speed is dependent upon the piston surface and the delivery entering the hydraulic cylinder. A hydraulic motor can be seen as an alternative pressure oil end user when rotational motion is required instead of translational working motion.

The hydraulic pump is the key component of a hydraulic system. Most of the mechanical energy supplied by the drive shaft is converted to hydraulic energy. The rest is lost to the environment in the form of heat [GROL10].

All hydraulic pumps operate using the same principle: Chambers are repeatedly filled and emptied with a pressure medium on an alternating basis. The filling process starts when the chamber volume is at its minimum. Once the maximum chamber volume has been reached (filled to the maximum), the filling procedure is ended by closing the chamber off from the filling line (tank side). The chamber volume decreases as the process continues, which builds up pressure. The energy from the volume flow leaving the pressure connection of the pump is available to operate hydraulic motors or hydraulic cylinders. These pressure oil end users convert the hydraulic energy back into mechanical energy. Electric motors are generally used to drive hydraulic pumps, but internal combustion engines are also used in some cases, e.g., diesel engines in construction machinery.

Hydraulic systems usually require high pressures with low deliveries rarely greater than 300 l/min. This cannot be achieved with centrifugal pumps. Hydraulic pumps operate almost exclusively using the displacement principle. This is based on expanding and contracting chambers. Machines that operate on this principle are also known as positive displacement machines. In positive displacement pumps, the medium is usually fed by either a piston, gears, a vane or a diaphragm [GROL10].

Most displacement units can be used as pumps or motors because the hydrostatic pumping process is reversible. Striving for the best possible solution in each case, however, means that, in practice, pumps and motors often have different designs. For the same reason, a number of displacement principles are applied for different application scenarios [MURR12].

The structures, functions and the calculations for the different displacement principles are described in detail in Chap. 5 of Volume 2 in this series.

10.2.5.3 Design Variants of Hydraulic Units

Described below in addition to a classic hydraulic unit are different design variants of energy-efficient hydraulic units and a compact pump unit.

A classic hydraulic unit uses a variable displacement pump. This pump is driven by a motor at a constant rotational speed independent of the operating state of the machine tool. Two pressure circuits (low pressure (LP) and high pressure (HP)) are frequently necessary for the different hydraulic functions. In such a case, two variable displacement pumps are installed on a common shaft (see ◻ Fig. 10.43, top left). When a hydraulic function is performed, the pump delivers at its maximum volume flow rate. When no hydraulic function is active, the pump pumps "on block" and pivots to minimal stroke (pressure hold function).

It is possible to increase the energy efficiency of a hydraulic unit with two pressure circuits by using a booster. Here the high pressure pump is replaced by a booster that is supplied by the low pressure pump and generates the high pressure required using a fixed gear ratio (see ◻ Fig. 10.43, center left). This optimization greatly reduces the power losses occurring from the high leakage rates in the high-pressure circuit. In comparison to a classic hydraulic unit, using a booster in the pressure hold function can reduce leakage in the high-pressure circuit by 94%. Leakage in the low pressure circuit is very low in both cases. Power consumption is reduced by 90%. Another benefit of a booster is its insensitivity to dirt and particle contamination [BREC13b; TRIE15].

The variable-speed pump drive (VSD) is another energy-efficient hydraulic unit. This system is fitted with an accumulator charging circuit that enables switching to a low circulation pressure when no hydraulic function is required. The main aim of such units is to reduce the power consumption during the pressure hold function. When the hydraulic circuit is in the no-load state, the motor equipped with a frequency converter runs at a very low speed. So, the pump maintains a low circulation pressure and delivers no volume flow (standby state). If the pressure in the high or low-pressure accumulator falls below the pressure switch point set, the speed increases to a set value and the pressure accumulators are filled again by the pump. Check valves prevent pressure equalization between the two circuits during operation and block the return flow towards the pump during the idle cycle. When the accumulators are filled, the pump switches back to standby [BREC13b; TRIE15].

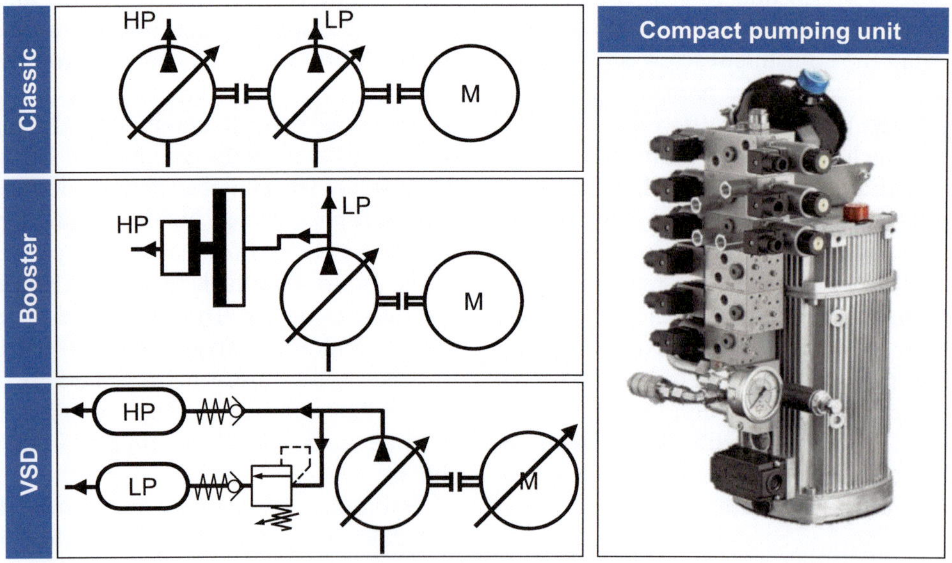

◻ **Fig. 10.43** Design variants of hydraulic units. *Sources* HAWE Hydraulik SE, [TRIE15]

Compact pumping units allow machine tools with few hydraulic functions the option of having a hydraulic unit requiring very little space with low energy expenditure. The compactness of the unit is achieved by submerging the motor in oil (submersible motor). The pump and motor are submerged in oil in the casing which also acts as the tank. Due to the improved heat dissipation from the motor via the hydraulic oil compared to a conventional air-cooled motor, it is possible to use a motor with a lower nominal power and smaller dimensions, depending on the application. The low energy expenditure is due to the fact that compact pumping units are used for short-time or intermittent operations. Depending on the requirement, an additional pressure accumulator, which maintains the operating pressure, can be used, and ensures the hydraulic functions are supplied with the necessary volume flow rate and pressure. For more information, refer to Chap. 5 of Volume 2 of this series.

10.3 Tool Systems

A tool system is understood to be the mechanical system comprising the tool interface, tool holder and tool. Depending on the machine type, tool systems can be divided into stationary systems for lathes, and rotating systems for drilling, milling and grinding machines.

The following chapters contain examples of constructive solutions for tool interfaces and for change and magazine equipment. Preceding the examples explained are specifications on the functions of holders and clamping devices as well as change and magazine equipment. Refer to the respective variants for the machine type for general descriptions of the functions of tool and workpiece carriers.

10.3.1 Tool Systems for Drilling and Milling Machines

Tool systems for drilling and milling machines are characterized by a mainly rotationally symmetric structure and are also adapted to their characteristic application scenario. Different interfaces have been developed with the introduction of automatic tool changers.

Interfaces can be divided into three types.

The first interface, between drive spindle and tool, is called the tool interface. It is used to exchange complete tools and in many cases is suitable for automated tool changes.

The second interface, between the tool interface and the interface to the cutting inserts, is for the modular structure of cutting tools. For this manufacturer-specific interface, screws are often selected as the connecting elements because no quick-change functionality is required here.

The third interface denotes the seat of the replaceable cutter elements (such as indexable inserts). These wearing parts can be replaced quickly and easily by a mechanical interface.

10.3.1.1 Mechanical Interface Between Spindle and Tool (Tool Interface)

In the pursuit of uniform drilling and milling tools that can be used universally on different machines, standardized interfaces have been developed, i.e., connection forms and clamping systems. Despite this standardization, there is still an array of manufacturer-specific variants.

Morse Taper to DIN 228

For simple drilling operations on the shop floor area without any automated tool changes, the morse taper as per DIN 228 [DIN87a; DIN87b] continues to be used. In the field of automatically exchangeable tool interface systems, different geometries have established themselves. These follow national and international standards.

The drilling spindles of vertical drilling machines and the quills of tailstocks have taper bores to accommodate the tool tapers in accordance with DIN 228 [DIN87a; DIN87b] (◘ Fig. 10.44, top left). For these morse tapers with a taper gradient of 1:20, the force and torque are transferred by a self-locking friction fit. For higher loads, variants with tightening thread or cotter are used. The tool is released from the spindle with jackscrews or a cotter driven into the end of the morse taper with a hammer. Drilling spindle heads as per DIN 55058 [DIN78] (◘ Fig. 10.44, bottom) are used in multi-spindle drilling machines to mount DIN 6327 adjuster sleeves [DIN85a; DIN85b; DIN85c; DIN81]. These heads enable tools to be adjusted to the required lengths. The holders presented are unsuitable for automatic tool changes, however (◘ Fig. 10.45).

Other tool interface types have become established on the market for the automated tool change of drilling and milling tools. The first interface developed in the chronology of automatically changeable tool interfaces is known as a steep taper.

Steep Taper (SK) to DIN ISO 7388

The tool steep tapers (taper 7:24) are non-self-locking and are chucked by the clamping system with the machine spindle. Different functional surfaces on the gripper flanges are used to hold and orient the tools in the gripper and magazine equipment. The common steep taper form as per DIN ISO 7388-1 [DIN14a] is form A with a trapezoid groove and orientation groove. For low loads, the torque is transferred by the friction

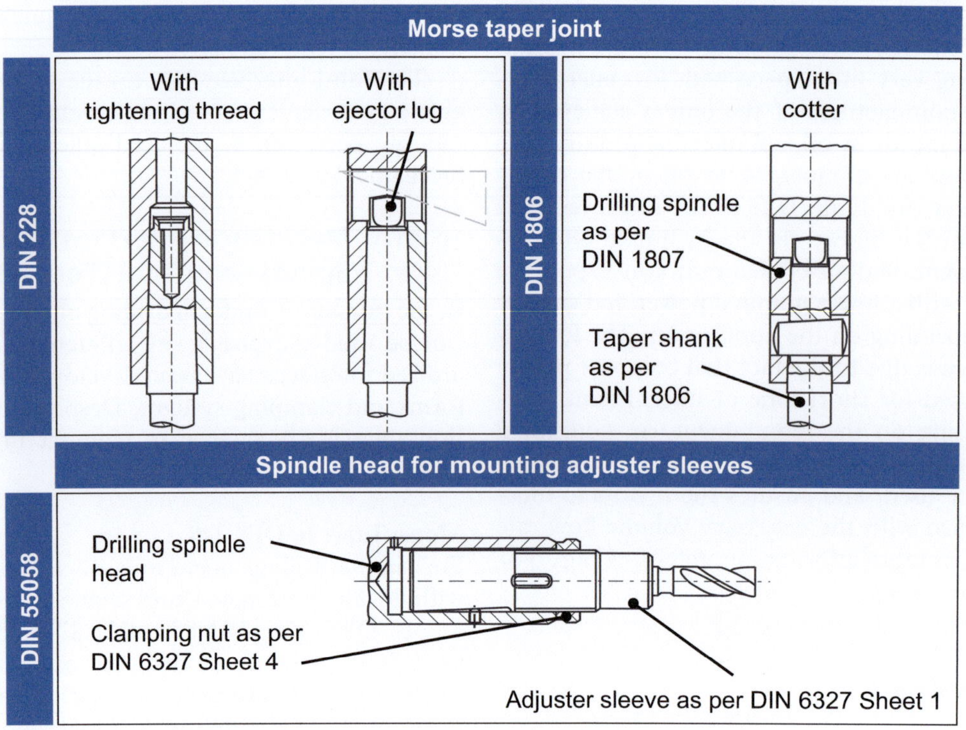

Fig. 10.44 Design variants of drilling spindle heads, according to [DIN87b; DIN87a; DIN64; DIN78]

fit of the taper. For large torques and impact loading of a steep taper, the driving keys take the main component of the torque transfer (Fig. 10.46). The driving keys on the spindle that engage into the driving slots of the tool holder have different widths to guarantee unique orientation for the tool. The tools complying to DIN ISO 7388-1 [DIN14a] have tightening bolts that comply with [DIN15b] and can be replaced very quickly and automatically using hydraulically operated collet chucks in most cases (see ▶ Sect. 10.3.1.2).

Hollow Taper Shank (HSK) to DIN 69893

The hollow taper shank interface (HSK) as per DIN 69893 [DIN11a] has enjoyed widespread usage in metal-cutting processes. It can be regarded as a further development of the steep taper. Its benefits pertaining to precision and stiffness, suitability for high rotational speeds and quick tool changes now mean it takes clear precedence over conventional steep tapers.

The hollow shank with a 1:10 taper is radially elastic. This enables, despite over-determination, the tool flange on the spindle to be brought to the flange contact surface. The form-fit torque transfer takes place at the shank end. The system is clamped from the inside by widening collet segments. Here too, functional surfaces on the gripper flanges are used for holding and orienting the tools. There are six different forms with different characteristics (Fig. 10.47). The width of the flange contact surface is a constructive difference. While types A, C and E (row on the left in Fig. 10.47) exhibit a narrow flange contact surface, types B, D and F, with the same flange face diameter, have the taper of the next smallest size and thus a wider flange contact surface (row on the right in Fig. 10.47). There are also differences in the coolant supply. Designs B and D use sunk bores via the flange while the cooling lubricant is transferred centrally to the tool for the other designs. Variants C and D have no gripper grooves and are only suited for manual tool changes. According to [DIN11a], the most common form of hollow taper shank is form A. Form E is suited to high-speed machining in particular.

Its flange contact surface and associated larger support diameter mean the hollow taper shank attains higher radial and axial stiffness compared to the steep taper. Bending stiffness in particular is much higher than for steep tapers and cylindrical shanks of comparable size (Fig. 10.94).

Due to the flange contact surface, the axial changing repeat accuracy of the hollow shank is reproducible to values below 1 μm. For the steep taper, axial position deviations in the tenths of millimeters can occur just because of the tolerance. The lower mass and shorter length of the HSK compared to the steep taper simplify the tool change and handling processes.

Polygonal Hollow Taper Shank with Flange Contact Surface (PSC) as per DIN ISO 26623

The polygonal hollow taper shank with flange contact surface (PSC) was developed and sold by Sandvik Co-

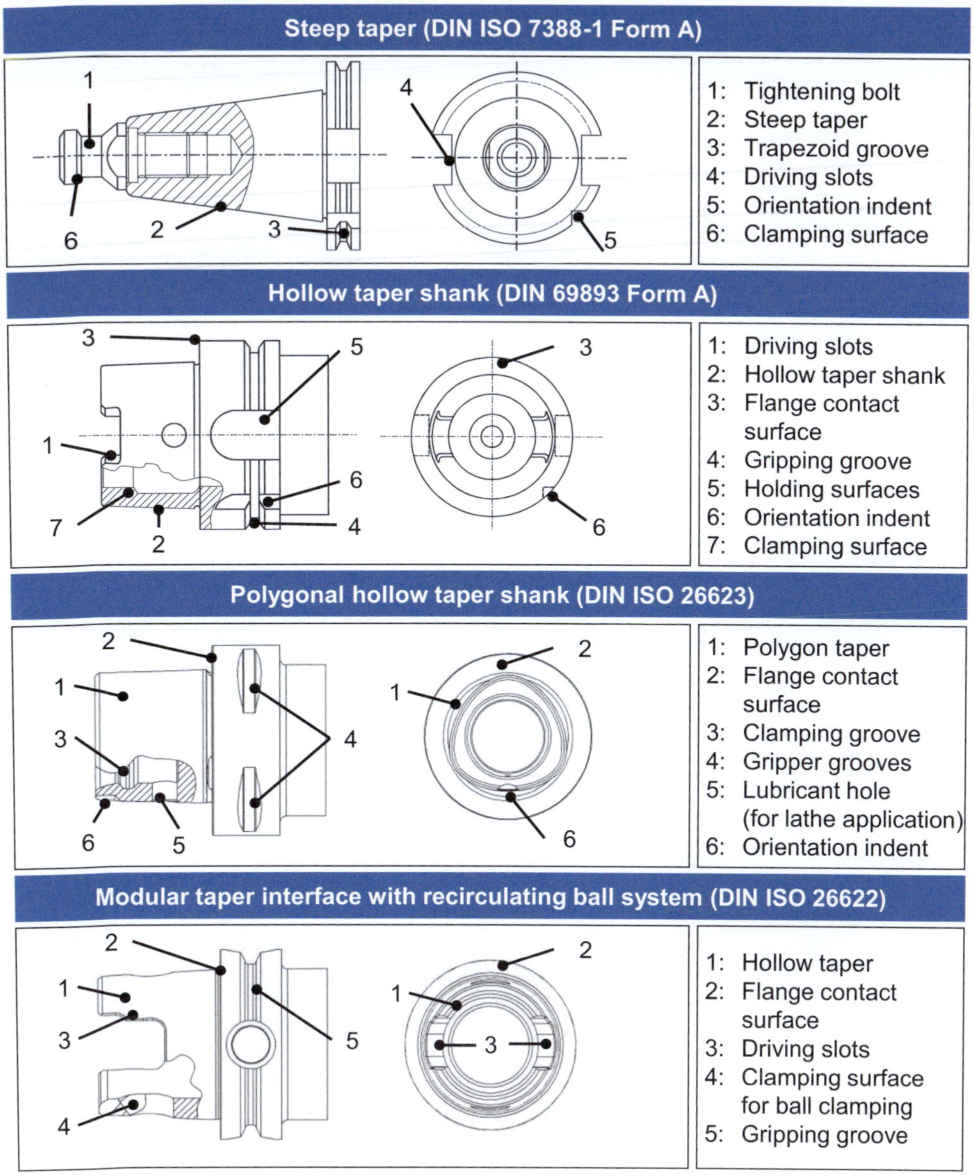

Fig. 10.45 Comparison of standardized tool interfaces for drilling and milling applications, according to [DIN14a; DIN11a; DIN10e; DIN13b]

romant under product name Capto. This tool interface system has since been standardized in DIN ISO 26623 [DIN10e].

The basic structure is similar to that of the HSK but the conical shank with a pitch of 1:20 has a 3-sided polygonal profile. This means the torques are transferred differently than on the HSK. That is, via the polygon and not the slot nuts. Another differentiating factor is how the force of the clamping system is introduced to the tool. While there is a clamping bevel on the HSK, the PSC is held by a clamping groove or a thread located in the center in the tool.

Modular Taper Interface with Ball Track System (TS) as per DIN ISO 26622

The now standardized TS system as per DIN ISO 26622 [DIN13b] from Kennametal is also based on a hollow taper for centering and a flange contact surface for axial positioning and increasing bending stiffness. The type of tool clamping is a fundamental differentiating factor. HSK and PSC use collets to introduce the axial clamping force into the tool. The TS system has recessed, inclined bores in the taper of the tool. Two balls are pressed radially outwards into the bores by the clamping system. This generates the clamping force required to chuck the tool.

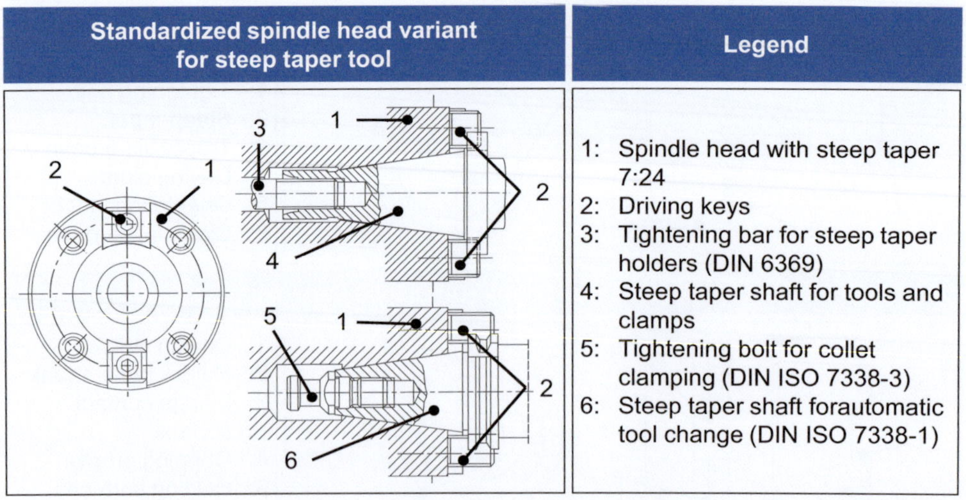

Fig. 10.46 Standardized spindle head variants for steep taper tools, according to [DIN87c]

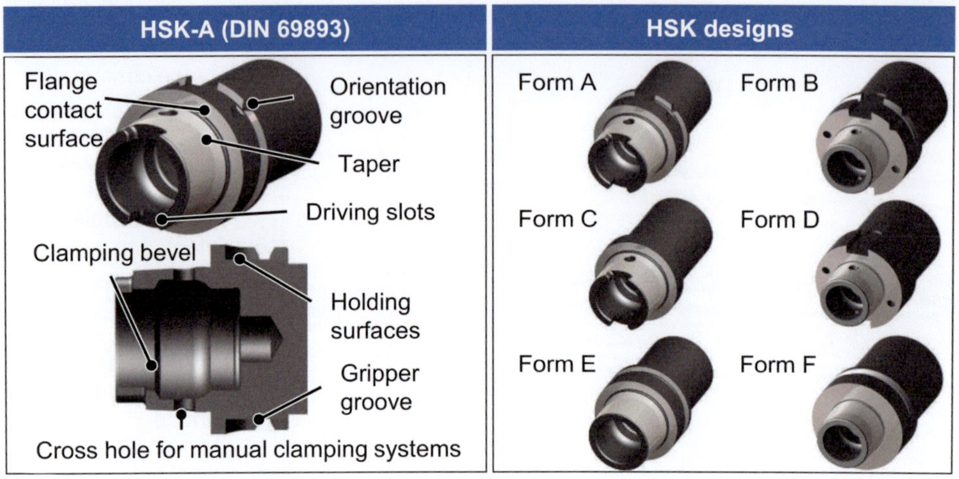

Fig. 10.47 Overview of designs of the hollow taper shank interface, according to [DIN11a; DIN13c; DIN12; DIN03]

10.3.1.2 Tool Clamping Systems

All the tool systems introduced here require an axial force to provide the functionality. Automatic as well as manually operated systems are used. Steep taper, HSK and PSC tools are usually clamped with collet tool clamps for quick and automatic clamping. The necessary actuation force is often provided by a spring assembly.

Figure 10.48 shows an example of a structural solution with release and clamping positions for steep taper clamping. In the clamping position, the disk spring assembly (9) presses onto the clamping sleeves (6) so that the spring force is transferred to the clamping ring (3) via the balls (5). The different inclination angles of the clamping sleeve and clamping ring at the contact point to the ball create a clamping force that is 3 to 3.5 times higher than normal spring compressors, that generate their lowest clamping force in the clamping position (longest spring travel). The clamping ring (3) is permanently joined to the tension rod (7). This means that the increased tensile force acts upon the tension rod and the collet holder (8). An overview of the clamping forces in the standards is in Table 10.3.

Hydraulic releasing units are popular for releasing/forcing out the tool. Pneumatic or electrical systems can also be used depending on the application. In the example in Fig. 10.48, the piston (1) is pressed from left to right by hydraulic pressure. The pressure bolts (2) then push the clamping sleeves to the right against the spring force. The solution shown in Fig. 10.48 has the benefits that the retraction force is 3 to 3.5 times higher than the spring force and only the spring force needs to be applied by the hydraulic system for releasing. Also, the direct docking of the cylinder at the spindle means the spindle bearings are not subjected to loading by the high tightening forces.

Compact clamping fixtures with as low a rotating mass as possible permit rotational speeds up to

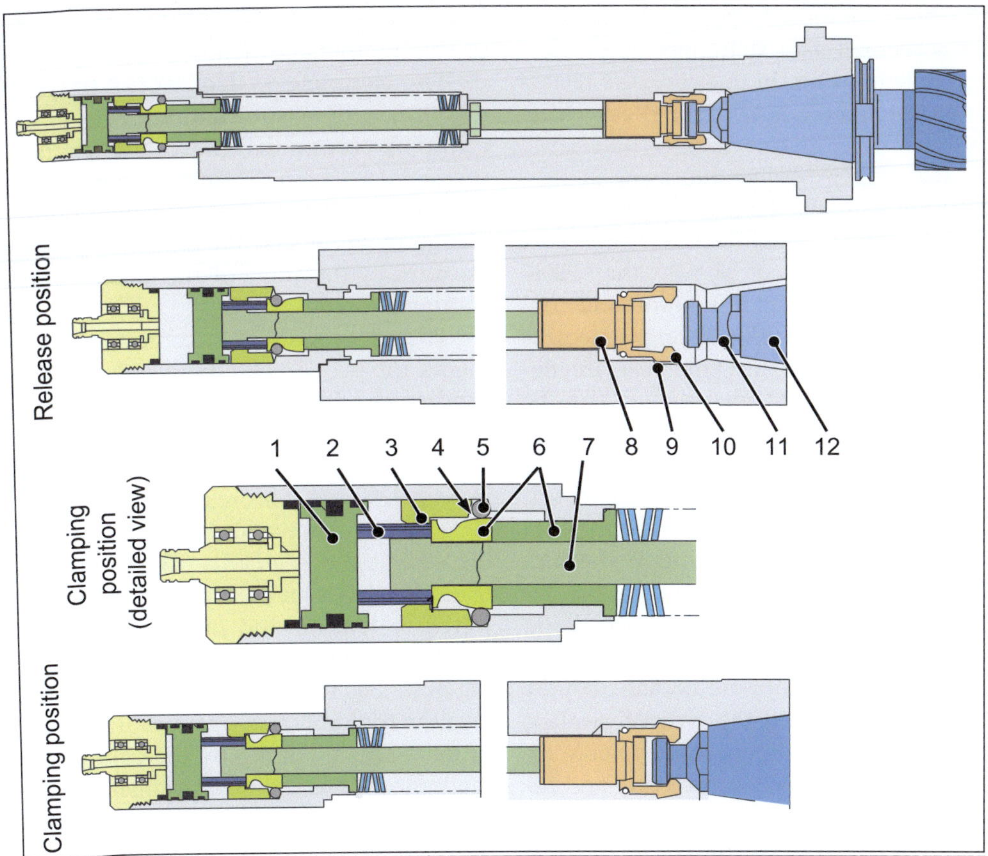

1: Piston
2: Pressure bolt
3: Clamping ring
4: Wedge
5: Balls
6: Clamping sleeves
7: Tension rod
8: Collet holder
9: Control edge
10: Collet chuck
11: Tension bolt
12: Steep taper

Fig. 10.48 Tool clamp with force-multiplying wedge gear. *Source* OTT-JAKOB SPANNTECHNIK

Table 10.3 Overview of standardized minimum clamping forces, in kN

Interface	Size 25	32	40	50	63	80	100	125	160
SK									
HSK-A/C	3	5	6.8	11	18	28	45	70	115
HSK-B			5	6.8	11	18	28	45	70
HSK-E	1.5	5	6.8	11	18				
HSK-F				6.8	11	18			
HSK-T		5	8	14	24	37	55	86	150
PSC		15	20	25	30	40			
TS		9–18	13–27	22–40	36–58	53–80	75–110		

45,000 l/min. For high rotational speeds, the hydraulic releasing unit is secured as a stationary, non-rotating unit on the spindle housing. In this case, the releasing forces act upon the spindle bearings and can easily result in overloading.

To use tools with inner cooling lubricant supply, and for taper cleaning with air, the tool clamps are fitted with 2-channel rotating unions.

The flange spindle depicted shows the HSK tool holder and the clamping set for automatic tool change with tension rod and the spring assembly (■ Fig. 10.49). The compressed air channels are designed such that, during a tool change, air can be used to clean chips from the holder taper. Cooling lubricant is supplied centrally by the rotating union via the tension rod all the way into the fitted clamping set. From here, the cooling lubricant is routed into the tool via a transfer pipe. Coolant supply through the machine spindle can, in addition to the central supply described, also uses channels recessed into the spindle at the front for other tool holders. Tools or tool holders must be equipped for the appropriate coolant supply on the spindle side depending on the type of coolant supply. Coolant supply outside the machine spindle requires a tool holder with stationary supply ring and rotary transmission.

The clamping mechanism for the hollow taper shank is shown in ■ Fig. 10.50. Key features are the clamping force effect radially from the inside to the outside and the positioning of the clamping bevel inside the hollow shank. The collet segments of the clamping system used to grip the clamping bevel and, due to its 30° inclined position, increase the axial retraction force. At high rotational speeds, the clamping force is increased, not reduced, by the centrifugal force of the clamping elements.

The clamping system works (depending on nominal size) with six or eight collet segments. During the clamping process that moves the tension rod backward, the segments keep moving radially outward on the bevels of the collet holder until they come to rest on the clamping shoulders. Pressing the collet chucks further outward by the angle of the tension taper builds up the clamping force.

The self-contained force flow runs via the collet elements, the two 30° clamping surfaces of the spindle and tool shank, the steep taper and the flange contact surface. The only function of the short disk spring assembly is, in the release process, to press the collet chucks radially inwards. This releases the tool for a tool change.

The PSC tool system is also pulled into the spindle using collet segments (■ Fig. 10.51). In contrast to usual HSK designs, the collet chucks here are not subjected to tension and instead are only used to transfer the pressure forces passed to the tool by the clamping bolt. The retraction force is thus introduced directly into the tool and not, as is usual for the HSK, with additional strengthening by the clamping bevels. If force multiplication is required, it must be accomplished with suitable components, such as a mechanical wedge gear downstream.

The TS tool system is fundamentally different from the previous systems (■ Fig. 10.52). Here, the axial force required is applied to the tool via two balls. They are each guided mechanically and radially in an inclined hole in the tool. The 3-point contact between the tool, fixed guide and clamping bolt means the tool is subjected to a radial and axial force component onto the shank.

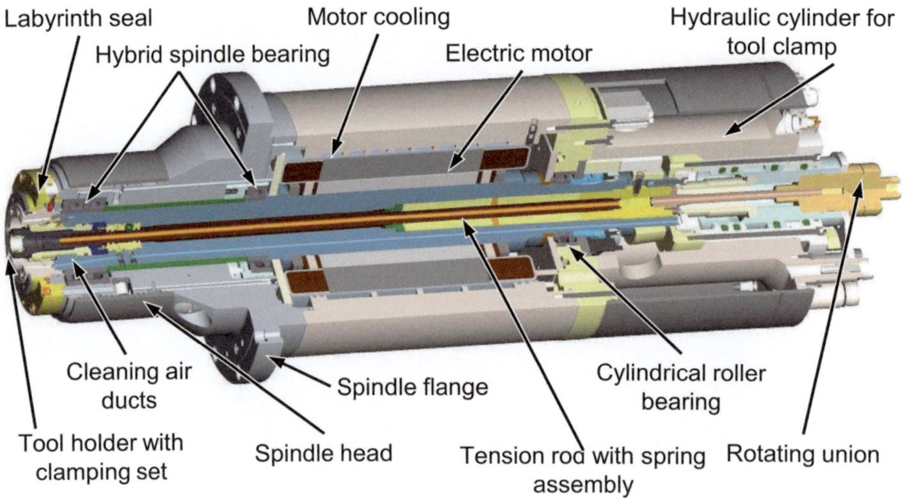

■ **Fig. 10.49** Tool holder and clamping device in a flange spindle. *Source* GF Machining Solutions

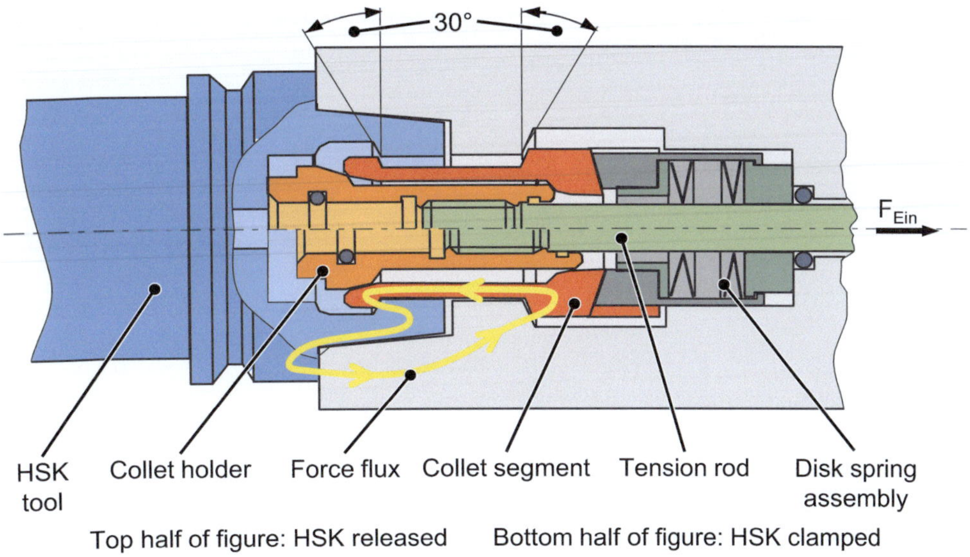

Fig. 10.50 Automatic HSK clamping for spindles. *Source* OTT-JAKOB SPANNTECHNIK

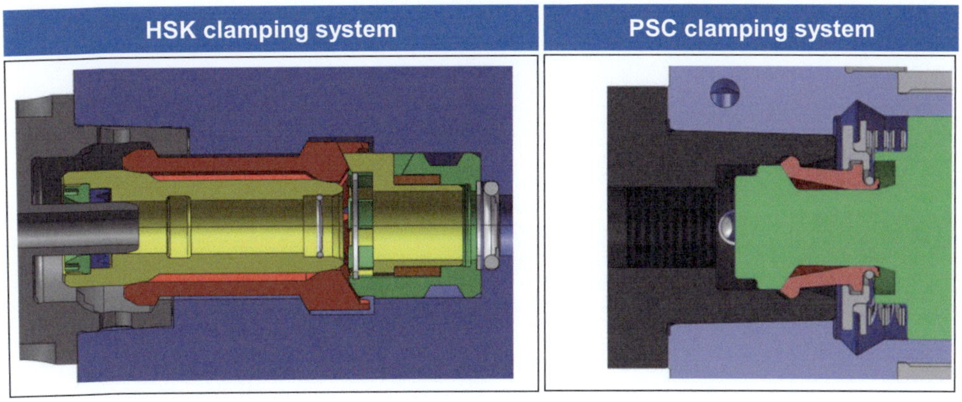

Fig. 10.51 Comparison of the designs of HSK and PSC clamping systems. *Sources* OTT-JAKOB SPANNTECHNIK, Sandvik Coromant

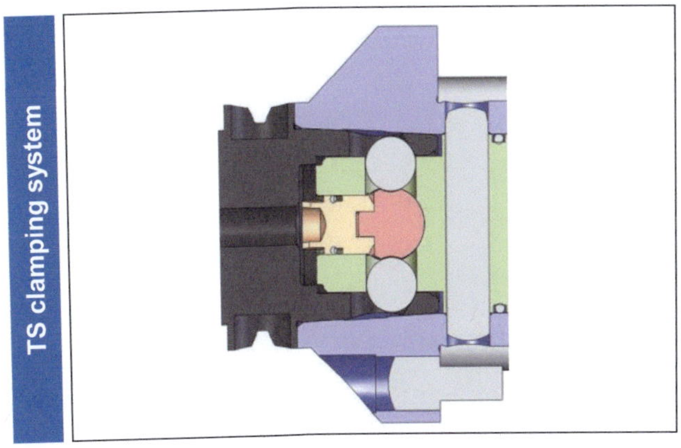

Fig. 10.52 Structure of the TS tool clamping system. *Source* Kennametal

10.3.1.3 Tool Types (Modular Tools/Tool Holders)

For drilling and milling tools, there are a number of standard variants having different interfaces that are specific to the function. An overview of conventional standard tools is shown in Figs. 10.53 and 10.54. These standard tools are clamped in tool holders. The interfaces of these holders on the machine side are usually standardized (HSK, PSC, SK or TS). A tool holder acts as an adapter. A second and possibly more interfaces connect to it.

Different clamping principles are applied to mount drilling and milling tools with a cylindrical shank. Thermal, hydraulic and purely mechanical clamping mechanisms are commonplace. Furthermore, additional equip-

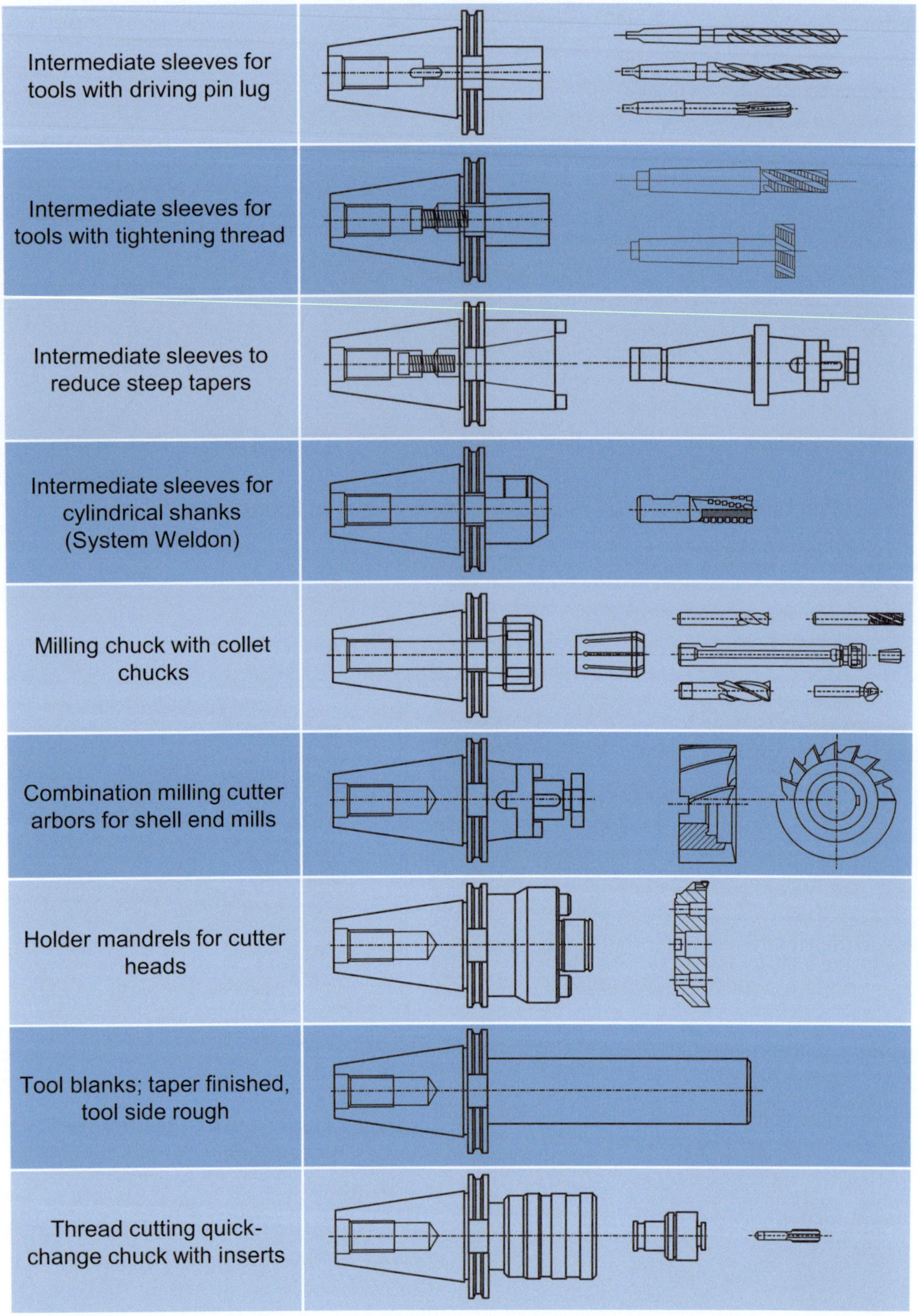

Fig. 10.53 Base adapters to DIN ISO 7388 [DIN14b] with different tool holders. *Source* Kelch

ment for heating or injecting is often required. Shell end mills and cutter heads are joined to a base holder by special milling cutter arbors or holder mandrels.

A base holder with a universal chuck is shown in Fig. 10.55. The hydraulic expansion chuck for shaft tools represents a completely enclosed chamber system that is resistant to dirt and coolant. A clamping bolt actuates a piston with stroke limiter. The piston applies hydraulic pressure to the oil-filled compartment around the clamping bush. To reach the full clamping force, the clamping bolt must be turned all the way. The thin-sided expansion bush, made of special high-speed steel with a hardness of 55 HRC, is pressed together and clamps the cylindrical tool. The stroke limiter installed stops over-expansion. The bush shrinks uniformly over its entire length, cylindrically and centrally with respect to the central axis. It has multiple flutes on the inner side to catch waste particles and chips.

Fig. 10.54 Different base adapters to DIN ISO 7388 [DIN14b]. *Source* W&F Werkzeugtechnik

The chamber system, filled with the hydraulic medium, has a damping effect on the tool chucked. At both ends, the expansion bush is joined tightly to the base mandrel without a sealing element. This metallic joint to the base body guarantees good stability and torsional stiffness. After the pressure is relieved with release of the clamping bolt, the expansion bush expands again. It moves back to its initial position and releases the tool shank.

With a maximum centering accuracy of 3 μm, this chuck has a high continuous running accuracy. When slitted intermediate bushes are used to clamp more than one shank diameter with one expansion chuck, the circular running accuracy is reduced to 5 μm. Many variants also have an additional, radially operated adjustment mechanism for the presetting of tool lengths down to μm accuracy (shown on the right in Fig. 10.55). Tools with smooth cylindrical shanks up to 32 mm diameter as well as with recesses (as per DIN 1835-1 [DIN99] forms B, C, D, E and DIN 6535 [DIN92] forms HB, HE) can be clamped directly and without intermediate sleeves. For stationary use of the expansion clamping tool, pressure can also be applied directly by a hydraulic unit. A variant of the chuck shown, with an axial oil supply, is used for this.

In addition to the complete tools where the tool body is joined directly to the base adapter, modular tool systems are used for NC technology. A large tool requirement, due to the variety of machining tasks, can be addressed in a cost-effective manner. One way of meeting such requirements is to use interchangeable tool modules. These can be assembled as required into individual tools. The tool stock can be reduced to a relatively few base holders, adapters and tools.

In addition to the suitability for different machining processes, modular tool systems differ mainly in the type of structural variants of sectioning points. They are available in different sizes. There is an array of structural solutions, that can be categorized by the joining or clamping principle. There can be indirect clamping via a bolt positioned axially or radially, direct screw fitting of the modules or a tensioning with screw-operated clamping wedges.

With all structural methods, precise axial position determination and unambiguous orientation of the cutter position can be achieved. This is in addition to centering that guarantees a changing repeat accuracy of less than 5 μm for the diameter. The basic design of a modular tool system based on the standardized hollow taper shank/steep taper is shown in Fig. 10.56. Different drilling, milling and boring tools can either be held directly or by an extended adapter in the base holder. Different base holders enable tools to be adapted to different holders in the machines.

The range of clamping devices shown includes all known elements for preferably rotating applications. These elements run from chucks for cylindrical shanks

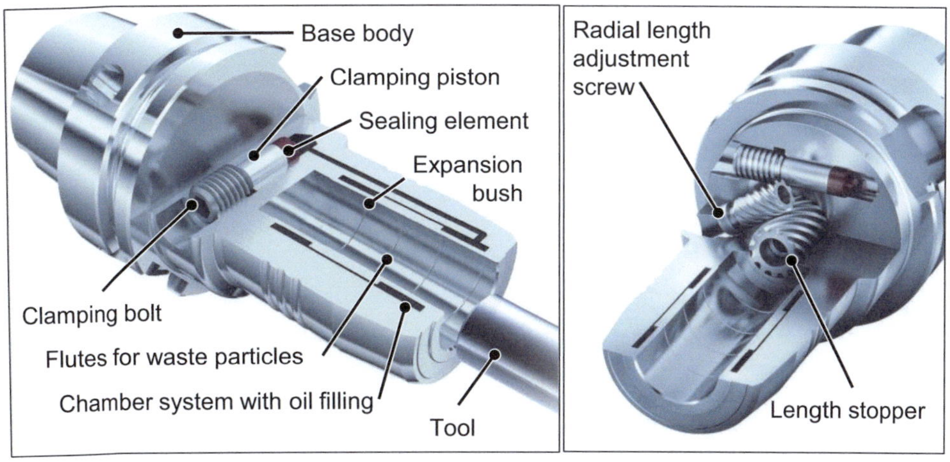

Fig. 10.55 TENDO hydraulic expansion chuck for manual clamping. *Source* Schunk

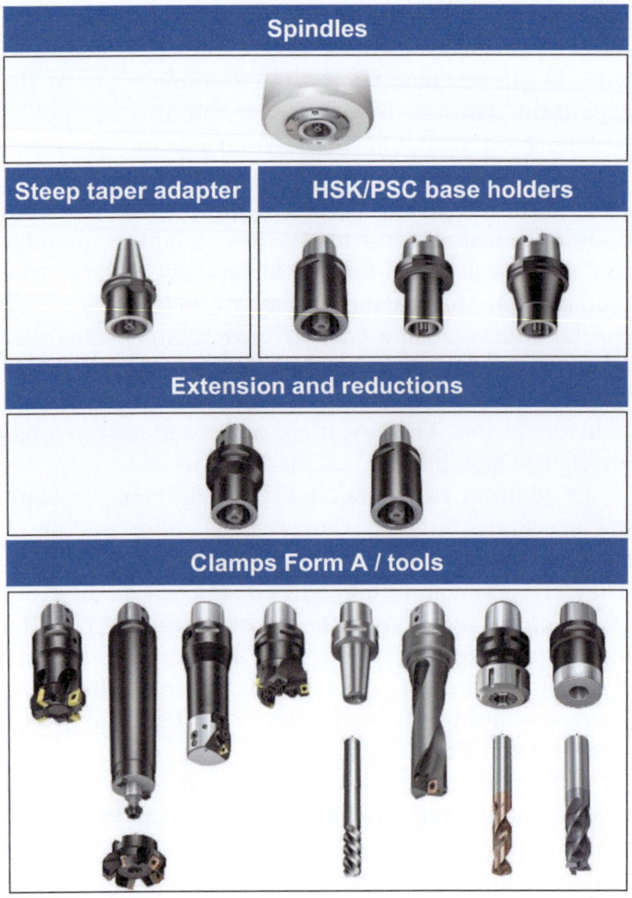

Fig. 10.56 Modular tool system for manual and automatic changing. *Source* Sandvik Coromant

and collet variants, hydraulic expansion chucks and thread cutting quick-change chucks, to tool holders for drills, intermediate holders for morse tapers and milling cutter arbors.

The core of this modular clamping system is a clamping cartridge. This is used in spindles and base holders directly or via the adapter flange (Fig. 10.57). In the clamping process, two clamping jaws are moved radially outward by a bolt with external and internal thread. Here, the jaws abut against the 30° shoulder of the hollow shank and generate axial and radial clamping forces. For the declamping process, an ejector pin is actuated at the same time that ejects the tool from the pretensioned joint (Fig. 10.58).

Combining different system elements give rise to modular systems for mainly rotating applications. These also permit very good machining results for large overhangs. After a reduction of the HSK nominal size, an extension tensioned by a clamping cartridge is held here. A milling cutter arbor is held by another clamping cartridge and then the milling head is mounted.

In addition to the applications shown above, there are also special tools such as the actuating tool in Fig. 10.59 for the combined machining of multiple surfaces. It comprises a seven-edge boring tool at the tip of the tool and a fine boring tool with an actuating system having a total of seven cutting edges for chamfering and boring a larger diameter. The fine boring tool is inserted into the base holder via a HSK 100 holder with a

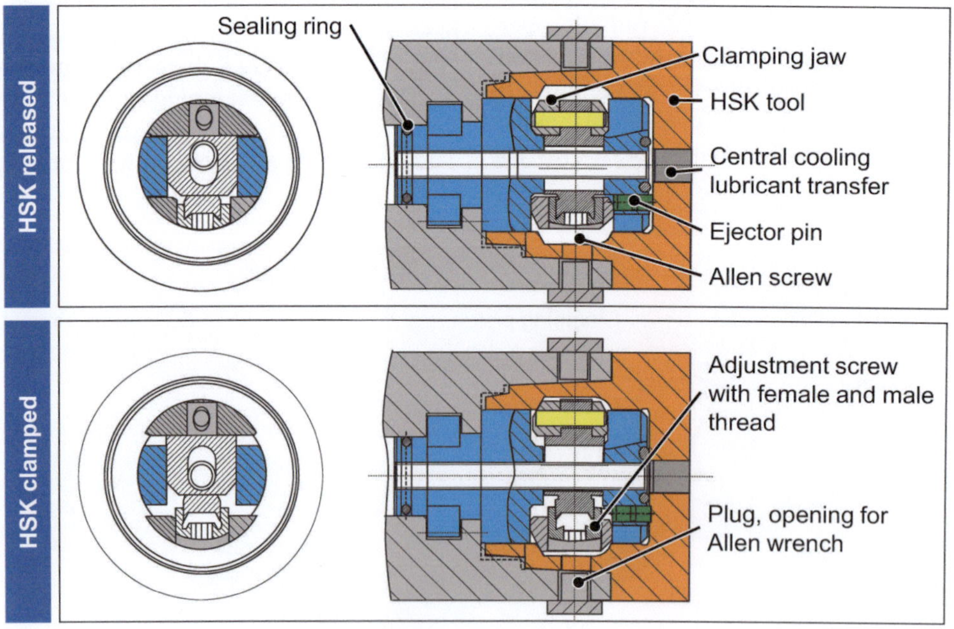

Fig. 10.57 Manual clamping system for hollow taper shank tools and minimum quantity lubrication (MQL). *Source* MAPAL

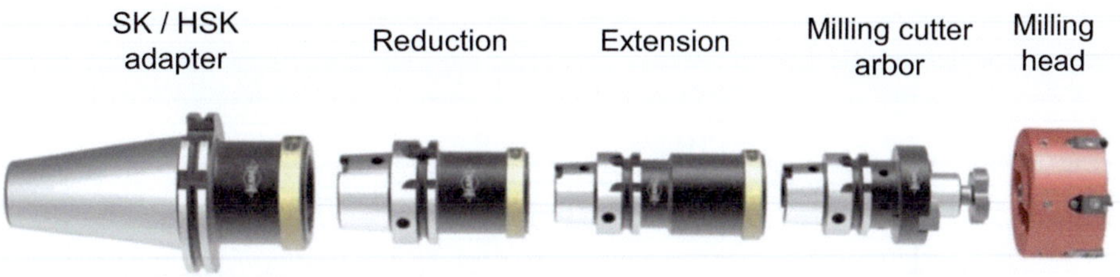

Fig. 10.58 Modular tool system. *Source* MAPAL

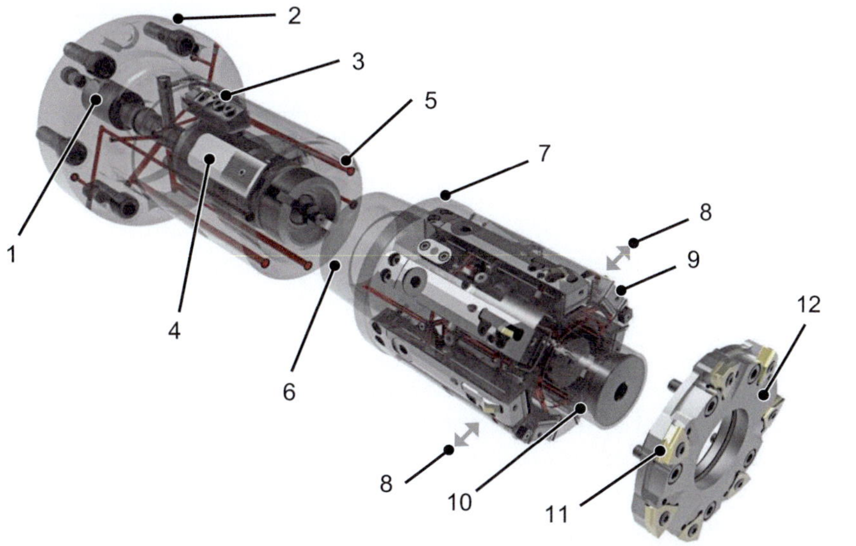

1: Tension rod and counter slide	7: Fine boring tool
2: Module 160 interface	8: Actuating movement
3: ISO cartridge with indexable insert for chamfering	9: Seven-edge actuating system
4: Cover for HSK clamping bolt	10: Concentricity check point
5: Rerouted coolant flow-through	11: Indexable insert
6: HSK-E 100 interface	12: Boring tool

Fig. 10.59 Modular actuating tool. *Source* MAPAL

manual clamping set, the base holder itself has a clamping level with ISO cartridges and indexable inserts.

10.3.1.4 Change Equipment and Magazine for Tools

The depiction of the individual components and functional processes of an extensive tool supply system, in Fig. 10.60, illustrates in general the handling, transport and magazine functions. Tool handling includes the approach and return movements of a tool by machine components, the actual tool change, tool preselection with corresponding positioning of the tools in the machine magazine and tool exchange. In each of these handling processes, the tools must be positioned, oriented, held, released, etc.

From the number of possible magazine modules, those that are to be assigned close to the spindle (machine magazines) are discussed in the following chapters.

An overview of possible magazine designs with moving and stationary tool slots is in Fig. 10.61. The storage of tools, that is, orderly keeping of geometrically determined bodies [VDI90b], can use different geometrical arrangements. The key characteristics of a magazine with moving tool slots are the positions of the tools in relation to the rotation axes, the number of rotation axes and the number of tool levels.

In addition to the magazine designs described, it is possible in the group of magazines with stationary tool slots to differentiate rails, pallets, cassettes and racks as

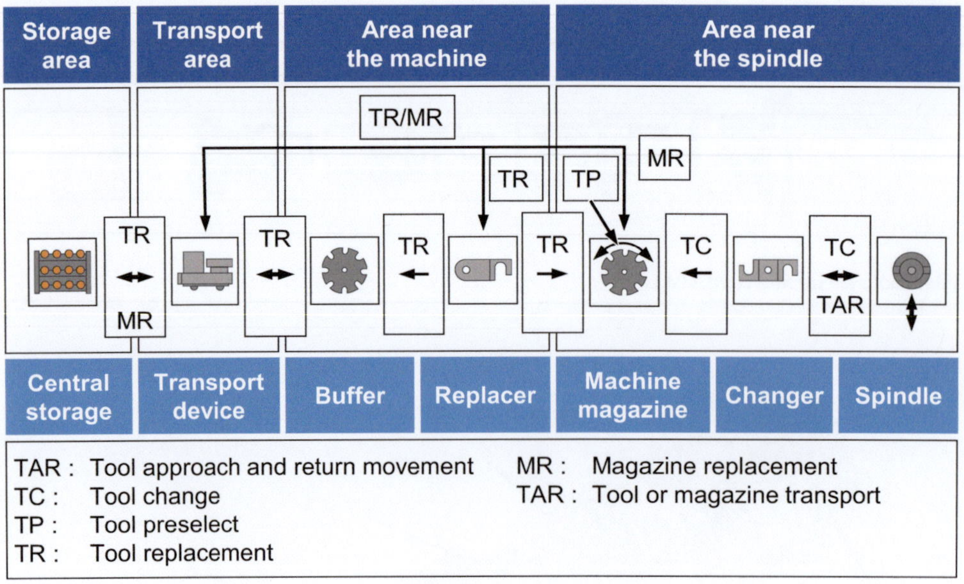

Fig. 10.60 Individual components and processes in a tool supply system

10

designs (Fig. 10.61, bottom) from the number of coordinate axes.

Components that can be assigned to the immediate machine area, as well as those that perform functions in the central storage area (see Fig. 10.60), belong to automatically operating tool supply systems. This chapter introduces change equipment and magazines close to the spindle. Refer to ▶ Chap. 9 for explanations extending beyond this area of equipment components, in particular the interlinking of multiple machines.

Tool supply systems close to the spindle can be of various forms depending on the different machine designs and sizes. The initial criteria in specifying the components for the tool supply are the number and position of working spindles and the number of axes of motion in the tool holder of the machine. In the development of an overall concept, the magazines and change equipment are then designed optimally depending on the machine design necessary for a particular production task.

In addition to the magazine design, the change principle and the gripper design are the main definition criteria for the automatic tool supply equipment (see Fig. 10.61). Differentiation can be made between the principles of changing with additional equipment, usually dual-arm grippers, and changing with combined movements of functional units in the machine and magazine (pick-up).

The main factor in selecting suitable change equipment is the chip-to-chip time attainable that is determined primarily by the change time. Depending on the designs of the machines and the tool change frequency, tool preselection and time for tool approach and return movements also have a bearing on the chip-to-chip time.

Selected tool change concepts for machining centers are described in detail below using examples.

The workspace of a high-speed machining center with a vertical spindle is shown in Fig. 10.62. The spindle is moved to the tool change position at the start of the tool change process. The chain-type magazine moves into the workspace. Here, the tool currently clamped in the spindle is placed in a free magazine slot. The chain-type magazine now makes the next tool available. After the clamping process, the chain-type magazine moves out of the workspace and machining can continue.

Other tool magazine and tool change methods are shown in Fig. 10.63. The tools are housed in a basket radially around the spindle. The basket is secured to the spindle axis. Every tool is held in the gripper groove by a separate arm. The arms are operated pneumatically. They switch from the magazine position in the basket downward for transfer into the spindle, and vice versa. At the start of the tool change process, the whole basket moves down in the spindle axis direction. The tool is pulled out of the spindle (Fig. 10.63a). The previous tool is retracted, and the next tool brought into position beneath the spindle (Fig. 10.63b and c). Then, the entire basket is moved up again and the new tool is moved into the spindle and clamped (Fig. 10.63d).

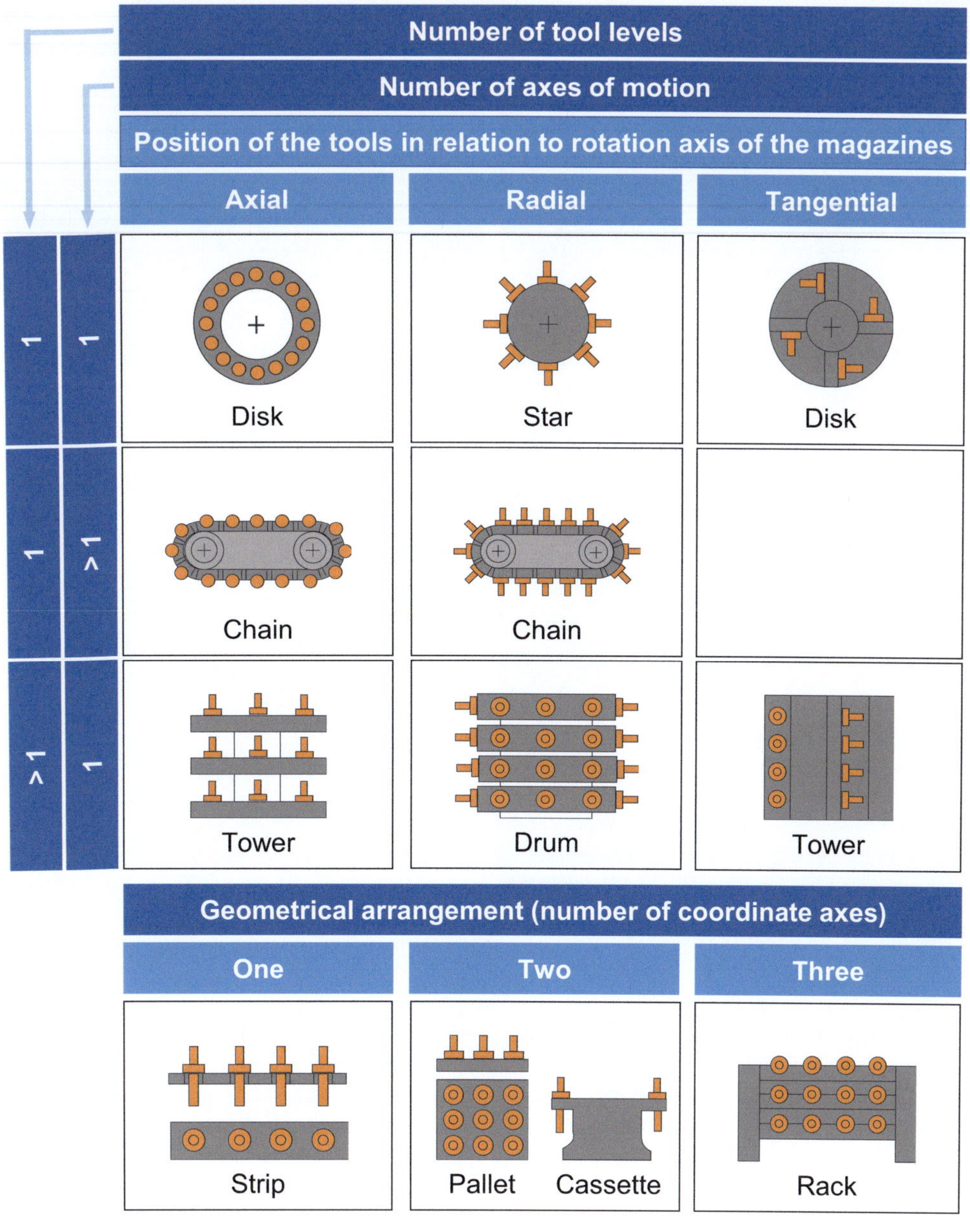

Fig. 10.61 Magazine designs with moving and stationary tool slots

This form of tool change is very quick and maintenance-free due to the pneumatic drive for the lift motions of the basket and individual arms. Furthermore, there is no traversing of the spindle to a certain tool change point where tool changing takes place. A downside is the additional weight on the spindle head as the complete and limited tool set must be moved with the change mechanism.

Using a magazine with fixed instead of moving tool slots provides a tool supply concept that is characterized in particular by the capability to simply extend the magazine capacity (Fig. 10.64). Pallets, cassettes or racks are used as magazine designs.

In these concepts, the handling devices have a portal design or are rack feeders for tool preselection and provision. They also bridge the large gap between the tool magazine and the spindle. Any tools can be removed during primary machining time and be provided for the dual-arm gripper. Magazine cassettes loaded with tools can be retooled manually or with industrial trucks.

For a tool change with a dual-arm gripper, the tool to change must be retrieved by the handling unit

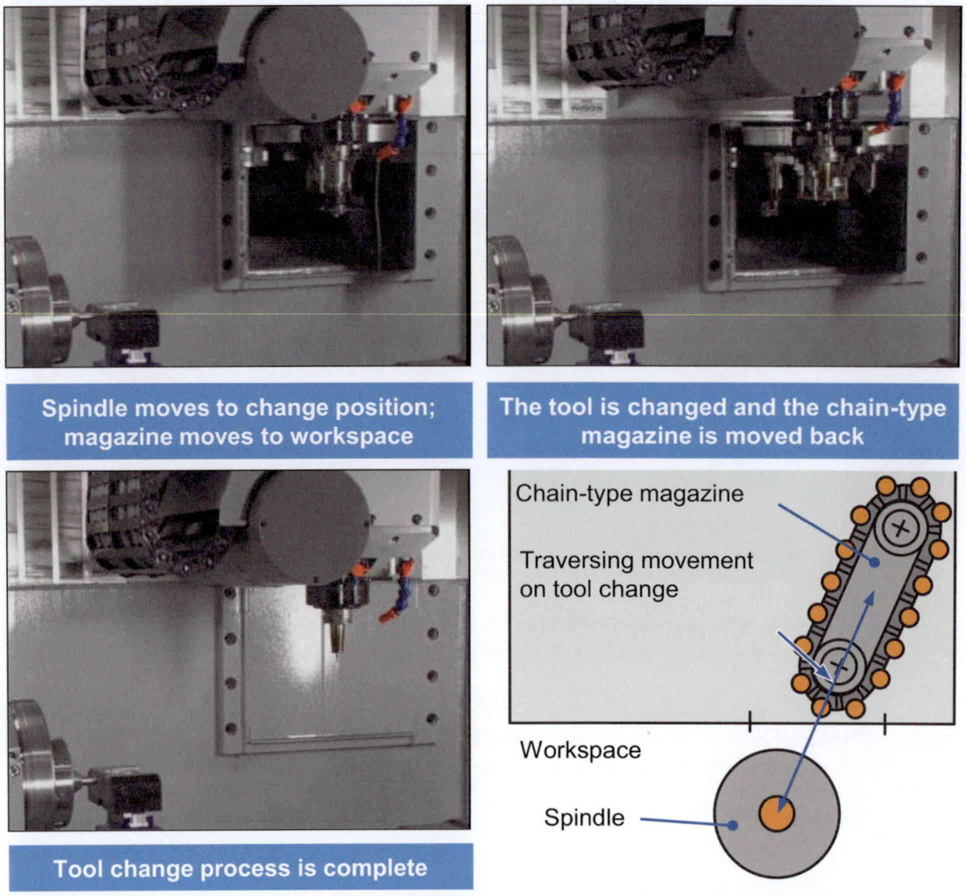

Fig. 10.62 Tool change with chain-type magazine. *Source* CHIRON Werke

from the cassette tool magazine and be provided in the transfer position near the spindle (Fig. 10.65). For this, the handling device swivels from its vertical position pointing downward to the horizontal position. The tool change can take place once a sliding door between the machining area and tool magazine has been opened. The dual-arm gripper pivots from its vertical position into the horizontal position, grips both tools, pulls the tools out of the spindle and handling unit with s stroke along parallel axes, rotates by 180° and loads them into the spindle and handling unit. The spindle and tool provisioning unit move out of the tool change position so that the dual-arm gripper can turn back into the vertical idle position. Once the sliding door is closed, machining can continue, while the tool provisioning unit puts the tool down and prepares the next tool change.

Tool changes can also be performed with combined spindle and magazine movements on machining centers with horizontal and vertical spindles. However, a longer tool change time is generally associated with the savings for a tool change unit. The preference for changing heavy tools or tool holders is this pick-up method because the maximum tool weight that can be handled by dual-arm grippers is 2×40 kg (this is dependent on the version).

Figure 10.66 shows the tool change process on a machining center with a disk-type tool magazine. The tool to be replaced is grabbed by the clamping elements of the magazine disk from the spindle following a translational movement (1) of the disk-type magazine. The tool is released when the column moves back while the tool collet chuck is opened at the same time (2). The new tool is brought into the spindle position by rotating the magazine disk (3) and held in the spindle by a backward movement of the column (4). The magazine, moving back to its initial position, releases the machine again (5). This change principle is also applied to machines with movable column design (three axes in the tool holder) where the magazine does not require an additional translation axis.

Another option for tool storage is the chain-type magazine (Fig. 10.67). Here, the tools are secured in holders that are fastened to a chain link. The chains

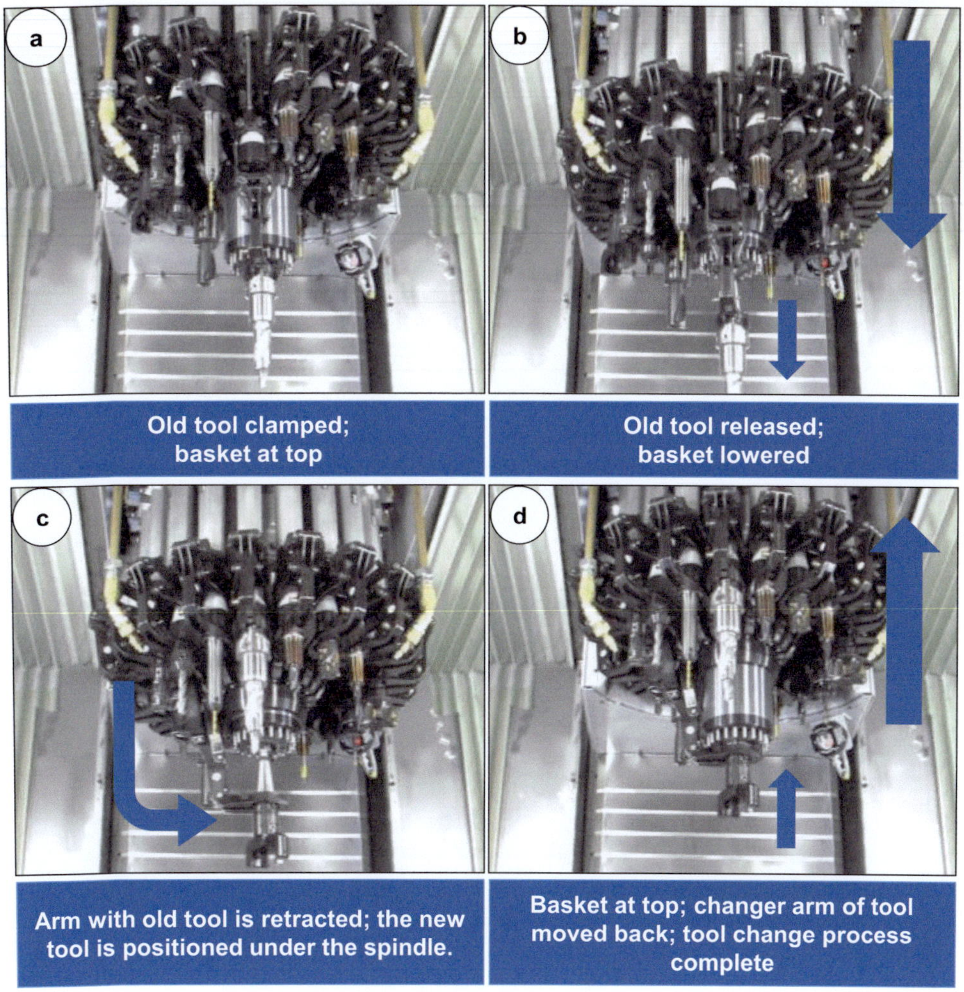

Fig. 10.63 Tool change with basket changer. *Source* Chiron

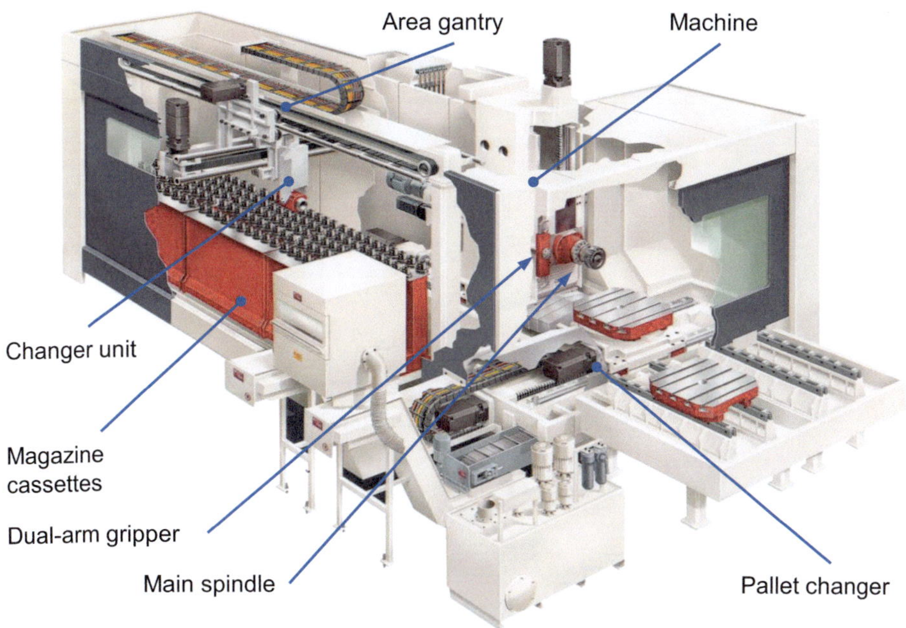

Fig. 10.64 Magazine cassettes with fixed tool slots on a cross-bed machining center. *Source* Zuse Hüller Hille

1 Single-arm gripper
2 Spindle in workspace
3 Handling device
4 Tool magazine

Fig. 10.65 Tool change with dual-arm gripper. *Source* Zuse Hüller Hille

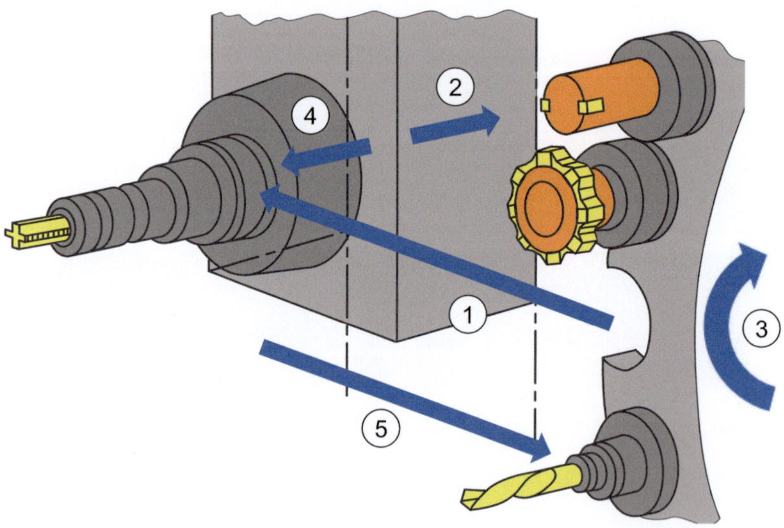

Fig. 10.66 Tool change with combined spindle and magazine movements, according to Zuse Hüller Hille

have a drive and make the tools available at the tool change and retooling position. As shown on the left in Fig. 10.67, chain-type magazines can also have a 3-dimensional arrangement such that the back and top of the machine can be used for the tool magazine. Routing the chains to follow a meandering path (Fig. 10.67, right), enables many tools to be stored when space is at a premium.

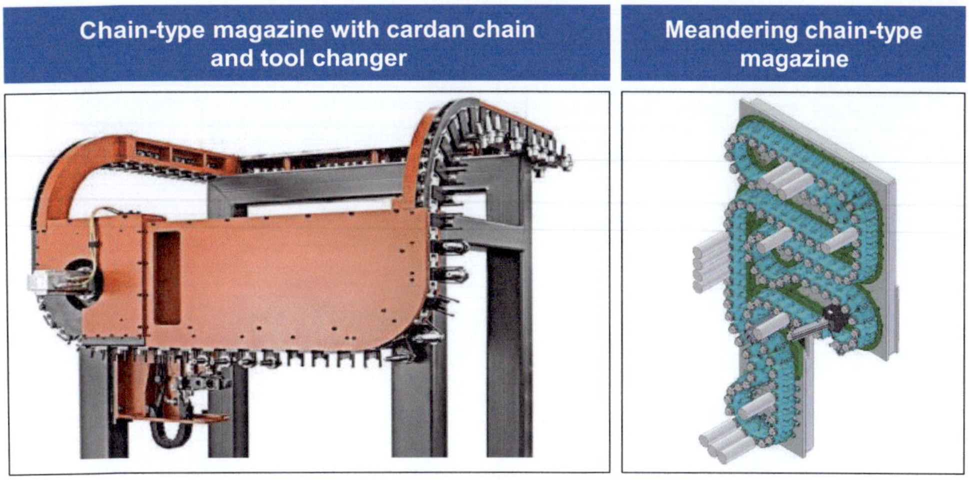

Fig. 10.67 Tool change with chain-type magazine. *Source* ktc Handhabungstechnik

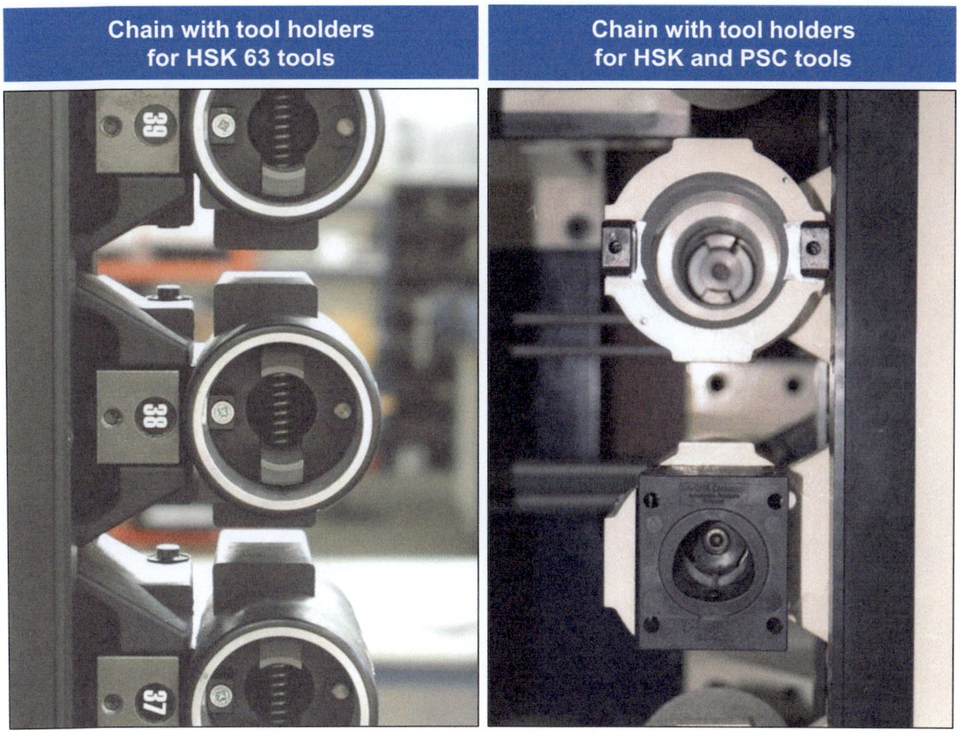

Fig. 10.68 Tool holders in a chain-type magazine. *Source* ktc Handhabungstechnik

Tools are usually locked in the magazine slots and are only released as required when at the tool change position. There is a variety of structural solutions here.

Applicable to all tool holders is the basic principle that, if the machine or provision equipment should fail, the tools must remain reliably clamped. Reliable clamping is normally guaranteed by springs and form-fit elements, while for declamping, either mechanical, hydraulic or pneumatic supplementary systems are necessary. Figure 10.68 provides a view into the tool holders of tool magazine chains for the HSK, steep taper and PSC tool holders.

A clamping mechanism for the HSK 63 is shown in Fig. 10.69. The spring tension ensures the collet chucks reliably hold the HSK shank at all times. The orientation of the HSK in the magazine is guaranteed by the slot nut. The tool can be pulled out of the holder using a gripper and be exchanged again. This is possible due to the unhindered access to the gripper groove of the HSK shank. An ejector pin is only in-

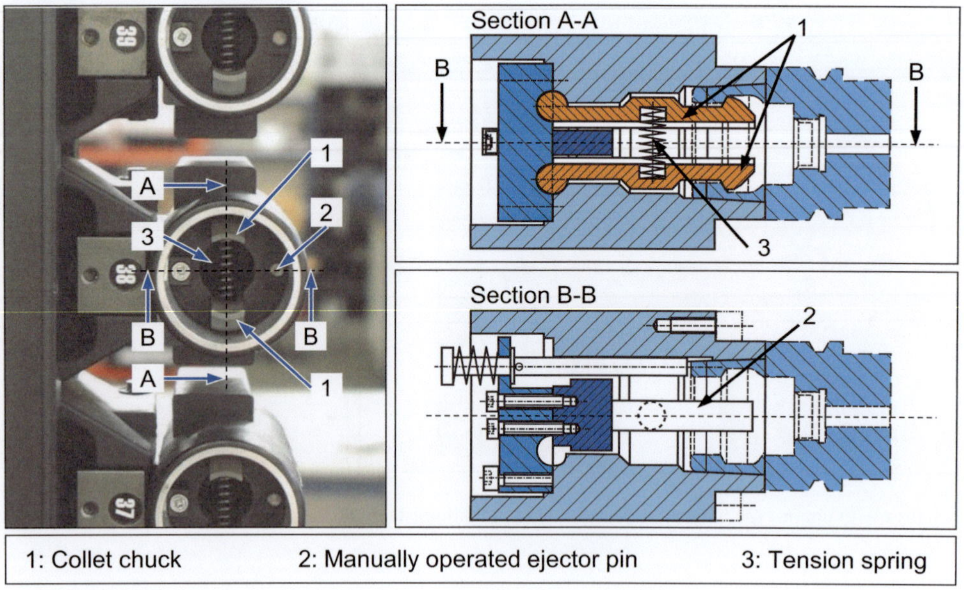

Fig. 10.69 HSK 63 tool holder in a chain-type magazine. *Source* ktc Handhabungstechnik

10

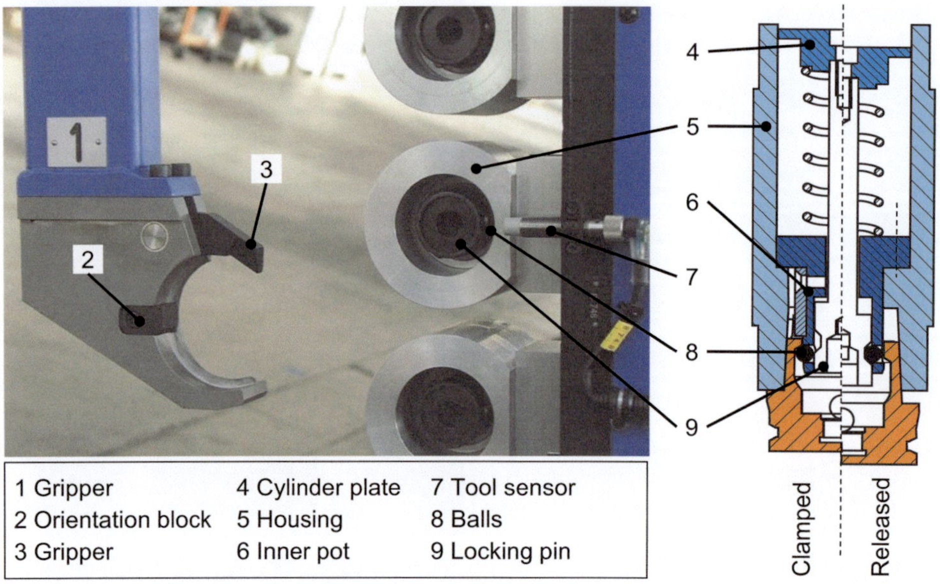

Fig. 10.70 Gripper and HSK 80 tool holder. *Source* ktc Handhabungstechnik

tegrated in the system (for ease of use) for manual removal of a tool from the holder, e.g., for loading the magazine or replacing worn tools.

The tool holder described clamps the tool with the springs. This only works up to a threshold load that cannot be very big due to removal and supply. For heavy tools, there is a risk of the tools working loose and falling out of the holder. Systems with locking mechanisms are better and more reliable (Fig. 10.70). In the example, the design has balls that move radially and that guarantee locked clamping on the inside of the HSK. The locking pin is retracted by the spring. This action presses the balls outward by the bevels of the pin. For a tool change, the gripper moves up to the tool and grips it in the HSK groove. The tool in the holder is released by a pneumatic cylinder that moves the cylinder plate, which moves the locking pin axially against the spring force (downward in the drawing). In this way, the balls are able to move into the groove of the locking pin when the HSK shank is removed, and this releases the tool. A constriction of the hole diameter at the outer edge of the inner pot ensures the balls do not fall out.

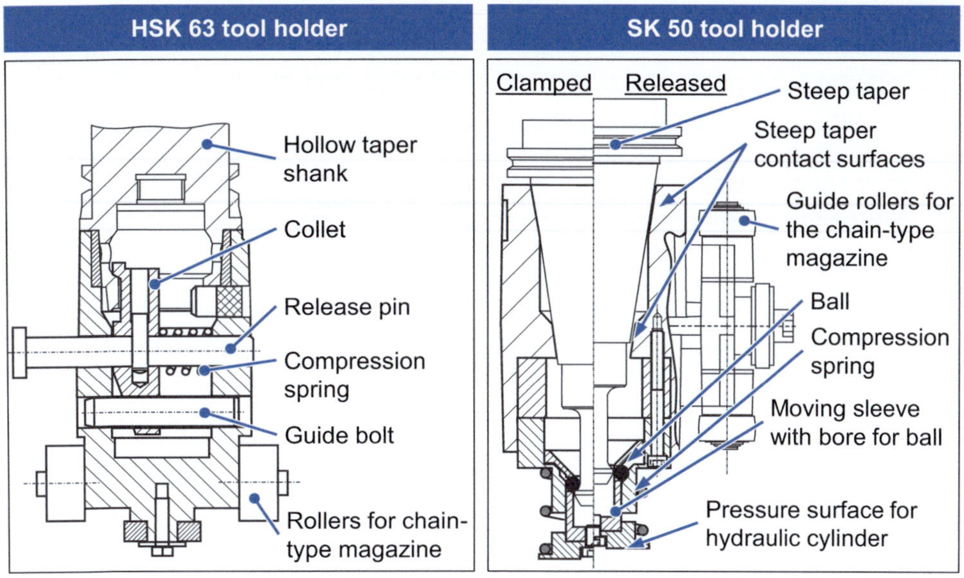

Fig. 10.71 HSK and SK tool holders. *Source* ktc Handhabungstechnik

In addition to the two systems described, there are a number of other ways to reliably store tools in chain-type magazines. Figure 10.71. The collet keeps the HSK shank reliably clamped. To unclamp, the release pin must be pressed against the spring force.

One principle for holding an SK 50 tool is shown on the right in Fig. 10.71. To manufacture the tool holder straightforwardly and cost-effectively, the taper is not continuous in the holder: The steep taper is only supported on two contact surfaces. The function of the clamping mechanism is similar to that of the hollow taper shank interface in Fig. 10.70. The balls grip from the outside into the groove of the tightening bolt. To release, the pressure cylinder must be actuated so that the balls can move radially outwards and release the shank.

If a tool is not retrieved from the magazine by a gripper but instead is held directly with the spindle using the pick-up method, simpler holding systems can be used (Fig. 10.72). The key difference is that, with the pick-up method, the gripper groove of the tool can be used for the magazining process. In the previous examples, it must be freely accessible for the gripper.

The majority of systems shown clamp the tool on the basis of the elastic retention force of the collet chucks (HSK 100, PSC 63, SK 40 and HSK 50). Sometimes specific actions need to be taken to guide the collets reliably over the longest shank diameter. For the PSC 63 collet at the front of the figure, rollers, for example, are used at the ends of the collet segments. Unlike these purely elastic systems, mounted collet segments that clamp the tool with a spring can also be used. This is the case, for example, for the holder of a PSC 63 tool (on the extreme right in Fig. 10.72) that has been specified to be as compact as possible.

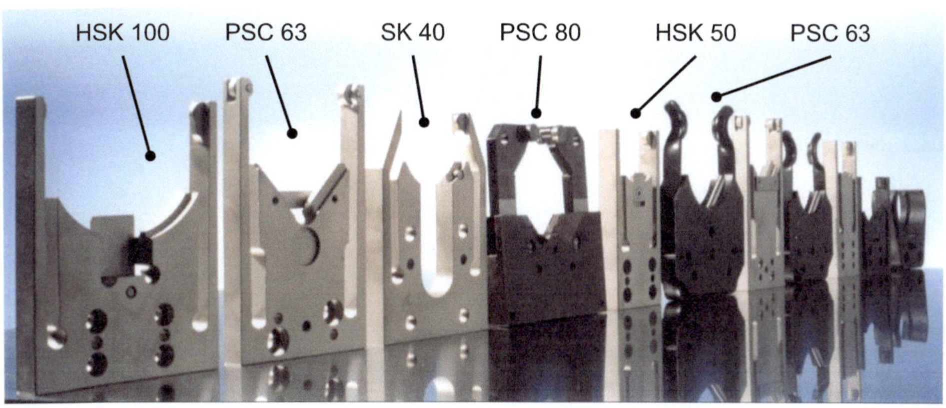

Fig. 10.72 Collets for different tools in the pick-up method. *Source* ktc Handhabungstechnik

10.3.1.5 Identification Systems for Tools

In automated manufacturing systems, the number of tools needed is sometimes very large meaning tool management requires suitable identification systems. Such systems are also required for the flexibly automated workpiece flow, for the coding of pallets, in particular. While sufficient space is available on pallets for code carriers, the sizes of code carriers for tools and tool modules must be minimized as much as possible.

Originally, coding screws or coding rings affixed to the tool shank or tool flange were mainly used for tool identification. Inductive switches can be used to check whether a bolt is present or not. The information storage options with these systems are limited, however. They also have large space requirements and cannot be used, for example, on cutting heads currently commonplace for lathe work. Similarly, adhesive bar code labels are only of limited use in automated systems due to the necessity for continuous updating. Optical sampling systems can only be used up to a point due to particle contamination in the machining area of the machine tools. However, on the plus side, these systems do not impact tool balancing and so are used in isolated cases. Figure 10.73 shows a tool with 2D matrix code on the tool flange.

The stable circuit designs, the use of RFID memory chips not entailing any wear at all and the high levels of interference immunity in inductive data transfer to and from the writing/reading head have proven themselves in practical industrial applications. In the EEPROM data carrier (a transponder) with a memory capacity of about 116 bytes [EUCH13], information is stored for over 20 years without a battery (Fig. 10.74, top right).

The writing/reading stations have different serial or bus interfaces. The 3964R protocol, for exam-

10

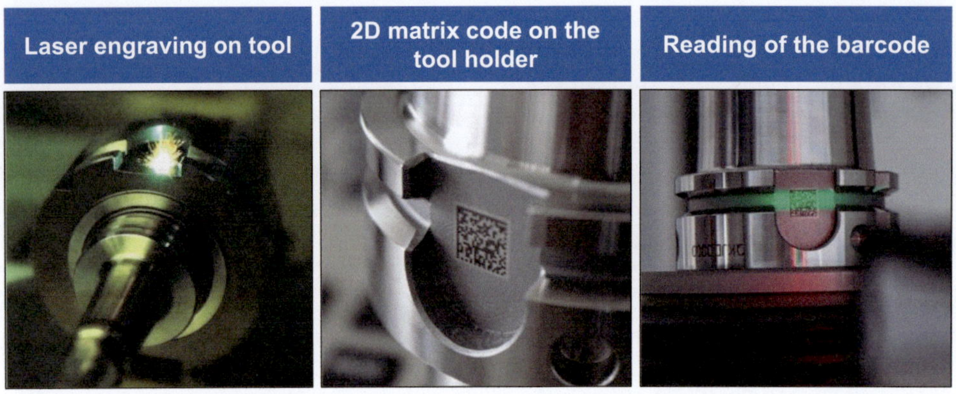

Fig. 10.73 Tool identification with 2D matrix code. *Source* MySolutions

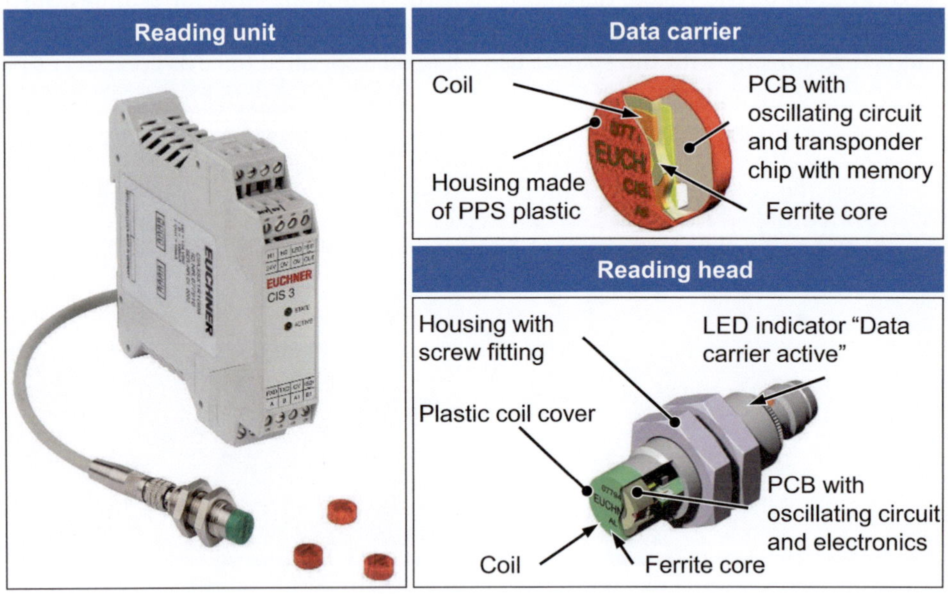

Fig. 10.74 Identification systems for tools. *Source* EUCHNER

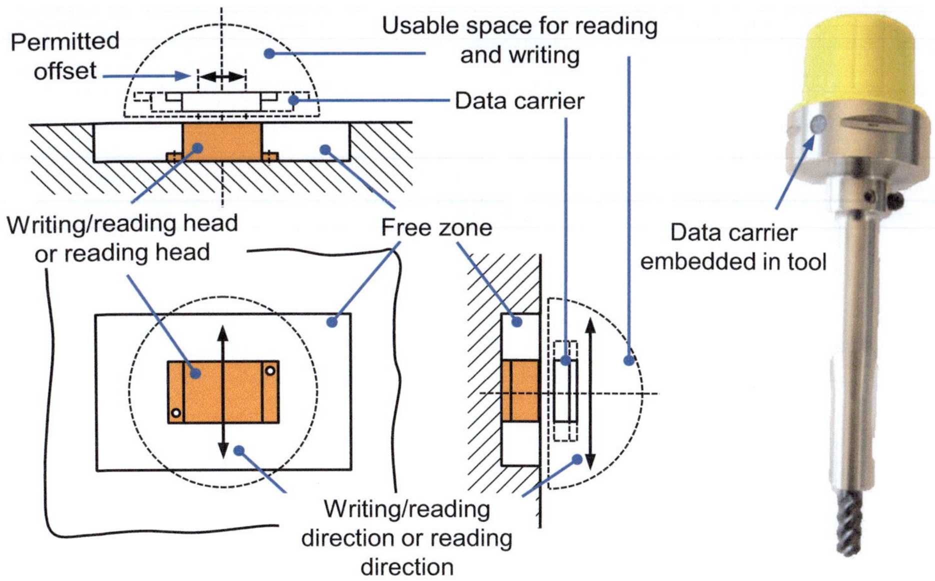

Fig. 10.75 Adaptation of identification systems on tools and depiction of the writing/reading process. *Sources* Balluff, SANDVIK COROMANT

ple, can be used for data transfer, enabling easy integration into the most popular PLCs. PC-based systems are used outside the system for data carrier programming. The data carriers can be very small in size (4 mm high with a 10 mm diameter) so they can be integrated relatively easily into tool holders (Fig. 10.75, right) [EUCH13]. The key factor for smooth data exchange between the writing/reading unit and the data carrier is the data carrier staying relatively close to the writing/reading head for a sufficiently long time (Fig. 10.75, left).

Data carriers having a rod antenna instead of the round antenna system shown in Fig. 10.75 are also suitable for dynamic read operations with relative speeds up to 25 mm/s. Note here however that the reading head and data carrier may only move in a certain direction relative to one another. The writing/reading process takes place as follows: The data carrier receives the energy signal of the transmitter, charges its supply voltage from this signal and sends its data in the direction of the writing/reading head. The head receives the data and exchanges information with the data carrier. The tool interface unit manages the reciprocal data transfer and acts as an interim buffer. It is the link between the higher-level system and the data carrier. The reliability of data transfer is guaranteed by check algorithms.

The high memory capacity of code carriers permits all data defining the tool (such as tool type, tool/cutting/correction dimensions, and reference and actual service lives), necessary for use in automated production, to be kept in the tool itself such that no information is lost if the master control computer fails.

10.3.2 Tool Systems for Lathes

In the production of turned parts, the geometric complexity, the required accuracy and the dimensions of the parts to make determine the machine type used and thus the accompanying tool system. Similarly, the degree of automation and flexibility in production play a role in the selection and design of the tool system. While on smaller manual workshop machines, the tools are joined to the tool carriage using simple tool holders, CNC lathes usually have a turret-type tool holder. Using the tool holders, the tools are either joined directly to the machine carriage or with corresponding adapters, such as cartridges. Nowadays, it is possible on many CNC lathes to also insert driven tools that can be used for additional milling or drilling operations. These tools are driven by an additional tool drive in the turret.

10.3.2.1 Tool Types in Lathes

Different types of tools are used for lathe work depending on the machine type and machining task. Cartridges with a rectangular or cylindrical shank are used in virtually all machine and tool carrier types. Tools with a quick-change tool interface can also be incorporated in most turrets of CNC machines. Driven tools for rotating applications can be inserted into special turrets with a tool drive. Figure 10.76 provides an overview of the different tool types.

In the classic basic form, a turning tool comprises a shank and a cutting part. Nowadays, cutting bodies are soldered to the shank only in rare cases. Rather, they are clamped in corresponding holders in the form of

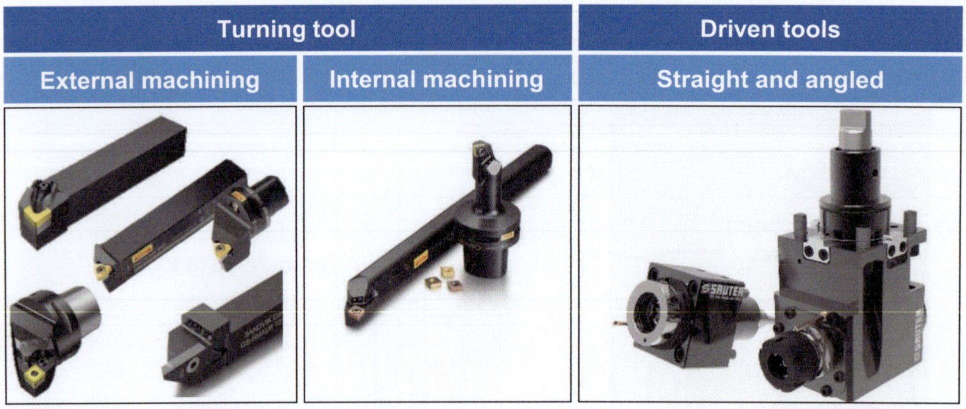

Fig. 10.76 Tool types in lathes. *Sources* SANDVIK COROMANT, Sauter Feinmechanik

indexable inserts. The standardized forms of cartridges with a rectangular shank are defined in DIN ISO 5610 [DIN15a]. External machining tools have quadratic or rectangular (a side ratio of 1:1.6) shank cross-sections. For internal machining, tools with a circular cross-section are used.

Indexable inserts made of carbide or oxide ceramics have become widely established. Worn cutting inserts can be used multiple times by fitting them rotated or indexed. Cutting inserts that have been exposed to high levels of wear or are damaged beyond repair are replaced. The function of a cartridge is to clamp the cutting insert at the same position all the time so that after the cutter is changed, no readjustment of the tool is necessary. Differentiation is made between cutting inserts with and without perforation and with and without chip groove, depending on the clamping system used. Lever, wedge, screw clamping and clamping finger systems are major structural variants of clamping systems.

Figure 10.77 shows the lever clamping of perforated cutting inserts. The clamping force is introduced by a clamping bolt that is easily accessible from the top and bottom of the cartridge. The head shape of the angle lever permits form-fit and force-fit transfer of the clamping force onto the indexable insert. The resultant clamping force has horizontal and vertical components. The force is aligned with the bisecting line of the two vertical support surfaces so that reliable clamping is guaranteed even when cutting is interrupted and feed directions change. At the same time, high positional accuracy can be attained regardless of the manufacturing tolerances of the indexable inserts.

The clamping mechanisms and indexable inserts are the same for the classic cartridges as for the turning tools with quick-change tool interface shown in Fig. 10.76.

In addition to conventional turning tools, i.e., stationary tools, driven tools are increasingly being used for drilling and milling operations on lathes. The drive power for the tools is delivered here via the turret. The complexity of the mechanics due to the relatively small space available for these tools is shown in Fig. 10.78 using a driven 90° angle head as an example.

The angle head has a cylindrical shank as the interface to the turret disk. The clamping teeth here are mounted on the two opposite sides. This universal shaft variant enables the spindle head to be clamped in the tool holder with an orientation offset by 180° so that different machining processes are possible with one spindle head. The alignment unit on the spindle head

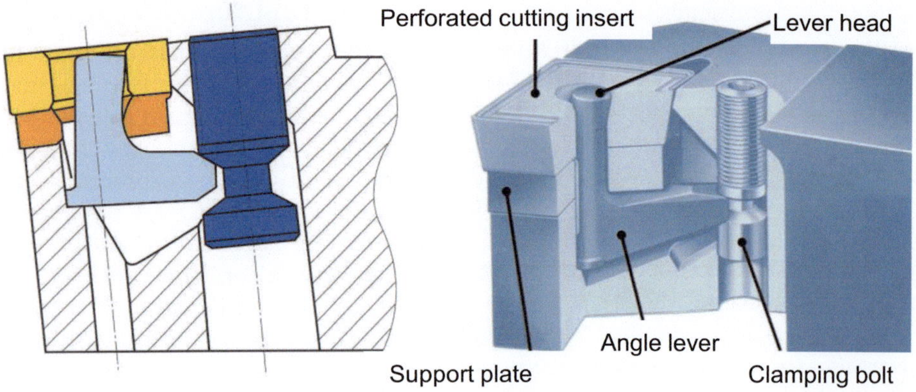

Fig. 10.77 Lever clamping of ISO perforated cutting inserts. *Sources* Walter Montanwerke, Hommel Hercules Werkzeughandel

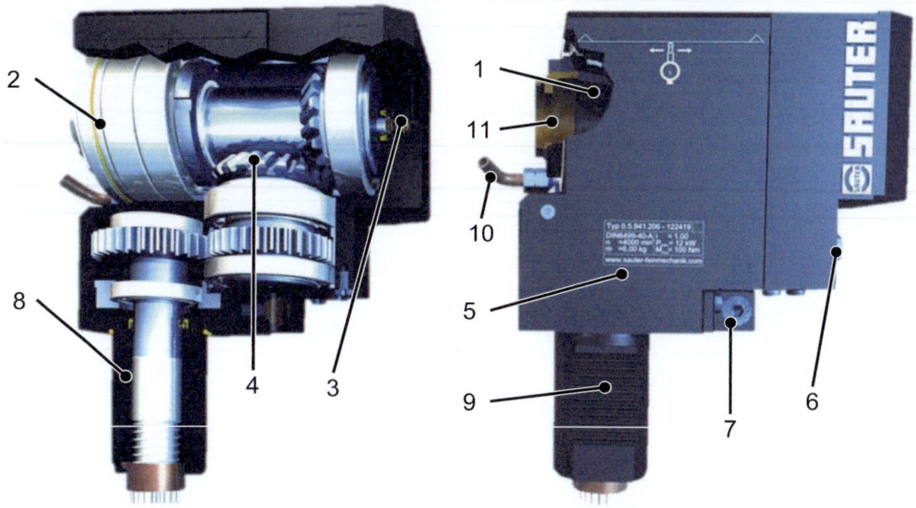

1: Tool interface, runout < 5 µm
2: Spindle bearing in O arrangement
3: Cooling lubricant supply (for internal cooling to 80 bar)
4: Cyclo-palloid bevel gears for good transfer of force
5: Nameplate with all technically relevant information
6: Separate, external cooling lubricant connector
7: Aligner for minimizing retooling time, improved handling, speedy alignment
8: Anti-twist lock
9: Universal shaft variant (geared on both sides)
10: External cooling lubricant supply
11: Internal clamping nut for highest stiffness and optimal leverage for the spindle bearing arrangement

Fig. 10.78 Driven 90° angle head with cylindrical shank. *Source* Sauter Feinmechanik

enables quick and straightforward alignment of the spindle head in the turret.

The geared coupling for the drive also has an axially movable sleeve so that a defined spindle orientation is maintained when the tool is not coupled. This reduces the coupling time for the drive because both partners in the coupling gearing are oriented identically and so no time-consuming search run is necessary.

The cooling lubricant can be supplied externally or through the tool. The spindle head can be fitted with different tool holders, such as collet chucks and an HSK or PSC interface. High stiffness and optimal leverage for the spindle bearing arrangement are attained with the internal clamping nut.

10.3.2.2 Turret Interface

The turret interface represents the link between the turret disk and the tool or tool holder. Different systems are used depending on the machine and application. Some of the systems have been standardized to satisfy the requirement for standardized tool holders. The tool interface systems for the cylindrical shank holder [DIN06] and prism holder [DIN87d] have been developed specially for lathe work (Fig. 10.79). The tool is clamped by a screw mechanism for these two interfaces. Another standardized system used to automatically or manually clamp tools with simple movements is HSK-T (turning) [DIN11b]. This is a modified design of the HSK already established in milling. The design is adapted to the requirements for turning. The PSC [DIN10e] and TS [DIN13a] systems, also known as standardized milling interfaces, can be used as turret interfaces in their original designs (see ▶ Sect. 10.3.1). There are also many manufacturer-specific solutions besides the standardized systems. In the bolt mounted turret (BMT) solution, the clamping force required is applied by a single bolt joint.

Tool Holders with Cylindrical Shank as per DIN 69880

The cylindrical shank [DIN06] (Fig. 10.79, top), standardized in multiple sizes, is preferred for use in smaller tools and machines. The tool shank is flattened on one side and has gear teeth. The tool holder is clamped by another geared clamping piece. This generates radial and axial clamping forces that contribute toward high clamping accuracy and high system stiffness. Exact positioning and high repeat accuracy are guaranteed by the design of the holder bore and geared clamping piece with minimum deviation from a circle. The system is normally used for horizontal lathes. The position of the turret in relation to the rotation axis, the machining type (inner or outer) and the direction

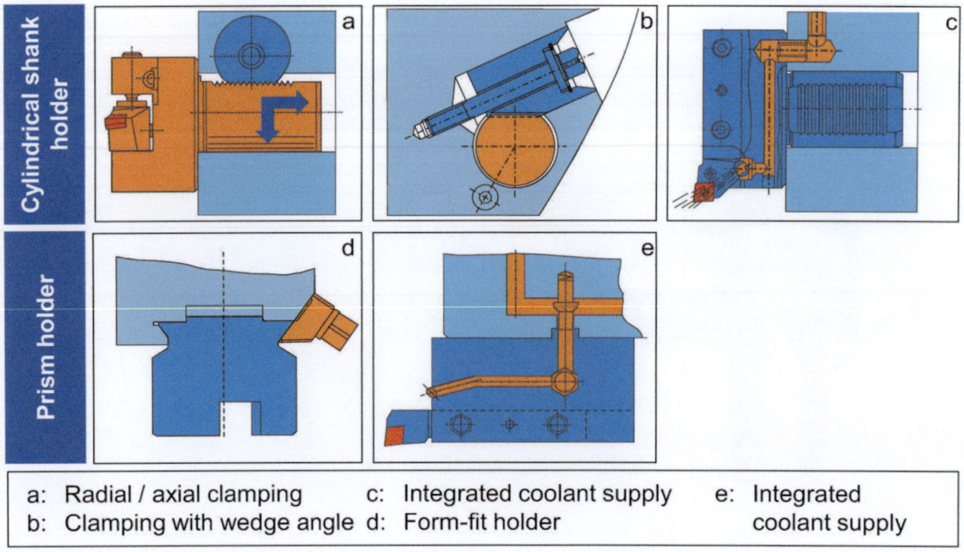

Fig. 10.79 Cylindrical shank and prism holders. *Sources* Heyligenstaedt, Hessapp

of spindle rotation determine the necessary design forms of turning tools and tool holders.

Prism Holder

The heavy prism holder tool system [DIN87d] (Fig. 10.79, bottom) is used for large vertical lathes. It is characterized by a dovetail prism. The classic form of the prism permits simple accommodating and clamping of the tool holder. In the longitudinal direction, the tool holder is fixed in place by a slot nut. The coolant is supplied through this nut.

Hollow Taper Shanks for Stationary Tools (HSK-T) to DIN 69863-7

The HSK-T [DIN11b] represents a modified version of the HSK-A familiar from milling (see ▶ Sect. 10.3.1.1). The major difference is the play at the driving slots. On the HSK-T, the dimensions and tolerances on tool and holder are changed such that there is smaller overall clearance. This is necessary because, in contrast to rotating applications for turning, it depends on precise angular positioning of the tool. Furthermore, the clamping force proposed in the standard has been increased to attain higher load capacity of the interface. Figure. 10.80 shows the system tolerance analysis. The tolerances between the driving keys of the HSK holders and the driving slots on the tools are shown on the left in the figure. The effects of the restricted tolerances of the HSK-T are on the right, showing the deviations of the tool cutter resulting from tool excursions for different projection lengths of the tools.

Bolt Mounted Turret (BMT)

The different manufacturer-specific bolt mounted turret systems (BMT) represent low-cost alternatives to standardized systems. On these, the joint between the

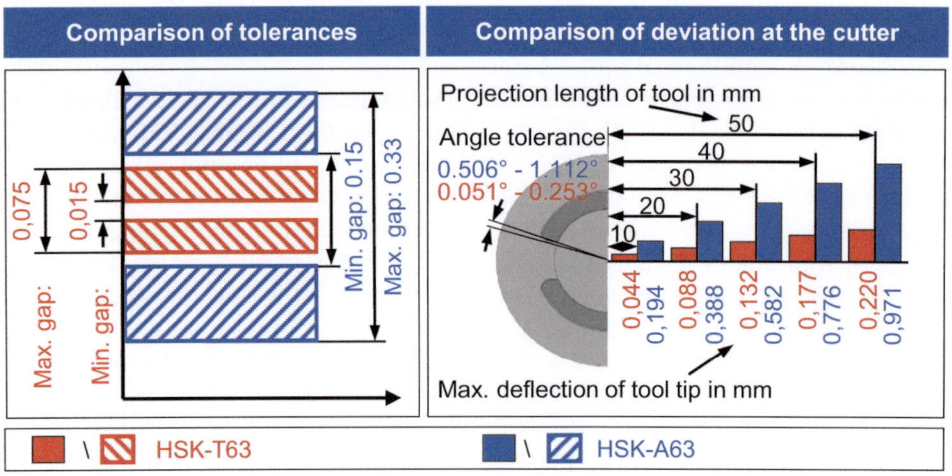

Fig. 10.80 Tolerances and resultant accuracy of the HSK-T

Fig. 10.81 BMT systems from different manufacturers. *Sources* ESA EPPINGER, INDEX, SANDVIK COROMANT

turret disk and tool is designed as a simple bolt joint. The alignment elements, different in each case and including pins and slot nuts, are used to position the tools. Figure 10.81 shows some design examples of BMT systems that have a quick-change tool interface for milling tools.

10.3.2.3 Tool Holders for Lathes

Selecting a certain tool holder for numerically controlled lathes means there is a direct impact on the flexibility of machine usage. Different tool holder designs result from the production tasks intended and the machine concept selected. The tool collision analysis plays a major part. This enables the important designs to be differentiated: linear, star, polygonal, drum, crown and disk-type turrets.

The turn mill center shown in Figs. 10.82 and 10.83 has one main and one counter spindle with the same design and performance. Two workpieces can be machined in parallel or one workpiece from all sides. For the latter, the workpiece is reclamped from one spindle to the other inside the machine, something that can be performed automatically using the traversable spindles. For finishing from all sides, the necessary tools are available in the star turrets. To create challenging tool contours, every multi-function unit also has driven tool spindles. The disk-type tool magazines loaded with hollow taper shank tools are positioned close to the workspace so that tool changes can be performed directly without a gripper and the fastest chip-to-chip times can be attained.

Shown in Fig. 10.83 is the parallel machining of the workpiece by turret 1 while the second multi-function unit is performing a tool change by swiveling the tool spindle unit.

Polygonal turrets are custom variants of star turrets that are used with a horizontal or vertical rotational axis. The polygonal turret shown in Fig. 10.84 is a 4-way turret head with prism holder. Two tool holders can be used on each side of the head. Indexing is by a hydraulically operated piston with Hirth spur gearing. Direction logic allows the position programmed to

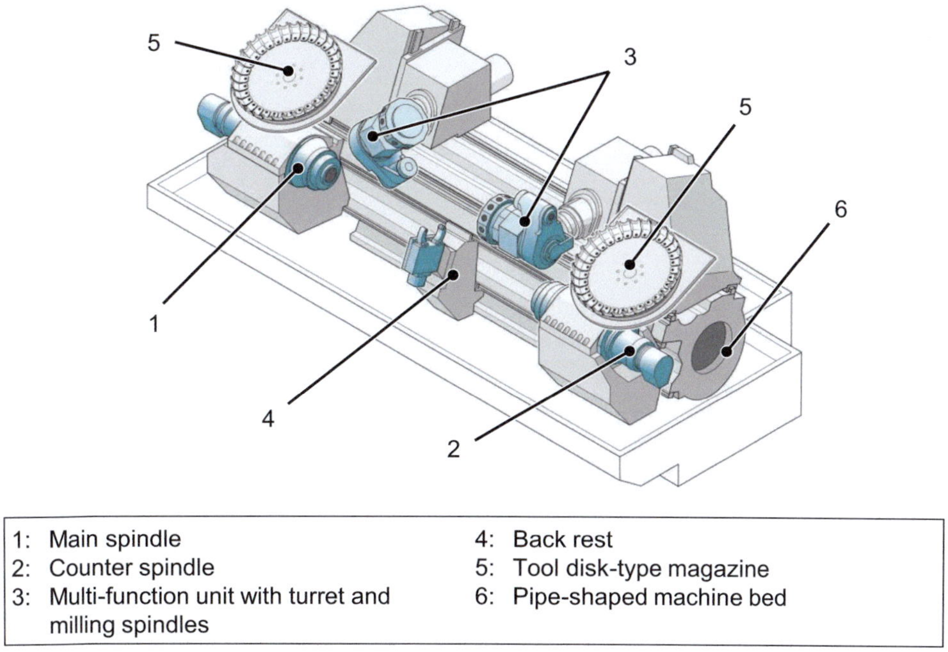

Fig. 10.82 Structure of a horizontal lathe. *Source* INDEX

1: Multi-function unit
2: Turret
3: Back rest
4: Counter spindle with workpiece
5: Driven tool spindle
6: Workpiece
7: Tool disk-type magazine

Fig. 10.83 Arrangement of two star turrets in a horizontal lathe. *Source* INDEX

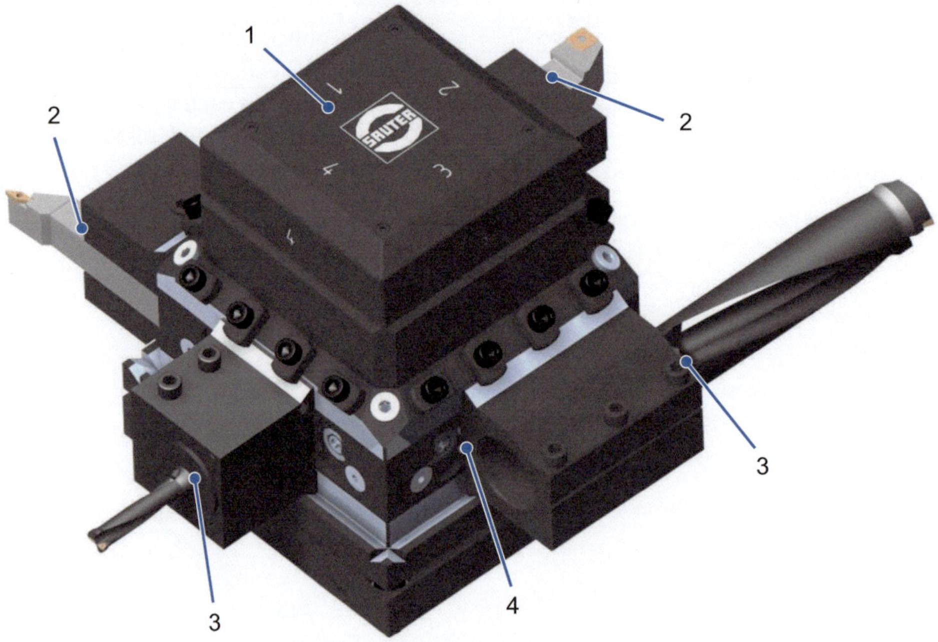

1: Main body
2: Tool for external machining
3: Tool for internal machining
4: Prism holder (DIN 69880)

Fig. 10.84 Four-way turret head with prism holder. *Source* Sauter Feinmechanik

be approached by the shortest path. The compact turret head is designed to accommodate high machining forces.

The number of tools necessary, the machine design and collision considerations can be crucial when selecting a crown-type turret. A variant with

12 slots for stationary and driven tools is shown in ▣ Fig. 10.85.

Disk-type turrets are characterized by high repeat positioning accuracy within a range of 1 to 2 arc seconds. The swivel time can be minimized with bidirectional rotation, meaning the swing angle is not greater than 180°. The diversity of variants is broad. There are disk-type turrets

- with or without tool drive
- for which all the tools are driven at the same time or only the tool in the working position
- for which the spindles of the driven tools can be actuated at their rotation angle (thread cutting)
- with two separate motors or only one motor for the tool and swivel drive.

While, for the first turret heads with multiple driven tools, all tools rotated simultaneously, in modern-day designs, only the active tool is driven, the others remain still. In this way, the heat influx, and the resultant thermo-elastic deformation of the turret head, are reduced to a minimum. Cylindrical shank holders compliant with DIN ISO 10889 [DIN06] have broadly established themselves as the interface between tool turret and tool holder.

Furthermore, there are disk-type turrets whose locking and control mechanisms are electromechanical, pneumatic or hydraulic. ▣ Figure 10.86 shows a disk-type turret with individually driven and differently aligned tools attached radially for large volume production. The tool holder of the turret enables tools for radial and axial machining to be used, with the capability of spindle positioning. Only one AC servo motor each is used for swiveling and the tool drive. The swivel drive can be locked either hydraulically or pneumatically.

If robustness and insensitivity to collisions are the priorities, a hydromechanical system with low oil volume and simple actuation with fast swiveling speed is used for the swivel drive and turret lock (▣ Fig. 10.87). An AC servo motor coupled to the turret is still used for the tool drive.

The shifting principle for such a hydraulically controlled disk-type turret with separate motors for swiveling and drive for the tools is shown in ▣ Fig. 10.88. The very compact hydraulic motor is controlled by the control valve, activated by the controller and by the shifting gate on the gear. This shifting gate ensures that the oil flow to the motor is stopped when the reference swivel position is reached.

This view shows the turret with the tool operating. The drive motor is connected either directly or via a toothed belt. When the tool holder disk needs to be turned, the tool coupling is moved inwards radially by hydraulics, so the tool is disconnected from the drive. At the same time, the lock of the tool holder disk (3-part Hirth serrations) is released by the control ring being moved to the right hydraulically. Sensors on the tool coupling and locking mechanism ensure that switching between operating modes is smooth.

As soon as the tool coupling and Hirth serration are fully released, the turret head is rotated to the next

1: Tool for external machining	2: Tool for internal machining	3: Driven tool

▣ **Fig. 10.85** Turret of a high-performance lathe, crown-type. *Source* INDEX

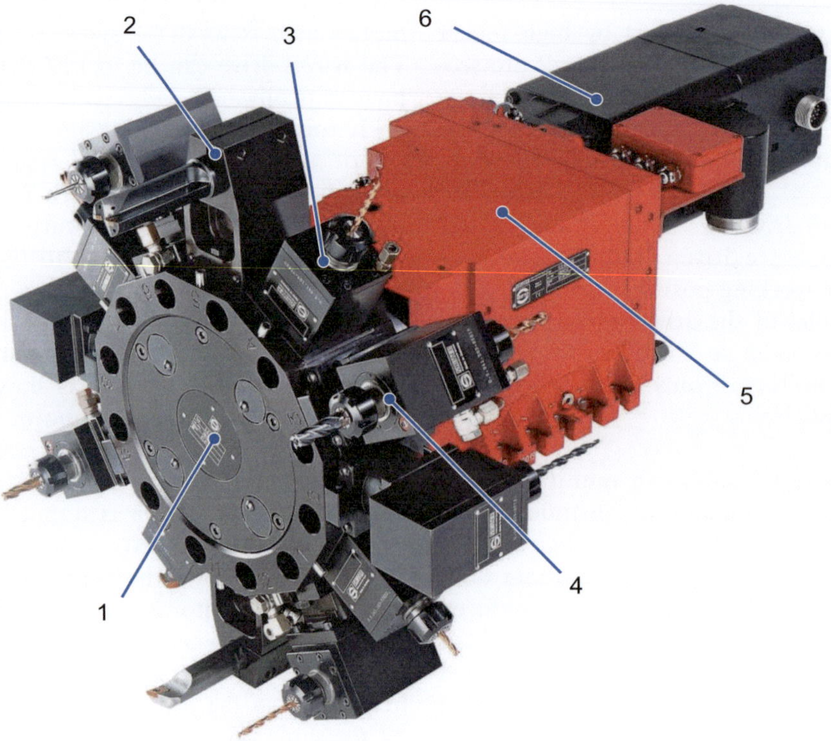

1: Turret disk	5: Turret housing
2: Fixed tool for turning	6: Drive motor
3: Spindle head for radial drilling	
4: Spindle head for axial machining (front and back)	

Fig. 10.86 View of a tool disk-type turret with AC servo motor. *Source* Sauter Feinmechanik

machining position. Twelve to 16 T-slot nuts spread around the circumference. These can slide into the T-slots when the torque load is too high and serve as overload protection in the event of a collision during the swivel process.

When the required swivel position is reached, the hydraulic motor stops and the next tool is joined to the drive by the tool coupling moving outward radially using hydraulics. The hydraulic supply lines shown inside the central shaft (there are a total of five lines, only two are shown in the section) are used to control the coupling process and to supply hydraulically operated tools such as grippers and steady rests.

10.3.2.4 Modular Tools

Modular tools have a second, quick change interface in addition to the turret interface. This has the benefit that the laborious alignment and clamping on the turret disk (different depending on the turret interface) are omitted for tool changes. Instead, only the cutting heads are replaced. Another benefit of these systems is that, depending on the requirement, just individual turret positions can be configured for quick change. The interfaces described in Sects. 10.3.1.1 and 10.3.2.2 (HSK-T, PSC and TS), and other manufacturer-specific systems, are used as quick-change tool interfaces.

The ABS system shown in Fig. 10.89 is an example of a modular tool system. It comprises a cylinder bore on the spindle side with orientation groove and clamping and taper bolts acting crosswise. On the tool side, there is the cylindrical holder with pendulum pin that moves crosswise. The structural axis offset of clamping and taper bolts relative to the pendulum pin means that when the clamping bolt is tightened, the cone surfaces come into contact and provide axial tensioning of the components. When cutting forces and torques occur, sufficient play between the positioning hole of the tool holder and the positioning pin on

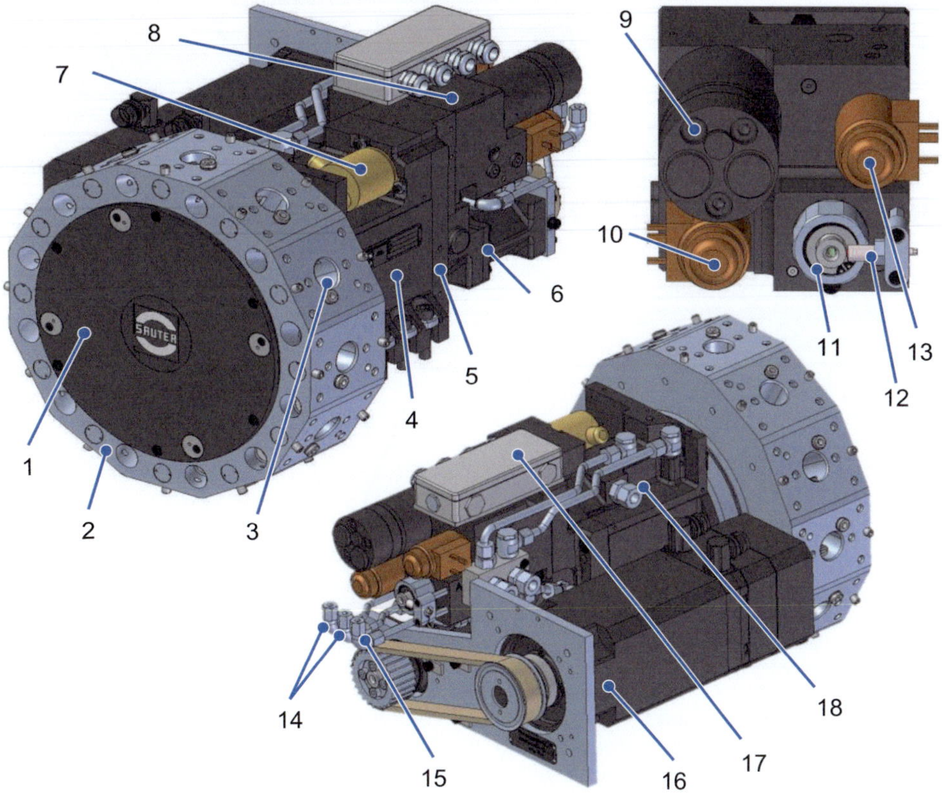

1: Cover plate	10: Hydraulic valve
2. Tool disk	11: Control valve
3: Tool holder	12: Proximity switch
4: Turret housing	13: Hydraulic valve
5: Bearing plate	14: Hydraulic connectors
6: Swivel drive	15: Connector for safety valve
7: Angle position encoder	16: Tool drive motor
8: Hydraulic swivel drive	17: Terminal box
9: Hydraulic motor	18: Connector for cooling lubricant valve

Fig. 10.87 View of a tool turret with a hydromechanical system. *Source* Sauter Feinmechanik

the tool side intentionally permits a minimal torsion. On tools requiring a precisely defined cutter position or transferring high torques, positioning and torque handling are by two driving keys on the holder or two driving slots on the tool. Precisely tailored sizing of the ABS system in conjunction with an extremely small fitting tolerance leads via minimal deformation to a self-centering 4-point system. This results in increased bending stiffness, higher change accuracy and beneficial damping behavior. The ABS system is also used as a direct holder in spindles.

The modular tool system shown in Fig. 10.90 is based on the PSC interface as the quick-change connection. It transfers torques without needing the usual gear teeth, wedges, balls or pins. The shanks of the tools are polygonal and chucked in the clamping unit with a high preload. The contact surfaces of the coupling (flange contact surface and cone) are precision-ground and ensure high repeat and change accuracies. The automatic clamping unit shown at the bottom of the figure is hydraulically operated. For the declamped state, pressure is applied to the front chamber and the rear chamber is depressurized. The tension bolt coupled via the pressure ring moves forward. The springs act via lever elements on the collet segments and cause them to butt against the tension bolt. This part of the system is similar to the inner clamping mechanism of the HSK. A tool can now be removed or inserted without collision. To chuck a tool, pressure is applied to the rear chamber and the front chamber is depressurized. This moves the hollow piston forwards and the wedge segments radially inwards. A tensile force directed backward is applied to the tension bolt. The abutting collet segments are pushed outward radially by the bevels on the tension bolt and clamp the tool.

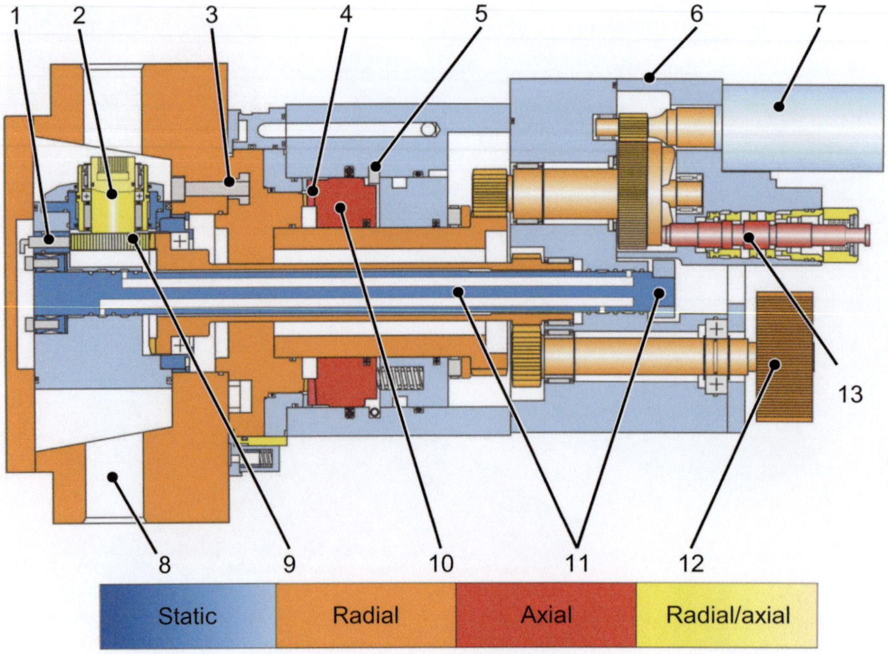

1: Position sensor	6: Swivel drive of tool disk
2: Gear head with tool head	7: Hydraulic motor
3: Twelve to 16 T-slot nuts spread around the circumference for overload protection in the event of a collision	8: Tool holder disk
	9: Crown gear for driving the gear head
	10: Control ring for turret fixing
4: Locking of tool holder disk with 3-part Hirth serration	11: Torque rod and supply for hydraulics
	12: Toothed pulley for tool drive
5: Sensor	13: Control valve

Fig. 10.88 Shifting principle of a tool disk-type turret. *Source* Sauter Feinmechanik

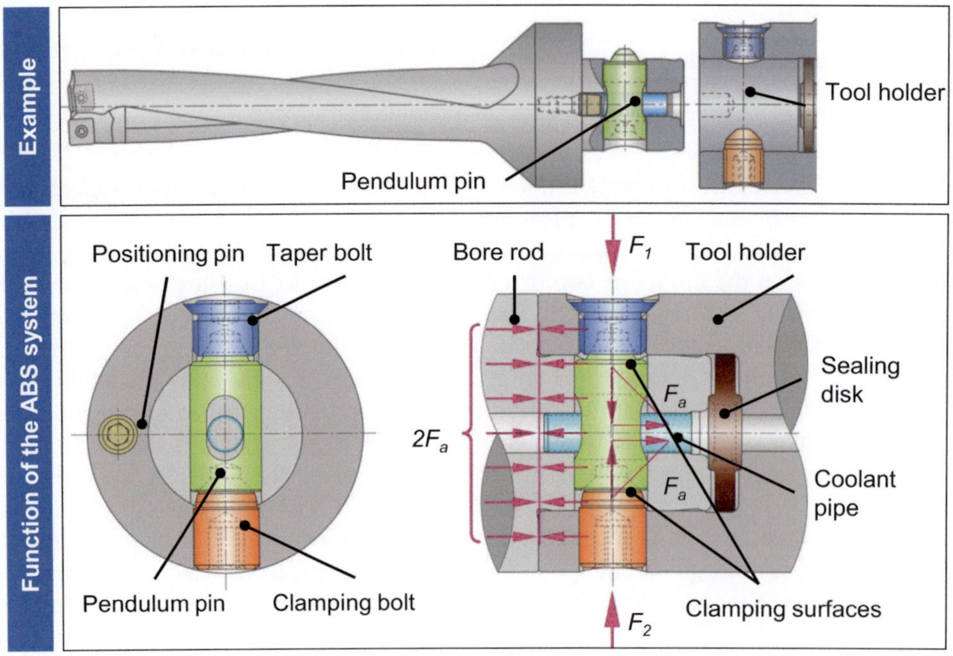

Fig. 10.89 Modular tool system with ABS clamping system. *Source* KOMET GROUP

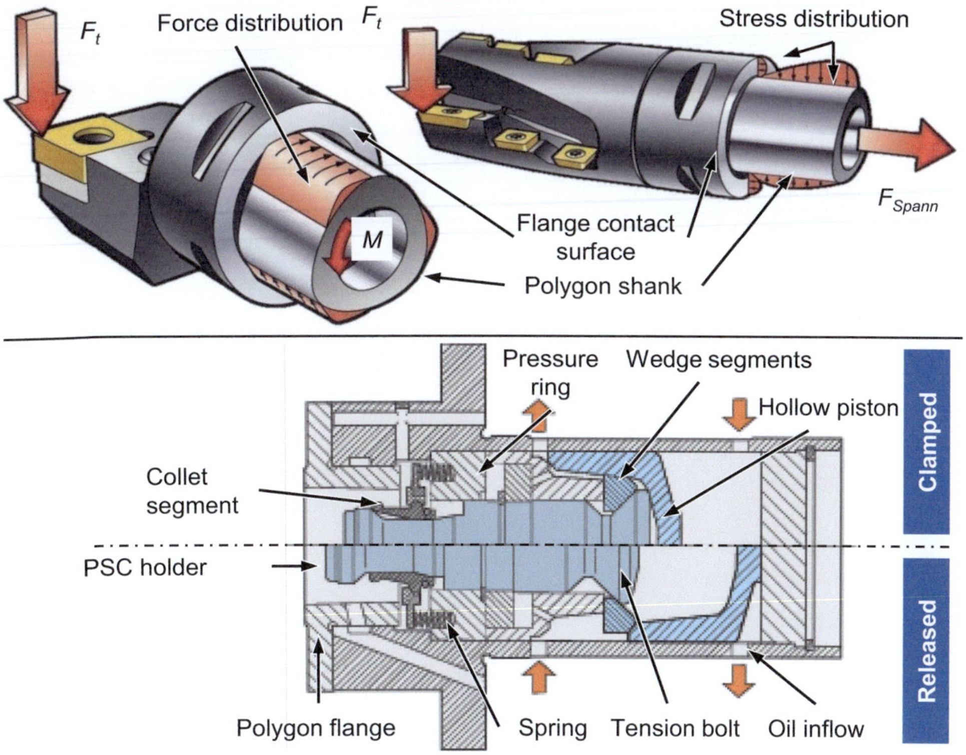

Fig. 10.90 Modular tool system with PSC interface. *Source* SANDVIK COROMAT

The combined tool interface (Fig. 10.91) merges a turning interface for heavy-duty machining and a milling interface in the Z-slide of a vertical lathe. The tension for large turning tools is shown on the left. If a milling tool is used, a cover protects the radial clamp from damage and dirt. The milling tool is clamped in the integrated milling spindle by the steep taper interface.

10.3.3 Mechanical Tool Interface Properties

The static behavior of tool interfaces when process forces are applied by the machining process has a strong influence on the machining result. The behavior under static loading represents an important criterion for assessing the tool interface properties regarding its performance and load capacity.

An important influencing factor on the behavior of tool interfaces is the retraction force used to chuck tools in the holder. In practice, force-multiplying clamping systems are commonly used. However, these do not permit direct measurement of the applied clamping force when it is clamped. Reference clamping systems without force multiplication are used for tests. Here, the clamping force can be determined with strain gauges on the tension rod used.

Depending on usage of the tool interface mainly the bending or torsion behavior of the interface is relevant. Figure 10.92 shows the characteristic bending and torsion curves that represent the behavior of the interfaces for radial and axial torque loads. These characteristics can be used to derive comparable parameters for different interfaces. The torsion curve shows the axial torque load starting from the origin to the torsion at the two end positions. The bending curve represents a load cycle with a radial load.

Bending

The parameters for the tool interface behavior under bending load are determined from the load range. Figure 10.93 shows a bending curve which is characteristic for tool interfaces with flange contact surface. In the first area (1), the bending curve is virtually linear, meaning bending stiffness k_b can be determined by the inverse of the gradient of an approximated straight line. In the first area, the bending torque is mainly taken up by the flange contact surface. At the end of linear area $M_{b,crit}$, the flange contact surface of the tool starts to lose contact with the spindle on one side. This means the tool interface stiffness drops in the subsequent transition area (2). Here the taper and clamping systems gradually start to absorb more bending torque until a new linear area arises (3). Here, the

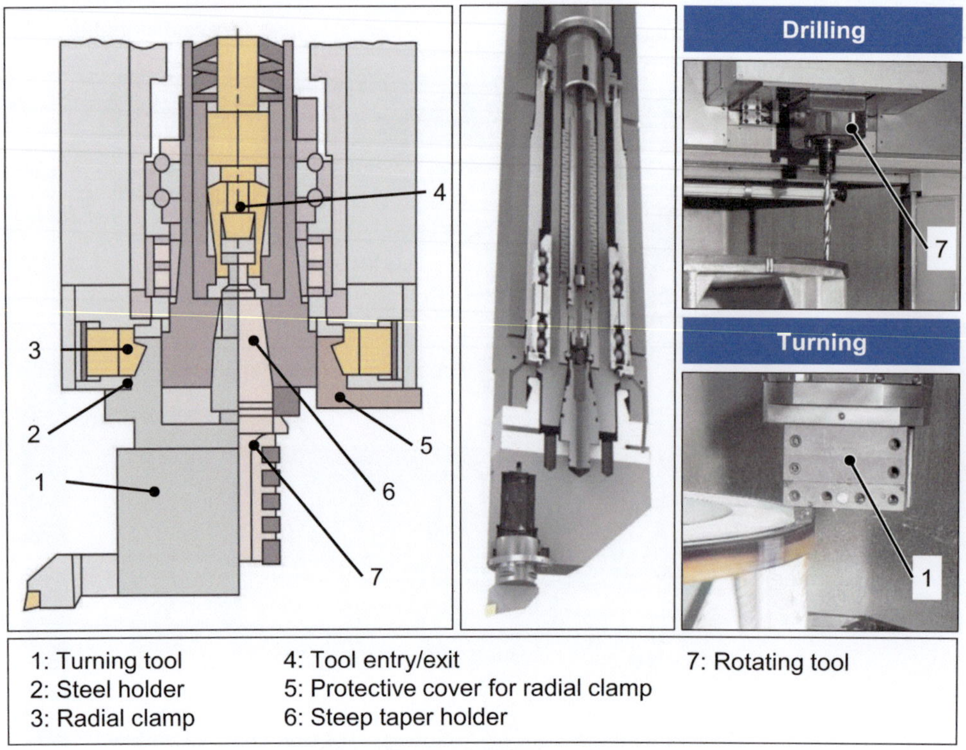

Fig. 10.91 Combined milling and turning interface. *Source* Starrag Technology (Scharmann division)

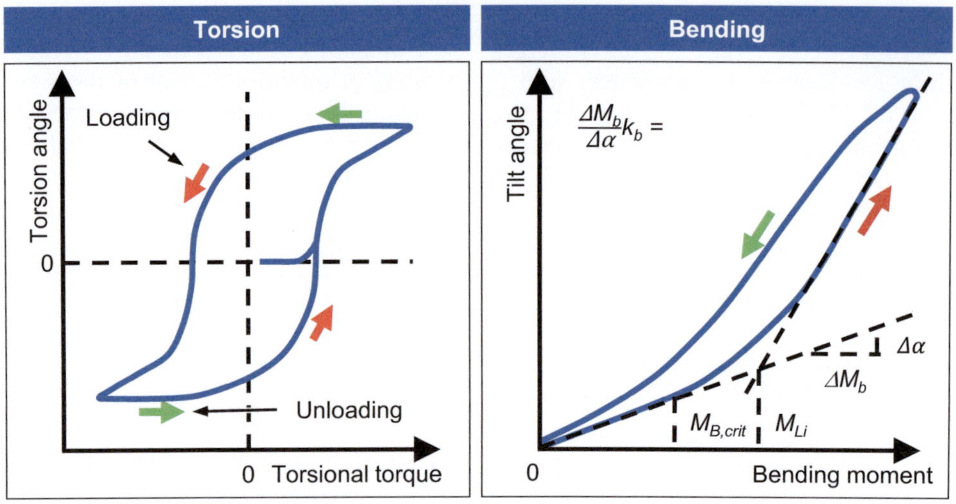

Fig. 10.92 Characteristic bending and torsion curves

clamping system and taper joint determine the stiffness. To achieve high tool interface stiffnesses, tool interfaces below $M_{b,crit}$ should be used [BREC15].

Lifting torque M_{Li} is determined by the intersection point of the straight lines that approximate the two linear areas of the bending curve in the first and third areas. Since the third area is strongly influenced by the clamping system and taper, this also applies to M_{Li}. The actual lifting point is independent of this and can be determined by approximation $M_{B,crit} \approx 0.7 * M_{Li}$ from the lifting torque [BREC15].

Figure 10.94 shows the load components of the bending curves of the HSK-A 63, PSC 63 and SK 40 tool interfaces. The top of the figure depicts the bending curves for the lowest retraction force in each case in accordance with the associated standard. The lower part has the same retraction force of 30 kN. Worthy of mention here is that the SK, because of the lack of a flange contact surface, has no kink in the bending curve and the lifting torque cannot be determined.

The bending stiffness before lifting of the flange contact surface is independent of the used clamping

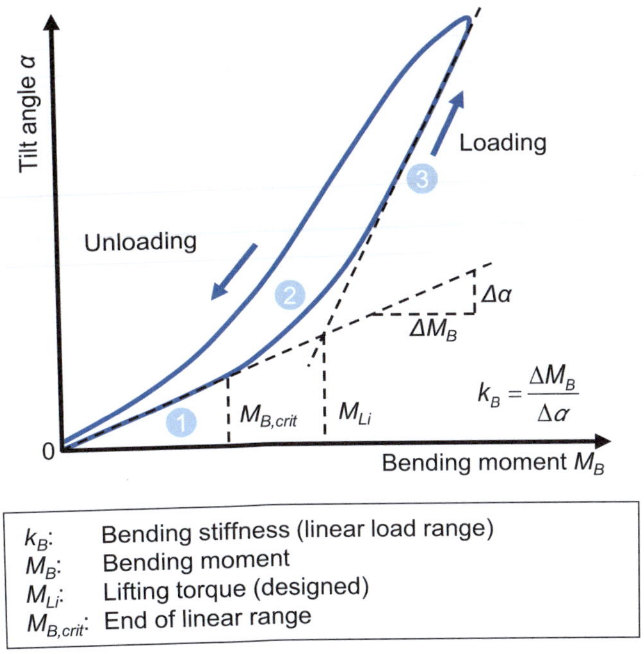

Fig. 10.93 Bending curve of an interface with a flange contact surface, according to [SCHU94; LEMB93]

system and can be only be influenced to a limited extent by the clamping force. Figure 10.95 shows that, from a certain clamping force, an increase in stiffness can no longer be attained, but the lifting point of the flange contact surface moves. This behavior is shown clearly in Figs. 10.96 and 10.97 for the HSK-A and PSC flange face contact interfaces.

Figure 10.98 shows the lifting torques of different HSK sizes depending on the clamping forces applied.

The lifting torques are generally greater than the torques occurring during operation. Tools with long overhangs and high cutting loads represent the only exceptions.

Torsion

The characteristic torsion behavior of a tool interface can be represented in the torsion curve, that shows relative torsion angle α_T between the tool and holder when torsional torque M_T is applied. The torsion curves for various tool interfaces differ greatly in part on the basis of different design features. Thus, a comparison using generally applicable parameters is not viable [BREC13a].

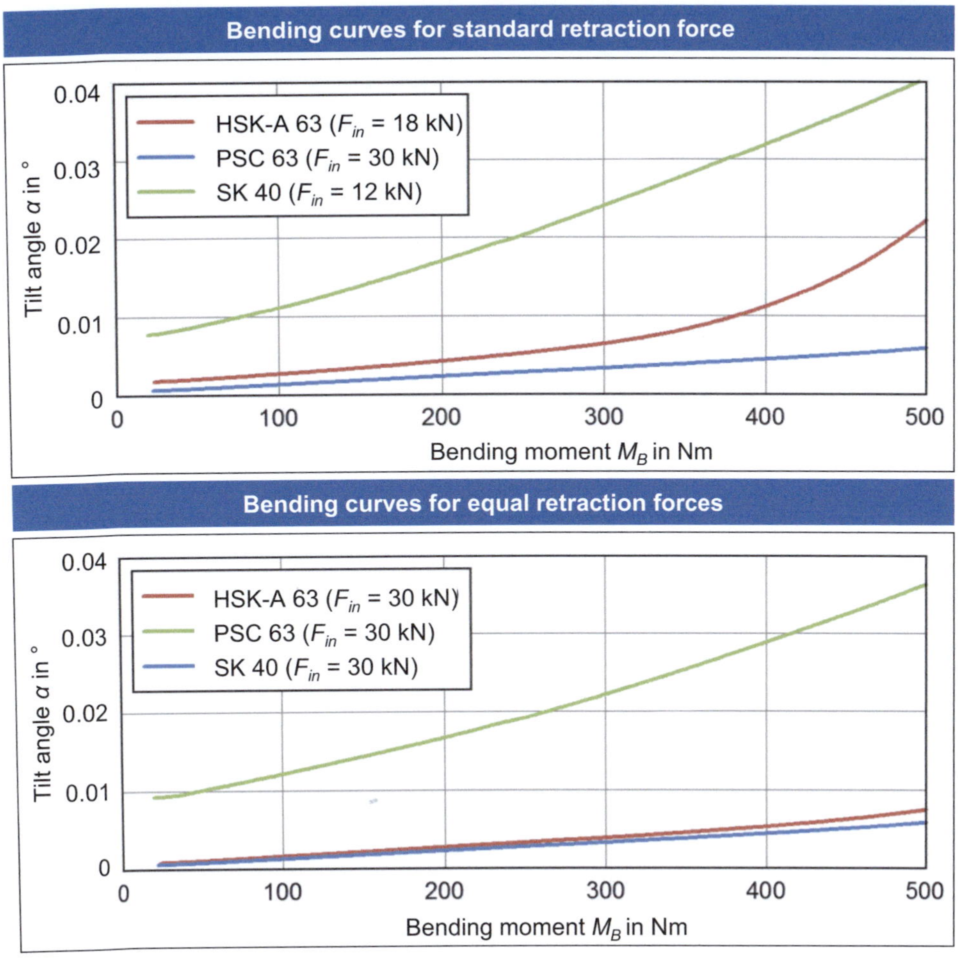

Fig. 10.94 Bending curves of standardized tool interfaces, according to [BREC15]

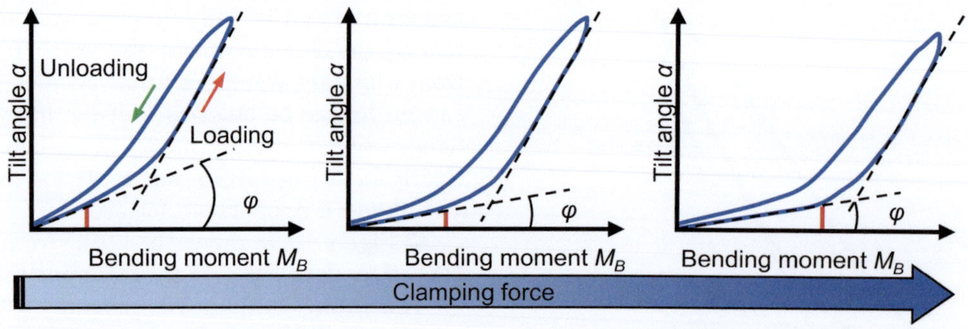

Fig. 10.95 Bending curve of a tool with a flange contact surface with an increase in clamping force

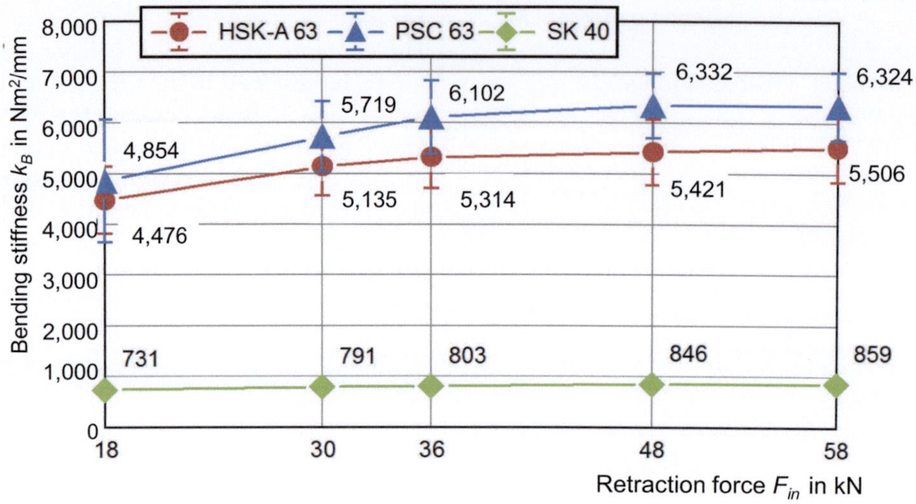

Fig. 10.96 Bending stiffnesses of standardized tool interfaces, according to [BREC15]

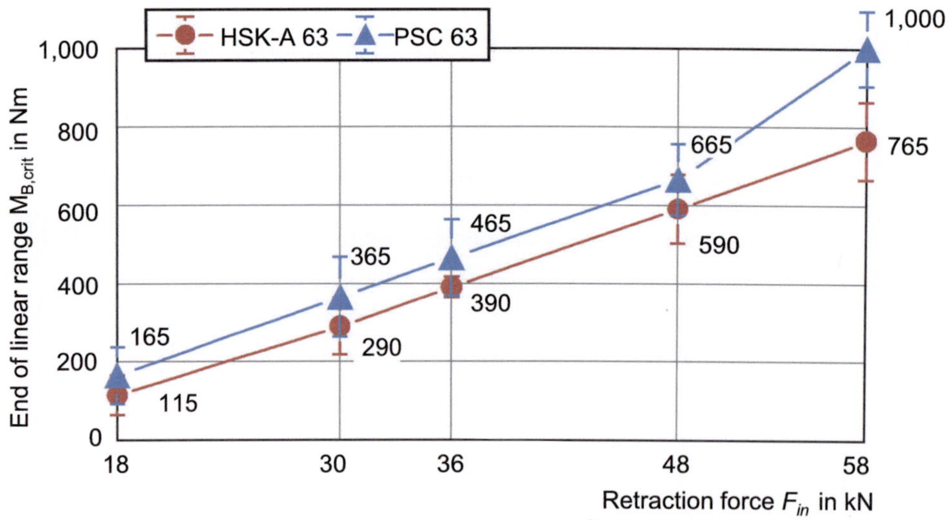

Fig. 10.97 End of the linear area $M_{B,crit}$ for flange face contact interfaces, according to [BREC16]

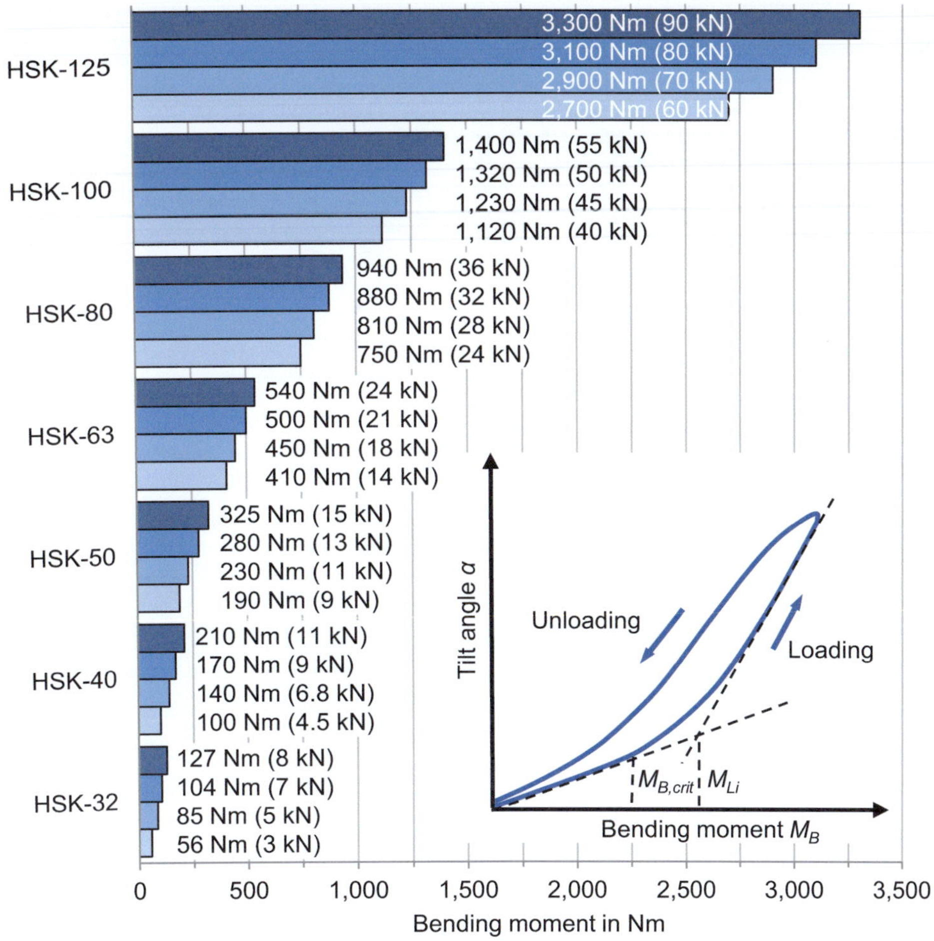

Fig. 10.98 Clamping force as a function of the lifting torque

Some tool interfaces exhibit pronounced torsional play on torsion loading. This play, for the HSK for example, can be attributed to the play between the driving keys and slots. The torsion curve for such a system is shown on the left in Fig. 10.99. When loading starts (1), the moment is transferred frictionally. There is negligible torsion due to the tangential-elastic behavior in the gap. When breakaway torque M_B is exceeded, the transition to sliding friction starts. When the moment M_R that can be transferred frictionally is reached, torsion occurs without a further load increase (2) until contact is established at the driver elements (3). Once contact has been established in full, linear load-deformation behavior is reached (4). The inverse of the gradient of the characteristic in this area results in the torsional stiffness k_T. For systems that exhibit no torsional play, the form fit exists when loading starts and the driver elements are not applied, meaning the linear load range is reached immediately [BREC16].

If the torque is reduced once the maximum torsional torque is reached, there is an elastic reversion of the torsion angle. When a sufficiently high torque in the opposite direction is reached, the tool in the holder starts to slide until it makes contact with the opposite driving pins [BREC16].

The torsion angle under load is dependent upon torsion play S, frictionally transferrable moment M_R and torsional stiffness k_T, as well as upon the curve profiles in the transition areas. A definition of torsion behavior with individual parameters only provides limited information [BREC16] (Figs. 10.100 and 10.101).

Fatigue Strength

For the torsional strength of the HSK interface, the failure moment with a static torsion load and the fatigue strength with a pulsating torsion load were determined. As the material used exerts a significant influence on the limiting load carrying capacity, three classes of materials used in practice were investigated as examples:

- 16MnCr5 (1.7131), 56 HRC
- 34CrNiMo6 (1.6582), 52 HRC
- X38CrMoV51 (1.2343), 52 HRC

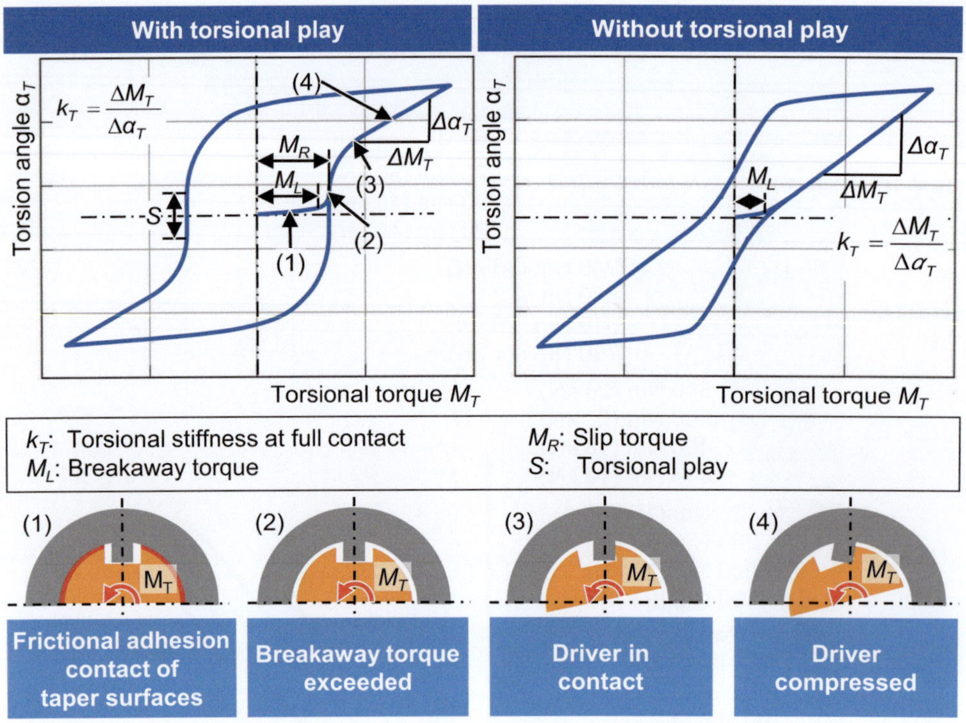

■ **Fig. 10.99** Torsion curves, according to [BREC16]

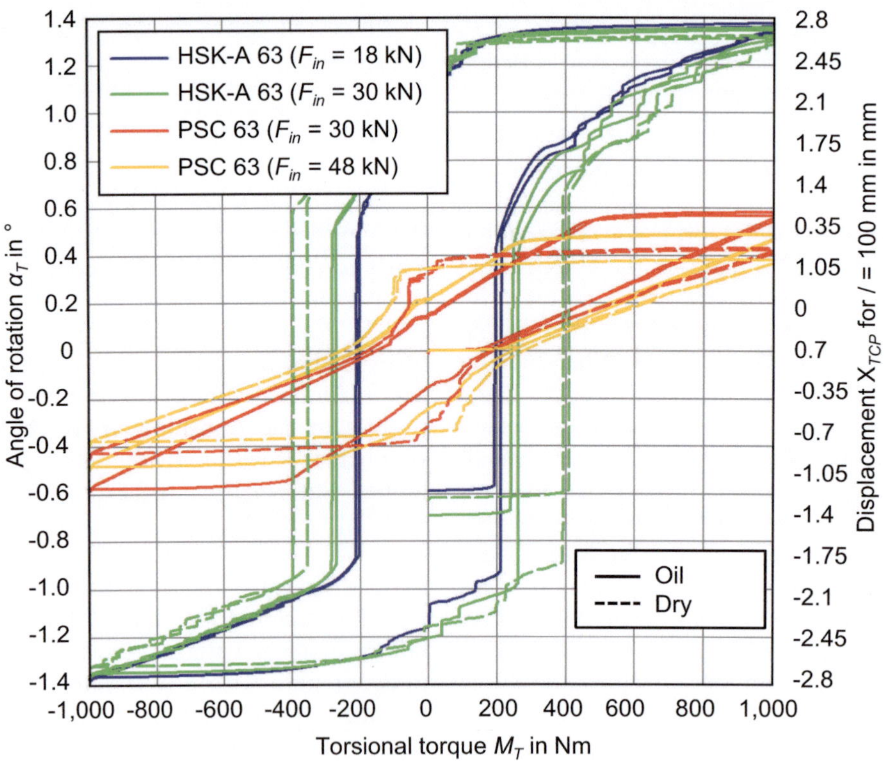

■ **Fig. 10.100** Influence of retraction force and friction on torsion behavior, according to [BREC16]

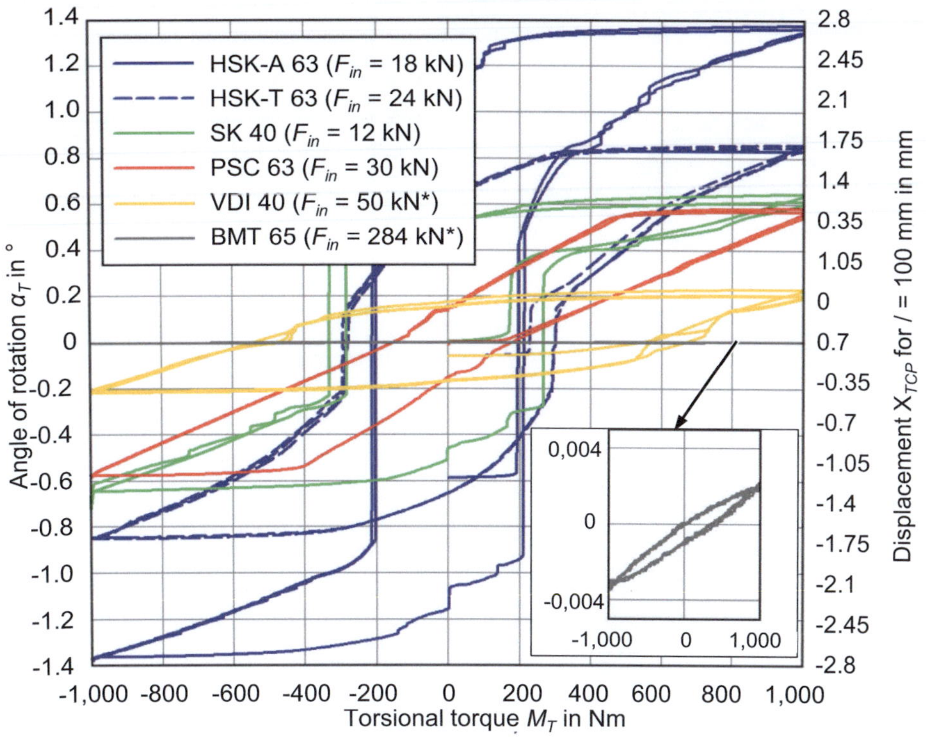

- Retraction force determined mathematically for M_{Re} = 50 Nm (VDI) / 120 Nm (BMT) and μ = 0.1

Fig. 10.101 Torsion curves of different interfaces for nominal clamping force, according to [BREC16]

The parameters specified correspond in each case to the first appearance of cracks in the tool holders tested. The typical failure pattern is a fracture in the loaded driving slot base. For statistical validation, all tests were performed several times on one batch with the same test tools. Clear differences in the limiting load carrying capacities can occur for nominally identical materials due to differences in the material microstructures, the metallurgical compositions and the hardening methods applied. The tests performed showed deviations of over 30%. The parameters given here can thus only serve as reference values.

The table in Fig. 10.102 provides an overview of the static torsion failure moments in relation to the HSK nominal size and the material. The case-hardened material 1.7131 was not investigated for nominal sizes less than 63 because its low side thicknesses in the shank area significantly increase the risk of through-hardening, which reduces the carrying capacity. For this reason, HSK nominal sizes less than 63 are not normally made of case-hardened steel. For interfaces made of the highly ductile tempering material 1.2343, no fracture could be detected. Instead, pronounced plastic deformation of the groove side surfaces occurred under high load. Since this deformation did not negatively impact the task of the groove, namely torque transfer, no failure threshold could be ascertained for this material. The values specified in the table thus represent the highest loads tested.

Static torsion tests do not fully characterize the load carrying capacity of the interface because, in the actual cutting process, an additional dynamic load occurs. The threshold of this additional load can only be determined with fatigue strength testing. On the basis of an actual milling process, a dynamic torsion load was superimposed on a static preload. The ratio of the dynamic to static component is 0.4. For fatigue strength, a minimum of 4×10^6 completed, non-destructive load cycles were assumed. The data shown in Fig. 10.102 provide the continuously applied mean dynamic moment versus the HSK nominal size and material. For interfaces made of material 1.2343, no failure could be verified (similar to the static failure moment). Plastic deformations of the groove side surfaces of the HSK appeared here, too.

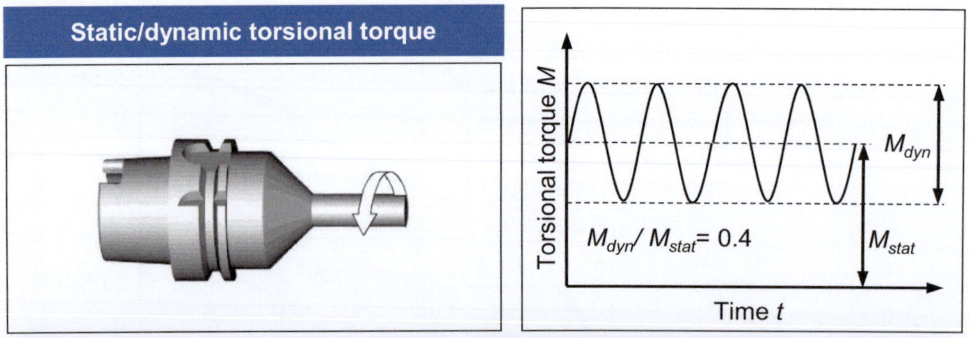

Size	Static torsional torque [Nm]			Average torsional torque [Nm]		
	1.7131	1.6582	1.2343	1.7131	1.6582	1.2343
32	—	400	> 600	—	350	> 500
40	—	880	> 880	—	800	> 800
50	—	2,270	> 3,000	—	1,400	> 1,800
63	2,600	3,500	> 3,500	1,600	2,000	> 2,500
80	4,400	7,000	> 10,000	3,300	4,000	> 5,200
100	9,300	14,200	> 18,000	6,000	7,500	> 10,000
125	15,000	—	—	> 11,500	12,900	—

Fig. 10.102 Static and dynamic torsional strength of the HSK

Threshold Rotational Speeds

The maximum possible rotational speed of the HSK is determined by the loss of the guide system in the front taper part and by reaching the fracture stress at the back clamping position. A number of factors have an influence on the threshold rotational speed. The length of the carrying holder taper, the radial oversize between the shank and spindle holder and the clamping system used thus have a key influence. So, for applications with high rotational speeds, determination of the threshold rotational speed is necessary on a case-by-case basis. Here, two basic criteria need to be checked:

- The stress maxima under the influence of rotational speed generally form in the area of the transition radius between the clamping bevel and internal shank diameter and in the base of the deep driving slot (Fig. 10.103).
- The maximum material-dependent reference stresses must not be exceeded.
- The precise radial guide of the tool in the front shank area near the flange contact surface must be ensured. No radial play is permitted in this area.

In general, a supporting holder over the entire taper length is preferred (Fig. 10.103, left). This prevents edge loading at high rotational speeds and necking in the taper center that lowers the amount of contact area. The tension increases in the shank, resulting from the unhindered expansion at the shank end in case of a widened taper of the spindles in the back area, significantly reduce the maximum rotational speed (Fig. 10.103, right).

In Fig. 10.104, the threshold rotational speeds of the HSK-63 with forms A and E were determined using finite element calculations. How the shank and holder fit give different maximum rotational speeds with different failure causes. It is important to note that much higher threshold rotational speeds can be reached in practice than those determined in a simulation.

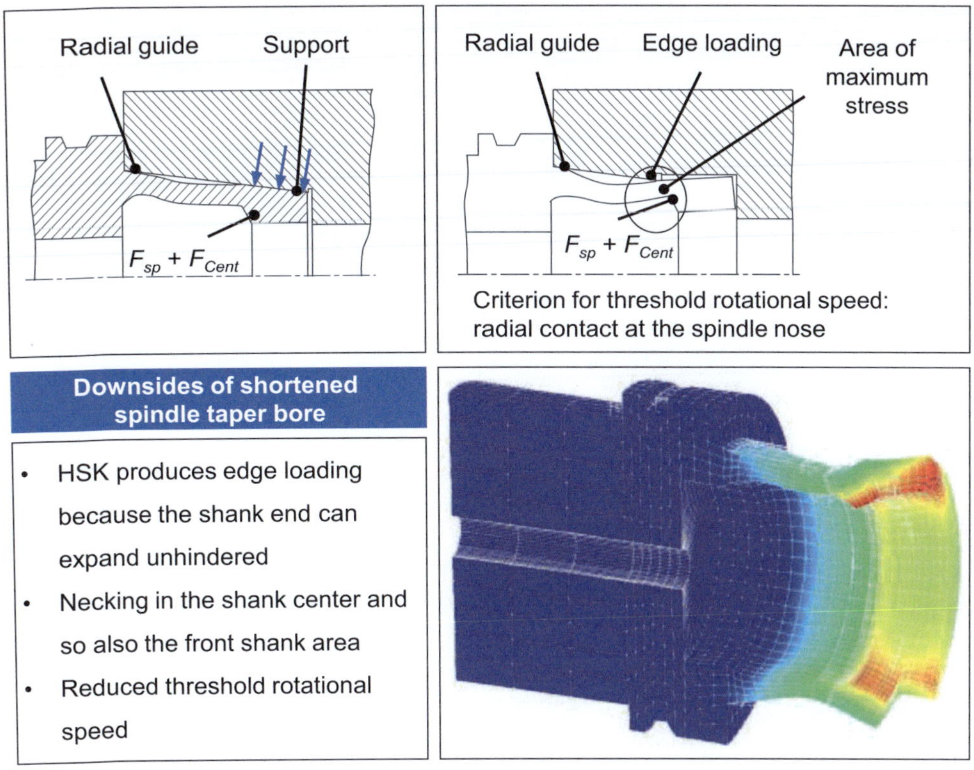

Fig. 10.103 Influence of the radial support on the threshold rotational speed

10.4 Workpiece Clamping and Supply Systems

Workpiece clamping and supply systems play an important role in automated production. The function of a workpiece clamp is to hold machined workpieces reliably and precisely in the machine. The tooling overhead and the forces acting upon the workpieces must be kept as low as possible. High levels of flexibility and adaptability to the most diverse components are, or a high degree of specialization is, required depending on the manufacturing task. The necessary set-up times must be kept as short as possible regarding the machining time of the machine tool. Supply systems are automated to provide workpieces for machining. Thus manual supply is omitted.

10.4.1 Functions of Workpiece Clamping and Supply Systems

In order to clamp a workpiece in a fixture or a tool in a machine tool with precision, the position of the element to clamp must be determined exactly. Thereby, body surfaces or center planes are the reference planes to be fixed. In the first case, the term "defining" is used, and, in the second, "centering". Defining and centering removes degrees of freedom from the workpiece or tool. The number and arrangement of centering and defining planes implemented in a fixture are specified by the workpiece geometry and machining task. It is not necessary for every machining task to define the position of the workpiece in all six possible degrees of freedom of a body in the space before the workpiece is fixed in place.

The relationship between the number of remaining degrees of freedom and the number of defining planes is specified at the top of Fig. 10.105. If machining is as shown in the bottom left of Fig. 10.105 for example, i.e., only one surface of the workpiece is machined (dimension *a*), the position (*x, z*) and orientation (φ_{xz}) of the workpiece have no significance on the defining plane. If the requirement, however, is to mill a continuous groove with clearances *a* and *b*, two defining planes are required. The position in the x-direction need not be defined precisely. For the milling of a groove with three clearances (*a, b, c*), three defining planes must be specified in the fixture. This fixes the workpiece fully in place in terms of position and orientation. All six degrees of freedom are removed from the workpiece here.

To avoid overdetermination, only six defining points should be selected in the fixture. These should be in line with the six possible degrees of freedom. As a plane is defined by three points, arrangements with three 2-dimensional defining planes are overdetermined ($3 \times 3 = 9$ degrees of freedom removed). The arrangement selected in Fig. 10.106 with six defining points

Form	Fit	Criterion	Threshold speed
HSK-A	Worst case	Radial play at the flange contact surface	22,500 l/min
	Best case	Exceedance of reference stress	27,500 l/min
HSK-E	Worst case	Radial play at the flange contact surface	25,000 l/min
	Best case	Radial play at the flange contact surface	37,500 l/min

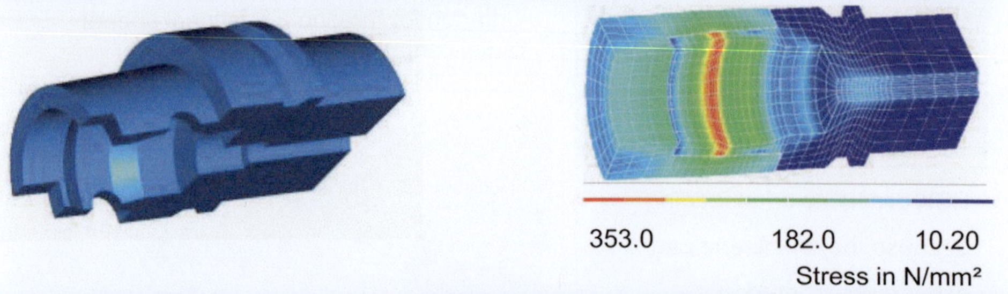

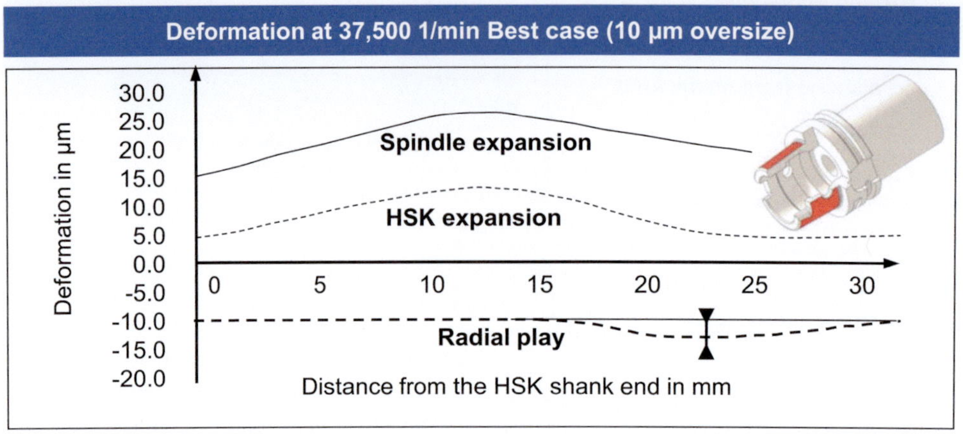

Fig. 10.104 Threshold rotational speeds of the HSK interface (HSK-63, forms A and E)

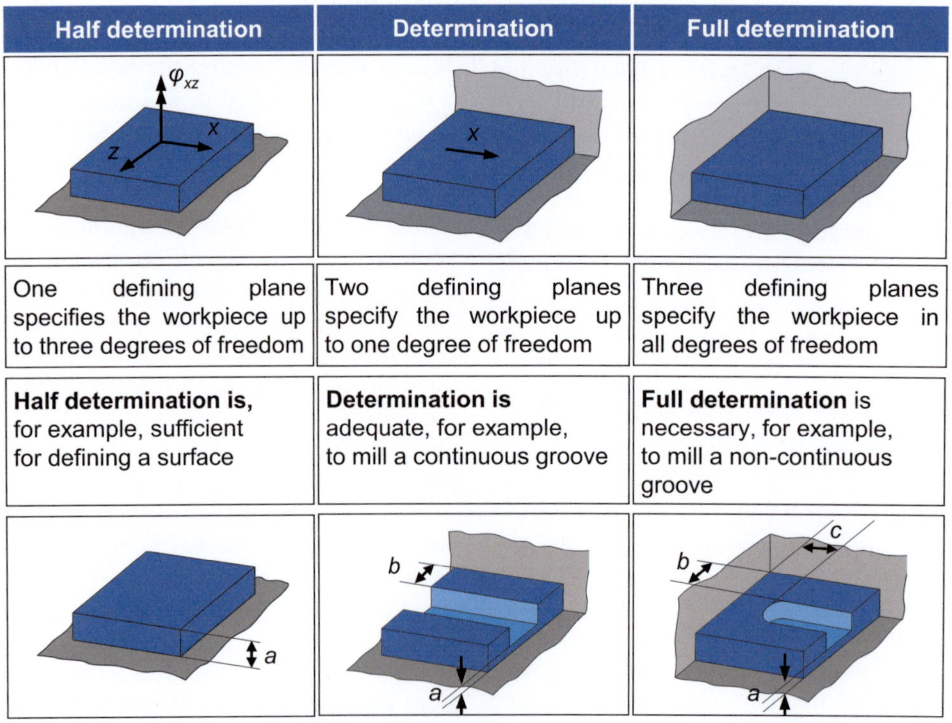

Fig. 10.105 Degrees of freedom and number of defining planes for different production tasks

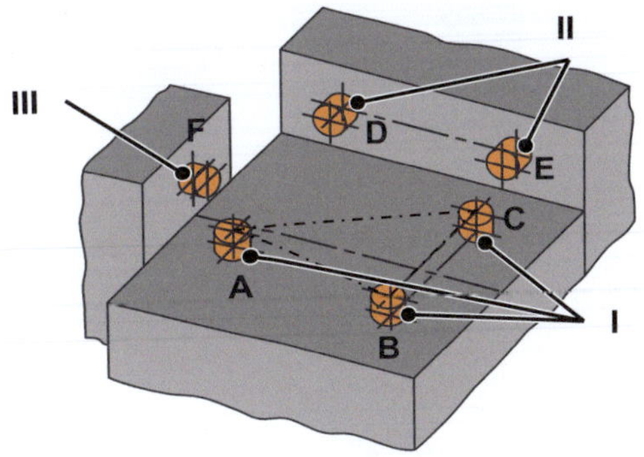

I. Defining plane:	3 defining points (A, B, C)
II. Defining plane:	2 defining points (D, E)
III. Defining plane:	1 defining point (F)
Total: 6 defining points for 6 degrees of freedom	

Fig. 10.106 Splitting of defining planes into defining points

is adequate for full determination. The defining points are fixed or moving or adjustable stops, locating pins, support bolts, support strips or support surfaces depending on the task at hand.

The second form of position determination is the fixation of the center planes of a body. The term used here is centering (Fig. 10.107). If the machining task is simply to make a radial hole in the center of a cylindrical workpiece having a diameter with a tolerance, one cylinder center plane must be fixed in position. On the left in Fig. 10.107, the vertical center plane is highlighted as the necessary reference plane. On the right are the workpiece support surfaces of the fixture (in the form of a prism) acting with a centering effect.

If the machining task, for example, is to make an axis-centric hole in a cylinder with a tolerance, the cylinder axis must be secured. The cylinder axis is the intersection line of two (preferably) orthogonal centering surfaces of the cylinder. This task is handled by the fixture in Fig. 10.108 by fixing the two center planes in position vertically one on top of each other. One center plane is fixed in place by the prismatic shape of the clamping jaws and the other center plane by the jaws that move synchronously towards each other [INST80].

10.4.1.1 Clamping

With position determination or centering, the workpiece or tool is moved to a defined position in relation to the machine tool or fixture. For clamping, this position is secured. This enables the absorption or transfer of forces and torques. The machining forces acting upon workpiece and tool during the machining process may not move the clamped workpiece or tool from its specified position, start to oscilate in an impermissible manner or deform it in such a way that workpiece accuracy and quality are no longer ensured. In fixtures and clamping devices for machine tools, the clamping forces are exerted on the workpiece and tool such that they act as normal and tangential forces. The ratio of clamping force to maximum possible retention force is given by the structure and configuration of the clamping elements. Similarly, different levels of automation can be provided depending on the design principle. Generation of all the force is made possible by using different physical effects (mechanical, hydraulic, pneumatic, etc.). Hydraulic clamping devices are growing in importance in NC production.

The two main methods of clamping, force-fit and form-fit, are shown in Fig. 10.109 using an example of workpiece clamping in a machine vise [VDI72]. Since, for form-fit clamping, the main cutting force acts directly against one clamping jaw, the clamping force

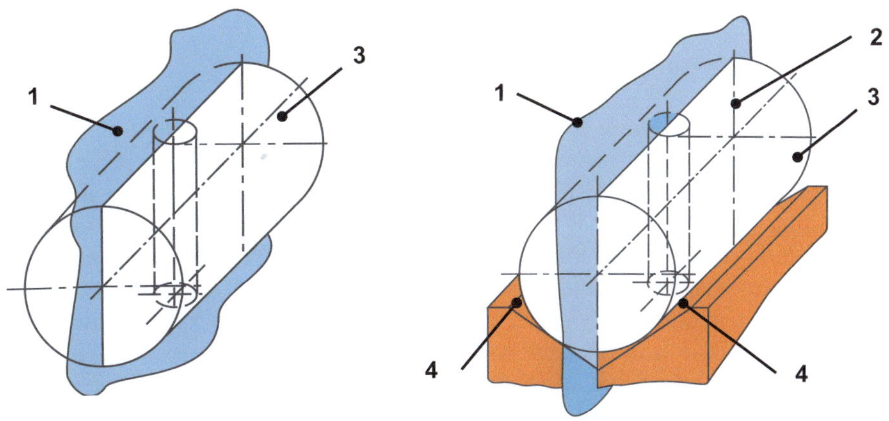

1: Center plane	3: Workpiece
2: Centering plane of the prism	4: Centering surfaces of the prism

Fig. 10.107 Half-centering of a cylindrical workpiece

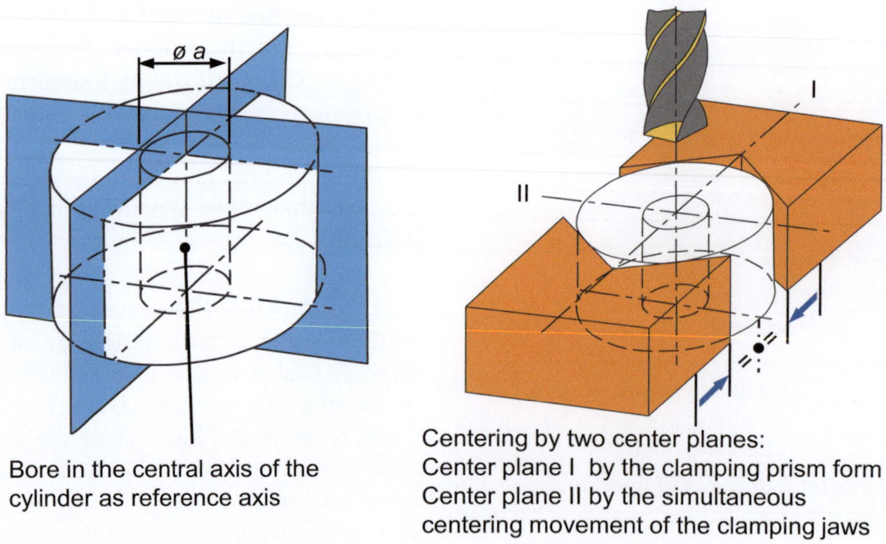

Fig. 10.108 Centering of a cylinder by two center planes

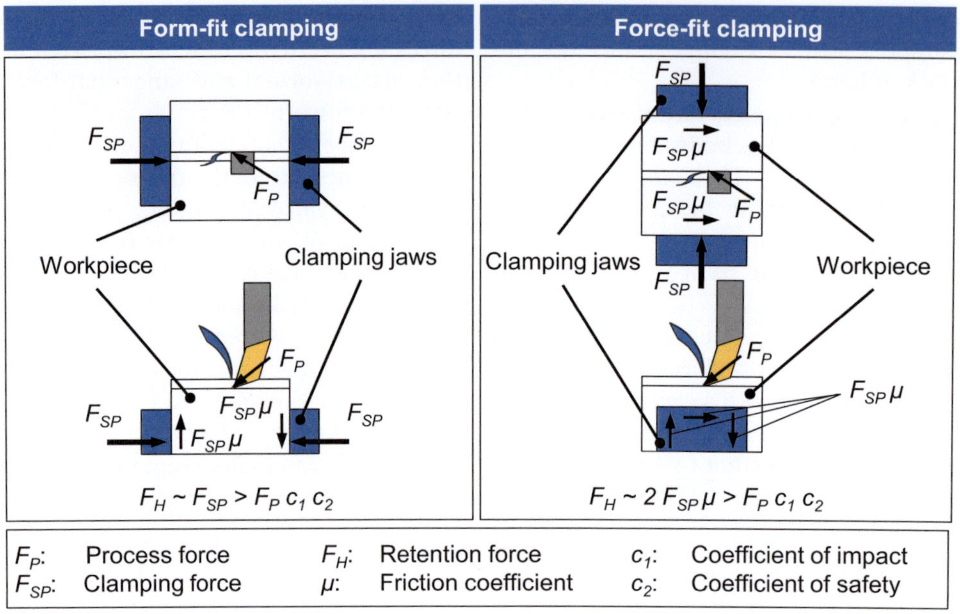

Fig. 10.109 Force-fit and form-fit clamping

need only be selected high enough to ensure the process forces cannot override the preload and the workpiece is reliably secured against falling out. For force-fit clamping, the workpiece is held just by the frictional forces at the two clamping jaws. Taking equal coefficients of impact and safety (*c1* and *c2*) for the computational estimation of the retention force for both cases gives, for example for the steel-steel material combination with a friction coefficient of $\mu = 0.1$, a clamping force necessary for force-fit clamping that is 5 times higher than for form-fit clamping. In selecting the clamping method, priority should be given to the form-fit connection due to the lower operating and clamping forces and the resulting lower level of workpiece deformation [THIE71]. For safety reasons, it is also important to ensure that the necessary clamping forces are maintained in full for the duration of the machining process. As a result, the fixture must have locking mechanisms or self-locking clamping elements. Corresponding considerations apply to tool clamping that is force-fit only for lower machining forces and force-fit and form-fit for heavy duty machining. In the latter case, the shaped elements are used for additional force absorption or represent safety elements in the event of overloading.

10.4.1.2 Workpiece Handling

Functional considerations for workpiece clamping apply in a similar manner to workpiece supply. The complex shapes of workpieces result in the need for interfaces between the workpiece and machine or handling devices to be flexible.

For automatic workpiece changes, a differentiation between two different change strategies can be made in line with selecting the location of change (◘ Fig. 10.110). Direct, automatic workpiece changes require flexible adaptation between workpiece and machine table as well as between workpiece and handling devices. For indirect workpiece changes using pallet systems, standardized interfaces between pallet and machine and between pallet and change equipment, are created.

10.4.2 Workpiece Clamping Systems and Fixtures on Drilling and Milling Machines

10.4.2.1 Workpiece Clamps and Fixtures

The function of workpiece clamping systems is to ensure the reliable holding of workpieces having different geometries. The higher level of design diversity of prismatic workpieces as compared to rotationally symmetric workpieces necessitates adaptable fixtures. In flexible NC production in particular, this means clamping systems differ greatly in terms of flexibility. The focus in this chapter is on fixtures that are considered universal or standard clamps. Workpiece-specific special fixtures are usually custom designs and are not discussed. Workpiece clamping fixtures can be divided into four groups:

- Simple clamping devices and positioning parts (outriggers, tensioning brackets, supports, and positioning and support elements)
- Standard fixtures (vises, drilling and milling devices, jaw chucks, swivel and tilting tables, indexing equipment)
- Modular and system fixtures
- Special fixtures

The manufacturing task often allows, in workshops in particular, the base bodies of fixtures and also workpieces to be clamped directly on the machine table. Clamping jaws, clamps or clamping supports are held on workpiece carriers using T-slots or threaded holes. A comprehensive range of standardized elements of assorted sizes enables flexible use but entails considerable manual tooling overhead. These standard clamping elements are only used in single-part and low-volume production. ◘ Figure 10.111 provides some examples.

The most frequently used standard clamping devices for smaller part dimensions are vises or machine vises, which are available in a wide range of different designs, sizes and accuracy classes. Vises are the preferred choice for accurate and quick clamping of prismatic workpieces and are in general-purpose use. By using custom chuck jaws that can be plugged or screwed on, the vises can be adapted to the workpiece geometries and materials (within certain limits). The

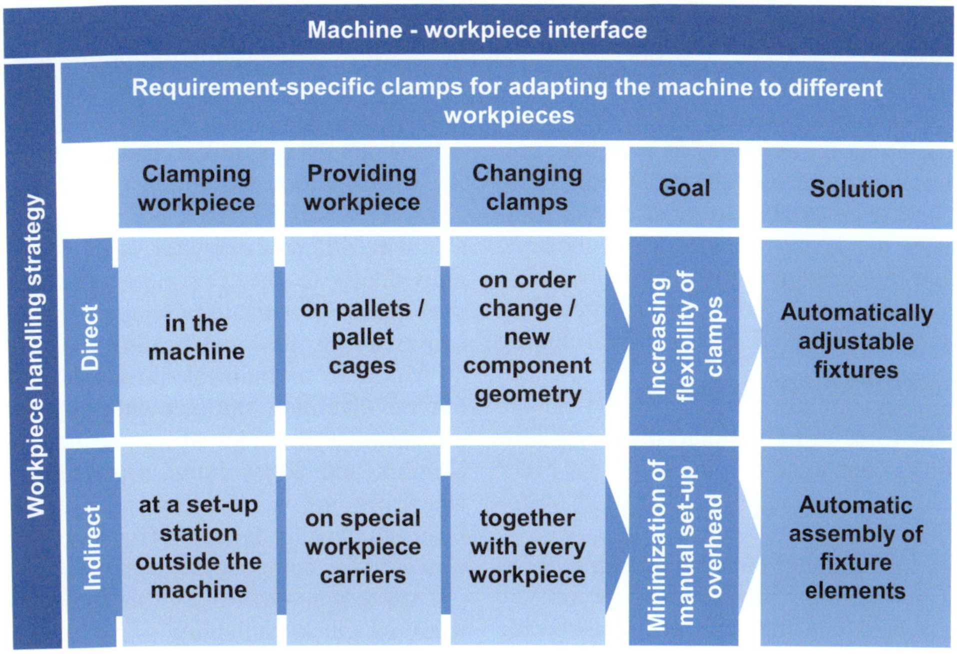

◘ Fig. 10.110 Workpiece clamping in flexibly automated production

Fig. 10.111 Clamps and clamping supports. *Source* AMF ANDREAS MAIER

chuck jaws can be actuated by hand or be moved with pneumatic, hydraulic or electrical energy.

A vise variant with mechanical drive and hydraulic force transmission is shown in Fig. 10.112. The mechanical preload is created by turning the hand crank. This turns the ball screw that moves the traversable jaw to contact against the workpiece. The operating moment is transferred by a coupling bolt, that disengages when a threshold torque is exceeded. After disengagement, the drive rotation is transferred to an inner spindle. The translational movement of this spindle actuates a pressure bolt. The axial movement of the pressure bolt creates (by displacing oil in the hydraulic unit) a high oil pressure that generates a high clamping force with the large cylinder surface. Additional safeguarding of the clamping force is ensured during the machining process by disk springs. Pressing the hand crank into the drive shaft unlocks an adjusting sleeve that can set a certain clamping force using the marking rings. This presetting facilitates constant clamping forces for every clamping process and consequently high repeat accuracy.

For drilling and milling circular workpieces in single-part and low-volume production, the use of quick-clamping devices can replace workpiece-specific fixtures in many cases. Figure 10.113 shows a manually operated vise and a hydraulically operated vise. The vises are mounted directly on the machine table. Turning the hexagon bolt with a wrench pulls the chuck head down. The chuck head comprises steel segments and elastic intermediate elements. Due to the conical inner surface of the chuck body, the chuck head moves radially, in addition to the axial movement, to clamp the workpiece. In the vice shown in the right part of the picture, the workpiece is clamped hydraulically. Applying pressure to the top piston surface moves the piston down and pulls the chuck head into the clamp taper of the chuck body. Chuck heads are available with different diameter variants or can be adapted individually to the workpiece geometry. Manually or pneumatically activated changing devices are used for conversion of the chuck heads [HAIN15].

Modular, mechanical fixtures provide virtually unlimited clamping options with regard to workpiece geometry. The range of applications comprises single-part and low-volume production of workpieces where it is not worth using workpiece-specific fixtures. From an existing range of (functional) elements, a fixture even for complicated parts can be assembled without any design overhead. Other benefits of modular fixtures are quick availability and reusability of the components.

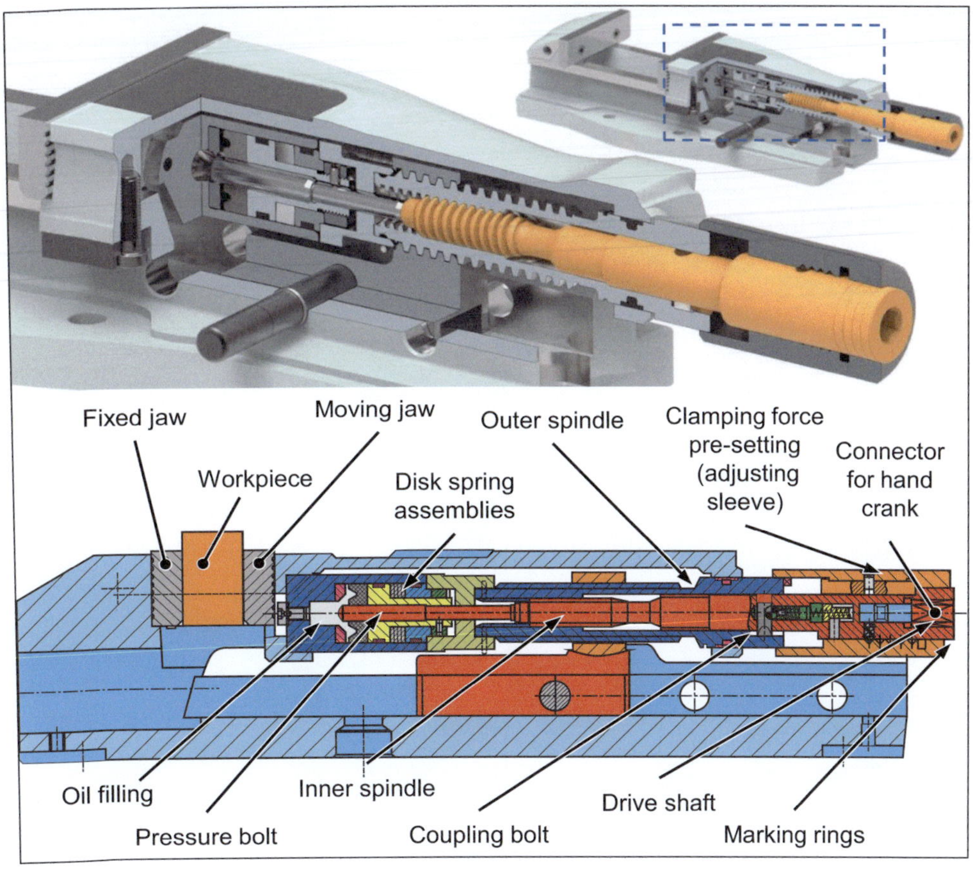

Fig. 10.112 Vise with mechanical-hydraulic drive. *Source* Röhm

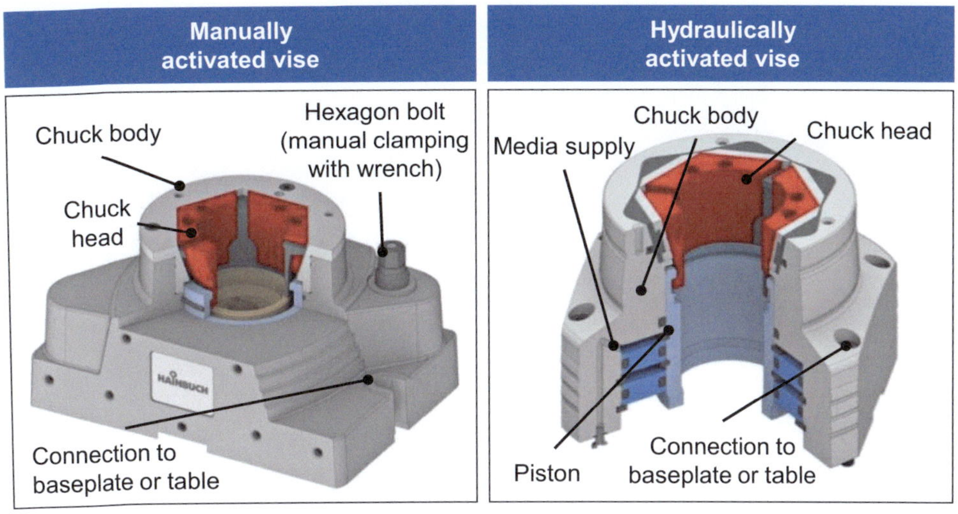

Fig. 10.113 Vises. *Source* Hainbuch Spanende Technik

A modular system is characterized by the structural design of the connection elements to hold the modular elements on the clamping plates. Slot and hole systems with different base raster plates are commonplace. A modular fixture with slot system is shown at the top in Fig. 10.114. The slot centering nuts ensure unique reference point specification at every slot crossing. The lower part of the figure shows the clamping options of the hole system. Here, the reference points are specified by the fit bores with threaded holes in the clamping plate. These bore/thread systems make it possible to combine the clamping and positioning functions in just one axis.

For the system shown in Fig. 10.115, any raster bore can be used (either fit bore or threaded bore).

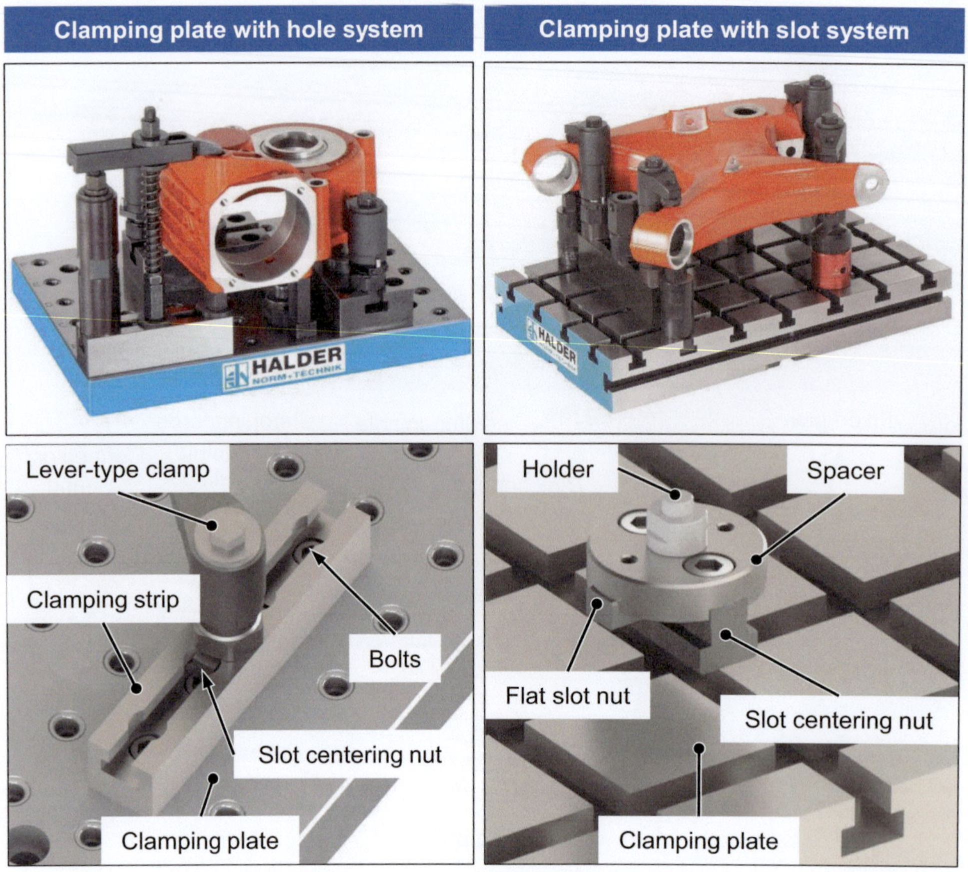

Fig. 10.114 Modular fixtures using the slot and hole systems. *Source* Erwin Halder

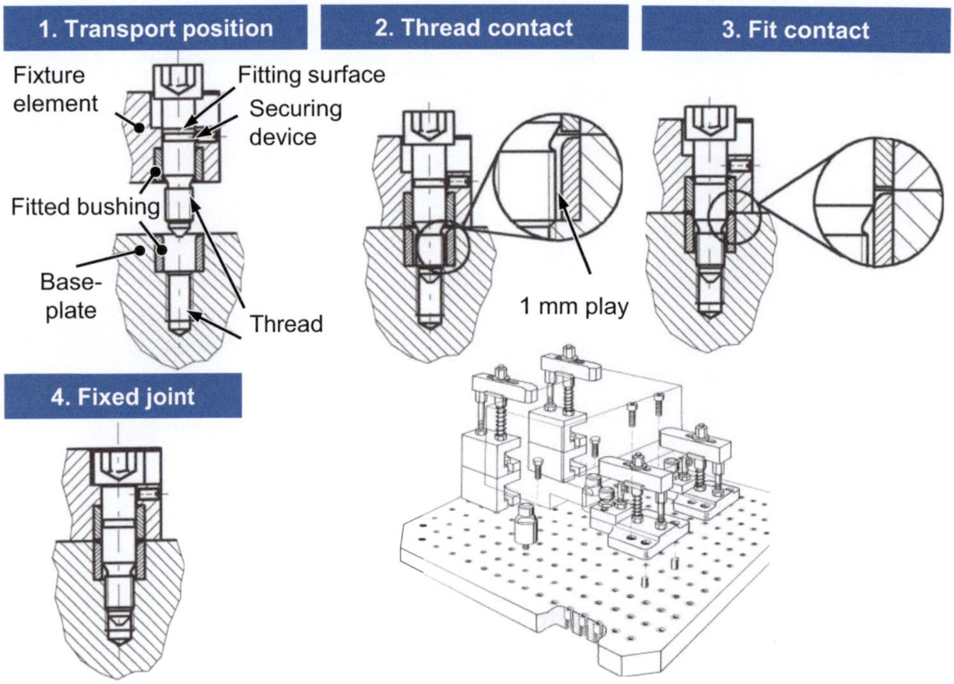

Fig. 10.115 Flexible definition and clamping system. *Source* Niedecker

A lower number of parts is required for the modular system. Furthermore, this system is suited to the robot-aided assembly of fixture elements. Shown on the right of the figure is the joining process in individual steps. The fixture elements involved in the centering and clamping tasks are also shown. After the bolt is positioned above the hole, it is fed axially until thread contact. There is then tightening to the fit contact and tight joining via the fitting pin of the bolt with the two fitting bushes of the baseplate and fixture element.

The clamping and support elements of the system shown in ◘ Fig. 10.116 have flattened balls that can move along all three spatial axes. These balls make possible distortion-free clamping. Distortion and displacement of the workpieces during the clamping process is not possible because there is a ball in the contact area as well as in the opposite clamping element of the units.

◘ Figure 10.117 shows the configuration options for a modular fixture system, comprising base, assembly, positioning and clamping elements. It also enables complex workpieces to be clamped securely and simply. A standout feature of this system is the combined fit bore and thread system.

High importance is attached to controllable, hydraulic clamping devices for lowering clamping overhead and for automatic workpiece changes. Sometimes, hydraulic elements are incorporated into the modular fixtures. Furthermore, clamping systems developed specially for the purpose of automatic workpiece chucking are available. Hydraulic elements are also suitable for adding on to multi-clamp fixtures where more than one small part is clamped on the workpiece carrier at one time.

In addition to generating the clamping force, maintaining it during the machining process is an important function of hydraulic clamping devices. The structure of a hydraulic swing clamp cylinder with mechanical lock and functional sequence is shown on the left in ◘ Fig. 10.118.

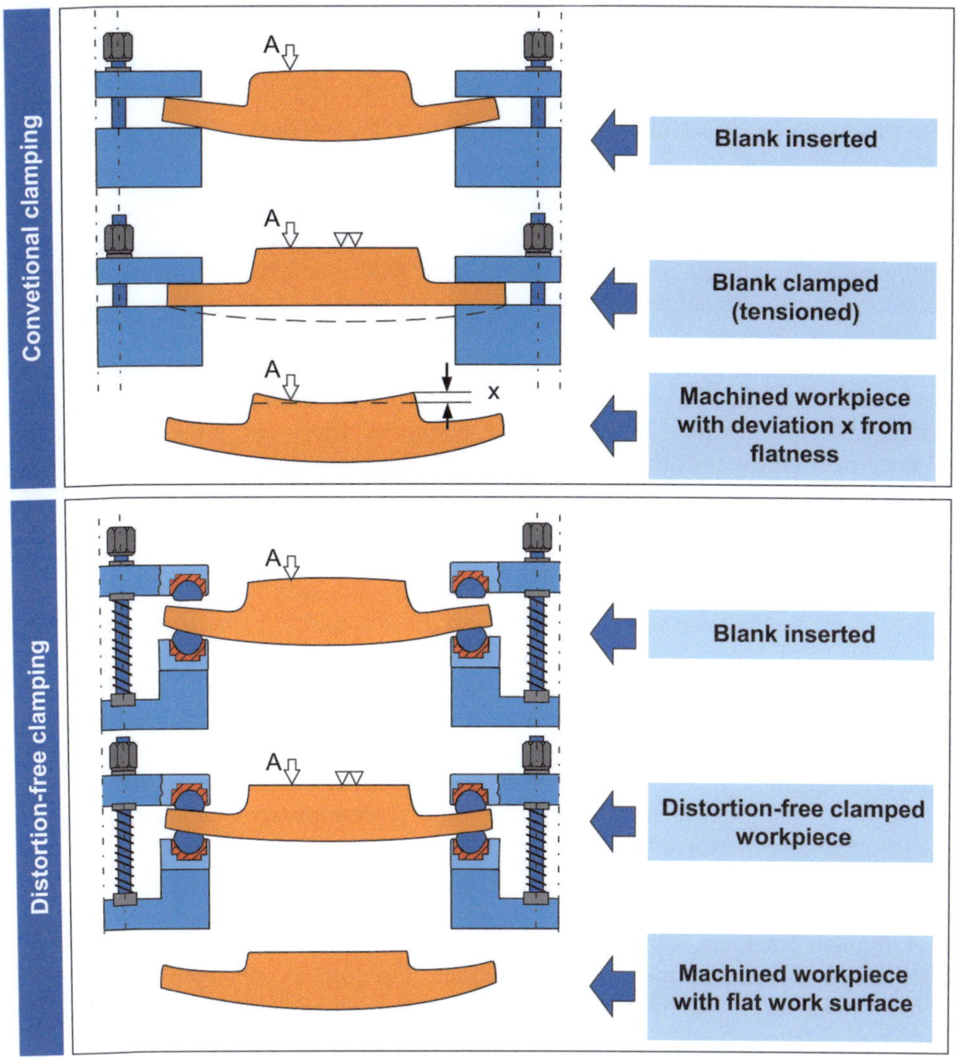

◘ **Fig. 10.116** Clamping of a severely distorted part, according to Niedecker

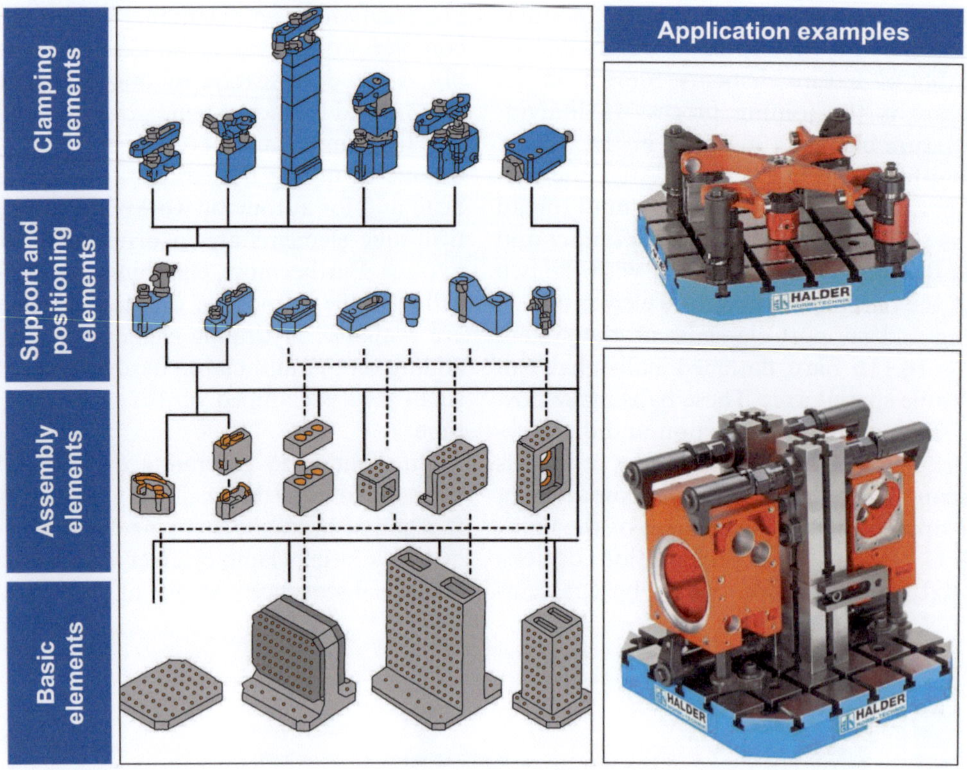

Fig. 10.117 Modular fixture system, according to AMF ANDREAS MAIER, Erwin Halder

Mechanical locks on these clamping elements function as follows. As soon as the clamping pressure is reached, the piston is mechanically locked by a hydraulically operated lock striker. The clamping force is maintained regardless of the hydraulic supply by self-locking of the striker elements. When exchange pallets are used, the workpiece can be changed outside the machine. The hydraulic unit is available at the clamping position and joined to the pallet. The device is disconnected from the pressure source during transportation to the machine and during the machining process. The functional sequence illustrates the hydraulically operated swiveling, clamping and locking (1, 2), depressurized operation (3) during the machining process in the phases (4) and (5) and hydraulic unlocking, releasing and swiveling back.

The magnetic clamping plate represents a flexible clamping device also suited to automatic workpiece changes. Here, the workpiece is an integral part of the magnetic circuit, i.e., the magnetic flux is closed by the workpiece. A distinction is made between three types: permanent magnetic clamping plates, electro-permanent magnetic clamping plates and electro-magnetic clamping plates. In the first type, the magnetic force is generated by permanent magnets. The permanent magnets are located inside the clamping plate and are swiveled by a control lever such that their magnetic flux passes through the workpiece (for clamping) or bypasses it (for releasing). They can be used everywhere as they are independent of an external power source. When the primary pole pitch is very tight and the magnetic field is at its lowest, they are particularly suitable for machining small or thin parts on eroding and milling machines [WALK04].

In contrast, electro-permanent magnetic clamping plates feature considerably higher holding forces. These clamping tools comprise a combination of permanent magnets and electromagnets. If clamped, the lines of flux of both magnets are aligned and flow through the workpiece (Fig. 10.119, top right). To release the workpiece, the electromagnet is switched on. Here, the lines of flux now acting against each other cancel out in the clamping range. The workpiece can be removed (Fig. 10.119, bottom right).

For electromagnetic clamping plates, the magnetic clamping force is generated by electromagnets. The electromagnets essentially comprise a steel core with coil and are supplied a constant direct current. A pole plate distributes the magnetic field of the steel cores. When switching it off, an AC voltage effective for a short time is usually used to reduce the residual magnetism in the workpiece. A special control unit is required for this. One downside of this variant is the heat build-up. As a result of this, high levels of accuracy are not possible (for grinding applications for example) [WALK04]. Heat generated in the coils can be dissipated using water cooling, however.

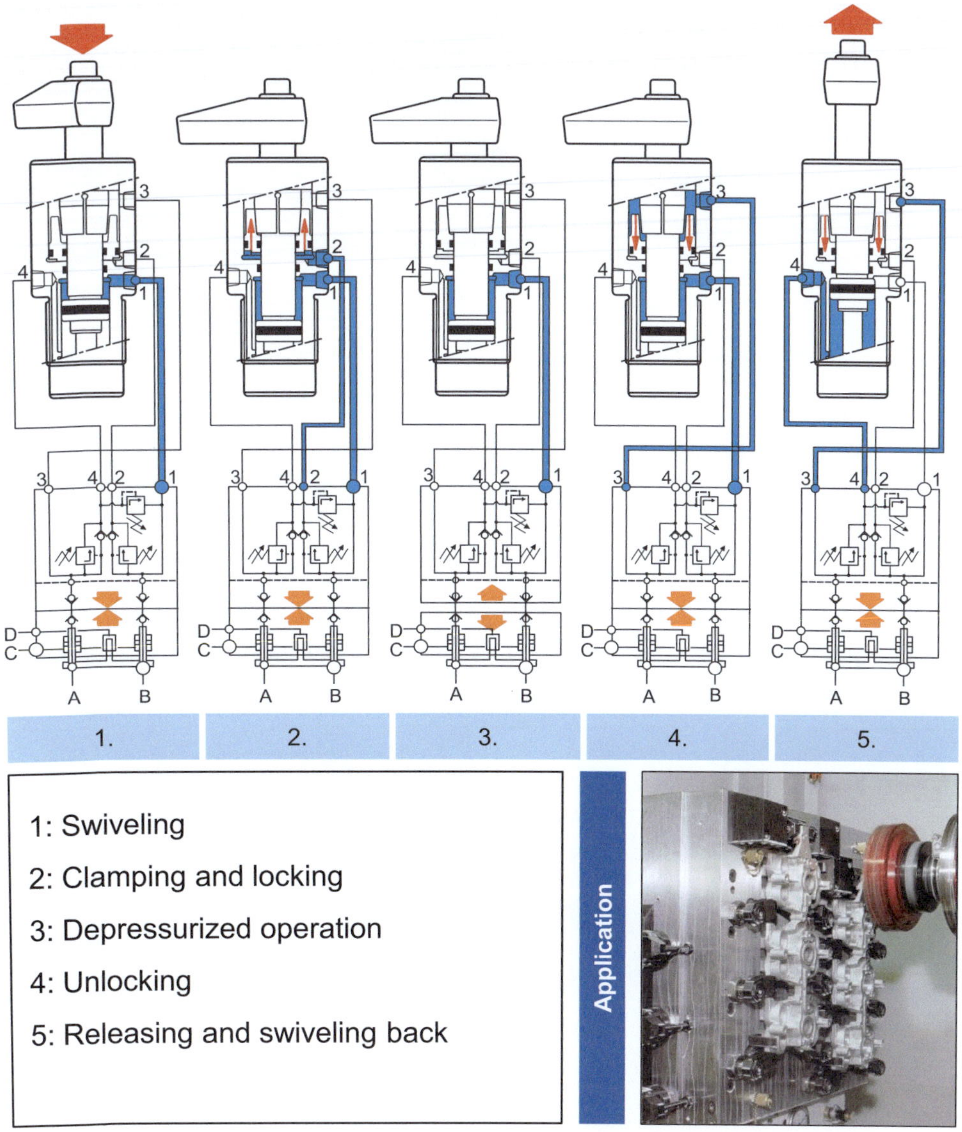

Fig. 10.118 Functional sequence of a hydraulic swing clamp cylinder, according to Enerpac, Zuse Hüller Hille

10.4.2.2 Workpiece Carriers and Pallets

Workpiece carriers include bottom plates, box tables, rotating tables, rotary-displacement tables and other bench variant forms. Workpiece carriers have already been introduced as equipment accessories under standard fixtures.

For flexible, automatic workpiece carrier systems, pallets that allow workpieces to be clamped during primary machining time are used. For universal, milling and drilling machines in particular, the use of different workpiece carriers is possible for task-adapted enhancement of the machine functions. Figure 10.120 shows machine tables for machining centers.

In addition to fixed machine tables driven by ball screw spindles, pinion racks or linear motors, different rotary tables can be used that have a linear axis of motion as well as a rotation axis. On rotary-swivel tables, the table plate can also perform a swivel motion while the rotary table can turn in complete revolutions (Fig. 10.121). On the 2-axis rotary-swivel table shown in the middle in the figure, the table plate swivels in the range of +91° to −121°. The axes are driven by worm gear motors on the NC rotary-swivel table shown on the left in the figure. NC rotary-swivel tables are also available with direct drives (Fig. 10.121, right). The direct drives on the B and C-axes allow high speeds and accelerations such as are necessary for 5-axis simultaneous high-speed machining. Rotary-swivel tables are also made as variants where a handwheel is used for manual swiveling.

Indexing equipment for the defined, repeated adjusting of component positions is available in an array of design variants. The automatic indexing device shown in Fig. 10.122 is used to position the face

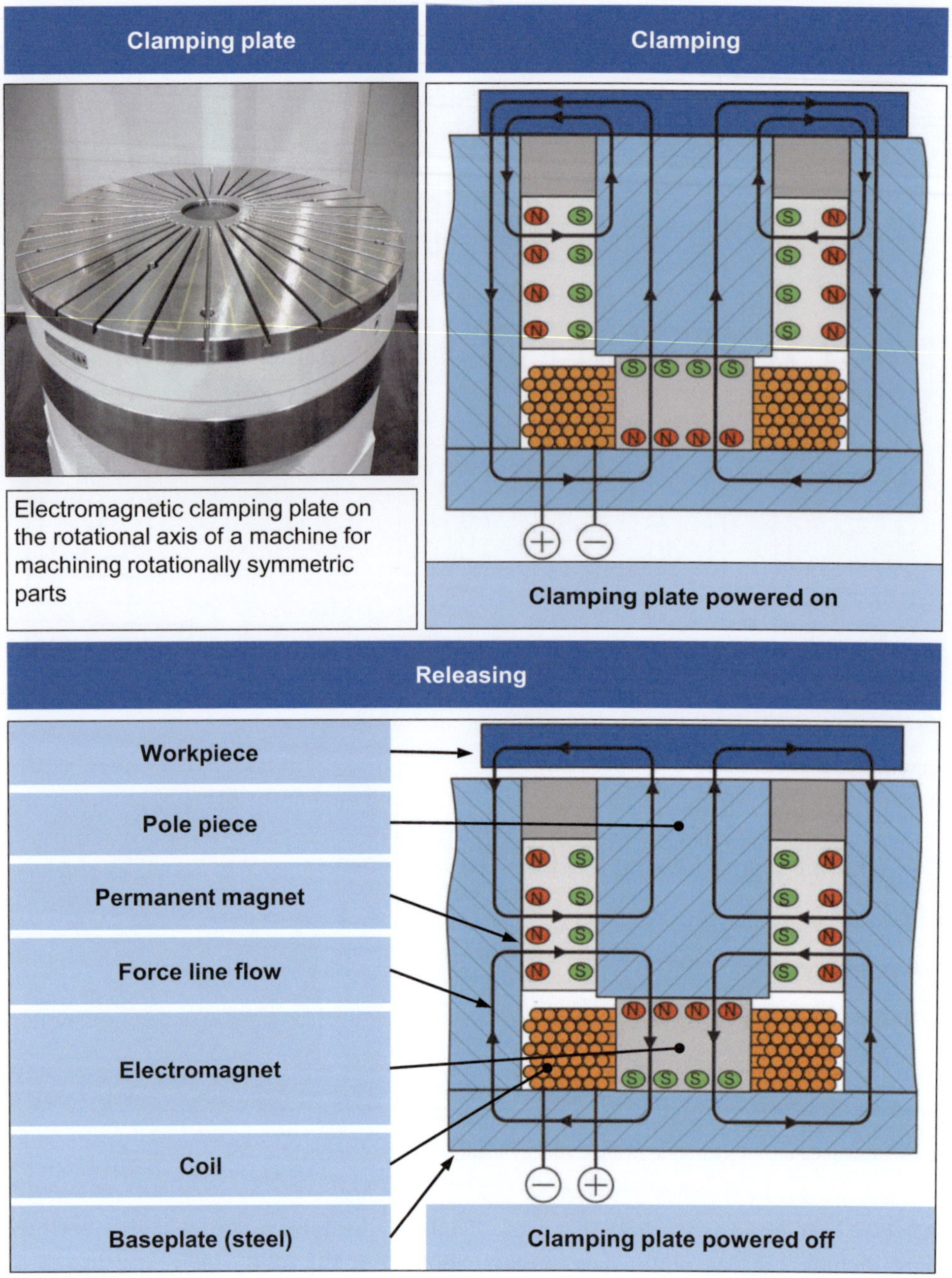

Fig. 10.119 Electro-permanent magnetic clamping plate, according to Starrag Technology (Scharmann division), Walker Magnetics

plate and its associated fixtures and workpieces at discrete angle settings of the rotational axis or to rotate it continuously. The housing and face plate each have a hardened and ground face gear. A third hydraulically operated face gear coupling locks the toothed rings of face plate and housing in the working position and centers both parts. With a build-up of pressure in the top chamber between the fixed piston and axially moving hydraulic cylinder, the cylinder and Hirth ring coupling, joined by bolts, are pushed up. The Hirth ring coupling decouples the housing and face plate. When the face gear coupling is lifted, the face plate can turn freely on rolling element bearings in the housing. They are multiple needle roller bearings that are compact and support the table on the inside and outside. An inductive proximity switch checks the position of the face gear ring. Contact seals protect the housing and face plates against the ingress of cooling water and chips. The non-lifting variant prevents the dangerous pump effect of the face plate. A servo motor, worm gear and spur gear realize the rotation drive for the face plate. NC indexing devices with this design are made as horizontal and vertical variants in sizes from 200 to 2500 mm. They are suitable for mid-duty to heavy-duty

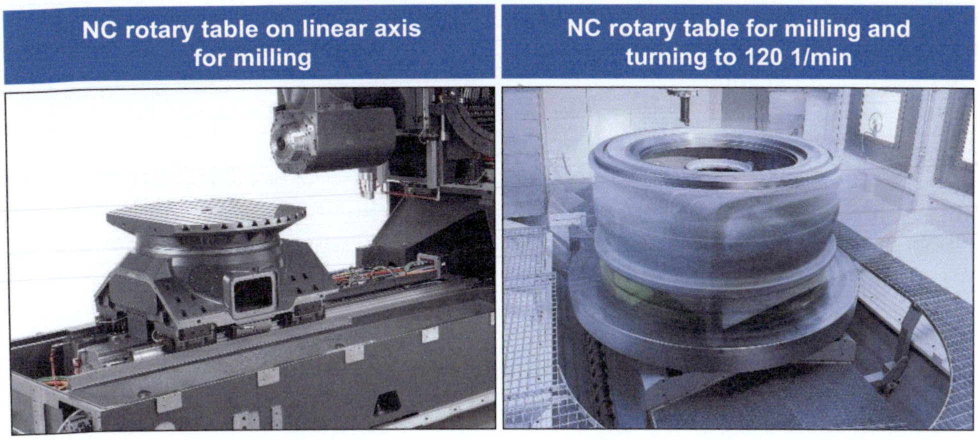

Fig. 10.120 Design variants of machine tables. *Source* DMG MORI

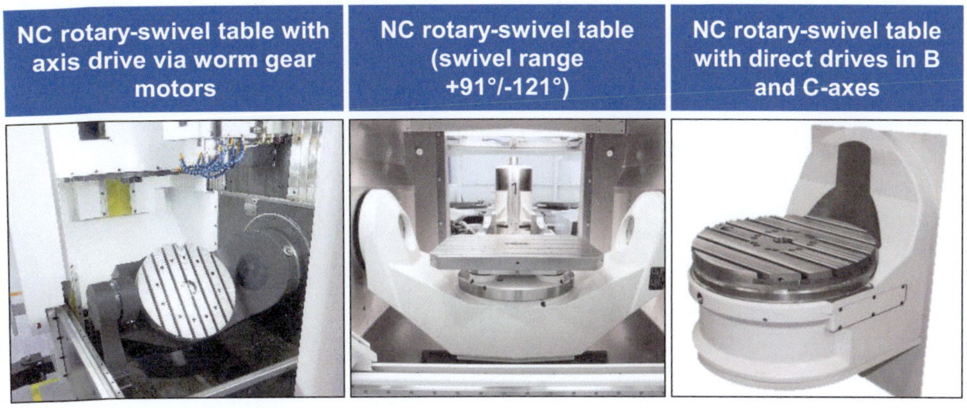

Fig. 10.121 Machine tables with swivel axis. *Source* GF Machining Solutions

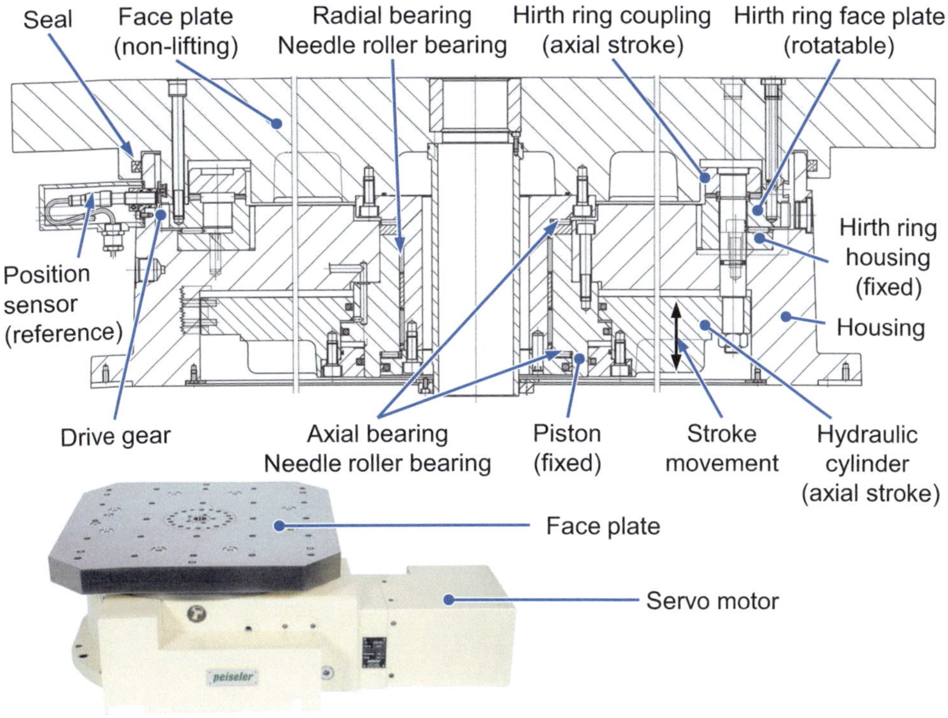

Fig. 10.122 Automatic indexing device with non-lifting face plate. *Source* Peiseler

cutting capacities and for positioning and circular machining [PEIS03].

The structural design of a machine table realized as an NC rotary table is shown in ◘ Fig. 10.123. A servo motor in conjunction with back gearing and a worm gear drive adjusted with no play provide the drive. High output torques permit circular milling with considerable machining forces. A hydraulic, axial clamp provides increased stiffness in the required position and prevents deformation and displacement of the workpiece chucking surface. A measuring system positioned centrally ensures highly precise positioning.

◘ Figure 10.124 shows another example of applying a worm gear drive in a vertical NC rotary table. The face plate is driven by a worm shaft and worm gear. The rotary table is mounted in axial-radial cylinder roller bearings that exhibit high stiffness. Clamping is provided by a hydraulic, radially acting bush that is subjected to pressure. Indirect measuring systems (on the worm shaft or drive motor), as well as direct measuring systems (on the indexing spindle), are used for exact reading of the angular position of the rotary table in NC operation.

◘ Figure 10.125 shows a compact rotary-swivel table. Both axis drives have numerical control. The servo motors initiate the rotational motion using worm gears. The fixed bearings for both axes are axial-radial cylinder roller bearings. Both axes can be clamped hydraulically to withstand large moments during the machining process. For clearance-free adjustment of the worm drives, the worms are each split into shank and hollow worms. The clearance is adjusted by turning the worm halves. Clamping between the shank and hollow worms is achieved with a clamping element.

Modern-day NC production is placing ever-stricter requirements on the exchangeability of workpiece carriers. A pallet forms an exchangeable part of the machine-table interface. Exchangeable pallets, like tools,

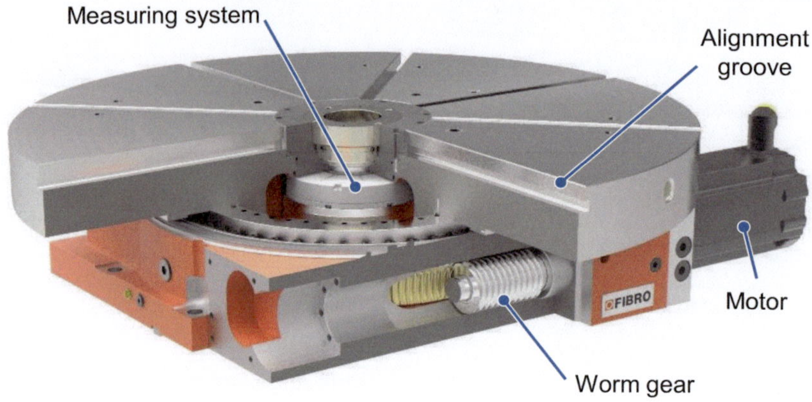

◘ **Fig. 10.123** NC rotary table for multi-sided machining in one clamping operation. *Source* FIBRO

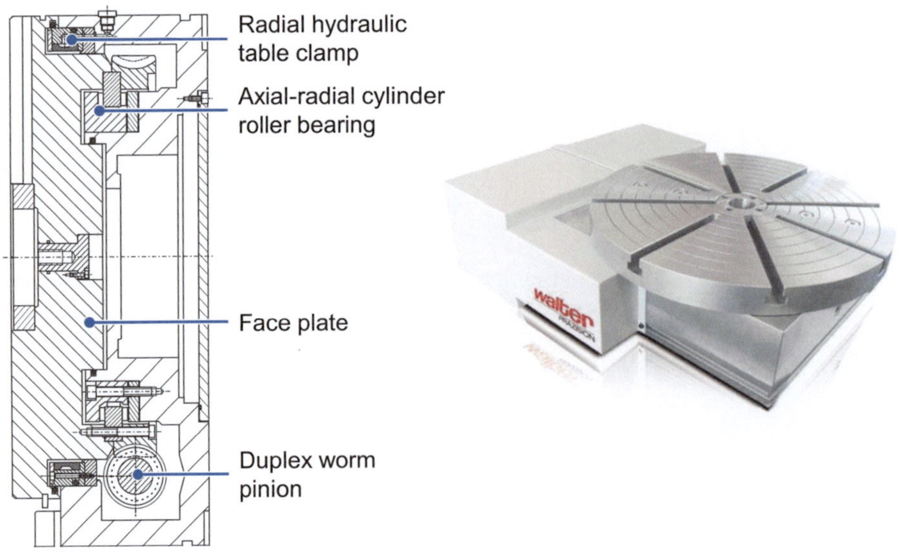

◘ **Fig. 10.124** Example of an NC rotary table. *Source* Walter Maschinenbau

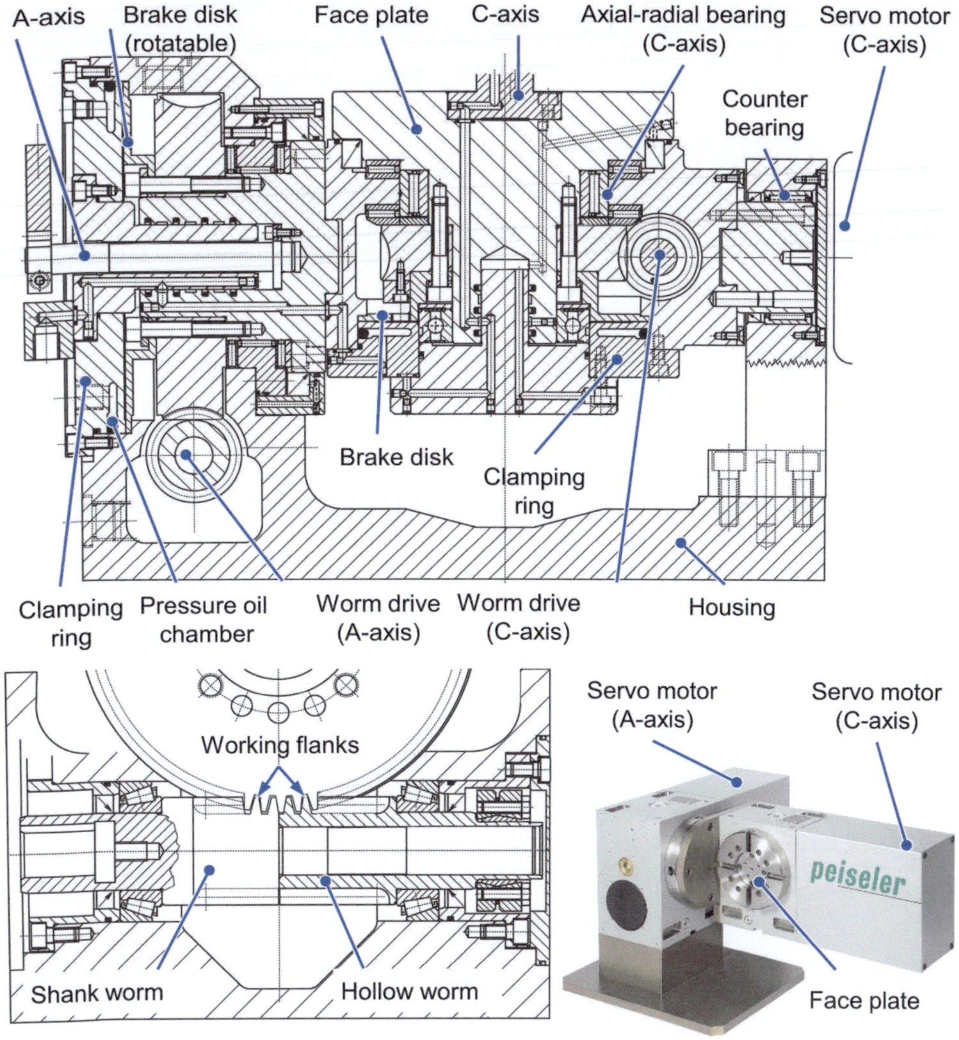

Fig. 10.125 Example of a 2-axis swivel unit. *Source* Peiseler

must necessarily have standardized interfaces to the machine. Standardized pallets are used in addition to some manufacturer-specific variants (Fig. 10.126). Two-part systems have also established themselves. In these, a clamping plate is added to a machine-specific pallet. A raster pattern of holes or parallel T-slots are added to square and rectangular pallets on the clamping side (shown with dot and dash lines in the figure). The transport guide surfaces and positioning/clamping surfaces for the machine are on the underside of the pallet. For pallets with nominal sizes greater than 800 mm, two dual-slots are provided to enable reliable pallet clamping [DIN85d; DIN87e].

Zero point clamping systems represent an enhancement to conventional clamping systems in terms of repeat accuracy of change processes and shortened set-up times. This type of clamping system is typically based on a pallet that can be positioned, aligned and clamped in a machine automatically using a special tool interface geometry and configuration. A system comprising clamping bolts, centering cones and a flange contact surface is the heart of the zero point clamping system introduced here. Combining these elements in a multiple arrangement as shown in Fig. 10.127 allows exact positioning of the pallet or workpiece on the machine table.

The clamping mechanism is usually operated hydraulically or pneumatically. With this mechanism, a spring assembly provides the necessary retention force should the power fail.

With this type of workpiece handling, it is possible to clamp components with high accuracy outside the machine and during primary machining time. Repeat accuracies less than 5 µm are certainly usual for zero point clamping systems.

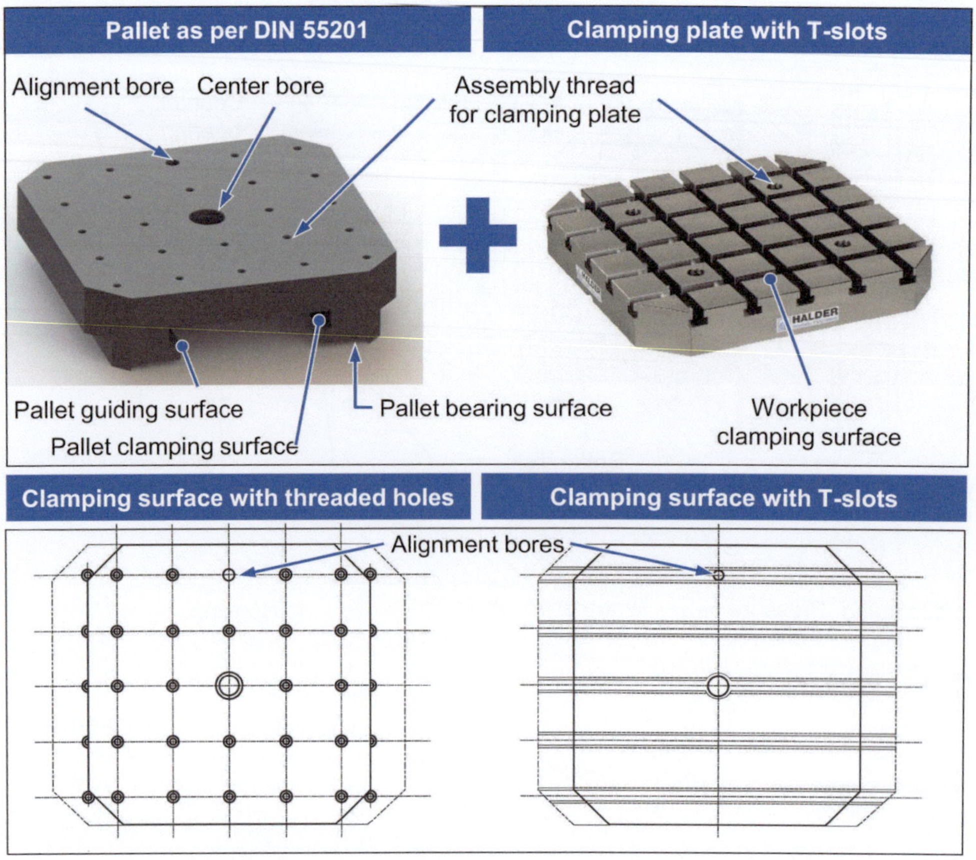

Fig. 10.126 Pallets for machine tools. *Sources* [DIN85d], Erwin Halder

Connection geometry	Tool interface design	Modules: Tool modules	Modules: Workpiece modules	Multiple arrangement
Connection geometry	Circle	• Milling heads • Rotating turrets • Hobbing units	• Swivel tables • Rotary tables • Swivel bridges	
	Row/ raster	• Multi-spindle heads • Rotating turrets • Tool holders • Spindle holders	• Clamping devices • Workpiece spindles	
	Rectangle	• Milling heads • Head turrets • Spindle holders	• Fixtures • Tables • Pallets	

Fig. 10.127 Zero point clamping system. *Source* Schunk

10.4.2.3 Pallet Changing Equipment and Magazines

Automatic workpiece changes by replacing clamping pallets is a common form of material handling in flexible production facilities. Machining centers for drilling and milling can be upgraded to flexible manufacturing cells using pallet changers and pallet magazines. In contrast to pallets, the transfer mechanism is not standardized. This means that different problems can occur when machines of different manufacturers are interlinked.

A form of automatic pallet change used for smaller pallets is a changer arm in analogy to the automatic tool change (◘ Fig. 10.128, right). To change the pallet, it must be moved into a transfer position. In this position, the double gripper fork of the swivel changer grabs both pallets. After the clamp is released, the dual-arm gripper lifts both pallets out of the holders and switches them with a 180° turn. This enables loading, unloading and setup of the pallets independently of the machining process. Pallet change times are 5 to 15 s depending on pallet size. There are 90° swivel changers with cross design having four pallets still available [HÜLL03a].

For greater workpiece weights, pallet changers are very often implemented as pallet displacement stations (◘ Fig. 10.128, left). The prerequisite for this type of automatic pallet changer is the ability for the workpiece carriage of the machine to move to two transfer positions. One station is required to provide a new pallet and the other to receive the completed pallet. A linear transport system with railed pallet carriage is, or floor conveyor transport systems are, used to supply pallets to the machine and transport them to the loading and unloading stations.

Linearly configured pallet magazines or magazines in rack form are usually used in multi-machine systems. If the requirement is to upgrade a machining center to an autonomous manufacturing cell, the pallets are often arranged in a circle or oval to form a pallet magazine (◘ Fig. 10.129). Six to eight pallets can be stored in multi-pallet magazines depending on the size. For a pallet change, a manipulator located in the center of the magazine moves underneath a pallet, lifts it, turns and places it down in the workspace. Multi-pallet magazines have a separate, rotatable loading and unloading station enabling loading and unloading to take place independently of the machining time. The version with two storage levels on top of each other is particularly space-saving and compact. It allows up to twelve pallets to be accommodated in the system [HÜLL03b].

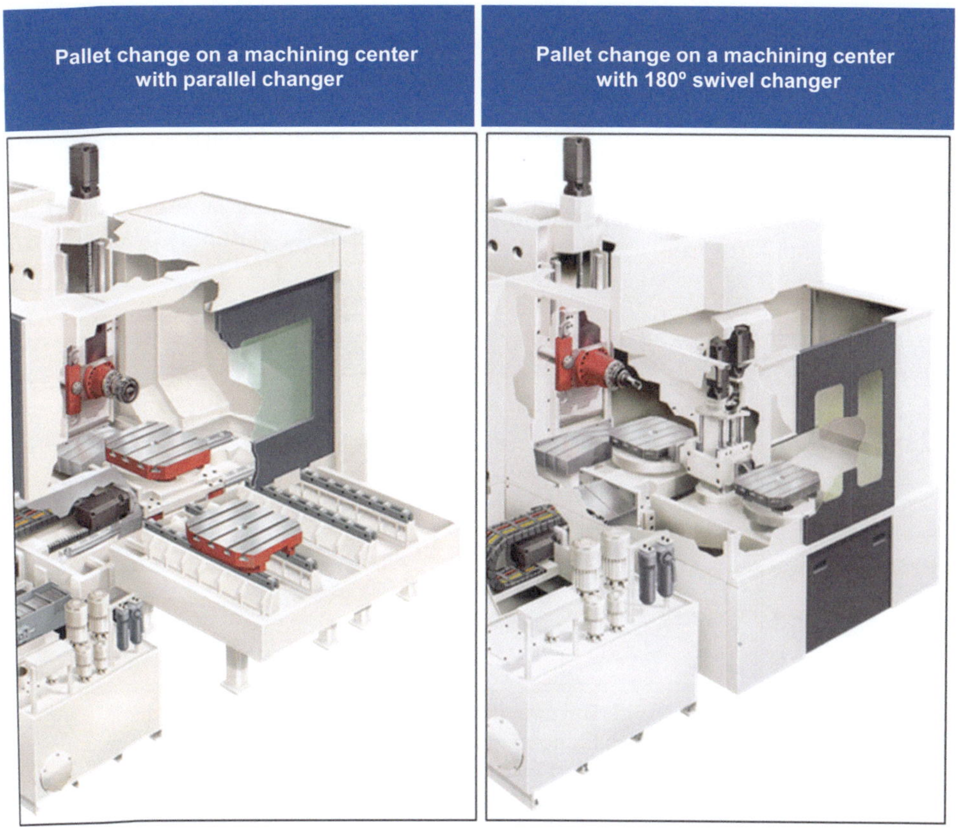

◘ **Fig. 10.128** Pallet changing equipment. *Source* Zuse Hüller Hille

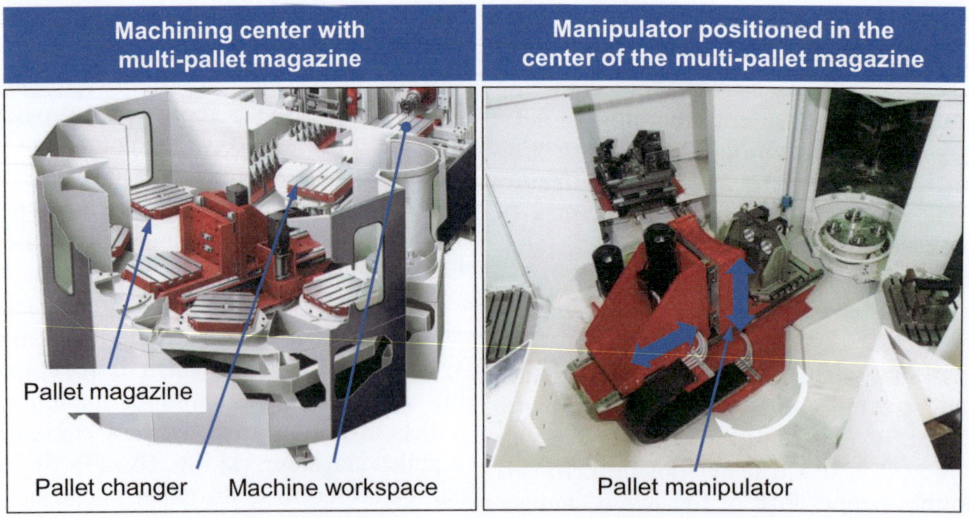

Fig. 10.129 Pallet storage system with central manipulator. *Source* Zuse Hüller Hille

10.4.3 Workpiece Clamping, Supporting and Supply on Lathes

10.4.3.1 Clamping Systems

The function of clamping systems on the working spindles of lathes is to align the workpiece axis and spindle rotation axis by means of centered clamping and to transfer the necessary machining forces and moments to the workpiece.

Turned workpieces are normally clamped in chucks of different sizes using diverse actuating mechanisms. Large workpieces are clamped on face plates and smaller workpieces and bar stock in collet chucks.

Spindle Head Design Variants

Spindle ends in lathes are usually short taper holders. There are different standards for securing the standardized workpiece chucks (Fig. 10.130). When chucks are not changed frequently, spindle heads with chuck screw threading from the front are used [DIN10b]. Similarly, chucks are held on spindle heads as per [DIN10c] via a short taper. In this case, the stud bolts secured in the chuck are screwed against the bayonet disk with flanged nuts. This principle is suitable when chucks are changed more frequently but requires a gap for the flanged nut (that is tightened manually) between the headstock and spindle head. Simple actuation of the locking mechanism is possible for the Camlock variant (as per [DIN10a]) by turning the eccentrically mounted clamping cams.

For holding rotationally symmetric workpieces (shafts and disks), centrally clamping three-jaw chucks with three clamping jaws offset by 120° to each other are used. For the clamping of multi-surface workpieces,

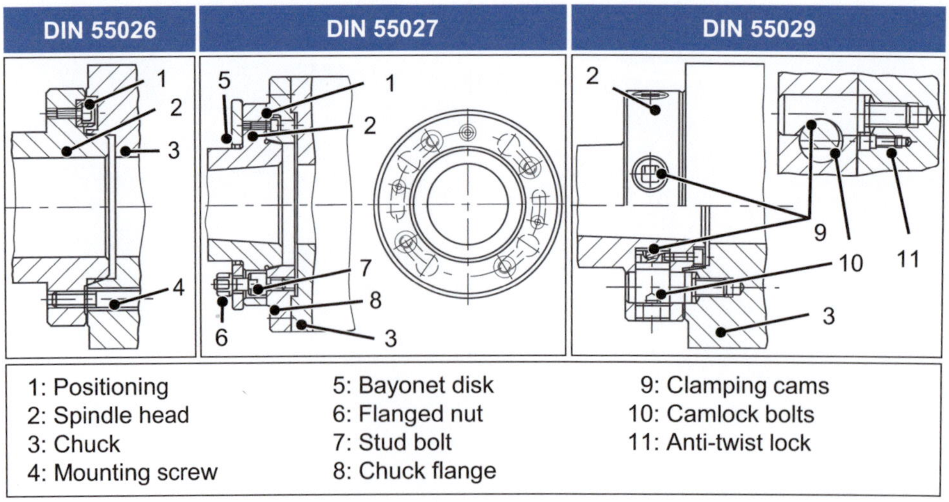

Fig. 10.130 Standardized spindle head variants for working spindles, according to [DIN80a; DIN80b; DIN80c]

four-jaw chucks offset by 90° are used. Three-jaw chucks are available for manual actuation (manual chucks) or as automatic chucks.

Figure 10.131 shows the ways to clamp workpieces with large and small diameters. Using stepped or gripper jaws enables the clamping diameter range to be aligned to the workpiece diameter.

Spiral chucks and wedge bar chucks are widely used among manual chucks [DIN10d]. Figure 10.132 shows a manual chuck working as per the wedge bar principle. A wedge bar movable crosswise is positioned in the chuck body behind every clamping jaw. It is engaged with the backward teeth of the clamping jaw by means of inclined module gear teeth. When the wedge bars are moved, the jaws perform a clamping stroke inwards or outwards depending on the direction of movement. The symmetric clamping by the exact synchronized speed of the three wedge bars is caused by the driver ring at the back. Three radial grooves accommodate the sliding blocks. Each block is mounted to a stud on the rear of the wedge bar. One of the three wedge bars is driven by the ball screw mounted in the chuck body. The end of the usable jaw stroke is marked by the indicator pin exiting. In such a case, there is a risk of the clamping teeth not being in contact, or not being adequately in contact, or there is not enough clamping stroke left for reliable retention of the workpiece. For difficult roughing work on untreated or pre-turned workpieces, hard stepped jaws with multiple teeth on the clamping surface are used. For the precision clamping of workpieces already machined having clamping surfaces that must not be damaged, soft clamping jaws with smooth clamping surfaces are used [FORK13].

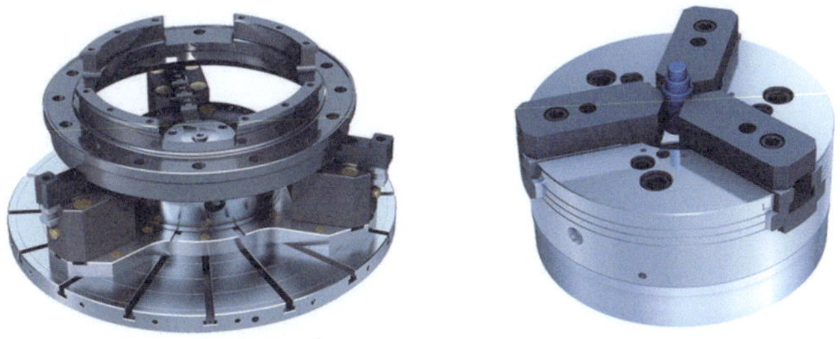

Fig. 10.131 Clamping large and small workpieces. *Source* Schunk

Automatically operated chucks are used for production machines. A clamping piston causes the clamping movement of the jaws using wedge or lever mechanisms. For actuation of the clamping piston, pneumatic/hydraulic cylinders or electromechanical power-clamping fixtures are used. These are affixed to the back end of the spindle. The pressure medium is transferred to the rotating cylinder/piston system by rotary joints. Hollow chucks or partially hollow chucks are used depending on the production task (Fig. 10.133).

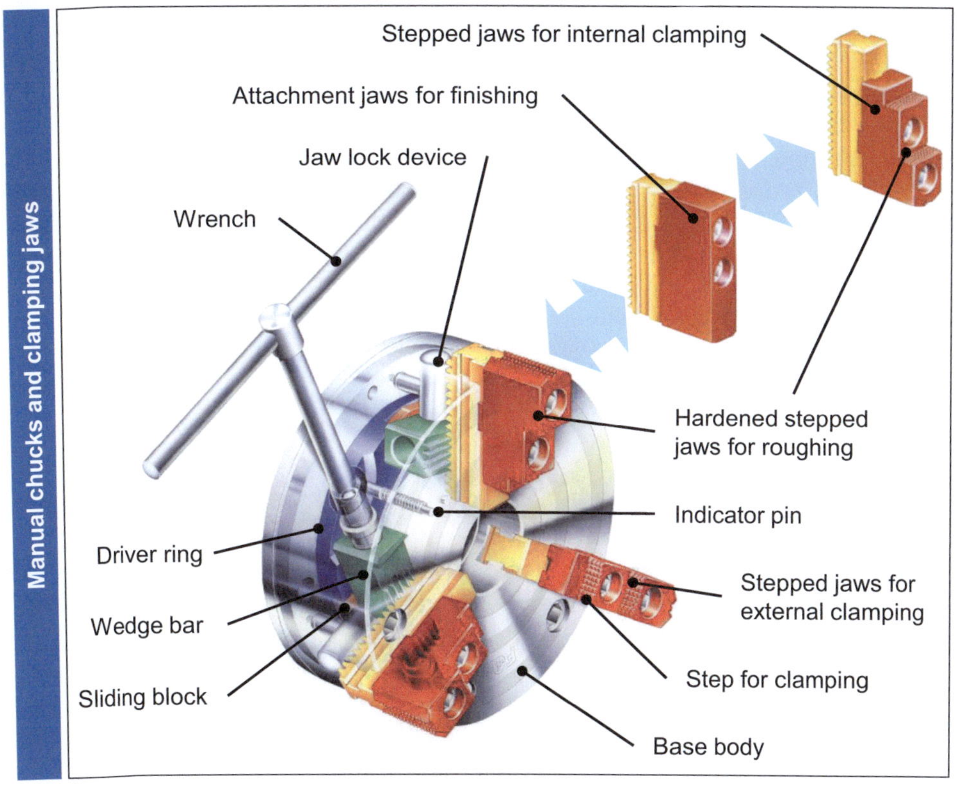

Fig. 10.132 Manual chuck as per the wedge bar principle. *Source* Forkardt

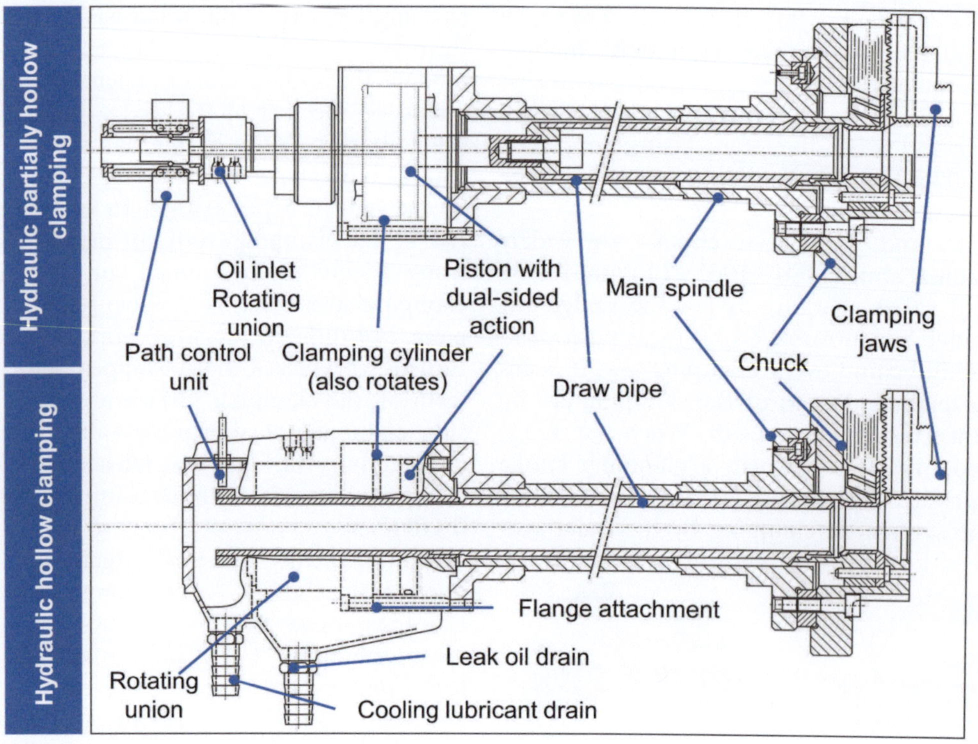

Fig. 10.133 Power-clamping fixtures for lathes. *Source* Forkardt

The pressure oil clamping cylinder for partially hollow clamping (Fig. 10.133, top) has a simple, short cylinder flange with attached from the back of the cylinder. It is relatively close to the back spindle bearing, something that has a positive effect on quietness at high rotational speeds. The pressure oil clamping cylinder for hollow clamping (Fig. 10.133, bottom) means bar stock can also be machined. The coolant, used in the cutting process and flowing backward through a spindle hole, is collected by the cylinder housing. A path control unit checking the position of the draw pipe, and so the positions of the clamping jaws, is common to both systems. Note that the axial movement of the draw pipe is different for different chucks. This means that, when chucks are replaced, the limit switches may need to be readjusted.

At high rotational speeds, the clamping jaws' centrifugal forces exert a major influence on the clamping force of the chuck. For external clamping, the clamping force can be reduced such that reliable machining of a workpiece is no longer guaranteed. To prevent this, centrifugal force compensation systems are used to compensate for the centrifugal force of the clamping jaws such that a high enough clamping force acts upon the workpiece in all speed ranges. In the example shown in Fig. 10.134, this compensation force is achieved by redirecting the centrifugal force of the compensation weight to the base jaw. Elastic expansion of the chuck body must also be factored into the designs of the masses and lever arms.

Diaphragm Clamping Technology

Figure 10.135 shows the principle of diaphragm clamping technology. The function of this chuck is based on the elastic deformation of a diaphragm. The diaphragm screwed on in the outer area of the chuck is deformed by a draw pipe. This closes the jaws in slanted positions and clamps the workpiece. Behind the diaphragm are heavy metal counterweights joined to the chucks. This compensates the centrifugal forces coming from the rotation. There are no sliding parts here, the reason why the system is broadly low-wear and insensitive to dirt [SMW13]. The clamping range is not very wide (tenths of millimeters).

Collet Chucks

Collet chucks are suitable for the automatic clamping of bar stock and prefabricated blanks. They are chucks made of steel jaws and have elastic intermediate elements (Fig. 10.136). The outer, conical chuck head is pulled into the taper of the chuck body by the draw pipe using a clamping cylinder. The chuck flange and base body are adapted to the machine and the chuck jaws to the workpiece respectively. The clamping jaws are manufactured such that they are machined collectively in the form of a closed ring and then separated

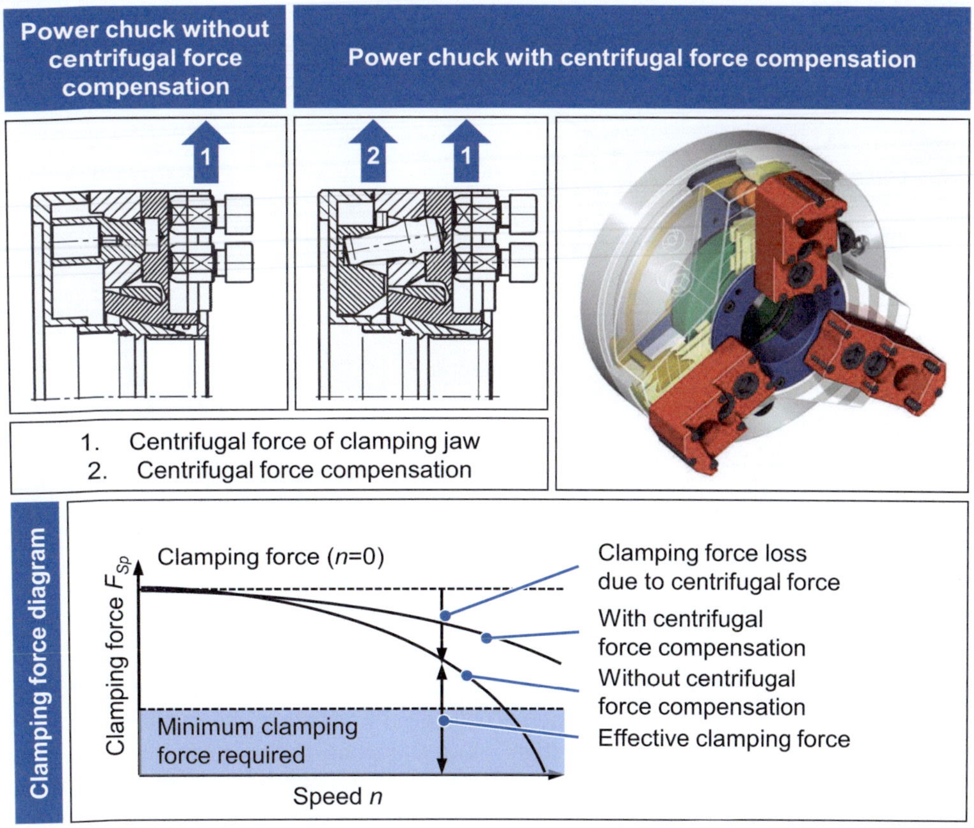

Fig. 10.134 Powered wedge hook chuck with centrifugal force compensation. *Source* Forkardt

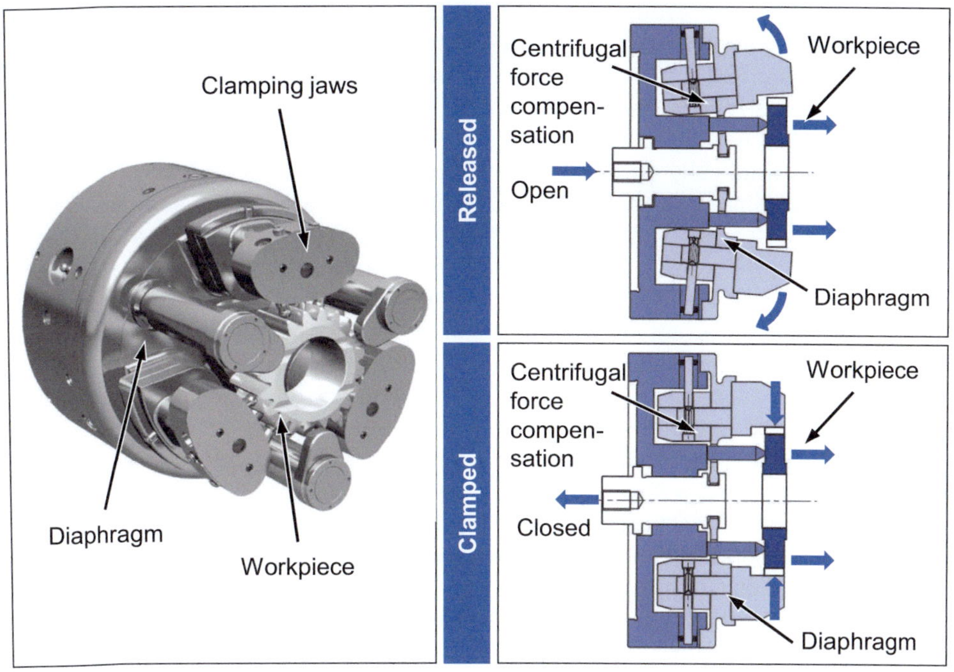

Fig. 10.135 Principle of the diaphragm clamping technology. *Source* SMW-Autoblok

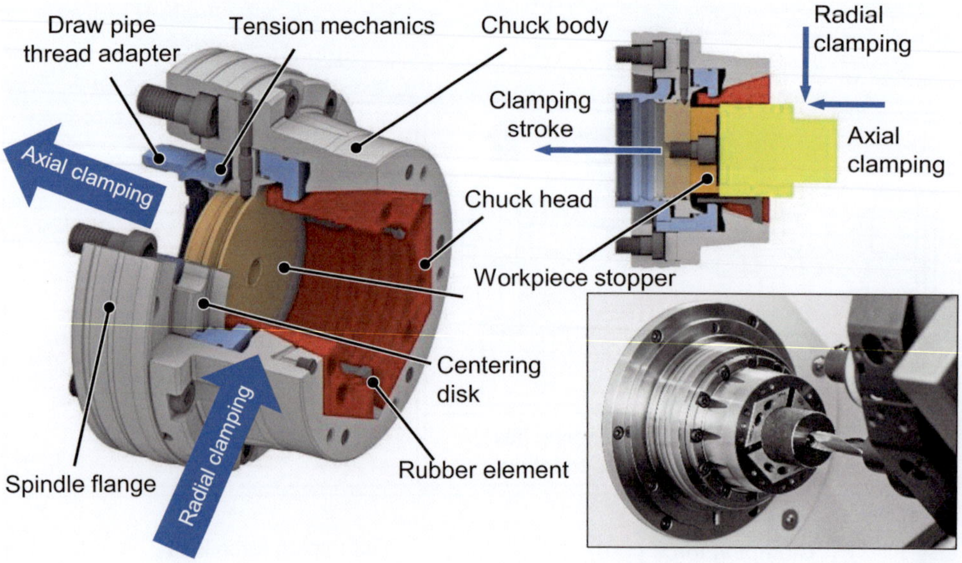

Fig. 10.136 Chuck with vulcanized steel segments. *Source* Hainbuch Spanende Technik

into individual jaws. Then, the steel segments are connected to each other elastically with vulcanized intermediate rubber layers [HAIN15]. The clamping range is less than 0.5 mm.

Collet chucks using a different design principle are also used for the mass production of bar stock. Collet chucks are tailored to specific workpiece diameters. Their clamping range is low (<0.3 mm) for this reason. With uniform clamping all the way around surfaces, they also permit the machining of workpieces with thin sides (such as pipes). Figure 10.137 shows a collet variant for which, unlike the variant in Fig. 10.136, the clamping sleeve is not pulled into the taper. Instead, the taper is pushed by the hydraulically operated pressure pipe over the elastic clamping sleeve to secure the workpiece.

Mandrels

Many design variants exist in the group of clamping tools for internal clamping, the mandrels. Given the short clamping stroke necessary for high clamping accuracy, mandrels must be tailored to the different workpiece diameters. A powered segment mandrel is shown in Fig. 10.138. The mandrel comprises a segmented steel bolt with elastic intermediate elements. The axial clamping stroke actuated by a tension bolt causes a radial expansion of the steel bolt on the chuck taper in addition to an axial motion. This means precise repeat positioning and reliable clamping of the workpiece are achieved.

Another variant is the hydraulic mandrel with synthetic clamping sleeve (Fig. 10.139). A piston applies pressure to the fluid grease or hydraulic oil. It expands the glass fiber or carbon fiber reinforced synthetic clamping sleeve. The maximum expansion can be 1% of the diameter of the expansion sleeve such that the low clamping range means only workpiece bores with tight tolerances can be clamped. The workpiece can be clamped directly by the expansion sleeve or indirectly by a cylindrical intermediate sleeve with slits on both sides. The intermediate sleeve can also be designed as an exchange part for tailoring to different clamping diameters. The design and fiber profile of the expansion sleeve permit cylindrical and circular expansion across the entire clamping range with a maximum workpiece diameter tolerance of 5 μm [RÖHM13].

Clamping Between Centers

Shaft parts are often clamped between centers and driven by face drivers. For a large working range, the front chuck shown in Fig. 10.140 has exchangeable cam cutters that dig themselves into the front surface of the workpiece and thus ensure form-fit transfer of torque. A hydraulic compensation system ensures uniform contact of the cam cutters, even when the front surfaces are highly uneven. For this, the four cutters are each mounted on pistons that in turn rest on oil columns that are connected. The oil is under a spring preload. The centering tip is also spring-mounted and performs the centering function [KARL14].

Face Plates

For the machining of large, non-circular or irregular workpieces, vertical lathes are often used. On these lathes, the workpieces are clamped on a face plate. A workpiece is aligned manually on the face plates, secured using four to eight individually adjustable

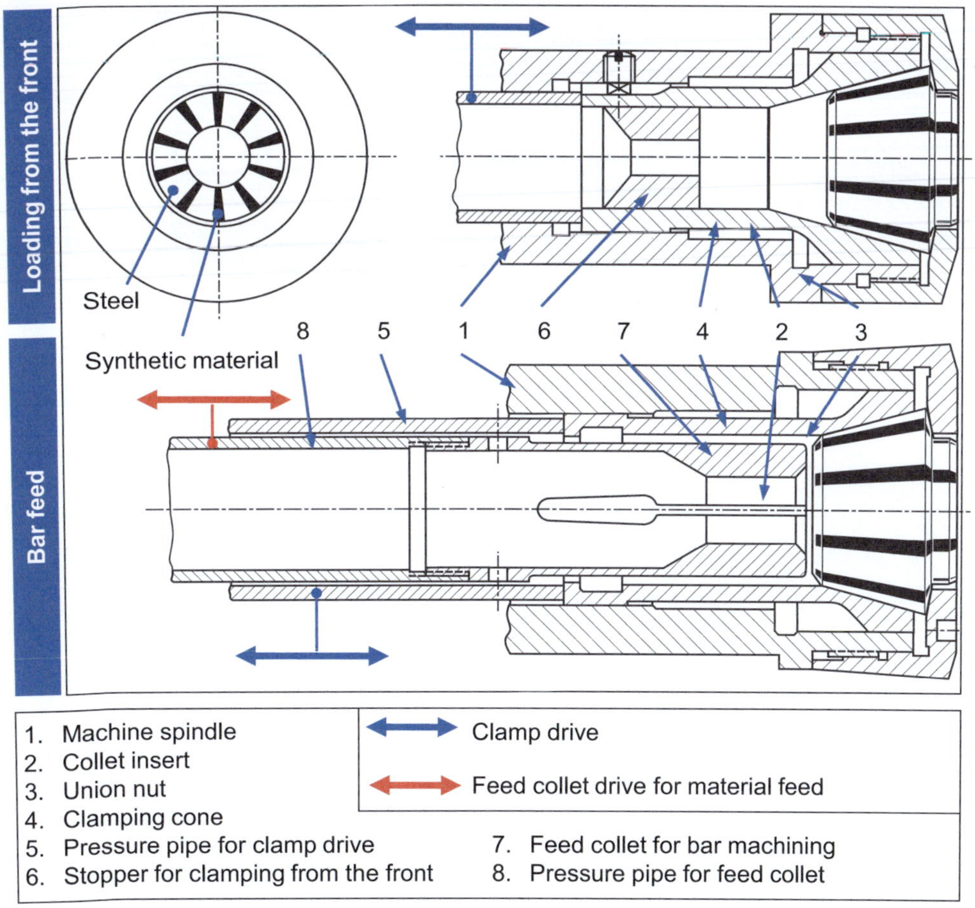

Fig. 10.137 Collet chucks with and without bar feed. *Source* Ortlieb Präzisionssysteme

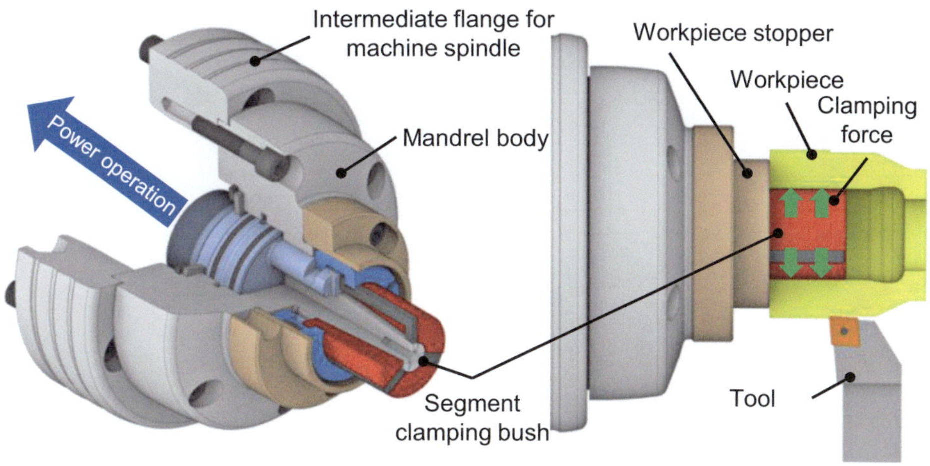

Fig. 10.138 Power operated segment mandrel. *Source* Hainbuch Spanende Technik

clamping jaws and, as required, with additional clamping jaws, clamps and clamping supports. In particular here, the centrifugal forces must be taken into account and imbalances compensated with corresponding counterweights. Heavy-duty lathes in this family are used to machine rotationally symmetric workpieces with a maximum circumferential diameter exceeding 12 m and a maximum workpiece weight of 350 tons. Face plates are also used as workpiece carriers for vertical lathes. Figure 10.141 shows a vertical lathe with face plate for parallel machining with two turning tools.

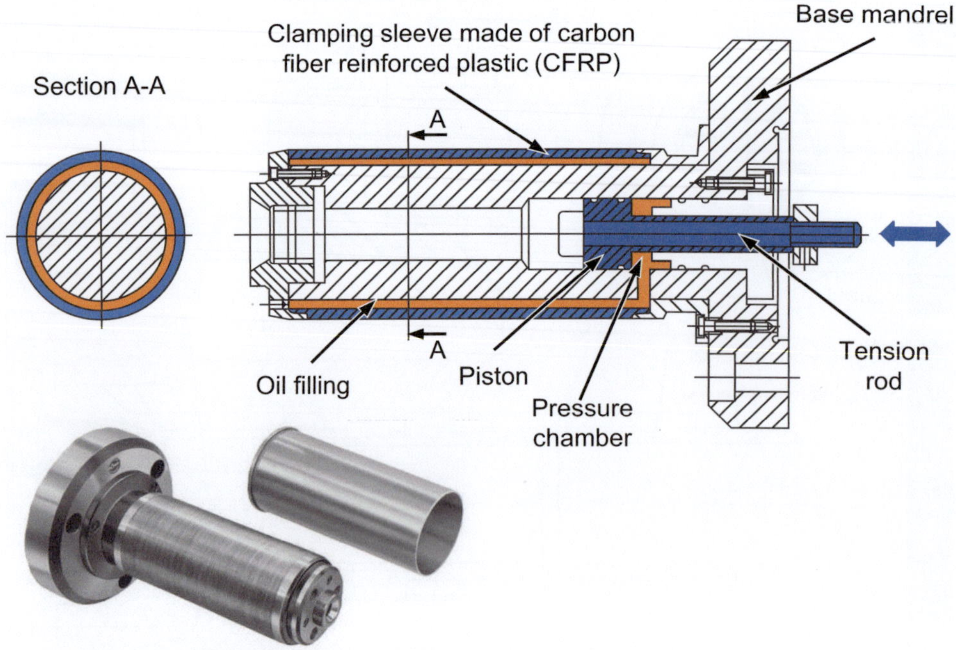

Fig. 10.139 Hydraulic mandrel. *Source* Röhm

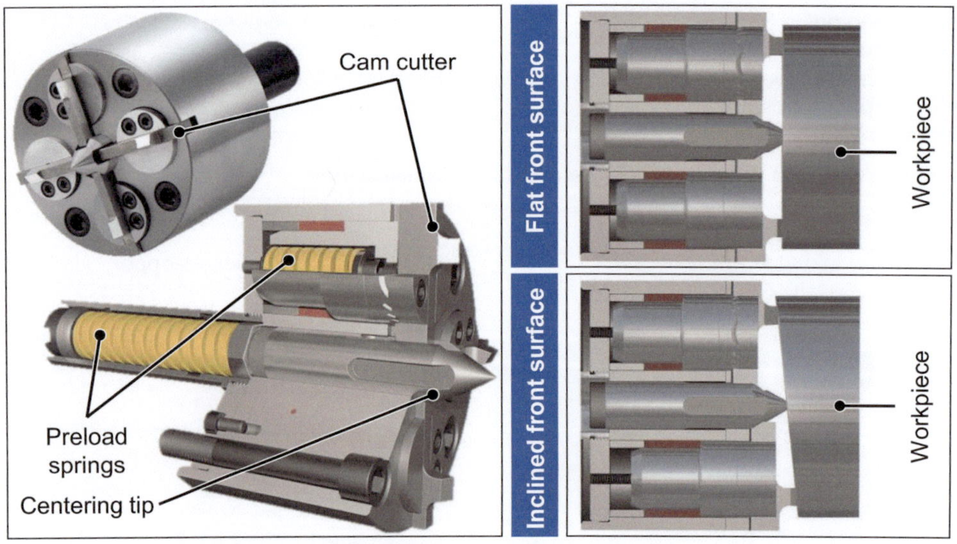

Fig. 10.140 Front chuck with hydraulic compensation system. *Source* Karl Bruckner

Integrated Machining of a Component

One variant for the integrated machining of shafts in one clamping operation is shown in Fig. 10.142. To turn on the clamping diameter, the workpiece is driven between centers by face drivers. The driving pins are pushed into the front side of the workpiece by the inner piston of the dual piston clamping cylinder. Uniform transfer of force to all driving pins is achieved by a ball joint. Then, higher torques are transferred to the workpiece for roughing by a shaft chuck. The front piston is used to actuate the chuck according to the wedge bar principle [SMW13].

10.4.3.2 Support of Long Workpieces

Long workpieces are held between centering tips or centering cones and supported by steady rests. A few representative application examples are introduced below.

Fixed or moving centering tips usually have a tip angle of 60°. They are hardened and ground. A moving tailstock center with triple bearing is shown at the top of Fig. 10.143. To drive the workpieces, a face driver can be inserted into the spindle that has a centering tip as well as a driving pin that engages into the front surface of the workpiece.

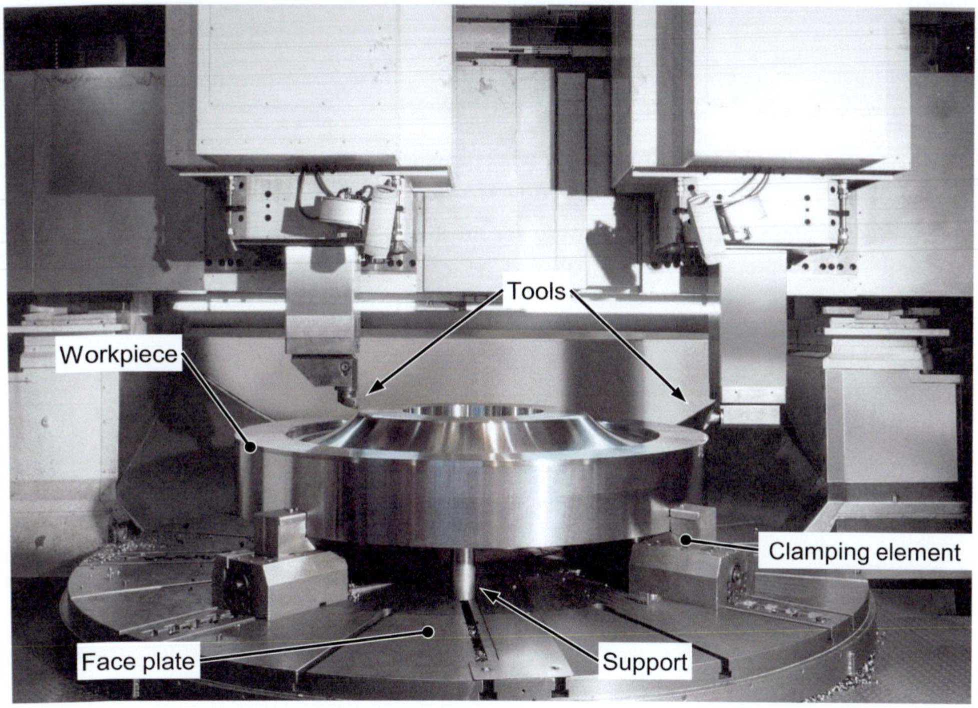

Fig. 10.141 Clamping of large workpieces on a face plate. *Source* Starrag Technology (Scharmann division)

For the support of long, rotating workpieces, steady rests or back rests are used. These prevent elastic deflection due to the dead weight and cutting forces. A steady rest has moving jaws that are adjusted concentrically to the diameter of the workpiece and to the correct rotation height. Back rests have rollers that support the workpiece and permit end face machining of the free end. For this, the workpiece is chucked at one end in the lathe chuck and the opposite end is supported by the back rest. The back rest is mounted on a separate carriage and can be moved along the workpiece axis. Workpieces are often first machined with a centering tip and only guided in the back rest for end machining (Fig. 10.144). If no separate carriage is provided, turret back rests can be used on 2-carriage lathes or turn mill centers that are mounted on the turret like a tool.

The back rest shown in Fig. 10.145 acts in a self-centering way via a gearbox. The actuating force can be applied hydraulically or pneumatically. The piston extending out causes the synchronous transfer of the movement to the two clamping arms via the bevels of the center piece to which the middle roller is attached. The clamping stroke is limited by the path controller and the clamping force using a safety check valve.

10.4.3.3 Change Equipment and Magazines for Workpieces

Equipment components for the automatic supply and unloading of workpieces for lathes range from simple loading aids to programmable robots. These will only be discussed here in part due to their many designs. Automatic change equipment that adapts to different workpiece shapes and permits unattended operation of a lathe is of interest.

Different workpiece supply concepts are used in line with the robot designs available. If there is enough room to do so, handling devices can be integrated into the workspace of a lathe (Fig. 10.146). Here, a 4-axis loading gantry moves the blanks from the workpiece carrier pallet system on the right of the machine to the workspace and the finished parts back again. For workpiece feed, several working spindles are integrated into the machine tool or external robots are used. For the CNC multi-spindle automatic lathe shown in Fig. 10.147, the 6-axis articulated arm robot mounted in the workspace loads and unloads the workpieces.

The workpiece loader in Fig. 10.147 is supplied with workpieces by the workpiece feed outside the machining area using the integrated palleting system in

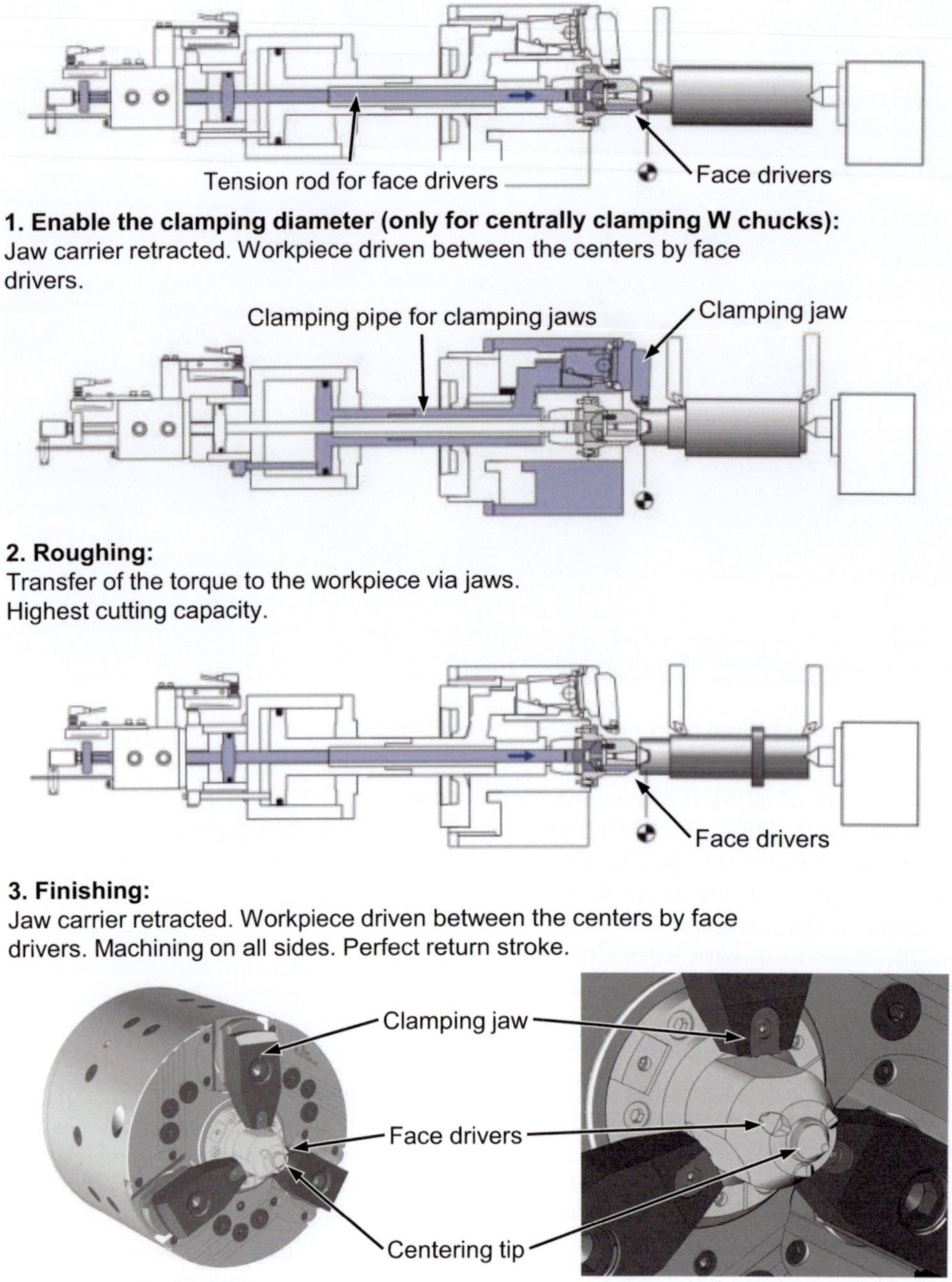

Fig. 10.142 Shaft chuck with retractable jaws and face drivers. *Source* SMW-Autoblok

Fig. 10.148. For workpiece changes, a gripper (with 3-point clamping) is mounted on each of two small area gantries. The finished part gripper takes a workpiece from the material lock and places it onto the finished part pallet. At the same time, the blank gripper takes a blank from the blank pallet and makes it available for machining. The two area gantries can each move towards the workpiece positions at the blank and finished part pallets. The pallets must always be in the correct position but need not be repositioned for every workpiece. Just as the workpiece magazine is emptied by removing blanks, the finished part pallets are filled and diverted in the direction of the material flow. There, they can be transported to other machines or for unloading.

The base material for turned parts are shaft and chuck blanks as well as material bars. The machining of bars is suitable for large volume production and requires high conversion effort on job changes. This effort can be reduced with the NC bar loading maga-

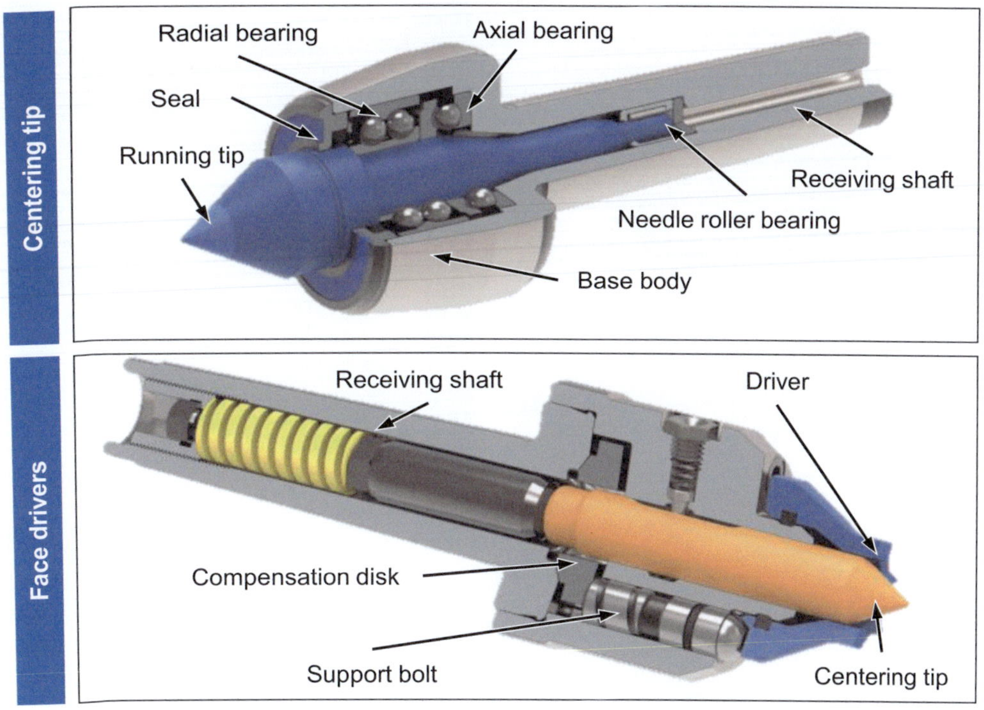

Fig. 10.143 Moving centering tip and face driver. *Source* Röhm

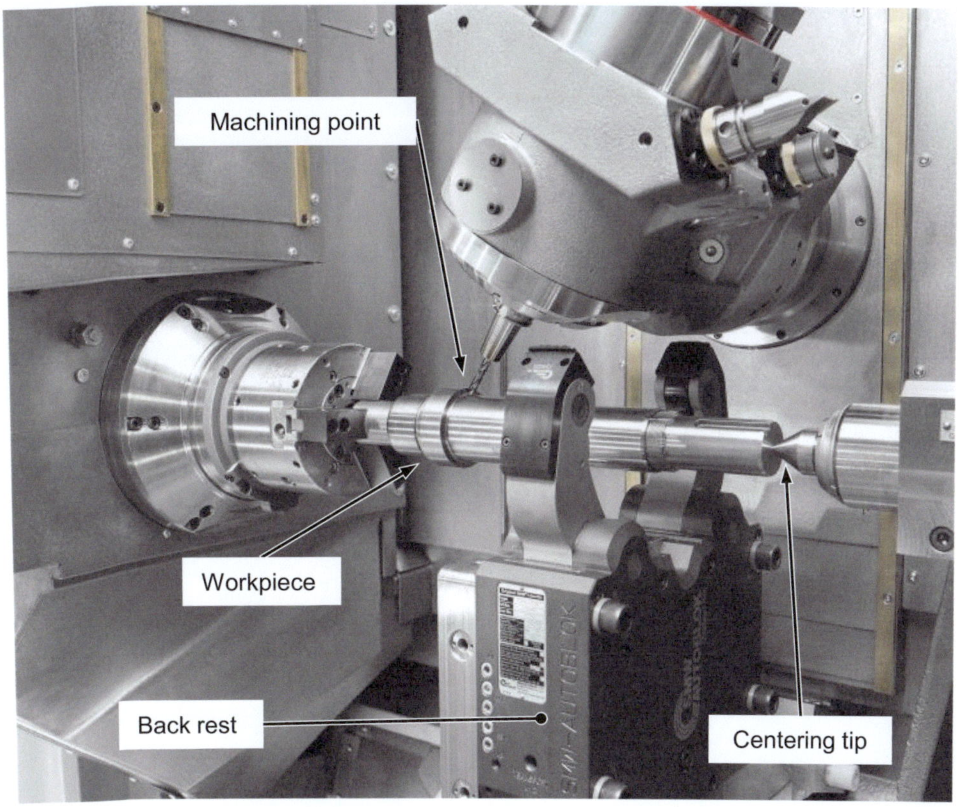

Fig. 10.144 Self-centering back rest in a turn mill center. *Source* INDEX

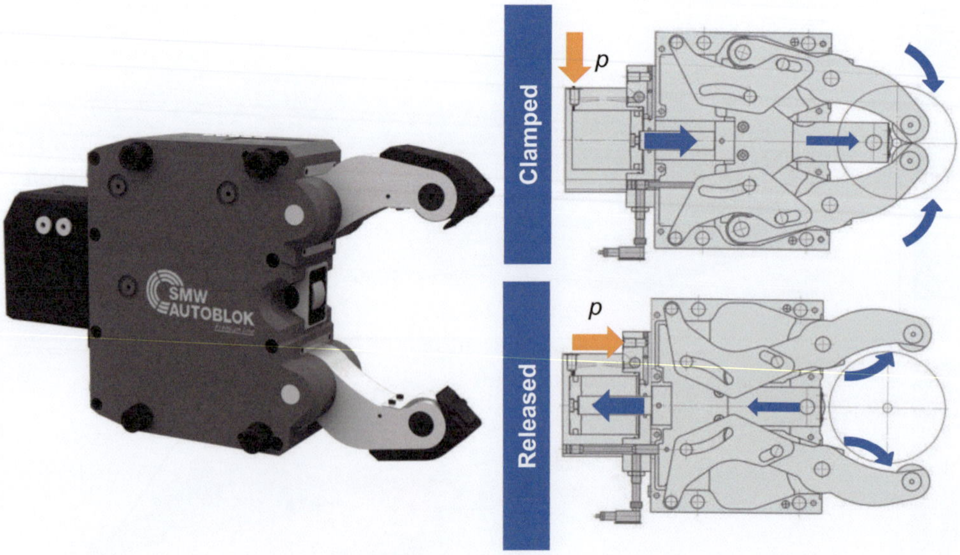

Fig. 10.145 Self-centering back rest. *Source* SMW-Autoblok

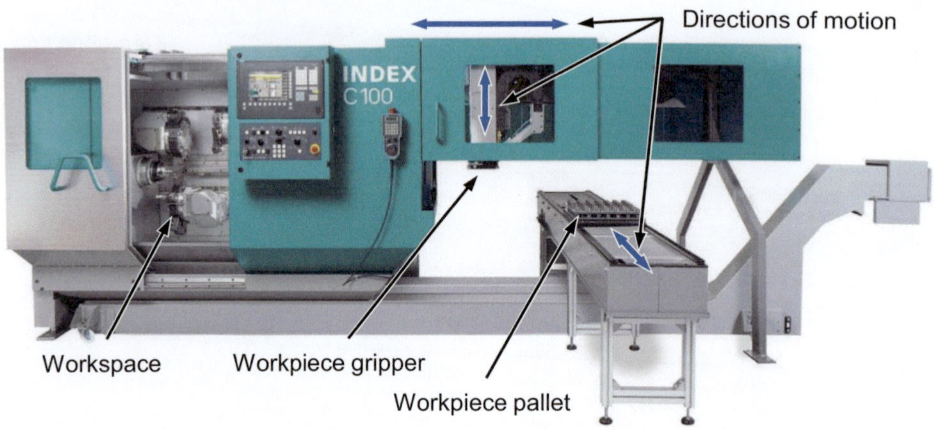

Fig. 10.146 Automatic workpiece feed with handling device. *Source* INDEX

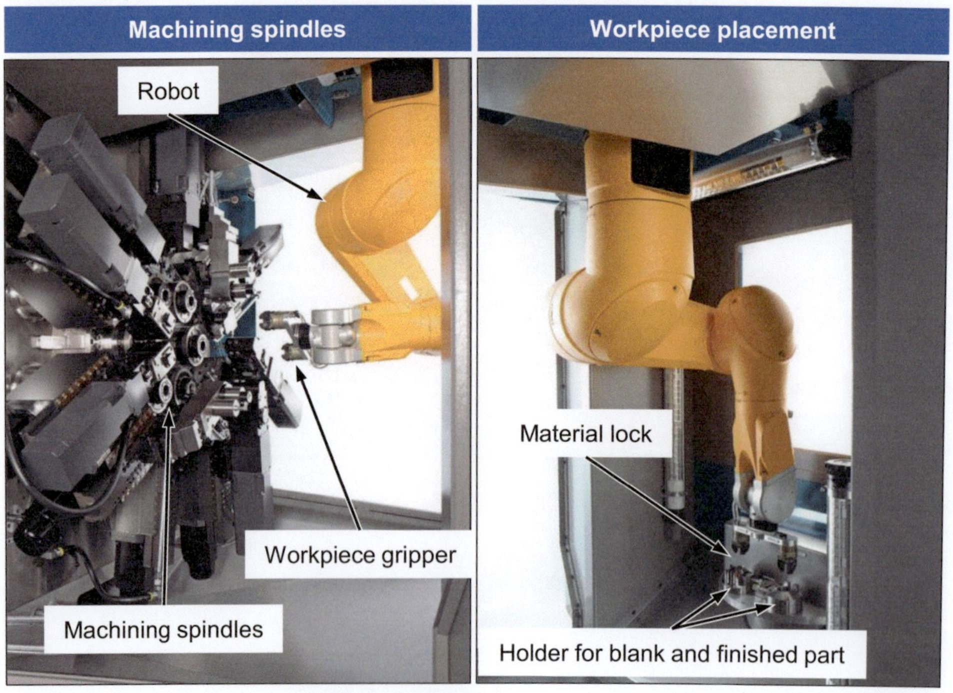

Fig. 10.147 Multi-spindle automatic lathe with automatic workpiece feed. *Source* INDEX

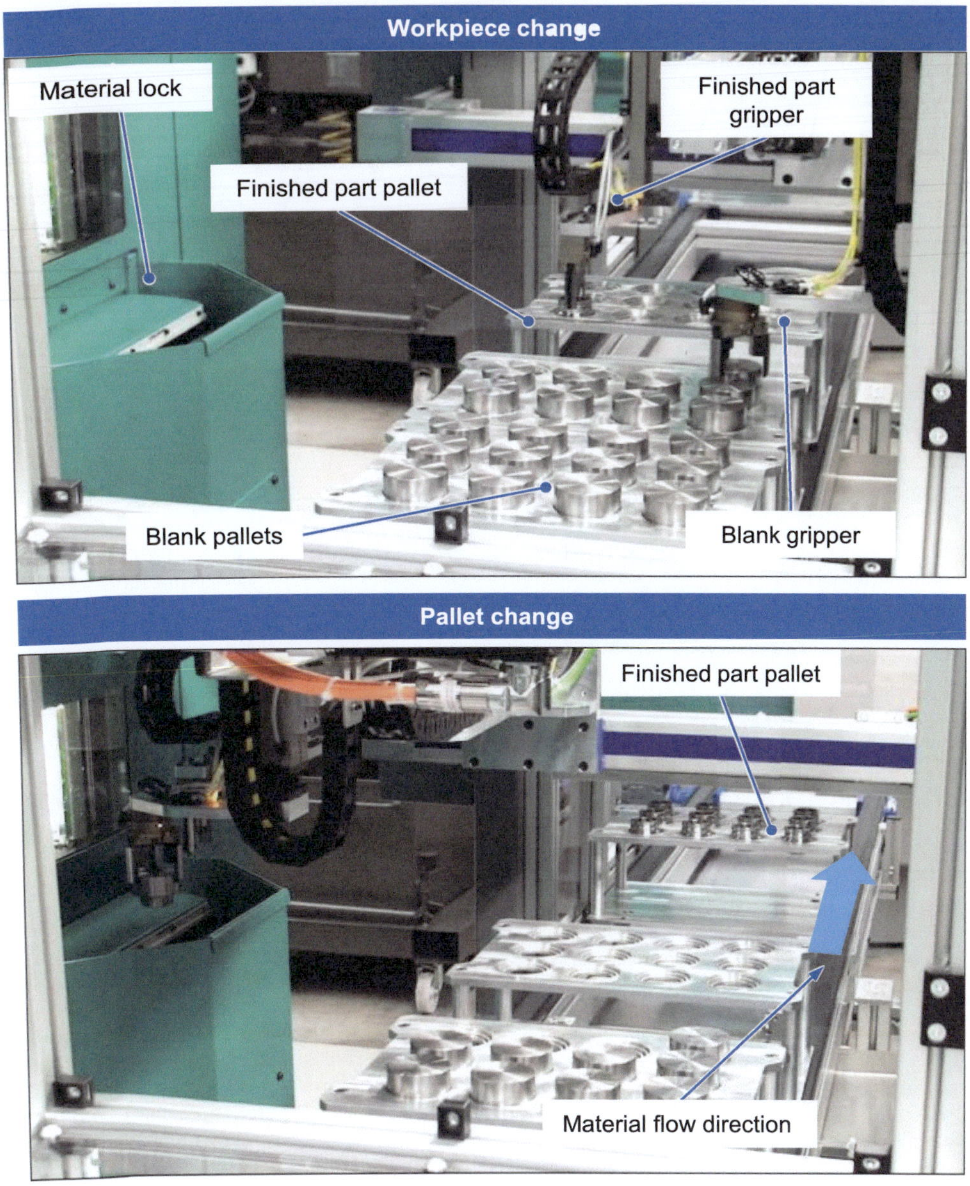

Fig. 10.148 Workpiece feed using pallets. *Source* INDEX

zine shown in Fig. 10.149. The material bars are guided by elastic rollers that automatically adapt to the material bar to be machined in each case. To prevent the material bars from oscillating, they are guided in the working spindle by an additional adapter sleeve. After a workpiece has been finish machined and is cut off, a bar up to 6 m in length is advanced section by section through the opened collet chuck by the telescopic feed. The material of the remaining piece of the material bar is fed in the area of the working spindle by the internal feed of the telescopic feed that also ejects the remaining piece. A new, single material bar is moved into the guide duct by the injector. This is where the feed guide and centering roll arms fold open, and the rollers adapt to the bar. The diameter-dependent clamping sleeves are not exchanged.

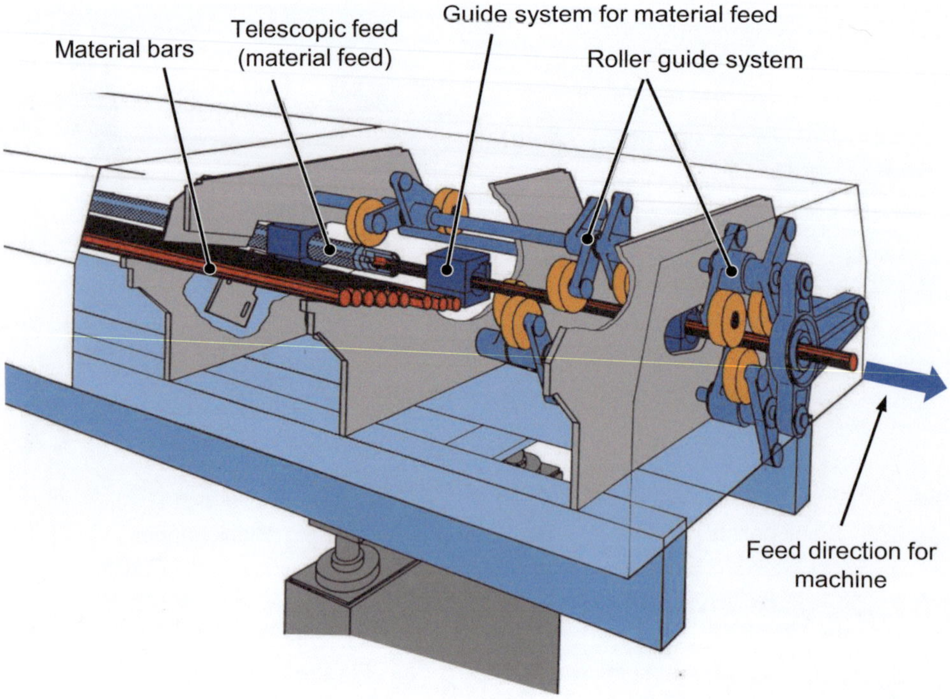

Fig. 10.149 Basic structure of an automatic bar loading magazine. *Source* Traub

10

References

[ABEL13] Abele, E., Sielaff, T., Beck M.: Schlussbericht zum Projekt Maxiem. Darmstadt (2013)

[BERU90] Richtlinie BGR 109 (April 1990). Richtlinien zur Vermeidung der Gefahren von Staubbränden und Staubexplosionen beim Schleifen, Bürsten und Polieren von Aluminium und seinen Legierungen

[BREC13a] Brecher, C., Müller, F., Wagner, P.: Statisches Verhalten von Werkzeugschnittstellen. VDI-Z **155**(1), 36–39 (2013)

[BREC13b] Brecher, C., Triebs, J., Jasper, D.: Energy efficient solutions for hydraulic units of machine tools. In: Nee, A.Y.C., Song, B., Ong, S.N. (eds.) Re-engineering manufacturing for sustainability, pp. 191–196. Springer, Singapore (2013)

[BREC15] Brecher, C., Müller, F., Wagner, P., Fey, M.: Analyse des Biegeverhaltens von Werkzeugschnittstellen. MM Maschinenmarkt **121**(30), 44–48 (2015)

[BREC16] Brecher, C., Müller, F., Wagner, P.: Werkzeugschnittstellen unter Torsionsbelastung. MM Maschinenmarkt **122**(14), 64–69 (2016)

[BUND06] Technische Regeln für Gefahrstoffe TRGS 900 (Oktober 2006). Grenzwerte in der Luft am Arbeitsplatz “Luftgrenzwerte”

[COMP17] Compatible Components Corporation: Urethane Vor-Spin Hydrocyclone. ▶ https://www.cccmix.com/urethane-vorspin-hydrocyclone/. [Stand: 01.12.2017]

[DENK11] Denkena, B., Möhring, H.-C., Hackelöer F., Hülsemeyer L., Dahlmann D., Augenstein, E., Nelles, J., Grigoleit, A.: Effiziente Fluidtechnik für Werkzeugmaschinen. wt Werkstattstechnik online **101**(5), 347–352 (2011)

[DGUV10] Information BGI/GUV-I 718 (November 2010). Minimalmengenschmierung in der spanenden Fertigung

[DIN03] Norm DIN 69893-6 (März 2003). Kegel-Hohlschäfte mit Plananlage – Teil 6: Kegel-Hohlschäfte Form F, Maße und Ausführung

[DIN06] Norm DIN ISO 10889-1 (Januar 2006). Werkzeughalter mit Zylinderschaft – Teil 1: Zylinderschaft, Aufnahmebohrung; Technische Lieferbedingungen

[DIN10a] Norm DIN ISO 702-2 (April 2010). Werkzeugmaschinen – Spindelköpfe und Drehfutter, Anschlussmaße – Teil 2: Kurzkegelaufnahme mit Camlock-Befestigung

[DIN10b] Norm DIN ISO 702-1 (April 2010). Werkzeugmaschinen – Spindelköpfe und Drehfutter, Anschlussmaße – Teil 1: Kurzkegelaufnahme mit Schrauben von vorne

[DIN10c] Norm DIN ISO 702-3 (April 2010). Werkzeugmaschinen – Spindelköpfe und Drehfutter, Anschlussmaße – Teil 3: Kurzkegelaufnahme mit Bajonett-Befestigung

[DIN10d] Norm DIN 6350-1 (Mai 2010). Drehfutter handbetätigt – Spannbacken zentrisch spannend – Teil1: Zylindrische Zentrieraufnahme

[DIN10e] Norm DIN ISO 26623-1 (November 2010). Polygonaler Hohlschaftkegel mit Plananlage – Teil 1: Maße und Bezeichnung von Schäften (ISO 26623-1:2008)

[DIN11a] Norm DIN 69893-1 (April 2011). Kegel-Hohlschäfte mit Plananlage – Teil 1: Kegel-Hohlschäfte Form A und Form C; Maße und Ausführung

[DIN11b] Norm DIN 69893-7 (April 2011). Kegel-Hohlschäfte mit Plananlage – Teil 7: Kegel-Hohlschäfte Form T für stehende Werkzeuge; Maße und Ausführung

[DIN12] Norm DIN 6989-5 (Oktober 2012). Kegel-Hohlschäfte mit Plananlage – Teil 5: Kegel-Hohlschäfte Form E; Maße und Ausführung

[DIN13a] Norm DIN 26622 (August 2013). Modulare Kegelschnittstelle mit Kugellaufbahnsystem – Teil 1: Maße und Bezeichnung von Schäften

[DIN13b] Norm DIN ISO 26622-1 (August 2013). Modulare Kegelschnittstelle mit Kugellaufbahnsystem – Teil

1: Maße und Bezeichnung von Schäften (ISO 26622-1:2008)
[DIN13c] Norm DIN 69893-2 (November 2013). Kegel-Hohlschäfte mit Plananlage – Teil 2: Kegel-Hohlschäfte Form B und Form D; Maße und Ausführung
[DIN13d] Norm DIN 51385 (Dezember 2013). Schmierstoffe – Bearbeitungsmedien für die Umformung und Zerspanung von Werkstoffen – Begriffe
[DIN14a] Norm DIN ISO 7388-1 (Juli 2014). Steilkegelschäfte für automatischen Werkzeugwechsel – Teil 1: Maße und Bezeichnungen von Schäften der Formen A, AD, AF, U, UD, UF (ISO 7388-1:2007)
[DIN14b] Norm DIN ISO 7388-1 (Juli 2014). Werkzeugschäfte mit Kegel 7/24 für automatischen Werkzeugwechsel – Teil 1: Maße und Bezeichnung von Schäften der Formen A, AD, AF, U, UD und UF
[DIN15a] Norm DIN ISO 5610-1 (Februar 2015). Klemmhalter mit Vierkannschaft für Wendeschneidplatten. Teil 1: Übersicht, Zuordnung und Bestimmung der Maße
[DIN15b] Norm DIN ISO 7388-3 (März 2015). Werkzeugschäfte mit Kegel 7/24 für automatischen Werkzeugwechsel – Teil3: Anzugsbolzen für Schäfte der Formen AC, AD, AF, UC, UD, UF, JD, und JF (ISO 7338-3:2013)
[DIN64] Norm DIN 1806 (Juni 1964). Kegelschäfte für Querkeilbefestigung
[DIN78] Norm DIN 55058 (September 1978). Spindelköpfe für Stellhülsen; Anschlußmaße
[DIN80a] Norm DIN 55026 (März 1980). Werkzeugmaschinen – Spindelköpfe mit Zentrierkegel und Flansch – Maße
[DIN80b] Norm DIN 55027 (März 1980). Werkzeugmaschinen – Spindelköpfe mit Zentrierkegel, Flansch und Bajonettscheibenbefestigung, Zubehör – Maße
[DIN81] Norm DIN 6327-4 (August 1981). Stellhülsen mit Werkzeugkegel, Nuttmuttern und Klemmmuttern
[DIN85a] Norm DIN 6327-3 (Juni 1985). Stellhülsen mit Werkzeugkegel – abgesetzte Bauart
[DIN85b] Norm DIN 6327-1 (Juni 1985). Stellhülsen mit Werkzeugkegel – kurze Bauart
[DIN85c] Norm DIN 6327-2 (Juni 1985). Stellhülsen mit Werkzeugkegel – lange Bauart
[DIN85d] Norm DIN 55201 Teil 1 (September 1985). Werkzeugmaschinen – Paletten für Werkzeugmaschinen – Nenngrößen bis 800mm – Form A
[DIN87a] Norm DIN 228-1 (März 1987). Morsekegel und Metrische Kegel; Kegelschäfte
[DIN87b] Norm DIN 228–2 (März 1987). Morsekegel und Metrische Kegel; Kegelhülsen
[DIN87c] Norm DIN 2079 (August 1987). Werkzeugmaschinen; Spindel mit Steigkegel 7:24
[DIN87d] Norm DIN 69881-1 (September 1987). Werkzeughalter mit Prisma; Prisma, Prismaaufnahme; technische Lieferbedingungen
[DIN87e] Norm DIN 55201 Teil 2 (Dezember 1987). Werkzeugmaschinen – Paletten für Werkzeugmaschinen – Nenngrößen ab 800mm – Form A
[DIN92] Norm DIN 6535 (Februar 1992). Zylinderschäfte für Spiralbohrer und Schaftfräser aus Hartmetall; Maße
[DIN99] Norm DIN 1835-1 (Januar 1999). Zylinderschäfte für Fräser – Teil 1: Maße
[ENER13] Energie-Loesungen: Digital-Scroll-Technologie. URL: ► http://www.energie-loesungen.de/energie-loesungen/encyclopedia/thema.php?keyID=441. [Stand: 15.01.2013]
[EUCH13] Euchner: Identsystem CIS. Leinfelden-Echterdingen, Januar 2013. Firmenschrift
[FORK13] Forkardt Deutschland: Betriebsanleitung Handspannfutter F+. Erkrath, März 2013. Firmenschrift
[GROL10] Grollius, H.-W.: Grundlagen der Hydraulik, 5th edn. Hanser, München (2010)
[HAIN15] Hainbuch: Katalog 2015. Marbach, 2015. Firmenschrift
[HAUP05] Richtlinie BGR 204 (August 2005). Umgang mit Magnesium
[HAUP06] Richtlinie BGR 143 (Januar 2006). Tätigkeit mit Kühlschmierstoffen
[HERS11] Herschel, D.: Druckflüssigkeiten. In: Will, D., Gebhardt, N., Ströhl, H. (eds.) Hydraulik, 5th edn. Springer, Berlin (2011)
[HIRS97] Hirsch, W., Hempel, D.: Untersuchungen zum Explosionsschutz beim Einsatz von Kühlschmierstoffen in Werkzeugmaschinen. (Reihe: PTB-Bericht). Bremerhaven: Wirtschaftsverl. NW (1997)
[HÜLL3a] Hüller Hille: Die CNC Zentren-Baureihe. Diedesheim, 2003. Firmenschrift
[HÜLL3b] Hüller Hille: Horizontal-Bearbeitungszentren. Diedesheim, 2003. Firmenschrift
[INST80] Institut für Fachschulwesen der DDR (eds.): Vorrichtungen. Verlag Technik, Berlin (1980)
[ISO12] Norm DIN ISO 1219-1: Fluidtechnik – Graphische Symbole und Schaltpläne – Teil 1: Graphische Symbole für konventionelle und datentechnische Anwendungen (2012)
[KARL14] Karl Bruckner: Stirnfutter HS. Weinstadt, 2014. Firmenschrift
[KLOC08] Klocke, F., König, W.: Fertigungsverfahren 1, 8th edn. Springer, Berlin (2008)
[KLOC98] Klocke, F.: Minimalmengen Kühlschmierung – Systeme, Medien, Einsatzmöglichkeiten. In: VDI-Bericht 1375. Tagungsband zu VDI-Fachtagung. Aachen, 30.–31. März 1998, pp. 337–369
[LEMB93] Lembke, D.: Untersuchung der Gestaltungsmöglichkeiten für die Schnittstelle Maschine/Werkzeug. Diss. RWTH Aachen (1993)
[LOCK84] Lock, F.: Konzeption und Entwicklung von Vorrichtungssystemen für die spanende Fertigung. Diss. RWTH Aachen (1984)
[MURR12] Murrenhoff, H., Linden, D.: Grundlagen der Fluidtechnik. (Reihe: Reihe Fluidtechnik. Bd. 2), 7th edn. IFAS, Aachen (2012)
[PEIS03] Peiseler: Automatisches Teilgerät. Remscheid, 2003. Firmenschrift
[PFLA98] Pflaumenbaum, B., Kleine, S.: BIA Report 1/98, Gefahrstoffliste 1998. St. Augustin: Hauptverband der gewerblichen Berufsgenossenschaften (HVBG) (1998)
[RITT10a] Rittal GmbH & Co. KG: Rittal ist Partner beim Themencluster Energieeffizienz Sparprogramm für Maschinenkühlung. Industrie Anzeiger **38**, 76 (2010)
[RITT10b] Rittal GmbH & Co. KG: Frequenzgeregelter Wärmetauscher mit 70 Prozent weniger Energieverbrauch. MAV Maschinen Anlagen Verfahren **9**, 122 (2010)
[RÖHM13] Röhm: Spanndorne. Sontheim/Brenz, 2013/2014. Firmenschrift
[ROTH10] Rothenbücher, S., Kuhrke, B.: Energiebündel auf dem Prüfstand. WB Werkstatt und Betrieb **9**, 130–137 (2010)
[SCHU94] Schubert, I.: Grenzlastverhalten von Schnittstellen zwischen Maschine und Werkzeug. Diss. RWTH Aachen (1994)
[SIEM15] Siemens Building Technologies: Handbuch Kältetechnik. ► https://www.cee.siemens.com/web/austria/de/in-

dustry/bt/support/Documents/5476_1811.pdf. [Stand: 16.07.2015]

[SMW13] SMW-Autoblok Spannsysteme: Katalog 13G. Meckenbeuren, 2013. Firmenschrift

[THIE71] Thiel, H.: Vorrichtungen – Gestalten, Bemessen, Bewerten. Hanser, München (1971)

[TOSH] Toshiba: Grundlagen der Invertertechnologie. ► http://www.toshiba-klima.net/images/content/file/Techniklexikon/INVERTER_allgemeine_Infos.pdf

[TRIE15] Triebs, J.B.: Analyse, Optimierung und Simulation des Energieumsatzes dezentraler Fluidkreisläufe von Werkzeugmaschinen. Dissertation RWTH Aachen (2015)

[VDI07] Richtlinie VDI 3397 (Mai 2007). Kühlschmierstoffe für spanende und umformende Fertigungsverfahren

[VDI72] VDI (eds.): VDI-Handbuch Betriebstechnik/Teil 3. Betriebsmittel. (Reihe: VDI-Handbuch Betriebstechnik. Bd. 3). VDI, Düsseldorf (1972)

[VDI90a] Richtlinie VDI 2860 (Mai 1990). Montage- und Handhabungstechnik

[VDI90b] Richtlinie VDI 2860 (Mai 1990). Montage- und Handhabungstechnik – Handhabungsfunktionen, Handhabungseinrichtungen; Begriffe, Definitionen, Symbole

[VDMA02] VDMA: Kühlschmierstoffe – Frische Luft am Arbeitsplatz. FV ALT, Frankfurt (2002)

[VDMA95] VDMA: Verpuffungen und Brände in Werkzeugmaschinen. Expertenbericht der Fachgemeinschaft Werkzeugmaschinen und Fertigungssysteme des VDMA. Frankfurt (1995)

[VDW96] Abschlussbericht 3002: Explosionsdruckentlastung von spanabhebenden Werkzeugmaschinen (1996)

[VDW98] Forschungsbericht VDW 3001/1: Untersuchungen zur Vermeidung von Brand- und Explosionsgefahren sowie zur Gewährleistung der Sicherheit des Bedienpersonals spanabhebender Werkzeugmaschinen (1998)

[WALK04] Walker Magnetics Europe: Magnetspanntechnik, 2004. Firmenschrift

[WECK] Weck, M., Marpert, M.: Einfluss des Kühlschmierstoffsystems auf die Gestaltung von Werkzeugmaschinen zur Magnesiumbearbeitung. In:

[WEIN99] Weinert, K.: Trockenbearbeitung und Minimalmengenkühlschmierung. (Reihe: VDI-Buch). Springer, Berlin (1999)

Design Examples of Selected Machine Tools

I thank my staff members Mathias Dehn M.Sc. and Tim Kaczinski M.Sc. for their contribution in revising this chapter.

C. Brecher and M. Weck, *Machine Tools Production Systems 1*,
Lecture Notes in Production Engineering, https://doi.org/10.1007/978-3-662-68120-6_11

Contents

C. Brecher and M. Weck, *Machine Tools Production Systems 1*,
Lecture Notes in Production Engineering, https://doi.org/10.1007/978-3-662-68120-6_11

The previous chapters have already dealt with an extremely wide range of machine tools and their functionalities. This chapter discusses three specific examples and provides an insight into the practical implementation of the design and structure of machine tools. The machine tools introduced include a Starrag Technology GmbH machining center, an Elb-Schliff Werkzeugmaschinen GmbH grinding machine, and an INDEX GmbH & Co. KG lathe.

11.1 Machining Center

11.1.1 Company Profile

Starrag Technology GmbH is a medium/large Swiss machine tool manufacturer producing high-quality and technologically advanced products for drilling, turning, grinding, and milling. With over 120 years of experience in mechanical engineering, the company has made a name for itself as a supplier to a wide range of international industrial sectors.

11

11.1.2 Overview

The ALPHA series from Starrag Technology GmbH is a modular machining center with a new concept in column/slant-bed design. Compared to conventional double-column or box-in-box designs, this concept allows the moved masses to be lighter with increased stiffness. Furthermore, the axis drives in this concept act on the center of mass in the main machine components allowing a smaller gap between the Z-guide system and the main spindle. The modularity of the system means a large number of different main spindles, tool supply systems and tables can be used.

Figure 11.1 shows the typical structure of an ALPHA machine with a horizontal spindle, rack-type magazine and setup area. The main spindle range covers horizontal spindles, motor spindles, a turning-drilling-milling head, quills, a face plate, a horizontal-vertical head, and a swivel spindle for simultaneous 5-axis machining. Tools are supplied from a standard chain magazine or a rack-type magazine in various sizes.

The table can be designed as a rotary table for additional turning operations. Several ALPHA machines can be combined to form a flexible production environment, a pallet system is available for this purpose.

11.1.3 General Description

As shown in Fig. 11.2 the machine consists of two separate beds, the table bed **7** with the rotary table 6, and the column bed **9**. Both beds are fitted to the floor with installation elements **10**. These elements allow the machine alignment to be adjusted. Due to its size, the machine is not inherently rigid. The bed stiffness comes from the foundation via the installation elements. The machine column **3**, which can move in the Z-direction, is located on the column bed. The slant-bed design of the bed can be seen in the front view. The spindle housing **2** is located on the machine column which can move in the Y-direction. The main spindle **1** is located in the

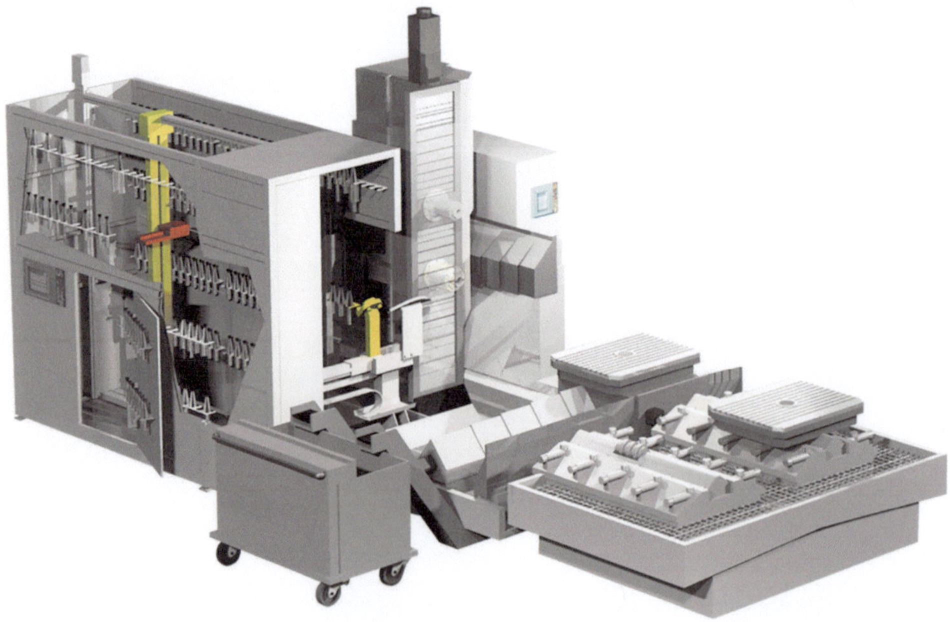

Fig. 11.1 Overview of the ALPHA series. *Source* Starrag Technology (Scharmann division)

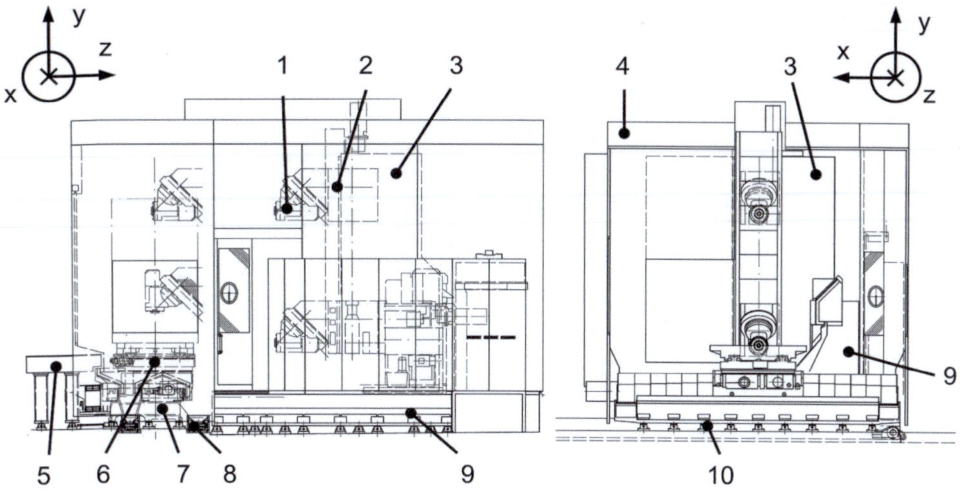

Fig. 11.2 Structure of the DST ALPHA with separate beds, according to Starrag Technology (Scharmann division)

spindle housing. A chip conveyor **8** is located under the rotary table. A feed station **5** is provided to allow the machine to be loaded quickly. The entire machine is surrounded by a safety enclosure **4**.

11.1.4 Machine Network

Several ALPHA machines can be connected to form a single machine network. Figure 11.3 shows a typical arrangement of two machines set up next to each other and connected by a pallet conveyor system **16**. Both machines have a coolant container **12**, an automatic tool changer with a tool storage area **11**, a chip conveyor **8** and a control panel **14**. The electronics for calculating the movement and for providing the electrical energy are located in the control cabinet **13**.

The machining area **15** is connected to the pallet conveyor system in such a way as to keep setup times to a minimum. Workpieces are set up on external pallets in the setup area **18** away from the machines. The entire system is controlled by the operator **17**.

11.1.5 Table Bed

Figure 11.4 shows the ALPHA machine table bed **7**. This is a welded construction located on the installation elements **10** as shown. The carriage **25**, forming the X-axis of the machine, is moved by a ball screw drive.

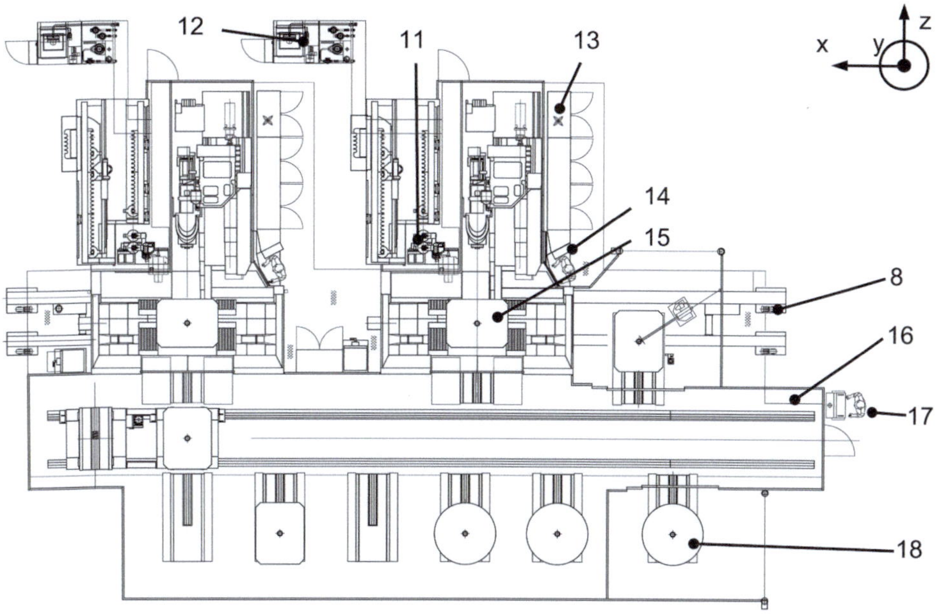

Fig. 11.3 DST ALPHA in a machine network, according to Starrag Technology (Scharmann division)

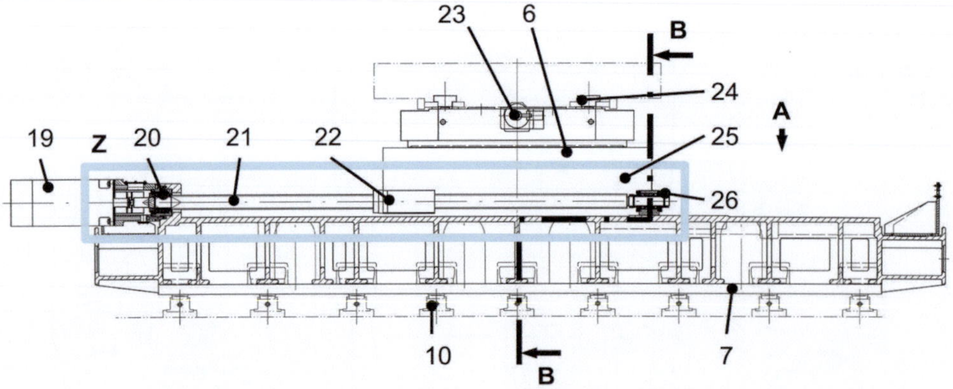

Fig. 11.4 Table bed in cross-section, according to Starrag Technology (Scharmann division)

This includes the motor **19**, the fixed bearing **20** and the directionally stable floating bearing **26**, the ball screw spindle **21** and the spindle nut **22**. The spindle nut is connected to the carriage and converts the spindle rotation to a translational movement of the carriage. A rotary table **6**, only indicated in the figure, is located on the carriage. The pallets are changed via the drive **23** and this action is guided by the guide systems **24**.

Table Bed: View A

The view from above the table bed in Fig. 11.5 shows the motor **19**, the ball screw drive spindle nut **22**, the ball screw spindle **21** and the directionally stable floating bearing **26**. The direct measuring system **27** captures the position of the carriage.

This view also shows the alignment of the guide rails **28**. The two lower guide rails are fixed on the table horizontally while the upper guide rail is offset at 90°. This concept with angled guide rails is repeated for the Y and Z-axes and utilizes the high tensile and compressive stiffness compared to the lateral stiffness of the linear guide systems in the main load direction.

Table Bed: Section B-B

The section shown in Fig. 11.6 shows the telescopic covers **29** over the axis and the ribbing **30** of the table bed. The arrangement of the guide rails **28** is as described above. The two guide rails on the left and in the center in this view are parallel to the foundation while the guide rail on the right is rotated by 90°. The ball screw spindle **21**, shown here without the bearings, is supported by the fixed bearing as well as the directionally stable floating bearing **26**. The installation elements **10** are mounted directly under the pockets in the bed structure.

Table Bed: Detail Z-Ball Screw Drive Bearing Arrangement

Figure 11.7 shows the detail of the ball screw drive bearing arrangement. The motor **19** is connected to the table bed via the flange **31**. The clutch **36** can be mounted via the access point **32**. The fixed bearing **20** consists of four angular contact ball bearings **35**. Three of these bearings are mounted in the same direction. The outer rings are fixed in position by the adjusting nut **34** and the bearing inner rings are fully tightened axially by the nut **33**.

The directionally stable floating bearing **26** consists of two sealed **38** cylindrical roller bearings **37**. This floating bearing arrangement ensures the achievement of a high critical whirling speed.

11.1.6 Column Bed

The column bed is designed as a new type of slant-bed to minimize the mass of the movable column. This

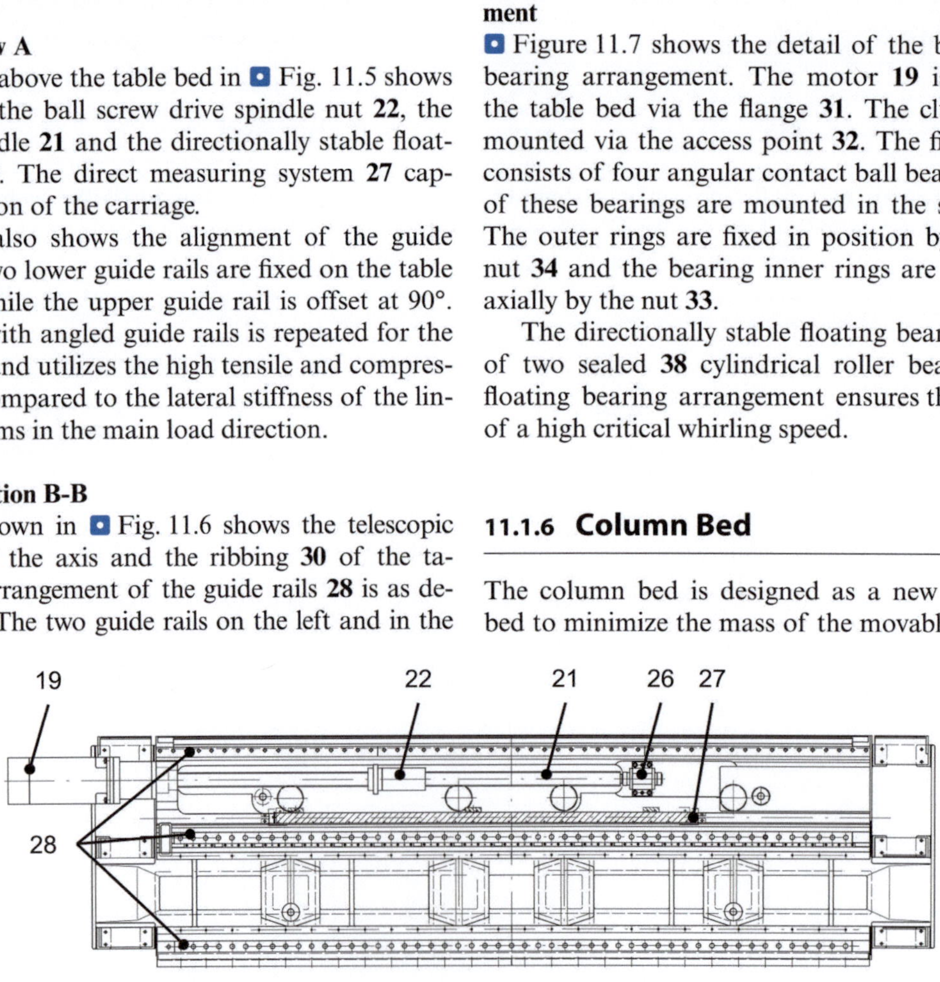

Fig. 11.5 Table bed, view A, according to Starrag Technology (Scharmann division)

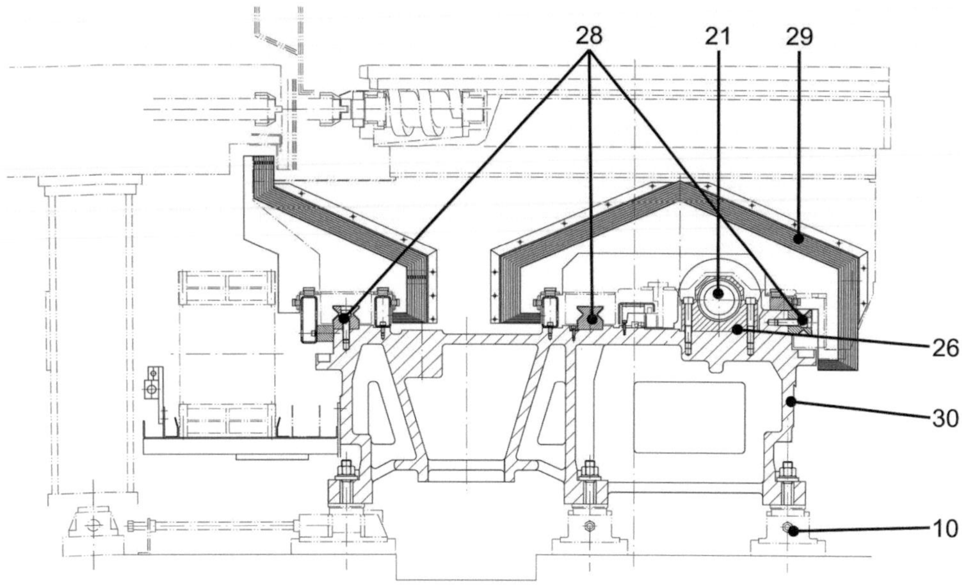

■ **Fig. 11.6** Table bed, section B-B, according to Starrag Technology (Scharmann division)

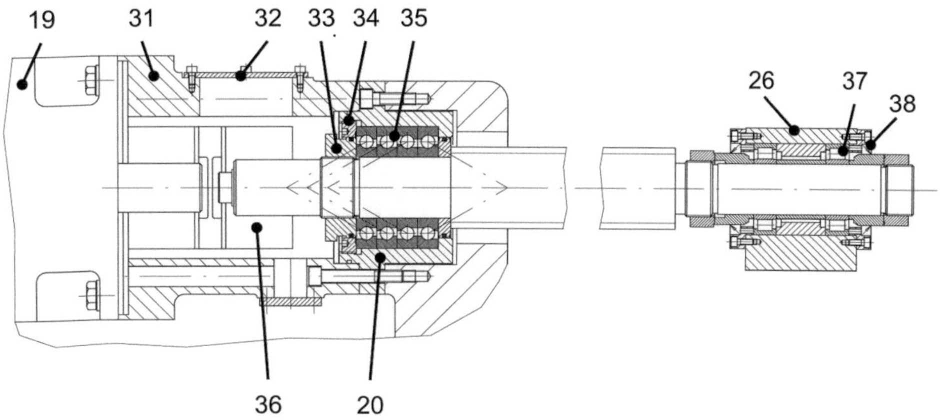

■ **Fig. 11.7** Table bed, detail Z, according to Starrag Technology (Scharmann division)

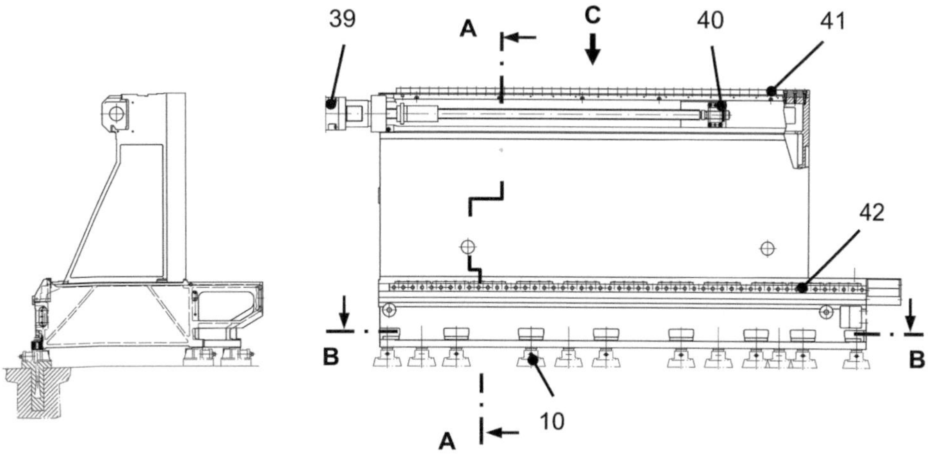

■ **Fig. 11.8** Column bed, according to Starrag Technology (Scharmann division)

bed is designed to be comparatively large in order to be able to absorb the process forces without a large lever arm. ◘ Figure 11.8 left, clearly shows this special design. Unlike the table bed, the guide systems are not in a single plane. Instead, the first guide system **41** is mounted at the highest point on the bed and is parallel to the ground. The second guide system **42** is located in the center of the design and is rotated by 90°. The drive **39** and the relevant bearing arrangement **40** are also located in the upper plane. As with the table bed, the column bed is mounted on installation elements **10**.

Column Bed: View C

The view in ◘ Fig. 11.9 shows the two guide rails **41** and **42** as well as the direct measuring system **43**. The measuring system measures the actual position of the column.

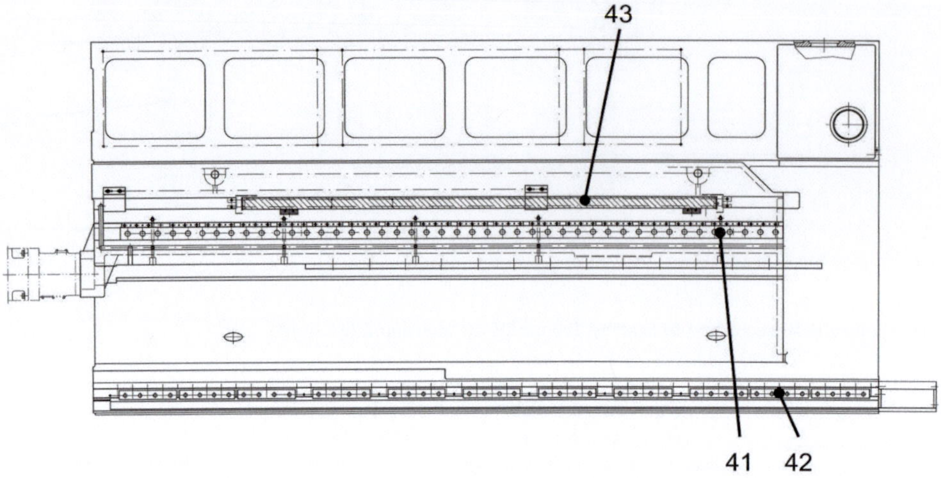

◘ **Fig. 11.9** Column bed: View C, according to Starrag Technology (Scharmann division)

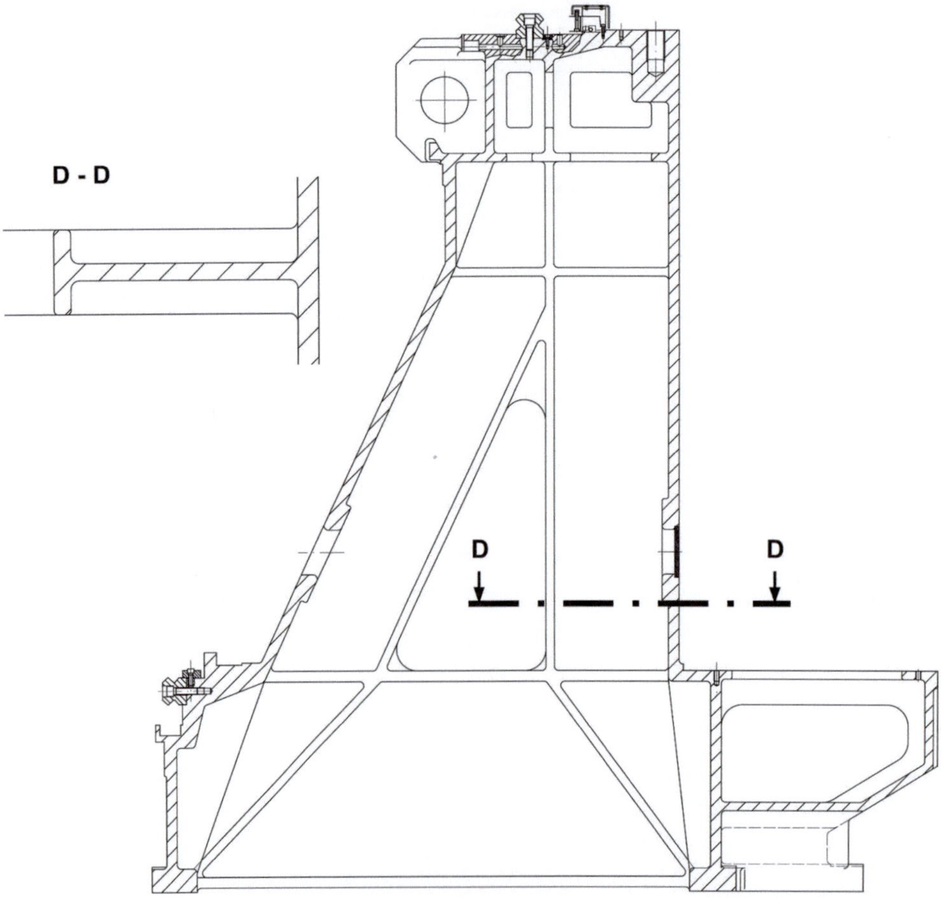

◘ **Fig. 11.10** Column bed: Section A-A and D-D, according to Starrag Technology (Scharmann division)

Column Bed: Section A-A

As can be seen in Fig. 11.10, the guide rails are screwed to the column bed. The ribbing of the column supports the guide rails. The transverse ribs can be seen in section D-D.

Column Bed: Section B-B

Section B-B of the column bed, Fig. 11.11, shows the pockets for connecting the installation elements as well as the ribbing of the bed.

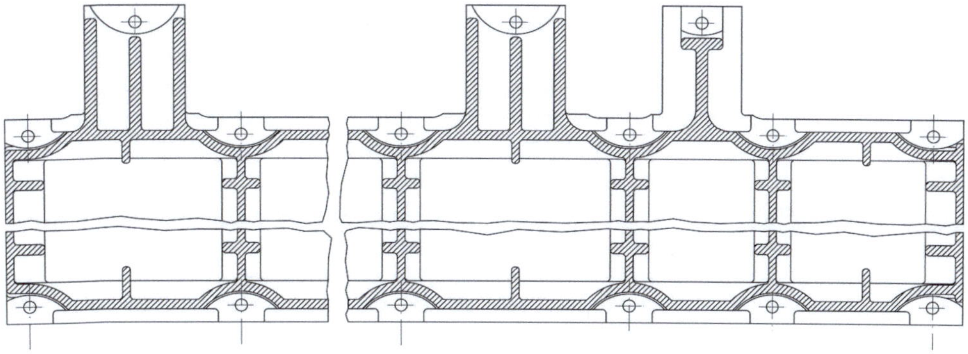

Fig. 11.11 Column bed: Section B-B, according to Starrag Technology (Scharmann division)

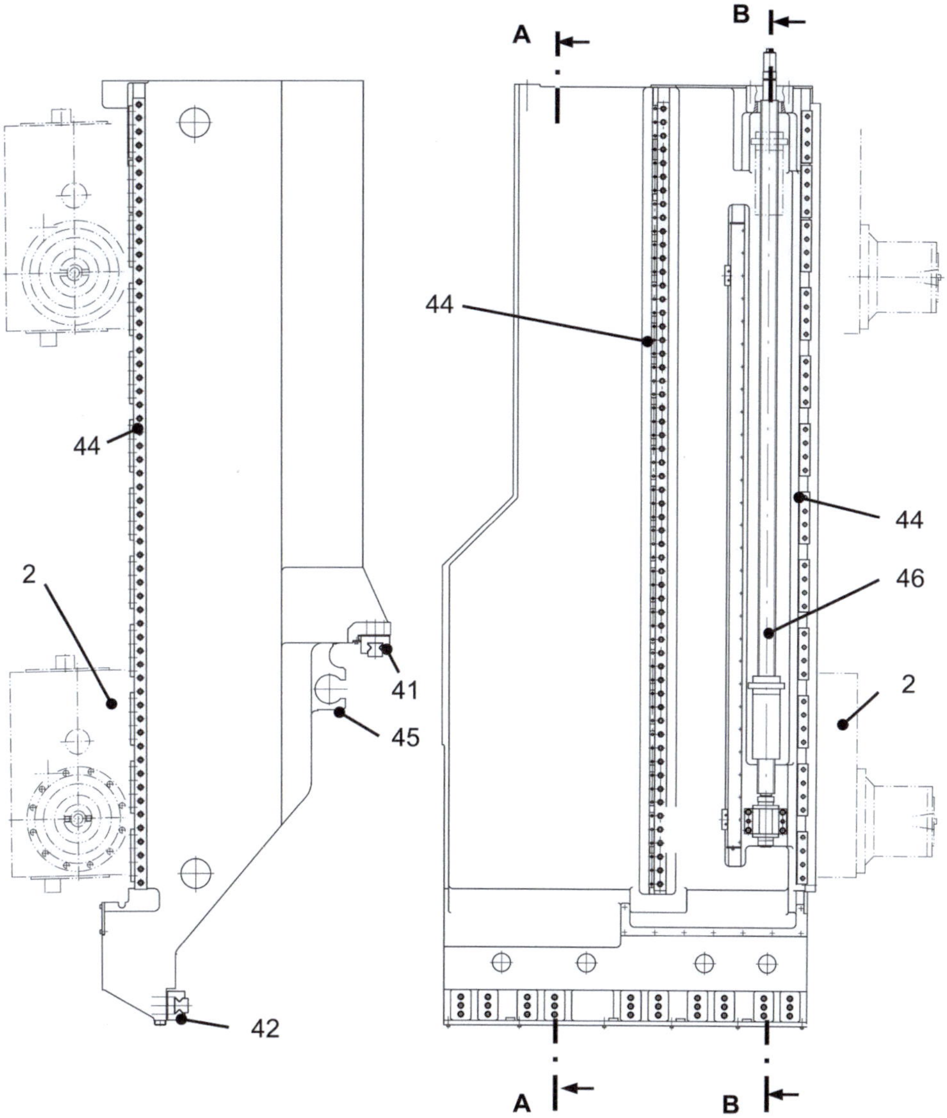

Fig. 11.12 Column, according to Starrag Technology (Scharmann division)

11.1.7 Column

▪ Figure 11.12 shows the movable column on the ALPHA machine. The spindle housing **2** is guided along the Y-axis by the guide systems **44** and is driven by the ball screw drive **46**. The spindle nut support **45** and the linear guide carriages **41**, **42** on the Z-axis are also shown.

Column: Section A-A and B-B

The sectional view of the column, ▪ Fig. 11.13, shows the ribbing that ensures high stiffness with a low mass. The greater wall thicknesses can be clearly seen in the more highly loaded areas, such as near the guide systems. The bearing arrangement on the ball screw drive spindle for the column box is a fixed/floating bearing arrangement. The directionally stable floating bearing **50** supports only radial forces and consists of two cylindrical roller bearings mounted at a distance apart. The fixed bearing **49** consists of four angular contact ball bearings. The ball screw drive spindle is driven by the motor **47** and the clutch **48**.

11.1.8 Horizontal-Vertical Head

▪ Figure 11.14 shows the structure of the horizontal-vertical head (HV head). This element allows the main spindle **1** to be moved in a horizontal position (left) and a vertical position (right) relative to the spindle housing **2**. To achieve this, the head is rotated by a swivel mechanism about the swivel axis **51**. The tool holder **52** is on the main spindle. Cooling lubricant is fed to the workspace via the cooling lubricant nozzles **53**.

Section of HV Head

▪ Figure 11.15 shows the HV head in section. The main spindle is driven from the drive shaft **54** via an interme-

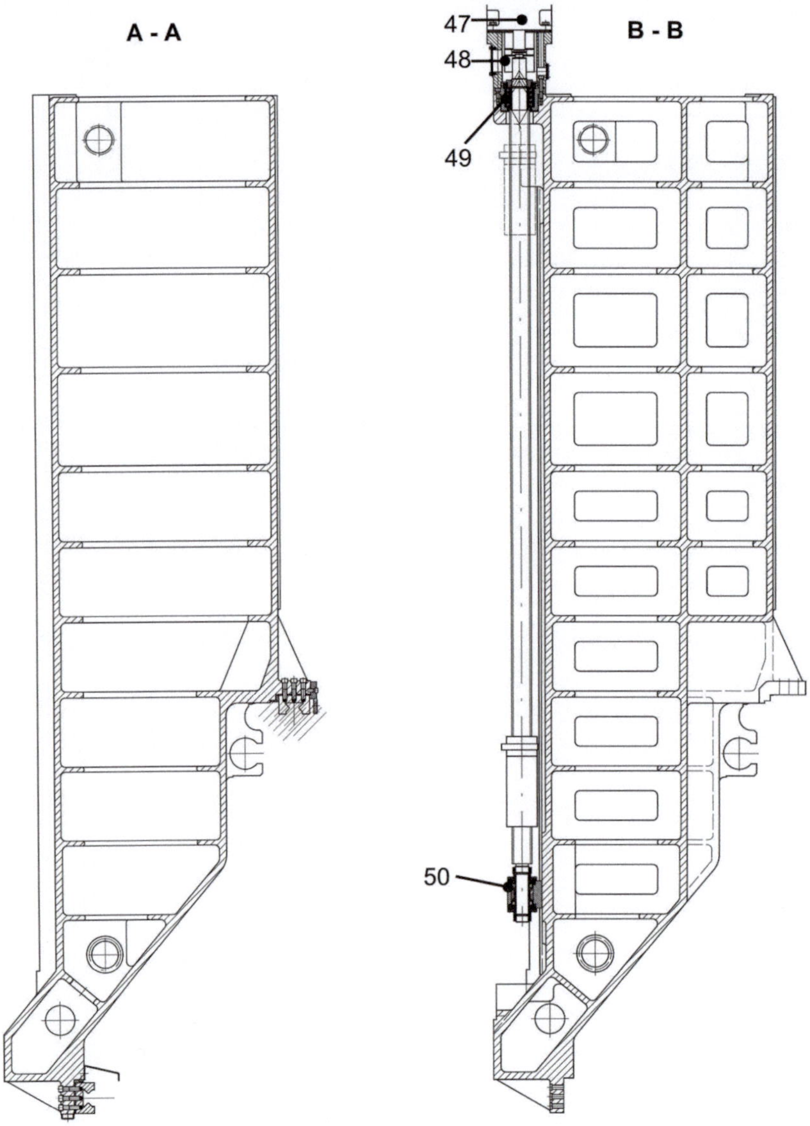

▪ **Fig. 11.13** Column section A-A and B-B, according to Starrag Technology (Scharmann division)

diate shaft **57** to the spindle shaft **61** using bevel gear pairs. The drive shaft and the intermediate shaft are supported in O arrangements (**55**). The gear teeth are oil-lubricated; the lubricating oil flows by gravity to the lowest point in the head (horizontal position: **62**, vertical position: **60**) and is extracted from this point. Depending on the position of the head, one of the two extraction lines **59** in the swivel housing **58** is connected to the extraction line in the head housing **56**, and the other is closed.

HV Head: Section A-A

The swivel component of the HV head shown in ▫ Fig. 11.16 is driven via a double-acting cylinder (see section B-B, ▫ Fig. 11.17). This moves the rack **65** which meshes with the pivot shaft **64**. The pivot shaft is hollow and, unlike the stationary housing, is supported radially by the needle roller bearing **63** and axially by the axial bearing **66**. The bearing housing **69**, which is screwed (**70**) to the swivel housing **58** and also swivels, is driven by the pivot shaft via slot nuts **73**.

Before the head can swivel, the clamping must be released. To achieve this, the clamping piston on the Hirth serration **72** must be pushed forward hydraulically so that the Hirth serration **68** is opened via the sleeve **67**. The head clamp condition is checked by the sensors **71**.

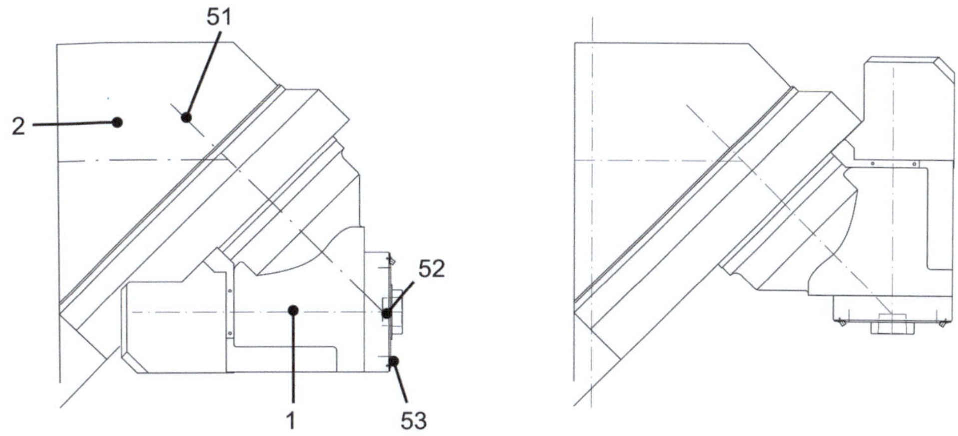

▫ **Fig. 11.14** Horizontal-vertical head, according to Starrag Technology (Scharmann division)

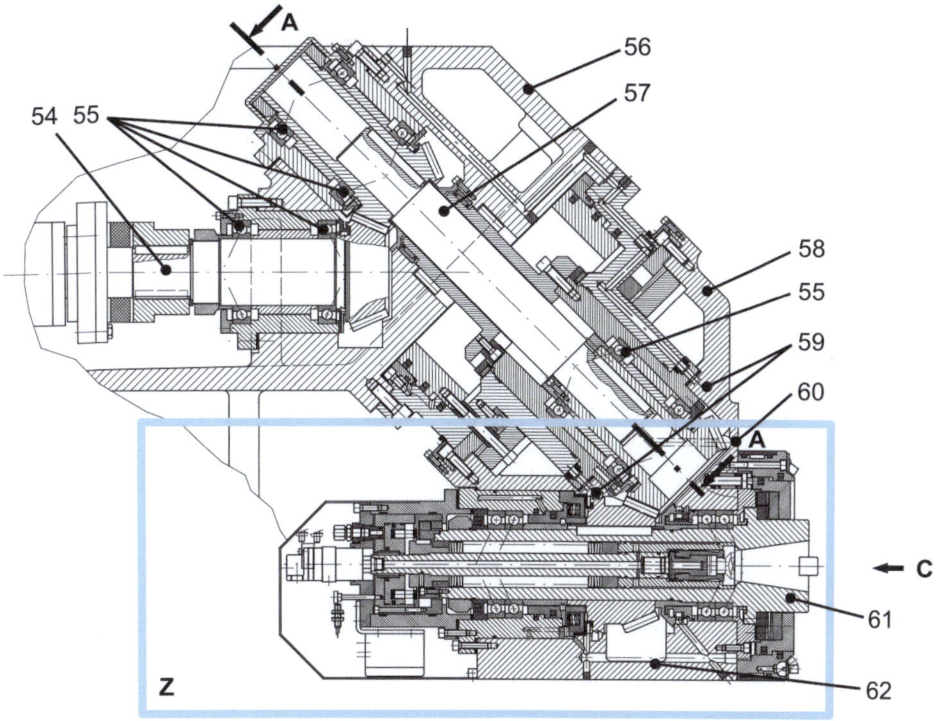

▫ **Fig. 11.15** Section of HV head, according to Starrag Technology (Scharmann division)

HV Head: Section B-B

Figure 11.17 shows the HV head's swivel mechanism. The rack **65** is moved via the double-acting hydraulic cylinder **74** (allows the cylinder to move in and out). The rack engages with the hollow pivot shaft **64** that encases the main drive shaft **57**. The movement of the rack leads to rotation of the pivot shaft. This rotation of the pivot shaft causes the main spindle to swivel about the swivel axis. The head clamping is monitored by a sensor **71**.

HV Head: View C and Cross-Sections D and E

Figure 11.18 shows details of the HV head. The cooling lubricant is fed through the swivel housing **58** into the attachment **75** and distributed via an annular groove **76** to the cooling lubricant nozzles **53**. The hydraulic line **77** on the spindle clamping is also guided through the swivel housing.

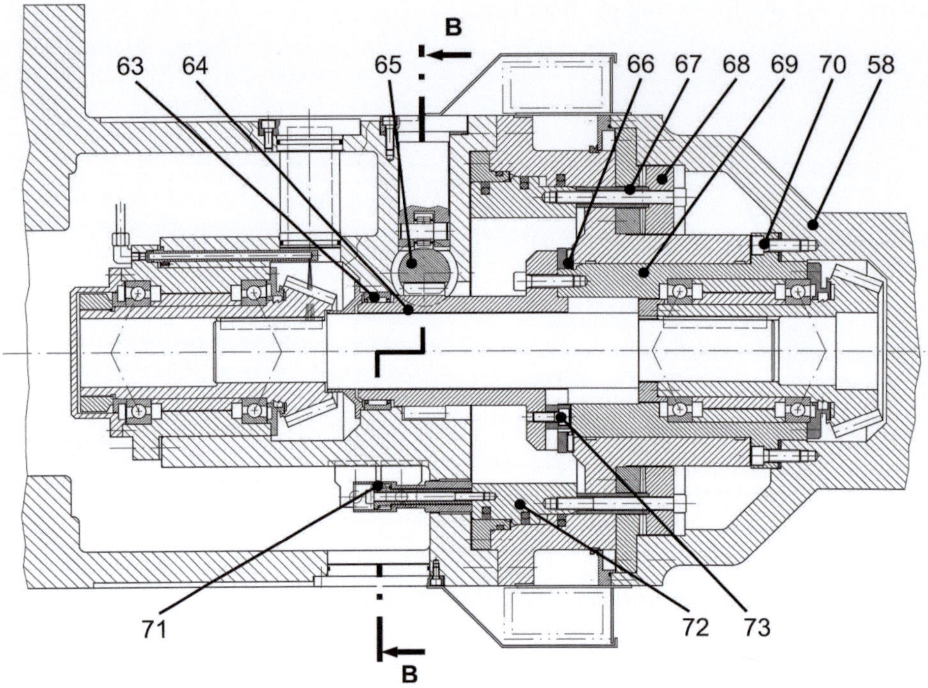

Fig. 11.16 HV head: Section A-A, according to Starrag Technology (Scharmann division)

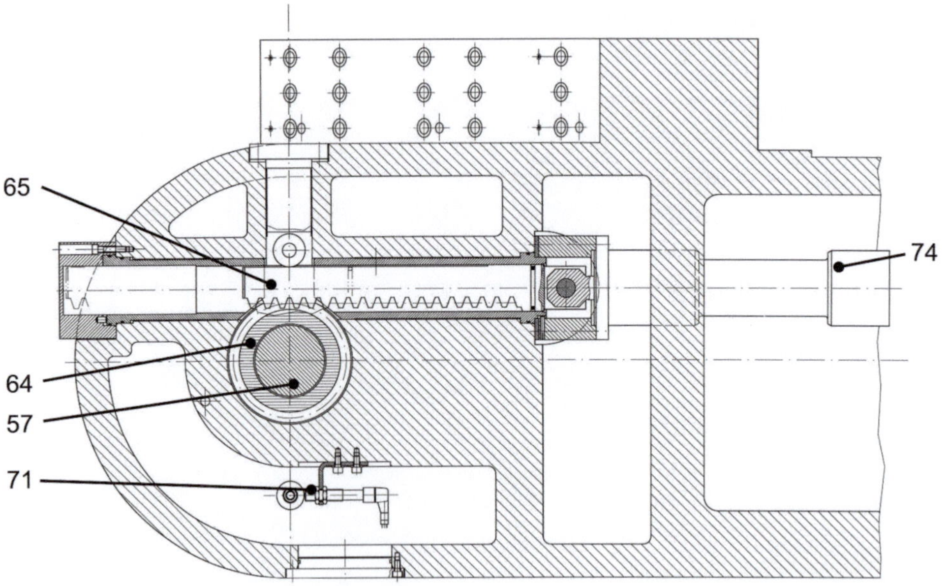

Fig. 11.17 HV head: Section B-B, according to Starrag Technology (Scharmann division)

HV Head: Detail Z

The spindle shaft **61** shown in ◻ Fig. 11.19 is driven via the bevel gear **87**. Here, the torque is transmitted from the bevel gear to the spindle shaft, which is supported by a tandem O arrangement via a key connection **86**. The gearing here is a cyclo-palloid gearing. The preload on the bearing arrangement can be adjusted during assembly via the shim **83**. The bevel wheel pairs are oil-lubricated and the angular contact ball bearings are greased. An air purge is introduced at points **85** and **88** to keep the lubrication media separate. This air is fed into the swivel component via a rotating union (not shown).

As turning operations on the rotating workpiece are possible with a high-speed machine table, the spindle shaft can be locked. To achieve this the Hirth serration **90** is locked at the front by the piston **89**. In addition, the internal bearing rings on the right-hand bearing package are displaced by the piston **84** such that the bearing arrangement is removed from the force flux and not damaged, for example by machining with an interrupted cut.

To release the tool, the tension rod **80** is pushed against the disk spring assembly **92** by the piston **81**. The system rests axially against the shoulders of the bayonet connection **82** to release the load on the spindle bearing. By displacing the tension rod **80**, the collet chucks **91** open and the tool is released. Cooling lubricant or compressed air is fed through the tension rod **80** for cleaning purposes. The connections for this are included in the rotating union **78** and the tool clamping is also checked here by the sensors **79**.

HV Head: Details X and Y

◻ Figure 11.20 shows X and Y details of the HV head. The rear head clamping is checked by the sensors **96** by measuring the position of a driving rod **93** that is moved together with the other components by the piston **84**

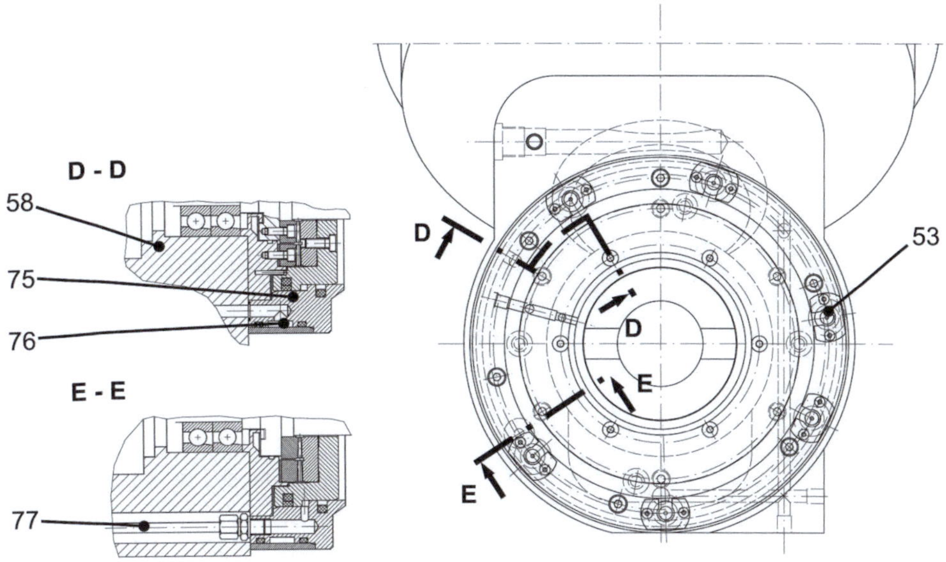

◻ **Fig. 11.18** HV head: View C, according to Starrag Technology (Scharmann division)

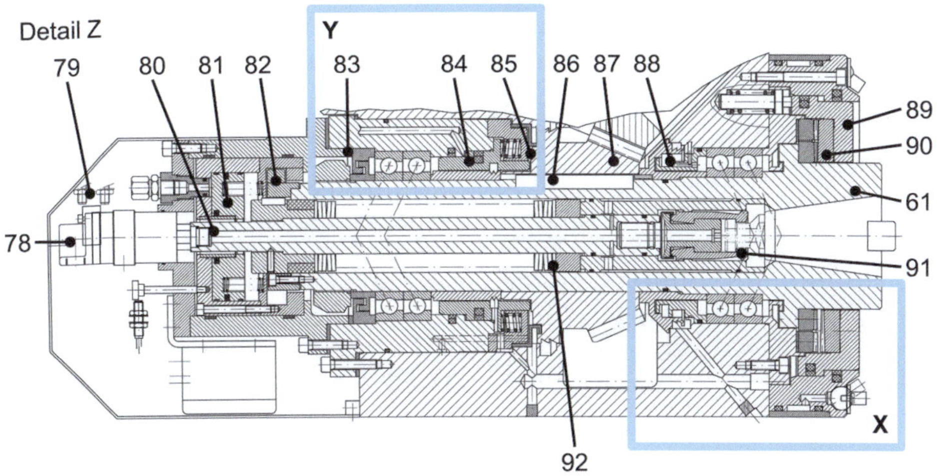

◻ **Fig. 11.19** HV head: Detail Z, according to Starrag Technology (Scharmann division)

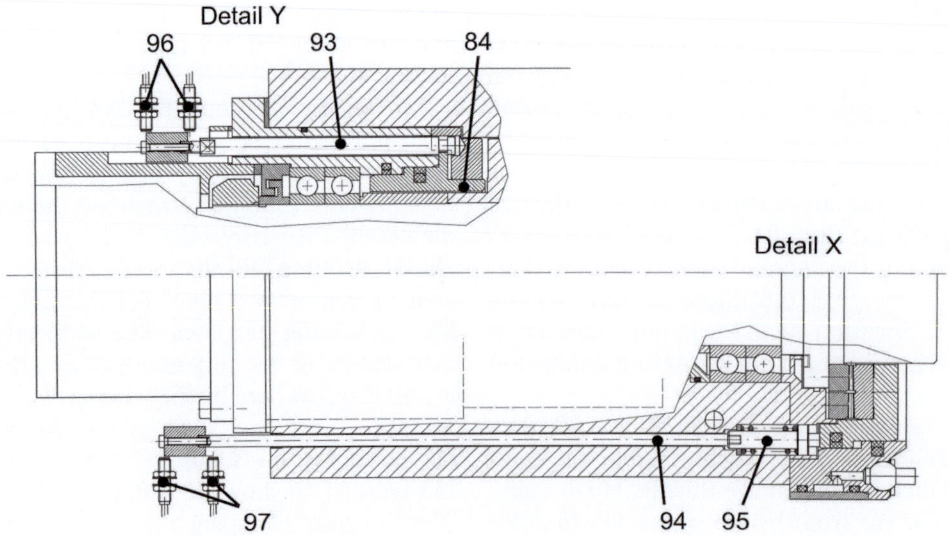

Fig. 11.20 HV head: Details X and Y, according to Starrag Technology (Scharmann division)

via a groove. An intermediate rod **94** is screwed into a spring bolt **95** and guided through the swivel housing to check the front spindle clamping. The clamping status is passed to the sensors **97** from the intermediate rod.

11.2 Grinding Machine

11.2.1 Company Profile

ELB-Schliff and aba Grinding Technologies are part of the Lapmaster Wolters Group, a machine tool corporation for technology-oriented system solutions in precision machining. Elb-Schliff Werkzeugmaschinen GmbH has been producing flat and profile grinding machines since 1948. The aba Grinding Technologies GmbH was established in 1898 under the name Messwerkzeugfabrik Alig & Baumgärtel Aschaffenburg and has been producing precision grinding machines since 1933. The production program covers high-accuracy flat, profile, and creep-feed grinding machines as well as rotary table machines and automated multi-technology production cells.

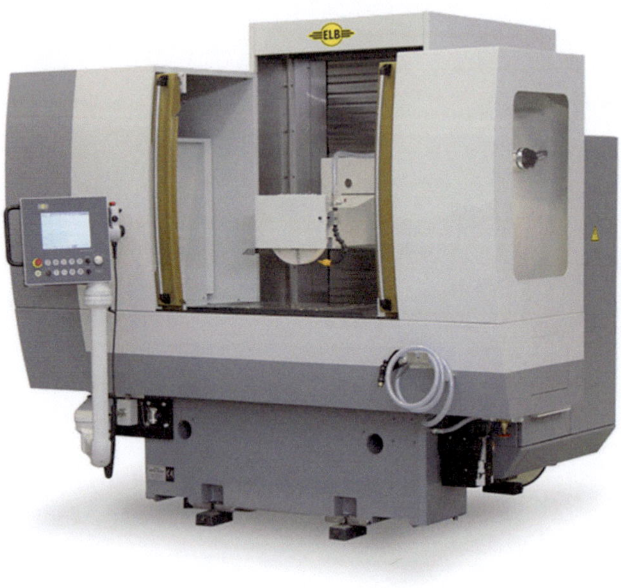

Fig. 11.21 Overview of the ELB-Schliff smartLine, according to Elb-Schliff

11.2.2 Overview

The ELB-Schliff smartLine (Fig. 11.21) concept covers the entire spectrum of traveling table grinding machines. Grinding processes from simple reciprocating flat grinding, speed-stroke grinding, slot and profile grinding, to creep-feed grinding can all be achieved. Many different combinations of components are possible due to the flexible modular system. The basic package for simple grinding applications serves as the base module that can be extended with additional components. These include a range of different table drives, guide systems, controls, and the appropriate accessories. The modular system comprises a wide range from components for use in the shop floor area, to high-technology solutions in automated production. The basic components of the ELB-Schliff smartLine are described below followed by some typical solutions.

11.2.3 General Description

Figure 11.22 shows the fundamental structure of the ELB smartLine Kombi. The machine consists of two separate beds: The front base **1**, on which the work table **2** is mounted and the rear base **3**. The front base can be modified as required. A range of different guide systems are available and the table length can also be varied. When installing the system, once the beds have been aligned, they are first glued and then cast together to maximize positional accuracy. The front base provides the table's traversing movement in the X-direction. The table drive **8** can be designed as a toothed belt, hydraulic, or ball screw drive depending on the requirements. The rear base is available in three versions having different lengths. This provides the column **4** with a range of different travel distances in the Z-direction. The column mounted on the rear base is also available in a range of different designs to suit the specific requirements. In addition to the choice

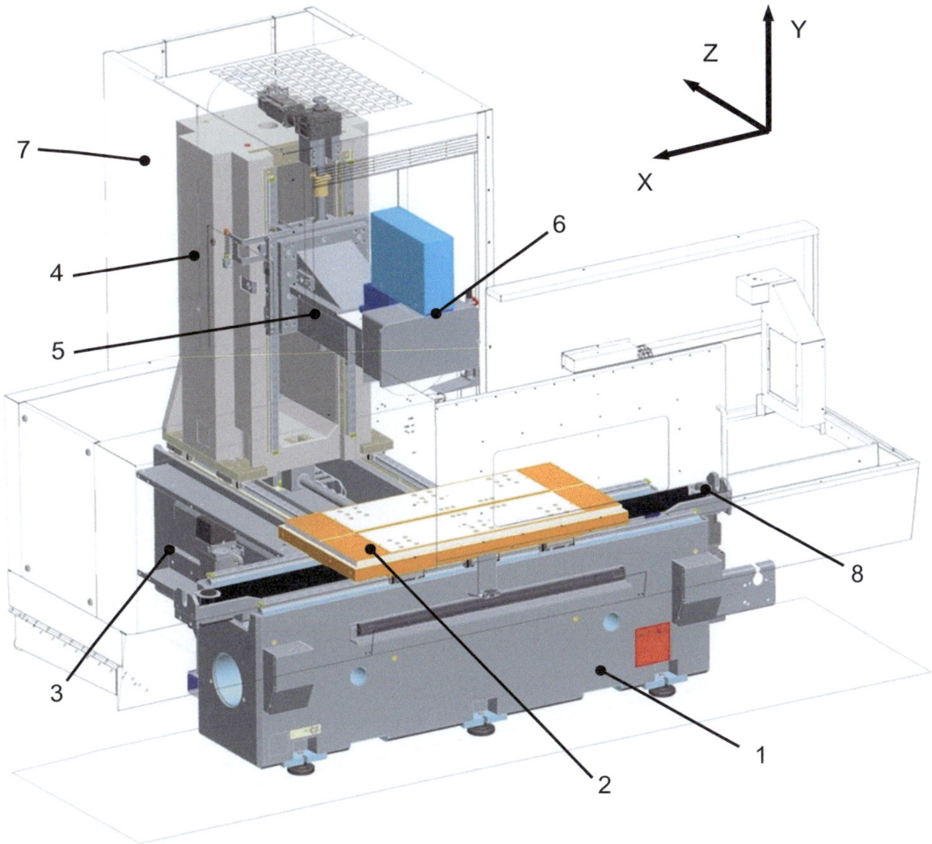

Fig. 11.22 Structure of the ELB smartLine Kombi, according to Elb-Schliff

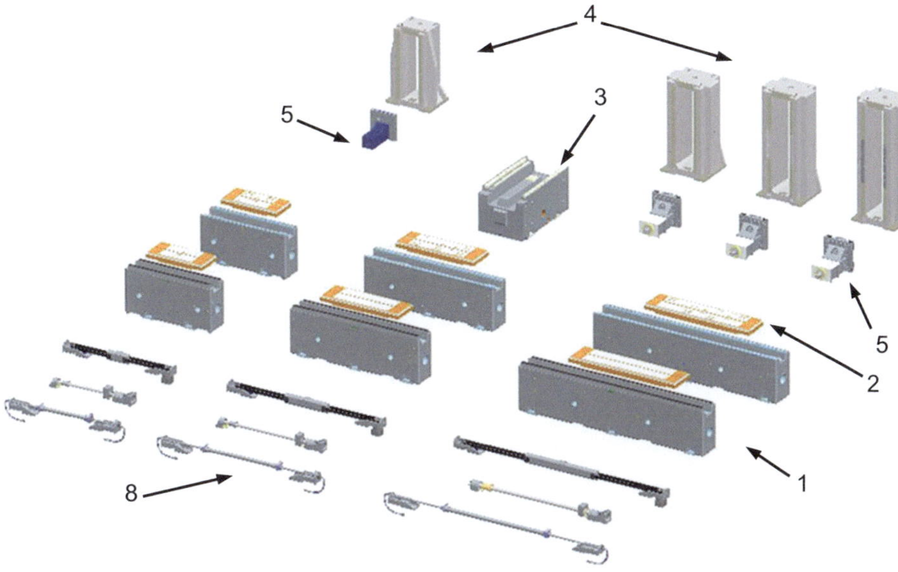

Fig. 11.23 Modular structure of the ELB smartLine Kombi, according to Elb-Schliff

of three different lengths, a stiffer column design is also available for applications with higher loads.

The spindle housing **5** with either a synchronous or asynchronous motor is attached to the column. This spindle design allows users to choose between an externally driven spindle or a motor spindle allowing spindle powers ranging from 5.5 kW to 52 kW. The grinding spindle is supported by bearings in the spindle housing and has a grinding wheel support at one end. The dresser **6** is also mounted on the spindle housing.

The entire machine is surrounded by a safety enclosure **7**.

The combinable components of the modular system are shown in ◘ Fig. 11.23. As explained above, different versions of front bases **1**, working tables **2**, rear bases **3**, columns **4**, spindle housings with spindle **5** and table drives **8** can be combined with each other. Control options of either NC (numerical controller) or PLC (programmable logic controller) are also available in addition to the machine's mechanical components.

11.2.4 Front Base and Table

11

The front base, as shown in ◘ Fig. 11.24, is made of a monolithic casting **9** of reaction resin concrete. Reaction resin concrete is a mixture of crushed hard stone in various grain sizes and a reaction resin. It has high vibration damping, thermal stability, resistance to media and excellent design flexibility. The stiffness of reaction resin concrete, which is related to the density, is comparable to cast materials.

Visible on the front stand are the transportation points **10** and the access to the front base fixators **11**. The fixators are arranged in the front base such that they are directly within the flux of forces of the frame components. ◘ Figure 11.24 shows a front base variant with linear guides. The guide rails **12** and the carriage **13** that supports the table are visible. Front bases with different lengths and guide rails set at different gaps can be used depending on the application.

The table is also available in a range of variants that differ in terms of the clamping option and type of guidance. One clamping option for the workpiece is standard magnetic clamping. Other table designs are available to suit the application, such as additional axes or T-slots.

There are two basic variants in terms of guideway systems:

- linear guides and
- V/flat guides.

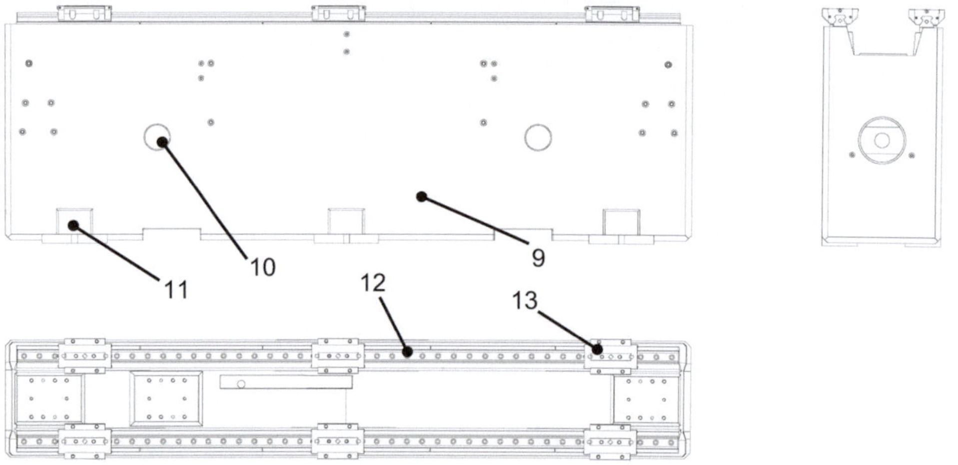

◘ **Fig. 11.24** Structure of the front base, according to Elb-Schliff

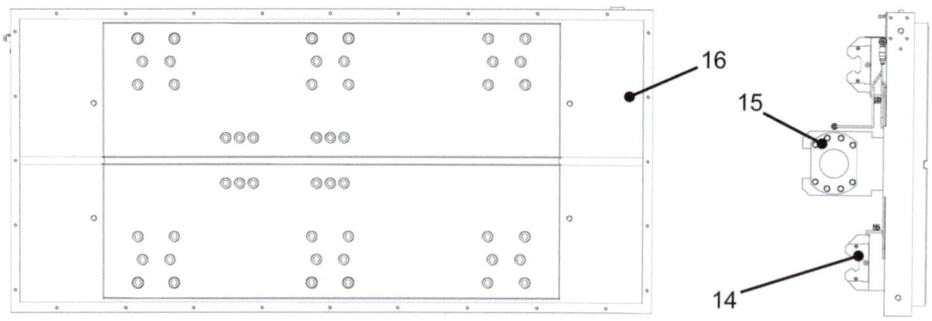

◘ **Fig. 11.25** Structure of the table with linear guides, according to Elb-Schliff

The strips for mounting the linear guide rails are cast when the front base is manufactured. For V/flat guides (see Fig. 11.26), it is also possible to form the guideway surfaces directly in the bed.

Figure 11.25 shows the structure of a table variant with linear guides as the guideway system. Profiled rail guide systems **14** are particularly suitable for creep-feed and profile grinding processes and for tasks with interpolations. The use of ultra-precision linear guides allows high-precision work. In this instance, the table is driven by a ball screw drive with the nut **15** mounted on the table. The workpiece can be fixed in position on the table using the clamping option **16**.

A table variant with the alternative guideway system of a V/flat guide is shown in Fig. 11.26. V/flat guides or coated plain guide systems with oil lubrication are particularly suitable for reciprocating grinding and for the production of optimum grinding patterns. The V guide system **17**, the flat guide system **18** and the clamping option **19** can be seen here. In this instance, the V guide system provides the guidance in the direction of movement. V/flat guides are particularly suitable for absorbing forces perpendicular to the sliding surfaces.

The infeed axis (Y-axis) and the feed axis (Z-axis) are usually designed with linear guide systems and ball screw drives. The drive and guide system of the longitudinal axis (X-axis) are chosen to meet the demands on the workpiece, the required accuracy and the grinding process. The choice of process parameters (infeed, workpiece speed, cutting speed and cutting volume) has a significant influence on the results.

The drives shown in Fig. 11.27 are available.

- The hydraulic cylinder is characterized by good damping. The damping characteristics of the hydraulic oil in the hydraulic cylinder ensure good grinding patterns in reciprocating surface grinding.
- Toothed belt drives are used if high oscillation speeds are required to machine the workpiece, e.g., in speed-stroke grinding.
- Ball screw drives offer the high axial stiffness typically required for creep-feed grinding processes. Ball screws using direct drives also offer clearance-free

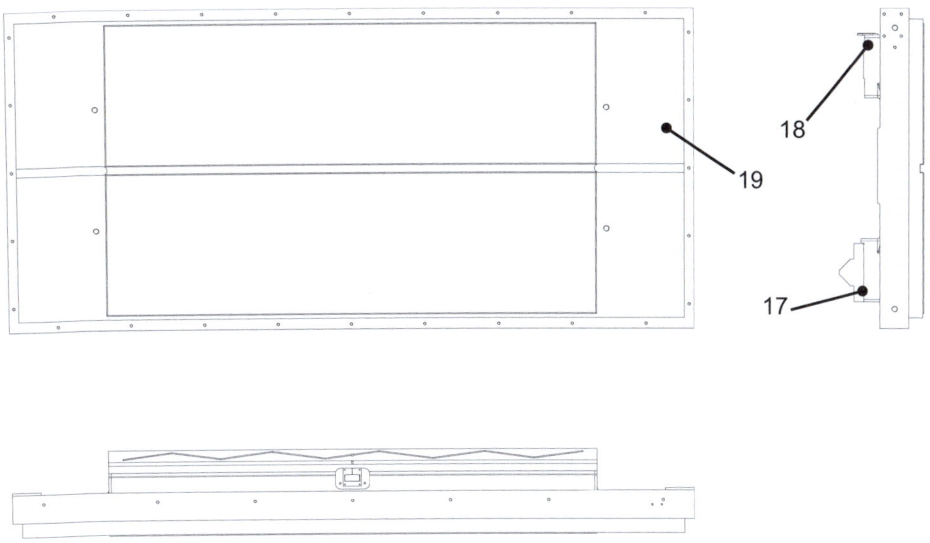

Fig. 11.26 Structure of the table with V/flat guide, according to Elb-Schliff

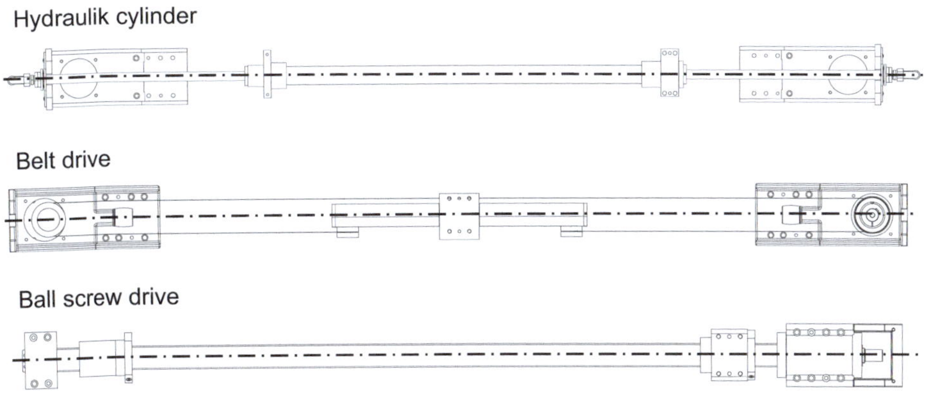

Fig. 11.27 Overview of table drive options, according to Elb-Schliff

positioning of the table in addition to the high stiffness. These features are necessary for profile grinding processes and interpolation tasks.

11.2.5 Rear Base

As with the front base, the rear base shown in ◘ Fig. 11.28 consists of reaction resin concrete and also has transportation points **20** and access points to the fixators **21**. The rear base has two guide rails **22**, four carriages **23** and a drive **24**, here a ball screw drive. This allows the vertical column installed on the rear base to be moved in the Z-direction. A scale **25** ensures extremely accurate positioning in the Z-direction. The rear base is available in three versions, ALPHA, BETA and GAMMA, each with a different guideway length. The available guideway lengths are 400 mm, 600 mm or 750 mm.

11.2.6 Column

The column is the base element of the grinding machine's Y-axis **26** (see ◘ Fig. 11.29). It has two guide rails **27** equipped with four carriages **28**. The spindle

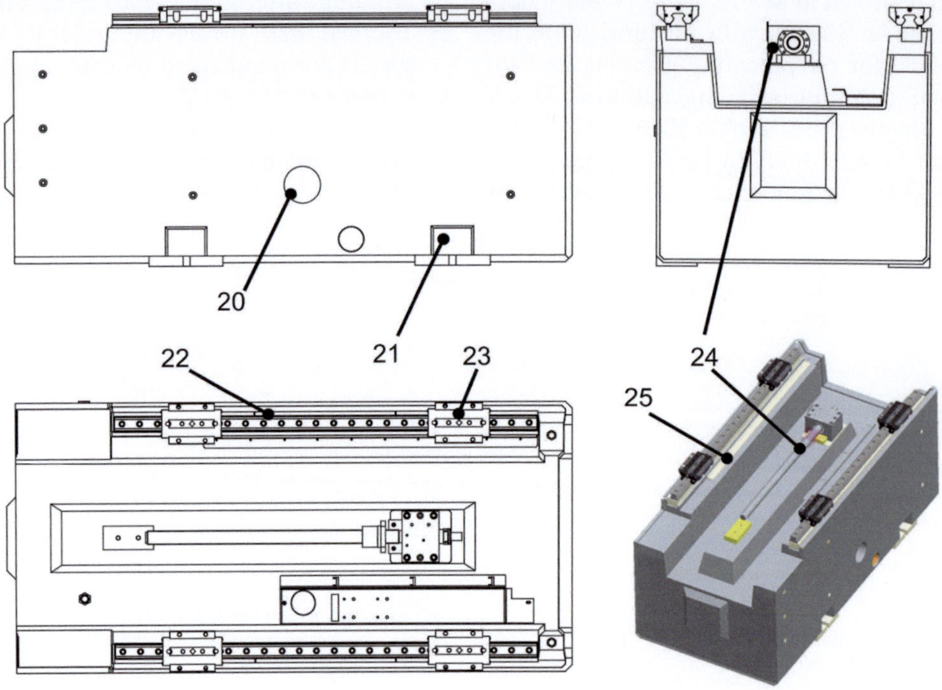

◘ **Fig. 11.28** Structure of the rear base, according to Elb-Schliff

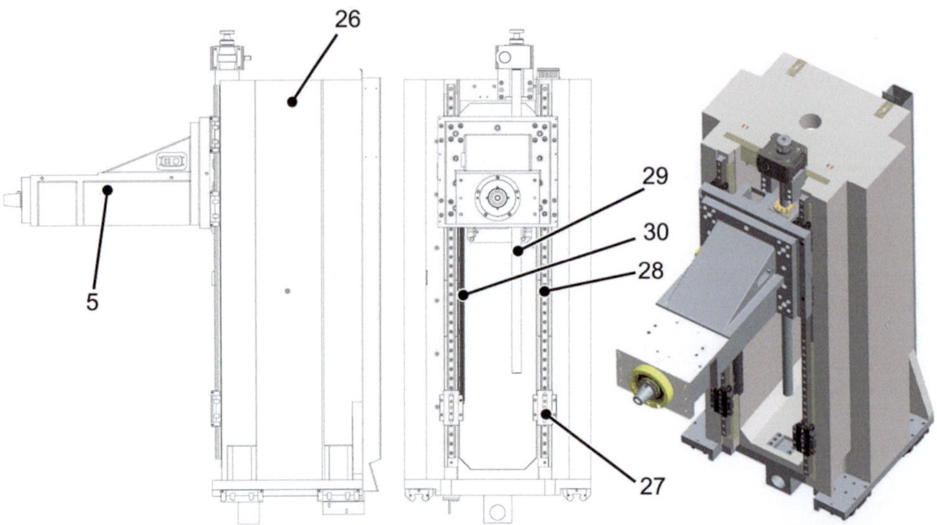

◘ **Fig. 11.29** Overview of the column with grinding spindle unit, according to Elb-Schliff

housing **5** is mounted on the carriages and is moved in the Y-direction by a ball screw drive **29**. The grinding spindle unit located in the spindle housing is driven by a synchronous motor. A Rotex GS clutch transfers the torque. A scale **30** ensures extremely accurate positioning in the Y-direction. A measuring head records the traverse position.

Figure 11.30 shows the column without the grinding spindle unit. The ball screw drive is driven by a worm gearbox **31**. The traversing movement in the Y-direction is achieved via the nut **32** fixed to the spindle housing. Guiding and driving of the column on the rear base in the Z-direction are realized by the four carriages **23** and the nut block **33** (with integral nut). The column is available in a range of different heights depending on the application area. A total of four different grinding heights are available (500 mm, 750 mm, 1000 mm and 1250 mm). As with the machine bed, the column is also made from reaction resin concrete and, if necessary, can be reinforced for the greater grinding heights by integrating a honeycomb reinforcement (see Fig. 11.31).

This is primarily used for grinding processes with greater spindle power such as creep-feed grinding processes requiring higher stiffness of the column. The stiffening elements are cast into the column with the structure remaining the same. Mounting ledges and the guide rails **27** are also cast in this way. The reinforced column is shown in Fig. 11.31. The reinforcement consists of a basic honeycomb construction **34** fixed to the baseplate **35**.

11.2.7 Spindle

The spindle for driving the tool can be adapted to meet specific requirements. External spindles or motor spindles may also be used. This allows spindle powers from 5.5 kW to 52 kW. The Tool change can be either manual or by an automatic wheel changer. Depending on the machining process, it is possible to provide internal cooling of the spindle and allow lubricants to pass through the spindle.

The cross-sectional drawing shown in Fig. 11.32 illustrates a grinding spindle unit **36** supported by bearings within the spindle housing **37**. The fixed bearing **39** is on the grinding wheel support side **38** of the spindle, while the floating bearing **40** is on the motor side. The fixed and floating bearing arrangements are both designed as bearing packages in an O arrangement. The grinding spindle has four or six bearings, depending on the application. A synchronous motor drives the grinding spindle via a Rotex GS clutch mounted on the end of the spindle **41**.

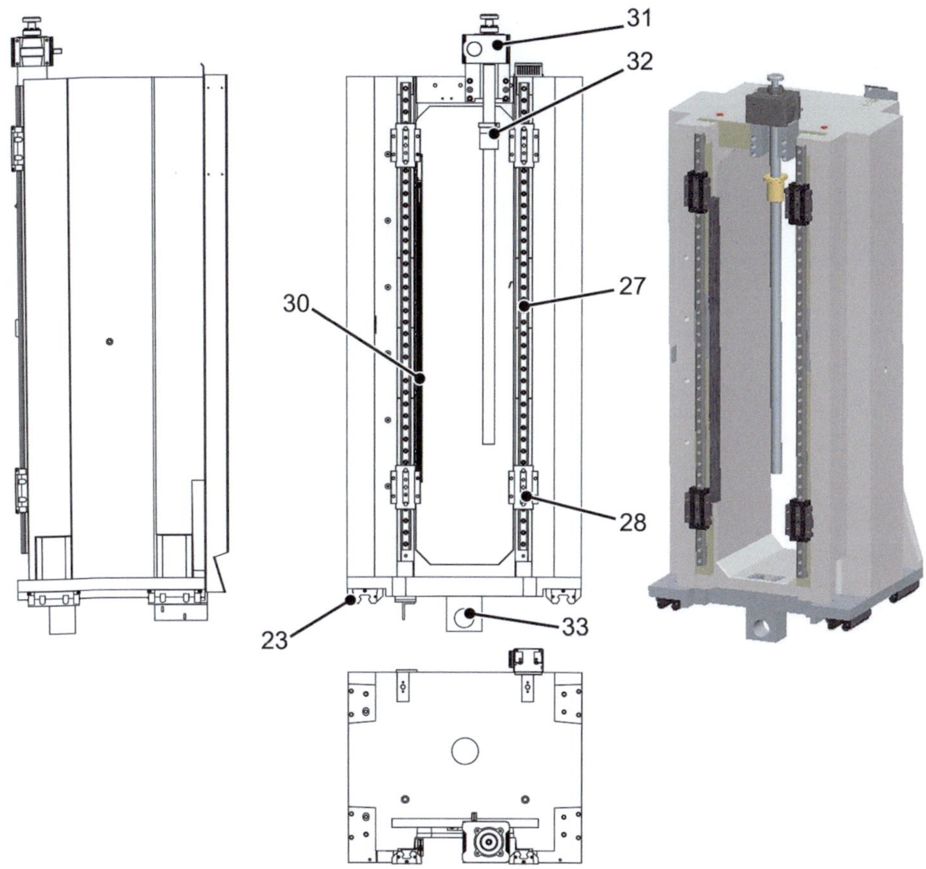

Fig. 11.30 Structure of the column, according to Elb-Schliff

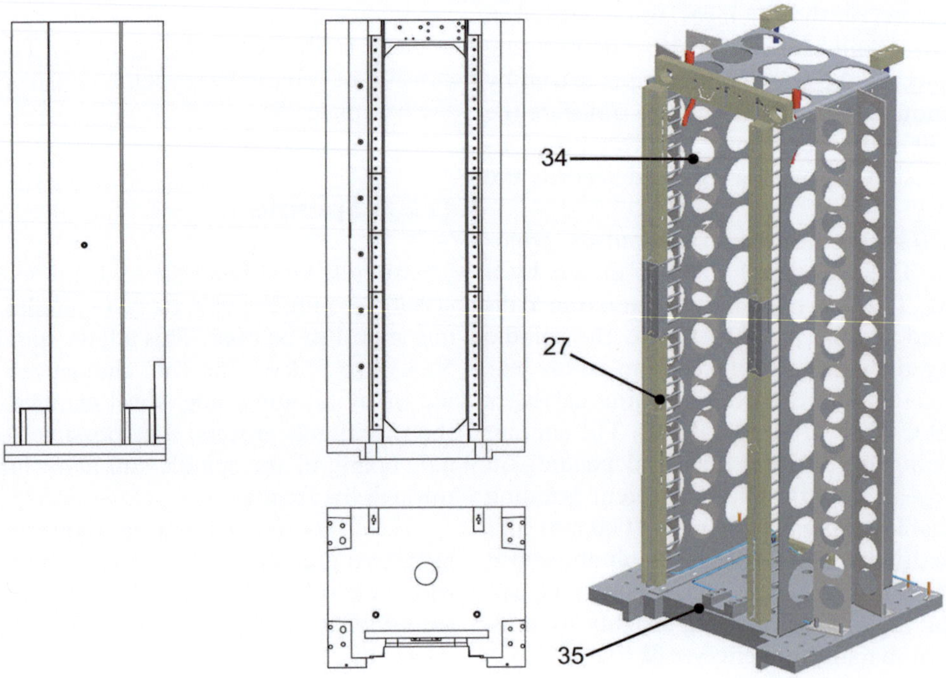

Fig. 11.31 Column reinforcement, according to Elb-Schliff

11

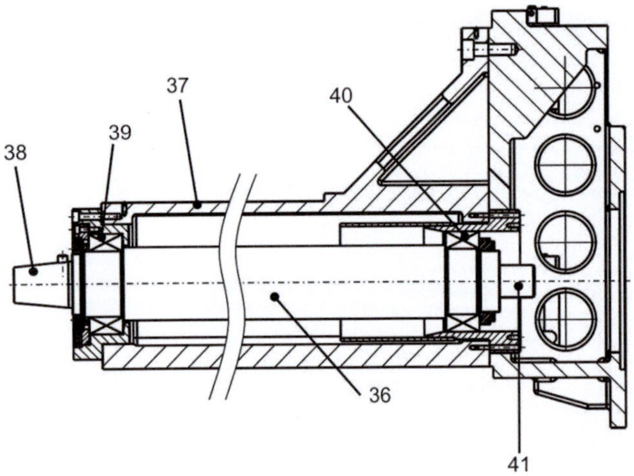

Fig. 11.32 Overview of the spindle housing with an externally driven spindle, according to Elb-Schliff

A motor spindle, as shown in Fig. 11.33, can be used as an alternative to an externally driven spindle. In the motor spindle, the rotor of the synchronous motor **42** is shrunk directly onto the shaft of the spindle. The stator of the synchronous motor **43** is located in the spindle housing. In this case, the spindle's fixed bearing **44** is also on the tool side with the floating bearing **45** on the opposite side; the two bearings are designed to form bearing packages in an O arrangement. The spindle is sealed on the tool side by an air purge seal **46**. The grinding wheel **47** can be automatically changed using a tension rod and a tool clamping system **48**. The spindle speed is measured by the rotary encoder **49**.

11.2.8 Dressers

Dressing ensures the grinding wheels are ready for use. It produces the desired grinding wheel profile (macrogeometry) and a topography (microgeometry) capable of cutting. A distinction can be made between systems with a stationary or a rotating dressing tool. Systems with a stationary tool include single-point, multi-stone, multi-point and folding dressers. Dressers with a rotating shape or profile roller are attached to the machine table to increase their accuracy. Rotating tools are also used in swivel dressers or roll dressers for continuous dressing. Examples of various dressing options are shown below.

Figure 11.34 shows a folding dresser with three single-point diamonds **50**. This dresser is mounted on the working table using the table mounting facility **51**. To prevent the dressing block and the machine components from colliding, the dressing block can be rotated to the side using the swivel axis **52** during processing. This rotation is provided by moving a rack **54** using a pneumatic or hydraulic system **53**. The rack rotates a pinion **55** located on the swivel axis.

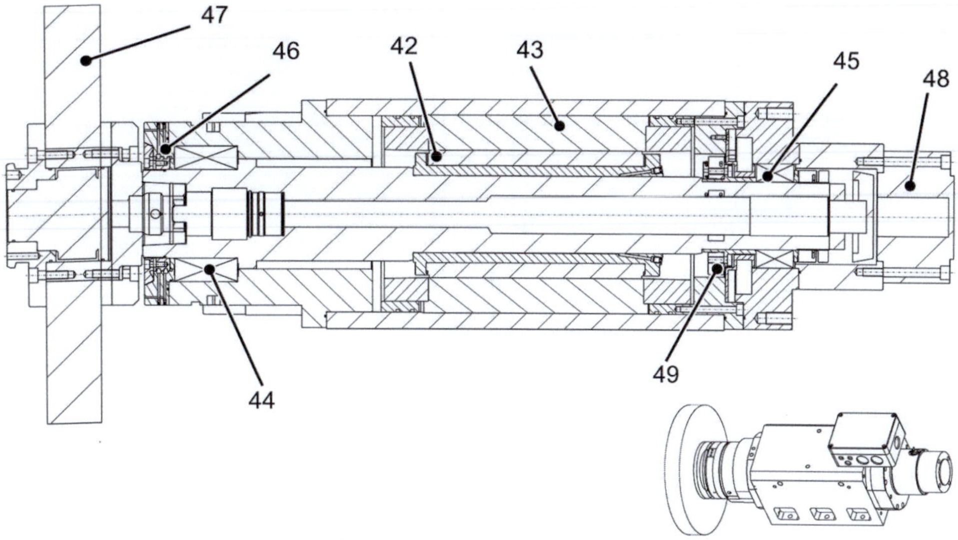

Fig. 11.33 Overview of a motor spindle with automatic wheel changer, according to Elb-Schliff

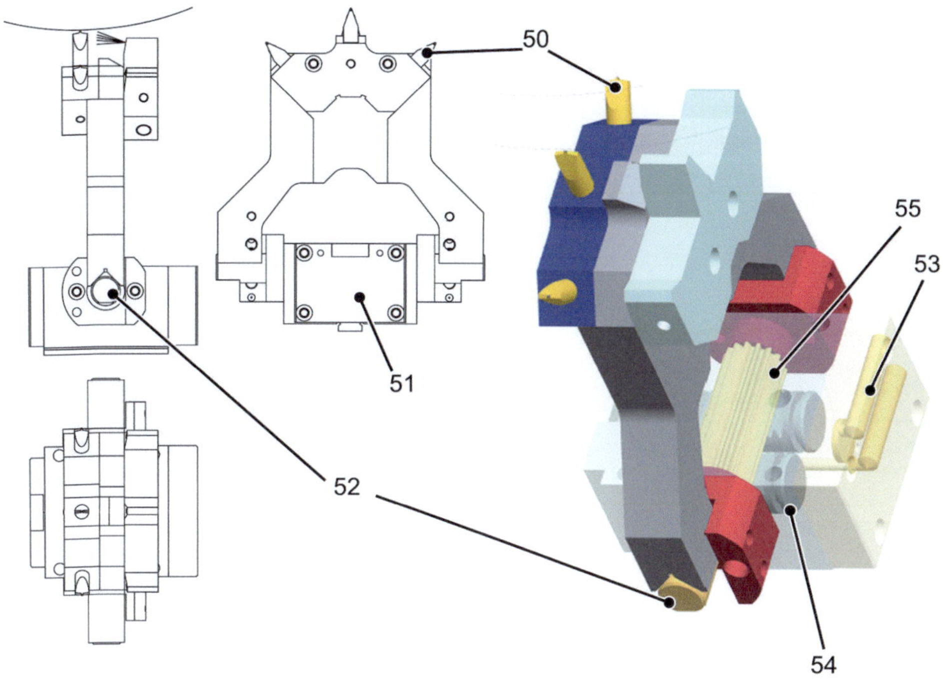

Fig. 11.34 Folding dresser structure, according to Elb-Schliff

One option for dressing using a rotating dressing tool is a table roll dresser (see Fig. 11.35). This element is mounted on the working table using the table mounting facility **56**. A shaft **58**, which is supported by two bearings, is driven by a motor **57**. The shaft acts as the support for the dressing tool (not shown). The fixed bearing **59** is located on the toothed belt **60** side. The toothed belt couples the shaft to the motor. The floating bearing **61** is located on the opposite side. The bearing arrangements are sealed by labyrinth bushings **62**.

Figure 11.36 shows a swivel dresser with rotating dressing tool **63**. The dressing tool holder **64** has a rotating bearing such that the required profile can be applied to the grinding wheel. Lubricant and coolant can be fed to the dressing process via the cooling lubricant supply **65**. The swivel dresser is mounted on the table using the

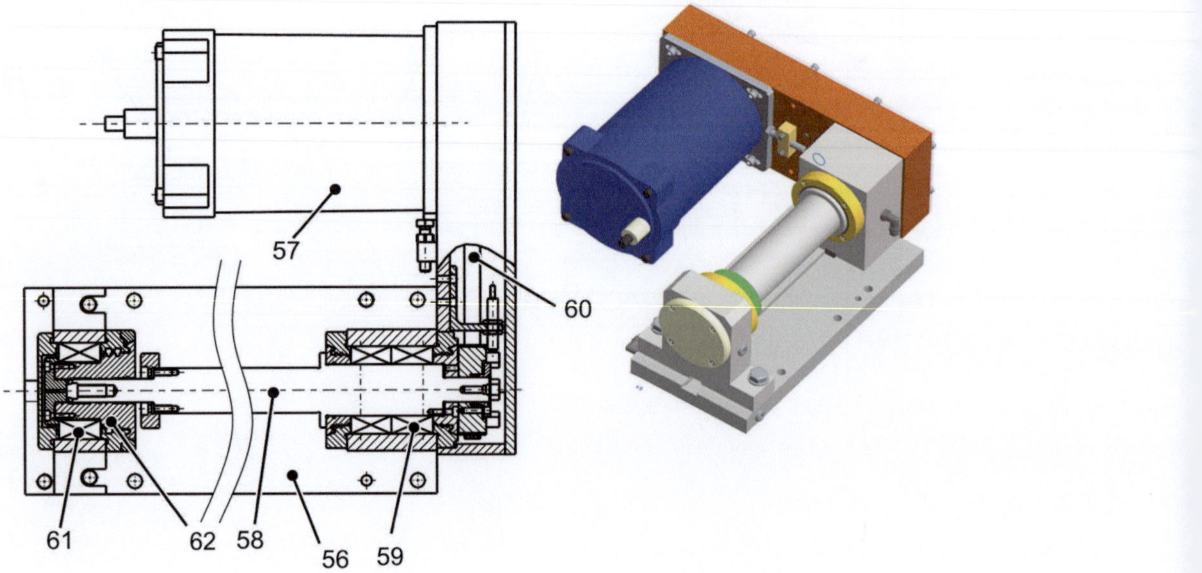

Fig. 11.35 Structure of a table roll dresser, according to Elb-Schliff

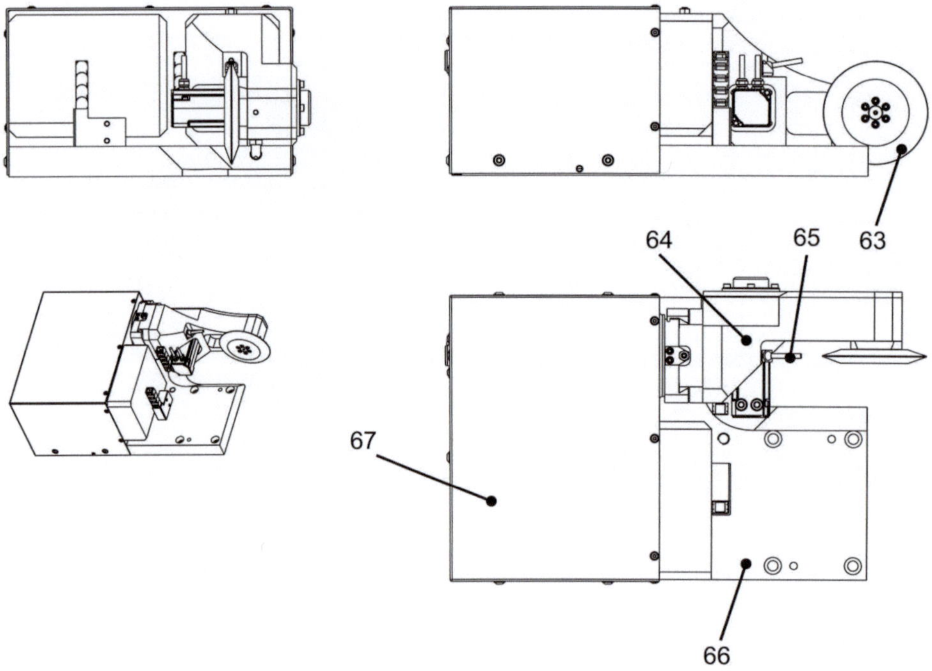

Fig. 11.36 Structure of a swivel dresser with rotating dressing tool, according to Elb-Schliff

table mounting facility **66**. A motor-driven toothed belt located in the housing **67** drives the dressing tool.

Dressing can be continuous (during primary machining time) or discontinuous (interval dressing) using profile rollers. Continuous dressing or dressing during primary machining time is also known as the CD process (continuous dressing). In the CD process, a dressing roller is brought up to the grinding tool during the grinding process. The dressing amount is compensated in the infeed axis. In continuous dressing, the profile rollers must be mounted directly on the spindle housing and have their own traverse axis. This has the benefit that profile changes due to wear are constantly compensated during the process increasing productivity. The CD process is predominantly used in high productivity creep-feed grinding systems used in large scale production. In addition to the benefit of high profile consistency coming from the continuous wear compensation, the CD process also offers very consistent cutting behavior as a quasi-constant micro-geometry is formed. The high procurement costs and the increased grinding wheel consumption leading to frequent tool changes should also be mentioned as disadvantages.

Figure 11.37 shows a corresponding roll dresser **68** together with the grinding spindle. The grinding wheel **69** is fixed to the spindle by the wheel flange **70**. The dressing tool **71** (e.g., a diamond roller) is located above the grinding wheel. The shaft **72**, which supports the dressing tool, is driven by a motor **73** via a belt.

11.2.9 Engineering Solutions and Production Systems

Standardized engineering solutions are available based on the basic smartLine modular system discussed above. These range from additional axes, special and additional spindles, tool and workpiece change systems to additional technologies such as milling and laser metal deposition (LMD). The production systems are usually optimized for specific component types and are linked to matched processing software. A wide range of measurement technology is also available for machine condition monitoring as well as component calibration and measurement. Two typical equipment variants are illustrated below.

The millGrind series, based on the smartLine design, is shown in Fig. 11.38. This series was specially developed for the turbine industry and the requirements for multi-technology machining with one clamping process. Depending on the requirements, it can be used with either a horizontal or swivel spindle **74**. The variant shown here also has an overhead roll dresser **75**. The rotary tilt table **76** comes as a 2 or 3-axis variant. This allows the machining center to be equipped with up to seven machining axes. A flexible interface to the clamping device, e.g., a zero-point clamping system, provides simple connection to workpiece change systems. A modular changer **77** for grinding wheels, dressing rollers, drilling and milling tools, and measuring probes is available on the tool side. Connecting a laser unit **78** facilitates other machining processes such as laser metal deposition (LMD).

When grinding with high stock removal rates, the grinding wheels are dressed either intermittently or continuously by an overhead roll dresser. For individual parts or small series, a table-mounted dresser is used for the axially interpolated generation of the profile via the infeed and transverse axes.

When combined with a rotary swivel table mounted on the working table, it is possible to machine work-

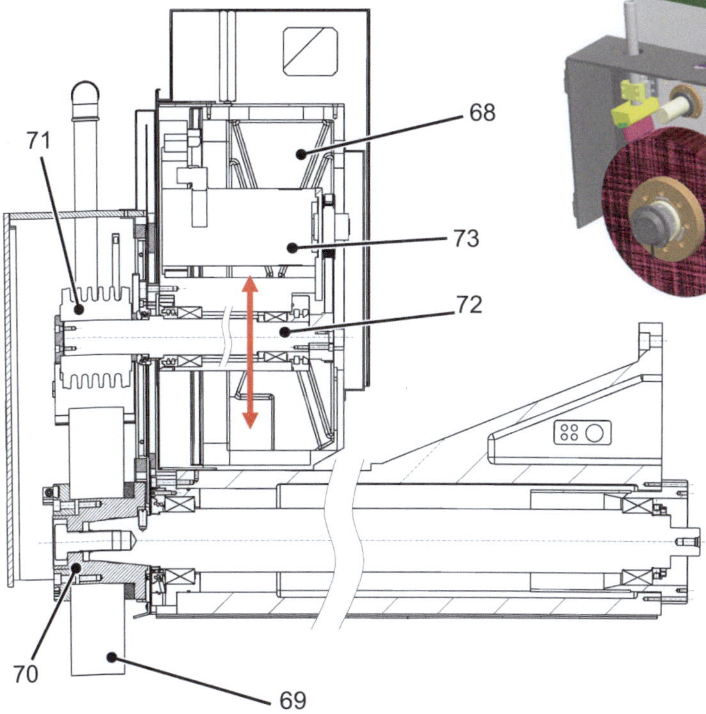

Fig. 11.37 Structure of a roll dresser for continuous dressing, according to Elb-Schliff

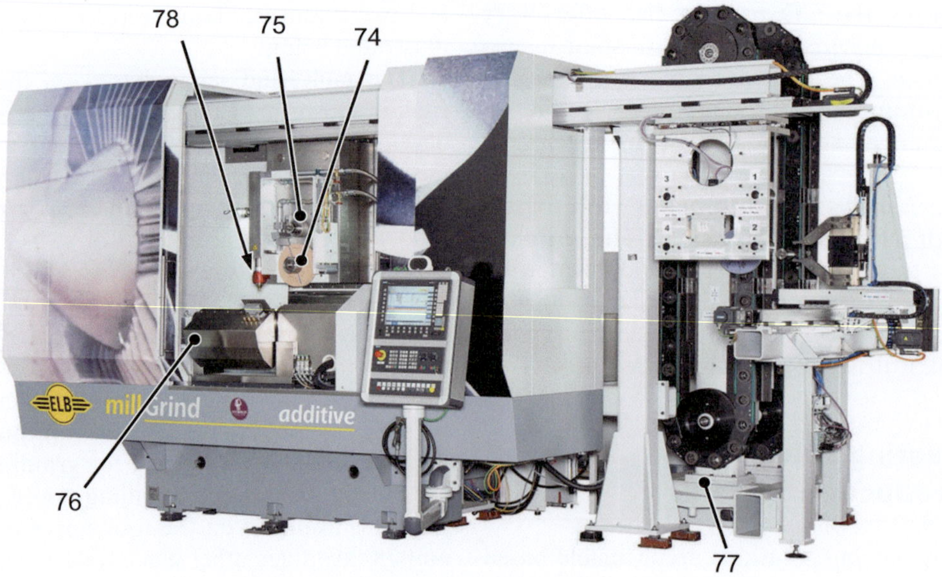

Fig. 11.38 Overview of the ELB-Schliff smartLine millGrind, according to Elb-Schliff

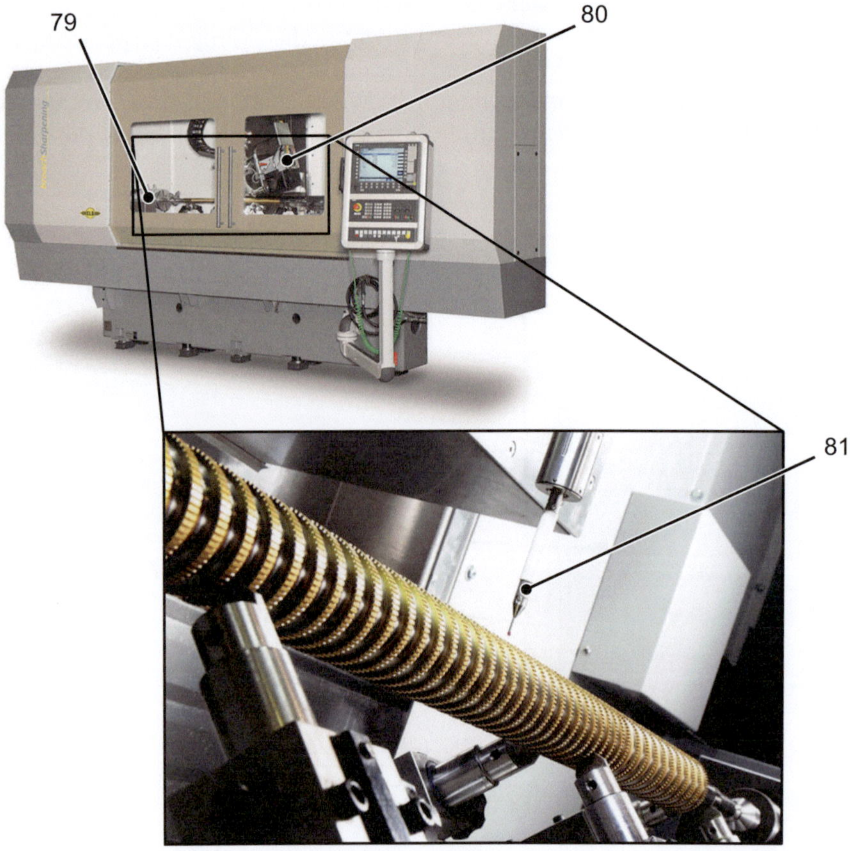

Fig. 11.39 Overview of the ELB-Schliff smartLine BGS, according to Elb-Schliff

pieces from multiple sides. The modern CNC control allows multiple axis interpolated grinding of complex shapes. Proven component-specific user cycles already exist for many applications.

The product range for the BG and BGS variants, (Fig. 11.39) includes machines for profile grinding and sharpening both round and flat broaching tools. The main markets are the automotive and aerospace industries.

Straight and helical-toothed broaching tools can be machined by integrating a precision rotary tilt table **79** into the smartLine on the machine table (A-axis) together with an optional swivel spindle **80**. Measuring probes **81** can capture, among other geometric properties, the tooth root and the cutting edge so that the tooth pitch can be determined. This series is suitable for many other grinding applications for rotationally symmetric workpieces.

11.3 Lathe

11.3.1 Company Profile

INDEX-Werke was founded in 1914 by Hermann Hahn who started by manufacturing automatic turret lathes in 1914. After the death of Eugen Hahn, the founder's son, 85% of the firm's shares were transferred to a foundation. In 1997, TRAUB Drehmaschinen GmbH & Co. KG, located in neighboring Reichenbach (Fils), was taken over and integrated into the INDEX Group. The company continues to run with two brands to this day.

Since 2002, INDEX has has been offering multifunctional production centers where various process technologies can be combined in a single machine. Having all the processing on one machine not only increases quality and accuracy but also reduces production costs by minimizing throughput times. The main customers of INDEX-Drehmaschinen include the automotive and automotive supply industries, mechanical engineering, electrical engineering, electronics industries, manufacturers in the fluid technology sector, as well as manufacturers of controls and instruments. Medical technology and the aerospace industry are considered as emerging growth markets.

11.3.2 Overview

The G250 machine is part of the INDEX RatioLine G series (Fig. 11.40). Its modular system design enables the creation of numerous machine configurations, which can be customized according to specific applications. These can be adapted to the relevant application. The turn mill center can be configured to accommodate both flexible machining of small/medium lot sizes and high-productivity manufacturing of large lot sizes.

11.3.3 General Description

Figures 11.41 and 11.42 depict the structure of the Index G250 RatioLine. The machine bed **1**, made of a single-piece gray cast iron, provides support for all the main machine tool components. A welded steel construction **0**, attached to the machine bed, serves as the support for the enclosure, auxiliary components, and acts as a closed sump to prevent fluids from escaping the machine. It also forms the connection between the machine bed and the foundation. The configuration shown has two rotating spindles. The main spindle **2** is mounted on the machine bed while the counter spindle **3** can be moved along the Z-axis. This setup enables simultaneous machining from two sides using both spindles. A motor milling spindle **7** mounted on a separate carriage **5** is also available for machining in

Fig. 11.40 Overview of the Index RatioLine G250, according to INDEX

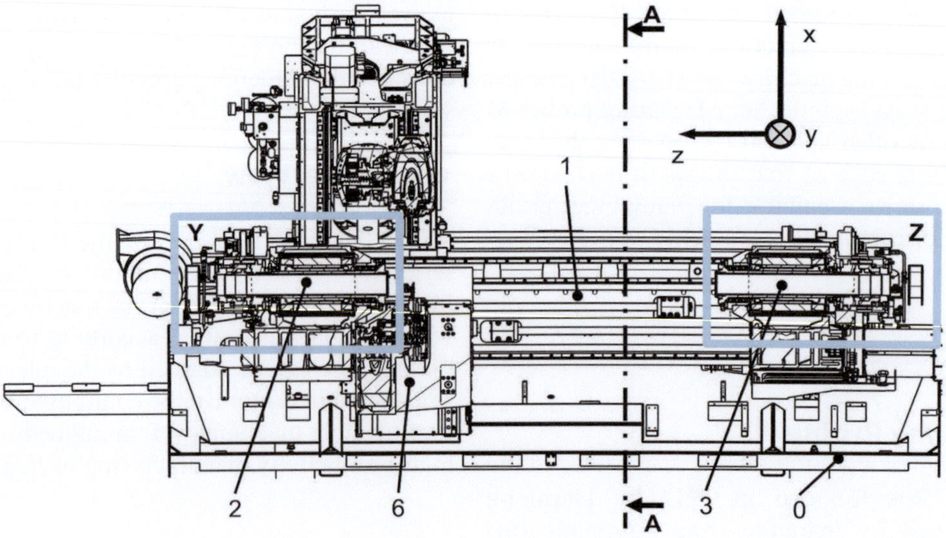

Fig. 11.41 Structure (front view) of the Index RatioLine G250, according to INDEX

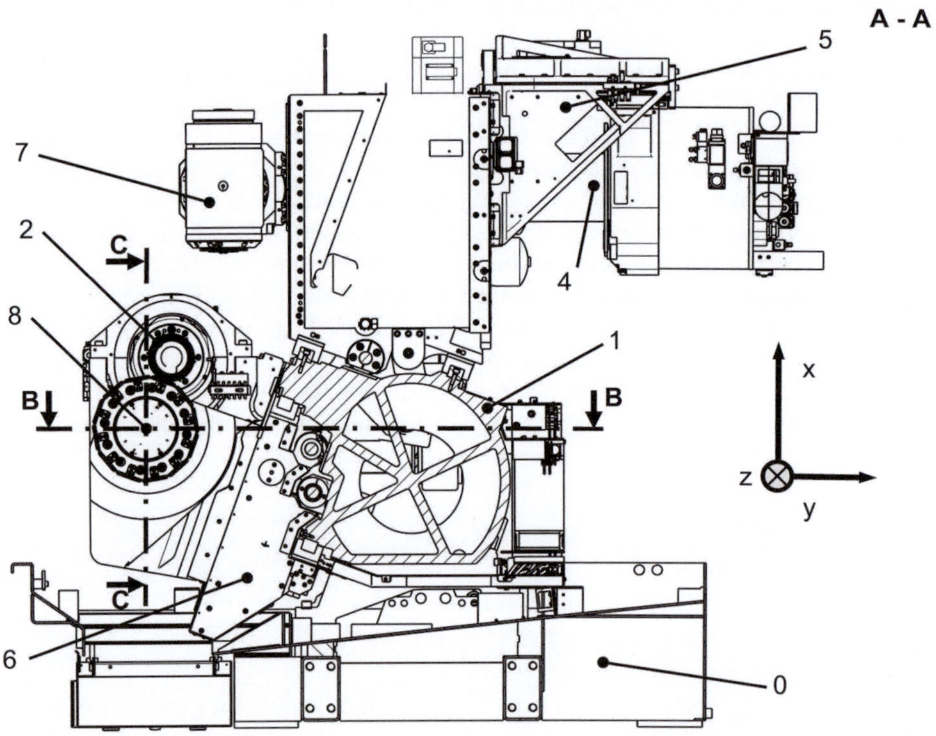

Fig. 11.42 Structure (side view) of the Index RatioLine G250, according to INDEX

addition to the tool carriage **6** with turret **8**. A special feature of the kinematics of the milling unit is the design with three traverse axes. The milling column, which moves in the Z-direction, supports a centrally guided vertical carriage **5** that can move in the X-direction. This vertical carriage further supports a hydrostatically guided Y-axis cantilever **4**. In other turn mill centers, the traversing movements of the X or Y-axis are often replaced by the simultaneous rotation of the rotating spindle.

11.3.4 Machine Bed

Figures 11.43 and 11.44 show the machine bed without the axis carriages. The Z-drive of the axis carriages uses servo motors **9**. These motors drive the ball screw **10** through a belt drive system. The ball screw drive shown in Fig. 11.43 produces the feed motion for the turret slide **6**. The long screw drives are equipped with a fixed bearing arrangement at both ends **11** to provide the feed system with high stiffness.

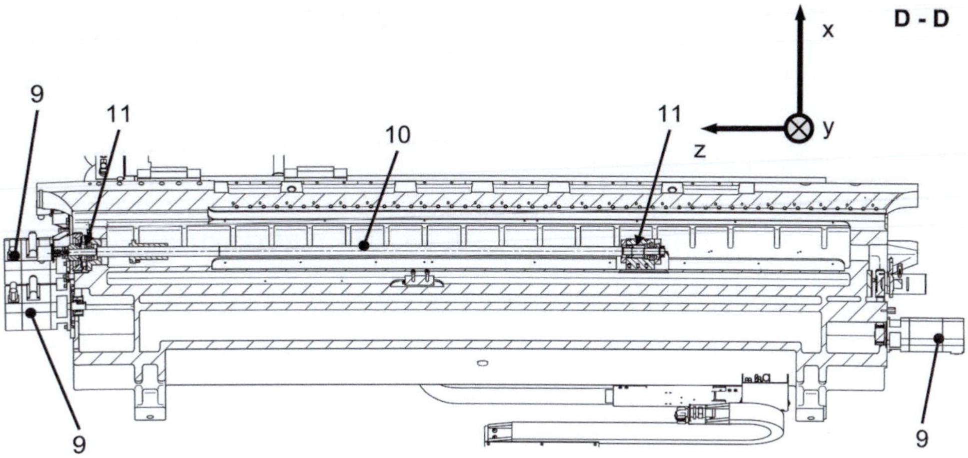

Fig. 11.43 Machine bed **1** (front view) on the Index RatioLine G250, according to INDEX

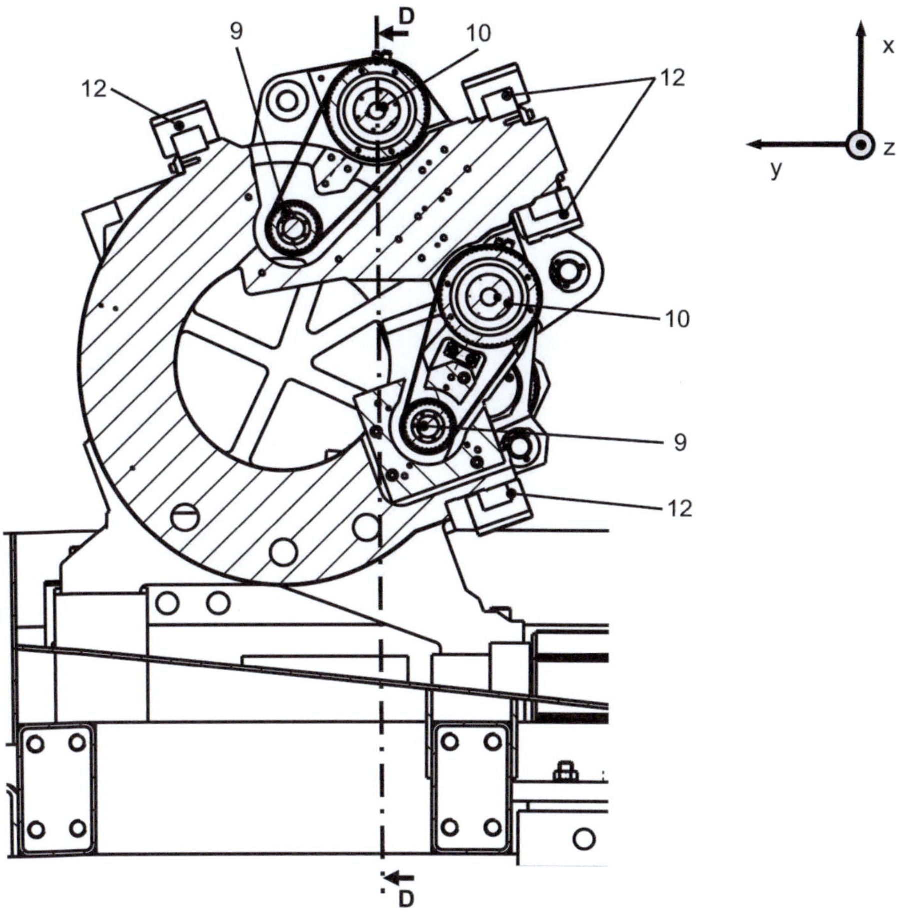

Fig. 11.44 Machine bed **1** (side view) on the Index RatioLine G250, according to INDEX

Figure 11.44 illustrates the configuration of the motors **9** and ball screw drives **10** as well as the gear ratio of the belt drives. Additionally, it showcases the striking layout of the linear rolling element guide systems **12**, designed as linear guide systems with balls as the rolling elements. The two upper guide systems **12** provide support for the milling column on the motor milling spindle. The turret slide and the counter spindle are mounted on the guide systems **12** on the right of the image.

11.3.5 Turret Slide

The turret slide, which is guided on the machine bed in the workpiece's axial direction (Z-direction), has an X-axis for the tools' radial infeed motion. The guide system here also consists of linear guide systems as can be seen in ◘ Fig. 11.45. A ball screw drive **13** provides the feed. The tool turret on the G250 provides 12 tool stations **14**. All of these are indirectly driven by one motor **20**. The figures 11.46, 11.47 and 11.48 show the drive train in detail. The torque is transfered via a two-speed bevel gearbox **16, 17**. The motor **20** drives the shaft **18** via the bevel gear pairing **16**. The shaft **18** within the turret is supported by rolling element bearings **15** so it can rotate and transmits the torque to the intermediate shaft **19** via the bevel wheel pair **17**. The intermediate shaft then drives the appropriate active tool. Coupling is achieved by radially swiveling the tools into a so-called pivot with cheeks in accordance with DIN 1809. The torque is transferred via a web, similar to an axial key, on the tool and a groove in the drive shaft. The unused tools remain undriven. The drive power at 25% operating time is 11 kW or 17 kW, with a maximum speed of 6000 min^{-1}.

11.3.6 Main and Counter Spindles

The structure of the main and counter spindles on the G250 is largely similar, as can be seen from ◘ Figs. 11.49 and 11.50. Depending on the selected design, the spindle bore has a diameter of 65 mm, 90 mm or 102 mm. Available chuck diameters range from 200 to 315 mm. The motor spindle torque is up to 610 Nm at 40% operating time. The output at this operating time is 60 kW. Both spindles have a resolution of 1 µ degree when in use as an active C-axis. Gear teeth allow the spindles to be locked for milling and drilling

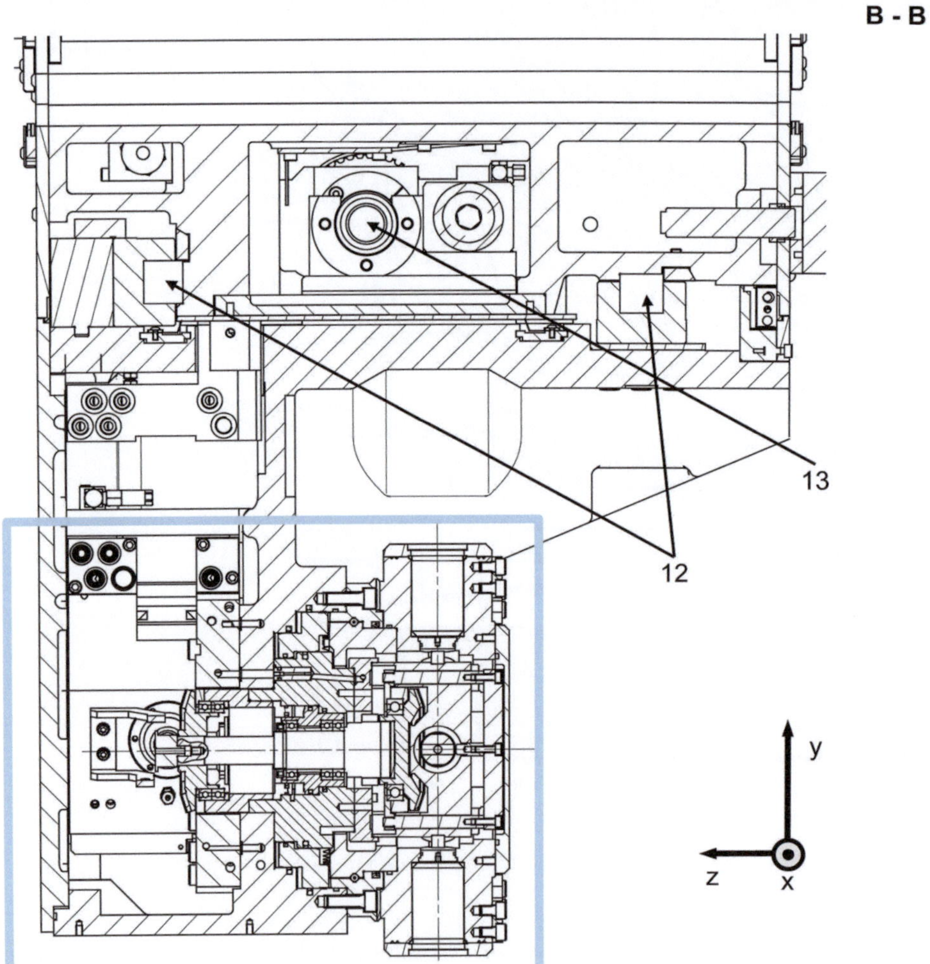

◘ **Fig. 11.45** Turret slide **6** (top view) on the Index RatioLine G250, according to INDEX

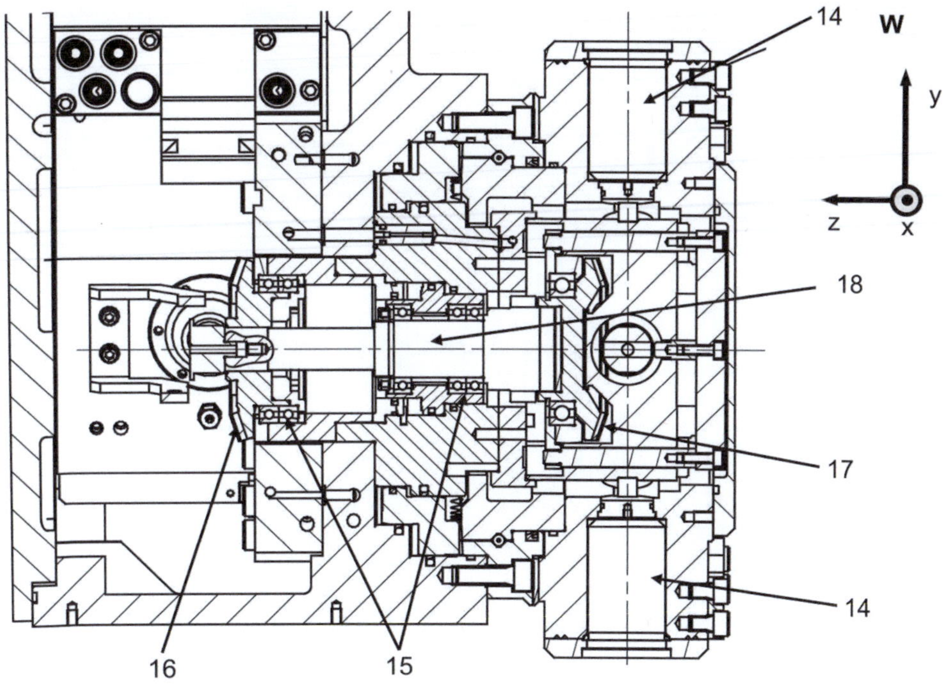

■ **Fig. 11.46** Turret (detail W) on the Index RatioLine G250, according to INDEX

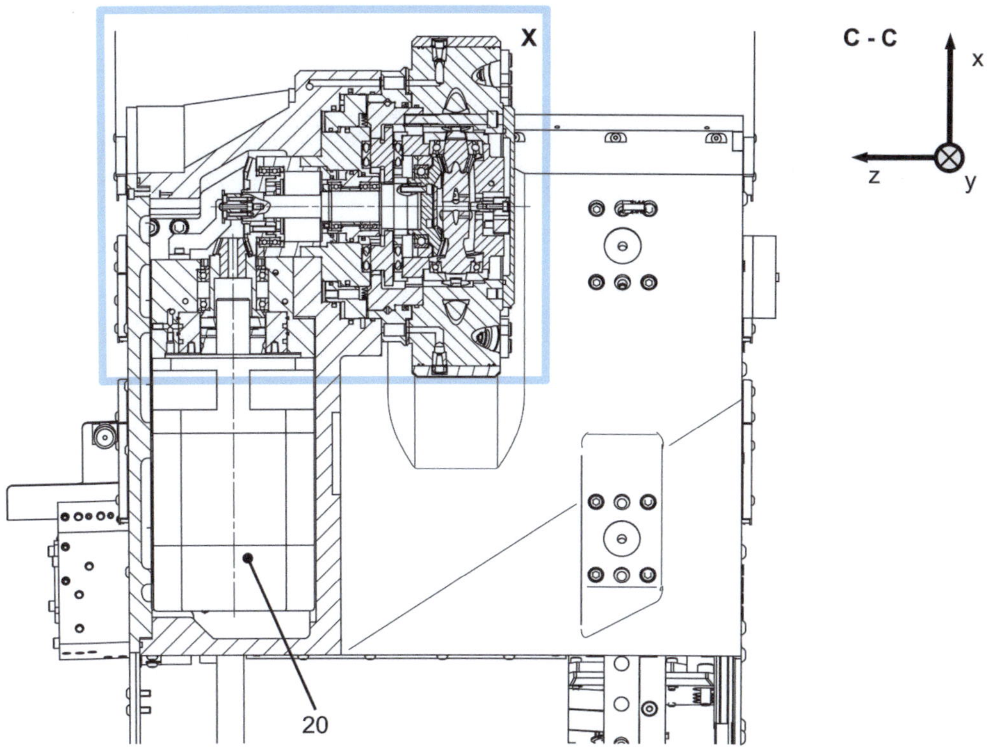

■ **Fig. 11.47** Turret slide **6** (front view) on the Index RatioLine G250, according to INDEX

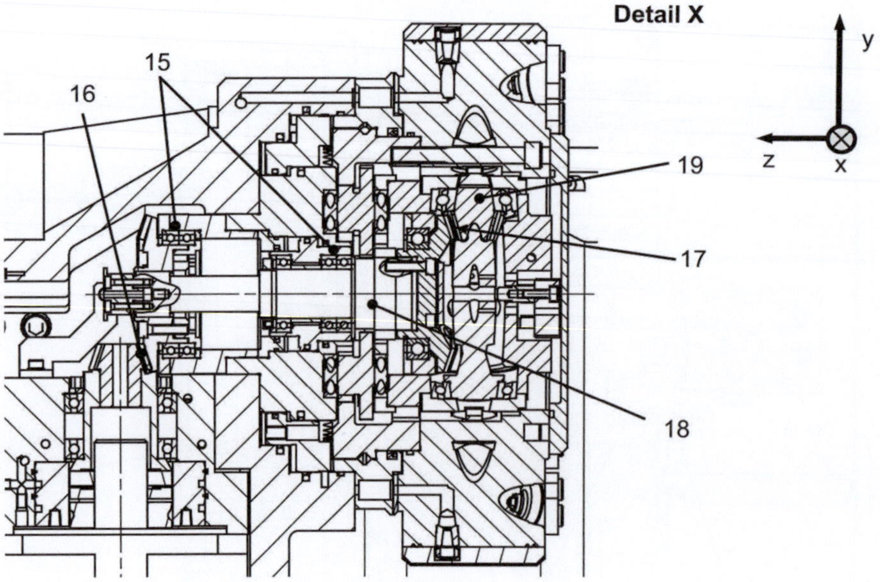

Fig. 11.48 Turret (detail X) on the Index RatioLine G250, according to INDEX

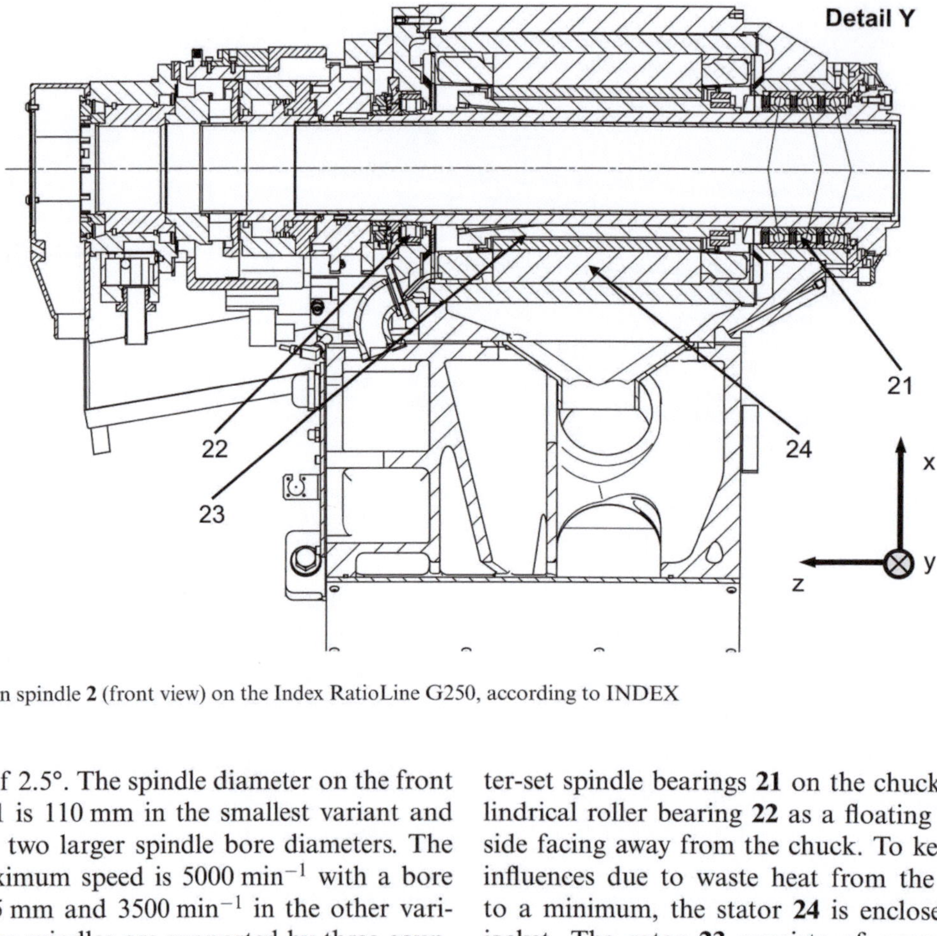

Fig. 11.49 Main spindle **2** (front view) on the Index RatioLine G250, according to INDEX

work in steps of 2.5°. The spindle diameter on the front bearing seat **21** is 110 mm in the smallest variant and 140 mm in the two larger spindle bore diameters. The achievable maximum speed is 5000 min^{-1} with a bore diameter of 65 mm and 3500 min^{-1} in the other variants. The motor spindles are supported by three counter-set spindle bearings **21** on the chuck side and a cylindrical roller bearing **22** as a floating bearing on the side facing away from the chuck. To keep temperature influences due to waste heat from the motor spindle to a minimum, the stator **24** is enclosed in a cooling jacket. The rotor **23** consists of permanent magnets and is force-fit to the spindle shaft.

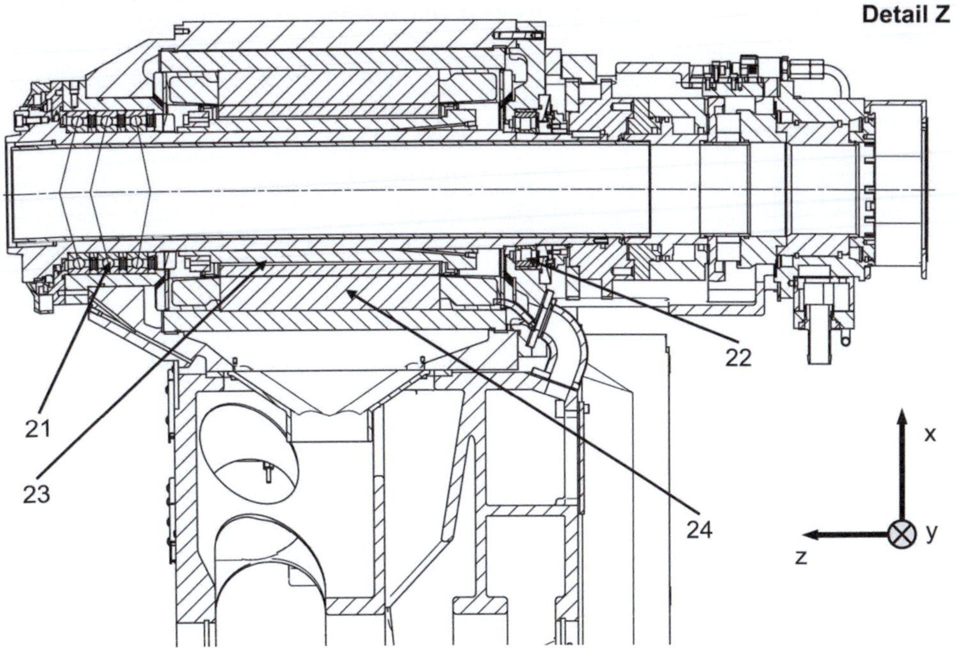

Fig. 11.50 Counter spindle **3** (front view) on the Index RatioLine G250, according to INDEX

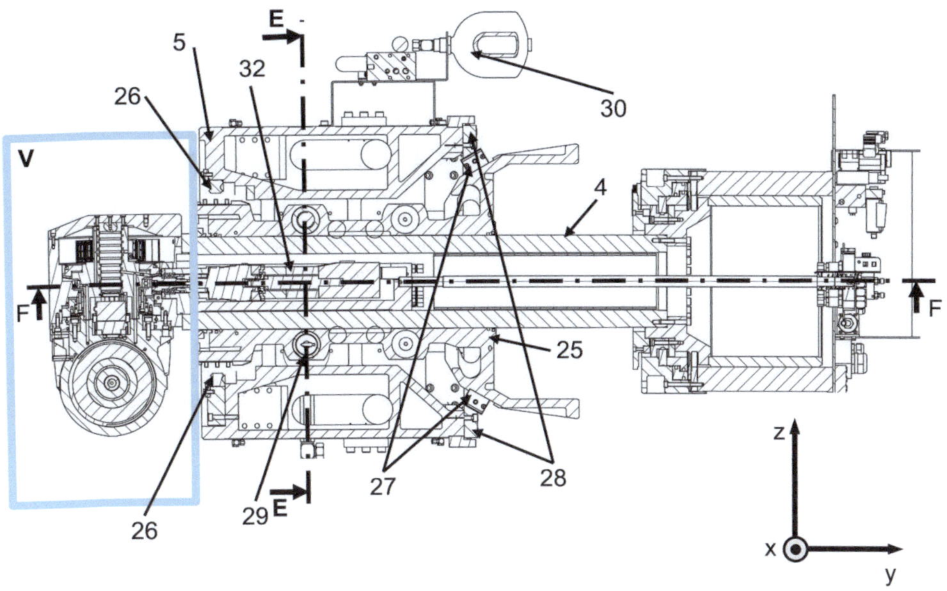

Fig. 11.51 Carriage for motor milling spindle **5** (top view) on the Index RatioLine G250, according to INDEX

11.3.7 Carriage for Motor Milling Spindle A and B-Axes

The carriage supporting the motor milling spindle **7** (Fig. 11.42) travels in the Z-direction on the linear guide systems **12** (Fig. 11.44) mounted on the upper side of the machine bed. A vertical carriage **25**, which can be moved in the X-direction, is integrated into the cross slides **5** on the motor milling spindle. The vertical carriage **25** is guided in the vertical X-axis by rolling element guide systems as shown in Fig. 11.51. Linear guide systems **26** are installed on the spindle side. Recirculating roller bearings **27** rolling on screw-mounted, hardened steel strips **28** provide the guide system on the rear of the vertical carriage. The counterweight for this vertical axis is achieved using a hydraulic cylinder **29** and a system of pressurized gas reservoirs **30**. A ball screw drive **33**, directly driven by a servo motor **34**, provides the feed motion, as shown in Fig. 11.53. The

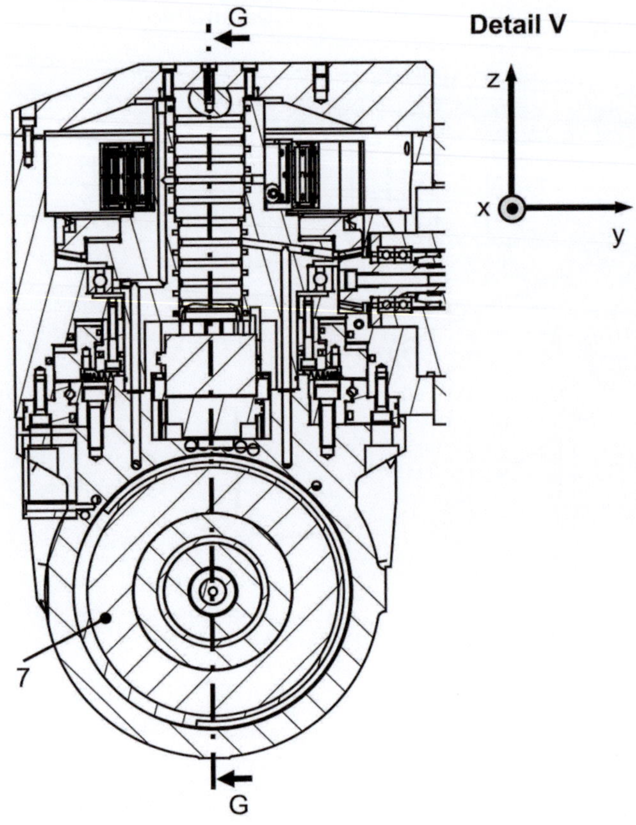

Fig. 11.52 Swivel head for motor milling spindle (detail V) on the Index RatioLine G250, according to INDEX

axial bearing **35** of the screw drive is mounted in an overhanging housing (Fig. 11.52).

The cylindrical cantilever on the Y-axis **41** is radially guided by two hydrostatic bearings **40** as shown in Fig. 11.54. These bearings allow the entire cantilever to rotate around the B-axis. The B-axis can be fixed in any angular position using a friction-locking clamping unit **42**. The axis is driven by a torque motor **43**, with the internal rotor flange-mounted directly to the cantilever on the Y-axis **41**. Unlike the A-axis, the B-axis is suitable for simultaneous machining. The projecting cantilever **36**, mounted on the vertical X-carriage **25** (see Fig. 11.51), supports the components of the feed axis for the Y-cantilever **4** and serves as the torque rod on the B-axis utilizing two linear guide systems **44**. A ball screw drive **39**, driven by a servo motor **37** via a belt drive **38**, provides the feed. Due to the short projection length, a locating/non-locating bearing arrangement is used with a needle roller/axial cylindrical roller bearing (see Figs. 11.55 and 11.56).

11.3.8 Motor Milling Spindle and A-Axis

The RatioLine G250 configuration shown here features an A-axis that allows the motor milling spindle to be swiveled by 180° about the X-axis. The axis designations for this machine configuration are illustrated in Fig. 11.58. While the motor milling spindle is

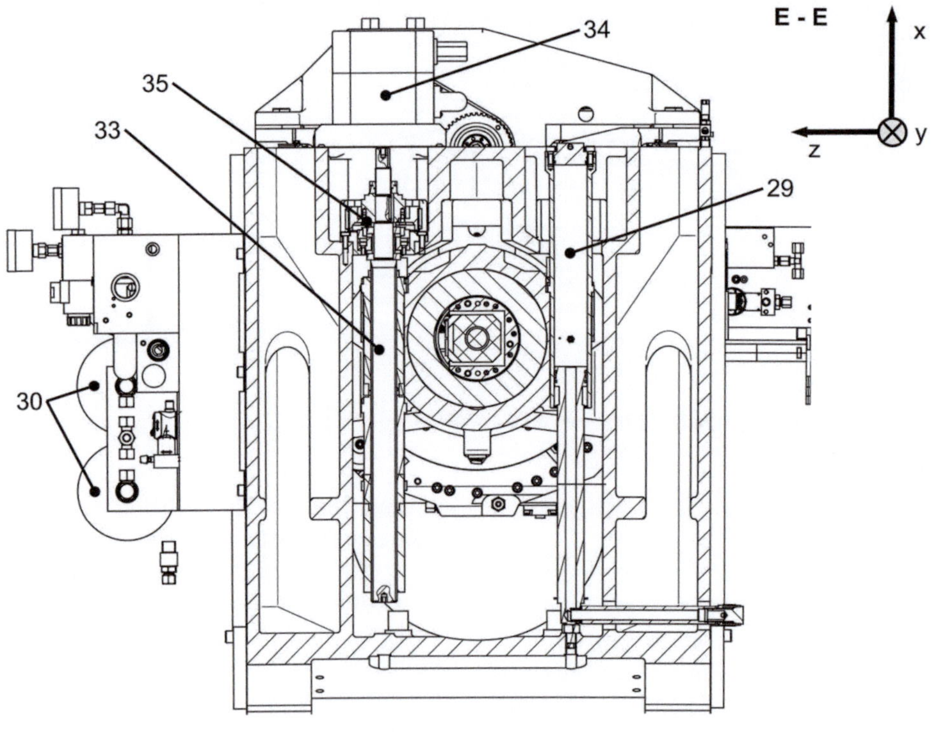

Fig. 11.53 Carriage for motor milling spindle **5** (section E) on the Index RatioLine G250, according to INDEX

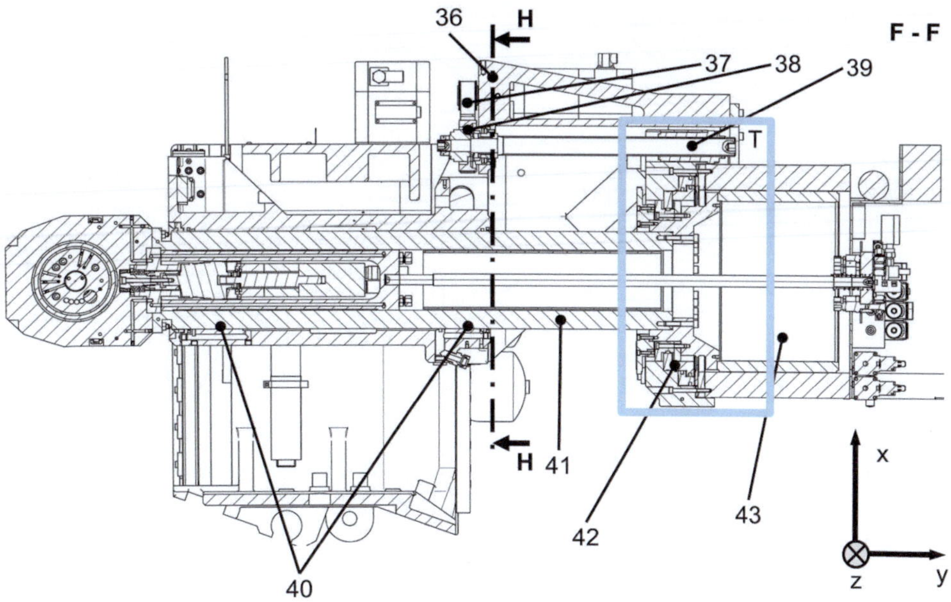

Fig. 11.54 Carriage for motor milling spindle **5** (side view) on the Index RatioLine G250, according to INDEX

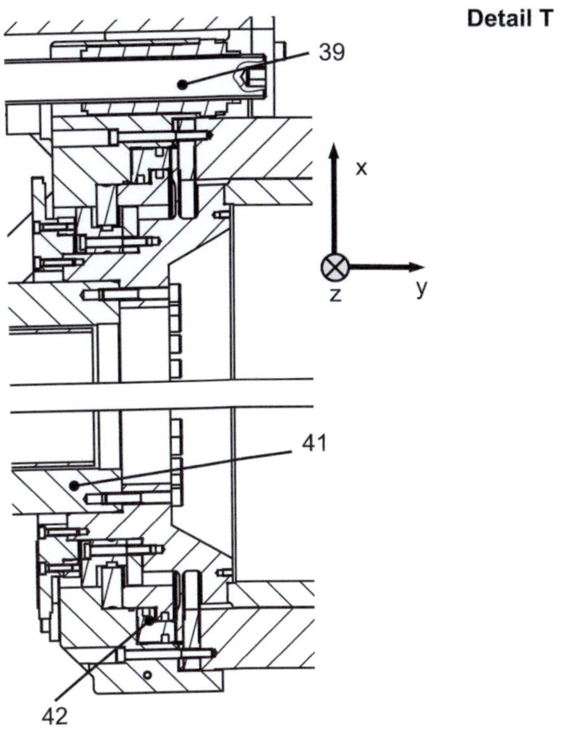

Fig. 11.55 Torque motor clamping unit (detailed view) on the Index RatioLine G250, according to INDEX

shown in a horizontal position in this specific illustration, it is in a vertical position in all other illustrations. Figure 11.57 provides a sectional view of the spindle **7** and the head of the Y-axis cantilever **4**. With this swivel arrangement, it is possible to make optimal use of the machining area and achieve short lever lengths during turning operations. The A-axis drive is powered by a servo motor with a planetary gear stage **30**, which swivels the motor spindle unit via a bevel gear stage **53**. The angular position can be locked in 5° increments using a hydraulically activated gear clamp **45** (Fig. 11.58). When not clamped, the motor spindle is supported by the crossed roller bearing **46** and the deep groove ball bearing **47**, allowing the spindle to rotate. The A-axis cannot be controlled while machining is in progress.

The motor milling spindle itself is driven by a servo motor with a continually excited rotor **52** that is directly integrated into the shaft. The stator **48** is located within the spindle housing and is liquid-cooled. The spindle shaft features a bearing arrangement comprising two duplex spindle bearing sets **50, 51** that are elastic adjusted. The two main bearings **50** on the toolholder **49** have a larger diameter than the two counter bearings **51** on the floating bearing side. The toolholder **49** consists of either an HSK-A63 or a Capto C6

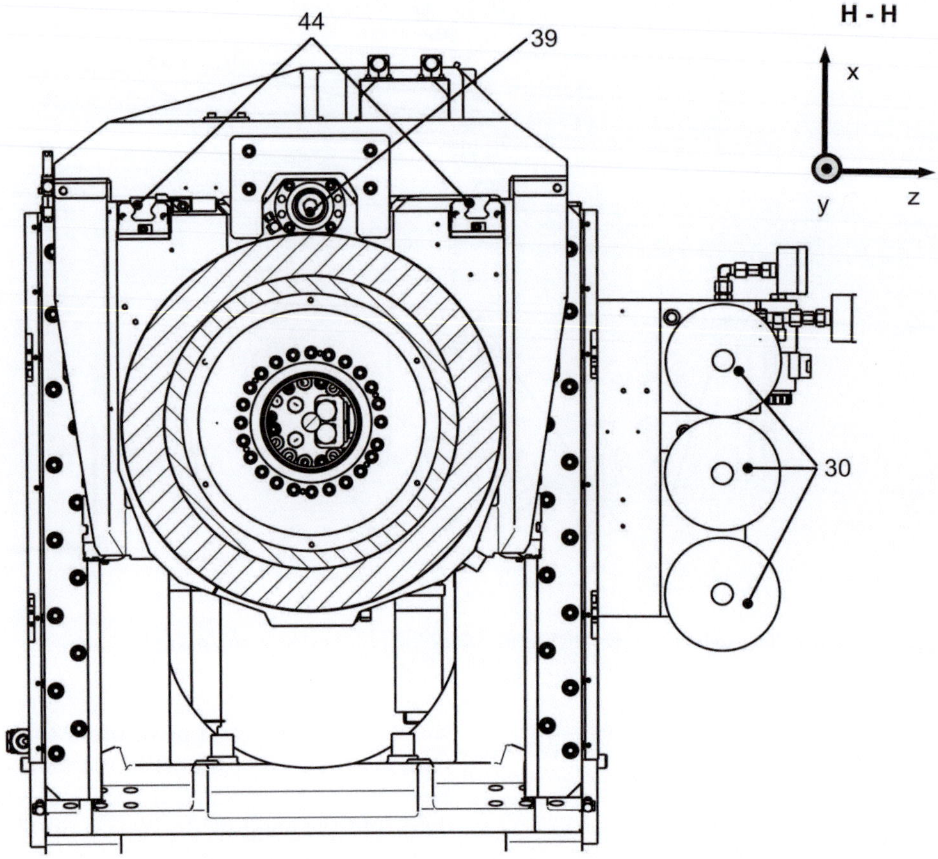

Fig. 11.56 Carriage for motor milling spindle **5** (rear view) on the Index RatioLine G250, according to INDEX

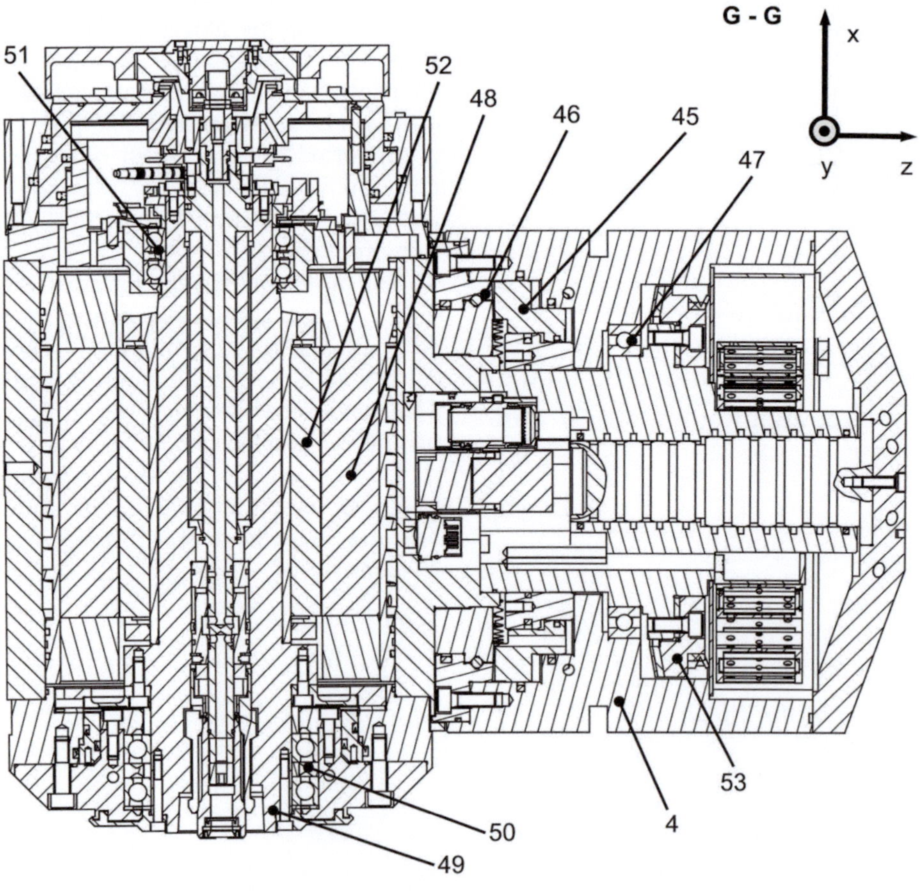

Fig. 11.57 Motor milling spindle (sectional view) on the Index RatioLine G250, according to INDEX

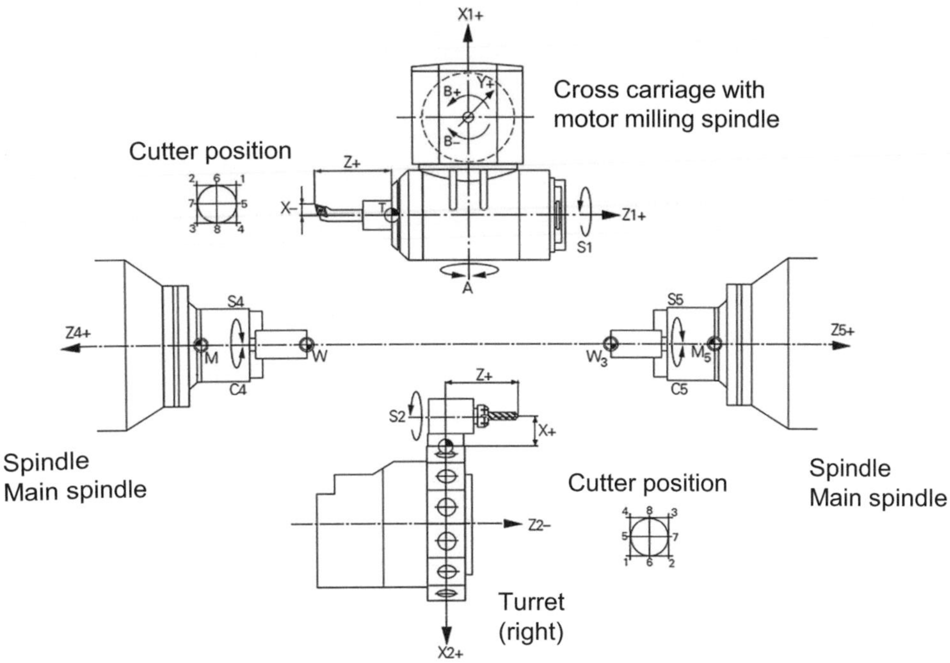

Fig. 11.58 Axis designations (front view) on the Index RatioLine G250, according to INDEX

design. The motor spindle has a maximum power output 27.5 kW, regardless of whether it operates at 100% or 25% operating time. The torque is 75 Nm at 100% and 115 Nm at 25% operating time. The spindle's maximum rotation speed is 12,000 min^{-1}, with the option to reach up to 18,000 min^{-1}.

Supplementary Information

C. Brecher and M. Weck, *Machine Tools Production Systems 1*, Lecture Notes in Production Engineering,
https://doi.org/10.1007/978-3-662-68120-6

Formula Directory

Formula Directory Chap. 3

Symbol	Unit	Description
A	m^2	Pressure surface
A_V	m^2	Valve cross-section
E	J	Work capacity
e	mm	Eccentricity
F	N	Straightening force
f	mm	Maximum deflection
F_b	N	Bending force
F_N	N	Nominal force
F_p	N	Pressing force
F_P	N	Stamping force
F_U	N	Forming force
g	m/s^2	Gravity
H	mm	Stroke
h	mm	Height
I_p	mm	Area of plastic deformation
J	kgm^2	Moment of inertia
l	mm	Length
l	mm	Connecting rod length
m	kg	Mass
M	Nm	Bending torque
m_O	kg	Upper ram mass
m_U	kg	Lower ram mass
p	N/m^2	Pressure
p_P	N/m^2	Pump pressure
r	mm	Eccentric radius
v	m/s	Rate of fall
α	°	Crank angle
$\dot{\varphi}_m$	s^{-1}	Mean forming speed
φ	-	Degree of deformation
ω	s^{-1}	Angular velocity

Formula Directory Chap. 4

Symbol	Unit	Description
$1-m$	[–]	Slope for cutting force
B	[mm]	Width
D	[mm]	Borehole diameter
F_c	[N]	Cutting force
F_f	[N]	Feed force
F_p	[N]	Passive force
H	[mm]	Height
R_a	[nm]	Average surface roughness
R_z	[nm]	Largest profile height
T	[mm]	Depth
a_p	[mm]	Cutting depth
b	[mm]	Cutting width
f	[mm]	Feed
h	[mm]	Cutting thickness
$k_{1.1}$	$[N/mm^2]$	Specific cutting force
l	[mm]	Drill depth
v_c	[m/s]	Cutting speed
κ	[°]	Tool cutting edge angle

Formula Directory Chap. 5

Symbol	Unit	Description
M	[mm]	Measurement range
R_a	[µm]	Arithmetical mean deviation of the profile
R_R	[µm]	Roughness of fracture surface
R_S	[µm]	Roughness of the cut surface
b_0	[mm]	Width of zone affected by jet
b_G	[mm]	Burr width
d_R	[mm]	Mean regulating wheel diameter
h_G	[mm]	burr height
h_R	[mm]	Fracture height
h_S	[mm]	Clean-cut part
m_i	[kg]	Mass
n_R	$[min^{-1}]$	Regulating wheel rotational speed
n_s	$[min^{-1}]$	Grinding wheel rotational speed
n_{WST}	$[min^{-1}]$	Work piece rotational speed
r_E	[mm]	Edge radius (pierce)
s	[mm]	Sheet thickness
u	[°]	Dimensional tolerance (perpendicularity and inclination)
v_f	[m/s]	Feed rate
v_L	[m/s]	longitudinal feed rate

Symbol	Unit	Description
v_{OSZ}	[m/s]	Oscillation speed
α	[°]	Clearance angle/spread of the 2 masses during balancing
β	[°]	Wedge angle/position of the compensation force with respect to the spindle/flank angle
γ	[°]	Cutting angle
Δs	[mm]	Free length on both sided in front of and behind the measurement length
ϕ	[°]	Angel between work piece and regulating wheel

Formula Directory Chap. 6

Symbol	Unit	Description
a	[kg/mol]	Atomic weight
F	[C/mol]	Faraday constant
H_2SO_4	–	Sulphuric acid
HCl	–	Hydrogen chloride
HNO_3	–	Nitric acid
i	[A]	Current
I	[A]	Working current
n	–	Solution valence
$NaCl$	–	Sodium chloride
$NaNO_3$	–	Sodium nitrate
p	[bar]	Pressure
t	[s]	Time
u	[V]	Voltage
U	[mm]	U-axis of the machine coordinate system (parallel to X-axis but independently movable)
V	[mm]	V-axis of the machine coordinate system (parallel to Y-axis but independently movable)
V	[mm³]	Removed volume during ECM
V_{Sp}	[m³*mol/C]	Material constant
W	[mm]	W-axis of the machine coordinate system (parallel to Z-axis but independently movable)
X	[mm]	X-axis of the machine coordinate system
Y	[mm]	Y-axis of the machine coordinate system
Z	[mm]	Z-axis of the machine coordinate system
ρ	[kg/m³]?	Density

Formula Directory Chap. 7

Symbol	Unit	Description
M_2	–	Beam quality
R_a	[µm	Average surface roughness
ϕ_A	[°]	Acceptance angle
ϕ_T	[°]	Threshold angle for total reflection

Formula Directory Chap. 8

Symbol	Unit	Description
E_{wi}	[mm]	Eccentricity during cut i
E	[mm]	Eccentricity
F_W	[N]	Rolling force
a_{ezi}	[mm]	Cutting width during cut i
d_s	[mm]	Shaping wheel diameter
m	[g]	Masse
p_z	[mm]	Tool pitch
v_c	[m/mm]	Cutting speed
v_0	[m/min]	Tool circumferential speed
v_f	[m/min]	Feed rate
α	[°]	Rolling contact angle
ξ_y	[rad]	Rolling angle
β	[°]	Helix angle
β_2	[°]	Helix angle of the gear to produce
γ_0	[°]	Pitch angle of the endless screw
η	[°]	Milling tool swing angle
Σ	[°]	Axis crossing angle
ϑ_y	[rad]	Arc segment

Formula Directory Chap. 10

Symbol	Unit	Description
F_A	[N]	Axial force
F_{in}	[kN]	Retraction force
F_{cent}	[N]	Centrifugal force of the clamping elements
F_{centri}	[N]	Centrifugal force of the drum
F_H	[N]	Retention force
F_P	[N]	Process force
F_{SP}	[N]	Clamping force
F_{sp}	[N]	Clamping force

Symbol	Unit	Description
F_{Spann}	[N]	Clamping force
F_T	[N]	Torsion force
M	[Nm]	Moment
M_{Li}	[Nm]	Lifting torque
M_{AN}	[Nm]	Tightening torque
M_B	[Nm]	Bending moment
$M_{B,crit}$	[Nm]	End of linear range
M_{dyn}	[Nm]	Dynamic torsion load
M_L	[Nm]	Breakaway torque
M_R	[Nm]	Slip torque
M_{stat}	[Nm]	Static torsion load
M_T	[Nm]	Torsional torque
a	[mm]	Clearance
a	[mm]	Diameter
b	[mm]	Clearance
c	[mm]	Clearance

Symbol	Unit	Description
c_1	–	Coefficient of impact
c_2	–	Coefficient of safety
k_b	[Nm^2/mm]	Bending stiffness
k_T	[Nm/rad]	Torsional stiffness at full contact
n	[min^{-1}]	Speed
p	[bar]	Hydraulic pressure
S	[μm]	Torsional play
x	[mm]	Direction in space
y	[mm]	Direction in space
z	[mm]	Direction in space
α	[°]	Tilt angle
α_T	[°]	Torsion angle
ϕ	[°]	Angle of inclination
ϕxz	[°]	Rotation axes
μ	–	Friction coefficient

Abbreviations

Abbreviations	Meaning
3D	Three-Dimensional
AC	Alternating Current
Al	Aluminum
Ar	Argon
BDC	Bottom Dead Center
BGR	Berufsgenossenschaftliche Regel (German Trade Association Regulations)
BMT	Bolt-Mounted-Turret
BPP	Beam Parameter Product
BSD	Ball Screw Drive
CAD	Computer-Aided Design
CAM	Computer-Aided Manufacturing
CAx	Computer-Aided Technologies
cbn	Cubic Boron Nitride
CD	Continuous Dressing (Kontinuierliches Abrichten)
CIM	Computer-Integrated Manufacturing
CM	Carbide Metal
CNC	Computerized Numerical Control
CO_2	Carbon Dioxide
CSP	Compact Strip Production
Cu	Copper
cw	Continuous Wave
DIN	Deutsches Institut für Normung (German Institute For Standardization)
DNC	Direct Numeric Control
EBM	Electron Beam Machining
ECM	Electrochemical Machining
EDM	Electrical Discharge Machining
EDP	Electronic Data Processing
EEC	European Economic Community
EEPROM	Electrically Erasable Programmable Read-Only Memory
EIA	Electronic Industries Association
EMD	European Machinery Directive
EMS	Electromagnetic Stirring
EN	Europäische Norm (European Standard)
Er	Erbium
FBG	Fiber Bragg Grating
Fe	Iron
FMS	Flexible Manufacturing System

Abbreviations	Meaning
GaAlAs	Gallium Aluminum Arsenide
GaAs	Gallium Arsenide
GFRP	Glass-Fiber Reinforced Plastic
H_2SO_4	Sulphuric Acid
HCl	Hydrogen Cloride
He	Helium
HF	High Frequency Excitation
HNO_3	Nitric Acid
HP	High-Precision
HRC	Hardness on Rockwell Scale
HSK	Hohl-Schaft-Kegel (Hollow Shaft Taper Interface)
HV head	Horizontal-Vertical Head
IHPF	Internal High-Pressure Forming Process
IMAS	Integrated Manufacturing And Assembly System
IR	Infrared
ISO	International Organization for Standardization
IT	ISO-Tolerance
LASER	Light Amplification by Stimulated Emission of Radiation
LCR	Liquid Core Reduction
LP	Low Pressure
MCD ara>	Monocrystalline Diamond
MIT	Massachusetts Institute of Technology
MOPA	Master Oscillator Power
MQL	Minimum Quantity Lubrication
MWC	Maximum Workplace Concentrations
N_2	Nitrogen
NaCl	Sodium Chloride
$NaNO_3$	Sodium Nitrate
NC	Numerical Control
Nd	Neodymium
O_2	Oxygen
PCD	Polycrystalline Diamond
PKM	Parallel Kinematics
PLC	Programmable Logic Controller
PSC	Polygonal Hollow Taper Shank With Flange Contact Surface
pw	Pulse Wave
RFID	Radio-Frequency-Identification

Abbreviations	Meaning
RRC	Reaction Resin Concrete
Si_3N_4	Silicon Nitride
SLDS	Single-Lip Drilling System
SLM	Selective Laser Melting
STL	Standard Tessellation Language
TDC	Top Dead Center
TEM	Thermal Deburring Method
TGC	Technical Guideline Concentrations
TS	Modular Taper Interface With Ball Track System

Abbreviations	Meaning
UP	Ultra-Precision
UV	Ultraviolet
VDI	Verein Deutscher Ingenieure (Association Of German Engineers)
VDMA	Verband Deutscher Maschinen- und Anlagenbau (Mechanical Engineering Industry Association)
VSD	Variable-Speed Pump Drive
WLAN	Wireless Local Area Network
YAG	Yttrium Aluminum Garnet
Yb	Ytterbium

Index

G

H

I

K

L

S

T

U

V

W

Y

MIX
Papier aus verantwortungsvollen Quellen
Paper from responsible sources
FSC® C105338

If you have any concerns about our products,
you can contact us on
ProductSafety@springernature.com

In case Publisher is established outside the EU,
the EU authorized representative is:
Springer Nature Customer Service Center GmbH
Europaplatz 3, 69115 Heidelberg, Germany

Printed by Libri Plureos GmbH
in Hamburg, Germany